Lecture Notes in Networks and Systems

Volume 164

The series "Lecture Notes in Networks and Systems" publishes the latest developments in Networks and Systems—quickly, informally and with high quality. Original research reported in proceedings and post-proceedings represents the core of LNNS.

Volumes published in LNNS embrace all aspects and subfields of, as well as new challenges in, Networks and Systems.

The series contains proceedings and edited volumes in systems and networks, spanning the areas of Cyber-Physical Systems, Autonomous Systems, Sensor Networks, Control Systems, Energy Systems, Automotive Systems, Biological Systems, Vehicular Networking and Connected Vehicles, Aerospace Systems, Automation, Manufacturing, Smart Grids, Nonlinear Systems, Power Systems, Robotics, Social Systems, Economic Systems and other. Of particular value to both the contributors and the readership are the short publication timeframe and the world-wide distribution and exposure which enable both a wide and rapid dissemination of research output.

The series covers the theory, applications, and perspectives on the state of the art and future developments relevant to systems and networks, decision making, control, complex processes and related areas, as embedded in the fields of interdisciplinary and applied sciences, engineering, computer science, physics, economics, social, and life sciences, as well as the paradigms and methodologies behind them.

Indexed by SCOPUS, INSPEC, WTI Frankfurt eG, zbMATH, SCImago.

All books published in the series are submitted for consideration in Web of Science.

More information about this series at http://www.springer.com/series/15179

João Manuel R. S. Tavares · Satyajit Chakrabarti ·
Abhishek Bhattacharya · Sujata Ghatak
Editors

Emerging Technologies in Data Mining and Information Security

Proceedings of IEMIS 2020, Volume 3

Springer

Editors
João Manuel R. S. Tavares
Faculdade de Engenharia da
Universidade do Porto
Porto, Portugal

Abhishek Bhattacharya
Institute of Engineering and Management
Kolkata, West Bengal, India

Satyajit Chakrabarti
Institute of Engineering and Management
Kolkata, West Bengal, India

Sujata Ghatak
Institute of Engineering and Management
Kolkata, West Bengal, India

ISSN 2367-3370 ISSN 2367-3389 (electronic)
Lecture Notes in Networks and Systems
ISBN 978-981-15-9773-2 ISBN 978-981-15-9774-9 (eBook)
https://doi.org/10.1007/978-981-15-9774-9

This Springer imprint is published by the registered company Springer Nature Singapore Pte Ltd.
The registered company address is: 152 Beach Road, #21-01/04 Gateway East, Singapore 189721, Singapore

Preface

This volume presents the proceedings of the 2nd International Conference on Emerging Technologies in Data Mining and Information Security (IEMIS2020), which took place in the Institute of Engineering and Management in Kolkata, India, from July 2 to 4, 2020. The volume appears in the series "Lecture Notes in Networks and Systems" (LNNS) published by Springer Nature, one of the largest and most prestigious scientific publishers, in the series which is one of the fastest-growing book series in their program. The LNNS is meant to include various high-quality and timely publications, primarily conference proceedings of relevant conference, congresses and symposia and also monographs, on the theory, applications and implementations of broadly perceived modern intelligent systems and intelligent computing, in their modern understanding, i.e., including tools and techniques of artificial intelligence (AI), computational intelligence (CI)—which includes data mining, information security, neural networks, fuzzy systems, evolutionary computing, as well as hybrid approaches that synergistically combine these areas—and also topics such as network security, cyber-intelligence, multi-agent systems, social intelligence, ambient intelligence, Web intelligence, computational neuroscience, artificial life, virtual worlds and societies, cognitive science and systems, perception and vision, DNA and immune-based systems, self-organizing and adaptive systems, e-learning and teaching, human-centered and human-centric computing, autonomous robotics, knowledge-based paradigms, learning paradigms, machine ethics, intelligent data analysis, various issues related to network security, big data, security and trust management, to just mention a few. These areas are at the forefront of science and technology and have been found useful and powerful in a wide variety of disciplines such as engineering, natural sciences, computer, computation and information sciences, ICT, economics, business, e-commerce, environment, health care, life science and social sciences. The LNNS book series is submitted for indexing in ISI Conference Proceedings Citation Index (now run by Clarivate), EI Compendex, DBLP, SCOPUS, Google Scholar and SpringerLink and many other indexing services around the world. IEMIS2020 is an annual conference series organized at the School of Information Technology, under the aegis of the Institute of Engineering and Management. Its idea came from the heritage of the other two cycles of events: IEMCON and UEMCON, which were organized by the Institute of Engineering and Management under the leadership of Prof. (Dr.) Satyajit Chakraborty.

In this volume of "Lecture Notes in Networks and Systems," we would like to present the results of studies on selected problems of data mining and information security. Security implementation is the contemporary answer to new challenges in threat evaluation of complex systems. Security approach in theory and engineering of complex systems (not only computer systems and networks) is based on multidisciplinary attitude to information theory, technology and maintenance of the systems working in real (and very often unfriendly) environments. Such a transformation has shaped natural evolution in topical range of subsequent IEMIS conferences, which can be seen over the recent years. Human factors likewise infest the best digital dangers. Work force administration and digital mindfulness are fundamental for accomplishing all-encompassing cybersecurity. This book will be of extraordinary incentive to a huge assortment of experts, scientists and understudies concentrating on the human part of the Internet and for the compelling assessment of safety efforts, interfaces, client-focused outline and plan for unique populaces, especially the elderly. We trust this book is instructive yet much more than it is provocative. We trust it moves driving per user to examine different inquiries, applications and potential arrangements in making sheltered and secure plans for all.

The Program Committee of IEMIS2020, its organizers and the editors of these proceedings would like to gratefully acknowledge the participation of all the reviewers who helped to refine contents of this volume and evaluated conference submissions. Our thanks go to Prof Dr. Zdzisław Pólkowski, Dr. Sushmita Mitra, Dr. Pabitra Mitra, Dr. Indrajit Bhattacharya, Dr. Siddhartha Bhattacharyya, Dr. Celia Shahnaz, Mr. Abhijan Bhattacharyya, Dr. Vincenzo Piuri, Dr. Supavadee Aramvith, Dr. Thinagaran Perumal, Dr. Asit Kumar Das, Prof. Tanupriya Choudhury, Dr. Shaikh Fattah and to our all session chairs.

Thanking all the authors who have chosen IEMIS2020 as the publication platform for their research, we would like to express our hope that their papers will help in further developments in design and analysis of engineering aspects of complex systems, being a valuable source material for scientists, researchers, practitioners and students who work in these areas.

Porto, Portugal — João Manuel R. S. Tavares
Kolkata, India — Satyajit Chakrabarti
Kolkata, India — Abhishek Bhattacharya
Kolkata, India — Sujata Ghatak

Contents

Layered Ensemble Learning for Effective Binary Classification 1
Md. Abulkalam Azad, Sadia Islam, Dewan Md. Farid, and Swakkhar Shatabda

A Preliminary Study of Knowledge Graphs and Their Construction 11
Sanju Tiwari, Devottam Gaurav, Animesh Srivastava, Chhiteesh Rai, and Kumar Abhishek

A Simple Ensemble Learning Algorithm For a Real Time High Dimensional Data 21
G. Jenilasree, N. Shenbagavadivu, and M. Bhuvaneswari

Promising Highest Resilient Network Using Preventive Route Selection Algorithm-IPS DSDV in MANET 33
Kanika Garg, Naveen Chauhan, Chiranjit Dutta, and Tanupriya Choudhury

An ANN Approach of Twisted Fiestel Block Ciphering 47
Sanjay Kumar Pal, Bimal Datta, and Amiya Karmakar

Musical Instrument Classification Based on Machine Learning Algorithm 57
Hasanuzzaman Anuz, Abu Kaisar Mohammad Masum, Sheikh Abujar, and Syed Akhter Hossain

Analytical Survey and Comparative Analysis of Socialized Network via Data Mining and Machine Learning Techniques 69
Akhilesh Kumar Singh, Manish Raj, Deepak Mangal, and Saurabh Singhal

Assessment of Severity Classification of Traffic Accidents on the Basis of K-Means Clustering and Adaptive Neuro-Fuzzy Inference System 75
Amrita Sarkar, Gadadhar Sahoo, and Umesh Chandra Sahoo

Design and Implementation of Login-Based Wi-Fi Hotspot Network for an University Campus . . . 87
Sayeed Islam and Md. Nurul Islam

A Framework to Capture the Shift in Dynamics of a Multi-phase Protest—A Case Study of Hong Kong Protests . . . 95
Aayushi Sanghi, Abhishek Sinha, Aditya Venkatesh, and Bhaskarjyoti Das

A Comparative Study to Analyze the Performance of Advanced Pattern Recognition Algorithms for Multi-Class Classification . . . 111
Ayan Chatterjee, Martin Wulf Gerdes, Andreas Prinz, and Santiago Martinez

ECG Signal Denoising Using a Novel Solution to the Heat Equation Through Wavelet Transform . . . 125
P. Upadhyay, S. K. Upadhyay, and K. K. Shukla

Design and Implementation of Chao-Cryptic Architecture on FPGA for Secure Audio Communication . . . 135
B. Sowmya, V. M. Padmapriya, R. Sivaraman, Amirtharajan Rengarajan, Sundararaman Rajagopalan, and Har Narayan Upadhyay

Diffusion + Confusion: Chaos-assisted Twofold Cryptosystem for Secure Video Transmission . . . 145
Marusani Sumanjali, R. Sivaraman, A. Sridevi, Amirtharajan Rengarajan, Sundararaman Rajagopalan, and Har Narayan Upadhyay

Chaos Blend LFSR—Duo Approach on FPGA for Medical Image Security . . . 155
N. Adithya, Hemanth Kumar Nalajala, R. Sivaraman, A. Sridevi, Amirtharajan Rengarajan, and Sundararaman Rajagopalan

Secured Routing System for Low Energy Networks . . . 165
Anup W. Burange and V. M. Deshmukh

Clinical Decision Support System Using Fuzzy Logic Programming and Data Analysis . . . 175
Sandip Paul, Kumar Sankar Ray, and Diganta Saha

Analysis and Review on Energy Efficient Cluster-Based Routing Protocol . . . 185
Kanu Patel and Hardik Modi

Analysis and Review on Service-Oriented-Based IoT Middleware . . . 197
Kanu Patel and Hardik Modi

Colon Cancer Classification Using Binary Particle Swarm Optimization and Logistic Regression . . . 211
Nimrita Koul and Sunilkumar S. Manvi

Accident Detection and Avoidance Using Spot Sensor Technology 219
Prathamesh R. Mahadik, Praj Shete, Shrenik Kabade, Shlok Gaikwad, and N. S. Shirsat

Classification of Focal and Non-focal EEG Signal for Epileptic Seizure Detection with Entropy Features Using KNN Classifier 229
N. Samreen Fatima, M. K. Mariam Bee, Abhishek Bhattacharya, and Soumi Dutta

Prediction of Future Career Path Using Different Machine Learning Models 241
Subhabrata Sengupta, Anish Banerjee, and Satyajit Chakrabarti

Holographic Renormalization with Machine Learning 253
Eric Howard

Optimization of Energy Consumption and Throughput in Cognitive Radio Network Using Swarm Intelligence Techniques 263
Ashim Jyoti Gogoi, Romen Gogoi, Anirban Gogoi, and Deepjyoti Talukdar

Estimation of Various Parameters for AES, DES, and RSA 275
Jyoti Agarwal, Manoj Kumar, and Atul Kumar Srivastava

Semi-automated Ontology Creation and Upgradation for Rail-Road Incidents: A Case of a Steel Plant in India 285
Anima Pramanik, Sobhan Sarkar, V. Sai Siddharth, J. Maiti, and Pabitra Mitra

Blockchain and Machine Learning: Background, Integration Challenges and Application Areas 295
Abhishek Goel, Bharat Bhushan, Bhavish Tyagi, Harshita Garg, and Siddharth Gautam

Various Types of Cybercrime and Its Affected Area 305
Harshit Yadav, Siddharth Gautam, Aniket Rana, Jatin Bhardwaj, and Nitin Tyagi

A Mathematical Recommendation Model to Rank Reviewers Based on Weighted Score for Online Review System 317
Runa Ganguli, Priyanka Banerjee, Sanjib Halder, and Soumya Sen

Virtual Numeric Authentication System Using Contour Detection of Color-Banded Fingertips 327
Aman Gurung, Siddharth Gautam, Tushar Garg, Yash Bhardwaj, and Harshul Gupta

Efficient Authentication Technique in Medical Application by Using Homomorphic Encryption on Finger Print and RFID Tags 341
Sonali S. Lunawat, Hitesh Bhandari, Madhuri Badole, and Ompriya Kale

Automated Waste Monitoring and Alert System Using IoT 349
Meegan Scariah, Kavya Ann Salbi, Rintu Ponnachan, Sruthi Anna Joseph, and Er. Jinu Thomas

English to Bengali Neural Machine Translation Using Global Attention Mechanism 359
Sheikh Abujar, Abu Kaisar Mohammad Masum, Abhishek Bhattacharya, Soumi Dutta, and Syed Akhter Hossain

Taxonomy of Cyber Security in Medical Science 371
Jatin Bhardwaj, Siddharth Gautam, Harshit Yadav, Nitin Tyagi, and Shafiqul Abidin

Tendulkar's Cat and Schrodinger's Bat—Knowledge-Enhanced Real-Word Error Correction 381
Shailesh Sridhar, E Saahitya, Satyam Shivam Sundaram, and Bhaskarjyoti Das

Multi Task Learning for Financial Forecasting 397
Deepali Prabhu, Arnav Chhabra, Saksham Goyal, and Bhaskarjyoti Das

Embedding-Assisted Auto-Tuned Community Detection 413
Akshara Subramaniasivam, Kaushik Ravichandran, Aishwarya Poomuttam Sreedas, and Bhaskarjyoti Das

Optimization of Coverage Hole Identification in 5G SON Using Data Mining 429
Biswadeep Deb Nandy and Kartick Chandra Mondal

A Machine Learning Based Heuristic to Predict the Efficacy of Online Sale 439
Aditya Vikram Singhania, Saronyo Lal Mukherjee, Ritajit Majumdar, Akash Mehta, Priyanka Banerjee, and Debasmita Bhoumik

Evaluation Metrics of Spatial and Spatiotemporal Data Mining Techniques 449
S. Sharmiladevi and S. Siva Sathya

ECG Assess Heartbeat rate, Classifying using BPNN while Watching Movie and send Movie Rating through Telegram 465
Sima Das and Aakashjit Bhattacharya

Software-Defined Networking with Multiple Routing Utilizing DFS to Improving QoS 475
F. M. Javed Mehedi Shamrat, Pronab Ghosh, Md. Dipu Sultan, Anup Majumder, and Imran Mahmud

Direct Farmer and Customer Commerce Using Data Mining 485
Siddharth Singh, Aditya Kumar, Aditya Chopra, and Mitchell D'silva

Automatic Question and Answer Generation from Text Using Neural Networks 497
Sonam Soni, Praveen Kumar, and Amal Saha

A Multi-Criteria Prioritization Based Slot Allocation Scheme in IEEE 802.15.4 for Wireless Body Area Networks 505
Sathi Roy and Chandreyee Chowdhury

A Secure Lightweight Authentication Protocol for Mobile Payment 515
Gungun Sharma and Mohona Ghosh

Create EC2 Instance, Install Splunk and Configure Using Ansible 529
Rishabh Shah and Vatsal Shah

Quantum Computation-Based Secured Key Distribution Protocols 539
Priyanka Jojan, Kapil Kumar Soni, and Akhtar Rasool

A Detailed Survey on Epilepsy Detection Using the Classifications of EEG Signals 555
M. K. Mariam Bee and K. Vidhya

A Comparative Study of Machine Learning and Deep Learning for Prediction of COVID-19 563
Shaswata Saha and Sarmistha Neogy

Epileptic Seizure Severity Analysis Using Artificial Neural Network and CANFIS Classification Approach 571
M. K. Mariam Bee and K. Vidhya

Inferring Political Preferences from Twitter 581
Mohd Zeeshan Ansari, Areesha Fatima Siddiqui, and Mohammad Anas

Implementation of an Intelligent Online Job Portal Using Machine Learning Algorithms 591
Zarrin Tasnim, F. M. Javed Mehedi Shamrat, Shaikh Muhammad Allayear, Khobayeb Ahmed, and Naimul Islam Nobel

A Review Paper on Reconfigurable Antenna Technique and Methodology 605
Mritunjay Srivastava and Abhijeet Kumar

An Exploratory Study for Process Optimization in IT Industry 617
H. Aditya Pai, Piyush Kumar Pareek, M. S. Narasimha Murthy, Sunanda Dixit, and Sneha Karamadi

Self-driving Car in Virtual Simulation: A Non-AI Approach 633
Debkumar Chowdhury, Sujoy Paul, Sreejit Bagchi, Trisha Das, and Bipasha Mukhopadhyay

Exploration Study to Study the Relationships Between Variables of Secure Development Lifecycle (SDL) 641
Sandhya Soman, Piyush Kumar Pareek, Sunanda Dixit, R. M. Chethana, and Veeranna Kotagi

Recommendation-Based Interactivity Through Cross Platform Using Big Data 651
G. K. Suhas, S. N. Devananda, R. Jagadeesh, Piyush Kumar Pareek, and Sunanda Dixit

A Comprehensive Survey: Toward Providing Personal Safety Services in Emergency Situations 661
Ramesh M. Kagalkar and Lokesh B. Bhajantri

Classification of Bharatnatyam and Kathak Dance Form Through Audio Signal 671
Sanghamitra Das, Suchibrota Dutta, Debanjan Banerjee, and Arijit Ghosal

SWOT Analysis of BYOD (Bring Your Own Device) 681
P. Soubhagyalakshmi and K. Satyanarayan Reddy

A Novel Approach for Improving Security in Software Development in Small Software Firms: A Literature Review 689
Prakash Chandra Behera, Chinmaya Dash, and Piyush Kumar Pareek

Literature Survey on Exploratory Analysis of Waste in IT Industries During Project Development Using Agile–Lean Practices 699
Chinmaya Dash, Prakash Chandra Behera, and Piyush Kumar Pareek

Classifying Medical Histology Images Using Computationally Efficient CNNs Through Distilling Knowledge 713
Aakash Garg, Karan Aggarwal, Manan Saxena, and Aruna Bhat

Analyzing the Spread of COVID-19 in India Through PageRank and Diffusion Techniques 723
Sawon Bhattacharya, S. Siva Sathya, and S. Sharmiladevi

Graph-Assisted Attention for Path Finding in Question Answering Task 735
M. Guruprasad, Jai Agarwal, T. N. Lokesh Kumar, and Bhaskarjyoti Das

Towards Open Ended and Free Form Visual Question Answering: Modeling VQA as a Factoid Question Answering Problem 749
Abhishek Narayanan, Abijna Rao, Abhishek Prasad, and S. Natarajan

Integrating Secured Crypto System with Cloud for Enhancing Cloud-Based Encrypted Data Sharing Services 761
Ravi Verma, Chetan Agrawal, and Neelesh Jain

Performance Analysis of UAV Relay in Vanets against Smart Jamming with Q-Learning Techniques 771
L. Rajesh, K. Bhoopathy Bagan, P. Thayal Sankar, and V. Suchitra

Recent Trends in IoT–Based Smart Healthcare Applying ML and DL 785
Sohail Saif, Mayurakshi Jana, and Suparna Biswas

House Price Prediction Using Machine Learning 799
Navia Kamal, Ekta Chaturvedi, Siddharth Gautam, and Shruti Bhalla

MatriVasha: A Multipurpose Comprehensive Database for Bangla Handwritten Compound Characters 813
Jannatul Ferdous, Suvrajit Karmaker, A. K. M. Shahariar Azad Rabby, and Syed Akhter Hossain

Study on Efficient DRLSE-Oriented Edge-Based Medical Image Segmentation of Cardiac Images 823
Pramit Brata Chanda and Subir Kumar Sarkar

Big Data Analysis Using R and Hadoop 833
Vikas Rao Vadi, Varun Tiwari, and Shafiqul Abidin

Operating Systems Support and Network Optimization View of Internet of Things 845
Vikas Kumar Agarwal, Nitish Kumar, Rajesh Singh, Anuradha Pathak, and Bharat Bhushan

Information and Data Security Model: Background, Risks, and Challenges 859
Bhuvan Taneja and Bharat Bhushan

Design and Implementation of Cryptographic Technique Employing Intricate Table and Rubik Rotation Algorithm 871
Rajdeep Chowdhury, Soma Sharma, Md Adil Alam, and Niharika

Impact of Routing Protocols and Artificial Intelligence Techniques on Ad hoc Networks in Critical Scenario 883
Meena Rao

Algorithm for Inference of Gene Regulatory Networks Using Partial Least Squares Regression and Mutual Information 891
Nimrita Koul and Sunilkumar S. Manvi

Analyzing Automatic Object Classification to Query-Based Image Search Engine 899
Mansi Mahendru and Sanjay Kumar Dubey

An Analytical Approach to Investigation of Asthma Diseases Using Soft Computing Techniques 909
Gautam Singh Yadav and Sanjay Kumar Dubey

Cataract Eye Detection Using Machine Learning Models 919
Ankush Shetty, Kunal Bathija, and Priya R. L.

Automated High Precision Path Annotations on Digital Map from Crowdsourced GPS Traces 931
Siddhartha Chatterjee, Saikat Samanta, and Ratna Mandal

Amplifier to Strengthen the Microvolt Signal for Biomedical Sensors 947
R. Chaitanya and R. Puviarasi

Automatic Street Light Controller Using LDR and Solar 955
C. H. Jyothi Priya and R. Puviarasi

EEG Processing for Computer-Aided Seizure Detection Using Artificial Neural Networks 963
P. Sai Ankith and M. K. Mariam Bee

An Automated Method for Epileptic Seizure Detection Using Support Vector Machine 973
P. Sai Ankith and M. K. Mariam Bee

Crop Production Prediction Using Machine Learning: An Indian Perspective 981
Ashi Agrawal, Deepak Arora, and Puneet Sharma

Author Index 991

Editors and Contributors

About the Editors

João Manuel R. S. Tavares is a Senior Researcher at the Instituto de Ciência e Inovação em Engenharia Mecânica e Engenharia Industrial (INEGI) and Associate Professor with Habilitation at the Department of Mechanical Engineering (DEMec) of the Faculdade de Engenharia da Universidade do Porto (FEUP). He graduated in Mechanical Engineering at the Universidade do Porto, Portugal, in 1992. He earned his M.Sc. degree and Ph.D. degree in Electrical and Computer Engineering from the Universidade do Porto in 1995 and 2001, respectively, and the Habilitation in Mechanical Engineering in 2015 from the same university. João Tavares is Co-Editor of more than 50 books and co-author of more than 35 book chapters and 650 articles in international and national journals and conferences. Further, he is author of 3 international as well as 3 national patents. He has been a committee member of several international and national journals and conferences and is the co-founding and co-series Editor of the book series "Lecture Notes in Computational Vision and Biomechanics" published by Springer. He is the founding Editor and Editor-in-Chief of the journal *Computer Methods in Biomechanics and Biomedical Engineering: Imaging and Visualization* published by Taylor & Francis, and Co-Founder and Co-Chair of the international conference series CompIMAGE, ECCOMAS VipIMAGE, ICCEBS, and BioDental. Additionally, he has been (co)supervisor of several M.Sc. and Ph.D. theses and supervisor of several postdoc projects. He has participated in many scientific projects both as a Researcher and as a Scientific Coordinator. His main research areas include computational vision, medical imaging, computational mechanics, scientific visualization, human–computer interaction, and new product development.

Dr. Satyajit Chakrabarti is Pro-Vice Chancellor, University of Engineering and Management, Kolkata and Jaipur Campus, India, and Director of Institute of Engineering and Management, IEM. As the Director of one of the most reputed organizations in Engineering and Management in Eastern India, he launched a

PGDM Programme to run AICTE approved Management courses, Toppers Academy to train students for certificate courses, and Software Development in the field of ERP solutions. Dr. Chakrabarti was Project Manager in TELUS, Vancouver, Canada, from February 2006 to September 2009, where he was intensively involved in planning, execution, monitoring, communicating with stakeholders, negotiating with vendors and cross-functional teams, and motivating members. He managed a team of 50 employees and projects with a combined budget of $3 million.

Abhishek Bhattacharya is an Assistant Professor of Computer Application Department at Institute of Engineering and Management, India. He did Masters in Computer Science from the Biju Patnaik University of Technology and completed Master of Technology in Computer science from BIT, Mesra. He remained associated as Visiting/Guest Faculty of several universities/institutes in India. He has 5 edited and 3 authored books to his credit. He has published many technical papers in various peer-reviewed journals and conference proceedings, both international and national. He has teaching and research experience of 13 years. His area of research is data mining, network security, mobile computing, and distributed computing. He is the reviewer of couple of the Journals of IGI Global, Inderscience, and Journal of Information Science Theory and Practice. He is member of IACSIT, UACEE, IAENG, ISOC, SDIWC, and ICSES; Technical Committee member and Advisory Board Member of various international conferences such as CICBA, CSI, FTNCT, ICIoTCT, ICCIDS, ICICC, ISETIST, and many more.

Sujata Ghatak is an Assistant Professor of Computer Application Department at the Institute of Engineering and Management, India. She received Masters in Computer Science and Master of Technology in Computer science from the Sikkim Manipal Institute of Technology (SMIT), Majhitar, Sikkim, in 2011 and 2013. She received the B.Sc. (Computer Science) degree from the University of Calcutta (CU), Kolkata, in 2008. She is currently pursuing the Ph.D. degree with the Department of Computer Science and Engineering, from Maulana Abul Kalam Azad University of Technology, West Bengal. Her research interests include image processing, pattern recognition, applied dynamical system in image processing, pattern recognition, information retrieval, natural language processing, and data science. She was the co-convener in IEMIS 2018. She is TPC member in various international conferences such as—International Conference on Smart Innovation, Ergonomics and Applied Human Factors (SEAHF 2019), Data Science and Machine learning and Application (DSMLA 2019), ARIAM-2019. She was also organizing Chair in CIPR 2019. She is the member of several technical functional bodies such as The Society of Digital Information and Wireless Communications (SDIWC), Internet Society as a Global Member (ISOC), and International Computer Science and Engineering Society (ICSES). She has published several papers in reputed journals and conferences.

Contributors

Kumar Abhishek NIT Patna, Patna, India

Shafiqul Abidin HMR Institute of Technology & Management, GGSIP University, New Delhi, Delhi, India

Sheikh Abujar Department of Computer Science and Engineering, Independent University, Dhaka, Bangladesh;
Department of Computer Science and Engineering, Daffodil International University, Dhaka, Bangladesh

N. Adithya Department of ECE, School of EEE, SASTRA Deemed University, Thanjavur, India

H. Aditya Pai Department of Computer Science and Engineering, K. S. Institute of Technology, Bangalore, India

Jai Agarwal PES University, Bengaluru, India

Jyoti Agarwal Amity University Tashkent, Tashkent, Uzbekistan

Vikas Kumar Agarwal Department of Computer Science and Engineering, RTC Institute of Technology, Ranchi, India

Karan Aggarwal Delhi Technological University, Shahbad Daulatpur, Delhi, India

Ashi Agrawal Department of Computer Science and Engineering, Amity University, Lucknow, India

Chetan Agrawal Department of Computer Science and Engineering, University Institute of Technology, Rajiv Gandhi Proudyogiki Vishwavidyalaya, Bhopal, India

Khobayeb Ahmed Department of Computer Science and Engineering, Daffodil International University, Dhaka, Bangladesh

Md Adil Alam JIS College of Engineering, Kalyani, India

Shaikh Muhammad Allayear Department of Multimedia and Creative Technology, Daffodil International University, Dhaka, Bangladesh

Mohammad Anas Department of Computer Engineering, Jamia Millia Islamia, New Delhi, India

Mohd Zeeshan Ansari Department of Computer Engineering, Jamia Millia Islamia, New Delhi, India

Hasanuzzaman Anuz Department of Computer Science and Engineering, Daffodil International University, Dhaka, Bangladesh

Deepak Arora Department of Computer Science and Engineering, Amity University, Lucknow, India

Md. Abulkalam Azad Department of Computer Science and Engineering, United International University, Dhaka, Bangladesh

Madhuri Badole Department of Computer Engineering, Pimpri Chinchwad College of Engineering and Research, Ravet, Pune, India

Sreejit Bagchi University of Engineering and Management, Kolkata, India

Anish Banerjee Department of Information Technology, Institute of Engineering and Management (IEM), Kolkata, India

Debanjan Banerjee Department of Management Information Systems, Sarva Siksha Mission Kolkata, Kolkata, West Bengal, India

Priyanka Banerjee Department of Computer Science, The Bhawanipur Education Society College, Kolkata, India

Kunal Bathija Department of Computer Engineering, Vivekanand Education Society's Institute of Technology, Chembur, Mumbai, Maharashtra, India

Prakash Chandra Behera St. Claret College, Bangalore, India;
Shri JJT University, Rajasthan, India

Lokesh B. Bhajantri Basaveshwar Engineering College (Autonomous), Affiliated to VTU, Belagavi, Bagalkot, Karnataka, India

Shruti Bhalla Department of Information Technology, HMR Institute of Technology and Management, Delhi, India

Hitesh Bhandari Tata Consultancy Services, Hinjewadi, Pune, India

Jatin Bhardwaj Branch of Information Technology, HMRITM, Delhi, India

Yash Bhardwaj Department of Information Technology, HMRITM, GGSIPU, New Delhi, India

Aruna Bhat Delhi Technological University, Shahbad Daulatpur, Delhi, India

Aakashjit Bhattacharya Department of Computer Science and Engineering, Maulana Abul Kalam Azad University of Technology, Kolkata, West Bengal, India

Abhishek Bhattacharya Department of Computer Science, School of Information Technology, Institute of Engineering and Management, Kolkata, India;
Institute of Engineering and Management, Saltlake, India

Sawon Bhattacharya Department of Computer Science, Pondicherry University, Puducherry, India

K. Bhoopathy Bagan Department of Electronics Engineering, Madras Institute of Technology, Anna University, Chennai, India

Debasmita Bhoumik Advanced Computing & Microelectronics Unit, Indian Statistical Institute, Kolkata, India

Bharat Bhushan Department of Computer Science and Engineering, Sharda University, Greater Noida, India

M. Bhuvaneswari Department of ECE, University College of Engineering, Anna University BIT Campus, Tiruchirappalli, Tamil Nadu, India

Suparna Biswas Department of Computer Science and Engineering, Maulana Abul Kalam Azad University of Technology, Kolkata, West Bengal, India

Anup W. Burange Department of Computer Science and Engineering, PRMIT&R, Badnera, Maharashtra, India

R. Chaitanya Saveetha School of Engineering, Chennai, India

Satyajit Chakrabarti Department of Computer Science and Engineering, Institute of Engineering and Management (IEM), Kolkata, India

Pramit Brata Chanda Department of Electronics and Tele Communication Engineering, Jadavpur University, Kolkata, India

Ayan Chatterjee Department of Information and Communication Technology, Centre for E-Health, University of Agder, Kristiansand, Norway

Siddhartha Chatterjee Department of Computer Science and Engineering, Gargi Memorial Institute of Technology, Kolkata, West Bengal, India

Ekta Chaturvedi Department of Information Technology, HMR Institute of Technology and Management, Delhi, India

Naveen Chauhan Department of Computer Science and Engineering, KIET Group of Institutions, Ghaziabad, Uttar Pradesh, India

R. M. Chethana Department of Computer Science and Engineering, East West College of Engineering, Bangalore, India

Arnav Chhabra PES University, Bengaluru, India

Aditya Chopra Department of Information Technology, Dwarkadas J. Sanghvi College of Engineering, Mumbai, India

Tanupriya Choudhury Department of Informatics School of Computer Science, UPES, Dehradun, Uttarakhand, India

Chandreyee Chowdhury Jadavpur University, Kolkata, India

Debkumar Chowdhury University of Engineering and Management, Kolkata, India

Rajdeep Chowdhury Academician and Author, Chinsurah, India

Mitchell D'silva Department of Information Technology, Dwarkadas J. Sanghvi College of Engineering, Mumbai, India

Bhaskarjyoti Das Computer Science and Engineering Department, PES University, Bengaluru, India

Sanghamitra Das Department of Information Technology, St. Thomas' College of Engineering and Technology, Kolkata, West Bengal, India

Sima Das Department of Computer Science and Engineering, Maulana Abul Kalam Azad University of Technology, Kolkata, West Bengal, India

Trisha Das University of Engineering and Management, Kolkata, India

Chinmaya Dash St. Claret College, Bangalore, India;
Shri JJT University, Rajasthan, India

Bimal Datta Department of Computer Science and Engineering, Budge Budge Institute of Technology, Kolkata, India

V. M. Deshmukh Department of Computer Science and Engineering, PRMIT&R, Badnera, Maharashtra, India

S. N. Devananda PES Institute of Technology and Management, Shivamogga, India

Md. Dipu Sultan Department of Computer Science and Engineering, Daffodil International University, Dhaka, Bangladesh

Sunanda Dixit Department of Computer Science and Engineering, BMS Institute of Technology and Management, Bangalore, India

Sanjay Kumar Dubey Department of Computer Science and Engineering, Amity University Uttar Pradesh, Noida, U.P., India

Chiranjit Dutta Department of Information Technology, SRM IST, Modinagar, Uttar Pradesh, India

Soumi Dutta Department of Computer Science, School of Information Technology, Institute of Engineering and Management, Kolkata, India;
Institute of Engineering and Management, Saltlake, India

Suchibrota Dutta Department of Information Technology and Mathematics, Royal Thimphu College, Thimphu, Bhutan

Dewan Md. Farid Department of Computer Science and Engineering, United International University, Dhaka, Bangladesh

Jannatul Ferdous Daffodil International University, Dhaka, Bangladesh

Shlok Gaikwad Pune Vidyarti Gruha College of Engineering, Pune, India

Runa Ganguli Department of Computer Science, The Bhawanipur Education Society College, Kolkata, India

Aakash Garg Delhi Technological University, Shahbad Daulatpur, Delhi, India

Harshita Garg HMRITM, Delhi, India

Kanika Garg Department of Information Technology, SRM IST, Modinagar, Uttar Pradesh, India

Tushar Garg Department of Information Technology, HMRITM, GGSIPU, New Delhi, India

Devottam Gaurav IIT Delhi, New Delhi, India

Siddharth Gautam Department of Information Technology, HMR Institute of Technology and Management, GGSIPU, Delhi, India

Martin Wulf Gerdes Department of Information and Communication Technology, Centre for E-Health, University of Agder, Kristiansand, Norway

Arijit Ghosal Department of Information Technology, St. Thomas' College of Engineering and Technology, Kolkata, West Bengal, India

Mohona Ghosh Department of Information Technology, Indira Gandhi Delhi Technical University for Women, New Delhi, Delhi, India

Pronab Ghosh Department of Computer Science and Engineering, Daffodil International University, Dhaka, Bangladesh

Abhishek Goel HMRITM, Delhi, India

Anirban Gogoi Department of Electronics and Communication Engineering, Kaziranga University, Assam, India

Ashim Jyoti Gogoi Department of Electronics and Communication Engineering, Kaziranga University, Assam, India

Romen Gogoi Department of Electronics and Communication Engineering, Kaziranga University, Assam, India

Saksham Goyal PES University, Bengaluru, India

Harshul Gupta Department of Information Technology, HMRITM, GGSIPU, New Delhi, India

Aman Gurung Department of Information Technology, HMRITM, GGSIPU, New Delhi, India

M. Guruprasad PES University, Bengaluru, India

Sanjib Halder Department of Computer Science, The Bhawanipur Education Society College, Kolkata, India

Syed Akhter Hossain Department of Computer Science and Engineering, University of Asia Pacific, Dhaka, Bangladesh;
Department of Computer Science and Engineering, Daffodil International University, Dhaka, Bangladesh

Eric Howard Department of Physics and Astronomy, Macquarie University, Sydney, NSW, Australia;
Centre for Quantum Dynamics, Griffith University, Brisbane, Australia

Sadia Islam Department of Computer Science and Engineering, United International University, Dhaka, Bangladesh

Sayeed Islam Department of Electrical and Electronics Engineering, Manarat International University Bangladesh, Ashulia, Khagan, Savar, India

R. Jagadeesh CMR University, Bengaluru, India

Neelesh Jain Department of Computer Science and Engineering, Sagar Institute of Research Technology and Science, Bhopal, India

Mayurakshi Jana Department of Computer Science, Bijay Krishna Girls' College, University of Calcutta, Howrah, WB, India

F. M. Javed Mehedi Shamrat Department of Software Engineering, Daffodil International University, Dhaka, Bangladesh

G. Jenilasree Department of Computer Applications, University College of Engineering, Anna University BIT Campus, Tiruchirappalli, Tamil Nadu, India

Er. Jinu Thomas Department of Computer Science and Engineering, Saintgits College of Engineering, Kottayam, Kerala, India

Priyanka Jojan Department of Computer Science and Engineering, Maulana Azad National Institute of Technology, Bhopal, India

Sruthi Anna Joseph Department of Computer Science and Engineering, Saintgits College of Engineering, Kottayam, Kerala, India

C. H. Jyothi Priya Saveetha School of Engineering, Chennai, India

Shrenik Kabade Pune Vidyarti Gruha College of Engineering, Pune, India

Ramesh M. Kagalkar KLE College of Engineering and Technology, Affiliated to VTU, Belagavi, Chikodi, Karnataka, India

Ompriya Kale Department of Computer Engineering, Pimpri Chinchwad College of Engineering and Research, Ravet, Pune, India

Navia Kamal Department of Information Technology, HMR Institute of Technology and Management, Delhi, India

Sneha Karamadi Department of Computer Science and Engineering, K. S. Institute of Technology, Bangalore, India

Amiya Karmakar Department of Computer Science and Engineering, Maulana Abul Kalam Azad University of Technology, Kalyani, West Bengal, India

Suvrajit Karmaker Daffodil International University, Dhaka, Bangladesh

Veeranna Kotagi Department of Computer Science and Engineering, East West College of Engineering, Bangalore, India

Nimrita Koul REVA University, Bangalore, India

Abhijeet Kumar Department of Electronics and Communication Engineering, HMR ITM, Delhi, India

Aditya Kumar Department of Information Technology, Dwarkadas J. Sanghvi College of Engineering, Mumbai, India

Manoj Kumar School of Computer Science, University of Petroleum and Energy Studies (UPES), Dehradun, India

Nitish Kumar Department of Computer Science and Engineering, RTC Institute of Technology, Ranchi, India

Praveen Kumar Amity University Uttar Pradesh, Noida, India

T. N. Lokesh Kumar PES University, Bengaluru, India

Sonali S. Lunawat Department of Computer Engineering, Pimpri Chinchwad College of Engineering and Research, Ravet, Pune, India

Prathamesh R. Mahadik Pune Vidyarti Gruha College of Engineering, Pune, India

Mansi Mahendru Department of Computer Science and Engineering, Amity University Uttar Pradesh, Noida, U.P., India

Imran Mahmud Department of Information Technology and Management, Daffodil International University, Dhaka, Bangladesh

J. Maiti Department of Industrial & Systems Engineering, IIT Kharagpur, Kharagpur, India

Ritajit Majumdar Advanced Computing & Microelectronics Unit, Indian Statistical Institute, Kolkata, India

Anup Majumder Department of Computer Science and Engineering, Jahangirnagar University, Dhaka, Bangladesh

Ratna Mandal Department of Computer Application, Institute of Engineering and Management, Kolkata, West Bengal, India

Deepak Mangal GLAU Mathura, Mathura, UP, India

Sunilkumar S. Manvi REVA University, Bangalore, India

M. K. Mariam Bee Department of Electronics and Communication Engineering, Saveetha School of Engineering (SIMATS), Chennai, Tamil Nadu, India

Santiago Martinez Department of Health and Nursing Science, Centre for E-Health, University of Agder, Kristiansand, Norway

Abu Kaisar Mohammad Masum Department of Computer Science and Engineering, Daffodil International University, Dhaka, Bangladesh

Akash Mehta Department of Computer Science, The Bhawanipur Education Society College, Kolkata, India

Pabitra Mitra Department of Computer Science & Engineering, IIT Kharagpur, Kharagpur, India

Hardik Modi Department of Electronic and Communication, CHARUSAT University, Changa, Gujarat, India

Kartick Chandra Mondal Department of Information Technology, Jadavpur University, Kolkata, India

Saronyo Lal Mukherjee Department of Computer Science, The Bhawanipur Education Society College, Kolkata, India

Bipasha Mukhopadhyay University of Engineering and Management, Kolkata, India

Hemanth Kumar Nalajala Department of ECE, School of EEE, SASTRA Deemed University, Thanjavur, India

Biswadeep Deb Nandy Samsung R&D Institute, Bengaluru, India

M. S. Narasimha Murthy Department of Computer Science and Engineering, East West Institute of Technology, Bangalore, India

Abhishek Narayanan Department of Computer Science and Engineering, PES University, Bangalore, India

S. Natarajan Department of Computer Science and Engineering, PES University, Bangalore, India

Sarmistha Neogy Jadavpur University, Kolkata, West Bengal, India

Niharika JIS College of Engineering, Kalyani, India

Naimul Islam Nobel Department of Software Engineering, Daffodil International University, Dhaka, Bangladesh

Md. Nurul Islam Department of Electrical and Electronics Engineering, Manarat International University Bangladesh, Ashulia, Khagan, Savar, India

V. M. Padmapriya Department of ECE, School of EEE, SASTRA Deemed University, Thanjavur, India

Sanjay Kumar Pal Department of Computer Science and Engineering, Maulana Abul Kalam Azad University of Technology, Kalyani, West Bengal, India

Piyush Kumar Pareek Department of Computer Science and Engineering, East West Institute of Technology, Bangalore, India

Kanu Patel Department of Computer Engineering, CHARUSAT University, Changa, Gujarat, India

Anuradha Pathak Department of Computer Science and Engineering, RTC Institute of Technology, Ranchi, India

Sandip Paul Electronics and Communication Sciences Unit, Indian Statistical Institute, Kolkata, India

Sujoy Paul University of Engineering and Management, Kolkata, India

Rintu Ponnachan Department of Computer Science and Engineering, Saintgits College of Engineering, Kottayam, Kerala, India

Aishwarya Poomuttam Sreedas Computer Science and Engineering Department, PES University, Bengaluru, India

Deepali Prabhu PES University, Bengaluru, India

Anima Pramanik Department of Industrial & Systems Engineering, IIT Kharagpur, Kharagpur, India

Abhishek Prasad Department of Computer Science and Engineering, PES University, Bangalore, India

Andreas Prinz Department of Information and Communication Technology, Centre for E-Health, University of Agder, Kristiansand, Norway

Priya R. L. Department of Computer Engineering, Vivekanand Education Society's Institute of Technology, Chembur, Mumbai, Maharashtra, India

R. Puviarasi Saveetha School of Engineering, Chennai, India;
Saveetha Institute of Medical and Technical Sciences, Kuthambakkam, India

A. K. M. Shahariar Azad Rabby Daffodil International University, Dhaka, Bangladesh

Chhiteesh Rai AKTU, Lucknow, India

Manish Raj GLAU Mathura, Mathura, UP, India

Sundararaman Rajagopalan Department of ECE, School of EEE, SASTRA Deemed University, Thanjavur, India

L. Rajesh Department of Electronics Engineering, Madras Institute of Technology, Anna University, Chennai, India

Aniket Rana Branch of Information Technology, HMRITM, Delhi, India

Abijna Rao Department of Computer Science and Engineering, PES University, Bangalore, India

Meena Rao Department of Electronics and Communication Engineering, Maharaja Surajmal Institute of Technology, New Delhi, India

Akhtar Rasool Department of Computer Science and Engineering, Maulana Azad National Institute of Technology, Bhopal, India

Kaushik Ravichandran Computer Science and Engineering Department, PES University, Bengaluru, India

Kumar Sankar Ray Electronics and Communication Sciences Unit, Indian Statistical Institute, Kolkata, India

Amirtharajan Rengarajan Department of ECE, School of EEE, SASTRA Deemed University, Thanjavur, India

Sathi Roy Asian International School, Howrah, West Bengal, India

E Saahitya PES University, Bengaluru, India

Amal Saha NetWest Markets, Gurugram, India

Diganta Saha Computer Science and Engineering Department, Jadavpur University, Kolkata, India

Shaswata Saha Jadavpur University, Kolkata, West Bengal, India

Gadadhar Sahoo Department of Computer Science and Engineering, Indian Institute of Technology (ISM), Dhanbad, India

Umesh Chandra Sahoo School of Infrastructure, Indian Institute of Technology, Bhubaneswar, India

P. Sai Ankith Department of Electronics and Communication Engineering, Saveetha School of Engineering (SIMATS), Chennai, Tamil Nadu, India

V. Sai Siddharth Department of Ocean Engineering and Naval Architecture, IIT Kharagpur, Kharagpur, India

Sohail Saif Department of Computer Science and Engineering, Maulana Abul Kalam Azad University of Technology, Kolkata, West Bengal, India

Kavya Ann Salbi Department of Computer Science and Engineering, Saintgits College of Engineering, Kottayam, Kerala, India

Saikat Samanta Department of Computer Science and Engineering, Bengal College of Engineering, Durgapur, West Bengal, India

N. Samreen Fatima Saveetha School of Engineering SIMATS, Chennai, India

Aayushi Sanghi PES University, Bengaluru, India

Amrita Sarkar Department of Computer Science and Engineering, Sarala Birla University, Ranchi, India

Sobhan Sarkar Management Science, Business School, The University of Edinburgh, Edinburgh, UK

Subir Kumar Sarkar Department of Electronics and Tele Communication Engineering, Jadavpur University, Kolkata, India

K. Satyanarayan Reddy ISE Department, Cambridge Institute of Technology, Bengaluru, India

Manan Saxena Delhi Technological University, Shahbad Daulatpur, Delhi, India

Meegan Scariah Department of Computer Science and Engineering, Saintgits College of Engineering, Kottayam, Kerala, India

Soumya Sen A.K. Choudhury School of I.T, University of Calcutta, Kolkata, India

Subhabrata Sengupta Department of Information Technology, Institute of Engineering and Management (IEM), Kolkata, India

Rishabh Shah IT Department, Birla Vishwakarma Mahavidyalaya, Anand, Gujarat, India

Vatsal Shah IT Department, Birla Vishwakarma Mahavidyalaya, Anand, Gujarat, India

Gungun Sharma Department of Information Technology, Indira Gandhi Delhi Technical University for Women, New Delhi, Delhi, India

Puneet Sharma Department of Computer Science and Engineering, Amity University, Lucknow, India

Soma Sharma JIS College of Engineering, Kalyani, India

S. Sharmiladevi Department of Computer Science, Pondicherry University, Puducherry, India

Swakkhar Shatabda Department of Computer Science and Engineering, United International University, Dhaka, Bangladesh

N. Shenbagavadivu Department of Computer Applications, University College of Engineering, Anna University BIT Campus, Tiruchirappalli, Tamil Nadu, India

Praj Shete Pune Vidyarti Gruha College of Engineering, Pune, India

Ankush Shetty Department of Computer Engineering, Vivekanand Education Society's Institute of Technology, Chembur, Mumbai, Maharashtra, India

N. S. Shirsat Pune Vidyarti Gruha College of Engineering, Pune, India

K. K. Shukla Department of Computer Science and Engineering, I.I.T.(B.H.U.), Varanasi, India

Areesha Fatima Siddiqui Department of Computer Engineering, Jamia Millia Islamia, New Delhi, India

Akhilesh Kumar Singh GLAU Mathura, Mathura, UP, India

Rajesh Singh Department of Computer Science and Engineering, RTC Institute of Technology, Ranchi, India

Siddharth Singh Department of Information Technology, Dwarkadas J. Sanghvi College of Engineering, Mumbai, India

Saurabh Singhal GLAU Mathura, Mathura, UP, India

Aditya Vikram Singhania Department of Computer Science, The Bhawanipur Education Society College, Kolkata, India

Abhishek Sinha PES University, Bengaluru, India

S. Siva Sathya Department of Computer Science, Pondicherry University, Puducherry, India

R. Sivaraman Department of ECE, School of EEE, SASTRA Deemed University, Thanjavur, India

Sandhya Soman Department of Computer Science, Shri JJT University, Jhunjhunu, Rajasthan, India;
Department of Computer Science, Kristu Jayanti College, Bangalore, India

Kapil Kumar Soni Department of Computer Science and Engineering, Maulana Azad National Institute of Technology, Bhopal, India

Sonam Soni Amity University Uttar Pradesh, Noida, India

P. Soubhagyalakshmi Visvesvaraya Technological University, Belagavi, India

B. Sowmya Department of ECE, School of EEE, SASTRA Deemed University, Thanjavur, India

A. Sridevi Department of ECE, School of EEE, SASTRA Deemed University, Thanjavur, India

Shailesh Sridhar PES University, Bengaluru, India

Animesh Srivastava Poornima University, Jaipur, India

Atul Kumar Srivastava Amity University Tashkent, Tashkent, Uzbekistan

Mritunjay Srivastava Department of Electronics and Communication Engineering, HMR ITM, Delhi, India

Akshara Subramaniasivam Computer Science and Engineering Department, PES University, Bengaluru, India

V. Suchitra Department of Electronics Engineering, Madras Institute of Technology, Anna University, Chennai, India

G. K. Suhas East West Institute of Technology, Bengaluru, India

Marusani Sumanjali Department of ECE, School of EEE, SASTRA Deemed University, Thanjavur, India

Satyam Shivam Sundaram PES University, Bengaluru, India

Deepjyoti Talukdar Department of Electrical Engineering, Kaziranga University, Assam, India

Bhuvan Taneja Department of Computer Science and Engineering, HMRITM, New Delhi, India

Zarrin Tasnim Department of Software Engineering, Daffodil International University, Dhaka, Bangladesh

P. Thayal Sankar Department of Electronics Engineering, Madras Institute of Technology, Anna University, Chennai, India

Sanju Tiwari Universidad Autonoma de Tamaulipas, Tamaulipas, Mexico

Varun Tiwari Trinity Institute of Professional Studies, GGSIP University, New Delhi, Delhi, India

Bhavish Tyagi HMRITM, Delhi, India

Nitin Tyagi Branch of Information Technology, HMRITM, Delhi, India

Har Narayan Upadhyay Department of ECE, School of EEE, SASTRA Deemed University, Thanjavur, India

P. Upadhyay DST-Centre for Interdisciplinary Mathematical Sciences (CIMS), Banaras Hindu University (B.H.U.), Varanasi, India

S. K. Upadhyay Department of Mathematical Sciences I.I.T. (B.H.U.), DST-Centre for Interdisciplinary Mathematical Sciences (CIMS), Banaras Hindu University (B.H.U.), Varanasi, India

Vikas Rao Vadi Don Bosco Technical School, Bosco Technical Training Society, New Delhi, Delhi, India

Aditya Venkatesh PES University, Bengaluru, India

Ravi Verma Department of Computer Science and Engineering, Radharaman Institute of Technology and Sciences, Bhopal, India

K. Vidhya Department of Electronics and Communication Engineering, Saveetha School of Engineering (SIMATS), Chennai, Tamilnadu, India

Gautam Singh Yadav Department of Computer Science and Engineering, ASET, Amity University, Noida, India

Harshit Yadav Branch of Information Technology, HMRITM, Delhi, India

Layered Ensemble Learning for Effective Binary Classification

Md. Abulkalam Azad, Sadia Islam, Dewan Md. Farid, and Swakkhar Shatabda

Abstract Often when a single learner fails to produce accurate results in prediction of class labels in supervised learning tasks, an ensemble of classifiers is applied. Ensemble classifiers have been quite successful in recent years to solve complex machine learning problems in different domains. In this paper, we present a novel multi-layered ensemble classifier for solving binary classification tasks. Our layered ensemble technique performs the classification in multiple stages or layers. In each layer a single classifier is applied that divides the instances into two sets according to the predicted labels. In the later stages or layers these subsets undergo further learning by another set of single classifiers and this process goes on until the classification performance is satisfactory or the subsets produced in these way are too small. The over-fitting issues are dealt by the termination criteria. On a selected benchmark datasets used to test our method, it performs significantly better compared to other state-of-the-art ensemble classifier.

Keywords Classification · Ensemble learning · Layered architecture · Supervised learning

Md. A. Azad (✉) · S. Islam · D. Md. Farid · S. Shatabda
Department of Computer Science and Engineering,
United International University, Dhaka, Bangladesh
e-mail: mazad141081@cse.uiu.ac.bd

S. Islam
e-mail: islamsadia88@gmail.com

D. Md. Farid
e-mail: dewanfarid@cse.uiu.ac.bd

S. Shatabda
e-mail: swakkhar@cse.uiu.ac.bd

J. Tavares et al. (eds.), *Emerging Technologies in Data Mining and Information Security*,
Lecture Notes in Networks and Systems 164,
https://doi.org/10.1007/978-981-15-9774-9_1

1 Introduction

Ensemble classification method is an way to enhance the performances of single classifiers to achieve more accuracy score or to classify instances more correctly [1]. Sometimes in many cases, when a single classifier fails to predict the class value of some instances, using a different classifier for those instances may result in better predicted output [2]. Therefore, instead of using a single classifier for a supervised learning task, combination of two or more classifiers is used to classify instances more correctly. This combination of classifiers has become an active area in machine learning research and been applied in various field with greater success. The combination may have different names and one of them is ensemble classifier [3]. In ensemble classification, the output of different single classifiers are combined in different ways in an aim to achieve better accuracy or classification power to classify more accurately than a single classifier. There are many existing ensemble classifiers such as AdaBoost, Bagging, Random forest, Voting Classifier etc. Each of these algorithms uses different strategies to combine multiple learners effectively.

Among all of the existing ensemble classifiers, the Voting classifier is one of the widely used classifiers in research field [4]. It mainly follows the strategy to combine the results of different single classifiers and to sum up with the final result by doing majority voting [2]. Majority voting is basically taking the result as final result which has got more vote as output when we have used single classifiers to classify. There are mainly two types of existing approaches to analyze the results of single classifiers to produce a final strong output. One is to hard way in which the result which has got more vote is being selected. The other one is soft way, in which the class that has got high probability for an instance is being selected as a final output of voting. However, this ensemble method uses the whole train set to train each of the single classifiers. Which means each of the single classifiers that has been designed to combine for better accuracy, will use the whole train set for each of them in the training phase. As a result, when the dataset is large, the computational cost of this method will be high. Among the other two most successful ensemble classifiers are Random Forest and Adaboost. Random Forest uses the bagging technique and randomly samples the feature space and creates a large number of decision trees and aggregate or combine the results to produce the final prediction. On the other hand, AdaBoost algorithm dynamically updates the weight distribution of the instances and in an iterative fashion learns classifiers on the whole dataset. All these algorithms are learned over the whole set of instances.

In this paper, we focus on the problem to reduce the computational cost along with the better accuracy but also to keep the algorithm as simple to use more. This paper presents an ensemble classifier algorithm, combining single classifiers in layers to produce result with better accuracy and less computational cost. In each layer, our proposed ensemble divides the set of instances into two subsets according to the binary labels that are predicted by a single classifier. In subsequent layers each of these subsets are further passed through a single classifier and this process goes recursively until a termination criteria is met. The algorithms terminates if the size of the sets are too small or satisfactory results are achieved. We have tested the performance of the ensemble classifier on several benchmark test datasets using cross fold validation. Multi-layered ensemble classifier significantly outperforms the competitor voting classifiers on all of these datasets.

Rest of the paper is organized as following: Sect. 2 describes the related work; Sect. 3 presents our layered ensemble classifier; Sect. 4 presents the experimental results and the paper concludes with a summary and indication of future work in Sect. 5.

2 Related Work

Ensemble classifier is playing an active role in the field of research. Many researchers are working in this field to find a better combination of classifiers in some way to construct a new classifier for predicting unknown instances more accurately [5]. The basic concept before designing an ensemble classifier refers three major sensitive points of ensemble classifier—combining classifiers in a proper way, incorporate weak learners to produce a strong classifier and including experts [5]. The existing state of the art methods for ensemble classifiers are bagging, boosting, voting etc. We briefly discuss these approaches below.

Bagging basically indicates a package of algorithms which converts a weak learner to a strong one. Bagging method includes AdaBoost algorithm which is the first and mostly used boosting algorithm [6]. AdaBoost basically follows the strategy of creating a set of weak learners and also maintains a weight with them to take decision giving more priority depending on the weight in next iterations [7].

In case of bagging method of ensemble, random forest is a well-known algorithm which has been introduced in 2001 and being used till now [8]. Random forest mainly adds a layer of randomness with the basic bagging algorithm. Random forest can be said as a user friendly layer algorithm as it requires only two parameters and also not sensitive to their values. This algorithm also gives the benefit of robustness against over-fitting [8].

If we focus on the Voting ensemble, we can find it as one of the most simple and widely used ensemble classifier. Majority voting algorithm is known as the easiest algorithm under some arguments [9]. This algorithm is also being used as support for other algorithms such as to handle missing values. In collaboration with AdaBoost algorithm, majority voting is used even to design a hybrid ensemble to detect credit card fraud [10]. Majority voting is mostly used to learn from the crowds [9]. However, when we analyze the performance of majority voting, we find that the computational cost of this algorithm is high as it uses the whole dataset for each of the basic single classifiers in the training phase.

Therefore, in this paper we have paid more attention on the fact to increase the performance of majority voting classifier by designing a new layered classifier which will work layer by layer to give a better prediction value with a comparatively low computational cost than majority voting classifier.

3 Layered Ensemble Classifier (LEC)

In this section, we present our layered ensemble classifier. In our proposed algorithm, we construct a multiple layer of single classifiers. In this paper, the single classifiers that we have used to combine in order to make an effective ensemble classifier are Support Vector Machine (SVM) [11, 12], K-nearest Neighbor (KNN) and Decision tree (DT). Support Vector Machines are maximum margin classifiers that separates the instances in the hyperplane using a margin of maximum width. Application of SVM with kernel functions have made it successful to solve non-linear problems as well. KNN depends on the neighborhood of an instance in the hyperplane and is a lazy instance based method. KNN is more widely used algorithm when the size of the train and test set is not very large [13]. In Decision tree algorithm, a tree is made where in every node of the tree indicates a probability or a decision to determine where to go next. Different scaled data can also be handled using DT algorithm [14]. In the construction procedure of LEC, we randomly choose a classifier out of the selected three classifiers in the first layer (layer 0) and train it with a train dataset. Now, we test this fitted model on the same train dataset and split the dataset into two sub-datasets according to the prediction of the model at layer 0. There will be two train sub-datasets in layer 1 as we are working on binary classification problem. Secondly, in layer 1, we again randomly choose a classifier in a same way as in the first layer and train it with one of the two datasets we get from layer 0 and the same action takes place for the other sub-dataset. Consequently, we get 4 smaller sub-datasets out of the two from the predictions of two models in layer 1. This process continues and the layer grows deeper until the stopping criteria meets at any layer. As soon as any stopping criterion is met, the process is eventually stopped at that moment. We discuss about stopping criteria later in this section. For example, see in Fig. 1, there is a train dataset with 14 instances in layer 0 and SVM classifier has been chosen randomly for this layer, and then train the classifier. Now testing the classifier on the same dataset, it predicts 9 instances as 'yes' out of 14 and rest of 5 instances as 'no' regardless of their actual class values. Here, the dataset is split up into two sub-datasets of sizes of 9 and 5 instances respectively according to the prediction and send them to the next layer for further layer generation. The same process takes place in layer 1 and so on to increase the number of layers until it meets the stopping criteria.

Fig. 1 Construction of LEC

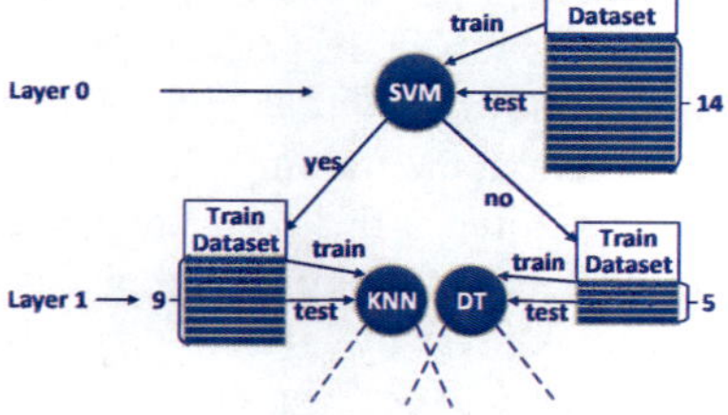

However, we set three different stopping criteria in our proposed algorithm to prevent from being an over-fitted model. First, we check the accuracy on train dataset of a classifier in each layer; if the accuracy goes beyond the threshold value (95 in our case) we immediately stop here. Secondly, if a classifier predicts the same class value for all the instances of current dataset which refers to that it is not possible to split out the current dataset according to our proposed method, we stop here. Finally, if any splitting sub-dataset has truly same value for all the instances, we do not let to generate this layer as this can reduce the accuracy of our proposed method. See the pseudo-code of our proposed criteria in Algorithm 1.

Algorithm 1 Construction of the LEC.

Input: D = {x1,x2,x3......xn}
Output: LEC
Method:

```
1: Classifier ← random classifier(SVM, KNN, DT)
2: Classifier ← fit(D)
3: LEC ← insert(Classifier)
4: Q ← LEC[0]
5: Data_Q ← D
6: while Q ≠ empty do
7:    node ← Q.pop()
8:    D ← Data_Q.pop()
9:    predictions ← node.predict(D)
10:    accuracy ← predictions' accuracy
11:    if accuracy < threshold_value then
12:       D1, D2 ← split D according to predictions
13:       if length(D1) or length(D2) = 0 then
14:          Go back to line 6
15:       end if
16:       if true labels of all instances are same then
17:          Go back to line 6
18:       end if
19:       Classifier ← random(SVM, KNN, DT)
20:       Classifier ← fit(D1)
21:       LEC ← append(Classifier)
22:       Q ← append(Classifier)
23:       Data_Q ← append(D1)
24:       Classifier ← random(SVM, KNN, DT)
25:       Classifier ← fit(D2)
26:       LEC ← append(Classifier)
27:       Q ← append(Classifier)
28:       Data_Q ← append(D2)
29:    end if
30: end while
31: return LEC
```

Algorithm 2 Prediction by the LEC.

Input: Test dataset D, the LEC and the classLabels
Output: predictions
Method:

```
N ← length of D
L ← length of LEC
for i = 1 to N do
   j = 1
   while j ≤ L do
      prediction ← LEC[j].predict(D[i])
      if prediction = classLabels[0] then
         j = 2 * j
      else
         j = 2 * j + 1
      end if
      predictions[i] ← prediction
   end while
end for
return predictions
```

The input for the algorithm is the train dataset, D and it produces the LEC as output which is actually an array of different single classifier models representing the trained LEC. In line 1, the algorithm randomly selects a single classifier and train the classifier on D in line 2. In 3 and 4, the classifier is inserted respectively into the LEC as layer 0 of our model and into the Q in order to generate further layer of the model. Also the D is pushed into the Data_Q in line 5 to maintain the right sequence of sub-data for a classifier in each layer. In line 9, it predicts on D it gets from line 8 by the classifier 'node' which is from line 7 and keeps result in predictions array. The accuracy of the classifier of this layer is calculated in line 10. However, it checks if the accuracy

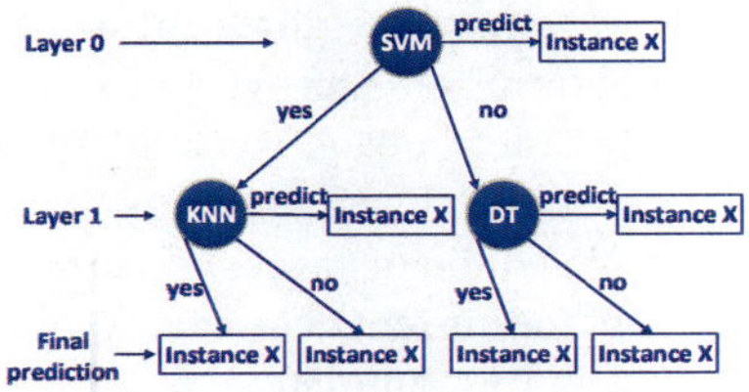

Fig. 2 Class label prediction of LEC

is more than the threshold value in line 11 which is our first stopping criteria. If the stopping condition is not met, the dataset D is split out into two sub-datasets, D1 and D2 respectively according to the predictions and from line 13–18, other two stopping conditions are checked. Then, from line 19–28, the operations are done for classifiers for the next layer in a same way as of line 1–5. Finally, Line 31 returns the trained LEC after completing all the calculation.

In fact, after the end of this process, we get our desired Layered Ensemble Classifier (LEC). Therefore, we test our model by predicting on a separate test dataset. The prediction takes place instance by instance as follows in Fig. 2. First, an instance comes to the classifier at layer 0, then according to prediction it goes to one of the classifiers in layer 1, and the final prediction is made here as it is the last layer for this example. However, if there are more layers in the model, the test instance has to go through all the layers to get the final prediction.

Now take a look at the pseudo-code of our prediction mechanism by LEC in algorithm 2. The input consists of the test dataset D, the LEC itself, and the array 'classLabels' which contains only two class values and the output is the prediction vector. In line 6, the i'th instance is predicted by j'th classifier of the LEC and depending on the prediction value, the classifier of the next layer will be chosen by changing the value of j in line 7–11. Hence, it changes the prediction of i'th instance layer to layer in line 12 but holds the value of final layer at the end. Finally, it returns the predictions of the dataset in line 15.

4 Experimental Results

In this section, we present the experimental results and necessary discussion.

4.1 Datasets

To test the LEC, we run our method on 10 binary datasets. We preprocessed these datasets by handling missing value, removing a column manually if it has nominal feature value. In fact, we avoid imbalanced datasets to test our proposed method.

Table 1 Datasets description

Dataset	Attributes	Instances
Australian	10	690
What makes A best seller item	42	337
Heart disease	14	303
Spectfheart	45	267
Bank loan modelling	14	5000
Spam	35	1200
Strategeion resume skills	224	619
Wdbc	30	569
Breast-cancer-wisconsin	9	699
Bupa	6	345

In the following Table 1, it shows the description of every dataset we use in this work. Most of the datasets were taken from Keel, kaggle and UCI Machine Learning Repository [15]. All of these datasets chosen are from binary classification task group and none of the features are nominal here.

4.2 Performance Measure

To measure the performance of the classifier, we have calculated the accuracy for the classifier. Accuracy refers to the percentage of correctly classified test instances. Accuracy mainly depends on two major factors which are True positive value (TP) and True negative value (TN). TP is the factor which indicates the correctly classified positive instances whereas TN refers to the negative instances that are correctly classified. Hence, according to [16] the equation to calculate accuracy is,

$$\text{Accuracy} = \frac{\text{TP+TN}}{\text{TN+TP+FN+FP}}$$

Here, FN and FP denotes false negatives and false positives. We have also considered three other measures: Sensitivity, Specificity and Precision. They are also defined in terms of discussed earlier.

$$\text{Sensitivity} = \frac{\text{TP}}{\text{TP} + \text{FN}} \quad \text{Specificity} = \frac{\text{TN}}{\text{TN} + \text{FP}} \quad \text{Precision} = \frac{\text{TP}}{\text{TP} + \text{FP}}$$

Here, all these measures have values in the range [0, 1]. The best or perfect classifier has the highest value 1 and the worst would have value of 0.

Table 2 Performance comparison of LEC and voting classifier

	Accuracy		Sensitivity		Specificity		Precision	
Dataset	LEC	Voting	LEC	Voting	LEC	Voting	LEC	Voting
Austrailian	**74.0579**	72.3188	**0.82**	0.81	**0.62**	0.61	**0.73**	0.72
What makes A best ...	**95.2941**	88.1299	1.00	1.00	**0.63**	0.07	**0.95**	0.88
Heart disease	**65.9945**	51.0655	**0.66**	0.41	**0.66**	0.60	**0.62**	0.46
Spectfheart	**76.4710**	71.5933	**0.33**	0.07	0.88	0.88	**0.41**	0.14
Bank loan modelling	**70.6000**	68.7400	**1.00**	0.95	0.00	0.06	0.71	0.71
Spam	**84.8333**	77.9166	**0.91**	0.77	**0.78**	0.77	**0.80**	0.77
Strategeion resume ...	**74.1162**	61.8581	**0.76**	0.68	**0.72**	0.55	**0.74**	0.61
Wdbc	93.8487	93.4979	0.95	0.96	**0.91**	0.88	**0.94**	0.93
Breast-cancer-wisconsin	**86.2826**	64.6793	0.91	0.91	**0.75**	0.12	**0.87**	0.66
Bupa	**67.5362**	66.3768	**0.51**	0.47	0.79	0.8	**0.64**	0.63

4.3 Performance Analysis

We have implemented our proposed method in scikit-learn and used KFold cross validation to get the result on each dataset with $k = 5$. Indeed, we compare our result with the Voting ensemble classifier as 'hard' voting and get different scenarios. Almost in all datasets, the LEC outperforms Voting classifier with very higher accuracy than the Voting classifier, and just in one dataset, both the algorithms perform similarly but still LEC is giving better performance. Since the LEC chooses classifier in each layer randomly, every-time we train the LEC it gives different results on the same dataset. Hence we train our model a few times on each dataset, get few results for each training, and choose the one which gives a comparable accuracy to Voting classifier. The performance comparison for both LEC and Voting ensemble classifier on the 10 datasets is given in Table 2. We note better performance of LEC over voting classifier in terms of the performance metrics as well. Moreover, LEC always needs less computational work than the Voting classifier. Because, unlike the Voting classifier, LEC trains all the classifiers on different sub-datasets of a single dataset whereas Voting classifier trains each classifier on the whole dataset for every classifiers.

5 Conclusion

In the results, we can see that LEC is giving better accuracy in all of the datasets compared to the Voting classifier and certainly the LEC needs less computational work than Voting. Therefore, we expect that the LEC will be used in many applications replacing the Voting classifier. In addition, we will try to improve the performance of this algorithm working with the stopping criteria to increase the accuracy by constructing the tree to proper layer and not to over-fit at the same time. More importantly, we want to replace choosing a classifier randomly on each layer by applying some methods in which the best classifier will be selected for each layer. Consequently, the LEC would need to train on a single dataset for a single time to get the best fitted LEC.

References

1. G.D. Thomas, *Ensemble learning* (2002)
2. D. Ruta, B. Gabrys, Classifier selection for majority voting. Inf. Fusion **6**(1), 63–81 (2005)
3. L.I. Kuncheva, C.J. Whitaker, Measures of diversity in classifier ensembles and their relationship with the ensemble accuracy. Mach. Learn. **51**(2), 181–207 (2003)
4. Y. Zhang, H. Zhang, J. Cai, B. Yang, A weighted voting classifier based on differential evolution, in *Abstract and Applied Analysis*, vol. 2014 (Hindawi, 2014)
5. T.G. Dietterich, Ensemble methods in machine learning, in *International Workshop on Multiple Classifier Systems*, pp. 1–15 (Springer, 2000)
6. Y. Freund, R.E. Schapire, et al., Experiments with a new boosting algorithm, in *Icml*, vol. 96, pp. 148–156 (Citeseer, 1996)
7. R. Wang, Adaboost for feature selection, classification and its relation with svm, a review. Phys. Procedia **25**, 800–807 (2012)
8. A. Liaw, M. Wiener et al., Classification and regression by randomforest. R News **2**(3), 18–22 (2002)
9. T. Tian, J. Zhu, Max-margin majority voting for learning from crowds, in *Advances in Neural Information Processing Systems*, pp. 1621–1629 (2015)
10. K. Randhawa, C.K. Loo, M. Seera, C.P. Lim, A.K. Nandi, Credit card fraud detection using AdaBoost and majority voting. IEEE Access **6**, 14277–14284 (2018)
11. K.-R. Muller, S. Mika, G. Ratsch, K. Tsuda, B. Scholkopf, An introduction to kernel-based learning algorithms. IEEE Trans. Neural Netw. **12**(2), 181–201 (2001)
12. W.S. Noble, What is a support vector machine? Nat. Biotechnol. **24**(12), 1565 (2006)
13. Q. Kuang, L. Zhao, A practical GPU based KNN algorithm, in *Proceedings. The 2009 International Symposium on Computer Science and Computational Technology (ISCSCI 2009)*, p. 151 (Citeseer, 2009)
14. M. Pal, P.M. Mather, An assessment of the effectiveness of decision tree methods for land cover classification. Remote Sens. Environ. **86**(4), 554–565 (2003)
15. A. Asuncion, D. Newman, *UCI Machine Learning Repository* (2007)
16. W. Zhu, Na. Zeng, N. Wang et al., Sensitivity, specificity, accuracy, associated confidence interval and roc analysis with practical sas implementations, in *NESUG Proceedings: Health Care and Life Sciences*, vol. 19, p. 67, Baltimore, Maryland(2010)

A Preliminary Study of Knowledge Graphs and Their Construction

Sanju Tiwari, Devottam Gaurav, Animesh Srivastava, Chhiteesh Rai, and Kumar Abhishek

Abstract Knowledge graphs are developed to extend and organize the Web of data application. It helps to store the data in structured form and also present the entities and their relations explicitly. An entity is considered as a knowledge graph unit such as a person, movie, and city or else that need to describe. All entities of knowledge graphs are presented with their attributes like name, age, nationality, birthdates, etc., and further connected with other entities during the construction of knowledge graph. Knowledge graph construction approaches and techniques play an important role to develop the knowledge graphs. This paper initially presents the study of some existing knowledge graphs like YAGO, DBPedia, ConceptNet, FrameNet, etc., and also focused on knowledge graph construction approaches and techniques. There are 21 knowledge graphs are analysed and presented at one place with their applications.

Keywords Knowledge graph · Knowledge extraction · Knowledge fusion · Triples · Ontology

---

S. Tiwari (✉)
Universidad Autonoma de Tamaulipas, Tamaulipas, Mexico
e-mail: tiwarisanju@ieee.org

D. Gaurav
IIT Delhi, New Delhi, India
e-mail: gauravpurusho@gmail.com

A. Srivastava
Poornima University, Jaipur, India
e-mail: er.animesh10@gmail.com

C. Rai
AKTU, Lucknow, India
e-mail: chhiteesh.rai@gmail.com

K. Abhishek
NIT Patna, Patna, India
e-mail: kumar.abhishek.cse@nitp.ac.in

J. Tavares et al. (eds.), *Emerging Technologies in Data Mining and Information Security*, Lecture Notes in Networks and Systems 164,
https://doi.org/10.1007/978-981-15-9774-9_2

1 Introduction

Knowledge graphs (KGs) have been presented as a common semantic network to organize and connect different entities and instances [1]. In 2012, Google confirmed the use of knowledge graph technology to enhance the present capability and efficiency of search engines in order to get better search quality and experience [2]. Traditional searching technology creates the Web pages index and performed keyword-based search from the sources. On the other hand, KGs extracts the whole knowledge to create a graph structure which is organized in nodes and edges to represent entities and relationships. As logical relation exists between knowledge in knowledge graphs, so it responses more efficiently and accurately for user's search. Knowledge graphs can be categorized into two different logical layers: pattern layer and data layer. Here, pattern layer has presented as a main layer of KGs and created above data layer to regulate the precise description of the data layer while the data layer gathers facts of knowledge in the graph database by using subject- predicate-object triples such as "**concept-property-value**" or "**concept-property-concept**" [3] as a basic unit. Concepts, properties are extracted to create the knowledge facts by the knowledge extraction and knowledge fusion process which help to remove redundancy and errors and enhance the structure by knowledge reasoning. KGs are growing continuously and most widely used in several fields of intelligent systems but still have problems in knowledge graph development and construction due to limited methodologies [4]. This paper provides a survey on existing KGs and also analyse the knowledge graph construction and techniques. To present the detailed description of KGs construction, a prescribed architecture has been followed with some relative techniques such as knowledge extraction or acquisition, knowledge fusion, knowledge retrieval, and knowledge storage. KGs construction is categorized into two parts: top-down approach and bottom-up approach. The top-down approach concentrating on knowledge schema like domain ontology and bottom-up approach concentrating on knowledge instances like Linked Open Datasets. In this paper, bottom-up approach has been discussed for knowledge construction as it is an iterative task to update the accurate knowledge. At the end of paper, KGs applications are discussed to highlight the uses of KGs in real world.

Rest of the paper is organized as follows: Sect. 2 presents a literature of existing knowledge graphs, Sect. 3 presents knowledge graph architecture and construction techniques. Section 4 highlights the problems of present knowledge graphs. Finally, we have drawn the conclusion in Sect. 5.

2 Literature Review

Several existing KGs are analysed in this section to understand their construction procedure and used techniques. We have analysed 21 knowledge graphs collectively in this study.

- ***WordNet***

WordNet [5] was generally accepted as a lexical database for the understanding of machine. At present, it is applied as ontology and a semantic network. In this knowledge base, 117,000 synset grouped as a synonym of the corresponding concept. There are various semantic relations to connect these synset with each other.

- ***HowNet***

HowNet [6] is considered as a common-sense knowledge base which opens inter-attribute relations and inter-conceptual relations of concepts in Chinese format and their English equivalents. This knowledge base has a significant concept "sememe" that is a minimal semantic unit. At present, there are 800 sememe in this knowledge graph.

- ***ConceptNet***

ConceptNet [7] is designed by a project, named by "Open Mind Common Sense (OMCS)". The aim of this project was to create and use a huge common-sense knowledge base by the support of several thousand people over the Web. It has acquired more than million facts of English from knowledge bases in different languages and over 15,000 contributors. This corpus builds a semantic network named as ConceptNet and used in common-sense reasoning, understanding, and natural language processing.

- ***FrameNet***

The FrameNet [8] was designed to maintain an English lexical database. It contains frames to describe a state, action, or situation and these frames more than 10,000 word senses. Most of them have annotated examples to present the usage and meaning of the entry. It also presents more than 170,000 sentences that are manually annotated and offered as a training dataset for several processes such as information extraction, machine interpretation, and sentiment analysis.

- ***KnowItAll***

KnowItAll [9] is designed as an information extraction system to automate the domain-independent, unsupervised tasks from the Web. It helps to extract the huge amount of facts such as academician, doctors, scientists, engineers, or politicians from the Web in a scalable way. To accomplish this task, it uses search engine which extract a huge amount of data from Web.

- ***Cyc***

Cyc [10] is the oldest knowledge graph to model the million amount of knowledge and started in the 1984 to interpret the human understandable information into machine understandable form. It is considered as a curated knowledge graph which is designed and maintained by CyCorp Inc1. It consists more than one million rules, common-sense ideas, and assertion defined by human.

- ***Freebase***

Freebase [11] has presented as a graph-shaped database for structured human knowledge. In this knowledgebase, knowledge is collected by crowd sourcing. There are millions of facts and more than ten millions of links among these facts in the current stored freebase.

- ***NELL***

It is a knowledgebase which is implemented by the ReadTheWeb Project. NELL (Never-ending language learner) [12] populating structured facts as a growing knowledge base. This knowledge base can extract 242,453 beliefs with an assessed precision of 74%. Hence, it has acquired over 50 million beliefs by understanding the web.

- ***Probase***

Probase [13] is designed to contain the probabilistic knowledge. This knowledge base stored more than 2.7 million facts which are extracted from a collection of 2.7 million facts. For modelling of uncertain and inconsistent facts, it uses probabilistic models to design a probabilistic hierarchy. Probase aims to interpret the human understanding in text by using the common-sense knowledge.

- ***YAGO***

YAGO [14, 15] is an extensible and lightweight knowledge base with high quality and coverage. It is extracted from DBpedia. YAGO is based on a model that is decidable, compatible, and extensible, with RDFS. It designs its classification from the lexical resource wordnet [5] and Wikipedia while DBPedia designs various associated knowledge graphs for all language category of Wikipedia. It contains more than 10 million concepts [cities, persons, organizations] and more than 120 million relations about these concepts.

- ***DBPedia***

DBpedia [16] as presented as a multi-lingual knowledge base. It extracts the structured facts from Wikipedia. DBPedia follows the crowd-sourced techniques to extract the structural knowledge. It has 24.9 million facts in 119 languages. This project aims to translate Wikipedia data into structural knowledge by using semantic Web technologies.

- ***Google's Knowledge Vault***

Knowledge vault (KV) automatically creates a Web-scale probabilistic knowledge base. It is a different project of Google which extract knowledge from heterogeneous sources like HTML tables, structured annotations, and text documents on the Web. It has three major components: Graph-based priors, extractors, and knowledge fusion. Like other knowledge bases, KV contains information as RDF triples. To present the probability of triple for its correctness, each triple is associated with a confidence value and only confident facts are stored into KV [17].

- ***Google Knowledge Graph***

Google using its knowledge graph[1] as a knowledge base to provide the semantic search facility to Google search engine. When user querying with Google's search engine about the subject it presents structured information by using Google Knowledge Graph. This knowledge graph has contained more than 3.5 billion concepts over 500 entities or objects.

- ***Yahoo Knowledge Graph***

The Yahoo knowledge graph [18] is a huge knowledge graph of all the objects and entities of Yahoo. It presents key information about objects (e.g. concepts, relationships, features, interlinking) and guides the knowledge-based applications. It acquired information about objects from difference heterogeneous sources. This knowledge graph is designed on both commercial sources as well as public data (Freebase and Wikipedia) for several domains. Yahoo knowledge graph uses wrappers for multiple sources and continuous examine the sources for updates. It contains 3.5 million concepts and 1.4 billion links between these concepts.

- ***Facebook Knowledge Graph***

Facebook knowledge graph [19] is a world largest graph for social entities such as locations, music, movies, and celebrities. This graph design drives for structure, correctness, coverage, and constant change of data. It contains data collected from over one billion users and also provides user-oriented query result as external data. Facebook knowledge graph provides semantic search for people, pages, check-ins, and places by using natural language.

- ***Satori***

Microsoft's Satori1 is similar to Google's knowledge graph. This knowledge graph is not publicly available but it is assumed that it contains 300 million facts and 800 million relations. It is using resource description framework for its data representation.

- ***Prospera***

PROSPERA [20] is a hadoop-based knowledge-extracting engine which integrates pattern-based collection of relational fact features with weighted MaxSat-based consistency reasoning to detect the most accepted correct facts.

- ***DeepDive***

DeepDive [21] is presented as a system that integrates machine learning concepts and database to help in the development of KBC systems. The main feature in DeepDive [22] is that machine learning and statistical inference are important tools to inject classical data issues in cleaning, extraction, and integration in a streamlined and more effective way.

- ***IBM***

IBM[2] designed its own knowledge graph based on Watson Discovery Services. IBM Watson utilizes the knowledge graph framework in two-folds: First, it permits to others to construct their knowledge graph with the predesigned knowledge graph as the core. Second, this knowledge graph directly boosts Watson Discovery, which attracts on using unstructured and structured knowledge to explore new, unexpected information and interlinked vertical services on top of discovery.

- ***eBay***

eBay[2] is developing a knowledge graph for its products to interpret semantic knowledge about entities, products, and their interlinking among them and with the external world. The facts of knowledge graph will also relate products to each other. This knowledge will be considered as a key input to analyse what a buyer is seeking for and seller is presenting and intelligently linking both.

- ***BING***

Bing[3] knowledge graph contains billions of real-world facts such as: things people, places, and local businesses. It can be used to attempt search within specified criteria.

After analysing these knowledge graphs, it is easy to identify the construction approaches and techniques of knowledge graphs for knowledge graph construction.

3 Knowledge Graph Construction Approaches and Techniques

Knowledge graph construction process follows a common architecture [4] which has some specific approaches and techniques. There are two main approaches to create knowledge graph based on ontology. One is top-down, and the other is bottom-up. Top-down approach means that ontology and schema should be defined, and then knowledge instances are added into knowledge base. This approach emphasizes the well-defined domain ontologies to represent the actual instances of knowledge graphs. The bottom-up approach extracts knowledge instances from the Linked Open Data (LOD) or other knowledge resources. After knowledge fusing the populated instances, the top-level ontologies are built by means of knowledge instances to create the whole KGs. This paper mainly discusses bottom-up approach and related techniques, has it is an iterative process to update the content which is divided into four main phases such as: Knowledge extraction, knowledge fusion, knowledge storage, knowledge retrieval, and visualization.

Knowledge extraction plays an important role to construct the KGs by using different approaches (natural language processing, text mining, and machine learning), types (entity, relation, attribute), data sources (structured, semi-structured, unstructured), and tools (StandfordNER [23], OpenNLP [24], AIDA [25], CiceroLite [26], etc.). Knowledge fusion works in an iterative manner which constantly performs ontology construction and evaluation of quality. Several approaches and methodologies [27, 28] are followed for ontology development and evaluation. Knowledge graph storage is still under development, generally used NoSQL databases for storage [25]. There are two main storage types: Graph database and RDF database [29]. Graph databases store knowledge graphs in the form of edges, nodes, and relations of graphs; they also considered as a significant method to store the knowledge graphs. On the other hand, RDF stores in a triple form subject-object-predicate as Triple-Store. The RDF-based storage provides query efficiency and merge-join of triples and indexing helps to improve the query efficiency. KG retrieval and visualization are current research topics, as knowledge retrieval performs semantic retrieval and provides the reasoning power. Query result and graph result are two forms to present the result of KG retrieval. For Quer result, SPARQL query provides result in different formats such as: JSON, JSON-LD, XML, JSON, TSV, CSV, HTML, RDF/XML, N3, and JSON-LD. A Graph result is also presented by different approaches [30] such as: RDF Gravity [31], IsaViz [32], Fenfire [33] and DBpedia Mobile [34]. Query result format provides fine-grained search result of KG while graph result provides a graphical visualization of KGs to explore and navigate the related knowledge.

Figure 1 shows the knowledge graph construction process in detail. Machine learning techniques [27, 35–38] have been successfully applied for the exploitation of different knowledge sources during the construction of knowledge graphs. These techniques are fully applied in entity extraction, relation extraction, and attribute

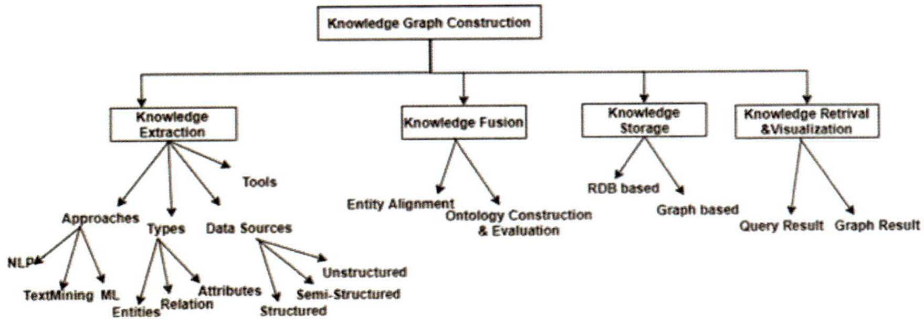

Fig. 1 Main phases of knowledge graph construction

extraction from different sources such as: unstructured, semi-structured and structured. For entity alignment, these techniques also play important role in knowledge fusion phase.

4 Knowledge Graph Applications

According to the literature, strongly structured knowledge is appropriate for the enrichment of several AI-based applications. Although it is still remains a challenge to integrate the symbolic knowledge and reasoning into real-world applications [39]. Currently, KGs are most widely applied to natural language understanding (NLU), query/answering (QA) system, smart search, and recommendation systems.

Knowledge-based natural language understanding (NLU) enhances the representation of language by injecting structured knowledge into semantic world. Chen et al. [40] introduced double graph random walks over two knowledge graphs. Question answering (QA) over knowledge graphs, answers natural language questions to perform semantic-based search and common-sense reasoning with the help of knowledge injection in the semantic environment. Recommender systems have also been widely adopted by collaborative searching that makes global use of general historical information. There are some other KGs applications [39] such as: Medical applications, mental healthcare, zero-shot image classification, sentiment analysis, and academic search engine. KGs help to organize medical reports by some common steps: retrieval, encoding, and paraphrasing. In mental healthcare, KGs provide a common understanding for mental disorder and mental conditions. KGs help in propagation and semantic representation of classes in zero-shot image classification. Sentiment analysis unified with the semantic-based concepts that can easily interpret the user's intention and sentiments. Knowledge graph helps to organize the academic research paper with semantic indexing and ranking which provides better search and interpret the meaning of query concept.

5 Conclusion

Knowledge graph presents a vast community of research and has a huge range of applications and methodologies. This paper conducted a survey of existing KGs and also focuses on their construction and applications in real world that will provide help to the beginners in future research. To sum up, the proposed survey analyses 21 existing KGs and also pointed out KG construction and their applications. For KG construction, there are four phases (Knowledge extraction, knowledge fusion, knowledge storage, knowledge retrieval, and visualization) have been discussed. Knowledge extraction and knowledge fusion are based on machine learning, NLP, and text mining techniques while RDF-based and graph databases are used for knowledge storage. This survey also discussed different tools for KGs retrieval and visualization that will help and inspire the researchers in the area of KGs development and construction.

Acknowledgements This work is not followed by any grant. No funding involve in this work.

References

1. Z.L. Xu, Y.P. Sheng, L.R. He, Y.F. Wang, Review on knowledge graph techniques. J. Univ. Electron. Sci. Technol. China **45**(4), 589–606 (2016)
2. S. Amit, Introducing the Knowledge Graph (Official Blog of Google, America, 2012)
3. W. Yuan, K. Zhang, Q. Dai, C. Peng, K. Zhao, Construction and application of knowledge graph in full-service unified data center of electric power system, in: IOP Conference Series: Materials Science and Engineering, vol. 452, No. 3 (IOP Publishing, 2018, December), p. 032065
4. Z. Zhao, S.K. Han, I.M. So, Architecture of knowledge graph construction techniques. Int. J. Pure Appl. Math. **118**(19), 1869–1883 (2018)
5. G.A. Miller, WordNet: a lexical database for English. Commun. ACM **38**(11), 39–41 (1995)
6. Z. Dong, Q. Dong, HowNet-a hybrid language and knowledge resource, in International conference on natural language processing and knowledge engineering, 2003. Proceedings (IEEE, 2003, October), pp. 820–824
7. H. Liu, P. Singh, ConceptNet—a practical commonsense reasoning tool-kit. BT Technol. J. **22**(4), 211–226 (2004)
8. C.F. Baker, C.J. Fillmore, J.B. Lowe, The berkeley framenet project, in *Proceedings of the 17th International Conference on Computational linguistics*, vol. 1 (Association for Computational Linguistics, 1998, August), pp. 86–90
9. O. Etzioni, M. Cafarella, D. Downey, A.M. Popescu, T. Shaked, S. Soderland, A. Yates, Unsupervised named-entity extraction from the web: An experimental study. Artif. Intell. **165**(1), 91–134 (2005)
10. C. Matuszek, M. Witbrock, J. Cabral, J. DeOliveira, An introduction to the syntax and content of Cyc. (UMBC Computer Science and Electrical Engineering Department Collection, 2006)
11. K. Bollacker, R. Cook, P. Tufts, Freebase: a shared database of structured general human knowledge, in *AAAI*, vol. 7 (2007, July), pp. 1962–1963
12. A. Carlson, J. Betteridge, B. Kisiel, B. Settles, E.R. Hruschka, T.M. Mitchell, Toward an architecture for never-ending language learning, in Twenty-Fourth AAAI Conference on Artificial Intelligence (2010, July)
13. W. Wu, H. Li, H. Wang, K.Q. Zhu, Probase: a probabilistic taxonomy for text understanding, in *Proceedings of the 2012 ACM SIGMOD International Conference on Management of Data* (2012, May), pp. 481–492
14. F.M. Suchanek, G. Kasneci, G. Weikum, Yago: a core of semantic knowledge, in *Proceedings of the 16th international conference on World Wide Web* (2007, May), pp. 697–706
15. J. Hoffart, F.M. Suchanek, K. Berberich, G. Weikum, YAGO2: a spatially and temporally enhanced knowledge base from Wikipedia. Artif. Intell. **194**, 28–61 (2013)
16. J. Lehmann, R. Isele, M. Jakob, A. Jentzsch, D. Kontokostas, P.N. Mendes, C. Bizer, DBpedia–a large-scale, multilingual knowledge base extracted from Wikipedia. Seman. Web **6**(2), 167–195 (2015)
17. X. Dong, E. Gabrilovich, G. Heitz, W. Horn, N. Lao, K. Murphy, W. Zhang, Knowledge vault: a web-scale approach to probabilistic knowledge fusion, in *Proceedings of the 20th ACM SIGKDD International Conference on Knowledge Discovery and Data Mining* (2014, August), pp. 601–610

18. N. Torzec, Yahoo's Knowledge Graph (2014). http://semtechbizsj2014.semanticweb.com/sessionPop.cfm?confid=82&proposalid=6452
19. S. Sengupta, Facebook unveils a new search tool. New York Times (2013)
20. N. Nakashole, M. Theobald, G. Weikum, Scalable knowledge harvesting with high precision and high recall, in *Proceedings of the Fourth ACM International Conference on Web Search and Data Mining*, (2011, February), pp. 227–236
21. C. De Sa, A. Ratner, C. Ré, J. Shin, F. Wang, S. Wu, C. Zhang, Deepdive: declarative knowledge base construction. ACM SIGMOD Rec.rd **45**(1), 60–67 (2016)
22. J. Shin, S. Wu, F. Wang, C,. De Sa, C. Zhang, C. Re, Incremental knowledge base construction using deepdive, in *Proceedings of the VLDB Endowment International Conference on Very Large Data Bases*, vol. 8, No. 11 (NIH Public Access, 2015, July), p. 1310
23. D. Rinser, D. Lange, F. Naumann, Cross-lingual entity matching and infobox alignment in Wikipedia. Inf. Syst. **38**(6), 887–907 (2013)
24. V. Khadilkar, M. Kantarcioglu, B. Thuraisingham, P. Castagna, Jena-HBase: a distributed, scalable and efficient RDF triple store, in *Proceedings of the 11th International Semantic Web Conference Posters & Demonstrations Track, ISWC-PD*, vol. 12 (2012, November), pp. 85–88
25. P. Cudre-Mauroux, I. Enchev, S. Fundatureanu, P. Groth, A. Haque, A. Harth, M. Wylot, NoSQL databases for RDF: an empirical evaluation, in *International Semantic Web Conference* (Springer, Berlin, Heidelberg, 2013, October), pp. 310–325
26. K. Sun, Y. Liu, Z. Guo, C. Wang, Visualization for knowledge graph based on education data. Int. J. Softw. Inf. **10**(3) (2016)
27. S. Mishra, S. Jain, Ontologies as a semantic model in IoT. Int. J. Comput. Appl. **42**(3), 233–243 (2018)
28. S. Mishra, S. Jain, An Intelligent Knowledge Treasure for Military Decision Support. Int. J. Web-Based Learn. Teach. Technol. (IJWLTT) **14**(3), 55–75 (2019)
29. D.C. Faye, O. Curé, G. Blin, A survey of RDF storage approaches (2012)
30. F. Ghorbel, F. Hamdi, N. Ellouze, E. Metais, F. Gargouri, Visualizing large-scale linked data with memo graph. Procedia Comput. Scie. **112**, 854–863 (2017)
31. S. Goyal, R. Westenthaler, RDF Gravity; Salzburg Research. http://semweb.salzburgresearch.at/apps/rdf-gravity/ (2004)
32. IsaViz: A visual authoring tool for RDF. http://www.w3.org/2001/11/IsaViz/ (2001–2006)
33. T. Hastrup, R. Cyganiak, U. Bojars, Browsing linked data with fenfire (2008)
34. C. Becker, C. Bizer, DBpedia Mobile: A Location-Enabled Linked Data Browser. Ldow **369** (2008)
35. D. Gaurav, S.M. Tiwari, A. Goyal, N. Gandhi, A. Abraham, Machine intelligence-based algorithms for spam filtering on document labeling. Soft Comput. 1–14 (2019)
36. D. Gaurav, J.K.P.S. Yadav, R.K. Kaliyar, A. Goyal, Detection of false positive situation in review mining, in Soft Computing and Signal Processing. Advances in Intelligent Systems and Computing, vol. 900 ed. by J. Wang, G. Reddy, V. Prasad, V. Reddy (Springer, Singapore, 2019)
37. M. Rahul, N. Kohli, R. Agarwal, S. Mishra, Facial expression recognition using geometric features and modified hidden Markov model. Int. J. Grid Util. Comput. **10**(5), 488–496 (2019)
38. S.P. Chatrati, G. Hossain, A. Goyal, A. Bhan, S. Bhattacharya, D. Gaurav, S.M. Tiwari, Smart home health monitoring system for predicting type 2 diabetes and hypertension. J. King Saud Univ.-Comput. Inf. Sci. (2020)
39. S. Ji, S. Pan, E. Cambria, P. Marttinen, P.S. Yu, A survey on knowledge graphs: representation, acquisition and applications. arXiv preprint arXiv:2002.00388 (2020)
40. Y.N. Chen, W.Y. Wang, A. Rudnicky, Jointly modeling inter-slot relations by random walk on knowledge graphs for unsupervised spoken language understanding, in *NAACL* (2015), pp. 619–629

A Simple Ensemble Learning Algorithm For a Real Time High Dimensional Data

G. Jenilasree, N. Shenbagavadivu, and M. Bhuvaneswari

Abstract High dimension data is growing rapidly, which challenges the existing data mining and learning systems. A stable performance can be achieved precisely in high dimension data using machine learning algorithm, ensemble learning and feature selection methods. The ensemble classifiers like Decision tree, Random forest, support vector machine, Navies Bayes, Linear Discriminant Analysis are applied as a self-contained module and the output label is reliably produced by the fusion of majority voting for data analysis. At present a lot of investigations for reduction of high dimensional data set are implemented, but the study result shows only a minor improvement in accuracy. In this paper, a wrapper method is properly used to improve certain key features such as random state, tune max depth, tune number samples split, and tuned loss function. The proposed work is carried out using real-time garbage data set collected from the municipal corporation of Tiruchirappalli, India and the selected sample is tested with FGDT (Forward Gradient Boost Decision Tree) shows improvement in accuracy.

Keywords Feature selection · Dimension reduction · Ensemble learning algorithm · Supervised machine learning algorithm

1 Introduction

Solid waste management (SWM) will remain an fundamental parameter in the coming decades to judge a civic body's remarkable accomplishment. Developing and densely populated countries like India face many fundamental problems to typically

G. Jenilasree (✉) · N. Shenbagavadivu
Department of Computer Applications, University College of Engineering, Anna University BIT Campus, Tiruchirappalli, Tamil Nadu 620024, India
e-mail: jenilasree@gmail.com

M. Bhuvaneswari
Department of ECE, University College of Engineering, Anna University BIT Campus, Tiruchirappalli, Tamil Nadu 620024, India

J. Tavares et al. (eds.), *Emerging Technologies in Data Mining and Information Security*, Lecture Notes in Networks and Systems 164,
https://doi.org/10.1007/978-981-15-9774-9_3

achieve the successful mission of smart cities and for effective solid waste management techniques. According to Swachh Bharat Mission (SBM) proposed by the Indian Government, the country should become clean and a garbage bin free cities by the ultimate end of 2019 [1] but unfortunately they cannot able to achieve due to the lack of information about the collection of garbage produced by the cities. It is accurately estimated that more than 6 million tons of garbage is collected in India per academic year and the wastages are dumped in a particular area. To know about the exact collection of garbage produced in Urban cities, a certain smart city like Tiruchirappalli is selected to know about the specific type of garbage collected and how it is segregated. The Trichy City Corporation launched a slew of initiatives to sustain the progress in making a city clean and green city for which segregation of waste at the reliable source for effective SWM is essential. Segregation of garbage is extremely essential as garbage can be recycled or reused or disposed of depending on its types [2]. The biodegradable waste could be efficiently converted into organic compost and sold to the local farmers. The non-biodegradable waste would be discharged to cement kilns to be used as a fuel in the kilns. The building debris will be crushed and used to lay roads, and the inert waste will be dumped. To overcome this time-consuming process a machine learning algorithm plays a vital role in the data mining field for both the independent To overcome this time-consuming process a machine learning algorithm plays a vital role in the data mining field for both the independent analysis and prediction process [3]. It is typically categorized into two division namely unsupervised learning and supervised learning. An unsupervised learning algorithm is focused on clustering methodology, while supervised learning is focused on classification [4]. In real-time data analysis, the significant challenge is to solve high dimensional dataset for this feature selection remain an effective and efficient method. These feature selection methodologies choose only a compact subset of relevant features and other features like irrelevant, noisy or redundant features are removed.

Feature selection can be carried out using various algorithms like Sequential Forward Selection (SFS), Sequential Backward Selection (SBS), Sequential Forward Floating Selection (SFFS), Sequential Backward Floating Selection (SBFS) and optimize selection (evolutionary) and so on. The possible combination of machine learning algorithms and feature selection algorithms doesn't yield the expected accuracy of the real-time garbage dataset of Trichy municipality. Due to that, a wrapper method is used to increase the level of accuracy in segregated wastages. A wrapper method is a set of a unique model that constructs a machine learning system that works both in parallel and sequential mode. For a given real-time application, the output obtained from the wrapper method is a combination of decision fusion strategy and the required outcome. There are many types of wrapper methods in which certain algorithms are used to compare the performance accuracy by implementing Ensemble Learning algorithms. The ensemble learning methods are classified into bagging, boosting and stacked methods. In this proposed paper, the boosting algorithm is properly used to efficiently convert weaker learners into stronger learners [5, 6]. The supreme aim of this proposed work in common is (1) to determine how the missing feature values are dealt and to check the performance by using an ensemble learning algorithm for the various Machine Learning Algorithms. (2) To reduce the computational time by using Feature Selection (SFFS). This creative process is performed by implementing an open-source tool Py charm and the paper is organized as naturally follows (1) Related work, (2) Various methods involved in this process, (3) Analysis of garbage data in India using ensemble learning algorithm, (4) Proposed Method, (5) Results and discussion, (6) Conclusion and (7) Future work.

2 Related Works

Pudil established the SFFS algorithm by a combination of sequential forward selection and sequential backward selection. But in this paper, the sequential forward floating selection method is used [7]. Freund and Schapire 1996 Proposed Ada Boost algorithm which is mainly used in Machine Learning Algorithm and it is an important ensemble method since it shows high accuracy and predictive process. The tree is grown by weighted data of the combination of Tree 1 and Tree 2 that act as a sequential method. The main difference is that the Ada boost is used to find the weaker learners and the gradient boost algorithm is used to find loss functions to improve learning [8]. Rattanawadee Panthong proposed a wrapper feature dimension selection that uses various types of data set for the analysis process. These wrapper methods are combined with ensemble Learning like gradient boost and Ada boost for Decision Tree (DT) and Navies Bayes (NB). The result shows that Sequential forward selection (SFS) based on the bagging algorithm using Decision Tree produces an accuracy of 89.60%. However, the loss function is very high [9]. Li gang Zhou1 describes Ada boost of the bankruptcy of missing data proposes a method for corporate bankruptcy prediction using two data sets with different sample sizes. It is mainly used to detect the missing values of data set accurately by using Ada boost ensemble learning algorithm. The result shows that Ada boost with imputation method provides better prediction with missing values [10].

3 Various Method Involved in This Process

There are three approaches used in this specific process to discover and figure out the exact accuracy using Wrapper method with Sequential Forward Floating Selection (SFFS), Supervised Learning method using Naive Bayes, Decision Tree, Random Forest, Support Vector Machine(SVM), Linear Discriminant Analysis (LDA) and Ensemble Learning algorithm by Ada Boost and Gradient Boost.

3.1 Wrapper Method

The wrapper method works on the fundamental principle of the greedy search algorithm. Justly compared to other modern techniques SFFS efficiently performs accurate feature selection [11]. SFFS remain a process in which it initially commences the process from an empty set and slowly it adds features selected by some assessment function. It on tops adds features into an empty set until there is no option to include possible features or improve their performance of estimation.

The supreme advantage of the SFFS algorithm is it is used to find out the effective capability of finding forward steps. Through which it makes better performance and outstanding accuracy compared favorably to Sequential feature selection (SFS) and Sequential backward selection (SBS) algorithms. Figure 1 represents the active process involved in the SFFS algorithm. Pseudo code (Table 1) carefully describes the possible inclusion and condition exclusion algorithm.

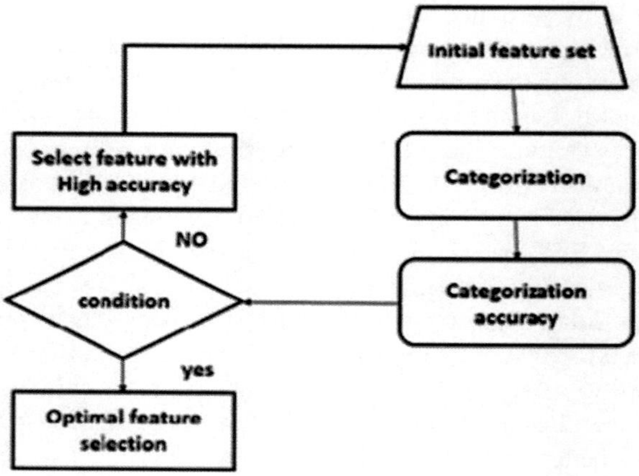

Fig. 1 Feature selection—flow chart

Table 1 Pseudo code

x + = arg max J (xk + x) where x ∈ Y − Xk x + = arg max J(xk+x), where x∈ Y− Xk Xk + 1 = Xk + x + Xk + 1=Xk+x+ ; k = k + 1 k=k+1 x − = arg max J (xk − x), where x ∈ Xk x− = arg max J(xk− x), where x ∈ Xk if J (xk − x)> J (xk − x) if J(xk− x)>J(xk− x) : Xk − 1 = Xk − x − Xk− 1=Xk− x− k = k − 1 k=k− 1

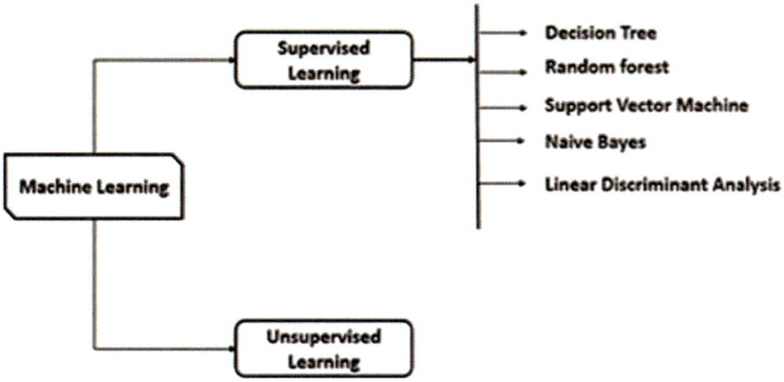

Fig. 2 Pictorial representation of machine learning algorithms

3.2 Machine Learning Algorithm

In Machine Learning Algorithms the trained models are exposed to current data to obtain accurate results. The intelligent machine undoubtedly learns from previous results for efficiently producing reliable, repeatable results for the given data set. Machine learning techniques are properly classified into two broad categories namely supervised learning and unsupervised learning. Supervised learning is used to classify the algorithm correctly whereas unsupervised learning, groups the given data set into a distinct group based on the features fed as input. Supervised learning consists precisely of Decision Tree (DT), Random Forest (RF), Support Vector Machine (SVM), linear discriminant analysis (LDA) and Naïve Bayes (NB) [12]. The decision tree wisely determines the predictive value based on the series of several factors into an account and classifies the given data set accurately with various branches in a single tree. Random forest algorithm could be used for both regression and classification kind of problems. The accuracy will be excellent when the sufficient number of trees is higher in RF. The support vector machine in common is to resolutely face challenges by finding the hyperplane, each data has an end dimensional space. Linear Discriminant Analysis (LDA) is used for pattern recognition. NB is a classification algorithm based on fundamental Bayes theorem. Figure 2 shows a pictorial representation of effective Machine Learning techniques and types.

3.3 Ensemble Learning Method

An ensemble learning algorithm represent a method to improve multiple models using the same learning algorithm. It has various methods like bagging, boosting. The significant drawback in learning is noise, variance, and bias. To overcome this practical drawback, the ensemble learning method is intentionally used. Bagging efficiently is a prevalent method used to positively enhance the stability and accuracy of the given data set. It sufficiently reduces overfitting. Boosting remain a method in which it consecutively builds on weaker learners into one stronger model. Boosting tries to reduce overfitting.

4 Proposed Method

The wrapper method, machine learning algorithm and effective SFFS methods are combined as a separate module and applied to the real-time data set and it is used to extract the features on the type of garbage that are collected most for a particular day from the given data sets. The wrapper method typically involves various specific types in which the proposed method properly includes a sequential forward floating selection algorithm (SFFS) since it efficiently is a family of greedy search algorithms and it enables the machine learning algorithm to train faster.

Figure 3 describes the process involved in the proposed models. The proposed scheme implements the real-time data set of garbage collection is obtained from Trichy municipal corporation, Tamil Nadu, India. The data set consists of absolute garbage collected in the city for a period from 2017 to 2018 for each day. The data is pre-processed using the SFFS module for feature selection. The sequential forward floating selection is used to reduce an initial d-dimensional feature space to the final k-dimensional feature subspace. It is performed on both train and test data sets along

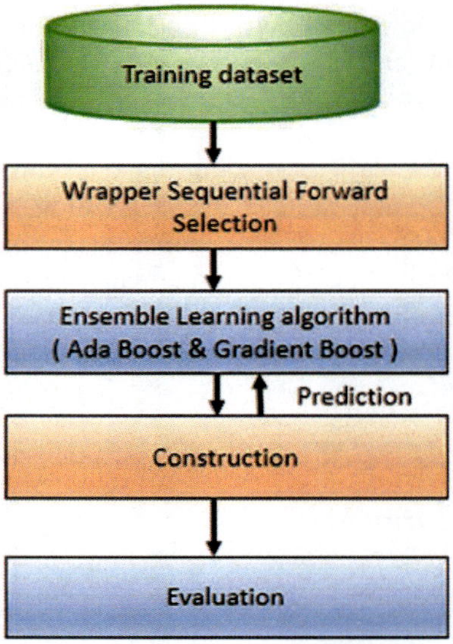

Fig. 3 Proposed method

with feature attributes by using the Gradient and Ada Boost algorithm. The primary aim of the forward floating selection algorithm is to improve computational efficiency along with accuracy in minimum time. At that time the selected feature is given to the ensemble learning algorithm using Ada boost and Gradient Boost. The Ada boost equation is represented as

$$(x) = (\theta m f m(x)) \quad (1)$$

where m limit is said to be from 0 to ∞. The gradient boost algorithm obtain the core objective of the supervised learning algorithm to find out the exact loss functions of the data set. The mean square of the loss function is defined as

$$\text{Accuracy} = \text{NTP} + \text{NTN}/\text{NP} + \text{NN} \quad (2)$$

$$\text{Sensitivity} = \text{NTP} + \text{NP} \quad (3)$$

$$\text{Specificity} = \text{NTN} + \text{NN} \quad (4)$$

where TP stands for True positive, TN stands for True Negative, N stands for a number. Thus, the evaluation process is done by feature subset using Decision Tree, Random Forest, Support Vector Machine, Naive Bayes, Linear Discriminant Analysis are accurately compared. The irrelevant features or noise is removed to reduce the standard error in this developed model (Table 2).

Table 2 Pesudo code 2

```
Input: Dataset D={(x1, y1...), (x2, y2...), ..............., (xn, yn...)}
Process:
Initialize feature set X 0 = ∅ , k = 0
x += arg max J (xk + x), where x ∈ Y − Xk x+= arg max J(xk+x), where x ∈ Y − Xk Xk+
1 = Xk + x + Xk+1=Xk+x+ k = k + 1 k=k+1
Dt= {bi I i=1,2,... .k; bi €X} where t = (0,1,2,3...)
T= Decision tree
Output: Decision tree Dt Initially it selects boosting algrm.
Build a boosting subset of samples from Dt. It selects randomly X features
Calculate loss function and weaker subsets
Resample using feature weights.
```

The above pseudo-code 2 properly tells about the working procedure and key parameter used for the algorithm. It initializes the algorithm was $\varnothing$ is said to be as null set and $k = 0$ where k is the size of the subset. Table 3 describes the various algorithm involved in this paper for the given garbage data set.

The various comparison chart for the efficient Ensemble Learning algorithms is carefully plotted for the wastage's collected for the year 2018. The performance anal-

Table 3 Search algorithms used in this study and their abbreviations

No.	Algorithms	Abbreviation
1.	SFFS +Ada boost +LDA	FALDA
2.	SFFS +Ada boost +RF	FARF
3.	SFFS +Ada boost+ DT	FADT
4.	SFFS +Ada boost+ SVM	FASVM
5.	SFFS +Ada boost+ NB	FANB
6.	SFFS +Gradient boost +LDA	FGLDA
7.	SFFS +Gradient boost +RF	FGRF
8.	SFFS +Gradient boost +DT	FGDT∗
9.	SFFS +Gradient boost +SVM	FGSVM
10.	SFFS +Gradient boost +NB	FGNB

∗ shows better performance

ysis of garbage collected is obtained by plotting the output value from the algorithms involved in it.

5 Analysis and Comparison Plot of Proposed Method

A powerful, free and open-source software tool Pycharm 3.7 for python Language is used to analyze data that includes the various supervised algorithm of the Machine Learning with the possible combination of wrapper method and SFFS. The performance is calculated by using recall and precision. The data sets are categorized into two sets namely train and test data set. For training 60% of the data set is used and the remaining data set is used for testing. Figure 4 represents the procedure followed by the system. At first, input data is properly fed into the module and it selects SFFS for dimension reduction as well as feature selection. The data is re-processed before properly feeding into the machine learning algorithms for missing values. After this process, it selects the supervised algorithm and boosting algorithm for further analysis.

As new data are updated, the model performs all the above-mentioned operations without the need to rebuild the system. In this paper, it uses 7 model data sets for feature extractions and accurate calculation. The confusion matrix is shown on Table 4

Table 5 describes the accuracies obtained from various supervised machine learning algorithms along with the Ensemble Learning algorithm and wrapper feature selection methods. It shows FGDT provides the highest accuracy of 85% on average.

Table 6 describes the combination of sequential forward selection along with gradient boost algorithm and supervised machine learning. The main advantage of implementing the Gradient Boost algorithm is to find an existing weak learner algorithm

Table 4 Confusion matrix

Dataset	Features	FALDA	FARF	FADT	FASVM	FANB
Municipal	28	67	76	**83**	64	68
Household	14	78	71	**85**	59	70
Hazardous	17	84	**77**	80	89	86
Industrial	28	73	81	**79**	78	75
Biomedical waste	19	86	**91**	93	87	78
Bio-degradable	20	79	85	**87**	90	93
Non-bio-degradable	24	88	84	**91**	86	79
Average	153	79	82	**85**	62	78

Table 5 Performance accuracies for sequential feature selection with Ada boost by LDA, RF, DT, SVM, NB

	Predicted class		
Actual class		Yes	No
	Yes	Positive	Negative
	No	Positive	Negative

and loss functions. When the boosting method is combined with Sequential Forward Floating Selection and supervised learning algorithm, the performance increases through an accuracy of about 94% in FGDT.

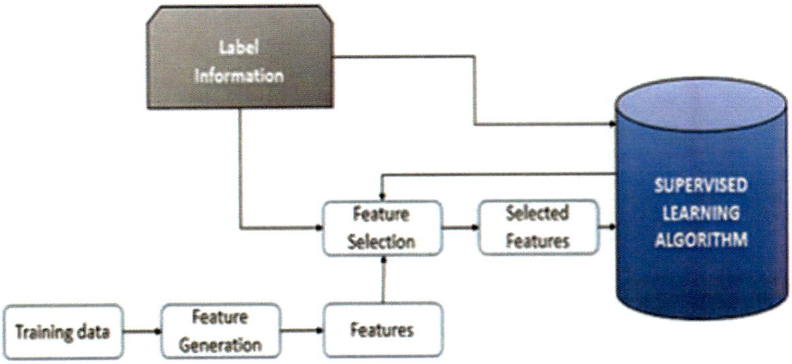

Fig. 4 Framework of supervised feature selection

Table 6 Performance accuracies for proposed method of sequential feature selection with gradient boost by LDA, RF, DT, SVM, NB

Dataset	Features	FGLDA	FGRF	FGDT	FGSVM	FGNB
Municipal	28	78	81	**94**	75	79
Household	14	86	82	**96**	68	82
Hazardous	17	93	**97**	92	88	91
Industrial	28	83	91	**95**	85	81
Bio-medical waste	19	94	**95**	92	90	89
Bio-degradable	20	92	90	**98**	92	87
Non-bio-degradable	24	86	89	**94**	79	86
Average	153	87	89	**94**	82	85

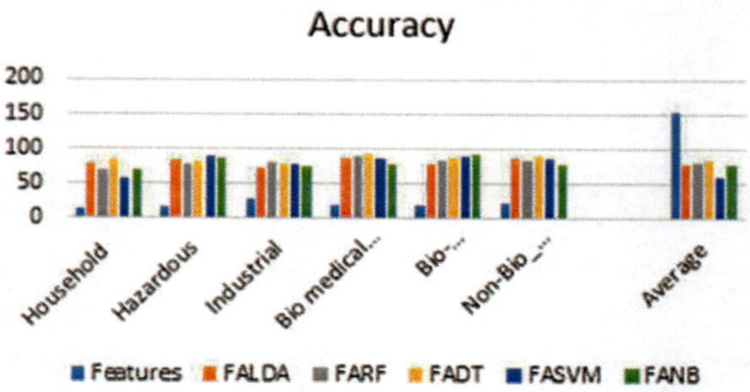

Fig. 5 Accuracy of the various dataset using Ada boost

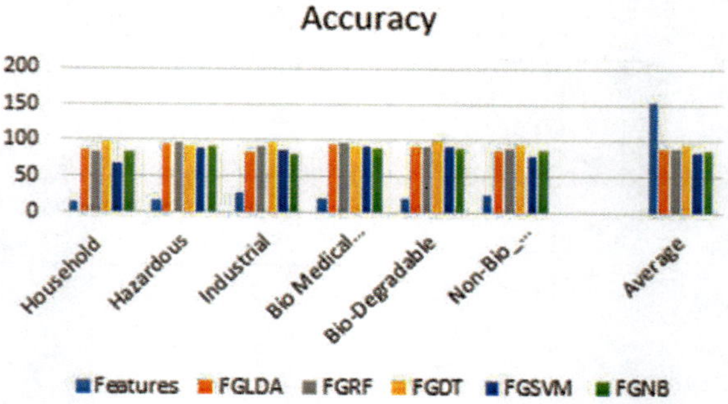

Fig. 6 Accuracy of the various dataset using Gradient boost

6 Results and Discussion

This model is constructed utilizing only two methods of ensemble learning algorithm i.e. Ada boost and Gradient Boost with the combination of Sequential Forward floating selection. The essential purpose of using SFFS is to identify which type of features is collected more in garbage's, during a particular month per year. The purpose of using two types of boosting Algorithms is to find out the exact accuracy of the learning algorithm for the given data set. By comparing Tables 4 and 5, it interprets the accuracy by gradient boost algorithm along with decision tree is more when compared to Ada boost. Therefore, with these results, we can find out which type of garbage is collected more in every city, and we can predict it for landfilling and other processes like the production of cobalt gas, manure, recycling, etc. Thereby public health hazards can be controlled. Figure 5 shows a pictorial representation of accuracy obtained using various algorithms using the Ada boost and Fig. 6 shows accuracy obtained using the Gradient boost algorithm.

7 Conclusion

In this paper, it possesses an exact reason for choosing garbage collected per year as a data set to know how much amount of garbage is collected and which types of features are collected from garbage frequently. Till now, study and performance for garbage data set have not been done. But designing smart garbage bins using IoT is constructed in various ways to prevent the overflow of garbage. Currently, the government municipal corporation of India announced to establish a city as a smart city. In that, one of the contributing factors in common is to instantly become bin free city. The previous corporation announced to collect the garbage from door-to-door from their homes to prevent the accumulation of wastage from dump yards. The Garbage data set includes Household, Industrial (electrical and electronic wastages), Hazardous, Biomedical Waste, Bio-degradable and Non-bio degradable. The Sequential forward feature selection exactly tells about which type of features of the garbage are rapidly increasing in a particular range. The direct results are generated using an SFFS based Ensemble Learning Algorithm for Garbage Data Analysis with five base learners (LDA, RF, DT, NB, SVM). With the help of results, it concludes that FGDT(proposed method) shows the best accuracy of (94%) with features of (kitchen waste, green waste, paper) for Biodegradable (Electrical applicants, mobile phones, chemical acids.) for Industrial (fertilizers, paints, construction waste) for Hazardous (pharmaceutical drugs, injection, plastics & cotton) for biomedical waste (glass and woods, aluminum types, computer hardware) for Non-bio degradable (sewage sludge, food waste, residual waste) for households. In the foreseeable future, with the help of this result, we can able to identify how much amount of garbages can be recycled periodically in various fields like bio gas, cobalt gas, recycling of papers & plastics, industrial recycling, etc. and improve the air quality of the whole region.

8 Future Work

Forthcoming work is to accurately compare stotastic gradient methods implementing the Ensemble Learning algorithm combined with tuned Light GBM and Stacked algorithm for better performance of accuracy, prediction, and risk assessment for real time various data sets and it is compared to the existing algorithms.

References

1. Rockefeller Foundation. *Solid Waste Management in India: An Assessment of Resource Recovery and Environmental Impact* (2019). Available from: http://icrier.org/pdf/4-Jan-2019/Utkarsh_Patel-SWM_%20in_India.pdf
2. R. Shrivastava, India's Trash Bomb:80% of 1.5 lakh metric tonnedaily garbage remains exposed,untreated. *India Today* (2019). Available from: https://www.indiatoday.in/india/story/india-s-trash-bomb-80-of-1-5-lakh-metric-tonne-daily-garbage-remains-exposed-untreated-1571769-2019-07-21
3. T. Komamizu, J. Nakazawa, T. Amagasa, H. Kitagawa, Analytical toolbox for smart city applications: garbage collection log use case, in *IEEE International Conference on Big Data (Big Data)* (Boston, MA, USA, IEEE, 2017), pp. 4105–4110, 11–14 Dec 2017
4. F. Pedregosa, G. Varoquaux, A. Gramfort, V. Michel, B. Thirion, scikit-learn: Machine learning in python. J. Mach. Learn. Res. **12**, 2825–2830 (2011)
5. P. Li, C.J.C. Burges, Q. Wu, Mcrank: learning to rank using multiple classification and gradient boosting, in *Advances in Neural Information Processing Systems*, vol 7 (Vancouver, British Columbia, Canada, 2007), pp. 845–845, 3–6 Dec 2007
6. T. Chen, C.G. Xgboost, XGBoost: a scalable tree boosting system, in *The 22nd ACM SIGKDD International Conference*, ed. by T. Chen et al. (2016), pp. 785–794. Available from: https://www.researchgate.net/publication/310824798_XGBoost_A_Scalable_Tree_Boosting_System
7. P. Somol, J. Novovicova, P. Pudil, Efficient feature subset selection and subset size pptimization, in *Pattern Recognition Recent Advances*, vol 1, ed. by A. Herout (Intech Open, United Kingdom, 2010)
8. H. Liu, S.M. Chen, Multi-level fusion of classifiers through fuzzy ensemble learning, in *11th International Symposium on Computational Intelligence and Design (ISCID)* (Hangzhou, China, China. IEEE, 2019), pp. 324–330, 8-9 Dec 2018
9. R. Panthong, A. Srivihok, Wrapper feature subset selection for dimension reduction based on ensemble learning algorithm. Procedia Comput. Sci. **72**, 162–169, in *The Third information Systems International Conference* (Elsevier, 2015)
10. L. Zhou, K.K. Lai, AdaBoost models for corporate bankruptcy prediction with missing data. J. Comput. Econ. **50**(1), 69–94 (2016)
11. F. Chang, C.Y. Guo, X.R. Lin, C.J. Lu, Tree decomposition for large-scale SVM problems, in *International Conference on Technologies and Applications of Artificial Intelligence* (Hsinchu, Taiwan, IEEE, 2010), pp. 2855–2892, 18–20 Nov 2010
12. J. Morgenstern, S. Samadi, M. Singh, U. Tantipongpipat, S. Vempala, *Fair Dimensionality Reduction and Iterative Rounding for SDPs*, vol 3, pp. 3855–3858, Arxiv 2019

Promising Highest Resilient Network Using Preventive Route Selection Algorithm-IPS DSDV in MANET

Kanika Garg, Naveen Chauhan, Chiranjit Dutta, and Tanupriya Choudhury

Abstract A primary issue which has been evolved in wireless ad hoc network is how to conserve or preserve energy while transmitting the data. In order to overcome this challenge, a mechanism has been proposed which is deployed with Destination Sequenced Distance Vector (DSDV) routing protocol known as Intelligent Promising Support DSDV (IPS DSDV) that promises the highest resilient network. The obstacle of early termination of nodes due to limited power in battery has been faced while carrying out the execution of DSDV routing protocol. The utmost noteworthy point which has been noticed while implementing DSDV routing protocol is that no criteria have been integrated to save and utilize energy in a well-mannered way in a cluster of ubiquitous devices network. In this study, we have proposed a preventive route selection algorithm with the objective of promising the highest resilient network. The proposed algorithm is implemented using the concept of DSDV in order to consume the energy of all nodes in a better way for enhancing the course of life of mobile ad hoc network (MANET). From the result, it can be stated that the main strategy behind the proposed mechanism is load balancing and even distribution of load by providing the optimal routes in order to achieve the resilient and robust network.

Keywords DSDV · MANET · Routing protocols · Energy-aware protocols

---

K. Garg (✉) · C. Dutta
Department of Information Technology, SRM IST, NCR Campus, Modinagar, Uttar Pradesh, India
e-mail: kanikagarg.kg@gmail.com

C. Dutta
e-mail: cse.chiranjit@gmail.com

N. Chauhan
Department of Computer Science and Engineering, KIET Group of Institutions, Ghaziabad, Uttar Pradesh, India
e-mail: naveen.chauhan@kiet.edu

T. Choudhury
Department of Informatics School of Computer Science, UPES, Dehradun, Uttarakhand, India
e-mail: tanupriya1986@gmail.com

J. Tavares et al. (eds.), *Emerging Technologies in Data Mining and Information Security*, Lecture Notes in Networks and Systems 164,
https://doi.org/10.1007/978-981-15-9774-9_4

1 Introduction

At the initial level, users are not having knowledge about the computing work being done known as omnipresent computing [1]. As there is no conceptualization of fixed base stations or central controller in the basic structure of mobile ad hoc networks, it means its topology is dynamic in nature not fixed [2]. Because of its prime feature, i.e., dynamicity in ad hoc networks, it has the potential to generate or demolish its own network anywhere anytime. Because of successful transmission or communication of data, this wireless network has gained a lot of attention. Since there is no fixed access point or central controller point in wireless ad hoc network, so this vigorous feature is depicting its powerfulness in comparison with infrastructure network as it is enabling the nodes to maintain connectivity [3]. As today's generation is solely dependent on Internet-enabled devices, so maintenance of minimum energy in all nodes of a network is the most urgent need [4]. In ubiquitous infrastructure less network, if any of the node dies early because of deficiency of power or energy, then it will not be possible for other nodes to perform communication, as a result the network will get disconnected. As there is unexpected growth in users who are using Internet on daily basis, so this scenario is calling for a robust system where services provided by wireless multimedia can extend point to point communication. There are two classes of wireless matrix, namely everlasting or fixed network and short-term network [5]. A network in which successful communication among the devices is done with the help of an access point (AP) is termed as static or fixed network. An AP is a central hub point through which all devices connect with each other within the defined network.

For the entire network, an AP is presenting itself as a central controller point. Short-term network is a grid which is not in the form of fixed mode [6], and its development depends on flexible adjustability of nodes. There are three categories of life cycle of MANETS. The third generation depicts today's scenarios which consist of Bluetooth sensors and routing topologies [7].

The organization of the paper is as follows: In Sect. 2, theoretical background about computing sector of wireless network and DSDV routing protocol has been presented. In Sect. 3, the problem statement of this paper has been defined. In Sect. 4, there is description about the simulation environment that is taken to examine different scenarios. In Sect. 5, the algorithm of proposed method is described, and the implemented results are explained.

2 Theoretical Background

The promotion in computing sector of advancement of mobile wireless network is escalating its section by utilizing its significance in every field due to speedy advancement in communication sector through wireless mode and transmission of data technologies. For victorious data transmission to take place in ad hoc network and the speedy widening of cellular and mobile devices like smart phones, tablet computers, smart watch, pocket PC, etc., or for the swiftly progress and advancement of mobile ad hoc networks, it opens the door for researcher to enhance the study for optimal routing protocol [8]. However, various studies have been done by many researchers, and in [9, 10] the challenges are addressed with respect to the requirement of efficient routing protocol.

At this moment of requirement of high computational power, there is an urgent need for decision-making process in which a node should be so capable of taking accurate decision whenever it is required. This situation will depict the intelligent behavior of nodes in an ad hoc network, although the problem of fast response time and fast computation time can be rectified by cloud-based computing which in turn can make the smart nodes. Most of the multimedia smart devices which are totally based on battery power or where energy has always been a chief QoS attribute are resource constrained. Each of the necessary task like collection of information and data, interaction with communication world has to be done over the entire network. The prime intention of ad hoc network is to distribute systematic information through smart devices with the help distinct routing algorithms [11, 12]. The prime features of nodes in mobile ad hoc network like flexible adaption behavior, high mobility and dynamic topological behavior of network present a necessary consequence of concern for ad hoc network where nodes are employed to provide connectivity for longer duration of time [13, 14]. Battery power or energy of a wireless node is one of the precious resources that must be carried out or executed in an actual way in order to demolish untimely termination of communication [15]. DSDV adds two features to distance vector routing protocol: First one is sequence number to avoid loops and second one is damping to hold advertisements for changes in short duration. In this protocol, each node maintains a routing table, and this routing information must be updated periodically through the network [16].

3 Problem Statement

In a mobile ad hoc network, wireless node works on the principle of radio waves to maintain linkage for longer period. In last couple of years, a quick growth can be seen in wireless ad hoc network. The open issues with respect to the highly mobile ad hoc network are how to maintain the energy in wireless nodes. Productivity in energy parameter can be defined as the period during which a node or device can maintain a minimum level of energy so that proper execution and performance of the network [17] can take place. The degree of power expenditure is mainly evaluated when a process of communication takes place from one end to another. The process of computation and dispatching of message drains out some amount of energy from each node, and when the node is in an inactive mode or idle mode, some amount of energy will be released. Drastic destruction of the network can happen because of inappropriate closing of few end points due to depletion of power. Unequal dispersion of data traffic in the network can lead to the central subject for the accessible of

Table 1 Parameters list

Parameters	Considered values
Channel type	Wireless
Transmission class	Two ray ground
Network	Physical/wireless
Category of queue	Droptail, CMU priority
Length of queue	50
Antenna type	Omni antenna
Data rate	11 Mb
Basic rate	1 Mb
Transmission radius	250
Interface scale	550

energy. The connectivity of wireless ad hoc network may completely be destroyed or the entire communication from origin point to destination point can be shut down on the account of early termination of wireless nodes. Unequal or irregular distribution of information traffic or data in the network is one of the primary reasons for the utilization of energy in larger amount while using DSDV in a network. In order to utilize battery power or energy in wireless nodes in mobile ad hoc network, an upgradation in traditional routing protocols is the most basic requirement. The prime criterion of energy familiar routing protocols which are handling the major issue of energy conservation is to keep down communication overhead and minimize the utilization of energy consumption.

4 Simulation Parameters

In this section, we have described various parameters taken to perform simulation in different scenarios. Here in this work, we used network simulator-2 [18, 19] with following attributes

Primary parameters for network setup are indicated in Table 1, which will carry out the operation of simulation of proposed algorithm in mobile ad hoc network.

At the initial level, energy level of each node is considered as 20 points and the communication range parameters are illustrated in Table 2. In IPS DSDV, strategy 4.2 units is set as the threshold energy level.

5 Implementation

An innovative algorithm is implemented to strengthen the overall life duration of the wireless network in an optimistic manner which will decline the energy utilization

Table 2 Parameters for communication setup

Parameters	Considered values
Gt/Gr	1/1
Antenna length	1.0
Frequency	2.472e9
Data rate	11 Mb
CPThresh/CSThresh	10.0/5.011e−12
RXThresh	5.82e−09
Node energy	20
Energy threshold level	4.2
Transmission power	0.9
Receiving power	0.8

of nodes by distributing the data traffic in a proper way (load balancing). Distinct parameters like network throughput, performance of network, normalized routing load, loss of packet, success rate and the chief criterion of this study are consumption of power or energy by nodes and remaining energy for different network magnitude is taken into account for evaluation of routing protocols of MANET.

Proposed Destination Sequenced Distance Vector Algorithm

DSDV smart energy saver is an accordance that controls the pitfalls of the primary DSDV and sets out a systematic routing agreement for a random automatic wireless network that may oversee battery power demands.

(A) **Falling of same RREQ packet** [6]

This step states that same RREQ packet will be discarded by an intermediate node instead of forwarding it. There is no use to forward the same multiple packets of RREQ in the network. By doing this, some amount of energy can be conserved.

(B) **Evaluation of power level at each node** [6]

Before forwarding the data to next node, the battery power of one and all intermediate nodes will be examined.

(C) **Comparison of two values** [6]

At this point before transmitting the data packet to next node, calculated power level will be compared with threshold value. If the former value is less than later value, then the packet will be simply dropped, and a notification message will be sent to source node about the rejection of packet. If the type of the packet is normal data packet, then by adopting the path discovery procedure it can follow another suitable route, and if it is an emergency packet type, then the same path will be available. When this method will be implemented, it may be possible that the shortest path criteria will not be selected by the algorithm but the main effort of boosting up the connectivity and lifetime of wireless network will be done.

Algorithm of IPS DSDV

In this strategy, the main goal is to drop non-priority packets when the battery power of node is less than the threshold value. In this strategy, the traffic is routed through multiple paths so that the energy consumption can be distributed intelligently. The proposed algorithm will decide whether to drop or not RREQ packet by intermediate node which is as follows.

```
Step 1. [Check packet type]
        if (PT_Type == TCP || PT_Type==RTR)
Step 2. To differentiate between emergency and normal packets an emergency sign must be
marked on the packet.
Step 3. // if the calculated value of energy is less than the pre-defined threshold and an
emergency mark is not present on the packet then the route will be in down state but it remains
active for high priority and control packets and other routes can be selected for other data packets.
                if (high priority != true &&calenergyval<energy threshold level)
                        discard the current packet
                        notify the source station using ICMP source quench message
                exit (0)
//Identifying routing loop
Step 4. //If packet_origin=src
  if (src == current_nodeaddress &&cmh->num_forwards() == 0)
        Include IPV4 Header
        exit;
  //check for routing loop
  else if(origin address == address of current node)
    discard the packet.
  //node ready to forward the packet, Check the TTL. If the value of TTL is zero then packet will
be dropped
  else
    if(TTL == 0)
          drop the packet
          exit;
  //if it is a broadcast packet
  if(destination address== ip broadcast)
    if(source address == current address)
      sendOutBCastPkt() function
  else
    forwardPacket() function;
```

IPS DSDV Performance Metrics

Estimation of quantifiable metrics fundamentally requires assessment of quality of routing protocols. Following metrics are considered to judge routing protocols.

Throughput—This parameter can be defined as the amount of data or information successfully transferred from one location to another location in a given period of time.

Throughput of both routing protocols DSDV and IPS DSDV is shown in Fig. 1 for 50 nodes scenario. The values of throughput for DSDV and IPS DSDV are almost equal. From graph, it is clearly noticeable that the value of throughput for node number 10 of both routing protocols is approximately 180 units. For node number 20, the throughput value is 19 for DSDV and IPS DSDV. Hence, a decision can be formed that both routing protocols have almost same throughput.

Network Performance—Network performance examines the standard or quality of the network. Basically, the quality of services offered by the network can be measured by the network administrator.

The result of network performance for both routing protocols DSDV and IPS DSDV is illustrated in Fig. 2. This parameter considers the success rate of total packets delivered for appropriate installation of mobile ad hoc network.

In Fig. 2, it is clearly visible that network performance of IPS DSDV is more than that of DSDV routing protocol. From figure, it is clear that for node 20 scenario the value of network performance of IPS DSDV is above 92% and in DSDV for same

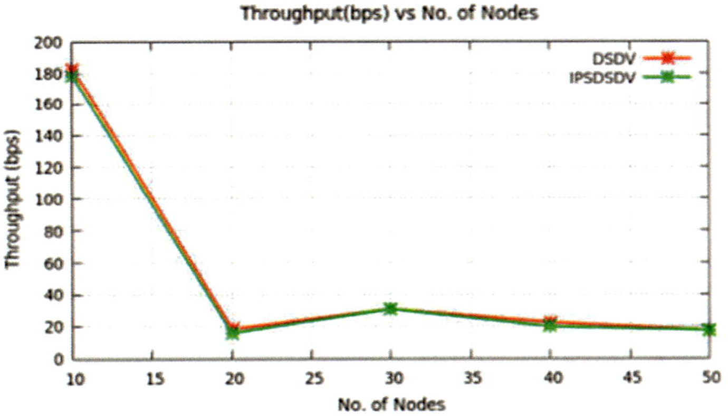

Fig. 1 Throughput of DSDV and IPS DSDV

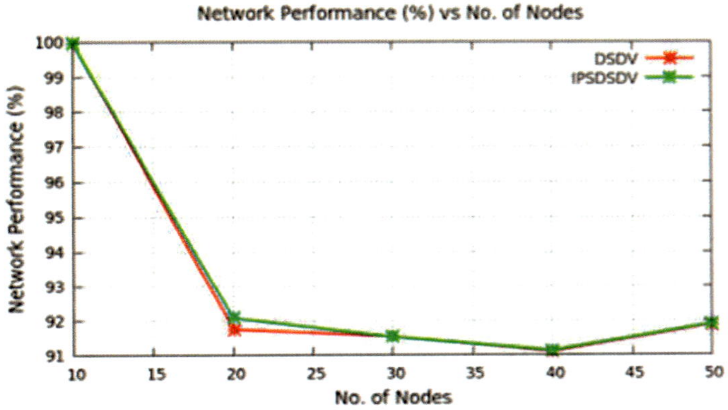

Fig. 2 Network performances of DSDV and IPS DSDV

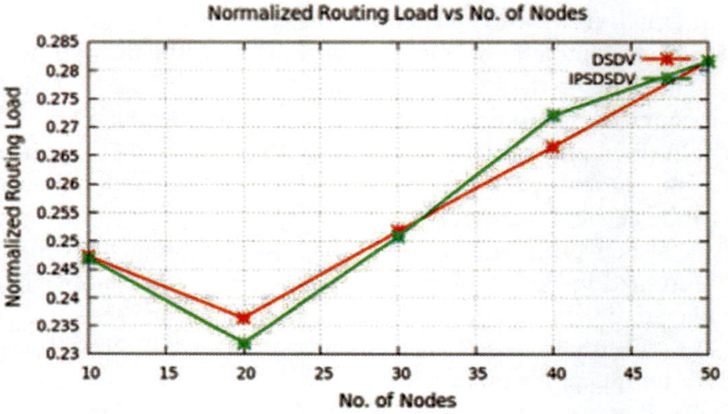

Fig. 3 Normalized routing load of DSDV and IPSD SDV

scenario the value is below 92%. Beyond 30 nodes network scenario, the values of network performance for both routing protocols are almost same. IPS DSDV provides better result in terms of network performance for small-scale networks consisting of 30 nodes. After that the overall performance is good but is equal to DSDV.

Normalized Routing Load—This parameter fundamentally narrates how well managed the routing is. It can be defined as general amount of routing package send-off to the network with respect to the received number of packets [20].

Figure 3 plotted the result of normalized routing load for IPS DSDV and DSDV. Above figure indicates that till 30 nodes scenario the normalized routing load is less in IPS DSDV comparatively to DSDV routing protocol. This can be related to the result of network performance described in Fig. 2 in which conclusion had been made that IPS DSDV performs better in small-scale networks consisting of 30 nodes. Also, in case of normalized routing load, IPS DSDV is having less routing load in 30 nodes scenario. Hence, normalized routing load is inversely proportional to network performance.

Packet Loss—Packet loss means one or more packets fail to reach their destination point.

In Fig. 4, network performance of two distinct routing protocols IPS DSDV and DSDV is examined based on percentage loss of packets with respect to successful delivery of packet. From the observation, in network with ten nodes, the percentage value of packet loss is equal in both routing protocols. There is high packet loss in both routing protocols at a scenario of 20 and 30 nodes. As the node's density increases, slight increment in packet loss percentage in DSDV routing protocol is obtained, while in IPS DSDV packet loss is less comparatively. After 30 nodes scenario, a slight decrement in value of packet loss in IPS DSDV can be noticed.

I

Energy Consumption—Energy consumption means the entire battery power dissipate by a node while transmitting a data packet from one point to another point.

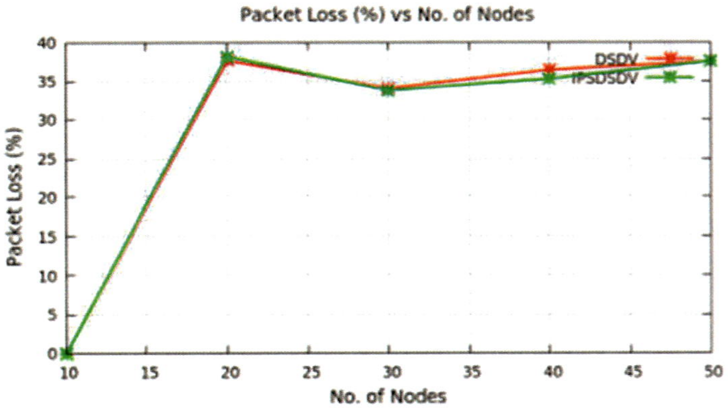

Fig. 4 Packet loss of DSDV and IPS DSDV

Average energy consumption of IPS DSDV and DSDV is depicted in Fig. 5. Energy consumed by IPS DSDV is more than that of DSDV routing protocol because of even load distribution among all nodes in a network. The main property of IPS DSDV is that to retain energy level among all nodes in a network, it takes all nodes to participate in data communication activity so that data is evenly distributed. It leads to more consumption of energy in a network as all nodes are participating in transmission activity in an equal manner. Figure 5 shows a network scenario of 50 nodes with an interval of 10. It is clearly shown from the result that across the whole scenario IPS DSDV is having more average energy consumption in comparison with DSDV routing protocol.

Remaining Energy—The parameter remaining energy or battery power defines to examine how much the energy remains in the node before further progressing the data packet.

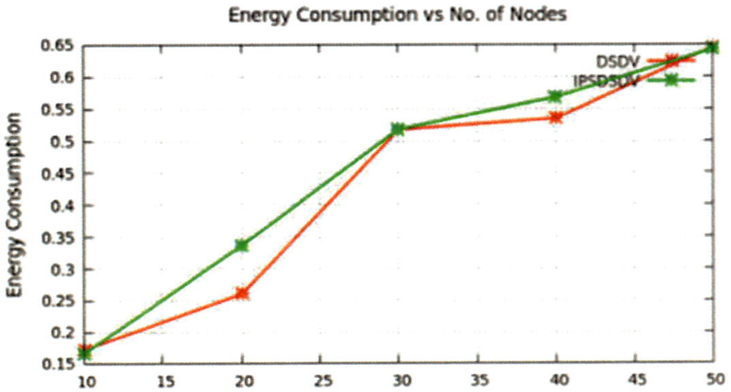

Fig. 5 Energy consumption of DSDV and IPS DSDV

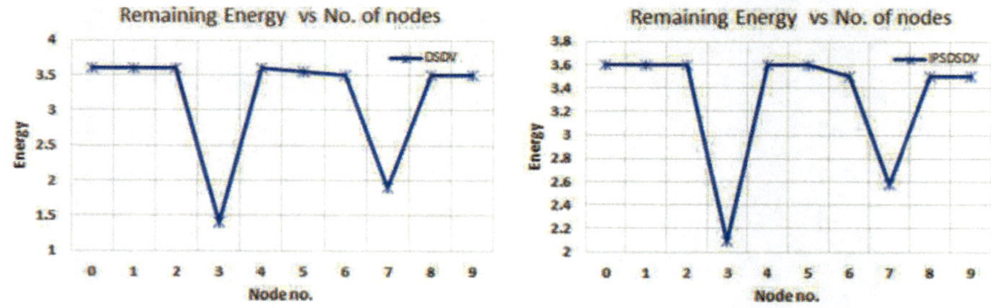

Fig. 6 Energy consumption on a network density of ten nodes

Energies to be left for ten nodes scenario of DSDV and IPS DSDV protocol are depicted in Fig. 6. In the figure, it is clearly shown that in DSDV routing protocol out of ten nodes, two nodes have less than two units of energy, means at any moment these two nodes can be terminated from the network, whereas in case of IPS DSDV according to figure out of ten nodes only one node is having minimum energy whose value is two units. This leads to the amount of energy left in nodes in IPS DSDV is better than that of DSDV because of proper dispersal of data packets throughout the network.

For 20 nodes environment, the result of remaining energy metric for DSDV and IPS DSDV routing protocol is plotted in Fig. 7. A great variance can be seen in the amount of energy left in both routing protocols. It is clearly visible from the figure that in DSDV routing protocol out of 20 nodes, six nodes are in condition of early termination from the network as the value of energy left in these nodes is less than two units. But in case of IPS DSDV, out of 20 nodes all nodes have more than two units of energy. The minimum amount of energy present in IPS DSDV result is 2.2 units of node number 7. The main intention, i.e., conservation of energy can be achieved by implementing IPS DSDV routing protocol.

Energies to be left for 30 nodes network scenario for DSDV and IPS DSDV routing protocols are graphed in Fig. 8. According to figure for DSDV routing protocol, out of 30 nodes seven nodes have 0 units of energies. While in IPS DSDV, the least amount of energy left in nodes is four units which clearly depicts that in IPS DSDV

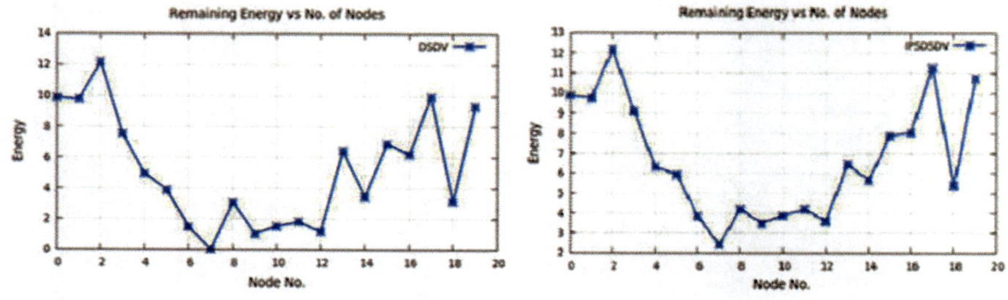

Fig. 7 Energy consumption on a network density of 20 nodes

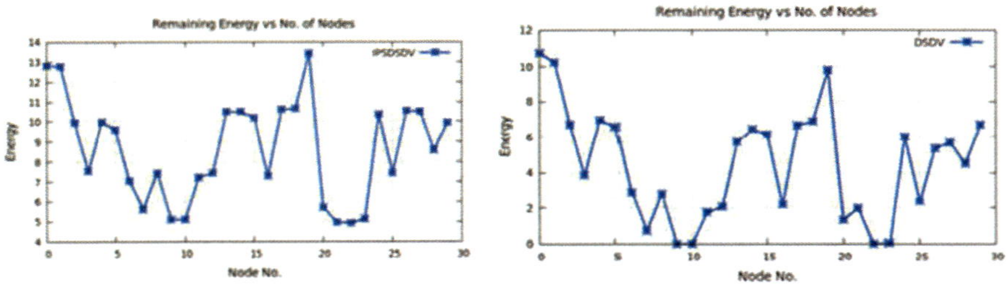

Fig. 8 Energy consumption on a network density of 30 nodes

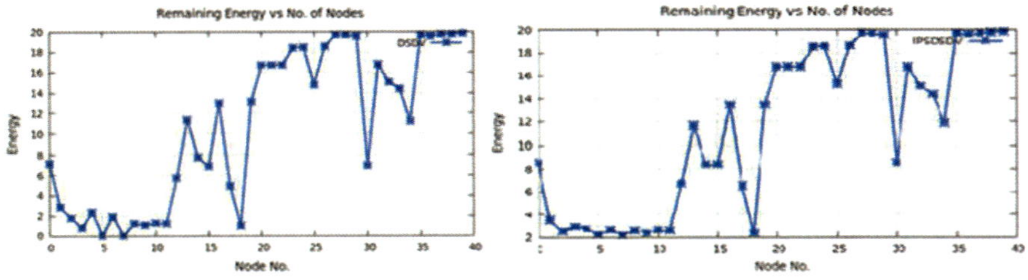

Fig. 9 Energy consumption on a network density of 40 nodes

the nodes have more amount of energy due to properly distribution of total load. The highest amount of energy left in IPS DSDV is 13.5 units by node number 19, whereas in DSDV the highest amount of energy left is 10.5 by node number 0.

Figure 9 depicts the result of remaining energies of DSDV and IPS DSDV for 40 nodes network environment. It is undoubtedly observable from the figure that there are certain nodes in the network which are in state of die out soon. In total, there are 40 nodes out of which nine nodes are in state of untimely termination (as their remaining energy value is less than 2 units) from network while in IPS DSDV none of the nodes from 40 nodes has less than two units of energy. The minimum amount of remaining energy in any node of IPS DSDV routing protocol is two units.

For 50 nodes network scenario, result for remaining energy for DSDV and IPS DSDV routing protocols is illustrated in Fig. 10. In DSDV, out of 50 nodes 11 nodes have been terminated means their residual energies are of 0 units while in IPS DSDV out of total 50 nodes eight nodes are in condition of termination. The highest amount of energy left in IPS DSDV is 13 units by node number 48 as shown in figure but the highest amount of energy left in DSDV is 12 units by the same node.

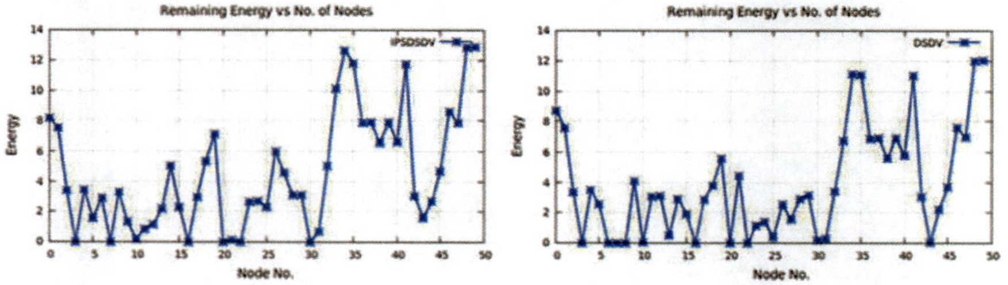

Fig. 10 Energy consumption on a network density of ten nodes

6 Conclusion

The main criteria behind the IPS strategy is to operate on the concept of making the distribution of load stable and in the interest to maintain the longer connectivity of the network, least level of energy or power is a must requirement. Therefore, from the effect of simulation results, one can observe very clearly that IPS intensifies the execution or throughput of the network. If the predefined value of threshold energy (least level of energy) matches with the value of energy of a node, then if it receives a high exigency packet then this data packet will be given priority for transmission and the normal data packet will be rejected by that node. Hence, by acquiring this innovative mechanism integration of network will be accessible by peak priority data for longer span of time, and it can be concluded from result part that after performing simulation on different scenarios of network all nodes consist of sufficient level of energy to extend the life expectancy of the network as in comparison with the conventional routing protocol in which some of the nodes die soon due to lack of energy.

References

1. M.M. Ajmal, S.A. Madani, T. Maqsood, K. Bilal, B. Nazir, K. Hayat, Coordinated opportunistic routing protocol for wireless mesh networks. Comput. Electr. Eng. **39**(8), 2442–2453 (2013)
2. S. Singha, B. Jana, S.H. Jana, N.K. Mandal, A survey to analyse routing algorithms for opportunistic network. Procedia Comput. Sci. **171**, 2501–2511 (2020)
3. P. Parvathi, Comparative analysis of CBRP, AODV, DSDV, routing protocols in mobile Adhoc networks, in International Conference on Computing, Communication and Applications (2012)
4. R. Agrawal, Performance of routing strategy (bit error based) in fading environments for mobile Adhoc networks, in IEEE International Conference on Personal Wireless Communications (2005), pp. 550–554
5. V. Jha, K. Khetarpal, M. Sharm, A survey of nature inspired routing algorithms for MANETs, in 3rd International Conference on Electronics Computer Technology, vol. 6 (2011)
6. S.V. Kanika, Implementation & analysis of IPS-AODV for emergency scenario in MANET. Int. J. Comput. Sci. Inf. Technol. **7**(4), 1783–1787 (2016)

7. M. Malathi, S. Jayashri, Robust against route failure using power proficient reliable routing in MANET. Alexandria Eng. J. **57**(1), 11–21 (2018)
8. M.E. Rouidi, H. Moudini, H. Mouncif, A. Merbouha, A balanced energy consumption in mobile Ad hoc network. Procedia Comput. Sci. **151**, 1182–1187 (2019)
9. G. Kaur, P. Thakur, Routing proctocols in MANET: an overview, in 2nd International Conference on Intelligent Computing, Instrumentation and Control, vol. 1 (2019)
10. S. Preethi, B. Ramachandran, Energy efficient routing protocols for mobile Adhoc networks, in International Conference on Emerging Trends in Networks and Computer Communications (2011)
11. C.T. Calafate, J. Oliver, J.C. Cano, P. Manzoni, M.P. Malumbres, A distributed admission control system for MANET environments supporting multipath routing protocols. Microprocess. Microsyst. **31**(4), 236–251 (2007)
12. L. Yu, B. Nazir, Y. Wang, Intelligent power monitoring of building equipment based on Internet of Things technology. Comput. Commun. **157**, 76–84 (2020)
13. S.S. Joshi, S.R. Biradar, Communication framework for jointly addressing issues of routing overhead and energy drainage in MANET. Procedia Comput. Sci. **89**, 57–63 (2016)
14. S. Vinay, S.B. Honnalli, G. Varaprasad, Multipath source routing protocol for mobile Adhoc networks with performance effective analysis, in 2018 Second International Conference on Inventive Communication and Computational Technologies (ICICCT) (2018)
15. S.S. Basurra, M.D. Vos, J. Padget, Y. Ji, T. Lewis, S. Armour, Energy efficient zone based routing protocol for MANETs. Ad hoc Netw. **25**(Part A), 16–37 (2015)
16. B.N. Jagdale, P. Patil, P. Lahane, D. Javale , Analysis and comparison of distance vector, DSDV and AODV protocol of MANET. Int. J. Distrib. Parallel Syst. **3**(2), 121–131 (2012)
17. M. Maiti, S. Mukherjee, P.K. Thakurta, An energy efficient teaching learning-based optimization approach for common content distribution in mobile adhoc networks. Comput. Electr. Eng. **72**, 296–306 (2018)
18. R. Kourdy, S. Yazdanpanah, M.R. Rad, Using the NS-2 network simulator for evaluating multi-protocol label switching in network-on-chip, in Second International Conference on Computer Research and Development (2010)
19. J.L. Font, P. Inigo, M. Dominguez, J.L. Sevillano, C. Amaya, Analysis of source code metrics from ns-2 and ns-3 network simulators. Simul. Model. Pract. Theor. **19**(5), 1330–1346 (2011)
20. Y. Jahir, M. Atiquzzaman, H. Refai, A. Paranjothi, P.G. Presti, Routing protocols and architecture for disaster area network: a survey, Ad Hoc Netw. **82**, 1–14 (2019)

An ANN Approach of Twisted Fiestel Block Ciphering

Sanjay Kumar Pal, Bimal Datta, and Amiya Karmakar

Abstract The twenty-first century is a digital era when bulk volume of data is engendering per second over the cyber world which is somehow an indispensability of digital life. So, for the purport of its auspice from unauthorized betokens, sundry encryption and decryption techniques were engendered which are predicated on different aspects of bulwarking the data. One of them is the Feistel cipher algorithm which is predicated on symmetric key cryptography where a single key is utilized for both encryption and decryption of the text. In this method, there is immensely colossal drawback that the Feistel cipher only transmutes a component of the internal state in each round, i.e., after user feeds the data, only half part of the main text is manipulated in each round which as a result is bit insecure as the left half is the pristine text for the very first round. So, to surmount this issue we have proposed one algorithm predicated on Feistel cipher which transmutes the entire internal state in each round and makes the data entered by user more secure. Not only that, our method is more time-efficient than that of the traditional Feistel cipher. In this paper, we introduced an incipient convoluted structure where our main motive is to make an algorithm which is more expeditious, lighter and time preserver than the traditional Feistel cipher.

Keywords Artificial neural networks · Block cipher · Cryptography · Decryption · Encryption · Symmetric key · Tree parity machine

S. K. Pal (✉) · A. Karmakar
Department of Computer Science and Engineering, Maulana Abul Kalam Azad University of Technology, Kalyani, West Bengal, India
e-mail: sarbojay@gmail.com

A. Karmakar
e-mail: amiya.karmakar@rediffmail.com

B. Datta
Department of Computer Science and Engineering, Budge Budge Institute of Technology, Kolkata, India
e-mail: bkd_hetc@yahoo.co.in

J. Tavares et al. (eds.), *Emerging Technologies in Data Mining and Information Security*, Lecture Notes in Networks and Systems 164,
https://doi.org/10.1007/978-981-15-9774-9_5

1 Introduction

Cryptographic methods and techniques were employed in numerous numbers of applications in today's Internet-driven communications. The cryptographic techniques are applied in modern network communications for several reasons like security of the data and efficiency of communication. Each node in the network performs distinct tasks independently and integrated in a manner such that if some nodes are not available then others can perform the task to improve the security of the information [1].

Cryptographic encryption and decryption are major processes [2]. In this processes, key generation is a crucial step. The security of an algorithm is based on the efficiency and potency of the key [3]. Confidential information needs high privacy and security from unknown attacks [4]. The integrity of data, data privacy from the sender side as well as the non-repudiation of information should be well maintained.

On the contrary, in [5], they suggested hierarchical quadrature amplitude modulation (HQAM) for better protection of high-priority data during image transmission. Nevertheless, no channel model such as AWGN is considered in this work.

Here, only salt and pepper noise is taken into account. A raised cosine filter is used in the AWGN channel introduced in [6]. Here, authors evaluate the performance of the communication channel through the bit error rate (BER). In [7], M array QAM is used in AWGN channel for evaluating the polar coding technique. In this method, the characteristic of the band limited and channels inter symbol interference not considered.

Two major categories of cryptographic methods are, namely the symmetric key encryption and asymmetric key encryption. In this paper, the symmetric key methodology is utilized, where a homogeneous key (i.e., same key is utilized for encryption and decryption process) is utilized in between the communicating entities involved. Key generation is one of the major tasks that ascertains the dynamism and security of the encrypted data, how well it is not affected to the assailments and its performance in sundry parameters shows its efficiency. In asymmetric key encryption two keys, public and private keys are utilized between the sender and receiver to encrypt the data [8].

The asymmetric key cryptographic technique was commenced from the modern cryptography era. Precisely verbalizing, this type of techniques was first verbalized by the Diffie–Hellmann key exchange technique during 1970. In Diffe–Hellmann key exchange technique, both the parties, i.e., sender and receiver jointly establish a shared secret key over a channel which is not secure and without having prior understanding. In this method, actual secret key is never shared [8].

In this paper, we have utilized Feistel block cipher technique as foundation; this method is predicated on symmetric key algorithm which denotes that it utilizes the same key for both encryption and decryption process. This method utilizes an XOR operation for encryption and decryption. We have amended the Feistel cipher by amending its efficiency and not only that but with it engenders a different secret key in every moment for same data, so detection of the secret key is scarcely infeasible [9].

2 Related Work

There are various number of encryption and decryption algorithm available, and it is pretty tough to say one the most reliable because as it varies according to conditions. An algorithm which is suited for one situation may not be suitable for another. Besides these facts, there are certain parameters which help in selecting the correct algorithm. The basic factors upon which the algorithm should be chosen are its execution time, the level of security it provides, ease of implementation and complexity of the algorithm. As the structure of Feistel cipher is not much secure to do encryption and decryption as different cryptanalysis attacks can be possible on this current weak structure of Feistel cipher [10]. Using DES algorithm [1] for encryption and decryption by improving the size of input length, i.e., bigger block size will increase the security against brute force attack but also the efficiency of encryption and decryption will be affected. In the paper, "Distributing the Encryption and Decryption of a Block Cipher" [11], the author talks about how to securely and efficiently sharing the text between the both sender and receiver using block cipher. Also, an ultra-light weighted block cipher technique has been constructed for better security of data transfer in the paper "An Ultra-Lightweight Block Cipher" [12]. One more modified Feistel Cipher was presented in paper "A Modified Feistel Cipher Involving Substitution, shifting of rows, mixing of columns, XOR operation with a Key and Shuffling Cipher", based on matrix and other substitution methods by which the security was considered but complexity did not clearly analyzed [13]. The new algorithm given in this paper is compared with the traditional Feistel cipher. With same complexity that of Feistel cipher but upgradation in security level by introducing the change in the structure of encryption and decryption with better efficiency than traditional Feistel cipher; it, as a result, increased high randomness in cipher text. Last but not least, the individual and average execution time for encryption and decryption are better than the traditional Feistel cipher which gives the reason for the proposed technique to be most elegant.

3 Terminology

This section describes the basic terminologies used throughout this paper.

3.1 Artificial Neural Network

Artificial neural networks (ANN) are unique systems of computing and simulating the approach of human brain functionalities and signal processing techniques. The biological neurons structure of the human brain is represented using the concept of graph in ANN [5]. The human brain is built up integrating 86 billion nerve cells called neurons. In artificial neural network, each neuron is represented by node. All neurons are connected by an edge or link, and it has some weight; one side of link shows the output of one neuron and other side of link shows input of another neuron. The input layer nodes of artificial neural network take input data and then process it. The outcomes of the input layer are then forwarded to the hidden layers nodes; finally we get the desired output from the output layer node [6].

3.2 Block Cipher

Block ciphers generally encrypt a group of plaintext and generate cipher text of same size group and decrypt using same concept. Block cipher uses deterministic algorithm with symmetric key in the cryptographic process. Typically, block sizes used are 64 bits, 128 bits, etc. [11].

3.3 Fiestel Block Cipher

Feistel block cipher was visualized by the eminent German scientist, physicist and cryptographer Horst Fiestel. Feistel block cipher is a symmetric structure encryption process, and it is also known as a Feistel network. Feistel cipher consists of multiple round processing of plain text. Each round of the Fiestel cipher consists of substitution step followed by permutation step. The advantage of Fiestel cipher is the process of encryption and decryption which is nearly similar, therefore the process of implementation is simple and time effective. A Feistel cipher divides the current internal state of the cipher into two parts and operates only on a single part in each round of encryption or decryption. After each round, the left and right parts switch over the sides [14].

3.4 Secret Key

The cryptography is the process of encryption and then decryption using some standard algorithm and key. The key may be private or public, and the key is generated using some key generator or key may be decided confidentially. When similar key is used by the sender and receiver for encryption and decryption process, the process is called symmetric key cryptography; the key is secret for this case, and the sender and receiver only know the key. In asymmetric key cryptography, the sender and receiver use secret key as well as public key for encryption and decryption process [15].

3.5 *Symmetric Key*

In symmetric key cryptosystem, same key is used by the sender and receiver for encryption and decryption process; the key is extremely secret in this method, other than sender and receiver nobody knows the key [15]. In private key cryptography, the sender and receiver share secret key confidentially and separately before communication over the public channel. Therefore, both the parties must have a copy of the secret key [4].

3.6 *Tree Parity Machine*

Tree parity machine (TPM) is a unique model of multilayer feed forward neural network. In this model, the connections between the nodes never form circuits. TPM is having one output neuron, H hidden neurons and HxN input neurons. TPM is the oldest model of artificial neural networks. In this method, the data moves from the input neurons to the hidden neurons to the output neuron [7].

4 TPM Approach for Secret Key Generation

The secret key is generated using tree parity machine (TPM) based on ANN. Here, user input passes through various layers, in first layer the input string is converted into a string of ASCII values and further divided into two halves. The next layer is a hidden layer, where we pass both the halves through sigma (σ), and the output produced by sigma (σ) will be sent to the final layer. This layer will be through an output which will be passed through tau (τ). We will match the value of tau (τ) for both the halves, and if the value of tau is equal for both the halves then we can say that input value is synchronized. After that, the synchronized input value will be treated further accordingly as per rules in order to generate the secret key. The complexity of the secret key generation was calculated as $O(n)$, [9].

5 Encryption Technique

In encryption technique, at first, have the plain text that needs to be encrypted, then using it, generate a secret key based on TPM. Now count the total length of input data, and if data length is even then divide it into two halves. If not even then append single space at end in order to make it even and then divide it accordingly. Now we have two equal parts of input data, say left part and right part. Convert each character of the input data with its ASCII value (now we can say that we have two arrays of ASCII values, left and right). Take the right part (r) and secret key value and put it in the function together. The output of the function needs to be XORed with the left parts (l) which as a result will give us the new left part, say L. Now do AND operation between the new left part (L) and the right part (r) which as a result will give us the new right part, say R. After this, now interchange the positions of L and R with each other. After this particular interchange, the first round of encryption process is completed. We may continue this process multiple times means multiple rounds (more the rounds higher the security). In the last round, the process will be same except the interchange. Finally, concatenating the left part and right part of data, convert it back into characters from ASCII value which will give us the cipher text which we can send to a recipient.

5.1 Encryption Algorithm

Step 1: Accept a plain text.
Step 2: Generate the Secret key using TPM, say K, [9].
Step 3: Count the total length of input data, say I.
Step 4: If input length (I) is even number, then divide it into two equal parts else add single space at the end of plaintext in order to make i even and then convert all into their respective ASCII values.
Step 5: Divide it into two equal parts, i.e., left half (l) and right half (r), now pass the values of r and K through a function f, f (r_{i}, K_i).
Step 6: Perform XOR operation with the output value of f and l to make new left half, i.e., L.
Step 7: Perform AND operation with L and r to make new right half, i.e., R.
Step 8: Swap the L and R and concatenate them (one round of encryption completed).
Step 9: Repeat steps 5 to step 8 to for (n − 1)th times, where n is total number of rounds.
Step 10: Repeat steps 5 to step 7 to for (n)th round and concatenate L with R (i.e., L + R) and convert it back into character values from ASCII values in order to generate cipher text, C.

6 Decryption Technique

At first, the cipher text needs to be decrypted which is sent by the sender. Then use the same secret key K and send to the receiver which is used during encryption. Similarly, as encryption, convert each character of the input data with its ASCII code and divide it into two equal halves (now we can say that we have two arrays of ASCII codes, left and right). Take the right half and K and put it in the function together. The output of the function needs to be XORed with the left half which as a result will give us the new left half, say L'. Now do AND operation between the L and the right half R which as a result will give us the new right half, say R'. After this, now interchange the positions of L' and R' with each other. Continue this process to $(n - 1)$ rounds, in the last round, the process will be same except the interchange, which means same number of rounds used for encryption needs to be processed for decryption. Finally, concatenating the L' and R' of data and converting back it into characters from ASCII values will give us the plain text. This is how communication takes place between sender and receiver over an open network in a secure way using proposed algorithm.

6.1 Decryption Algorithm

Step 1: Take cipher text, C.
Step 2: Take the same secret key in K which was being used during encryption.
Step 3: Convert each character of E into ASCII values.
Step 4: Divide it into two equal parts, i.e., L and R, then take pass the values of R and K through the function f, $f\ (R_i,\ K_i)$.
Step 5: Perform XOR operation with the output value of f and L to make new left half, i.e., L'.
Step 6: Perform AND operation with L' and the R to make new right half, i.e., R'.
Step 7: Swapping L' and R' will complete one round of decryption.
Step 8: Repeat steps 4–7 till $(n - 1)$th number of rounds, where value of n is same as used in encryption process.
Step 9: Repeat steps 4–6 for nth round and concatenate L' and R' (i.e., $L' + R'$).
Step 10: Treat result values as ASCII codes and convert it into characters in order to have the plain text.

7 Complexity and Experimental Result Analysis

We have calculated the complexity of the new algorithm; the complexity of the new algorithm including encryption and decryption process is $O(n)$. We recorded the individual time and the average time taken for different set of input values in Table 1 by the proposed rule and traditional Feistel cipher, with deference to the

Table 1 Experimental results of Fiestel cipher versus proposed cipher algorithms (time taken in microsecond)

No. of round	Feistel cipher	Proposed cipher
300	0.212999	0.17173
800	0.288738	0.210213
1200	0.262082	0.251693
1600	0.271724	0.257714
2000	0.377291	0.273611
2400	1.619082	0.512803
2800	0.669622	0.435288
3200	0.427197	0.493809
3800	0.372097	0.340981
4800	0.422328	0.374096
5200	0.327538	0.304479
5800	0.382442	0.287669
6200	0.912706	0.284549
6800	0.411378	0.375512
7300	0.896183	0.372337

visual examinations, one will find that the outputs of proposed rule are far lesser than that the traditional Feistel cipher. Hence, we can verbalize that the proposed rule is bit more expeditious in the encryption and decryption processes than that of traditional Feistel cipher. All the visual examinations are recorded in same environment, i.e., each readings recorded in same machine.

The hardware used for experiment is Intel i3 six generation processor, and the fortified system is a 64-bit windows machine with 8 GB RAM capability [16]. The programming language is used for implementation, and the algorithm is Python (Fig. 1).

The experimental results are presented in tabular form as well as in graphical form to make a better visual understanding for observation of the execution values of both the algorithms with different lengths of input values. In Table 1, the first column, i.e., "Round", represents the input length of the message, the second column shows execution time taken by traditional Feistel cipher method, and the third column shows execution time taken by the proposed cipher method.

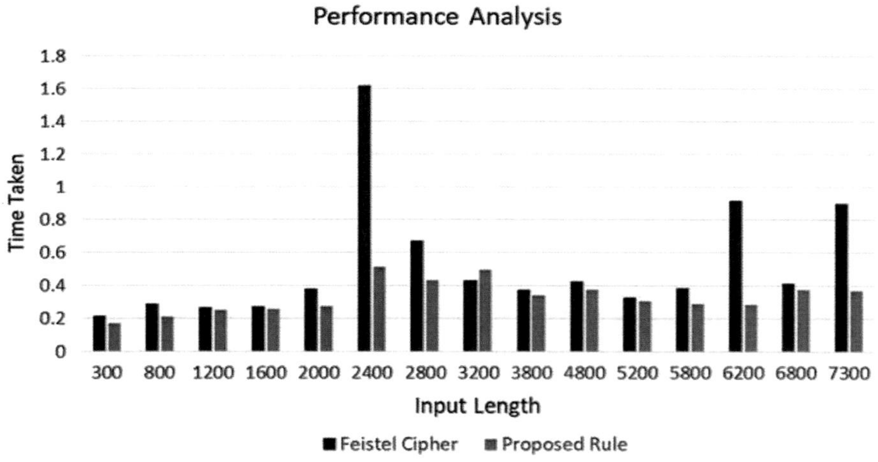

Fig. 1 Comparative chart of the experimental result shown in Table 1

References

1. B.F. Cruz et al., Expanded 128-bit data encryption standard. Int. J. Comput. Sci. Mob. Comput. **6**(8), 133–142 (2017)
2. K.P. Sanjay et al., An encryption technique based upon encoded multiplier with controlled generation of random numbers. Int. J. Comput. Netw. Inf. Secur. **7**(10), 50–57 (2015)
3. K.P. Sanjay et al., Revolutionary change in cryptography. Invertis J. Renew. Energy **9**(2), 43–54 (2019)
4. K. Sarita, A research paper on cryptography encryption and compression techniques. Int. J. Eng. Comput. Sci. **6**(4), 20915–20919 (2017)
5. N.E.K. Mohamed et al., Artificial neural networks in data mining. IOSR J. Comput. Eng. **18**(6), 55–59 (2016)
6. M.P. Jonathan et al., Security evaluation of tree parity re-keying ma-chine implementations utilizing side-channel emissions. EURASIP J. Inf. Secur. **3**(1), 1–16 (2018)
7. S.K. Pal et al., Cryptography based on RGB color channels using ANNs. Int. J. Comput. Netw. Inf. Secur. **5**(5), 60–69 (2018)
8. S. Karthik, A. Muruganandam, Data encryption and decryption by using triple DES and performance analysis of crypto system. Int. J. Sci. Eng. Res. **2**(11), 24–31 (2014)
9. K.P. Sanjay et. al., An TPM based approach for generation of secret key. Int. J. Comput. Netw. Inf. Secur. **11**(10), 45–50 (2019)
10. Amandeep et al.: Research problems in block cipher cryptanalysis: an experimental analysis. Int. J. Innov. Technol. Exploring Eng. **8**(8), 323–327 (2019)
11. K.M. Martin et al., Distributing the encryption and decryption of a block cipher. Des. Codes Crypt. **36**(3), 263–287 (2005) (USA)
12. A. Bogdanov et al., PRESENT: an ultra-lightweight block cipher, in Cryptographic Hardware and Embedded Systems-CHES 2007, Switzerland (2007), pp 450–466
13. V.U.K. Sastry, K.K. Anup, A modified feistel cipher involving substitution, shifting of rows, mixing of columns, XOR operation with a key and shuffling cipher. Int. J. Adv. Comput. Sci. Appl. **3**(8), 23–29 (2012)
14. S. Vidhya et al., Format preserving encryption using feistel cipher, in *IJCA Proceedings on International Conference on Research Trends in Computer Technologies*, vol. 1 issue No. (1), (2013) 5–8

15. S.A. Dar, RSA algorithm key generation extension. Int. J. Mod. Trends Eng. Res. **5**(1), 73–75 (2018)
16. K.P. Sanjay et al., Application of Cosmos's law of merge and split for data encryption. Int. J. Comput. Netw. Inf. Secur. **5**(10), 11–20 (2017)

Musical Instrument Classification Based on Machine Learning Algorithm

Hasanuzzaman Anuz, Abu Kaisar Mohammad Masum, Sheikh Abujar, and Syed Akhter Hossain

Abstract Musical instrument classification from an audio file is a very interesting and important topic in machine learning. In this paper, we represent a method to classify a musical instrument from a single audio file of a specific instrument. We focus on classifying six musical instruments that are very popular for Indian subcontinent, basically used to folk songs. It is also helpful for music genre classifiers. A fairly small dataset contains 600 audio files from harmonium, flute, monochord (ektara), cylindrical wooden drum (dhol), tawala, and violin that are classified using MFCC and various types of classifier. MFCCs are based on signal disintegration with the help of a filter bank. The great things of MFCCs over spectrogram is that they try to model the way perceive like frequency. To classify musical instruments, we are used as the k-nearest neighbor and support vector machine classifier with RBF kernel which provides optimum classification ability. A very high accuracy is achieved (97%) on the test set of our generated dataset.

Keywords SVM · KNN · Kernel · MFCCs · Feature extraction

H. Anuz (✉) · A. K. M. Masum · S. Abujar · S. A. Hossain
Department of Computer Science and Engineering, Daffodil International University, Dhaka 1212, Bangladesh
e-mail: hasanuzzaman15-10298@diu.edu.bd

A. K. M. Masum
e-mail: mohammad15-6759@diu.edu.bd

S. Abujar
e-mail: sheikh.cse@diu.edu.bd

S. A. Hossain
e-mail: aktarhossain@daffodilvarsity.edu.bd

J. Tavares et al. (eds.), *Emerging Technologies in Data Mining and Information Security*, Lecture Notes in Networks and Systems 164,
https://doi.org/10.1007/978-981-15-9774-9_6

1 Introduction

Musical instrument classification is a popular topic and problem of musical information retrieval (MIR) section. In MIR, most of the music is in digital format then we use mp3. There are different types of features [1] used to classify musical instruments LPC, LPCC, LSF, DWT, MFCC, etc. We use MFCCs feature extraction method for capturing textural or timbre [2] aspects from audio sounds. The music can be in various textures like monophonic, polyphonic, homophonic, and bi-phonic. We use monophony that deals with only one instrument being played at a time. Sound of same musical instrument has the same feature. Timbre is important in music when you want to differentiate a piece of tone, music.

Suppose you play violin and piano at the same frequency or pitch and same rhythm [3], it is difficult to classify which one is violin or piano from waveform, to classify we should need a tumbrel aspect of audio. Each and every instrument has its unique timbre, even two violins can sound totally different from each other. A professional or beginner musician plays an instrument in a different timbre. In general, musical instruments, timbre, are divided into four grouped by their sound generated source (1). Metal. 2. Air. (3) Lather. and (4) Wire. Timbre mainly describes harmonic content of a sound when an instrument generates a sound and this sound is actually a combination of pitches or harmonics. Many methods are available to successfully extract features from sound samples [4]. In [5], it compares several features that are used to classify. Feature selection is the key problem because it is closely tied to the perception of sound and its physiological effects.

In this work, as a feature extraction method we use Mel frequency cepstral coefficient (MFCC). To find out similar sounds, features extracted by using the same method are compared with each other based on a classifier algorithm named K-nearest neighbor (KNN) and support vector machine (SVM). Although our dataset contains 600 samples of 6 instruments using KNN and SVM with RBF kernel, we get high accuracy.

2 Related Work

Previous research on this topic focuses on features to accurately identify instruments. Guignard et al. [6] build machine learning models to construct classifiers from a monopoly audio. Their dataset contains 10 musical instrument samples that are basically used in rock music and audio file in mp3 format. They extracted features using discrete Fourier transform of the normalized signal. For regularization, use of tenfold cross-validation trains on 90% and 10% tests. Three models are applied PCA, K-means, and SVM with various kernels.

Singh et al. [7] made a technique to construct a classifier. The audio aspects firstly preprocessed into its visual representation (e.g., graph or image) because it is difficult to differentiate between similar sounding instruments like violin, piano. In music recommendation, it is easy to suggest if the listener prefers a particular instrument. Every audio convert into an image using Librosa audio analysis package. Mel spectrogram and MFCC were used for feature extraction and also used weighted average of prediction. CNN classification gives high accuracy. In classification problem, applying SVM classifier is a great idea which can give us a better result.

Azarloo et al. [8] build a model of musical instruments that can perform automatically on every possible combination of instrument. K-nearest neighbors and multi-level perceptron is used to classify. Four types of features are uses temporal, cepstral, spectral, and perceptual and finally use UTA algorithm for average feature from all feature into one. After the feature selection, UTA algorithm accuracy increased. Multi-layer perceptron classifier average accuracy is better than the KNN classifier.

3 Methodology

The first step in audio data is preprocess the signals. We start from a waveform then we extract MFCCs, and we pass MFCCs directly on our machine learning model two models we use SVM and KNN.

(A) Feature Extraction

MFCCs are the important aspect for standard features from all feature sets. MFCCs are a way of that can capture the information tumbrel quality of a sound [9]. For extracting MFCC, we perform a Fourier transform. And we goes time domain into frequency domain. The great things of MFCCs over spectrogram are that they approximate the human auditory system. The Mel scale divides the uniform frequency into filter banks of different bandwidth. MFCCs are used in speech recognition and different types of applications that are audio similarity. We will give basic information about MFCC [10] and then describe why we do things. The relationship between frequency and Mel scale is given below:

$$M(f) = 1125 \ln\left(1 + \frac{f}{700}\right)$$

Figure 1 presents the block diagram for extracting MFCC features.

We will now describe why each step is needed. Basically, an audio signal changes continuously then we need a short timescale where audio will not change as much. By sampling, we slice an audio several piece. We apply FFT [11] to convert audio signals into frequency domain from time domain then get a spectrum. Next step is to calculate power of the spectrum. The Mel scale tells us exactly how to space our filter banks and how wide to make them.

$$\text{Mel frequency } (m) = 1125 \log\left(1 + \frac{f}{700}\right) \tag{1}$$

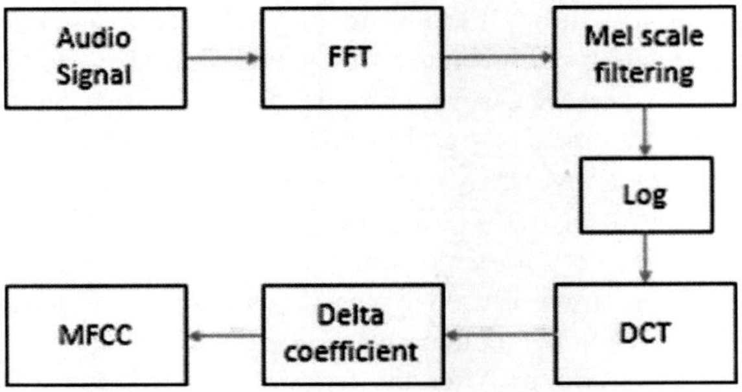

Fig. 1 MFCC feature extraction

$$\text{Mel to original frequency } (f) = 700\left(\exp \frac{m}{1125} - 1\right) \quad (2)$$

When we get the filter bank energies, then we apply logarithm of the energy. We use cepstral mean subtraction by logarithm, also known as a channel normalization method. In filter bank, energies are correlating with each other DCT is used to de-correlates the energies. DCT [12] is used to convert the data into the summation of a series of cosine waves oscillating at different frequencies. Deltas and double deltas are a differential and acceleration coefficient. The MFCCs' vectors describe only the power spectral envelope of a signal frame but delta cepstral features capture dynamic information and greatly improve the accuracy. Delta coefficient formula:

$$d_t = \frac{\sum_{n=1}^{N} n(c_{t+1} - c_{t-n})}{2\sum_{n=1}^{N} n^2}$$

MFCC captures the timbre aspect of sound, and its feature is also frequency domain. MFCC coefficient range is 13–40.

Finally, spectrogram is fundamental for performing machine learning application on audio. Whole preprocessing pipeline for audio data for machine learning is based on spectrogram. Spectrogram is an input for our machine learning model.

(B) **SVM with RBF kernel**

SVM is most probably very useful algorithm for classification problem. SVM is used many place, many problem in different way. For this purpose, many optimization methods are use and also have many applications. Before apply this method, first we know the core concept.

In Fig. 2 shows that, round shape points are in class 1 and star shape points are in class 2. Now we calculate the distance in between each and every point from one class to another class and try to find out which points are closes one to another between two classes. If we try to separate two classes by drawing a line called decision boundary, then we can clearly notify the closest points from the line are (4,2), (4,6), and (6,6) called as support vector. On the left-hand side of the red line points (4,2) and (4,4) in class 1 are the closest points from the line and their distance is $d1$. As we can

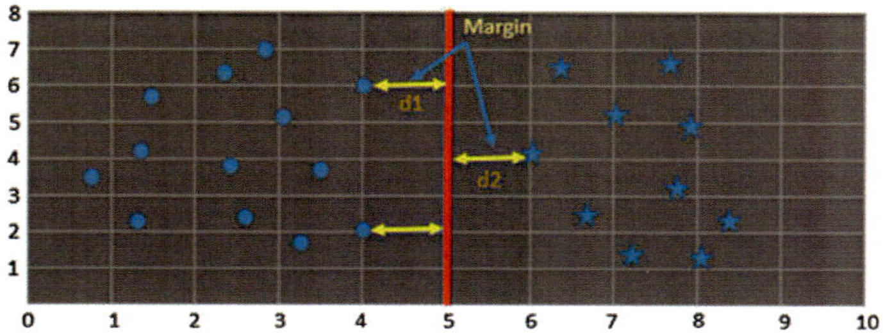

Fig. 2 SVM

say that minimum distance is $d1$ from the line to class 1 points. Same as in class 2, minimum distance of line is $d2$. Now total distance between two classes is $D = d1 + d2$ called as margin. Our aim is to maximize this margin (D). Because it might be chance overlapping when points of two classes stayed at closely and also misclassify one class by another class. Support vector points help us to correct the decision boundary and margin both. Rest of the points are not more useful in classification. If we move our support vectors then decision boundary moves with vectors, and the value of margin will be changed. If the value of margin changes, it will affect our classification. We should maximize the value of margin to properly set the decision boundary into points. Support vector machine helps us to find the decision boundary for the maximum value of margin. Now we work a kernel function known as radial basis function (RBF) kernel. Formula of finding every feature data using this kernel is:

$$F_i = \mathrm{e}^{\frac{-||x - x_i||}{2\sigma^2}}$$

Let us consider Table 1 is our dataset. Now we use kernel function for point (4, 5) and try to find out new feature point. Our input feature is $X_1 = 4$ and $X_2 = 5$. If we apply our dataset to the kernel input, our feature will be 3 because no of training data is 3. Our new feature is given below:

$$\sigma^2 = 1, \gamma = \frac{1}{2\sigma^2}$$

$$f_1 = \mathrm{e}^{\frac{-||x - x_1||}{2\sigma^2}} = \mathrm{e}^{\frac{-(4-2)^2+(5-3)^2}{2\sigma^2}} \approx 0$$

Table 1 Imaginary dataset

X_1	X_2
2	3
4	5
6	7

Fig. 3 KNN

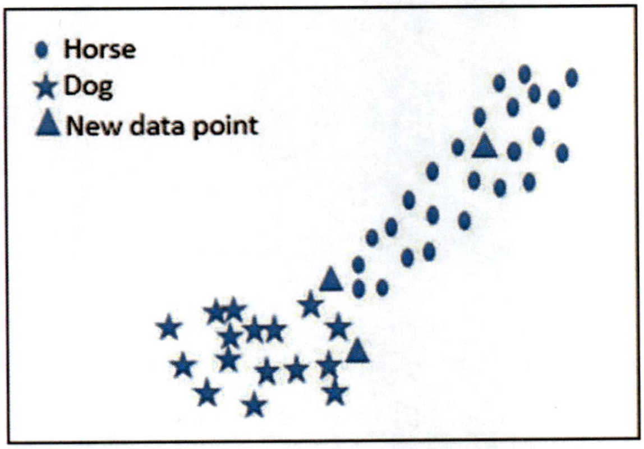

$$f_2 = e^{\frac{-||x-x_2||}{2\sigma^2}} = e^{\frac{-(4-4)^2+(5-5)^2}{2\sigma^2}} \approx 1$$

$$f_3 = e^{\frac{-||x-x_3||}{2\sigma^2}} = e^{\frac{-(4-6)^2+(5-7)^2}{2\sigma^2}} \approx 0$$

Our new feature vector is $(f_1, f_2, f_3) \approx (0, 1, 0)$
RBF kernel uses two parameters.

1. Gamma = how spread the region is
2. C = Penalty for misclassification

(C) **KNN**

KNN is a very simple algorithm but if we apply the KNN properly, it will be very fruitful to us. It is best known though example! Imagine we had some imaginary data on dogs and horses with height and weights as shown in Fig. 3.

Calculate the distance from new data point in your data. Sort the points in your data by increasing distance from new point. From the sorting data, first collect number of (K) points. In k-number data points, new point will be the class which contains most of the points. Before implementing the KNN, we need to preprocess the data. Here we used standard scale function and transformer class. The idea behind standard scalar is that normalization or standardization. Normalization means its scaled our dataset and the value range 0–1. Standardization is little bit different scaled the data over the standard deviation. In case of multivariate data, this done feature-wise. Given the scattering of the data points, every value in the samples will have the average subtraction value, and then subtraction value divided by standard deviance. How to calculate scaling [13]

(i) Standardization: $z \frac{x-\mu}{\sigma}$
(ii) With mean: $\mu = \frac{1}{N}\sum_{i=1}^{n}(X_i)$
(iii) Standard deviation: $\sigma = \sqrt{\frac{1}{N}\sum_{i=1}^{n}(X_i - \mu)}$

(iv) Min–Max Scaling:

$$X_{\text{norm}} = \frac{x - x_{\text{min}}}{x_{\text{max}} - x_{\text{min}}}$$

For choosing, the appropriate value of K is difficult without any technique. We need to apply the elbow method because using the elbow method, we can get the best value of k by checking low error rate. A small of K-means that K leads to a sensitive model and a large value makes it computationally expensive. Data scientists usually choose an odd number if the number of classes is 2.

4 Results

Out of 600 audio files 75% are randomly assigned training and 25% test. Scikit-learn is used for split the dataset. For feature extraction purposes, we use MFCCs. Figures show how we convert an audio file into its Mel frequency spectrum.

In Fig. 5, we show the waveform of our input audio file. Real world sound waves are not simple as the sine waves. This is like a massy sound that there is a lot of like complexity in it. Actually it turns out the nature has given us like an incredible way of knowing quite a lot about complex sounds and that is given though a Fourier transform.

In Fig. 6, we see the power spectrum of our signal by using FFT. Basically, the power spectrum gives us the magnitude as a function of a frequency. When we do Fourier transform basically, we move from the time domain toward the frequency

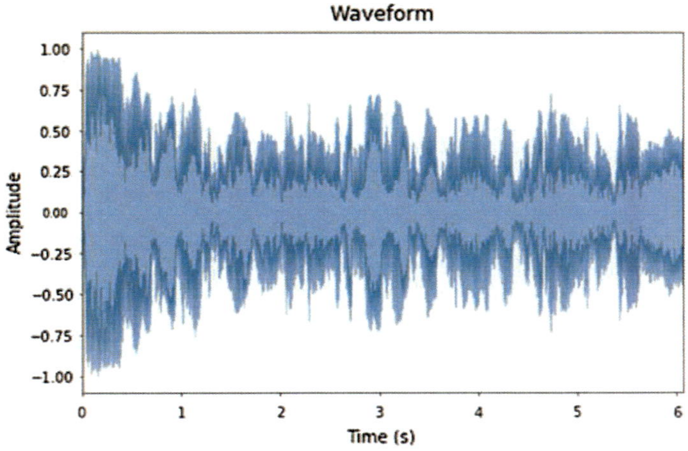

Fig. 5 Waveform

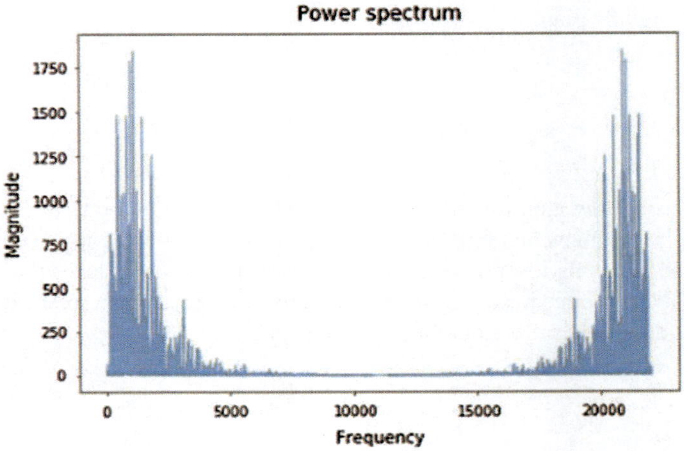

Fig. 6 Spectrogram

domain. Because this happens we lose information about time so it is as if this is a power spectrum. Power spectrum is telling us that difference frequencies have different power but throughout all of the sound here so its a snapshot, it is static.

In Fig. 7, we see short-time Fourier transform(STFT) or DFT. It computes several Fourier transform at different intervals and in doing so preservers information about time. At the spectrogram of the sound file, we can see we are back in time on the x-axis but we also have frequency on the y-axis and we have a third axis basically given by the color and the color is telling us how much a given frequency is present

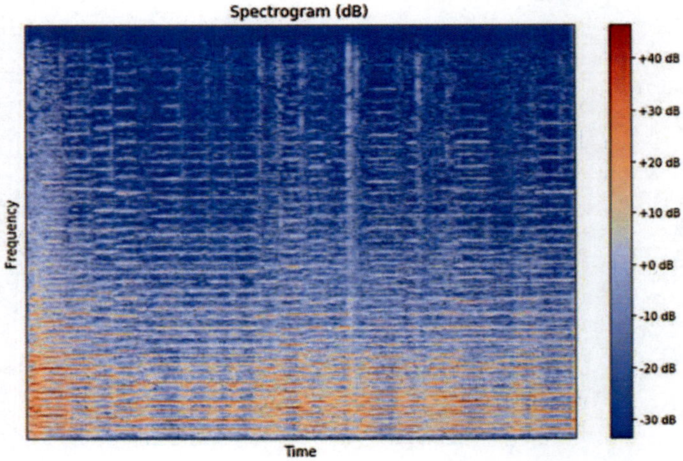

Fig. 7 STFT

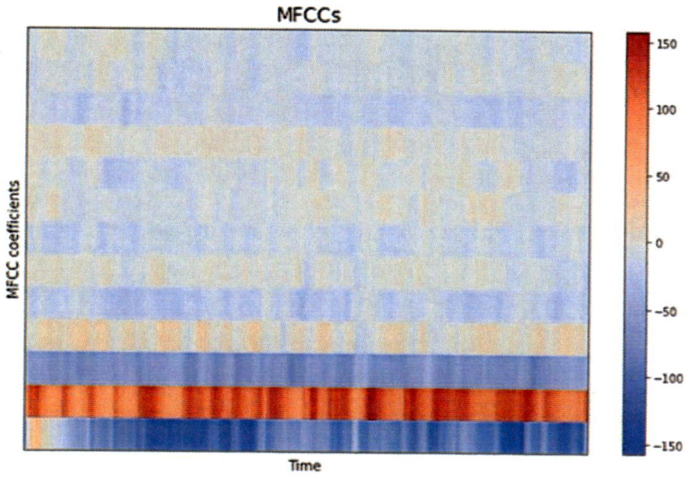

Fig. 8 MFCCs

in the sound at a given time. Mel frequency cepstral coefficients are shown in Fig. 8. On the y-axis, we have the different MFCC coefficients and thirteen coefficients are here. Value of the coefficient is represented with color by the color bar. As usual, the more red the value will bigger itself.

In Table 2, we show the SVM model performance for every instruments. Here accuracy of flute is very high for every steps of accuracy. Also, average recall is good for each instruments.

In Table 3, we show the KNN model performance for every instruments. Here accuracy of Ektara is very high for every steps of accuracy. Average precision is good for each instruments.

In Table 4, we show the accuracy of our applied models. Two algorithms we use (i) SVM and (ii) KNN. SVM is always best for the classification problem that provides best accuracy result for our audio samples.

Table 2 SVM class performance

Class	Recall	Precision	$F1$-score
Harmonium	1	0.961	0.980
Flute	1	1	1
Ektara	1	0.961	0.980
Dhol	0.88	1	0.936
Tawala	1	0.961	0.980
Violin	0.96	96	0.96

Table 3 KNN class performance

Class	Recall	Precision	$F1$-score
Harmonium	0.96	1	0.979
Flute	0.958	1	0.978
Ektara	1	1	1
Dhol	0.84	1	0.913
Tawala	1	0.892	0.943
Violin	1	0.892	0.943

Table 4 Model accuracy

Algorithm	Accuracy
SVM	97
KNN	96

5 Conclusion

The model can classify musical instruments from a monophonic audio. We sampled audio at 20,500 per second and converted into frequency domain. MFCC is used for feature extraction. Two different classifiers we use are KNN and SVM. Study of MFCC, KNN, and SVM in our system has been done by using some reference. Six musical instruments contain our dataset. By using MFCC, we get a better performance, and SVM gives high accuracy compared to KNN. Future work is for this work and uses more features such as discrete wavelet transform (DWT) for polyphonic audio.

Acknowledgements We would like to thank DIU NLP and Machine Learning Lab, CSE Department.

References

1. S. A. Alim, N.K.A. Rashid, Some Commonly Used Speech Feature Extraction Algorithms, in *From Natural to Artificial Intelligence—Algorithms and Applications*, ed. by R. Lopez-Ruiz (IntechOpen) (2018, December 12). https://doi.org/10.5772/intechopen.80419. Available from https://www.intechopen.com/books/from-natural-to-artificial-intelligence-algorithms-and-applications/some-commonly-used-speech-feature-extraction-algorithms
2. S. McAdams, Musical timbre perception. Psychol Music 35–67 (2013)
3. J. Burred, A. Roebel, X. Rodet, An Accurate Timbre Model for Musical Instruments and its. in Application to Classification Workshop on Learning the Semantics of Audio Signals (Athens, Greece, pp. 1–1, December 2006). https://hal.archives-ouvertes.fr/hal-01161413
4. G. Agostini, M. Longari, E. Pollastri, Musical instrument timbres classification with spectral features. EURASIP J. Adv. Signal Process. **2003**(1), 943279 (2003)

5. A. Eronen, “Comparison of features for musical instrument recognition, in *Proceedings of the 2001 IEEE Workshop on the Applications of Signal Processing to Audio and Acoustics (Cat. No. 01TH8575)* (IEEE, 2001), pp. 19–22
6. L. Guignard, G. Kehoe, *Learning Instrument Identification* (Stanford University CS229 Machine Learning Autumn, 2015)
7. P. Singh, D. Bachhav, O. Joshi, N. Patil, Musical instrument recognition using CNN and SVM. Int. Res. J. Eng. Technol. (IRJET) **6**(3) (2019)
8. A. Azarloo, F. Farokhi, *Automatic Musical Instrument Recognition Using K-NN and MLP Neural Networks* (Faculty of Electrical and Electronics Engineering Islamic Azad)
9. https://hal.archives-ouvertes.fr/hal-02270931 Contributor: PatrickSusini patrick.susini@ircam.fr Submitted on: Monday (2019 September 2). (2:19:52 PM)
10. http://practicalcryptography.com/miscellaneous/machine-learning/guide-mel-frequency-cepstral-coefficients-mfccs/
11. U. Oberst, The fast fourier transform. SIAM J. Control Optim. **46**(2), 496–540 (2007)
12. http://datagenetics.com/blog/november32012/index.html
13. https://sebastianraschka.com/Articles/2014_about_feature_scaling.html#bottom-up-approach

Analytical Survey and Comparative Analysis of Socialized Network via Data Mining and Machine Learning Techniques

Akhilesh Kumar Singh, Manish Raj, Deepak Mangal, and Saurabh Singhal

Abstract We are aware that almost all the complex systems that exist in real world can be framed as a network commonly termed as socialized network. The most important feature of real system that can be represented in this socialized network is clustering or community structure. The proposed research throws attention on the analytical comparison and theoretical review of socialized network with respect to analysis and explorations of big data. The prime focus in this work is to discuss the issues related to socialized network such as exploring and finding communities or cluster, searching based on personalization and individuality, anomaly identification, and other relevant issues. As per the various surveys, we know that community identification is very common problem in socialized network so detection of communities plays an important task in various scientific domain and analysis. The inclination of comparison in this paper is toward improvisation and optimization of mentioned issues on technological advancement of genetic algorithm, machine learning, and data mining. Several different properties of socialized network such as embeddedness and triadic closure are also reviewed.

Keywords Big data · Community detection · Clustering · Genetic algorithm

A. K. Singh (✉) · M. Raj · D. Mangal · S. Singhal
GLAU Mathura, Mathura, UP, India
e-mail: akhileshkr.singh@gla.ac.in

M. Raj
e-mail: manish.raj@gla.ac.in

D. Mangal
e-mail: deepak.mangal@gla.ac.in

S. Singhal
e-mail: saurabh.singhal@gla.ac.in

J. Tavares et al. (eds.), *Emerging Technologies in Data Mining and Information Security*, Lecture Notes in Networks and Systems 164,
https://doi.org/10.1007/978-981-15-9774-9_7

1 Introduction

The main concept of socialized network lies in the billion of nodes which are considered to be ubiquitous across varieties of domain. These are sometime referred to be structureless and forms different topology of graph. The main point of view in these socialized networks is community structure. The main attention of this proposed work is toward the identification, exploration, and analysis of various challenges and issues of social connectivity and network that can be categorized as exploring and finding communities or cluster, searching based on personalization and individuality, anomaly identification along with balancing of network structure and cut space problem. As the concept of exploring community detection explains data from various socialized active Web sites, Web services and cellular devices are stored and various processing based on predefined attributes are exercised on these gathered data. These refined processed socialized data are again forwarded for analysis on the basis of various parameters like connectivity, efficiency, number of nodes, etc.

The process flow of the evolutionary traditional algorithm defines that the process start with basic step of selection in the evolutionary algorithm in which the optimized best plotted algorithm and individual is selected for the subsequent processing and rest are dropped and discarded or can be used in the further round of iterations. If any next individual is found that is more best fitted, the current selected individual cannot survive and hence discarded. This repetition of selection is continued until the criteria of finding best individual are met. The concept of socialized networking involves the collaboration within the general populated advancements. The incorporation of quantitative and subjective associations along with collaborative behavior and other individuals within set of family or friends, group of companies, and co-employees and co-workers. The expansion of network can be handled through various paths inside socially active destinations that can be fulfilled via of directly or indirectly connected network that exists within a network. Socialized network is designed by socially active researchers to human interpersonal that organize mathematically and physician-based complicated system hypothesis and also in recent time the researchers are in the examination of data or Internet-powered socialized network. The chance of getting two networks expanding their feature of availability by exchanging thoughts of their own area important to one another must be incorporated also. Hence, various ideas and suggestions are flooded inside a socially networked platform Web site. The massive expansion of socialized connectivity, interconnection, and intercohesion inside data is also increasing. This diversification convinced people to generate, interpret, and process the recent finding of information and knowledge. Among all, major of the data are secondary, non-primitive, derived and accommodates inside the internetwork, there is always a continuous pressure of choosing how to collect, disseminate, along with follow up of the vast accumulated amount of information.

Vast information is continuously analyzed and utilized using the concepts of velocity, i.e., rate at which information is generated, value is finding of valuable processed data, volume referring to the measure of the information produced, veracity means governance and variety refers diversification and bifurcation of information spread. The current informal data communication social networking platform Web sites have caught numerous amount of data coordinated with day by day cooperation of each and every individual that associates and supports vast client populace [1].

Generally, to generate solution of complicated problems, traditional evolutionary algo imitate problem of biological imitation in the form of applications of computer field. Various mechanisms like random sampling and finding one best solution of problem are deployed and hence evolutionary algorithms can be considered as the optimized solution for various social connectivity and networking issues because of its numerous benefits. Out of its several importance, one major beneficial situation of evolutionary algo is its adaptability gains to even complicated situations and problems as it is fit to achieve the target focus. Further, much optimized improvement is achievable with evolutionary algo.

The outline of this paper is as follows: in Sect. 2, various challenges of socialized network are summarized. In Sect. 3, related work and comparative analysis are explained. In Sect. 4, based on the results, proposed application is concluded.

2 Various Challenges of Socialized Network

Considering current era, socialized platform and network sites proven to be the latest and most extensively accepted and widely used platform for social connection establishment, socialized gathering, and exchange of personal information in a fastest manner. Various social connectivity and platform and Web sites like Orkut, Instagram, Facebook, and applications of Google like YouTube allow having the above-said functionalities. As the consumption of the Internet is increasing day by day, a majority of population is now involved in these socialized sites as per their requirement and interest. These social connectivity network platforms are considered to be useful as per their interest but several challenges are need to be issued during the usage of these platforms. Some common challenges are listed below in regard with the social connectivity and networking platforms.

Seed Selection: The particular vertex selection among the collection of present vertices enhances the probable outcome of the connectivity within the socialized network. The problem of identifying and exploring the group of end users that influences the other users within a socialized network is NP hard, thus seed selection is sometimes performed on the static network sometimes on evolutionary and greedy algo.

Signed Network: The measurement of relation factor between end user on socialized network is combination of friendly and antagonistic result. Thus to get better knowledge on socialized network, the best efficient tool is signed network. The graph of signed network is often signified as a triangle which is considered as balanced and imbalanced. It will be in balanced state only if the product of signs is positive otherwise it will be treated as imbalanced. If all the conditions of structural balancing do not fulfill, the graph will be treated as unbalanced.

Cluster and Community Detection: Clustering is the way of making a collection of same items together in structures known as clusters. The hierarchy of cluster is formed in hierarchy clustering. A cluster identification and recognition depend on the condition of intra-association joins and togetherness between associations joins. A community can be well defined as the collection of entities closer to each other with respect to others in the complete datasets. Within the community, the individual interacts and exchanges information with each other more frequently than from the individual of outside group. Thus, cohesiveness within the community is very high. Community detection is a performance enhancement strategy within the socialized network for searching and detecting maximum modular density partitions inside the socialized network. If the nature of selection feature used varies, number of the clusters varies. Also, all clusters are not necessarily of same size.

Anomaly Detection: It is a self-configured rearrange able inconsistent detection mechanism which follows naturally inspired mechanisms like ant colony optimization and insect colony optimization that identifies the anomaly on the basis of collaborative effort of nodes that uses non-global information.

Other parameter includes individual search, security analysis, and routing.

3 Related Work

Various researchers have proposed many works that include algorithm to handle large or very large-scaled signed networks. The concept of multi-evolutionary algorithm is also introduced. Various researches have been introduced in the field of social connectivity, seed selection, and personalized search, and some researchers have highlighted the cluster detection as well as community detection work. Genetic algorithms and swarm optimization were also introduced by various researchers. The study of various fitness functions is also introduced in the social networking area. The community score fitness function is also introduced. Algorithms like MEA-SN, MOEA-SA, NMI, PSO, GA-SA, and IEDA are also highlighted by many prior authors. Some paper includes the community detection algorithm which handles large-scale signed networks. Some detect clusters via study of various fitness functions. Some undergo community detection with the use of cat swarm optimization. Some includes normalization of datasets, while some uses adjacency graph representation, some focuses on security issue while some focuses on multi-evolutionary algorithms. Some introduces the concepts of vertex properties and topological structure in social networking and social connectivity area.

Main of the focus is also given by various researchers toward anomaly detection and searching cut spaces. Works like EDA, DIAMOND, and Edge set encoding are also discussed prior. Issues like finding a littlest subset of hubs, search based on individual interest, vertex-based encoding, indirect and direct merged representation of communities, seed node determination, utilizing general reason registering on designs handling unit, and multiple decisions making are also proposed. The concept and issue of clustering, community detection, gradually reducing random variation, evolutionary algorithm, genetic algorithm, and social networking were seriously examined during the last decade. The wellness works in evolutionary methodology toward network identification in the complex system are given by Jora and Chira [2] in the year 2016. Jami and Reddy [3] in the year of 2016 proposed the community detection based on evolutionary algorithm. Lighari and Hussain [12] introduced the composite model for clustering analysis and rule-based for big data security using apache spark in 2017. Yan et al. [4] introduced a novel-based bi-target display for structural balancing and a multi-objective discrete PSO. Weskida and Michal-ski [5] expressed developmental calculation for seed choice in a social network concerned about the potential for self-sorting out irregularity discovery on a system. In [6], there is a HONIED: Hive Oversight for Network Intrusion Early Warning using DIAMoND—a honey bee enhanced strategy for completely dispersed digital protection. Balegh and Farzi [11] proposed the algorithm for recognition of community in social networks in the year of 2017 by overcoming the disadvantages of k-means clustering and also by utilizing kmeans++ and k-Rank-D algorithm. Liu et al. [7] proposed the paper on community detection in the year 2014.

There is also lot of research work proposed in the area of advancing the bi-target demonstrate, partitioning the hub, enhancing the bi-objective model, amplifying the spread of impact, focusing on greedy algorithm for seed node detection, and merging genetic learning and probabilistic learning models and vertex-based encoding for dense network.

This subsequent part mainly focuses on the prior researches performed in the field of community detection, social connectivity, and more sophisticated problems of socio-networking like detection of anomaly, detection and identification of seed nodes, individual searching, cut space issue, and more detail of structure balancing. Hence, digging in deep research indicates here that traditional evolutionary algorithms perform as a way to deal and resolve several and major issues of socio-connectivity and networking.

Few prior researches have been performed on genetic algorithm with discussion on the connection determination in which the system can be utilized to locate the close ideal arrangement. These works have special attention toward friendship selection in social IoT. Some research has also shown inclination toward particle swarm optimizer in socialized network to find structural balance. Basically the focus is toward dual objective display and multi-objective PSO. Multi-objective discrete particle swarm optimizer upgrades the bi-objective model. Some authors have also proposed a mixed framework model based on rules and clustering for the security enhancement of big data. They use the concept of big data tools, apache spark, and kddcup99. In this procedure, it is been perceived that the grouping goes about as the normalize for the anomalies in dataset, in light of the fact that subsequent to clustering, less time is required to work the rest of the dataset. There is also some work on firefly heuristic surprise maximization to detect graph community: Zachary's Karate Club (ZKC), Relaxed Caveman Model (RCM), Power Law Cluster Graph Model (PLCGM) [16], and Erdos-Renyi Model (ERM) graph. They used Naïve firefly algorithm to verify the optimality of their proposed model. During our exploration, we also found that some of the authors have also work on representation of spanning tree for cut space searching to signify edge set. They have coordinated well-known maximum cut problem that is performed instead of vertex-based encoding.

4 Conclusion and Future Work

The proposed paper focuses on several issues and challenges of socio-connectivity and networking. This paper also emphasizes on various survey and research performed in the field of social networking. Several algorithmic tools mentioned here are described in detail through which these challenges can be overcome. Several algorithms including GA, CSO, PSO, evolutionary algorithms, community detection, seed node selection, DIAMOND, EDA, MEA-SN, MOEA-SA, NMI, PSO, GA-SA ,and IEDA are analyzed.

Research work from various researchers have been studied, analyzed, and compared based on dataset, technology used, and outcomes that have been discussed. It has been observed during study of paper that evolutionary algorithms prove to be more efficient and effective in clustering as well as other socio-networking issues. It also focuses on survey which can be improvised in future to detect the issues of overlapping community detection challenge.

References

1. F. Moradi, T. Olovsson, P. Tsigas, An evaluation of community detection algorithms on large-scale email traffic, in *SEA* (Springer, Berlin/Heidelberg, 2012), pp. 283–294
2. H. Ishikawa et al., *Social Big Data Mining* (Taylor & Francis Group, CRC 2015)
3. N.F. Samatova, W. Hendrix, J. Jenkins, K. Padmanabhan, A. Chakraborty, in N.F. Samatova , W. Hendrix, J. Jenkins, K. Padmanabhan, A. Chakraborty, ed. by *Practical Graph Mining with R* (Chapman and Hall/CRC: CRC Press, 2013)
4. M. Chen et al., *Big Data Related Technologies, Challenges and Future Prospects* (Springer, Berlin, 2014)
5. J. Moorthy et al., Big data prospects and challenges. Makers (2015)
6. E.H. Alkmanash et al., *Annotation of Smart Cities Twitter Microcontents for Enhanced Citizen's Engagement* (IEEE, 2019)
7. S. Bi et al., The impudence of online information on investing decisions of reward-based crowdfunding (2017)
8. I. Paul et al., *Understanding Big Data Analytics for Enterprise Class Hadoop and Streaming Data* (McGraw Hill, 2011)
9. P.K. Gopalan, D.M. Blei, Efficient discovery of overlapping communities in massive networks. Proc. Natl. Acad. Sci. **110**, 14534–14539 (2013)
10. http://www.businessinsider.in/Social-Big-Data-The-User-Data-Collected-By-Each-Of-The-Worlds-Largest-Social-Networks-And-What-It-Means/28969179.cms

Assessment of Severity Classification of Traffic Accidents on the Basis of K-Means Clustering and Adaptive Neuro-Fuzzy Inference System

Amrita Sarkar, Gadadhar Sahoo, and Umesh Chandra Sahoo

Abstract Nowadays, significant advancement has been done for vehicles and urban roads for the purpose of safety, but still many accidents of high severities happen in urban roads. Factors related to the human being have uncertainty and are complex in nature and it is complicated to separate every feature since these show irregularity. Driver's age, gender, driving experience, and education may be selected as human factors. Determination of the significant contributing factors is the primary task in road accident analysis. In view of that, this paper deals with the model of Adaptive Neuro-Fuzzy Inference System (ANFIS) is applied in predicting severity of accidents applying in urban roads. K-means clustering was applied, and the mean of silhouette value was selected as an indicator to identify the number of subgroups of the numerical variable for ANFIS model. Classification accuracy, specificity sensitivity, and precision values were being calculated for evaluating performance of ANFIS method. Here, the best-fit model was also selected according to mean square errors (MSE)). The classification accuracy reached 76.92% which reflects the high fitness due to the good classification accuracy which concludes that the model is suitable for prediction of accident severity as well as other classification-related applications.

Keywords Accident severity · Driver characteristics · K-means clustering · Silhouette value · ANFIS

A. Sarkar (✉)
Department of Computer Science and Engineering, Sarala Birla University, Ranchi, India
e-mail: amrita.sarkar@sbu.ac.in

G. Sahoo
Department of Computer Science and Engineering, Indian Institute of Technology (ISM), Dhanbad, India
e-mail: gsahoo@bitmesra.ac.in

U. C. Sahoo
School of Infrastructure, Indian Institute of Technology, Bhubaneswar, India
e-mail: ucsahoo@iitbbs.ac.in

J. Tavares et al. (eds.), *Emerging Technologies in Data Mining and Information Security*, Lecture Notes in Networks and Systems 164,
https://doi.org/10.1007/978-981-15-9774-9_8

1 Introduction

Continuous investigation of significant factors contributing to accidents allows researchers to perform computations to predict accident severity [5]. While investigating the accident severity, the relationship among risk factors and severity of accident may not be recognized correctly because of existence of a lot of factors contributing to intensities of accident severity. Those factors are human characteristics, vehicle-related factors, environment factors, and geometrical design related variables [1]. There remains substantial importance for accident severity models in predicting urban road accidents. In most cases, the fatalities are used to denote the severity. Traffic-related and geometry-related data are found to be mostly used in accident severity classification, number of accident predictions, and black spots identification methods. But the main cause of accidents is driver characteristics as shown in Fig. 1. Hence, this study has formulated a database consisting of information about each driver.

There are several methods applied for data modeling. Statistical method is a common method employed for accident data analysis. Road accident prediction is a real-life problem and it happens in a real-time environment. Hence, it cannot give an estimation of the importance of the problem using crisp functions. Fuzzy logic creates complex nonlinear relationships between features and makes a model using those features to evaluate the accident prediction. Whenever fuzzy logic is used as a single method, it will be unsuccessful to explain the specificity of the user. At that time, ANFIS will be perfect to fulfill this condition using the special characteristics of the learning capabilities.

In this analysis, *k*-means clustering was applied and the mean of silhouette value was selected as an indicator to identify the number of subgroups of the numerical variable for ANFIS model. Finally, four-cluster method has been found suitable for selecting the number of subgroup in this research.

In this paper, model has been built using four features related to the driver characteristics to classify accident severity. The classification accuracy reached 76.92% which reflect the high fitness due to the more or less good classification accuracy.

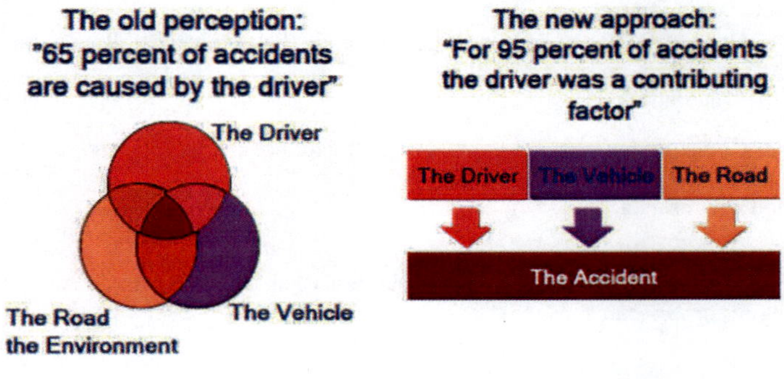

Fig. 1 Old and new percentage of road accidents in the developing countries [7]

2 Traffic Accident Data

In this research, traffic accident data have been collected from the year 2008–2012 on Acharya Jagadish Chandra Bose Road, Kolkata, India. Accident dataset consists of four features and one hundred thirty-three situations. While some of the parameters considered are numeric, rest are categorical in nature [9]. Here, only driver characteristics are taken as features for accident severity method. This dataset contains two classes represented as fatal accident and non-fatal accident. The data applied in this chapter relate to the driver characteristics are:

- Driver's age (DA)
- Driver's traffic education (DTE)
- Driving experience (DE)
- Driver's gender (DG).

3 Methodology

The first step of modeling the dataset by ANFIS is the input and output determination. In this study, driver's age (DA), driver's traffic education (DTE), driving experience (DE), and driver's gender (DG) are represented as inputs and the 'accident type' is represented as the output. Therefore, this ANFIS model comprises of four inputs and one output.

Here, accidents have been classified into two categories, i.e., fatal and non-fatal. Driver's gender and driver's traffic education are considered as categorical variables. Another two variables, driver's age, driving experience, are considered as numerical variables.

In this study, several ANFIS models were generated to find the suitable representation of prediction for severity of accident using effects of driver characteristics on the severity of accidents. The primary task concerning the structuring of the ANFIS model is establishing the optimal number of subgroups of input variables represent as fuzzy sets. The *k*-means clustering has been engaged, and the mean of silhouette value has been adopted as standard to find out the number of subgroups of the numerical variable. Finally, four subgroups have been chosen for each input numerical variable for ANFIS model comprising four driver characteristics. Additionally, three subgroups for each input numerical variable have been determined for ANFIS model including four driver characteristics and predicted accident index. For binary variable, two subgroups have also been defined for each fuzzy input.

3.1 Adaptive Neuro-Fuzzy Inference System (ANFIS)

Fuzzy logic generates a model using the knowledge of human being and creates the if-then rules, but this model does not always give enough results using this individual method. This statement also can be true for ANN model. ANN model only applies the numbers as an input rather than linguistic variables. This insufficiency can be defeated by integrating the beat characteristics of the two methods [3].

ANFIS method was discovered by Jang in 1993. Basically, this is a hybrid method which comprises artificial neural networks (ANN) and Sugeno-type fuzzy inference systems (FIS). ANFIS uses the merits of the two methods in a single structure and at the same time defeats their own limitations. ANFIS model is built as more methodical and less dependent on expert knowledge by using its structure. ANFIS method has non-rigidity and adaptive capacity. Based on this behavior of ANFIS, it proves much more efficiency for dealing with doubtful human natures.

The structure ANFIS method is shown in Fig. 2. In this architecture, a number of nodes were present at each layer which is identified by node function. Here, inputs were passed within the nodes from previous layers to present layers. To demonstrate the method of ANFIS, one output $f(x, y)$ and two inputs (x, y) were employed in this method. The two fuzzy if–then rules are stated below using first-order two-rule Sugeno fuzzy inference [2]:

$$\text{Rule 1: If } x \text{ is } B_1 \text{ and } y \text{ is } D_1, \text{ then } f_1 = p_1x + q_1y + r_1 \tag{1}$$

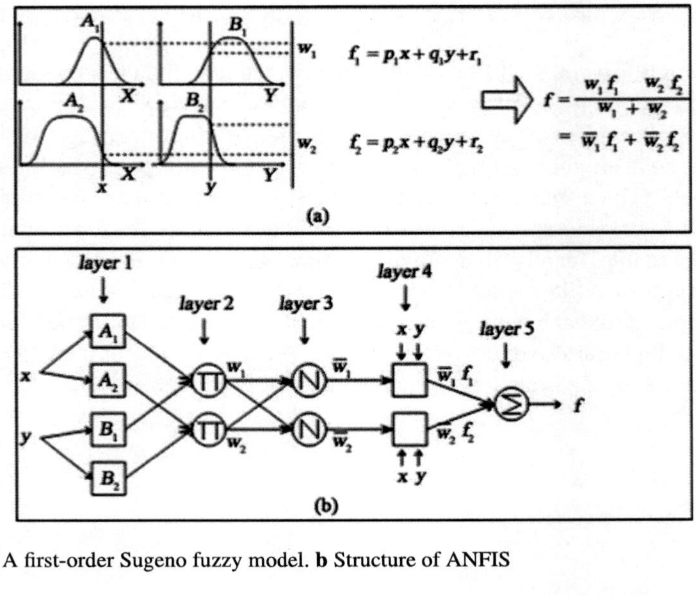

Fig. 2 **a** A first-order Sugeno fuzzy model. **b** Structure of ANFIS

$$\text{Rule 2: If } x \text{ is } B_2 \text{ and } y \text{ is } D_2, \text{ then } f_2 = p_2x + q_2y + r_2 \tag{2}$$

where B and D are the fuzzy sets in the antecedents, f_i is a crisp function in the consequent, x and y are the two input variables of model.

Usually, $f(x, y)$ is a first-order polynomial and it implies output of each ith rule ith and p_i, q_i and r_i are the consequent parameters.

Figure 2a demonstrates method of the fuzzy reasoning included with Sugeno fuzzy model for first order. Figure 2b displays the structure of ANFIS. Functions of node, belong to the similar function family which are in the similar layer, are explained below. Here O_i^j is represented as the output in layer j of ith node [6].

In this research, type 3 FIS invented by Takagi and Sugeno (1983) was used. Output of every rule was computed using input variables by adding a constant in this method. The weighted average of output has been calculated as final output of every rule [10].

Every layer in the architecture of ANFIS is described below [6]:

Layer 1: Each node i. In this layer has the capacity of adaption with a node function described as:

$$O_i^1 - \mu_{A_i}(x) \tag{3}$$

where μ_{A_i} is the membership function of A_i, x is the input for node i and A_i is the linguistic variable combined with the node function. Usually, $\mu_{A_i}(x)$ is chosen as

$$\mu_{A_i}(x) = \frac{1}{1 + \left[\left(\frac{x - c_i}{a_i}\right)^2\right]^{b_i}} \tag{4}$$

where x is the input and $\{a_i, b_i, c_i\}$ is the premise parameter set;

$$\mu_{A_i}(x) = \exp\left\{-\left(\frac{x - c_i}{a_i}\right)^2\right\} \tag{5}$$

where $\{c_i, a_i\}$ the set of parameters and it transforms the profile of the membership function. Parameters are represented as 'premise parameters' in this layer.

Layer 2: Each node in this layer is denoted as fixed node which calculates the firing strength w_i of a rule. Here, each node in output layer is estimated by the multiplication of the entire incoming signals toward it and is derived by,

$$O_i^2 = w_i = \mu_{Ai}(x).\mu_{Bi}(y) \quad \text{for } i = 1, 2 \tag{6}$$

Layer 3: The concept is to represent every node as fixed node in this layer. The firing strength of all rules is the summation of the ratio of ith node to the firing strength of ith rule. The normalized firing strength of the ith node is the output here and is given by,

$$O_i^3 = \bar{w}_i = \frac{w_i}{w_1 + w_2} \quad \text{for} \quad i = 1, 2 \tag{7}$$

The firing strength is denoted by the output signal w_i. Here in the fuzzy inference system, number of nodes for this layer are equated with number of fuzzy 'if-then' rules.

Layer 4: In this layer, every node is an adaptive node labeled by a square, with a node function given by,

$$O_i^4 = \bar{w}_i f_i = \bar{w}_i(p_i x + q_i y + r_i) \quad \text{for} \quad i = 1, 2 \tag{8}$$

where w_i is the output of Layer 3 and $\{p_i, q_i, r_i\}$ is the parameter set of consequent and f_i is the output of the rule ith.

Layer 5: This layer contains a single fixed node and it computes the total output by aggregation of all incoming signals, i.e.,

$$O_i^5 = \text{Total output} = \sum_i \bar{w}_i.f_i = \frac{\sum_i w_i f_i}{\sum_i w_i} \tag{9}$$

Figure 2b illustrates that the total output f can be stated as a linear combination of the consequent parameters with the given the values of the premise parameters. On the basis of the above figure, ANFIS applied a hybrid-learning algorithm combining two methods of backpropagation and least-squares, for learning the FIS and updating the parameters. The total output was describes as a linear combination of the consequent parameters, when the values of the premise parameters are fixed:

$$f_{\text{out}} = \bar{w}_1 f_1 + \bar{w}_2 f_2 = \frac{w_1}{w_1 + w_2} f_1 + \frac{w_2}{w_1 + w_2} f_2 \tag{10}$$

$$f_{\text{out}} = (\bar{w}_1 x)p_1 + (\bar{w}_1 y)q_1 + (\bar{w}_1)r_1 + (\bar{w}_2 x)p_2 + (\bar{w}_2 y)q_2 + (\bar{w}_2)r_2 \tag{11}$$

where $p_1, q_1, r_1, p_2, q_2, r_2$ are consequent parameters

3.2 Silhouette Analysis on K-Means Clustering

K-means clustering is one kind of segmentation method. This algorithm divides data into k significant clusters and sends back the index of the cluster where every observation has been allocated. K-means represents every observation in dataset as a sample holding a location in space. It creates the segment on the basis of two rules. First rule is the distances between the samples of same clusters should be less, and second rule is the distances between samples of other clusters will be more as much as possible. There are five different distance measures used for distances measurement. Any distance measure can be selected depending on the types of data.

Every cluster in the segment is comprised by samples and centroid. The centroid for every cluster is denoted as the central point. K-means calculates centroid of cluster for every measure of distance, to reduce the sum of distance [4].

K-means algorithm applies an iteration method that reduces the sum of distances from every sample to its centroid of cluster. This algorithm moves sample between clusters until the sum cannot be decreased further. The ultimate goal of this algorithm is the making of the set of well-separated clusters. To estimate whether the clusters are well-separated or not, one can create a silhouette plot employing the output of cluster indices obtained by K-means.

The silhouette value for every sample is the measurement similarity between samples of its own cluster, comparing with the samples within other clusters [8]. The silhouette value for the ith sample, S_i, is described as

$$S_i = (b_i - a_i)/\max(a_i, b_i) \tag{12}$$

where a_i is the average distance in the same cluster from the ith sample to another sample and b_i is the minimum average distance from the ith sample to sample s in a different cluster.

The value of silhouette varies from -1 to $+1$. A high silhouette value signifies that the sample is well fitted to its own cluster, and inadequately fitted to adjacent clusters. The clustering method is referred as suitable method if maximum objects contains a high silhouette value. If many points comprise low or negative silhouette value, in that case the clustering method may contain either a lot of or a small number of clusters. A silhouette value of zero specifies that the sample is on or very close to the decision boundary between two adjacent clusters and negative values shows for the samples allocated to the wrong cluster. The silhouette clustering method can be applied with any distance measurement method [8].

3.3 Description of Model

The first step of modeling the dataset by ANFIS is the input and output determination. In this study, driver's age (DA), driver's traffic education (DTE), driving experience (DE), and driver's gender (DG) are represented as inputs and the 'accident type' is represented as the output. Therefore, this ANFIS model comprises of four inputs and one output.

Here, accidents have been classified into two categories, i.e., fatal and non-fatal. Driver's gender and driver's traffic education are considered as categorical variables. Another two variables, driver's age, and driving experience, are considered as numerical variables.

In this study, several ANFIS models have been generated to find the suitable model of accident severity prediction using the effects of driver characteristics on the severity of accidents. The primary task in the generation of the ANFIS model is to select of the optimal number of subgroups of input variables represent as fuzzy sets. The *k*-means clustering has been applied, and the mean of silhouette value has been chosen as the criterion to select the number of subgroups of the numerical variable. Finally, four subgroups have been chosen for each input numerical variable for ANFIS model. For binary variable, two subgroups have also been defined for each fuzzy input.

The entire dataset with one hundred and thirty-three collected samples has been grouped into training and testing subsets randomly. Individually, these subsets are employed for learning and validating the model. All the training and testing data used as input variables are normalized for better generality. The building process consists of five steps.

Step I: The first step determines the optimal number of fuzzy sets for input variables of training data. Here, the mean of silhouette value is applied as the standard to quantify the number of clusters by employing *k*-means clustering method [11]. Finally, four clusters or subgroups are determined for each numerical fuzzy input for the first model.
In this research, the silhouette analysis is employed to decide an optimal value for the number of clusters or subgroups of input variables represent as fuzzy sets. The silhouette plot shown in Table 1 represents the *n* number of clusters value. In this table, silhouette values have been found to be below average, and wide fluctuations in the size of the silhouette plots using two and five number of clusters are observed. Therefore, two and five number of clusters is not suitable for selecting as the number of subgroups of numerical variables. Silhouette analysis is more complicated to decide in between 3 and 4 cluster numbers where silhouette values above average are reported.
In addition, from Fig. 3, thickness of the cluster size in the silhouette plot can be visualized. One big cluster can be observed due to the partitioning of the subclusters in silhouette plot using three clusters. More or less of similar thickness of clusters in the plot is observed with four clusters. It can be also verified from the scatter plot shown in Fig. 3. Finally, four clusters have found suitable for selecting the number of subgroups for ANFIS model in this research. As a result, four subgroups have been applied to the fuzzy inputs of driver's age (DA) and driver's traffic education (DTE) to implement ANFIS method.

Step II: In this step, preparation of parameters in antecedent and consequence has been done

Table 1 Represents the *n* number of clusters and corresponding mean silhouette value

Number of clusters (*n*)	Mean silhouette value
2	0.5029
3	0.5579
4	0.6578
5	0.4249

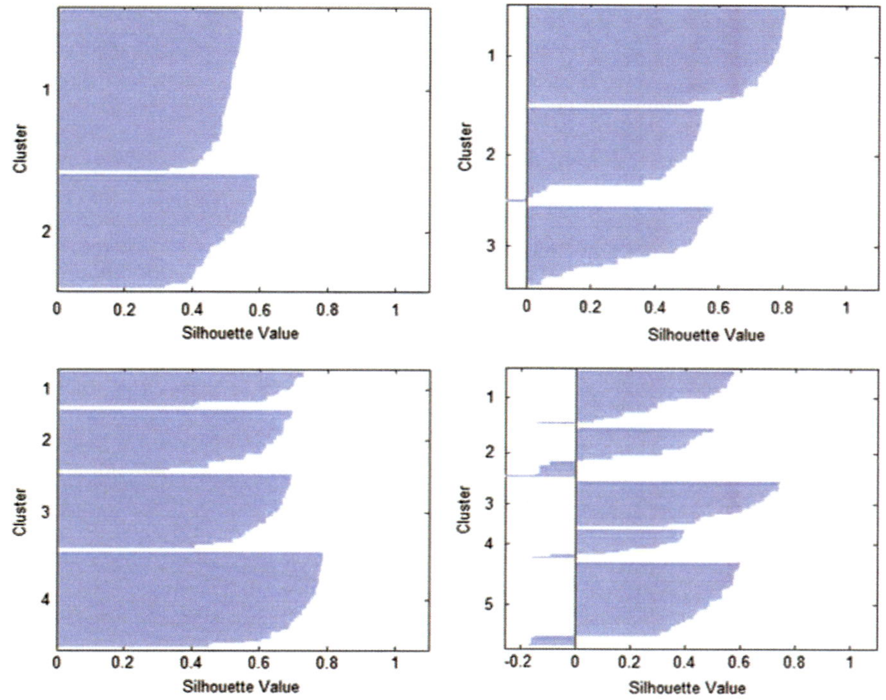

Fig. 3 Thickness of the silhouette plot for the cluster sizes 2, 3, 4, and 5

Step III: In this step, first k-means clustering method has been executed then develops the fuzzy inference system. By this method, sixty-four effective fuzzy rules were framed by first method and one hundred and eight fuzzy rules were formulated by the second method.

Step IV: In this steps, training of the model by backpropagation method has been done. After some iterations, optimal network with minimum error is obtained.

Step V: Examination of the optimal model has been done in the last step

In the present study, different models of ANFIS have been examined and the outline of the methods and outputs are represented in Tables 2 and 3. In these models:

1. For every model, Gaussian membership function was applied.
2. For every model, backpropagation algorithm was employed.
3. Three sets of training and testing data were applicable in this method. Those divisions are 70–30%, 60–40%, and 50–50%.

Table 2 Measuring error of ANFIS model using only driver characteristics

Model	Number of features	Number of membership function	Input membership function	Split data	MSE
ANFIS 1	4	4,4,2,2	Gaussian	70_30	0.1729
ANFIS 2	4	4,4,2,2	Gaussian	60_40	0.1735
ANFIS 3	4	4,4,2,2	Gaussian	50_50	0.1835

Table 3 Sensitivity, specificity, precision values, and classification accuracies obtained from ANFIS model using only driver characteristics

Model	Number of membership function	Input membership function	Split data	Sensitivity	Specificity	Precision	Accuracy
ANFIS 1	4,4,2,2	Gaussian	70_30	75.000	80.000	85.714	76.923
ANFIS 2	4,4,2,2	Gaussian	60_40	88.889	25.000	72.727	69.231
ANFIS 3	4,4,2,2	Gaussian	50_50	81.818	0.000	81.818	69.231

4 Results and Discussion

4.1 Performance Evaluation of Model

Once the model has been fit, the process of assessment of the model begins. Then various methods are used for giving judgment for the suitability of the model. It can also be applied on the dataset that are employed for computation of accuracy prediction. Each of these evaluative methods is being measured as a view of the efficiency of the model in allotting instances. The error measuring methods like mean square error (MSE), confusion matrix, and classification accuracy measures have been applied to minimize the errors and maximize of the performance in this research.

4.2 Discussion on Model

This research is basically a two-class problem, i.e., fatal and non-fatal, indicating severity of traffic accident. Traffic accident dataset has been classified using accident prediction model where four driver characteristics (used as inputs). The structure of the model is based on ANFIS. In this model, driver's gender and driver's traffic education are divided into two subgroups that implies two subgroups applied for these inputs to execute ANFIS model specially at the time of execution of fuzzy logic. Other two inputs, i.e., driver's age and driving experience, are defined by four subgroups using k-means clustering based silhouette value analysis. Three sets of training and testing data have been selected for execution of proposed model. In this analysis, ANFIS1 model determined the minimum MSE (0.1729) value that reflects the model comprising 70%-30% division of training and testing data are suitable for accident severity prediction. ANFIS1 model also computed 76.923% accuracy which reflect the high fitness due to the good value of classification accuracy. The outline of the methods and outputs are presented in Tables 2 and 3.

The effect of the total four features on the network outcome is so high that cannot be omitted to reduce network complexity. Therefore, the selections of these four variables have been a desired collection. Comparing the network outcomes, it is observed that ANFIS 1 model with Gaussian function gives the least MSE error and more classification accuracy.

5 Conclusion

An effort has been made to quantify the optimal number of fuzzy sets for input variables to develop the ANFIS model for road accident severity classification using driver related dataset. In this study, mean of silhouette value is applied as the standard to quantify the number of clusters of the numerical variable by employing *k*-means clustering method. In this study, a model of accident severity prediction has been developed using four driver's related features. This model resulted in a classification accuracy of 76.923%. The results recommend that the model is very much suitable for prediction of accident severity classification, emphasizing the application of driver characteristics as dataset.

References

1. L. Chang, F. Mannering, Analysis of injury severity and vehicle occupancy in truck- and non-truck-involved accidents. Accid. Anal. Prev. **31**(5), 579–592 (1999). https://doi.org/10.1016/s0001-4575(99)00014-7
2. J.S.R. Jang, Input selection for ANFIS learning, in *Proceedings of the Fifth IEEE International Conference on Fuzzy Systems*, vol. 2 (IEEE, New Orleans, 1996), pp. 1493–1499
3. J.S.R. Jang, ANFIS: Adaptive-network-based fuzzy inference system. IEEE Trans. Syst. Man Cybern. **23**(3), 665–685 (1993). https://doi.org/10.1109/21.256541
4. Mathworks Documentation Center, http://in.mathworks.com/help/stats/k-means-clustering.html. Last accessed 15 June 2019
5. F.R. Moghaddam, S. Afandizadeh, M. Ziyadi, Prediction of accident severity using artificial neural networks. Int. J. Civ. Eng. **9**(1), 41–48 (2011)
6. T. Mote, S. Lokhande, Temperature control system using ANFIS. Int. J. Soft Comput. Eng. **2**(1), 156–161 (2012)

7. Road Accidents in India Issues & Dimensions, Ministry of Road Transport & Highways Government of India, http://www.unescap.org/sites/default/files/2.12.India_.pdf. Last accessed 15 June 2019
8. P.J. Rousseeuw, Silhouettes: a graphical aid to the interpretation and validation of cluster analysis. Comput. Appl. Math. **20**, 53–65 (1987). https://doi.org/10.1016/0377-0427(87)90125-7
9. A. Sarkar, G. Sahoo, U.C. Sahoo, Feature selection in accident data: an analysis of its application in classification algorithms. Int. J. Data Anal. Tech. Strat. (IJDATS) **8**(2), 108–121 (2016). https://doi.org/10.1504/ijdats.2016.077484
10. T. Takagi, M. Sugeno, Derivation of fuzzy control rules from human operator's control action, in *Proceedings of the Symposium on Fuzzy Information, Knowledge Representation and Decision Analysis* (Marseillles, France, 1983), pp. 55–60
11. L. Zheng, M. Xianghai, An approach to predict road accident frequencies: application of fuzzy neural network, in 3rd International Conference on Road Safety and Simulation, Indianapolis, USA (2011)

Design and Implementation of Login-Based Wi-Fi Hotspot Network for an University Campus

Sayeed Islam and Md. Nurul Islam

Abstract In recent years, wireless hotspots have been recognized as a promising network platform to extend network connectivity in public places, which provides local coverage for Internet applications on the move in urban areas and rural environments. Wireless hotspot-based technology has popped up in corporate offices, coffeehouses, airports, restaurants, and bookstores worldwide. Wi-Fi technology has become the mobile connectivity mechanism of choice for everyone for all their features and ease of use. With the ongoing demand of secure Internet access in educational institutions due to modern standard of education and freedoms offered by the Internet, many educational institutes are keenly stepping forward to set up secured and stable wired or Wi-Fi network campus for their students. In this project work, we will propose a Wi-Fi hotspot network design for a university campus where each user will have a personal user ID and password.

Keywords Hotspot · Radius · Freeradius · MikroTik HotSpot Gateway · Unifi AP

1 Introduction

Wireless LAN technology has been reknown as a promising networking technology to extend network connectivity outside private networks. Wireless technologies offer a very cost-effective and promising solution to avail wireless Internet at public venues [1]. As per the current trend, many of the students are more likely to carry several devices with them. As we know these devices consume more and more bandwidth,

---

S. Islam · Md. Nurul Islam (✉)
Department of Electrical and Electronics Engineering, Manarat International University Bangladesh, Ashulia, Khagan, Savar, India
e-mail: shuvoa3@gmail.com

S. Islam
e-mail: Sayeed_nsu@live.com

J. Tavares et al. (eds.), *Emerging Technologies in Data Mining and Information Security*, Lecture Notes in Networks and Systems 164,
https://doi.org/10.1007/978-981-15-9774-9_9

educational institutions are struggling to keep up with the bandwidth demand. Educational institutions have serious needs for several security parameters like user access control, control over bandwidth consumption, and differentiated policy control for management faculty's access and student's access. These institutions with Internet facilities require continuous Internet bandwidth to access internet from class room and other places inside the campus. Designing a complete network with different solution providers would cost more and would be very difficult for IT admin to manage technical issues. Education institutes require cost-efficient manageable solution which can offer offers all the advance technologies and features in single platform with single point of contact for technical support [2].

But, what are the Wi-Fi challenges that school, university, and college face? A Wi-Fi solution for education must guarantee information security, especially that of students, staff, and the educational institution. A Wi-Fi solution for education must be capable of keeping hundreds of end-devices connected simultaneously. A Wi-Fi solution for an educational institutions must provide reliable connectivity throughout the entire campus. A Wi-Fi solution for education must have an optimal price-to-performance ratio to meet budgetary needs [3].

By addressing those major challenges and concerns in this paper, we have proposed a Wi-Fi hotspot network for a university campus that will include MikroTik HotSpot Gateway to provide network access for clients, FreeRADIUS for centralized authorization, authentication, and accounting. Also, we will include UniFi AP-AC-LR wireless access points for client access to Wi-Fi network. End users will have own user IP and password for authentication. The administrator will have more control over user authentication, authorization, and accounting. Moreover, the design will also provide a flexible way of monitoring and managing network devices.

2 System Model

2.1 MikroTik HotSpot System

The MikroTik HotSpot provides network access for end users using wireless or wired network connectivity. This product of MikroTik ensures features for user authentication using local client database, or RADIUS server, accounting using local database, or RADIUS server, and walled garden system (accessing some webpages without authorization) [4].

2.2 Introduction to Hotspot

Hotspot is a means that authorizes end users to access Internet. It does not ensure data encryption. To log in, users can use almost any web browser (either HTTP or HTTPS over encrypted TLS/SSL protocol), so they are not required to use additional software. The gateway is accounting the uptime and amount of traffic each of its clients have used and also can send this information to a RADIUS server. The hotspot system may limit each user's bandwidth, total amount of traffic, uptime, etc. [5].

The HotSpot gateway is intended to provide authentication within a local network, but may also be used to authorize access from public networks to access local resources. Configuring walled garden feature, it is possible to allow users to access some webpages without the need of prior authentication.

2.3 Description

MikroTik HotSpot system must have at least two network interfaces: one is HotSpot interface, that is used to connect HotSpot clients, two LAN/WAN interface that is used to access network resources. For example, DNS and RADIUS server(s) should be accessible. Figure 1 shows a sample HotSpot setup.

The HotSpot interface must have an IP address assigned to it. Physical network connection needs to be established between the HotSpot user's end device and the hotspot gateway. It can be wired or wireless.

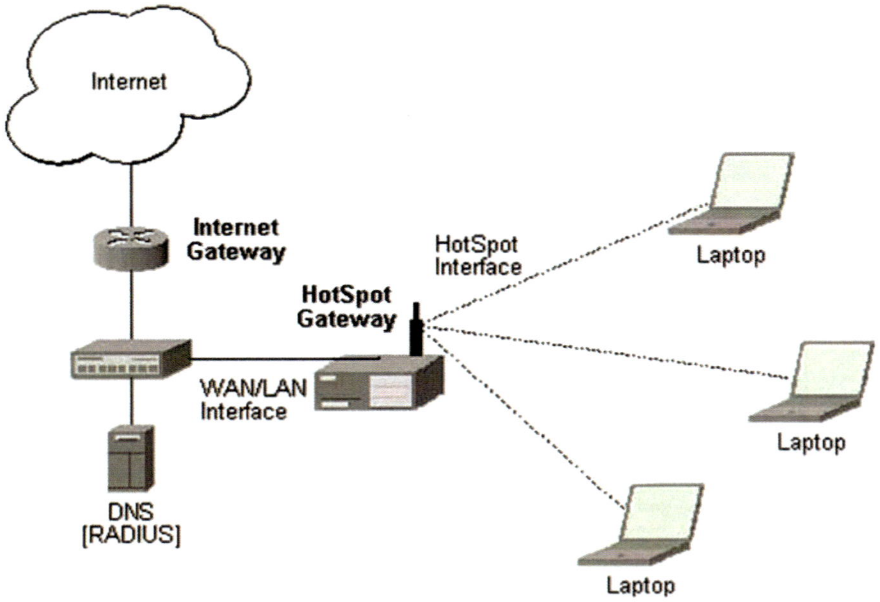

Fig. 1 A sample hotspot network

2.4 What Is RADIUS?

RADIUS stands for "Remote Authentication Dial In User Service" is a network protocol—a system that defines rules and conventions for communication between network devices—for remote user authentication and accounting. Commonly used by Internet Service Providers, cellular network providers, and corporate and educational networks, the RADIUS protocol serves three primary functions: Authenticates users or devices before allowing them access to a network, authorizes those users or devices for specific network services and accounts for and tracks the usage of those services [6].

3 Design and Implementation

First of all, the design we are going to describe is a Wi-Fi hotspot solution that is being proposed and implemented by us from Link3 Technologies Ltd. for South East University. The university has one main campus located at Banani and one permanent campus located at Tejgaon. The main campus has five buildings located few distance away from each other. And the permanent campus has two buildings which are also few distance apart from each other.

The entire campus (both) will be connected to the Internet via an ISP using one or more uplink. In this case, we have connected the main campus via two uplink and the permanent campus is connected to the Internet via one uplink.

We will use MikroTik router as edge router for the campus as well as for hotspot gateway. A Cisco switch will be used as core switch which will act as a distribution point of the hotspot LAN network. Remote buildings will be connected via fiber channel using a Cisco switch. A radius server will be connected to the MikroTik routed for user authentication, authorization, and accounting. The same server will also be used as a UniFi controller to monitor and configure UniFi Aps from a single dashboard.

3.1 Core Network Design

The main campus is connected to the Internet via ISP over iBGP routing. We have implemented one active and one standby link for main campus, which are connected to two different router of same ISP via fiber channel. The permanent campus is connected to the ISP with single link via fiber channel. The main campus and permanent campus is connected to each other via a layer 2 VPN channel. We have use MikroTik as an edge router for each campus which will also be used as a hotspot gateway for the Wi-Fi solution. A Cisco core switch is used in each campus as a distribution point for whole network. A radius server is connected to the MikroTik

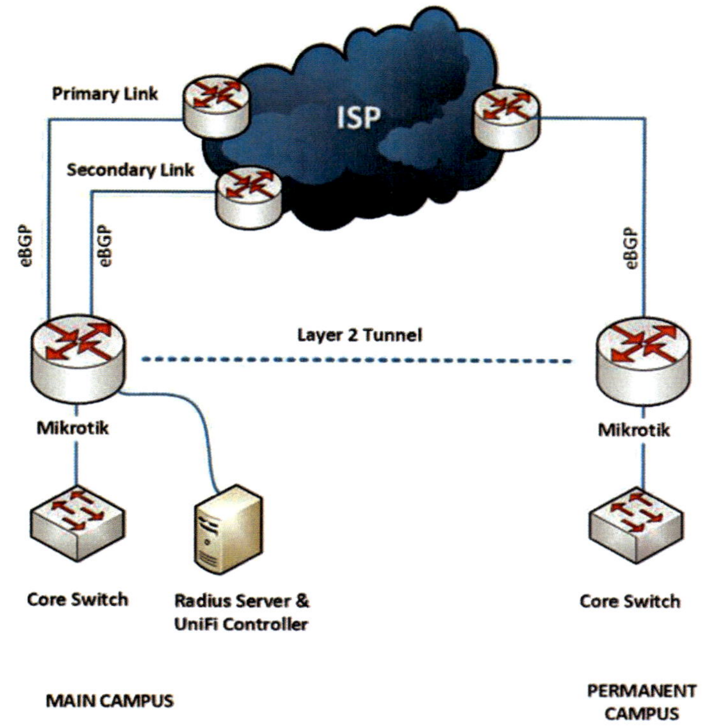

Fig. 2 Network diagram of university core network

router in main campus for user authentication, authorization, and accounting. All users in main campus and permanent campus will be authenticated at radius server. The radius server hardware will also be used as Unifi controller for monitoring and configuring Unifi wireless access points from a single dashboard. Figure 2 shows network diagram of university core network.

3.2 Distribution Network Design

In distribution layer, network is distributed all over the campus from the core switch. The access switch in each building is connected to the core switch over fiber channel. Since each switch port is an Ethernet port, we have placed two media converter between the two end of fiber channel to convert Ethernet to fiber and from fiber to Ethernet. According to client's requirement, we have placed one or more wireless access points (UniFi AP-AC-LR) in several floor of the building. Since UniFi AP-AC-LR is power over Ethernet (POE) enabled we have used Cisco switch that support

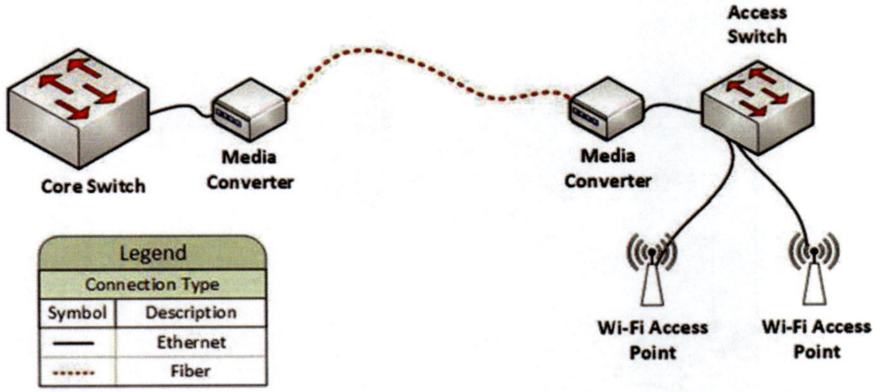

Fig. 3 Network diagram of university distribution network

POE. So, UniFi AP-AC-LR does not require any power adapter. Figure 3 shows network diagram of university distribution network.

4 Result and Discussion

We have also deployed network statistics monitoring functionality that the administrator can use to monitor LAN traffic. Figure 4 shows network statistics to monitor/analyze LAN traffic (Fig. 5).

The Unifi controller is configured to monitor and configure Unifi access points centrally. Figure 6 shows Unifi controller dashboard. Figure 7 shows Unifi controller statistics.

5 Conclusion

Nowadays, Wi-Fi technology has a positive impact on learning. For educational purposes, students and teachers are increasingly using Wi-Fi enabled devices in university campus. Wi-Fi connectivity at university campus leverage Wi-Fi enabled devices to connect to the Internet. As more and more students use mobile devices on campus to do things like, working on homework, checking their social feeds, and watching YouTube, IT teams need to rethink how to deliver and manage the growing demand for faster and more reliable campus Wi-Fi performance. In this project work, we have introduced a Wi-Fi network that is much more appropriate for university campuses. The wireless access points (Unifi AP) we used in this design covers larger areas than regular access points. Since these access points are POE enabled we do not need to have additional power adopter that makes it so easy for the IT persons

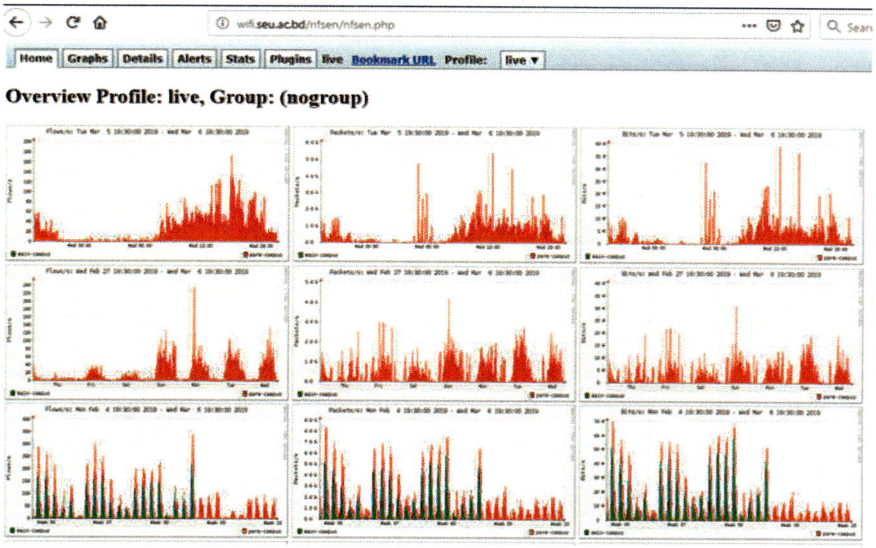

Fig. 4 Network statistics to monitor/analyze LAN traffic

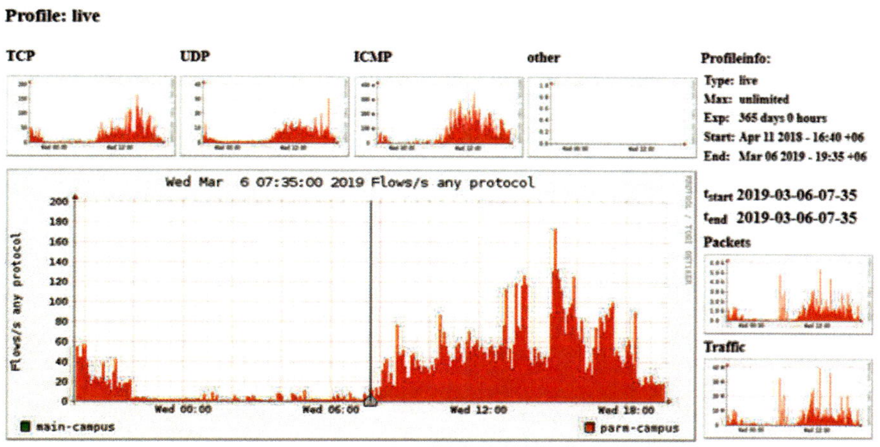

Fig. 5 Network statistics to monitor/analyze WAN traffic

to manage. We have also configured Unifi controller server to monitor and manage wireless access points from a single dashboard. All the connected Wi-Fi users can also be monitored from this Unifi controller.

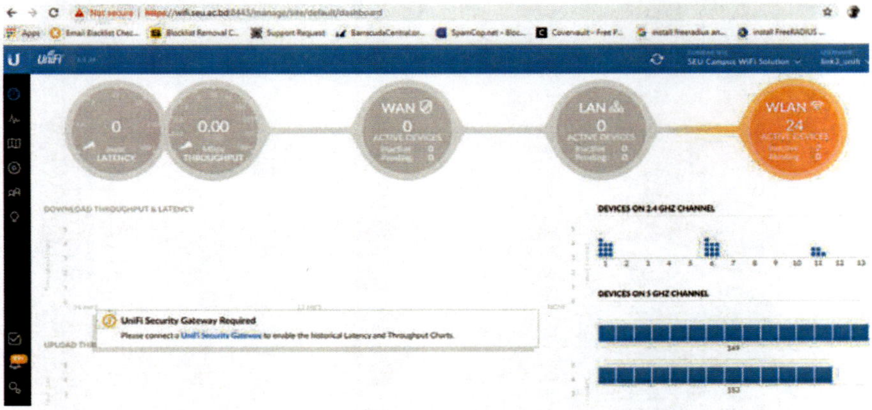

Fig. 6 Unifi controller dashboard

Fig. 7 Unifi controller statistics

References

1. H. Jorge, J.-C. Cano, C.T. Calafate, P. Manzoni, A wireless mesh network-based system for hotspots deployment and management, in International Conference on Networking and Services (June 2007), p 1
2. R. Kumar, M. Noura, *Wifi Hotspot Solution for Single & Multi-Location University Campuses* (24 Online Info Tech, 2017), p. 1
3. Ligowave, *Wifi Solutions for Schools, Universities and Colleges* (Ligowave, 2019), p. 1
4. A. DeKok, *FreeRADIUS* (Wikipedia), p. 1
5. J. Salazar, C. Fernando, *HotSpot Gateway* (MikroTik, July 4), p. 1
6. A. DeKok, *The Freeradius Technical Guide* (FreeRadius, January), p. 1

A Framework to Capture the Shift in Dynamics of a Multi-phase Protest—A Case Study of Hong Kong Protests

Aayushi Sanghi, Abhishek Sinha, Aditya Venkatesh, and Bhaskarjyoti Das

Abstract There has been a shift in protest dynamics in the 2019 Hong Kong protests, when compared to the Hong Kong Umbrella Protests of 2014. The intensity, magnitude and characteristics of the movement in 2019 were perceptibly different. This work presents a framework to capture such a shift using techniques of social network analysis, data science and natural language processing (NLP). This work models and analyzes the underlying social network structure on Twitter to decipher how things changed in 2019—from a social network analysis and NLP view. The social structure analysis indicates the diffused nature of leadership in 2019 versus the central nature of leadership in 2014. It also shows that the relative number of advocates versus supporters changed drastically in 2019. Brokerage analysis indicates the significant roles leaders played in 2014 as compared to the 2019 protest. The topic modelling analysis, along with sentiment analysis of trending hashtags, reveal how the themes of protest changed in 2019. In 2014, the theme evolved from civil disobedience to call for democracy. In 2019, it evolved towards the fight for freedom. Emotion has been investigated at various levels of granularity, i.e. across all tweets, emojis, hashtags, and for each central topic in 2014 versus 2019. The 2019 protest was characterized much more by anger, fear and anticipation. There was a larger sense of hopelessness and despair in 2019. Today social movements have moved online, and happen in phases. The proposed framework successfully deciphers the dynamics behind the two phases of Hong Kong protests and can be reused for similar analysis in future.

A. Sanghi (✉) · A. Sinha · A. Venkatesh · B. Das
PES University, Bengaluru, India
e-mail: aayu.sanghi@gmail.com

A. Sinha
e-mail: sinha.abhish3k@gmail.com

A. Venkatesh
e-mail: adityavenky97@gmail.com

B. Das
e-mail: bhaskarjyoti01@gmail.com

J. Tavares et al. (eds.), *Emerging Technologies in Data Mining and Information Security*, Lecture Notes in Networks and Systems 164,
https://doi.org/10.1007/978-981-15-9774-9_10

Keywords Framework · Multi-phase social movements · Hong Kong protests · Leadership pattern · Social network analysis · Brokerage analysis · Topic model · Emotion analysis

1 Introduction

Ever since people have started moving online, social media has occupied a place of prominence in social movements. This is proved beyond doubt in various social movements in the past decade i.e. Egypt in 2011, Iran in 2009, Turkey in 2013, Ukraine in 2013. As a result, organizers of social protests today actively use social media to enable collective discussion as well as mobilization. These protests have some common characteristics:

- They bear an active footprint on social media.
- Social media is actively used to organize protest as well as mobilize support.
- There are two primary stakeholders i.e. government and public.

It has been observed that in various social movements that governments succumb to pubic pressure once a critical mass is generated. Among the protest participants who actively use social media, often there are two categories of participants:

- **Advocates** who initiate and actively guide others and as a result incur high cost. Often they are perceived as leaders and get targeted by the authorities.
- **Supporters** who support and amplify the protest. They essentially perform a maintenance function, they are mostly followers and do not play a hyperactive role like the leaders.

Ever since China took over Hong Kong from Britain in 1997 with the proposal of "one country, two systems", protests were witnessed in Hong Kong. The most prominent ones are the 2014 and 2019 protests. The 2014 Hong Kong protest, known as the "Occupy Movement", was led by the "Occupy Trio" i.e. sociologist Benny Tai Yiuting, academic Chan Kin-man, and reverend Chu Yiu-ming. There were also some prominent student leaders, and the most prominent amongst them was Joshua Wong Chi-fung. The 2014 protest started as a 3-day sit-in but got extended into a 79 day occupy movement at prominent locations in Hong Kong. It was called "The Umbrella Movement" as protesters carried umbrellas to protect themselves. This 2014 protest was finally crushed and the leaders were jailed. The Occupy Movement abruptly ended after 79 days and failed to garner any concession from the local as well as mainland Chinese government.

The 2019 protest started when an Extradition Bill (to hand over fugitives to mainland China) was formally proposed on April 3, 2019. On April 28, 2019, tens of thousands of protesters marched calling for the withdrawal of this bill. On May 31, this bill was amended by the government, but was not withdrawn leading to a historic protest March on June 16 with more than 2 million protesters. From this point onward, violence took over and on August 12, Hong Kong airport was shut down. On

Sept 2, a large scale student boycott happened at the start of the new academic year, and on Sept 4, authorities announced withdrawal of the controversial bill. However, disobedience refused to die down completely and on Oct 1, the first case of a protester being shot was seen.

The 2019 Hong Kong protest is suspected to have its roots in the 2014 protest, though it was different in mood, scale, and in its ability to achieve success. There are many who participated in both 2014 as well as 2019 protests that yielded completely different results. In this perspective, this research primarily focuses on understanding the difference between these two related but widely different social movements based on available social media data.

2 Existing Work

Today Twitter is steadily replacing [7] the print and television media as the source of news and can be called the microphone [23] for the masses. This has led to numerous attempts to analyze Twitter data sets. Since any Twitter data set is typically large, frameworks such as MapReduce [30] have been used to collect, analyze, and visualize the data.

2.1 Analysis of Public Events

Hua et al. [15] presented a framework to correlate surge in tweet volume with events in a civil unrest and extracted the corresponding event topics from the tweet texts. Tumasjan et al. [32] used Twitter data to predict outcome of 2009 federal election in Germany. Kirilenko et al. [20] used one year worth of tweets to analyze public opinion about climate change issues across various parts of the world. Ackora et al. [4] used a time dimension analysis of Twitter data to identify breakpoints in public opinion. Politicians recently have started using Twitter effectively for the same reason. Usage of Twitter for European parliamentary election [13] as well as the last US election [26] has been extensively analyzed. Twitter is now indisputably the mirror of the public minds in spite of increasing level of misuse.

2.2 Twitter as an Organizing and Mobilization Tool

Proskurnia et al. [27] analyzed over 100 campaigns about climate change over social media platforms to present a comprehensive analysis framework covering categories of protests, actors, actions, tweets etc. The 2011 "Occupy Wall Street" movement in USA has been extensively organized through Twitter [33] covering aspects such as forming public opinion, sharing views, mobilization and movement logistics. On the contrary, Wilson et al. [36] has shown that in 2011 Tahrir Square movement in Cairo, Twitter played a key role in connecting and motivating the protesters. Not only has Twitter become integral to present day social movements, Twitter data is also increasingly used to analyze and understand such movements. Earl et al. [11] analyzed the tweets during the protests at G20 meeting venue at Pittsburgh in 2009. They examined the tweets to analyze how the protesters used Twitter to share information. Thus in a modern protest, Twitter is successfully addresses the information asymmetry between the protesters and law enforcement agencies. Bastos et al. [5] analyzed geocodes of tweets in 2013 Brazil Vinegar protests. They concluded that hashtagged messages from actual protest locations are used to connect sympathetic social media users from distant geographic locations. Ozturkcan et al. [24] analyzed the 2013 Gezi park protest in Istanbul that originally started as a small scale protest, and evolved into a large protest spread across multiple regions. Since this protest did not have sufficient media coverage, the protesters effectively used Twitter to spread information and create awareness.

2.3 Emotion Analysis of Social Movements Using Twitter Data

Sentiment and emotion analysis are often attempted, but given the cryptic short forms used on Twitter, along with short tweets, it remains a challenging [18] task. Dini et al. [10] discusses this challenge and talks about the three main approaches i.e. supervised machine learning, lexicon based approach, and a hybrid approach combining both. The model based approach is limited by the tweets they are trained on. Hence, in this work, a lexicon based approach [22] is employed. dos Costa et al. [29] analyzed 42 million tweets during the summer of 2013 in Brazil. They concluded that the peaks in Twitter activity coincided with the protest events and words in the tweets accurately represented the emotional characteristics of such protests. Ahmed et al. [3] analyzed Twitter data during the 2013 Nirbhaya gangrape protest at Delhi. Twitter was extensively utilized in organizing this protest. The authors looked at the tweets from Twitter as well as the stories from leading newspapers in India during the period when this protest happened. With the help of psycholinguistic lexicon provided by LIWC [25], the work analyzed the emotion before, during and after the protest from online and offline media. They concluded that there was a significant correlation between the emotion in the offline and online world.

2.4 Analysis of Underlying Network

Huberman et al. [16] studied the social interaction on Twitter to conclude that the underlying implicit network of friends and followers drive the interactions. Kwak et al. [21] pointed out that Twitter ceased to be social media alone, and is now a social network. Welch et al. [34] investigated Twitter beyond the follower relationships and focused on implicit relationships such as "retweets" and "mentions" to build two implicit sub-graphs and investigated users from the perspective of topical relevance. Garcia et al. [14] used tweet data to analyze the 2011–12 Chilean student movement. They did this analysis by constructing the underlying student network from tweet data, and then analyzing the topology of this network. Tinati et al. [31] analyzed tweets and defined role classification of users such as idea starter, amplifier, commentator, curator, and viewer. Khan et al. [19] used centrality analysis of the underlying Twitter social network to identify opinion leaders. Abascal-Mena et al. [1] used methods of social network analysis to analyze the meaning of a public movement. In their subsequent work [2], they extended the same method to detect socio-semantic communities from tweets. Recently, Chaudhuri et al. [9] studied the same Nirbhaya protest at Delhi. The finding of their study was the new paradigm of such social movements i.e. no specific interest groups were leading this protest but it was spontaneous and horizontal in nature, enabled by social media.

2.5 Hong Kong Protests

The Hong Kong protests, which are the subject of our work, have also attracted some attention from researchers. The 2014 protest (Umbrella Movement) has been subjected to research, whereas the recent 2019 movement is yet to see any scholarly work. Flowerdew [12] discussed the 2014 movement from a critical discourse perspective. Hones et al. [17] analyzed the critical role videos played on social media to build momentum, in 2014. Wetzstein [35] analyzed the visuals in tweets during the 2014 protest, to conclude that they provided very important "on the ground" perspective to people. Since Hong Kong had one of world's highest Facebook penetration rates, Chan [8] analyzed Facebook data to conclude that network size and connections with political actors influenced the participants directly. Since the 2019 Hong Kong protest is yet to die down completely, various websites and newspapers are still the best source of information about this. Recently, Purbrick [28] has published a report analyzing the historical events behind this 2019 protest. Any data driven academic work on the 2019 Hong Kong protest is yet to be published at the time of this work.

3 Data Set

Data collection was divided into two phases i.e. 2019 protest and 2014 protest. It was done using two methods i.e. Tweepy and Gephi's Twitter Streaming Importer plugin. In Phase 1, with 2019 data, Tweepy and Gephi's Twitter Streaming API were used to collect tweets through the months of November 2019 to the beginning of February 2020. We had collected around 33,000 profiles by the end of this period.

Phase 2 of data collection was to collect data from the 2014 protests. The Twitter API does not allow one to access tweets which were sent out earlier than one week from the date of collection. However, one can access old tweets of a particular account, i.e. by specifying the user's Twitter handle. Hence Twitter handles which were active/important during the 2014 Umbrella protests were manually found. Since the Twitter API only allows access to the 3200 most recent tweets of a user, some prolific users who were also active in 2014, could not be included. We were able to re-create the 2014 network, with 35,000 nodes using this method.

4 Methodology

To understand the shift in the protest dynamics, the proposed framework considered the two phases of the Hong Kong protest i.e. 2014 and 2019. The exact same steps were carried out for both. The key steps in the framework are as follows:

1. **Analysis of the network parameters**: From the set of tweets extracted, the nodes and edges were determined. Each node was each unique user. Each edge in the network was of one of the following types:

    a. Retweet: when a user retweets another user's tweet
    b. Mention: a user mentions another user in his/her tweet
    c. Quote: when a user quotes another users tweet adding or replying to the particular tweet.

    The first level of pre-processing was generation of both network graphs by applying the Fruchterman Reingold algorithm for the layout. Since the graph was too dense for a visual analyssi, an in-depth analysis was done using different attributes of each network i.e. namely the number of triangles, betweenness centrality, closeness centrality, average clustering co-efficient of each independent network (2014 and 2019).
2. **Contrasting the brokerage in both the movements**: Brokerage has been defined as the normalized value of the number of times a user has retweeted. The assumption here is that if a user is retweeting relatively more relating to the movement, then they are trying to be a broker for the movement, and spread the message of the movement. Hence the distribution of brokerage across both networks were calculated and contrasted. Brokerage lies between 0 and 1.
3. **Identifying the leadership patterns in both protests**: It was expected that network analysis and brokerage analysis together would provide an insight into the leadership patterns of both protests i.e. whether it was centralized or decentralized, and how it affected the final outcome in terms of diffusion and scale of the protest.
4. **Identifying the main themes in the protest**: This is done by LDA [6] topic modeling. Top topics were identified for both 2014 and 2019 protests. Similarly the main hashtags were also identified for each.

5. **Contrasting the emotions of the 2014 Umbrella Movement with the 2019 Protests**: The aim here was to bring out the subtle contrasts in negative sentiment in the protests—as both protests brought a wave of negativity which manifested in different ways.
   a. First the most trending hashtags and their attached emotions to contrast the mood which represented these protests online were found.
   b. Emotion analysis was then performed on all the tweets to find the distribution of six emotions i.e. anger, fear, anticipation, surprise, sadness, and disgust. Both, 2014 and 2019 protests came with a surge of negative sentiment in a slightly different way.
   c. An emoji analysis was also done to see which emojis represented each movement. Emojis became popular (and available) in the beginning of 2014, and they are a very important point of emotional analysis in the present day.
   d. All tweets were converted to fastText word vectors with a dimension of 300, and tweets who had a similarity of >0.8 with the respective topics, were segregated. These tweets were then analysed for the sentiments they represented by employing the NRC [22] lexicon.

5 Results and Analysis

5.1 Comparing the Underlying Protest Network—2019 Versus 2014

This analysis below goes to prove the contrast in leadership styles between both protests. In 2014, the leadership was centralized and people mostly amplified the voices of the recognized leaders. In 2019 protest, people mostly tweeted their own opinions and sometimes amplified the voices of a few prominent people as no concrete leaders were identified.

5.1.1 Clustering Coefficient

In network analysis, clustering coefficient measures the degree to which nodes are connected together. Figure 1 shows the distribution of the same in year 2014 and 2019. There is a stark difference in the number of close-knit communities/clusters between both networks, and hence the difference in clustering co-efficient between 2014 and 2019. 2019 has 25,344 triangles, whereas 2014 has 17061 triangles in the network. The average clustering co-efficient for 2019 (and ongoing) is 0.302, whereas for the 2014 Umbrella Movement, it is 0.149 (as shown in Table 1). This is representative of the actual dynamics of both movements. 2014 movement had leaders who were at the epicenter of the movement and the communities formed

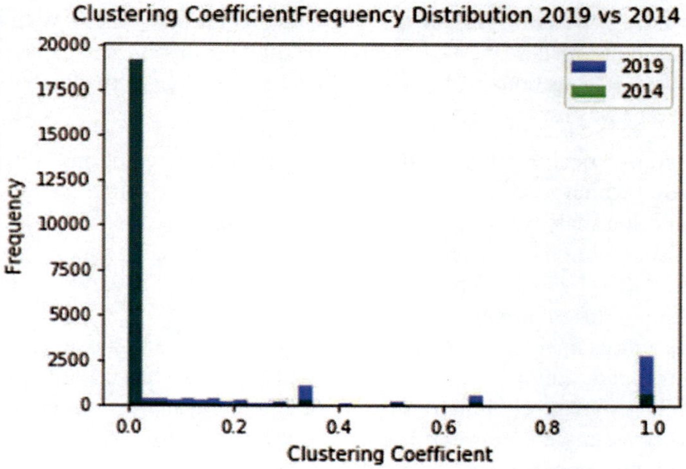

Fig. 1 Comparison of clustering co-efficient distributions

Table 1 Comparison of relevant graph attributes

Attribute	2014	2019
Number of triangles	17,069	25,344
Average clustering co-efficient	0.149	0.302
Average closeness centrality	0.403	0.300

around them. Moreover, the leadership was concentrated amongst a few people, hence the clusters were less dense than the ones in 2019 (Fig. 1).

In the 2019 protests, the leadership scenario is a lot more haphazard. There are no distinct leaders identified, hence there are a lot more important ego-centers with their own closely knit networks. These networks are weakly connected amongst each other, but there is a high level of cohesion within networks themselves (people generally tend to retweet the more prominent people within their own groups, as opposed to a small cohort of people who lead the movement globally). Therefore, in 2019, due to a more decentralized leadership, clusters formed around the more relevant/important/active members—hence the higher number of triads and higher clustering co-efficient in this network.

5.1.2 Closeness Centrality

The closeness centrality of a graph indicates how close a node is to the rest of the network. The closeness centrality distribution (as shown in Fig. 2) of the 2014 data shows that it has a lot more egos with higher closeness centrality values, whereas

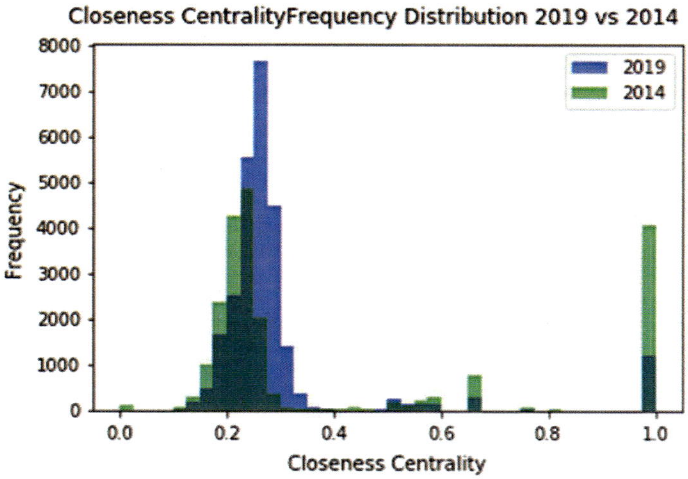

Fig. 2 Comparison of closeness centrality distributions

2019 has more egos with lower centrality values. This once again re-enforces the situation on the ground—the 2014 protest was characterized by a small number of very important egos around whom the network was centered. This means that fewer nodes were "far away" from the network. The leadership was concentrated and centralized.

This is in contrast with the 2019–2020 protests, where the leadership is more erratic and distributed, so all nodes are not close to each other. The closeness centrality values are more concentrated on the lower side when compared to the 2014 protests. There are very few nodes with higher centrality values. This also means that this network is characterized by a larger number of weak links—which is to be expected from a decentralized network, and which has a positive effect on information dissemination—as weak links aid information flow through a network. The above leadership patterns are also proved by the investigation of the relative percentage of roles played by Twitter users i.e. initiator (author of seed tweets), amplifier (author of retweets), curator (forwards a seed tweet by adding his own text). It is found that 2019 protest has more initiators than the other two categories compared to the 2014 protest.

5.1.3 Distribution of Brokerage

The 2014 protests showed significantly higher values of brokerage when compared to 2019 (as shown in Fig. 3). This means that a significant number of nodes played a higher role in amplifying the main voices of the movement when compared to 2019.

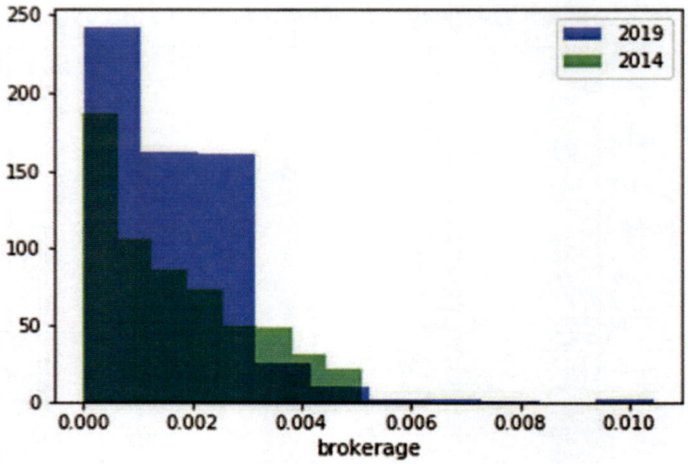

Fig. 3 Comparison of brokerage distribution across the two protests

5.2 *Main Themes in the 2014 and 2019 Protests*

5.2.1 LDA Topic Model Analysis

The top three identified topics for 2019 protest are the following:

1. **2019 Topic 1—police brutality state terror stand with Hong Kong**: This topic focuses on the fear and anticipation people felt due to the acts of law enforcement agencies. The sentiment of the people towards the police was never a central topic in 2014, but in 2020, it seemed a lot more prevalent. This sentiment was mostly driven by fear and anger.
2. **2019 Topic 2—anti chinazi china nazi boycott ccp boycott China**: This topic focuses on the view the protesters currently have of the Chinese government. Their view is very aggressive, and they are clearly not satisfied with the current government.
3. **2019 Topic 3—students sing gather tonight liberty slogan**: This topic focuses on the fact that a lot of students united against the government during these protests, and would plan to gather together and sing songs and slogans on Twitter. This third topic has significantly higher values of "sadness" because it also represents solidarity with each other in times of distress.

The top three identified topics for 2014 protests are the following:

1. **2014 Topic 1—umbrella revolution occupy protest civil disobedience**: This topic focuses on the way protesters had decided to approach the revolution—through civil disobedience. Students and other civilians protested outside government headquarters, and some began to occupy several major city intersections.

Protesters blocked both east–west arterial routes in northern Hong Kong Island near Admiralty.

2. **2014 Topic 2—hkdemocracy protest china democracy support**: The 2014 protests were triggered by students disillusioned by Beijing, and in particular, the pronouncement of the NPCSC on 31 August 2014—that said Hong Kong had to accept an electoral process in which up to three candidates would be pre-approved by the Election Committee (an electoral college widely seen to be loyal to Beijing authorities.) Therefore the biggest goal of this movement was to attain democracy, and that was also the most trending topic online.
3. **2014 Topic 3—help sign petition online protest thank you**: Petitions like "Condemn HK Police for Excessive Force Against Citizens and Call for Independent Inquiry", and "Act Now to Protect Hong Kong's Freedom and Democracy" were afloat all across Twitter. The protesters clearly tried to garner national and international attention through petitions, as it is one of the popular topics for 2014.

5.2.2 Analysis of Hashtags

At the most preliminary level, the primary hashtags used are analyzed. Hashtags are generally curated by central figures in the protest, to capture the entire mood of the protest in a few catchy words or phrases—hence it is a good way to analyse the kind of emotion it elicited from protesters. Table 2 shows the top hashtags for both the protests.

5.3 Comparing the Underlying Emotions—2019 Versus 2014

During both protests, there was a surge of negative emotion. However, the mood and theme of this negative emotion were completely different, and stemmed from different causes.

Table 2 Comparison of top hashtags

2014	2019
Occupy central	Hong Kong protest
Umbrella revolution	Stand with Hong Kong
Occupy HK	Fight for freedom
Umbrella movement	5 demands not one less
HK student strike	Hong Kong policestate

5.3.1 Analysis of the Primary Emotions

To distinguish between the negative emotion and mood of the protest, we considered the six emotion types i.e. anger, fear, anticipation, surprise, sadness, and disgust. We analysed the tweets for both years, to see how these emotions were distributed amongst these six buckets. We see that for 2019, anger, fear, and disgust occupy a visibly larger proportion of negative feelings when compared to 2014 (as shown in Fig. 4). In 2014, people dealt with a lot more anticipation when compared to 2019, indicating that protest culture was still newer and the mood was a lot more sombre than it is currently. While 2014 was rooted in dissatisfaction with the government and the electoral process, with people camping out on the streets, the 2019 protests seems to be characterized by more violence, fear, disgust, and resolve.

5.3.2 Analysis of the Emojis

As the last layer of analysis, the top emojis being used in each protest were extracted. Emojis became popular on Twitter in the beginning of 2014. They can display a wide range of complex emotions and actions. The top emojis used in 2019 were the angry face, the crying face, and the hands joined in prayer (as shown in Fig. 5). These are indicative of fear, resentment, anger, disappointment, and solidarity. In 2014, people mostly used Umbrellas (to represent the "Umbrella Movement"). While some emojis with negative connotations were used, the use of them was nowhere as wide-spread and intense as it is in the 2019–2020 protests.

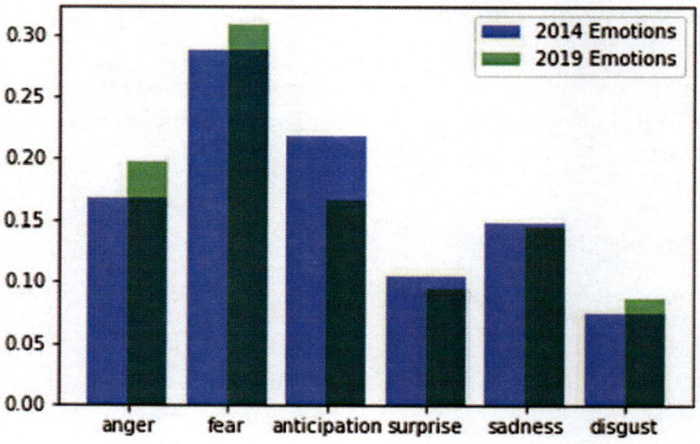

Fig. 4 Comparison of negative emotions

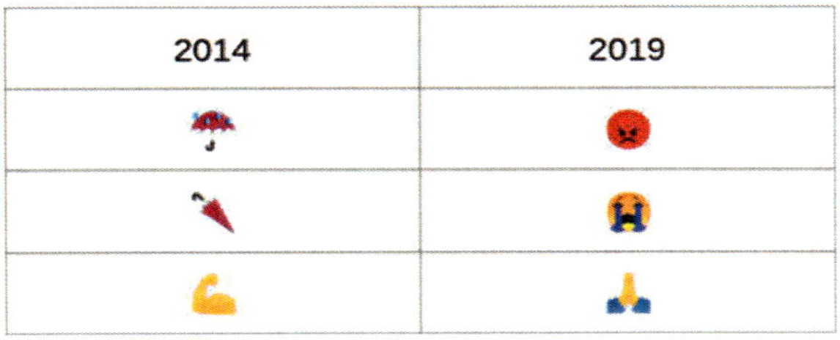

Fig. 5 Comparison of top emojis across the two protests

5.3.3 Emotions in the Main Hashtags

The hashtags used to represent the 2019 protests are indicative of fear, anger and violence. They are a lot more aggressive than the hashtags used to represent the 2014 protests. The 2014 protests incited a lot less anger and disgust than the current protests, as is clear from the wording of hashtags. 2019 shows the desire to fight the system, stand in solidarity with those protesting on the ground, and be firm about all the demands without any concessions.

5.3.4 The Emotions Around the Main Topics

On analyzing the emotions behind all topics in 2019 as shown in Table 3, it can be concluded that the central topics of discussion are mostly driven by fear and anger, closely followed by sadness. On analysing the emotions of the 2014 protest, we see that the percentage of "anticipation" and "sadness" are significantly higher than those in 2019. This is because the 2014 protests never got as violent as the 2019 protests did, and while the central topics did portray fear, they also portrayed a lot of sadness and anticipation. However in 2019, the central topics were driven by large percentages of fear and anger because the situation on ground was a lot more drastic and tense than it was during the Umbrella Revolution. While the central topics for both protests were backed by fear, there is a stark difference in the other emotions which dominated the topics in each era.

Table 3 Comparison of topic wise emotions in 2014 and 2019

	2019			2014		
Emotion	Topic 1 (%)	Topic 2 (%)	Topic 3 (%)	Topic 1 (%)	Topic 2 (%)	Topic 3 (%)
Anger	21	21	20	17	18	19
Fear	38	39	30	31	33	34
Anticipation	12	10	16	21	20	19
Surprise	8	7	10	11	10	10
Sadness	13	12	14	14	13	13
Disgust	8	10	9	6	6	6

6 Conclusion and Future Works

Though it is a well-accepted fact that social media like Twitter is an enabler in a modern social movement, any attempt to understand it in a thread-bare fashion is always an intriguing task. There has been attempts to understand social events using social media data. However, this work is a first attempt to deliver a framework to understand the reasons for the difference in dynamics and scale of two movements that happened in phases, used Twitter extensively, had common players and are somewhat related. This research has compared the 2014 and 2019 Hong Kong protest from different perspectives such as structure of the underlying implicit network, leadership patterns, leaders versus brokers, protest themes and the dominant emotional tones. The framework successfully explains why the 2019 Hong Kong protest has become far more successful than the 2014 protest which failed to extract any concessions from authorities. Today with society moving online, social movement happen in phases and leave an indelible footprint on social media. The framework proposed can be successfully used to explain any shift or difference in the dynamics in any multi-phase social movement.

As a next step, we plan to extend this comparison further into several directions. Firstly, we plan to identify those aspects of the underlying network's structure that apparently helped the diffusion process better in 2019. As the presence of weak ties help in diffusion, we plan to do an analysis of tie strength and percentage of weak versus strong ties in the underlying network in 2014 versus 2019. Secondly, during this phase of work, the Twitter participants were not explicitly classified into all possible categories such as government, individual, advocates, amplifiers etc. We plan to do those classifications to enable a more fine-grained analysis. We plan to enhance the analysis from the perspectives of roles played in the virtual world and influence exercised in the protests.

References

1. R. Abascal-Mena, R. Lema, F. Sèdes, From tweet to graph: social network analysis for semantic information extraction, in *2014 IEEE Eighth International Conference on Research Challenges in Information Science (RCIS)*, pp. 1–10 (IEEE, 2014)
2. R. Abascal-Mena, R. Lema, F. Sèdes, Detecting sociosemantic communities by applying social network analysis in tweets. Soc. Netw. Anal. Min. **5**(1), 38 (2015)
3. S. Ahmed, K. Jaidka, Protests against# delhigangrape on twitter: analyzing India's arab spring. eJournal eDemocracy Open Gov. **5**(1), 28–58 (2013)
4. C.G. Akcora, M.A. Bayir, M. Demirbas, H. Ferhatosmanoglu, Identifying breakpoints in public opinion, in *Proceedings of the first workshop on social media analytics*, pp. 62–66 (2010)
5. M.T. Bastos, R. Recuero, G. Zago, Taking tweets to the streets: A spatial analysis of the vinegar protests in Brazil. First Monday **19**(3) (2014)
6. D.M. Blei, A.Y. Ng, M.I. Jordan, Latent dirichlet allocation. J. Mach. Learn. Res. **3**, 993–1022 (2003)
7. M. Broersma, T. Graham, Twitter as a news source: how dutch and british newspapers used tweets in their news coverage, 2007–2011. Journalism Pract. **7**(4), 446–464 (2013)

8. M. Chan, Social network sites and political engagement: exploring the impact of facebook connections and uses on political protest and participation. Mass Commun. Soc. **19**(4), 430–451 (2016)
9. S. Chaudhuri, S. Fitzgerald, Rape protests in india and the birth of a new repertoire. Soc. Mov. Stud. **14**(5), 622–628 (2015)
10. L. Dini, A. Bittar, Emotion analysis on twitter: the hidden challenge, in *Proceedings of the Tenth International Conference on Language Resources and Evaluation (LREC'16)*, pp. 3953–3958 (2016)
11. J. Earl, H. McKee Hurwitz, A. Mejia Mesinas, M. Tolan, A. Arlotti, This protest will be tweeted: twitter and protest policing during the pittsburgh g20. Inf. Commun. Soc. **16**(4), 459–478 (2013)
12. J. Flowerdew, Understanding the hong kong umbrella movement: a critical discourse historiographical approach. Discourse Soc. **28**(5), 453–472 (2017)
13. A. Frame, A. Mercier, G. Brachotte, C. Thimm, *Tweets from the Campaign Trail*, vol. 11 (Peter Lang, 2016)
14. García, C., Chauveau, P., Ledezma, J., Pinto, M.: What can social media teach us about protests? analyzing the chilean 2011-12 student movement's network evolution through twitter data. arXiv preprint arXiv:1308.2451 (2013)
15. T. Hua, C.T. Lu, N. Ramakrishnan, F. Chen, J. Arredondo, D. Mares, K. Summers, Analyzing civil unrest through social media. Computer **46**(12), 80–84 (2013)
16. B.A. Huberman, D.M. Romero, F. Wu, *Social Networks that Matter: Twitter Under the Microscope*. arXiv preprint arXiv:0812.1045 (2008)
17. R.H. Jones, N.C.H. Li, Evidentiary video and "professional vision" in the Hong Kong umbrella movement. J. Language Polit. **15**(5), 569–591 (2016)
18. K. Kenyon-Dean, E. Ahmed, S. Fujimoto, J. Georges-Filteau, C. Glasz, B. Kaur, A. Lalande, S. Bhander, R. Belfer, N. Kanagasabai, et al., Sentiment analysis: it's complicated!, in *Proceedings of the 2018 Conference of the North American Chapter of the Association for Computational Linguistics: Human Language Technologies*, vol 1 (Long Papers), pp. 1886–1895 (2018)
19. N.S. Khan, M. Ata, Q. Rajput, Identification of opinion leaders in social network, in *2015 International Conference on Information and Communication Technologies (ICICT)*, pp. 1–6 (IEEE, 2015)
20. A.P. Kirilenko, S.O. Stepchenkova, Public microblogging on climate change: one year of twitter worldwide. Glob. Environ. Change **26**, 171–182 (2014)
21. H. Kwak, C. Lee, H. Park, S. Moon, What is twitter, a social network or a news media?, in *Proceedings of the 19th International Conference on World Wide Web*, pp. 591–600 (ACM, 2010)
22. S.M. Mohammad, P.D. Turney, *Nrc emotion lexicon* (National Research Council, Canada, 2013)
23. D. Murthy, Twitter: microphone for the masses? Media Cult. Soc. **33**(5), 779–789 (2011)
24. S. Ozturkcan, N. Kasap, M. Cevik, T. Zaman, An analysis of the gezi park social movement tweets. Aslib J. Inf. Manage. (2017)
25. J.W. Pennebaker, R.J. Booth, M.E. Francis, Linguistic inquiry and word count: Liwc [computer software], vol 135 (liwc. net, Austin, TX, 2007)
26. L. Perry, P. Joyce, Tweeting on the campaign trail: the when, how, and what of donald trump's tweets, in *The Role of Twitter in the 2016 US Election*, pp. 63–74 (Springer, 2018)
27. J. Proskurnia, R. Mavlyutov, R. Prokofyev, K. Aberer, P. Cudré-Mauroux, Analyzing large-scale public campaigns on twitter, in *International Conference on Social Informatics*, pp. 225–243 (Springer, 2016)
28. M. Purbrick, A report of the 2019 Hong Kong protests. Asian Aff. **50**(4), 465–487 (2019)
29. J.M. dos Reis Costa, R. Rotabi, E.L. Murnane, T. Choudhury, It is not only about grievances: emotional dynamics in social media during the Brazilian protests, in *Ninth International AAAI Conference on Web and Social Media* (2015)
30. A. Sechelea, T. Do Huu, E. Zimos, N. Deligiannis, Twitter data clustering and visualization, in *2016 23rd International Conference on Telecommunications (ICT)*, pp. 1–5 (IEEE, 2016)

31. R. Tinati, L. Carr, W. Hall, J. Bentwood, Identifying communicator roles in twitter, in *Proceedings of the 21st International Conference on World Wide Web*, pp. 1161–1168 (2012)
32. A. Tumasjan, T.O. Sprenger, P.G. Sandner, I.M. Welpe, Predicting elections with twitter: what 140 characters reveal about political sentiment, in *Fourth International AAAI Conference on Weblogs and Social Media* (2010)
33. Z. Wang, K. Caskey et al., # occupywallstreet: an analysis of twitter usage during a protest movement. Soc. Netw. **5**(04), 101 (2016)
34. M.J. Welch, U. Schonfeld, D. He, J. Cho, Topical semantics of twitter links, in *Proceedings of the Fourth ACM International Conference on Web Search and Data Mining*, pp. 327–336 (ACM, 2011)
35. I. Wetzstein, The visual discourse of protest movements on twitter: the case of Hong Kong 2014. Media Commun. **5**(4), 26–36 (2017)
36. C. Wilson, A. Dunn, The arab springl digital media in the egyptian revolution: descriptive analysis from the tahrir data set. Int. J. Commun. **5**, 25 (2011)

A Comparative Study to Analyze the Performance of Advanced Pattern Recognition Algorithms for Multi-Class Classification

Ayan Chatterjee, Martin Wulf Gerdes, Andreas Prinz, and Santiago Martinez

Abstract This study aims to implement the following four advanced pattern recognition algorithms, such as "optimal Bayesian classifier," "anti-Bayesian classifier," "decision trees (DTs)," and "dependence trees (DepTs)" on both artificial and real datasets for multi-class classification. Then, we calculated the performance of individual algorithms on both real and artificial data for comparison. In Sect. 1, a brief introduction is given about the study. In the second section, the different types of datasets used in this study are discussed. In the third section, we compared the classification accuracies of Bayesian and anti-Bayesian methods for both the artificial and real-life datasets. In the fourth section, a comparison between the classification accuracy of DT and DepT classification methods for both the artificial and real-life datasets is discussed. In the fifth section, a comparison between the classification accuracy of the four algorithms, such as (a) Bayes, (b) anti-Bayes, (c) DTs, and (d) DepTs for both the artificial and real datasets is explained. We used 5-fold cross-validation to determine the classification accuracy of individual, machine learning-based, advanced pattern recognition (PR) models.

Keywords Bayesian · Anti-Bayesian · Decision tree · Dependence tree

A. Chatterjee (✉) · M. W. Gerdes · A. Prinz
Department of Information and Communication Technology, Centre for E-Health, University of Agder, 4604 Kristiansand, Norway
e-mail: ayan.chatterjee@uia.no

M. W. Gerdes
e-mail: martin.gerdes@uia.no

A. Prinz
e-mail: andreas.prinz@uia.no

S. Martinez
Department of Health and Nursing Science, Centre for E-Health, University of Agder, 4604 Kristiansand, Norway
e-mail: santiago.martinez@uia.no

J. Tavares et al. (eds.), *Emerging Technologies in Data Mining and Information Security*, Lecture Notes in Networks and Systems 164,
https://doi.org/10.1007/978-981-15-9774-9_11

1 Introduction

Pattern recognition [1, 2] identifies patterns with the help of either machine learning or deep learning algorithms. Pattern recognition can be specified as the classification of data based on acquired knowledge or statistical information separated from patterns and their interpretation. In a regular pattern recognition application, the raw data is processed and transformed into a flexible form for a machine to use. Pattern recognition encompasses classification and cluster of patterns. Pattern recognition algorithms aim to deliver a sensible answer for all possible data and to classify input data into objects or classes based on specific features. In machine learning [3–5], the entire dataset is divided into two categories, one which is used to train the model, i.e., a training set and the other that is used to test the model, i.e., testing set [6, 7]. The performance of the individual model has been evaluated with an accuracy score and visualized in a confusion matrix table. A confusion matrix is a table that is generally used to explain the performance of a classification model (or "classifier") on a set of test data for which the correct values are known. The confusion matrix itself is relatively simple to understand, but the related terminology can be confusing. We have summarized our identified research questions for this study as follows—(a) How to develop algorithms for Bayesian, anti-Bayesian, decision tree, and dependence tree? (b) How to generate simulated data? (c) How to train and test different pattern recognition models with real and artificial datasets?, and (d) What is the accuracy score of different pattern recognition models under different training and testing environments? In this study, we have focused only on the machine learning-based pattern recognition (PR) algorithms.

2 Datasets and Data Pre-processing

In the study, we have used two types of dataset—(a) real glass identification dataset, and (b) artificially generated binary dataset.

2.1 Real Dataset

The glass identification dataset was used to classify the type of glasses, with the following given features as specified in the actual order: (1) ID, (2) RI: Refractive index, (3) Na: Sodium, (4) Mg: Magnesium, (5) Al: Aluminum, (6) Si: Silicon, (7) K: Potassium, (8) Ca: Calcium, (9) Ba: Barium, (10) Fe: Iron, and (11) Class (6 possible types). This dataset can be found at the "UCI Machine Learning Repository (https://archive.ics.uci.edu/ml/machine-learning-databases/glass/)" [8].

Data pre-processing steps on real glass data [7, 9, 10]:

Step-1: Load the dataset with Python library (pandas/NumPy).
Step-2: Remove **Id: Number** feature.
Step-3: Check if any missing value is found (if true then perform cleaning)
Step-4: Shuffle the cleaned data to return a random sample of items from an axis of object.
Step-5: Partition the data for training and testing (80:20) after random data selection strategy.
Step-6: Train a machine learning model (algorithm) with the training dataset and perform testing on the trained model with testing dataset following 5-fold cross-validation, to determine the classification accuracy of individual model.
Step-7: Repeat the Step-6 for five times and then take the average of classification accuracy for individual model.

The shape of the real glass data is (214, 11) distributed among six classes as follows: class—1 (70, 11), class–2 (76, 11), class—3 (17, 11), class—5 (13, 11), class—6 (9, 11), class—7 (29, 11). In this study, we have used only class-1, class-2, class-3, and class-7 for multi-class classification (as this combination produced the highest volume of data) using Bayesian, anti-Bayesian, decision tree, and dependence tree. For training, testing, and classification using the DT and DepT, we have used binary features for the real glass data accepting a threshold mechanism. We have used **mean** value as a threshold from each feature column to generate binary data from real glass dataset. Values greater than the threshold are converted to **1** otherwise, converted **0**.

2.2 *Artificial Dataset with DepT*

The artificial binary dataset is generated randomly based on the probability theory. Generated artificial data has ten-dimensional feature space (including index which is removed later in data pre-processing) with four classes. We have assumed that the vector components follow a dependence tree structure [2] between the various features. This **DepT** is arbitrarily assigned and is unknown to the classification algorithm for both training and testing. Values (0|1) for individual features (10-d) under four different classes are generated randomly following the below rule:

Rule: For class $j = 1$ to 4 and for feature indices $i = 1$–10, we have assigned randomly the value $V_{i,j} = \Pr[\mathbf{Xi = 0|W = Wj}]$ following a **DepT** structure, where **P** is the probability. Based on the distribution, we have modeled four classes. The generation of each feature point is dependent on the other feature point. They are unknown to the pattern recognition (**PR**) system but known to the data generation program.

Example For class $W1$, $p1 = [X1 = 0|X3 = 0;\ W1] = 0.7$, $\mathrm{p}1 = [X1 = 0|X3 = 1;\ W1] = 0.15$, and for the class $W2$, $q1 = [X1 = 0|X3 = 0;\ W2] = 0.9$, $q1 = [X1 = 0|X3 = 1;\ W2] = 0.27$, where **p, q** are the probabilities related to classes $W1$ and

$W2$, respectively, and X are the features. Each generated feature point is dependent to another.

Assumed dependence tree: Our assumed decision tree has a feature vector [$X0$, $X1, X2, X3, X4, X5, X6, X7, X8, X9$]. We have generated the root node ($X3$) first and followed by other nodes following the above rule. Node $X1, X5, X7$ depends on $X3$. Node $X6, X9$ depends on $X1$. Node $X2$ depends on $X5$. Node $X0$ depends on $X8$, and node $X4$, $X8$ depends on $X7$. We have generated total 8000 artificial data samples for four classes (2000 data sample/class) based on the feature dependency. We have performed classification assuming that all the random variables are dependent based on the **DepT**.

2.3 *Artificial Dataset with Feature Independency*

The artificial dataset is generated randomly based on the probability theory following the rule: $P[Xi = 0|W = W1] = p_i$, $[X_i = 1|W = W1] = (1 - p_i)$, and $[X_i = 0|W = W2] = q_i$, $[Xi = 0|W = W2] = (1 - q_i)$, where p, q are the probabilities related to classes **W1** and **W2**, respectively and X are the features. Each generated feature point is independent to another. They are known to the pattern recognition (**PR**) system but known to the data generation program. We have generated total 8000 artificial data samples for four classes (2000 data sample/class) based on the feature independency. We have performed a classification assuming that all the random variables are independent following a binary distribution [1, 2, 11].

2.4 *Bayes Classification Rule for Binary Data*

Binary data does not follow Gaussian or normal distribution [4, 12, 13], and it has different classification rule [1–5]:

$$G(X) = \sum_{i=1}^{d} X_i[\log\left(\frac{1-pi}{1-qi}\right) - \log\left(\frac{pi}{qi}\right)] + \log\left(\frac{P1}{P2}\right)$$
$$+ \sum_{i=1}^{d} \log\left(\frac{pi}{qi}\right), \text{ if } G(X) > 0 \text{ then } X \in W1 - \text{class, else} X \in W2 - \text{class.}$$

where

$$P(X|W2) = \prod_{i=1}^{d} q_i^{(1-X_i)}.(1-q_i)^{(X_i)}$$

Provided, X is data sample and d is total number of features.

2.5 Training and Testing Strategy

With regard to training and testing, we have used 5-fold cross-validation for granulation by removing little randomness to evaluate machine learning classification models on the limited data samples both real and artificial [3–5, 11].

2.6 Adopted Multi-Class Classification Technique

We have adopted binary class comparison method to calculate the final best class. The resultant confusion matrix (**CM**) has helped us to evaluate classification model accuracy.

3 Bayesian Versus Anti-Bayesian

Bayesian classification is based on Bayes' Theorem [12] that includes two types of probabilities—posterior probability $[P(H/X)]$ and prior probability $[P(H)]$, provided X is data tuple and H is some hypothesis or a class in a classification problem. The ultimate target is to create a classifier or a simple discrimination function ($[P(H/X)] = P(X/H) * P(H)/P(X)$) based on the training data to classify provided testing data among trained classes. $P(X/H)$ is called class conditional distribution. A good discrimination function gives good accuracy. A general Bayesian classifier can be represented as below:

$G(X) = (P(X/W1)\, P(W1)/P(X)) - (P(X/W2)\, P(W2)/P(X))$, if $G(X) > 0$, then data tuple **X** will belong to class **W1** else class **W2**. Bayesian is optimal but costly and hard. We have implemented both optimal Bayesian and Naive Bayesian following a Gaussian distribution of provided real glass data. The derived classifier is quadratic in nature and is equated as below:

$G(X) = X^T \cdot A \cdot X + B \cdot X + C$, if $G(X) > 0$, then data tuple **X** will belong to class **W1** else class **W2**. Provided A, B, C are constants.

$$A = (\sum_B^{-1} - \sum_A^{-1})$$

$$B = -\left(2 * \left(M_B^T \cdot \sum_B^{-1} - M_A^T \cdot \sum_A^{-1}\right)\right)$$

$$C = \left(M_B^T . \sum_B^{-1} . M_B - M_A^T . \sum_A^{-1} . M_A + \ln |\sum_B| - \ln |\sum_A| \right)$$

Our data **X** has a feature dimension d = 9 and the distribution follows normal or Gaussian following the central limit theorem. "μ" is the mean and "σ^2" is the variance of the distribution. "$\sum$" is the co-variance matrix ($d * d$). The only difference between Naive Bayes and optimal Bayes classifiers is the co-variance matrix used. For Naive Bayes, $\sigma_i^2 = E\ [x_i - \mu_i]^2$ (diagonal matrix) and for optimal Bayes, $\sigma_i \sigma_j = E\ [(x_i - \mu_i)\ (x_j - \mu_j)]$ (full matrix).

The steps are explained as below:

Step-1: Load the real glass dataset and perform pre-processing as explained in Sect. 2.1.
Step-2: Partition the random data into 5-folds (4-folds for training and 1-fold for testing). For each fold, repeat Step-3 to Step-6.
Step-3: Partition the training data into four sets those belonging to four classes.
Step-4: Calculate the normal distribution for each set (μ and σ):

$$\mu = \frac{1}{N} \sum_i^{N_i} X_i, \text{ and}$$

$$\sum^{+} = \frac{1}{N-1} \sum_i^{N_i} (X_i - \mu_i)(X_i - \mu_i)^T,$$

where X is given feature vector and N is total number of data sample.
Step-5: Form the classification equation as demonstrated above (Sect. 3).
Step-6: Perform binary class comparison for classification of test sample.
Step-7: After 5-fold cross-validation, calculate average testing accuracy.

Note: We have used pseudo-determinant and pseudo-inverse [14] for co-variance [15] calculation to avoid determinant of a singular matrix (0), and the logarithm of **0** (undefined).

Anti-Bayesian statistical pattern recognition [1, 2] is another classification technique. We have used the same approach in real glass dataset assuming Gaussian distribution, where classification is performed by quantile statistics (pattern recognition based on the two points rather than mean).

The steps for anti-Bayesian are explained as below:

Step-1: Load the real glass dataset and perform pre-processing as explained in Sect. 2.1.
Step-2: Partition the random data into 5-folds (4-folds for training and 1-fold for testing). For each fold, repeat Step 3 to Step 6.
Step-3: Partition the training data into four sets those belonging to four classes.

Step-4: Find two points (1/3rd percentile: a_i) and (2/3rd percentile: b_i) for each class considering Gaussian assumption.
Step-5: Perform classification of test sample (x) based on below rule:
if $x* < b1$ and $x* > a1$, $x* \in W1$ and if $x* < b2$ and $x* > a2$, $x* \in W2$. If, $b1 < x* < a2$, then decision is based on the following: $\|x* - b1\|$ and $\|x* - a2\|$ (This process is performed for each features of the test sample for future majority vote calculation).
Step-6: Perform binary class comparison for classification of test sample following majority voting scheme.
Step-7: After 5-fold cross-validation, calculate average testing accuracy.

3.1 Real Glass Data

The used real glass dataset is small. Optimal Bayes gives the highest accuracy than Naive Bayes and anti-Bayesian. In anti-Bayesian with Gaussian assumption and followed by quantile, statistics produces some information loss (Fig. 1).

Optimal Bayes	Class-1	Class-2	Class-3	Class-7	Accuracy
Class-1	37	0	6	10	61% (rounded)
Class-2	13	26	7	17	
Class-3	1	0	13	0	
Class-7	0	0	6	16	

Naïve Bayes	Class-1	Class-2	Class-3	Class-7	Accuracy
Class-1	17	40	3	9	60% (rounded)
Class-2	2	62	0	12	
Class-3	1	8	8	1	
Class-7	0	1	0	27	

Anti-Bayes	Class-1	Class-2	Class-3	Class-7	Accuracy
Class-1	49	11	2	6	42% (rounded)
Class-2	52	13	2	9	
Class-3	12	2	0	3	
Class-7	6	5	1	17	

Fig. 1 Confusion matrix (with 5-folds CV) and average accuracy comparison in between optimal Bayes, Naive Bayes, and anti-Bayes for real glass dataset (4-class)

3.2 *Artificial Dataset*

The artificial dataset (8000×11) is generated and classified as described in Sects. 2.2 and 2.3. As depicted in Fig. 2, Bayes-I represents Bayesian classification based on the independent random variables and Bayes-II represents Bayesian classification based on the dependent random variables (based on the **DepT** that has been inferred). For random data, we have assumed a binary distribution instead of Gaussian distribution for the features. It is not possible to perform anti-Bayesian classification over artificial binary data. The binary data is non-Gaussian and its histogram will produce only two columns (0 and 1). Over binary data, quantile points will be hardly able to classify features or attributes.

4 Decision Tree Versus DepT

A decision tree (DT) decides based on a tree [16] representation with a set of if-else rule. Each node in the tree specifies a test. It is used for approximating discrete valued target function (non-Gaussian). Tree is created with recursion based on apriori data. Figure 7 represents a sample decision tree created from real glass data after rendering to binary and following the algorithm as described below.

The steps for decision tree (DT) are explained as below

Step-1: Load the real glass dataset and perform pre-processing as explained in Sect. 2.1 (Therefore, rendered to binary data based on the threshold).

Step-2: Calculate the entropy for entire dataset.

Step-3: For every feature, do perform the following steps:

1. calculate entropy for all categorical values

Bayes - I	Class-1	Class-2	Class-3	Class-4	Accuracy
Class-1	2000	0	0	0	25% (rounded)
Class-2	2000	0	0	0	
Class-3	2000	0	0	0	
Class-4	2000	0	0	0	

Bayes - II	Class-1	Class-2	Class-3	Class-4	Accuracy
Class-1	1998	0	0	0	25% (rounded)
Class-2	1999	0	0	0	
Class-3	2000	0	0	0	
Class-4	1998	0	0	0	

Fig. 2 Confusion matrix (for 5-folds) and average accuracy of Bayes for artificial dataset (4-class)

2. take average information entropy for the current attribute
3. calculate gain for the current attribute

Step-4: Select the highest gain attribute.
Step-5: Repeat until we get the desired tree.

A dependence tree model [2] contains a dependence in between nodes (features) which is unknown to the pattern recognition (**PR**) system. Generated each feature point is dependent on another (parent node). Data generation with DepT is explained at Sect. 2.2. Probability of every node depends on its parent.

The steps for dependence tree (DepT) are explained as below

Step-1: Load the real glass dataset or artificially generated dataset and perform pre-processing as explained in Section-2.1 [Therefore, rendered to binary databased on the threshold].
Step-2: Create a fully connected undirected graph (G). $G = (V, E, W)$.
V = set of vertices (random variables)
E = set of edges (connecting any two vertices V_i and V_j)
W = set of weights associated to each edge
Step-3: Calculate weight (W) in graph (edges) following "**Expected Mutual Information Measure.**"
Step-4: Root probability calculation for competing classes (such as, $W1$, $W2$) for test sample and followed by, calculate $Pa(V)$ = **probability for first and second order marginals.**
Step-5: Class with better $Pa(V)$, will be the fittest.

4.1 Real Glass Data After Rendered to Binary

The used glass data sample size is small. Therefore, obtaining a good accuracy with this dataset is challenging. Binary conversion from real number has a chance to lose information. Training with more data can increase the prediction accuracy. The rendered binary data used to train **DepT** is not generated based on a **DepT** and thereby, accuracy is low (Fig. 3).

4.2 Artificial Data

Increasing the volume of the training data with the artificial data generation technique based on the dependence tree (Sect. 2.2), we have increased the classification accuracy both in **DT** and **DepT** as depicted (Fig. 4).

DT	Class-1	Class-2	Class-3	Class-7	Accuracy
Class-1	59	19	10	3	71% (rounded)
Class-2	7	50	5	3	
Class-3	0	4	2	0	
Class-7	2	3	0	23	

DepT	Class-1	Class-2	Class-3	Class-7	Accuracy
Class-1	32	15	15	7	39% (rounded)
Class-2	23	21	20	11	
Class-3	7	0	7	3	
Class-7	10	2	3	14	

Fig. 3 Confusion matrix (for 5-folds) and average accuracy comparison in between DT and DepT for real glass dataset (4-class)

DT-Artificial	Class-1	Class-2	Class-3	Class-4	Accuracy
Class-1	1806	136	212	64	79% (rounded)
Class-2	77	1496	288	112	
Class-3	94	242	1339	161	
Class-4	22	124	159	1663	

DepT-Artificial	Class-1	Class-2	Class-3	Class-4	Accuracy
Class-1	1408	82	260	248	51% (rounded)
Class-2	562	239	488	710	
Class-3	317	126	1132	425	
Class-4	113	18	602	1265	

Fig. 4 Confusion matrix (for 5-folds) and average accuracy comparison in between DT and DepT for artificial dataset (4-class)

5 Accuracy Comparison

The accuracy comparison between optimal Bayes, Naive Bayes, anti-Bayes, DT, and DepT on real glass dataset is depicted in Fig. 5, and the same comparison on artificial data is illustrated in Fig. 6. In Sect. 5.3, accuracy of individual methods is explained.

5.1 Real Glass Data

See Fig. 5.

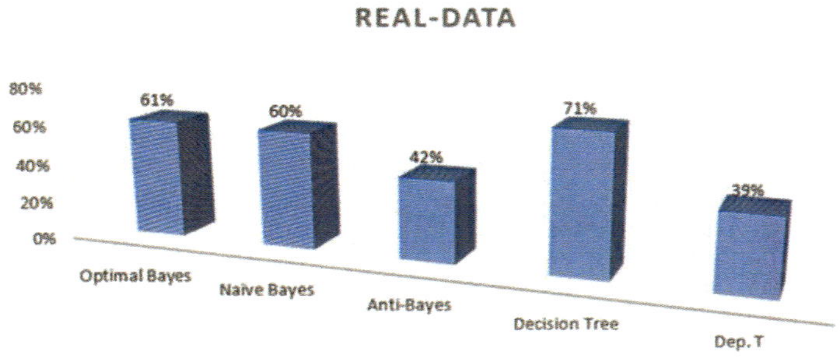

Fig. 5 Accuracy comparison for real data

Fig. 6 Accuracy comparison for artificial data

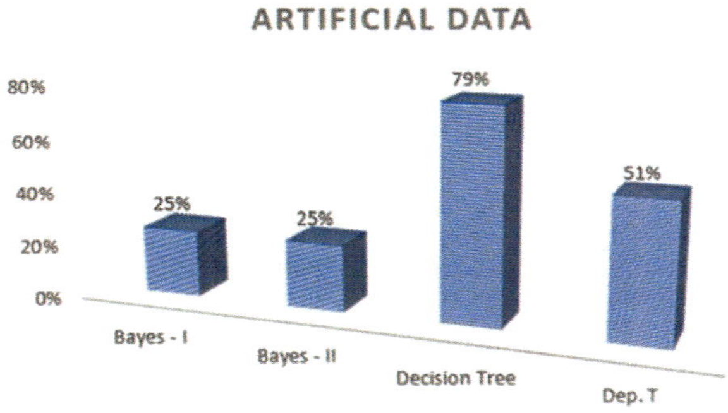

5.2 *Artificial Data*

See Fig. 6.

5.3 *Discussion on Performance Analysis*

The provided real glass data is minimal for training a pattern recognition model to achieve higher accuracy [6]. Some of the features have mostly zero values, but we have considered those features as well in our training else data volume would have become even small. Therefore, it hits the accuracy calculation. Instead of real-inverse and determinant, we have calculated approximated value (pseudo-determinant and pseudo-inverse) in the co-variance calculation for Bayesian.

As depicted in Fig. 5, it seems that Bayes has performed well with a limited set of real data. Anti-Bayesian produces less accuracy on limited data as compared to Bayes. We have made a Gaussian assumption for anti-Bayesian, and there might be some information loss due to quantile calculation and missing-out border point data. Histogram might produce better approximation, but we have not tried it here.

The decision tree has produced the most significant accuracy with this data after rendered them to binary. Binary conversion from real number has a chance to lose information. Training with more data can increase the prediction accuracy. The binary data supplied to DepT is not generated based on a dependence tree, and thereby, accuracy is low for DepT.

As depicted in Fig. 6, Bayes-I represents Bayesian classification based on the independent random variables, and Bayes-II represents Bayesian classification based on the dependent random variables (based on the DepT that has been inferred). Generated simulated data is binary, and they are not following Gaussian distribution. Artificial data for Bayes has been generated as specified in Sects. 2.2 and 2.3. The performance of Bayes with binary data is not suitable due to randomness in data. On the other hand, it is not possible to perform anti-Bayesian classification over artificial binary data. The binary data is non-Gaussian, and its histogram will produce only two bins (0 and 1). Over binary data, quantile points will hardly be able to classify features. Artificial data generated by feature dependence (DepT) has helped to increase the accuracy of DT and DepT as compared to the real glass data.

6 Conclusion

In Fig. 7, a sample decision tree is depicted (drawn with Grpahviz) with binary decision making on every branch using an information gain approach. Calculated entropy is displayed at each winning node. Leaf nodes are the classes. This paper gives a noble concept on how to use Bayesian and anti-Bayesian for multi-class classification on both artificial and real datasets! It appears that if we can reduce certain assumed randomness in artificial data generation, then we can increase the accuracy of the classifiers. More data in training will increase the accuracy of the classifier. For real-life data, we must select more robust dataset for efficient training. To reproduce the results as discussed from Sects. 3–5, Python codebase has been uploaded in GitHub repository (https://github.com/ayan1c2/AdvancedPatternRecognitionCode.git).

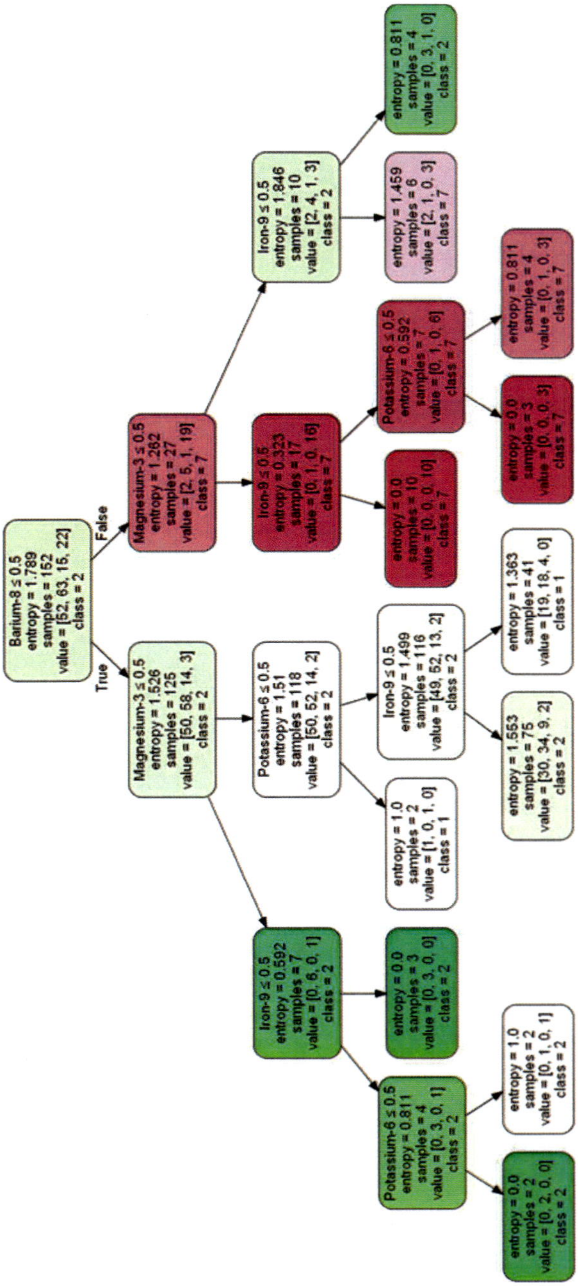

Fig. 7 A sample decision tree generated with information gain

Acknowledgements Thanks to Dr. John Oommen (Chancellor's Professor, School of Computer Science, Carleton University, Ottawa, Ontario K1S 5B6, Canada) for teaching us the course on "Advanced Pattern Recognition" as a part of PhD coursework, and reviewing the article which is written based on the assignment of the coursework.

References

1. A. Thomas, B.J. Oommen, The fundamental theory of optimal "Anti-Bayesian" parametric pattern classification using order statistics criteria. Pattern Recogn. **46**(1), 376–388 (2013)
2. B.J. Oommen, A. Thomas, "Anti-Bayesian" parametric pattern classification using order statistics criteria for some members of the exponential family. Pattern Recogn. **47**(1), 40–55 (2014)
3. E. Alpaydin, *Introduction to Machine Learning* (MIT Press, 2020) London
4. T.M. Mitchell, *Machine Learning* (1997)
5. D. Michie, D.J. Spiegelhalter, C.C. Taylor, Machine learning. Neural Stat. Classif. **13**, 1–298 (1994)
6. S. Visa, B. Ramsay, A.L. Ralescu, E. Van Der Knaap, Confusion matrix-based feature selection, in *MAICS*, vol. 710 (2011) pp. 120–127
7. S.B. Kotsiantis, D. Kanellopoulos, P.E. Pintelas, Data preprocessing for supervised leaning. Int. J. Comput. Sci. **1**(2), 111–117 (2006)
8. A. Asuncion, D. Newman, UCI machine learning repository (2007)
9. A. Chatterjee, M.W. Gerdes, S.G. Martinez, Identification of risk factors associated with obesity and overweight—a machine learning overview. Sensors **20**(9), 2734 (2020). https://doi.org/10.3390/s20092734
10. A. Chatterjee, M.W. Gerdes, S.G. Martinez, Statistical explorations and univariate timeseries analysis on COVID-19 datasets to understand the trend of disease spreading and death. Sensors **20**(11), 3089 (2020)
11. F. Pedregosa et al.: Scikit-learn: machine learning in python. J. Mach. Learn. Res. 2825–2830 (2011)
12. H. Zhang, The optimality of naive Bayes, in AA, vol. 1(2), (2004) pp. 3
13. Y.L. Tong, *The Multivariate Normal Distribution* (Springer Science & Business Media, 2012)
14. T. Górecki, M. Łuczak, Linear discriminant analysis with a generalization of the Moore-Penrose pseudoinverse. Int. J. Appl. Math. Comput. Sci. **23**(2), 463–471 (2013)
15. J. Baik, B.A. Gérard, P. Sandrine, Phase transition of the largest eigenvalue for nonnull complex sample covariance matrices. Ann. Probab. **33**(5), 1643–1697 (2005)
16. S.R. Safavian, D. Landgrebe, A survey of decision tree classifier methodology. IEEE Trans Syst Man Cybern. **21**(3), 660–674 (1991)

ECG Signal Denoising Using a Novel Solution to the Heat Equation Through Wavelet Transform

P. Upadhyay, S. K. Upadhyay, and K. K. Shukla

Abstract This paper proposes a wavelet transformation solution to the Heat equation. This method, associated with wavelet obtained by taking derivative of the Gaussian kernel, has been applied to denoise ECG signals. The proposed method is tested by experimenting on different signals of the MIT-BIH Arrhythmia database corrupted by power line interference, electromyographic (EMG) noises and synthetic interference. It is observed that the proposed method gives appreciable results in the cases of EMG noise, power line interference and synthetic interference. The proposed method is compared with some of the recenetly reported methods. We found that the proposed method outperfoms the recently reported methods.

Keywords ECG signals · Denoising · Heat equation · Wavelets · Initial conditions · Gaussian Kernel

P. Upadhyay (✉) · S. K. Upadhyay
DST-Centre for Interdisciplinary Mathematical Sciences (CIMS),
Banaras Hindu University (B.H.U.), Varanasi, India
e-mail: prateep.upadhyay1@bhu.ac.in

S. K. Upadhyay
e-mail: sk_upadhyay2001@yahoo.com

S. K. Upadhyay
Department of Mathematical Sciences I.I.T. (B.H.U.), DST-Centre for Interdisciplinary Mathematical Sciences (CIMS), Banaras Hindu University (B.H.U.), Varanasi, India

K. K. Shukla
Department of Computer Science and Engineering, I.I.T.(B.H.U.),Varanasi, India
e-mail: kkshukla.cse@iitbhu.ac.in

J. Tavares et al. (eds.), *Emerging Technologies in Data Mining and Information Security*, Lecture Notes in Networks and Systems 164,
https://doi.org/10.1007/978-981-15-9774-9_12

1 Introduction

ECG signals are the reflections of the heart activities. A number of intertwined and complex chemical, electrical and mechanical processes of the heart are related to ECG signals. ECG signals also equip us with useful information about other body functionings e.g. nervous system or circulation [4].

ECG signals are low-frequency band signals [8] which are noise-prone and weak electrical interpretation of the activities of heart [1, 15]. Path deformities and electrical disturbances introduce noise in the ECG signals. ECG signals are chiefly affected by baseline wander, power line interference, muscle noise etc. [16].

An approach of ECG signal denoising based on adaptive dual threshold filter (ADTF) and the discrete wavelet transform (DWT) has been proposed by [8]. Hesar and Mohebbi [6] used marginalized particle extended Kalman filter with an extended particle weighting strategy for ECG signal denoising. Hesar and Mohebbi [5] introduced a model-based Bayesian denoising framework associated with a marginalized particle extended Kalman filter (MP-EKF), variational mode decomposition (VMD) and a novel fuzzy-based adaptive particle weighting strategy (New MP-EKF). A method based on the combination of modified frequency slice wavelet transform (MFSWT) and convolutional neural network (CNN) to denoise wearable ECG signals has been proposed by Zhao et al. [17]. A convex optimization approach that used linear time-invariant filtering with sparsity has been proposed by Wang et al. [14] to denoise ECG signals corrupted chiefly by baseline wander noise. Chiang et al. [2] has introduced a denoising autoencoder (DAE) using fully convolutional network (FCN) to denoise ECG signals. A guided filter based multi-lead model for ECG signal denoising has been applied by [7]. Nasser mourad [11] introduced a two-stage ECG signal denoising method which is helpful in eliminating Gaussian white noise present in the ECG signals. The first stage of this method uses majorization-minimization (MM) technique whereas the second stage utilizes singular spectrum analysis (SSA) technique. Jilong wang et al. introduced adversarial denoising method to denoise ECG signals. This method is based on a new loss function [13].

In this paper, a novel solution to the Heat equation has been proposed by changing the initial conditions. The solution thus obtained is in the form of wavelet transformation. We apply this solution to different signals of the MIT-BIH arrhythmia database of Physionet, as reported in [8]. The results obtained, give superior results to the reported results of [2, 7, 11, 13, 14, 17].

In our present work, we have used the definitions and some of the properties of Fourier transform, inverse Fourier transform, convolution theorem, wavelets and continuous wavelet transform from [3].

1.1 Proposed Solution of the Heat Equation

In this paper, the noisy ECG signal is modeled using the Heat equation. Here, the amplitude of the signal corresponds to the temperature along a thin rod, given by the function $v(X, T)$, where X is the distance along the rod and T is time.

The one-dimensional Heat equation with no sources or sinks involved can be written as

$$\frac{\partial v}{\partial T} = \underline{a}^2 \kappa \frac{\partial^2 v}{\partial X^2}, \text{ for } -\infty < X < \infty, \tag{1}$$

where our assumptions are

$v(X, 0) = \underline{a}\, g(X)$ and $v(X, T) \to 0$ as $X \to \pm\infty$, $\underline{a} \in \mathbb{R}$, $\underline{a} \neq 0$ [3].

κ is any arbitrary constant known as thermal diffusivity in (1). Thermal diffusivity describes the rate of temperature spread through a material [10]. The thermal diffusivity term present in the Heat equation helps in better analysis of noise present across consecutive points of ECG signals.

On application of Fourier transform on both sides of (1) we obtain,

$$\tilde{\mathscr{F}}\left(\frac{\partial v}{\partial T}\right) = \underline{a}^2 \kappa \tilde{\mathscr{F}}\left(\frac{\partial^2 v}{\partial X^2}\right). \tag{2}$$

According to the definition of Fourier transform,

$$\tilde{\mathscr{F}}(v) = \hat{v}(\underline{\omega}, T) = \int_{-\infty}^{\infty} v(X, T) \mathrm{e}^{-i\underline{\omega}X} \mathrm{d}X. \tag{3}$$

(3) can further be written as

$$\frac{\partial}{\partial T} \int_{-\infty}^{\infty} v(X, T) \mathrm{e}^{-i\underline{\omega}X} \mathrm{d}X = \frac{\partial \hat{v}(\underline{\omega}, T)}{\partial T}. \tag{4}$$

We obtain,

$$\tilde{\mathscr{F}}\left(\frac{\partial v}{\partial X}\right) = \int_{-\infty}^{\infty} \frac{\partial v(X, T)}{\partial X} \mathrm{e}^{-i\underline{\omega}X} \mathrm{d}X. \tag{5}$$

Applying integration by parts we get,

$$\int_{-\infty}^{\infty} \frac{\partial v(X, T)}{\partial X} \mathrm{e}^{-i\underline{\omega}X} \mathrm{d}X = i\underline{\omega}\hat{v}(\underline{\omega}, T). \tag{6}$$

(2) gives,

$$\tilde{\mathscr{F}}\left(\frac{\partial^2 v}{\partial X^2}\right) = -\underline{\omega}^2 \tilde{\mathscr{F}}(v)$$

$$\tilde{\mathscr{F}}\left(\frac{\partial^2 v}{\partial X^2}\right) = -\underline{\omega}^2 \hat{v}(\underline{\omega}, T). \tag{7}$$

Substituting the values of (4) and (7) in (2), we get

$$\frac{\partial \hat{v}(\underline{\omega}, T)}{\partial T} + \underline{a}^2 \kappa \underline{\omega}^2 \hat{v}(\underline{\omega}, T) = 0. \tag{8}$$

solution of (8) gives,

$$\hat{v}(\underline{\omega}, T) = d e^{-\kappa \underline{a}^2 \underline{\omega}^2 T}. \tag{9}$$

In (9), d is an arbitrary constant with respect to the partial differential variable T. Thus d can be expressed as a function of $\underline{\omega}$.

Therefore we have

$$\hat{v}(\underline{\omega}, T) = d(\underline{\omega}) e^{-\kappa \underline{\omega}^2 T} \tag{10}$$

at $T = 0$; $\hat{v}(\underline{\omega}, 0) = d(\underline{\omega})$

Alternatively, we can say that the Fourier transform of $v(X, 0)$ is $d(\underline{\omega})$.

Therefore we write

$$d(\underline{\omega}) = \int_{-\infty}^{\infty} v(X, 0) \mathrm{e}^{-i\underline{\omega} X} \mathrm{d}X$$

$$= \int_{-\infty}^{\infty} \underline{a}\, g(X) \mathrm{e}^{-i\underline{\omega} X} \mathrm{d}X$$

$$d(\underline{\omega}) = \underline{a}\, \hat{g}(\underline{\omega}). \tag{11}$$

We thus obtain

$$d(\underline{\omega}) = \underline{a}\, \hat{g}(\underline{\omega}). \tag{12}$$

Substitution of the value of $d(\underline{\omega})$ from (12) in (10) gives

$$\hat{v}(\underline{\omega}, T) = \hat{g}(\underline{\omega}) \mathrm{e}^{-\kappa \underline{a}^2 \underline{\omega}^2 T}. \tag{13}$$

We assume

$$\hat{h}(\underline{\omega}) = \mathrm{e}^{-\kappa \underline{a}^2 \underline{\omega}^2 T}. \tag{14}$$

The definition of inversion formula of Fourier transform [3] gives,

$$h(X) = \int_{-\infty}^{\infty} \mathrm{e}^{i\underline{\omega}X} \mathrm{e}^{-\kappa \underline{a}^2 \underline{\omega}^2 T} d\underline{\omega}. \tag{15}$$

Simplifying the integral of (15) we get

$$h(X) = \frac{1}{\underline{a}}\sqrt{\frac{\pi}{\kappa T}}\ \mathrm{e}^{\frac{-X^2}{2\underline{a}^2\kappa T}}$$

$$v(X, T) = \widetilde{\mathscr{F}}^{-1}(\hat{g}(\underline{\omega})\hat{h}(\underline{\omega})). \tag{16}$$

From (16) and definition of convolution theorem [3], we have,

$$v(X, T) = \widetilde{\mathscr{F}}^{-1}(\hat{g}(\underline{\omega})\hat{h}(\underline{\omega})) = (g * h)(X). \tag{17}$$

From (17) and definition of the convolution of two functions [3] we get

$$v(X, T) = \int_{-\infty}^{\infty} g(y)\sqrt{\frac{\pi}{\kappa T}}\ \mathrm{e}^{\frac{-(X-y)^2}{2\underline{a}^2\kappa T}}\,\mathrm{d}y \tag{18}$$

$$= \sqrt{\frac{\pi}{\kappa T}}\left[\int_{-\infty}^{\infty} g(y)\mathrm{e}^{\frac{-(X-y)^2}{2\underline{a}^2\kappa T}}\,\mathrm{d}y\right]$$

$$= \sqrt{\frac{\pi}{\kappa T}}\left[\int_{-\infty}^{\infty} g(y)\mathrm{e}^{-\left(\frac{X-y}{\underline{a}}\right)^2\frac{1}{2\kappa T}}\,\mathrm{d}y\right].$$

Differentiating with respect to X we get

$$v'(X, T) = \sqrt{\frac{\pi}{\kappa T}}\int_{-\infty}^{\infty} g(y)\mathrm{e}^{-\left(\frac{X-y}{\underline{a}}\right)^2\frac{1}{2\kappa T}}\left(-\left(\frac{X-y}{\underline{a}\kappa T}\right)\right). \tag{19}$$

Now, if we take

$$h(X) = \mathrm{e}^{-\left(\frac{X}{\underline{a}}\right)^2 \frac{1}{2\kappa T}} \left(-\left(\frac{X}{\underline{a}\kappa T}\right)\right). \quad (20)$$

We get

$$v'(X, T) = \sqrt{\frac{\pi}{\kappa T}} \left[\int_{-\infty}^{\infty} g(y) h\left(\frac{X - y}{\underline{a}}\right) \mathrm{d}y \right]. \quad (21)$$

Thus we can write the solution of (1) in the form of continuous wavelet transform with respect to the wavelet h as obtained from (20).

Therefore we obtain,

$$v'(X, T) = \mathscr{W}_h[g](\underline{a}, y). \quad (22)$$

The proposed method has been implemented in MATLAB®. After getting the noisy ECG signal, we applied the Heat equation to it and got the wavelet denoised derivative of the noisy ECG signal where the wavelet is mentioned in (20). The denoised ECG signal is thus finally obtained with the help of the wavelet denoised derivative of the noisy ECG signal. We calculate MSE value for the denoised ECG signal. If the MSE value is not within the acceptable range ($\leq 10^{-5}$), we repeat the same procedure with the noisy ECG signal. The denoising performance of the proposed method with different types of noise that affect ECG signals viz power line interference, baseline wander, abrupt baseline drift, electromyographic (EMG) noise, composite noise, white Gaussian noise (WGN), pink noise and brown noise respectively [6, 9] and the flowchart of the proposed method are depicted in Fig. 1a, b respectively.

The wavelet in (20) is known as derivative of the Gaussian kernel. The mother wavelet of (20) is given in (23).

$$\psi(t) = te^{-t^2}; \quad -\infty < t < \infty \quad (23)$$

2 Results

The proposed solution of the Heat equation has been tested for three types of noises viz, the power line interference, the EMG noise and the synthetic interference. The powerline interference noise is simulated as given in [8] whereas EMG interference and synthetic interference are simulated as given in [12]. The quantitive comparison of the proposed methods with the reported results of [2, 7, 11, 13, 14, 17] have been done in terms of the mean square error (MSE), the root mean square error (RMSE) the

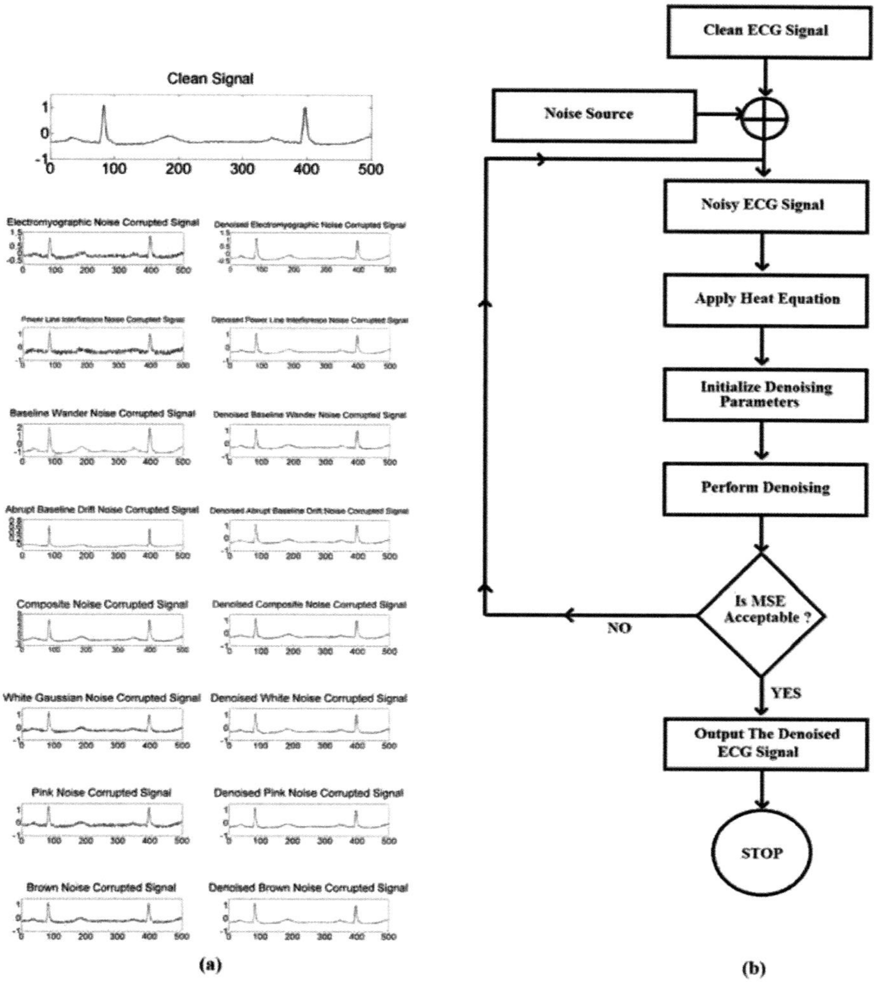

Fig. 1 **a** Proposed method denoising results with different kinds of noises for MIT BIH signal 101, **b** flowchart of the proposed method

percent root mean square difference (PRD), the signal-to-noise ratio output (SNR_{out}) and the signal-to-noise ratio improvement (SNR_{imp}) respectively.

$$\text{MSE} = \frac{1}{n}\sum_{i=1}^{i=n}(x(i) - y(i))^2.$$

$$\text{RMSE} = \sqrt{\frac{1}{n}\sum_{i=1}^{i=n}(x(i) - y(i))^2}.$$

$$\text{PRD} = 100\sqrt{\frac{\sum_{i=1}^{i=n}(x(i) - y(i))^2}{\sum_{i=1}^{i=n}(x(i))^2}}.$$

$$\text{SNR}_{\text{out}} = 10\log_{10}\left(\frac{\sum_{i=1}^{i=n}(x(i))^2}{\sum_{i=1}^{i=n}(y(i) - x(i))^2}\right).$$

$$\text{SNR}_{\text{imp}} = 10\log_{10}\left(\frac{\sum_{i=1}^{i=n}(\Delta(i) - x(i))^2}{\sum_{i=1}^{i=n}(y(i) - x(i))^2}\right).$$

where $x(i)$, $\Delta(i)$, $y(i)$ and n denote original signal, noisy signal, denoised signal and length of ECG signal respectively [8].

3 Discussion

In this paper, the novel and effective solution of the Heat equation has been applied to denoise several MIT-BIH signals. The motivation and aim of the proposed method are to improve the filtering of ECG signals and get better denoised ECG signals. To make thermal diffusivity term in the Heat equation capable of capturing noise at different points of the signal, we have taken the product of squared dilation term and thermal diffusivity in (1) along with the presence of dilation factor in the initial conditions. This framework first, captures the effect of noise in terms of positive and negative signal dilation equally well. Second, this mathematical construction also captures the dilation in signal due to noise at the very beginning. Further, due to the product of thermal diffusivity term along with the squared dilation term, this methodology also helps in analyzing the amount of noise present at different consecutive points of the ECG signals differently.

Table 1 clearly reflects the denoising superiority of the proposed method in the case of powerline interference when compared to the reported methods of [2, 7, 11, 13, 14, 17] in terms of SNR_{out} and MSE for the signal 115 of MIT-BIH. The comparison of results of proposed method with the results of the aforesaid methods with respect to SNR_{out} and MSE for the MIT-BIH signal 115 in the case of EMG interference is given in Table 2. It is evident that the proposed method outperforms all the above reported methods. Tables 3 and 4 are respectively the SNR_{imp} and PRD comparisons of the above mentioned methods with the proposed method for different MIT-BIH signals corrupted by 5 dB synthetic interference. Tables 3 and 4, clearly reflect the superiority of the denoising performance of the proposed method in comparison to all the above reported methods in terms of PRD and SNR_{imp} values.

Table 1 Correction results of power line interference of MIT-BIH signal 115

Method name	SNR_{out}	MSE
Adversarial denoising	30.71	0.051
Group sparsity	20.23	0.046
MFSWT + CNN	18.92	0.019
FCN	20.11	0.00202
Sparsity	14.33	0.008
Multi lead model based	18.76	0.042
Proposed method	**75.81**	**0.0000022**

Table 2 Correction results of EMG interference of MIT-BIH signal 115

Method Name	SNR_{out}	MSE
Adversarial denoising	25.46	0.29
Group sparsity based method	15.92	0.081
MFSWT + CNN	23.58	0.029
FCN	19.60	0.0039
Sparsity	12.85	0.0068
Multi lead model based	12.09	0.057
Proposed method	**73.41**	**0.0000014**

Table 3 SNR improved comparison of some of the MIT BIH signals

MIT BIH signal	Adversarial denoising	Group sparsity	MFSWT + CNN	FCN	Sparsity	Multi lead model based	Proposed method
100	20.81	12.01	15.66	23.38	16.22	14.11	**85.16**
101	26.88	11.82	12.03	21.46	17.51	16.86	**81.46**
103	25.31	10.56	14.48	23.79	14.21	12.99	**86.14**
113	28.11	9.81	15.58	25,93	18.29	11.02	**89.96**
115	23.63	11.98	15.45	23.31	14.75	18.49	**82.75**
117	21.02	10.88	16.88	23.12	12.82	16.53	**87.78**
119	22.81	10.75	12.35	24.88	15.77	17.12	**85.29**
122	25.69	11.54	12.87	21.53	16.68	17.59	**86.25**

Table 4 PRD comparison of some of the MIT BIH signals

MIT BIH signal	Adversarial denoising	Group sparsity based	MFSWT+ CNN	FCN	Sparsity	Multi lead model based	Proposed method
100	20.11	15.86	20.51	12.34	17.06	14.20	**6.42**
103	20.98	12.11	22.54	11.78	18.42	16.11	**6.14**
105	23.75	14.89	18.79	10.92	16.55	15.86	**5.49**
115	27.68	15.01	19.05	11.39	17.98	18.81	**5.19**

4 Conclusion

This paper shows the application of the novel solution of the Heat equation to denoise several ECG signals from the MIT-BIH database. We have applied the proposed method for denoising the ECG signals corrupted by power line interferences, EMG interferences and synthetic interferences. It was found that the proposed method outperforms the recently reported methods [2, 7, 11, 13, 14, 17]. We also found that apart from outperforming all the above reported results, the proposed method gives much superior results in the case of filtering of ECG signals.

References

1. S. Ari, M.K. Das, A. Chacko, ECG signal enhancement using S-transform. Comput. Biol. Med. **43**(6), 649–660 (2013)
2. H.T. Chiang, Y.Y. Hsieh, S.W. Fu, K.H. Hung, Y. Tsao, S.Y. Chien, Noise reduction in ECG signals using fully convolutional denoising autoencoders. IEEE Access **18**(7), 60806–60813 (2019)
3. L. Debnath, *Wavelet Transforms and Their Applications* (Birkhäuser, Boston, 2002), pp. 63–371
4. A. Gacek, W. Pedrycz, *ECG Signal Processing Classification and Interpretation: A Comprehensive Framework of Computational Intelligence* (Springer, Berlin, 2012)
5. H.D. Hesar, M. Mohebbi, An adaptive particle weighting strategy for ECG denoising using marginalized particle extended Kalman filter: An evaluation in arrhythmia contexts. IEEE J. Biomed. Health Inform. **21**(6), 1581–1592 (2017)
6. H.D. Hesar, M. Mohebbi, ECG denoising using marginalized particle extended kalman filter with an automatic particle weighting strategy. IEEE J. Biomed. Health Inform. **21**(3), 635–644 (2016)
7. H. Hao, M. Liu, P. Xiong, H. Du, H. Zhang, F. Lin, Z. Hou, X. Liu, Multi-lead model-based ECG signal denoising by guided filter. Eng. Appl. Artif. Intell. **1**(79), 34–44 (2019)
8. W. Jenkal, R. Latif, A. Toumanari, A. Dliou, O. El B'charri, F.M. Maoulainine, An efficient algorithm of ECG signal denoising using the adaptive dual threshold filter and the discrete wavelet transform. Biocybern. Biomed. Eng. **36**(3), 499–508 (2016)
9. A.S. Khaing, Z.M. Naing, Quantitative investigation of digital filters in electrocardiogram with simulated noises. Int. J. Inf. Electron. Eng. **1**(3), 210–216 (2011)
10. R. Mishra, J. Militky, *Nanotechnology in Textiles* (Woodhead Publishing, Cambridge, 2018)
11. N. Mourad, ECG denoising algorithm based on group sparsity and singular spectrum analysis. Biomed. Signal Process. Control **1**(50), 62–71 (2019)
12. J. Wang, Y. Ye, X. Pan, X. Gao, Parallel-type fractional zero-phase filtering for ECG signal denoising. Biomed. Signal Process. Control **18**, 36–41 (2015)
13. J. Wang, R. Li, R. Li, K. Li, H. Zeng, G. Xie, L. Liu, Adversarial de-noising of electrocardiogram. Neurocomputing **15**(349), 212–224 (2019)
14. X. Wang, Y. Zhou, M. Shu, Y. Wang, A. Dong, ECG baseline wander correction and denoising based on sparsity. IEEE Access **4**(7), 31573–31585 (2019)
15. Z. Wang, F. Wan, C.M. Wong, L. Zhang, Adaptive Fourier decomposition based ECG denoising. Comput. Biol. Med. **1**(77), 195–205 (2016)
16. S.K. Yadav, R. Sinha, P.K. Bora, Electrocardiogram signal denoising using non-local wavelet transform domain filtering. IET Signal Process. **9**(1), 88–96 (2015)
17. Z. Zhao, C. Liu, Y. Li, Y. Li, J. Wang, B.S. Lin, J. Li, Noise rejection for wearable ECGs using modified frequency slice wavelet transform and convolutional neural networks. IEEE Access **21**(7), 34060–34067 (2019)

Design and Implementation of Chao-Cryptic Architecture on FPGA for Secure Audio Communication

B. Sowmya, V. M. Padmapriya, R. Sivaraman, Amirtharajan Rengarajan, Sundararaman Rajagopalan, and Har Narayan Upadhyay

Abstract The emerging growth in communication arena firmly demands a secure channel and needless to mention that security and privacy of the user are the top priorities in audio communication. To transmit audio signals safely, cryptography is being adopted widely. Chaos-based cryptography is one of the effective strategies that attract researchers to develop more techniques. Characteristics of chaotic systems such as ergodicity and stochasticity are the main motivators for using them in encryption schemes. This work presents an field programmable gate array (FPGA) governed chaotic audio encryption scheme based on Lorenz attractor. This scheme of encryption undergoes two consecutive steps of operations, namely confusion by linear feedback shift register (LFSR) and diffusion through the random numbers generated in X, Y, and Z planes of the chaotic attractor. A study has also been carried out to analyze the best diffusion from the obtained results. Evaluation of the proposed scheme includes entropy, correlation, histogram, PSNR, and MSE. Further, NIST SP 800-22 batteries of test have been performed to ensure that the encrypted audio data possesses more randomness. Moreover, this work consumed 15,422 logic elements and 14.22 mW of dynamic power to accomplish the design. Noticeably, the proposed work has achieved 10.346 Mbps as a throughput to encrypt $2^{16} \times 16$ bit samples of audio data.

Keywords Cryptography · Audio encryption · FPGA · Lorenz attractor

1 Introduction

As the usage of communication devices has been accelerated, the need for secure communication has become a crucial task in day-to-day life. Audio encryption is the

B. Sowmya · V. M. Padmapriya · R. Sivaraman · A. Rengarajan · S. Rajagopalan (✉) · H. N. Upadhyay
Department of ECE, School of EEE, SASTRA Deemed University, Thanjavur 613401, India
e-mail: amir@ece.sastra.edu

J. Tavares et al. (eds.), *Emerging Technologies in Data Mining and Information Security*, Lecture Notes in Networks and Systems 164,
https://doi.org/10.1007/978-981-15-9774-9_13

process of encoding information in order to protect the data from eavesdropping or hackers and can provide access to only authorized users. The key used for encrypting the data plays a crucial role and determines the strength of the encryption algorithm [1]. In symmetric encryption schemes, a key used for encryption and decryption is the same, and only the intended users know it to secure the data. The most sensitive applications (such as FBI, military, some financial applications, etc.) use either advanced encryption standard (AES) or data encryption standard (DES). DES was developed in the nineteen seventies but has been changed by AES in many applications due to its complexity. However, audio signals have redundancy, and correlation among adjacent pixels was very high as it deals with a bulk amount of data [2]. Hence, AES alone has not been a suitable agent for bulk data encryption. To achieve adequate encryption, new algorithms must be proposed for secure audio communication.

Lokireddi Phani Kumar and A. K Gupta proposed one such work, through AES and sophisticated cryptographic algorithm for secure offline speech communication on FPGA [3]. Munawar A. Riyadi et al. proposed a 128-bit chaotic encryption method for voice communication using FPGA, where cipher feedback has been employed, and the channel was tested for four scenarios: fully secure channel, decrypted unsecure channel, eavesdropping and unsecure messages [4]. Mohammed F. Tolba experimented with hardware speech encryption using Nose Hoover chaotic generator, dynamic shift, and bit permutation, where the results have been analyzed using various perceptual and statistical tests [5]. Ahemed G. Radwan et al. focused on the reduction of area and delay of encryption architecture by using a modified chaotic logistic map and implemented the same on Xilinx virtex-5 xc5lx50T FPGA [6]. Salwa K. Abd-El-Hafiz et al. proposed an FPGA design of a bit permutation speech encryption system, which has been achieved using pipelining the encryption stages. It has been implemented using Xilinx Kintex 7 FPGA, and the results were analyzed using various standard analyses such as entropy, mean squared error and correlation [7]. Sudip Mondal and R. K. Sharma used AES standard for voice encryption, wherein the system consists of micro-blaze, UART control and 128-bit key encryption and GPIO using FPGA and Nexys4 DDR board for secure transmission through Pmod RF module [8].

Heba M. Yassin et al. have been experimented speech encryption using the chaotic generator and S-box table. It was implemented on FPGA, where the dynamic S-box depends on DNA module and chaotic system for increasing security [9]. Abdulaziz H. Elsafty et al. focused on generating FPGA-based speech encryption using S-box and memristor chaotic circuit. The values of S-box depend on the memristor chaotic circuit, and various analysis metrics have been used to ensure its effectiveness [10]. This proposed work aims to provide another solution for audio encryption through chaotic Lorenz attractor. In this work, confusion and diffusion operations have been performed to yield secure encryption. The design has been developed using Verilog HDL and has been implemented on Intel Cyclone IV E FPGA.

1.1 Preliminaries—Lorenz Attractor

In the scientific field of dynamical frameworks, an attractor is a set of numerical values toward which a system tends to evolve. The Lorenz attractor equations are dynamic three-dimensional system with a chaotic flow. In famous, the 'butterfly impact' comes from this present reality ramifications of the Lorenz attractor. The state of Lorenz attractor itself, when plotted, appears like a butterfly. They are very sensitive to the initial conditions. Therefore, when these are represented graphically, they show how a dynamic system develops in an exceedingly non-repeatable, complex pattern. The equations for Lorenz attractor are given below [11]:

$$\begin{aligned} x_{n+1} &= \sigma(y_n - x_n) \\ y_{n+1} &= x_n(r - z_n) - y_n \\ z_{n+1} &= x_n . y_n - \beta . z_n \end{aligned} \tag{1}$$

where the initial seed values of $X(n)$ = 32'b00111101110011001100110011001101, $Y(n)$ = 32'b01000000101000000000000000000000, $Z(n)$ = 32'b00000000000000000000000000011001. Also, β, σ, and r referred to as control parameters. All β, σ, and $r >$ zero, but usually, $\sigma = 10$, $\beta = 8/3$, and r is modified. For $r = 28$, the system shows chaotic behavior. Lorenz attractor equations are with initial conditions as mentioned above are plotted as X–Y, X–Z, and Y–Z plane in Fig. 1. This attractor is so sensitive to initial conditions and system parameters, which is the primary requirement in cryptography to offer the avalanche effect. The generated chaotic sequences can be utilized in cryptographic algorithms.

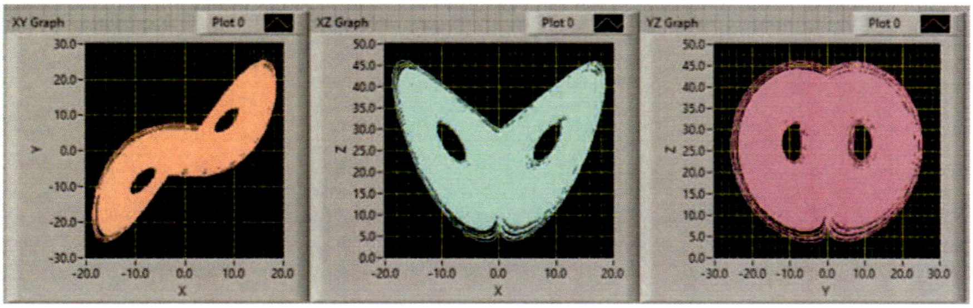

Fig. 1 Lorenz attractor response diagram for X, Y, and Z planes

2 Proposed Methodology

The algorithm presented in this paper is to design a Chao-cryptic chip for secure audio encryption during which it provides high reliability than many other encryption schemes, where the encryption keys are generated employing a Lorenz attractor. Figure 2 shows the block diagram of the propose encryption architecture. This scheme of encryption undergoes two consecutive steps of operations, namely, confusion, and diffusion part. In the confusion part, a random generator, i.e., 16-bit LFSR is used, and the output from the 16-bit LFSR is mod, and output is given to the 16-bit circular bit plane shift register. The input to the circular bit plane register is the digitized audio sample data from MATLAB 2016b. Moreover, to increase the security of the signal, the confused data is XORed with the key generated using the Lorenz attractor. Chaoic system has been constructed using the on-chip add, subtract, multiply and division modules of Cyclone IV E FPGA. Here, the bit position from the 32-bit key can be selected using a selector; it can be either LSB, MSB, or the mid-16-bit data.

2.1 Encryption Procedure

Step 1: The audio signal sample.wav files are taken as an input and digitized using MATLAB 2016b platform

Step 2: LFSR with tap positions at 16, 14, 12, and 11 with an initial seed value is used for generating random numbers for specifying the number of bits to be shifted in circular shift operation.

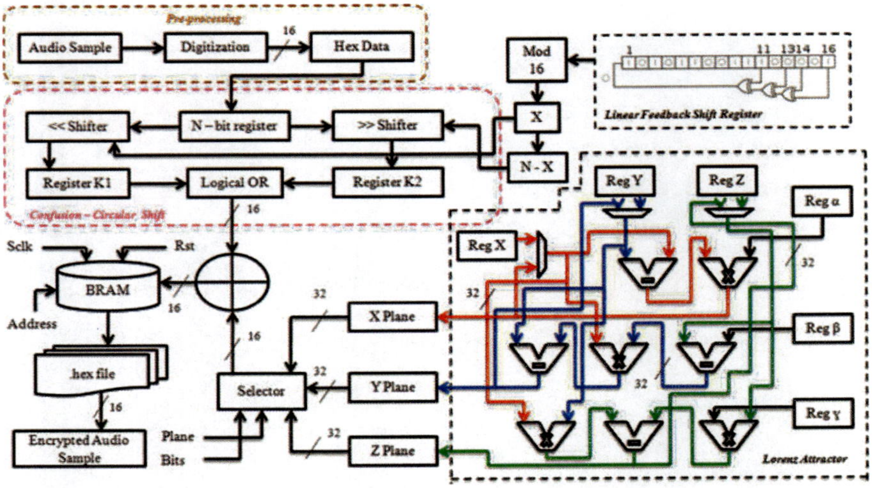

Fig. 2 Block diagram of the proposed encryption scheme

Step 3: 16-bit circular shift has been done with audio data as input and N bit generated from the earlier step. The circular shift accomplishes the confusion process, which breaks the dependency among the audio signal as follows: $c_{\text{out}} = ((\text{input} \ll n)|(\text{input} \gg (N - n)))$. Where input is the audio data, 'n' is the bits to be shifted, and 'N' is the 16 bits of an audio sample.

Step 4: Key generation for diffusion has been performed by Lorenz chaotic system with proper initial conditions and system parameters mentioned in Sect. 1.1. The selector has been used to select the plane and the 16 its keys from X, Y, and Z planes of Lorenz attractor through 2-bit multiplexers.

Step 5: The generated keys are XORed with the output of the 16-bit circular shift register and were stored to block random access memory (BRAM)available in FPGA.

Step 6: The values stored in BRAM are processed in MATLAB 2016b to construct the encrypted audio sample.

Step 7: The original audio data has been retrieved back by performing the reverse process at the receiver end.

3 Results and Discussion

This section deals with the performance analysis of the encrypted audio signal. The results were proposed based on the proposed algorithm, which has been tested on different audio samples taken from automatic speaker verification (ASV),and the data has been sampled at the rate of 8 kHz. Further, $2^{16} \times 16$ bits of digitized samples were coded into the target FPGA through Verilog HDL. To evaluate the scheme, entropy, correlation, histogram, MSE, PSNR, and NIST SP 800-22 tests have been used. MATLAB 2016b platform has been used for performance analysis, and the pictorial representation of encrypted and decrypted audio samples has been shown in Fig. 3.

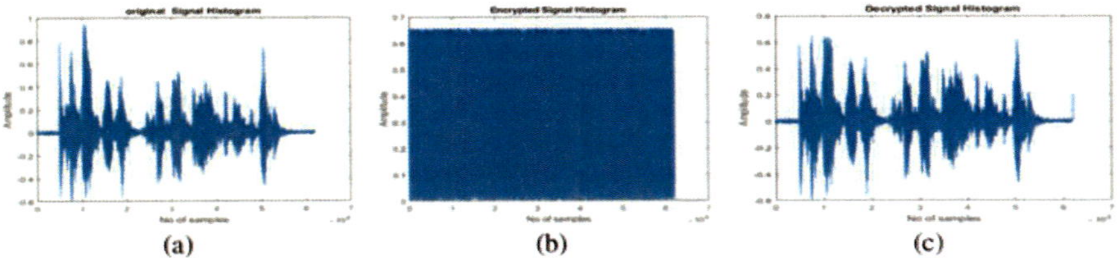

Fig. 3 Audio sample: **a** original, **b** encrypted and **c** decrypted

Table 1 Results of entropy and correlation analyses

Encrypted audio sample	Entropy	Correlation	MSE	PSNR
Audio sample (XOR(XYZ)_LSB)	15.1056	0.0053	0.2426	−43.3969
Audio sample (XOR(XYZ)_MID)	15.1744	0.0010	0.2451	−43.3902
Audio sample (XOR(XYZ)_MSB)	15.1701	0.0047	0.2472	−43.3978
Audio sample (X_LSB)	15.0867	0.0042	0.2176	−43.3976
Audio sample (X_MID)	15.1746	0.0012	0.2454	−43.3556
Audio sample (X_MSB)	15.1710	0.0018	0.2456	−43.4020
Audio sample (Y_LSB)	15.1175	0.0042	0.2493	−43.3974
Audio sample (Y_MID)	15.1740	0.0060	0.2488	−43.3891
Audio sample (Y_MSB)	15.1702	0.0002	0.2478	−43.3784
Audio sample (Z_LSB)	15.1538	0.0091	0.2494	−43.3603
Audio sample (Z_MID)	15.1770	0.0025	0.2457	−43.4142
Audio sample (Z_MSB)	15.1698	0.0044	0.2474	−43.3818

3.1 *Entropy Analysis*

Entropy is the primary measure of randomness [12]. It focuses on the equidistribution of encrypted data with its probabilistic values. The entropy of the encrypted data is tabulated in Table 1. From the table, the entropy of the encrypted audio sample has achieved near 16 entropy, which confirms the uncertainty.

3.2 *Correlation Analysis*

Correlation is a statistical measure that has been used to calculate the data dependency of the audio samples. It has three coefficients, namely horizontal, vertical, and diagonal, to determine the data relationship between the adjacent values in all three directions. In this proposed scheme, correlation analysis has been performed by the procedure given in [12], and the average results have been presented in Table 1.

3.3 *Histogram Analysis*

A histogram is a unique analysis to investigate the randomness of the audio signal [12]. It represents the intensity levels of the audio concerning the samples. Histogram analysis has been performed for the proposed algorithm, and the results have been shown in Fig. 4a, b. From the figures, it has been understood that the original audio sample has an uneven intensity level where the uniform distribution of data has been achieved in the encrypted sample.

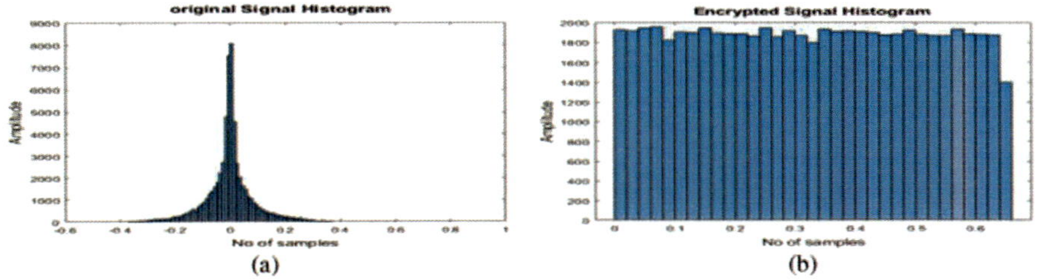

Fig. 4 Histogram analysis: **a** Original sample and **b** encrypted sample

3.4 Error Metric Analysis

Mean square error (MSE) and peak signal to noise ratio (PSNR) are the measure of avalanche effect, which gives difference between original and encrypted audio with the procedures given in [13]. Table 1 presents the results of MSE and PSNR where the values were less than 10 dB for MSE confirms the effective encryption.

3.5 NIST SP 800-22 Batteries of Test

NIST stands for National Institute of Standard and Technology released a test suite to evaluate the statistical characteristics of any random number generator [14]. NIST SP 800-22 tests have been performed for 993,280 encrypted audio bits and tabulated in Table 2.

From the table, it has been confirmed that the proposed audio encryption scheme has been passed all the tests in the NIST SP 800-22 test suites and it is cryptographically strong to resist attacks. Moreover, Tables 3 and 4 present the comparative study of the proposed scheme with [15, 16] in terms of correlation PSNR. The proposed design outperforms than the earlier work, which evidences its superiority.

Table 2 NIST SP 800-22 test results

Test parameters	*P*-values	Test parameters	*P*-values
frequency	0.066882	FFT	0.213309
Block frequency	0.350485	Non-overlapping template	0.991468
Cumulative sums-I	0.534146	Overlapping template	0.739918
Cumulative sums-II	0.739918	Approximate entropy	0.03013
Runs	0.008879	Serial-I	0.122325
Longest run	0.739918	Serial-II	0.213309
Rank	0.911413	Linear complexity	0.213309

P Value should be greater than 0.01 to pass the test

Table 3 Performance comparison-3

Reference	PSNR of encrypted audio data	Correlation coefficient
Ref. [15]	−37.26	0.0093
Proposed work (XOR(XYZ)_MID)	−43.3902	0.0010

Table 4 Performance comparison-4

Reference	PSNR of decrypted audio
Ref. [16]	60.2440
Proposed work	68.7880

Table 5 Results of hardware analysis

Target platform—Altera cyclone IV E EP4CE115F29C7 FPGA		
Throughput (Mbps)		10.346
Hardware footprint	Logic registers	7059
	Logic elements	15, 442
Power dissipation (mW)	Static power	104.67
	Dynamic power	14.22
	IO power	30.90

3.6 Hardware Analysis

It is the primary analysis that focuses on the hardware footprint of the proposed architecture. Hardware analysis includes the logic elements consumption such as Look Up Tables (LUTs), logic registers, and BRAM bits. In addition, power dissipation analysis has also been carried out through the power play power analyzer tool in the Quartus II tool. Hardware analysis results have been presented in Table 5, and it consumes less than 40% of its total register to accomplish the proposed encryption.

4 Conclusion

Enhanced audio encryption scheme has been proposed in this work through Chao-cryptic approach, wherein chaotic Lorenz attractor played a crucial role. Encryption of audio samples was achieved through two stages in which the first stage involves a confusion process attained using LFSR governed circular shift operation. Diffusion has been carried out using X, Y, Z, and the combination of all the three planes of Lorenz attractor generated pseudo-random sequences. The design has achieved a throughput of 10.346 Mbps with an operating frequency of 50 MHz. Further, efficiency of the proposed algorithm has been ensured through the various obtained standard metrics of encryption. In summary, the proposed work has yielded better performance than other diffusion techniques. The future work will be on developing a secure audio communication through DNA cryptography.

Acknowledgements The authors thank the Department of Science and Technology, New Delhi, for the FIST funding **(SR/FST/ET-II/2018/221)**. Also, the authors wish to thank the Intrusion Detection Laboratory at School of Electrical and Electronics Engineering, SASTRA Deemed University, for providing infrastructural support to carry out this research work. Further, the authors extend their sincere thanks to Mr R. Nagakrishnan, Research Scholar, School of EEE, SASTRA Deemed University, for his technical support.

References

1. N.K. Pareek, V. Patidar, K.K. Sud, Image encryption using chaotic logistic map. Image Vis. Comput. **24**, 926–934 (2006). 10.1016/j.imavis.2006.02.021
2. V. Patidar, N.K. Pareek, G. Purohit, K.K. Sud.: Modified substitution-diffusion image cipher using chaotic standard and logistic maps. Commun. Nonlinear Sci. Numer. Simul. **15**, 2755–2765 (2010). https://doi.org/10.1016/j.cnsns.2009.11.010
3. L.P. Kumar, A.K. Gupta., Implementation of speech encryption and decryption using advanced encryption standard, in 2016 IEEE International Conference on Recent Trends in Electronics, Information & Communication Technology (RTEICT), Bangalore (2016), pp. 1497–1501
4. M.A. Riyadi, N. Pandapotan, M.R.A. Khafid, T. Prakoso, FPGA-based 128-bit Chaotic encryption method for voice communication, in 2018 International Symposium on Electronics and Smart Devices (ISESD), Bandung (2018), pp. 1–5
5. M.F. Tolba, W.S. Sayed, A.G. Radwan, S.K. Abd-El-Hafiz, A.M. Soliman, Hardware speech encryption using a chaotic generator, dynamic shift and bit permutation, in 2018 30th International Conference on Microelectronics (ICM), Sousse, Tunisia (2018), pp. 100–103
6. M.F. Tolba et al., FPGA implementation of two fractional order chaotic systems. AEU Int. J. Electron. Commun. **78**, 162–172 (2017)
7. M.F. Tolba, W.S. Sayed, A.G. Radwan, S.K. Abd-El-Hafiz, A.M. Soliman, Permutation-only FPGA realization of real-time speech encryption, in 2018 25th IEEE International Conference on Electronics, Circuits and Systems (ICECS), Bordeaux (2018), pp. 93–96
8. H.M. Yassin et al., Speech encryption on FPGA using a chaotic generator and S-Box table, in 2019 Fourth International Conference on Advances in Computational Tools for Engineering Applications (ACTEA), Beirut, Lebanon (2019), pp. 1–4
9. S. Rajagopalan, S.R.H.N. Upadhyay, J.B.B. Rayappan, R. Amirtharajan, ONChip peripherals are ON for chaos—an image fused encryption. Microprocess. Microsyst. **61**, 257–278 (2018)
10. A.A. Abdo, S. Lian, I.A. Ismail, M. Amin, H. Diab, A cryptosystem based on elementary cellular automata. Commun. Nonlinear Sci. Numer. Simul. **18**, 136–147 (2013). https://doi.org/10.1016/j.cnsns.2012.05.023
11. Q. Zhang, L. Guo, X. Wei, Optik A novel image fusion encryption algorithm based on DNA sequence operation and hyper-chaotic system. Opt. Int. J. Light Electron. Opt. **124**, 3596–3600 (2013). https://doi.org/10.1016/j.ijleo.2012.11.018
12. X. Zhang, F. Han, Y. Niu Chaotic image encryption algorithm based on bit permutation and dynamic DNA encoding. Comput. Intell. Neurosci. **2017**(6919675), 11 (2017). https://doi.org/10.1155/2017/6919675
13. A.H. Elsafty, M.F. Tolba, L.A. Said, A.H. Madian, A.G. Radwan, FPGA speech encryption realization based on variable S-box and memristor chaotic circuit, in 2018 30th International Conference on Microelectronics (ICM), Sousse, Tunisia (2018), pp. 152–155

14. L.E. Bassham et al., A statistical test suite for random and pseudo-random number generators for cryptographic applications. Natl. Inst. Stand. Technol., Gaithersburg, MD, USA, Tech. Rep., No. April, (2010)
15. V.M. Padmapriya, M.Priyanka, K.S. Shruthy, S. Shanmukh, K. Thenmozhi, R. Amirtharajan, Chaos aided audio secure communication over SC-FDMA system, in 2019 International Conference on Vision Towards Emerging Trends in Communication and Networking (ViTECoN), Vellore, India (2019), pp. 1–5
16. P. Sathiyamurthi, S. Ramakrishnan, Speech encryption using chaotic shift keying for secured speech communication. J. Audio Speech Music Proc. **2017**, 20 (2017). https://doi.org/10.1186/s13636-017-0118-0

Diffusion + Confusion: Chaos-assisted Twofold Cryptosystem for Secure Video Transmission

Marusani Sumanjali, R. Sivaraman, A. Sridevi, Amirtharajan Rengarajan, Sundararaman Rajagopalan, and Har Narayan Upadhyay

Abstract Video broadcasting has attained high importance due to the popularization of networks. Video communication has lot of advantages but not limited to e-learning, surveillance, etc. Video encryption is necessary for securing video information from unauthorized users. To achieve this, an enhanced video encryption scheme has been proposed in this work which adopts two techniques such as confusion and diffusion. Sundarapandian–Pehlivan chaotic system (Akgul et al. Optik (Stuttg). 127:5491–5499 (2016), [9]) has been adopted as a keystream generator to accomplish the encryption process. This 3D chaotic system has been realized by RungeKutta fourth-order method through which the differential system has been modelled as a discrete system. LabVIEW 2013 environment has been used to develop the proposed cryptosystem in which the camera module has been interfaced to capture the real-time video. This scheme achieves maximum entropy of 7.94627514 and near-zero correlation for a test video which has 141 encrypted frames. Further, NIST SP 800-22 batteries of test has been carried out to ensure the statistical efficiency.

Keywords Runge Kutta fourth order method · Video encryption · NIST SP 800-22

1 Introduction

Due to the rapid growth in digital technology, data security plays a key role. The data is represented in different ways such as text, audio and video format [1]. Video is kind of information contains the frames of images. An authorized party can access the private video that has sensitive data. Due to the advancements in the field of business model, e-learning, online streaming, etc., protecting the video content became a challenging task [2]. In order to secure sensitive information, data encryption has been employed. Since the video has a large volume of data, traditional algorithms

M. Sumanjali · R. Sivaraman · A. Sridevi · A. Rengarajan · S. Rajagopalan (✉) · H. N. Upadhyay
Department of ECE, School of EEE, SASTRA Deemed University, Thanjavur 613401, India
e-mail: amir@ece.sastra.edu

J. Tavares et al. (eds.), *Emerging Technologies in Data Mining and Information Security*, Lecture Notes in Networks and Systems 164,
https://doi.org/10.1007/978-981-15-9774-9_14

such as DES, AES and GOST are not suitable due to their more rounds of operation that results in time consumption [3]. Also, the traditional algorithms fail to break the dependencies between the adjacent frames, which bring the vulnerability in terms of statistical attack. Hence, new methodologies have to be proposed to achieve high encryption. Hamidouche et al., proposed one such work for video encryption in which a scalable extension of High-Efficiency Video Coding (HEVC) has been used. This method concentrates only on the sensitive part of scalable HEVC parameters with a minimum delay and computational overheads. This method offers three types of video encryption through lower, higher and complete layers of HEVC. From the analyses, lower and complete later encryption schemes have been achieved more security [4]. Xu et al., suggested video encryption through chaos assisted random number generation. This work encrypts the data content of the H.264/AVC video [5].

Varalakshmi et al., also has performed selective encryption for scalable videos in Network Abstraction Layer (NAL) level. Proposed technique encrypts the data in three levels, namely intra-prediction mode, residual data and motion vector difference values. This work guarantees the robustness of the encrypted video during transmission [6]. Lian et al., proposed an encryption scheme using advanced video encoding in which the partial encryption has been adopted in prediction mode, residual data and motion vector difference levels. This work has been tested with known-plaintext attacks which ensure its attack resistant capability [7]. Chaudhary and Pandi developed a hybrid approach for video encryption to preserve the privacy of videos. It adopts symmetric and asymmetric cryptographic agents such as IDEA and RSA for encryption which brings the hybrid nature. The quality measures show the superiority of the algorithm in terms of security, and it has minimal time overhead which is to be improved [8].

2 Proposed Video Encryption Scheme

Proposed video encryption method contains four major blocks of operations, namely data extraction, cryptographic key generation, diffusion and confusion. The architectural view of the proposed methodology has been depicted in Fig. 1. The camera module has been interfaced with the LabVIEW environment, and the live video has been captured for encryption. Further, the captured video has been segmented as frames of size 256×256 and has been stored in the temporary buffers. Sundarapandian–Pehlivan 3D chaotic system has been the crucial part of this scheme through which the key generation has been performed. The 3D chaotic system has three differential equations which produce random numbers in all the three planes such as X, Y and Z. The random sequences generated from these planes have been utilised to perform encryption on the segmented frames.

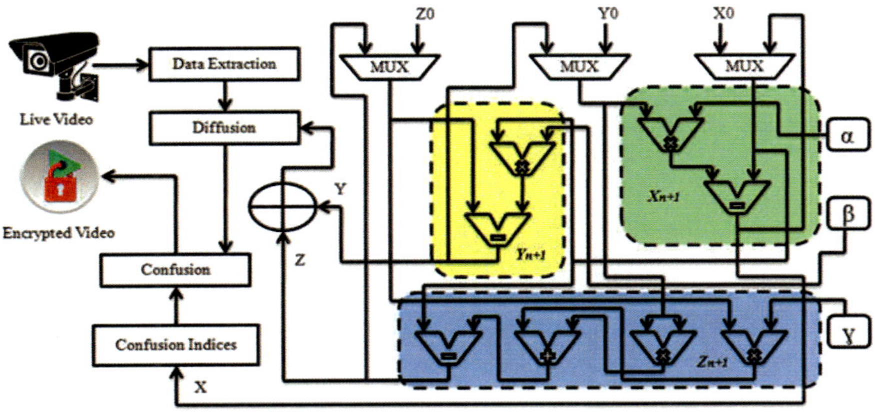

Fig. 1 Architectural diagram of the proposed scheme

2.1 Sundarapandian–Pehlivan 3D Chaotic System (SPCS)

SPCS is the three-dimensional chaotic system which has the properties such as ergodicity, stochasticity and extreme perturbations. SPCS has been the predominant chaotic agent for encryption [9]. SPCS is represented by using differential Eq. (1). To obtain chaotic behaviour, parameters were chosen as $a = 1.5$, $b = 0.4$, $c = 0.4$ and the initial conditions are determined as $x(0) = 0$, $y(0) = 0$, $z(0) = 0.1$.

$$\begin{aligned} \mathrm{d}x/\mathrm{d}t &= \alpha y - x \\ \mathrm{d}y/\mathrm{d}t &= -\beta x - z \\ \mathrm{d}z/\mathrm{d}t &= xz + xy^2 - x \end{aligned} \tag{1}$$

The differential equations can be solved in many methods, namely midpoint, Euler, Heun and fourth order of Runge–Kutta (RK4). Among all, RK4 has been considered as a superior method to produce accurate solutions. Equation (1) has been discretized using RK4 method, and it has been given in Eq. (2). In order to determine the term $y_{\lambda+1}$, the parameters k_1, k_2, k_3, k_4 must be calculated. A solution of each term depends on the previous terms. Hence, the equations were computed iteratively concerning the previous and current parameters with Δ_h distance. The next $y_{\lambda+1}$ values of the system are determined using the values of y_λ and Δ_h.

$$\begin{aligned} y\lambda + 1 &= y\lambda + \frac{1}{6}(k1 + 2k2 + 2k3 + k4)\Delta h, k1 = f(y\lambda) \\ k2 &= f\left(y\lambda + \frac{\Delta k}{2}k1\right), k3 = f\left(y\lambda + \frac{\Delta k}{2}k2\right), k4 = f(y\lambda + \Delta hk3) \end{aligned} \tag{2}$$

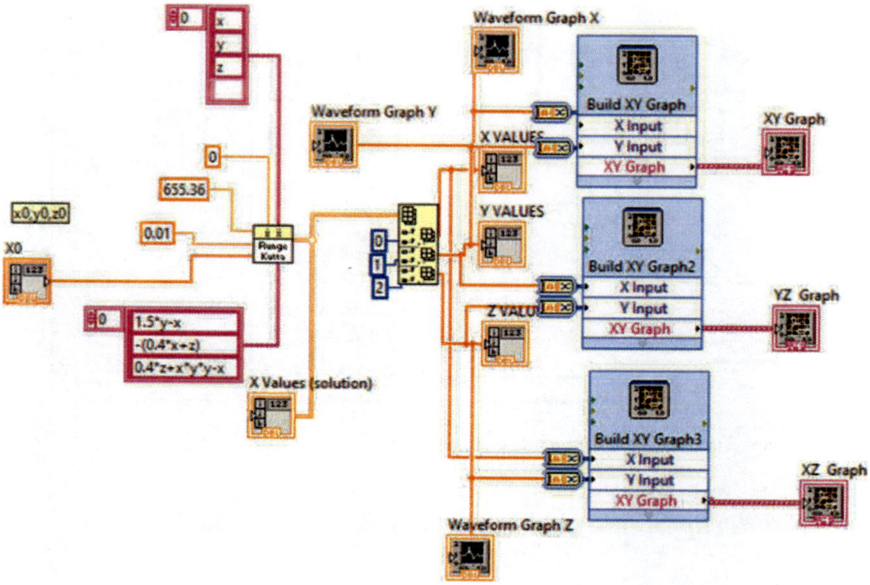

Fig. 2 SPCS using RK4 method in LabVIEW 2013

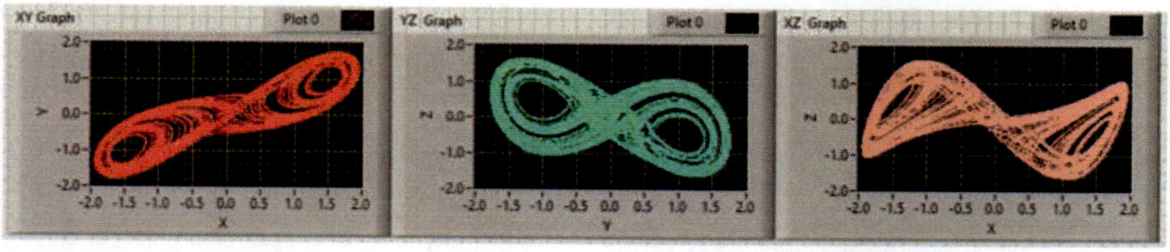

Fig. 3 Phase portraits of SPCS using RK4 method: *XY*, *YZ* and *XZ* planes

SPCS using RK4 has been developed through the LabVIEW 2013 graphical language tool, and the construction has been shown in Fig. 2. Phase portraits of SPCS have been presented in Fig. 3.

2.2 Encryption Procedure

Step 1: Live video is acquired by NI Acquisition Software.

Step 2: *IMAQ Create* function automatically creates a memory for storing image data.

Step 3: With the help of the display image control, the live video is displayed on the front panel, and the respective data is converted into array data using *IMAQ tools*.
Step 4: Video data has been segmented to N number of frames with a size of 256 $\times$ 256.
Step 5: The first frame has been chosen and converted into a 1D array of size 65,536.
Step 6: Data in the 1D array has been diffused by performing XOR operation with the random numbers generated from planes Y and Z of SPCS.
Step 7: Diffused data has been scrambled using the X plane values of SPCS as follows:

$$\text{Sort}(X_{\text{values}},\ \text{Indices})\text{-Sorting in Ascending order} \quad (3)$$

$$\text{Data}_{\text{confused}} = \text{Data}_{\text{diffused}}(\text{Indices}) \quad (4)$$

Step 8: Reconstruct the 1D array into 2D array of size 256 $\times$ 256 to generate a encrypted video frame.
Step 9: Repeat the steps 5 to 8 for N number of times to encrypt all the frames in the video.
Step 10: Combine all the frames to construct the encrypted video to analyse it.
Step 11: Decryption is the reverse process to retrieve back the original video.

3 Experimental Analysis

The proposed scheme has been developed using LabVIEW 2013 graphical language tool. NI acquisition software and Vision and Motion toolkit have been powerful tools used in this work. To test the proposed scheme, 4 s live colour video has been captured in the laboratory environment and segmented into 141 frames. Figure 4a–d shows the original and cipher data of frames 1 and 141, respectively. The encryption scheme has been evaluated through entropy, correlation and encryption quality analyses. In addition, NIST SP 800-22 tests have been employed to illustrate the strength of the algorithm towards randomness in encrypted data.

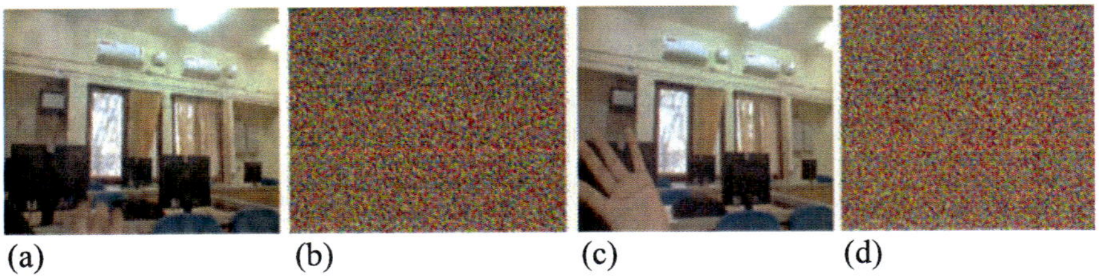

Fig. 4 Frame 1: **a** original data, **b** encrypted data, Frame 141: **c** original data, **d** encrypted data

Table 1 Results of entropy and correlation analyses

Entropy—original frames		**Entropy—encrypted frames**	
7.46097202		7.94627514	
Correlation—original frames		**Correlation—encrypted frames**	
Horizontal	0.96638484	Horizontal	0.03288562
Vertical	0.98296648	Vertical	0.02983350
Diagonal	0.95579468	Diagonal	0.03024162

3.1 Entropy Analysis

Entropy is the estimation of uncertainty which concentrates on the probability of distribution of video data. To achieve high entropy, the encryption must distribute the data uniformly [10, 11]. Uniformity has been verified through the following expression in which $P(X_i)$ is the probability of X with respect to the ith iteration, whereas N is the total number of iterations in the video data. Results of entropy analysis for the first 50 frames have been listed in Table 1. From Table 1, it has been identified that the proposed scheme has secured near eight entropy which confirms the amount of uncertainty produced by the algorithm.

3.2 Correlation Analysis

Correlation is an important metric to analyze the data dependency of a pixel with respect to its adjacent pixels in horizontal, vertical and diagonal directions [12–14]. Correlation value close to 1 illustrates the data has more dependency with its neighbours. A strong confusion mechanism can achieve near-zero correlation. In this work, correlation analysis has been performed between the first 50 frames of the encrypted video through the following equations, and the results have been tabulated in Table 1, which evidences the near-zero correlation. Figures 5a–c, 6a–c expresses the correlation of original and encrypted frames, respectively.

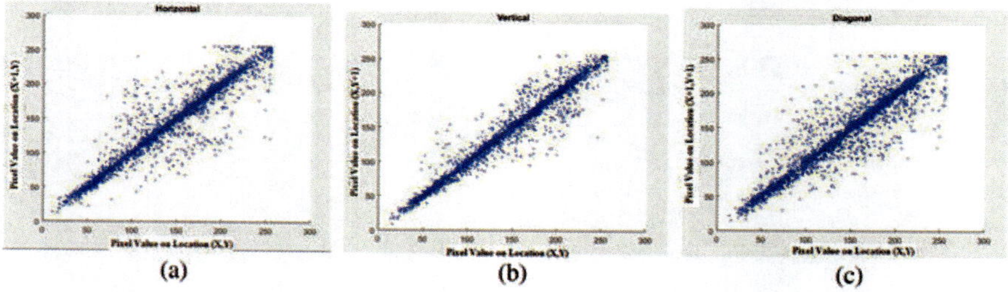

Fig. 5 Correlation analysis—original frame: **a** horizontal, **b** vertical and **c** diagonal

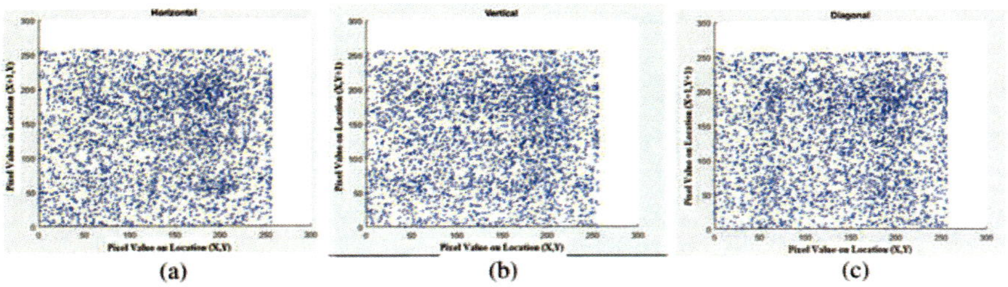

Fig. 6 Correlation analysis—encrypted frame: **a** horizontal, **b** vertical and **c** diagonal

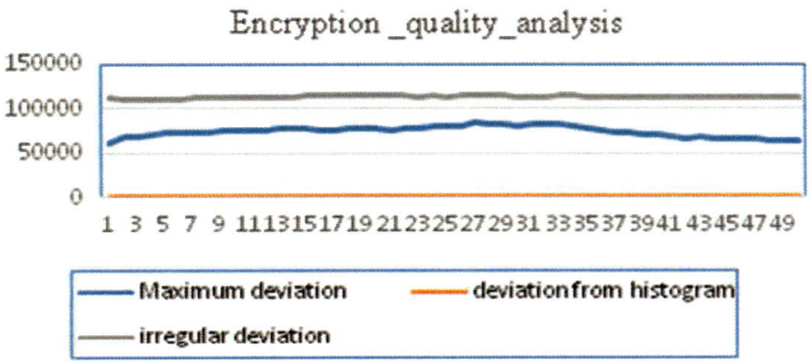

Fig. 7 Results of encryption quality analyses

3.3 *Encryption Quality Analysis*

To verify the quality of encryption, maximum deviation, irregular deviation and deviation from uniform histogram analyses [15] have been carried out for first 50 encrypted frames and reported in Fig. 7.

From Fig. 7, maximum deviation and irregular deviation have attained higher values which ensure the presence of high randomness in the encrypted frames. Also, deviation from the uniform histogram is very close to zero to evidence the encrypted video frames have a uniform distribution of values.

3.4 *NIST SP 800-22 Analyses*

National Institute of Standards and Technology (NIST) has released a Special Publication (SP) to investigate the statistical properties of Random Number Generators (RNGs) [16]. Since the encryption also exhibits randomness in its results, NIST tests can be applied for encryption to evaluate it. In this work, NIST was been performed for several encrypted frames of video and tabulated in Table 2. To conduct the test, $2^{16} \times 24$ bits have been taken from each frame.

Table 2 NIST SP 800-22 test results

Test parameters	P—values for 5 different frames				
	*F*1	*I*2	*I*3	*I*4	*I*5
Frequency	0.350485	0.350485	0.534146	0.350485	0.534146
Block frequency	0.350485	0.213309	0.020199	0.534146	0.122325
Cumulative sums-I	0.739918	0.739918	0.534146	0.739918	0.213309
Cumulative sums-II	0.739918	0.017192	0.017192	0.350485	0.213309
Longest run	0.534146	0.534146	0.350485	0.739918	0.534146
Rank	0.739918	0.534146	0.350485	0.739918	0.534146
FFT	0.739918	0.534146	0.911413	0.739918	0.739918
Non-overlapping template	0.066882	0.017912	0.534146	0.534146	0.018879
Overlapping template	0.350485	0.739918	0.122325	0.213309	0.534146
Approximate entropy	0.739918	0.125987	0.698745	0.265945	0.458986
Serial-I	0.350485	0.534146	0.213309	0.122325	0.911413
Serial-II	0.911413	0.739918	0.213309	0.739918	0.350485
Linear complexity	0.122325	0.739918	0.350485	0.350485	0.739918

NIST SP 800-22 has mentioned that the probability value for passing the test must be greater than 0.01. In Table 2, all the encrypted frames have been passed all the tests available in the NIST test suite, which proves the statistical efficiency of the proposed scheme.

4 Conclusion

In this work, video encryption through SPCS has been proposed using LabVIEW 2013. The 3D chaotic system has been realized through the RK4 method, and the random numbers generated from X, Y and Z planes have been utilized to achieve confusion and diffusion. Live video has been captured in the LabVIEW environment and segmented to frames of 256×256 size. Further, diffusion has been carried out using the Y and Z plane generated random sequences, whereas confusion has been accomplished through X plane sequences. Strength of the video encryption was tested through entropy, correlation, maximum deviation, irregular deviation and deviation from uniform histogram analyses. In addition, the statistical properties of the proposed scheme have been validated using NIST SP 800-22 batteries of tests. Future work will be on developing an FPGA prototype of the proposed work on NIs Compact RIO platform.

Acknowledgements The authors wish to thank SASTRA Deemed University for providing infrastructure through the Research & Modernization Fund (Ref. No.: R&M/0026/SEEE–010/2012–13) to carry out the research work.

References

1. Z. Li, X. Wang, Y. Lin, C. Cheng, RDEA: a novel video encryption algorithm, in *Advanced Multimedia and Ubiquitous Engineering*, ed. by J.H. Park, H.-C. Chao, H. Arabnia, N.Y. Yen. Lecture Notes in Electrical Engineering, vol 352 (Springer, Berlin, Heidelberg, 2015), pp. 183–189. https://doi.org/10.1007/978-3-662-47487-7_28
2. U. Andreas, A. Pommer, *Image and Video Encryption* (Springer, Boston, MA, 2005)
3. F. Liu, H. Koenig, A survey of video encryption algorithms. Comput. Secur. **29**, 3–15 (2010). https://doi.org/10.1016/j.cose.2009.06.004
4. W. Hamidouche, M. Farajallah, N. Sidaty, S. El Assad, O. Deforges, Real-time selective video encryption based on the chaos system in scalable HEVC extension. Signal Process. Image Commun. **58**, 73–86 (2017). https://doi.org/10.1016/j.image.2017.06.007
5. H. Xu, X. Tong, X. Meng, An efficient chaos pseudo-random number generator applied to video encryption. Optik (Stuttg). **127**, 9305–9319 (2016). doi:https://doi.org/10.1016/j.ijleo.2016.07.024
6. L.M. Varlakshmi, G.F. Sudha, G. Jaikishan, An efficient scalable video encryption scheme for real time applications. Procedia Eng. **30**, 852–860 (2012). https://doi.org/10.1016/j.proeng.2012.01.937
7. S. Lian, Z. Liu, Z. Ren, Z. Wang, Selective video encryption based on advanced video coding, in *Advances in Multimedia Information Processing—(PCM 2005*) ed. by Y.-S. Ho, H.-J. Kim (Springer, Berlin Heidelberg, 2005), pp. 281–290
8. K. Chaudhary, G. Pandi (Jain), Privacy preserving using video encryption technique—the hybrid approach, in *Computing and Network Sustainability*, ed. by H.R. Vishwakarma, S. Akashe (Springer, Singapore, 2017), pp. 29–37
9. A. Akgul, I. Moroz, I. Pehlivan, S. Vaidyanathan, A new four-scroll chaotic attractor and its engineering applications. Optik (Stuttg). **127**, 5491–5499 (2016). https://doi.org/10.1016/j.ijleo.2016.02.066
10. X. Zhang et al., Chaotic image encryption algorithm based on bit permutation and dynamic DNA encoding. Comput. Intell. Neurosci. (2017)
11. L. Liu, Q. Zhang, X. Wei, A RGB image encryption algorithm based on DNA encoding and chaos map. Comput. Electr. Eng. **38**(5), 1240–1248 (2012)
12. A. Belazi, A.A. Abd El-Latif, S. Belghith, A novel image encryption scheme based on substitution-permutation network and chaos. Signal Process. **128**, 155–170 (2016). https://doi.org/10.1016/j.sigpro.2016.03.021
13. S. Rajagopalan, R. Sivaraman, H.N. Upadhyay, J.B.B. Rayappan, R. Amirtharajan: ONChip peripherals are ON for chaos—an image fused encryption. Microprocess. Microsyst. **61**, 257–278 (2018)
14. M.A. Murillo-Escobar, C. Cruz-Hernández, F. Abundiz-Pérez, R.M. López-Gutiérrez, O.R. Acosta Del Campo, A RGB image encryption algorithm based on total plain image characteristics and chaos. Signal Process. **109**, 119–131 (2015). https://doi.org/10.1016/j.sigpro.2014.10.033
15. S. Rajagopalan, S. Rethinam, S. Janakiraman, H.N. Upadhyay, R. Amirtharajan, Cellular automata + LFSR + synthetic image: a trio approach to image encryption, in 2017 International Conference on Computer Communication and Informatics (ICCCI), Coimbatore (2017), pp. 1–6
16. L.E. Bassham et al., A statistical test suite for random and pseudo-random number generators for cryptographic applications. Natl. Inst. Stand. Technol., Gaithersburg, MD, USA, Tech. Rep., No. April (2010)

Chaos Blend LFSR—Duo Approach on FPGA for Medical Image Security

N. Adithya, Hemanth Kumar Nalajala, R. Sivaraman, A. Sridevi, Amirtharajan Rengarajan, and Sundararaman Rajagopalan

Abstract In recent years, encryption of medical images has been given much importance and several cryptographic algorithms and principles have been made and designed for secure communication. One such approach has been proposed in this work using pixel scrambling and diffusion techniques governed through linear feedback shift register (LFSR) and modified logistic map (MLM). The main advantage of this work is its concurrent encryption through which the scrambling and diffusion have been performed on the fly without any additional delays. The proposed design has been implemented on Intel Cyclone IV E FPGA using Quartus II Electronic Design Automation (EDA) tool. Further, the functional verification of LFSR and MLM has been carried out through ModelSim simulation. The design occupies 3580 logic elements and 135 logic registers to encrypt $256 \times 256 \times 16$ bits of DICOM image at 216 mS when it was operated with 50 MHz frequency. Also, this work achieves near-zero correlation, and the statistical properties of the encryption have been tested through standard metrics.

Keywords Encryption · DICOM · FPGA · Scrambling

1 Introduction

Information security is thus making a great impact on the data storage and secure transmission of data. Images are broadly used in numerous processes. Thus, the security of images is of primary concern, especially in various applications such as telemedicine [1]. Image encryption algorithms range from various spatial domain levels to highly reliable transform domain methods. Tradition encryption algorithms like DES, AES (private key cryptosystem), RSA and elliptic curve cryptography (public key cryptosystem) are not efficient for medical images due to its time

N. Adithya · H. K. Nalajala · R. Sivaraman · A. Sridevi · A. Rengarajan · S. Rajagopalan (✉)
Department of ECE, School of EEE, SASTRA Deemed University, Thanjavur 613401, India
e-mail: amir@ece.sastra.edu

J. Tavares et al. (eds.), *Emerging Technologies in Data Mining and Information Security*, Lecture Notes in Networks and Systems 164,
https://doi.org/10.1007/978-981-15-9774-9_15

complexity and fails to break the correlation between neighbourhood pixels [2]. Hence, new algorithms must be developed to achieve adequate security to protect the medical images.

One such work was suggested by Das et al., where MRI medical image has been encrypted using LFSR and logistic map. Further, the cipher image is encoded through DNA encoding to enhance security [3]. Prabhavathi et al., proposed a selective encryption scheme on a medical image which used Lorenz attractor for confusion and logistic map for diffusion. In addition, this method employs image steganography to conceal the patient information into the medical image [4]. Hua et al., developed a high-speed encryption method using pixel scrambling and adaptive XORing for diffusion. The design consumes two rounds of operation to achieve adequate encryption where the diffusion has been performed as two consecutive steps of bitwise XORing and modulo arithmetic operation [5]. Chen and Hu proposed a combined chaotic scheme for medical image encryption. Logistic—Sine chaotic system has been used for confusing the image plane, whereas the hyperchaotic system was utilized for diffusing the image [6]. Among the implementation platforms, FPGA is considered as predominant hardware platform which offers high prototyping capability and faster time to market because of its concurrent architecture [7]. Ravichandran et al., proposed one such work in which the DICOM image has been encrypted through pentalayer cryptosystem. LFSR, cellular automata (CA), Galois field (GF) product, PLL generated synthetic image, and the chaotic logistic map has been used in each layer of the design to perform encryption. This work occupies 2480 logic elements and dissipates 278.65 mW of power for Intel Cyclone II EP2C35F672C6 FPGA [8]. This proposed work also aims to provide one another solution for medical image encryption using FPGA where LFSR and MLM have been used for performing confusion and diffusion.

2 Proposed Encryption Architecture

The proposed architecture has confusion and diffusion units governed by pseudo-random key generators such as LFSR and MLM. Figure 1 shows the architecture of the proposed DICOM image encryption scheme in which pixel separation is the preprocessing to be done to extract the pixel data. Further, the extracted data has been stored in a 2D array of size 65,536 $\times$ 16 bits on FPGA. Each pixel has 16-bit depth, and it has been buffered into a temporary register to carry out the diffusion. The diffusion refers to a simple bitwise XOR operation between the output of MLM and the pixel value stored in the temporary register. The obtained diffused pixel has been scrambled using the random numbers generated through 16 bit LFSR. LFSR has been constructed using XNOR gates in a Fibonacci configuration in which the external feedback brings the randomness to LFSR. A block random access memory (BRAM) has been invoked on target FPGA to store the encrypted pixels.

BRAM has five ports, namely clock, Data_in, Data_out, Write_enable and address. 50 MHz of the operating clock has been connected to the clock terminal of

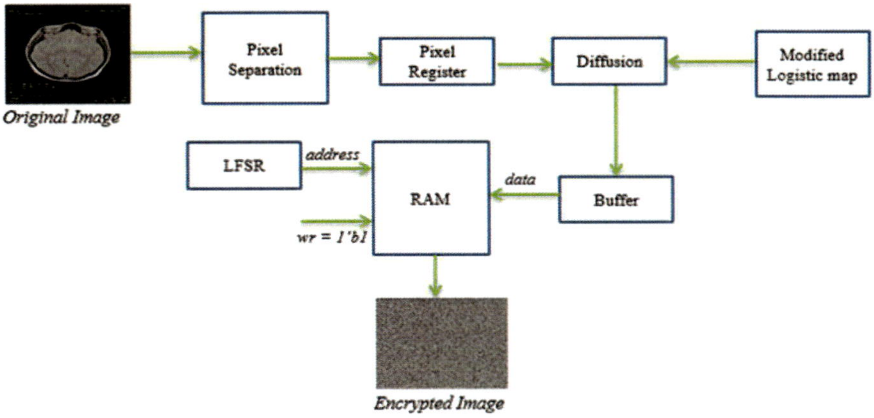

Fig. 1 Architecture of the proposed encrypted scheme

BRAM where the Write_enable has been kept as logic 1 to perform a write operation on it. Encrypted pixel has been applied to the Data_in port, whereas the address of the BRAM has been controlled using LFSR through which confusion has been achieved. In this work, confusion and diffusion are single-step operation to reduce the time of encryption. The following section explains the design of MLM on Cyclone IV E FPGA.

2.1 *Design of Modified Logistic Map*

Modified logistic map enhances the sensitivity and key space and also has a faster sequence generation rate that benefits to faster encryption. MLM has been mathematically expressed as [9]

$$X_{n+1} = 2\beta - x_n^2/\beta \tag{1}$$

In Eq. (1), X_{n+1} is the random numbers generated from MLM where $X(0) =$ 15 and 2, β are constants where $\beta = 9.8$ has been used in this work. The FPGA implementation of MLM has been shown in Fig. 2, which requires a subtractor, divider and two units of multiplier utilized from on-chip IP cores of Cyclone IV E FPGA to generate random numbers. Since the chaotic maps generate random numbers in floating points, it has been recommended by IEEE to use IEEE 754 float to binary conversion for realizing the chaotic maps on hardware. Hence, the initial value $X(0)$ and the constant β have been converted into single-precision binary values (32 bits) and applied to the MLM design for generating 65,536 iterations of 32-bit random numbers.

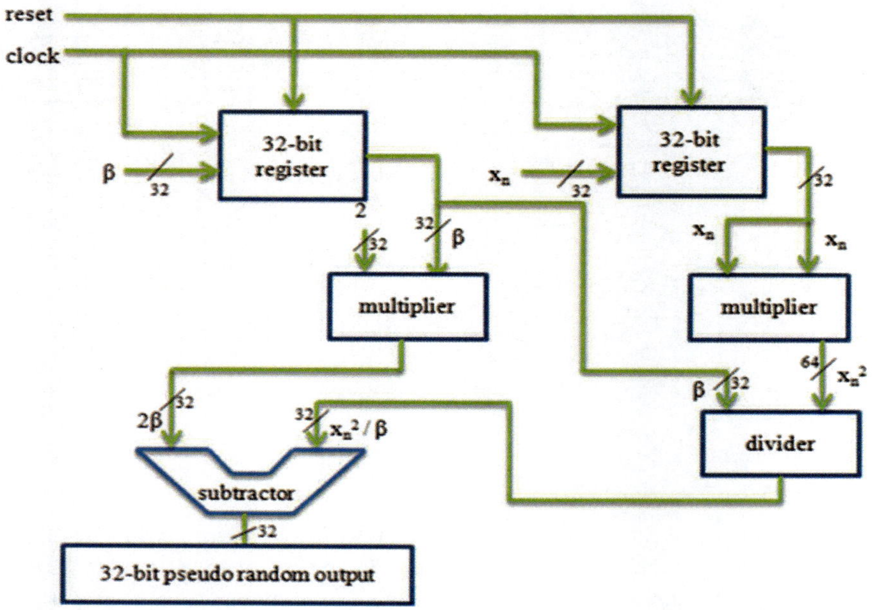

Fig. 2 FPGA realization of a modified logistic map

2.2 *Encryption Procedure*

The followings steps have to be followed to perform the encryption for 256 × 256 × 16 bits of DICOM image.

1. The pixel data is extracted from the image and stored in a temporary register.
2. The LFSR has been triggered by a seed value, and the random numbers were given to the address port of the BRAM. Figure 3 depicts the register transfer level (RTL) diagram of LFSR which has been realized on target FPGA.

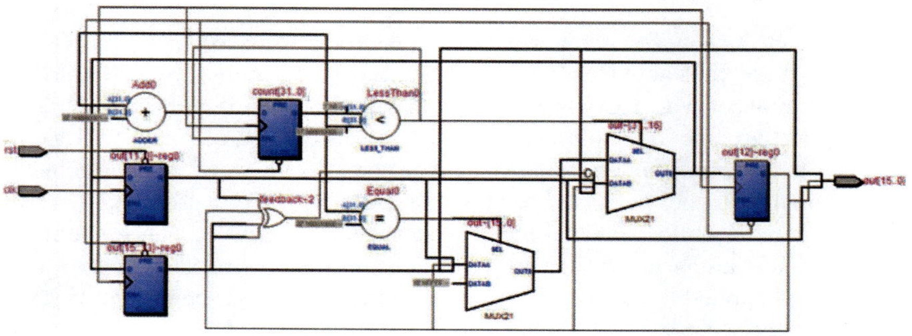

Fig. 3 RTL diagram of LFSR

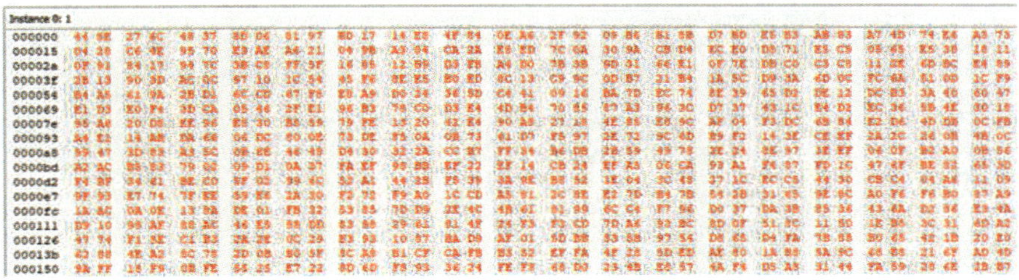

Fig. 4 BRAM snapshot of encrypted data

3. The image data is XORed with the modified logistic map (diffusion), and the output of the modified logistic map is fed back to the variable X_n for subsequent iterations.
4. The diffused output is assigned to the BRAM, and the output of the LFSR is given to the address of the BRAM.
5. Steps 3 and 4 have been continued until the 65536th pixel has been encrypted.
6. Finally, encrypted pixels were extracted from the BRAM to generate an encrypted image. Figure 4 presents the BRAM snapshot of the encrypted data.

3 Experimental Results and Analysis

The proposed work has been designed using Verilog HDL and implemented on Intel Cyclone IV EP4CE115F29C7 FPGA. Totally, 6 DICOM images have been taken for analysis. Original and encrypted images were presented in Figs. 5a–f and 6 a–f.

In addition, the functional simulations of LFSR and MLM have been presented in Figs. 7 and 8. The efficiency of the proposed design has been verified through hardware analysis and statistical analyses.

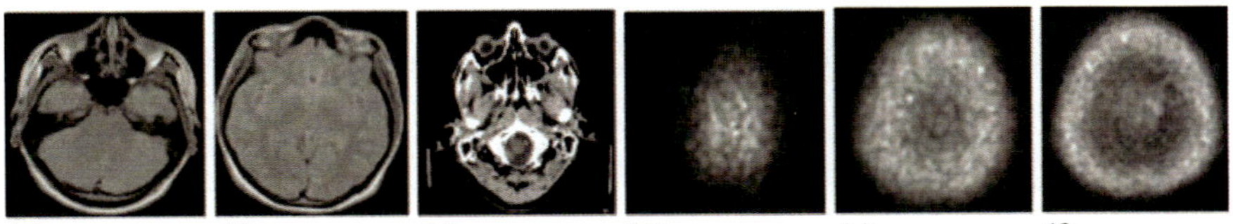

(a) (b) (c) (d) (e) (f)

Fig. 5 Original DICOM images: **a** D1, **b** D2, **c** D3, **d** D4, **e** D5 and **f** D6

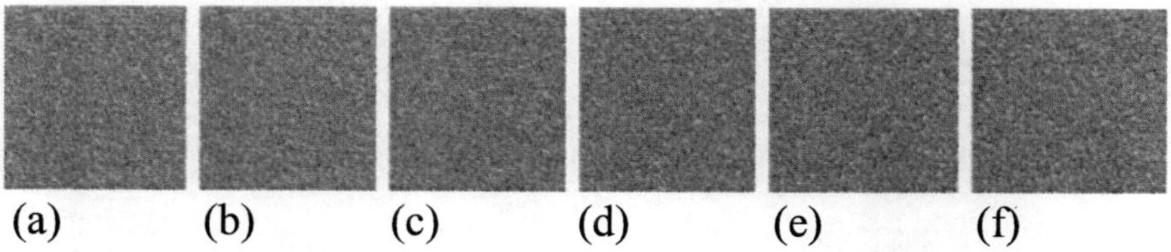

Fig. 6 Encrypted DICOM images: **a** ED1, **b** ED2, **c** ED3, **d** ED4, **e** ED5 and **f** ED6

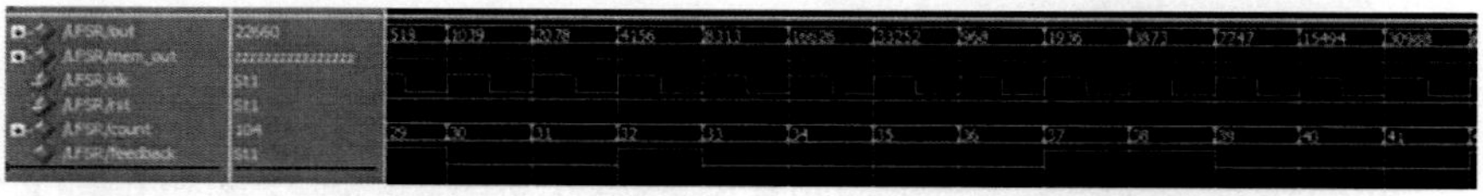

Fig. 7 Functionality verification of LFSR with the seed value of 16'hEFFF

Fig. 8 Functionality verification of MLM with the X (0) = 15 and $\beta = 9.8$

3.1 *Hardware Analysis*

To investigate the efficiency of the hardware, hardware analysis has been conducted. It includes the combinational logics such as logic elements, registers and power dissipation in terms of static, dynamic and I/O power. This analysis helps to understand the hardware utilization of the architecture, and the hardware analysis results have been tabulated in Table 1.

Table 1 Hardware analysis results of proposed encryption architecture

Target FPGA		Cyclone IV EP4CE115F29C7 FPGA
Hardware utility	Logic elements	3, 580/1, 14, 480 (3% Utilization)
	Combinational functions	3, 539/1, 14, 480 (3% utilization)
	Logic registers	135/1, 14, 480 (<1% utilization)
Power dissipation (mW)	I/O power	30.90
	Dynamic power	5.02
	Static power	98.80

Fig. 9 Timing analysis through zeroplus logic analyser

In addition, timing analysis has been carried out to determine the time taken form encryption through zeroplus logic analyzer, and the result has been shown in Fig. 9. From the figure, the proposed architecture consumes only 216 mS to encrypt a DICOM image when it was operated with 50 MHz of clock.

3.2 *Statistical Analyses*

3.2.1 Entropy and Correlation

These two metrics were considered as primary parameters which directly contribute to the evaluation of diffusion and confusion. As per the procedures in [10], entropy and correlation analyses have been performed and the results were listed in Table 2. From the table, it has been observed that the proposed scheme achieved near-zero correlation and the entropy close to 16 which confirms the uncertainty and uniform distribution in encrypted data.

Correlation has been calculated with respect to horizontal, vertical and diagonal direction of the image plane. Figure 10a–c depicts the correlation plot of ED2 image where the pixels are uniformly distributed throughout the plane which evidences the strong confusion being achieved in the proposed work.

Table 2 Results of entropy and correlation

Test images	Entropy	Horizontal	Vertical	Diagonal
ED1	15.1784	−0.0023	−0.0022	0.0050
ED2	15.1767	−0.0025	−0.0080	0.0009
ED3	15.1665	−0.0004	0.0023	0.0004
ED4	15.1760	−0.0016	−0.0007	0.0031
ED5	15.1741	0.0043	0.0024	−0.0059
ED6	15.1740	−0.0037	0.0017	0.0035

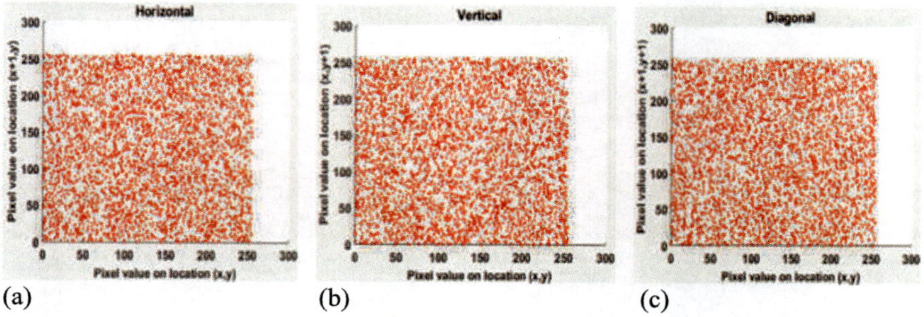

Fig. 10 Correlation analysis on ED2: **a** horizontal, **b** vertical and **c** diagonal

Table 3 Results of encryption quality measures

Test images	Maximum deviation	Deviation from uniform histogram	Irregular deviation
ED1	1.2888×10^5	0.0738	65,024
ED2	1.2892×10^5	0.0493	65,024
ED3	3.2130×10^5	0.0534	4873
ED4	1.1997×10^5	0.0496	65,024
ED5	1.1138×10^5	0.0541	65,024
ED6	1.0417×10^5	0.0539	65,024

3.2.2 Encryption Quality Measures

To distinguish the encrypted image from the original image, encryption quality analysis was applied where maximum deviation, deviation from the uniform histogram and irregular deviation have been computed [11]. The amount of pixel value deviation is the main concern of this analysis. Ideal histogram of the uniformly distributed data has been considered as the reference for conducting the test. Table 3 presents the results of this analysis, and it is inferred that the proposed encryption scheme has large deviations between original and encrypted images.

4 Conclusion

FPGA-governed DICOM image encryption has been proposed in this work where LFSR and MLM have been utilized for performing confusion and diffusion. The proposed design achieves on the fly encryption in which no additional delays in the encryption time due to the fair utilization of BRAM. It consumes 3580 logic elements on Intel Cyclone IV E FPGA which was operated at 50 MHz on board frequency. No design constrains of FPGA have been employed in this work to perform encryption. To validate the encryption, entropy, correlation and encryption quality analysis have been carried out. The future work will be on developing a colour DICOM image encryption on FPGA using chaotic attractors.

Acknowledgements The authors thank the Department of Science and Technology, New Delhi for the FIST funding **(SR/FST/ET-II/2018/221)**. Also, the Authors wish to thank the Intrusion Detection Lab at School of Electrical and Electronics Engineering, SASTRA Deemed University for providing infrastructural support to carry out this research work.

References

1. E.A. Krupinski, Medical Imaging, in *Handbook of Visual Display Technology* ed. by J. Chen, W. Cranton, M. Fihn (Springer International Publishing, Cham, 2016), pp. 545–558. https://doi.org/10.1007/978-3-319-14346-0_186
2. G. Ke, H. Wang, S. Zhou, H. Zhang, Encryption of medical image with most significant bit and high capacity in piecewise linear chaos graphics. Measurement **135**, 385–391 (2019). https://doi.org/10.1016/j.measurement.2018.11.074
3. A. Das, A.K. Das, N. Kar.: An Improved Chaos based medical image encryption using DNA encoding techniques, in *Advance Computer Intelligent Security Internet Things* ed. by A. Saha, N. Kar, S. Deb (Springer Singapore, Singapore 2020), pp. 207–220
4. K. Prabhavathi, C.P. Sathisha, K.M. Ravikumar, Region of interest based selective medical image encryption using multi Chaotic system, in 2017 International Conference on Electrical, Electronics, Communication, Computer, and Optimization Techniques (ICEECCOT), pp. 1–5. https://doi.org/10.1109/iceeccot.2017.8284614
5. Z. Hua, S. Yi, Y. Zhou.: Medical image encryption using high-speed scrambling and pixel adaptive diffusion. Signal Process. **144**, 134–144 (2018). https://doi.org/10.1016/j.sigpro.2017.10.004
6. X. Chen, C.-J. Hu, Adaptive medical image encryption algorithm based on multiple chaotic mapping. Saudi J. Biol. Sci. **24**, 1821–1827 (2017). https://doi.org/10.1016/j.sjbs.2017.11.023
7. A.J. Elbirt, W. Yip, B. Chetwynd, C. Paar, An FPGA-based performance evaluation of the AES block cipher candidate algorithm finalists. IEEE Trans. Very Large Scale Integr. Syst. **9**, 545–557 (2001). https://doi.org/10.1109/92.931230
8. D. Ravichandran, S. Rajagopalan, H.N. Upadhyay, J.B.B. Rayappan, R. Amirtharajan, Encrypted biography of biomedical image—a Pentalayer cryptosystem on FPGA. J. Signal Process. Syst. **91**, 475–501 (2019). https://doi.org/10.1007/s11265-018-1337-z
9. C. Han, OptikAn image encryption algorithm based on modified logistic chaotic map. Opt. Int. J. Light Electron Opt. **181**, 779–785 (2019). https://doi.org/10.1016/j.ijleo.2018.12.178
10. S. Arumugham, S. Rajagopalan, J.B.B. Rayappan, R. Amirtharajan, Networked medical data sharing on secure medium—a web publishing mode for DICOM viewer with three layer authentication. J. Biomed. Inform. **86**, 90–105 (2018). https://doi.org/10.1016/j.jbi.2018.08.010
11. S. Rajagopalan, R. Sivaraman, H.N. Upadhyay, J.B.B. Rayappan, R. Amirtharajan, ONChip peripherals are ON for chaos—an image fused encryption, Microprocess. Microsyst. **61**, 257–278 (2018). https://doi.org/10.1016/j.micpro.2018.06.011

Secured Routing System for Low Energy Networks

Anup W. Burange and V. M. Deshmukh

Abstract The Internet of things (IoT) is the network of different devices like sensors, actuators, and other different objects which have the capability to communicate without the need for human involvement. IoT mainly includes the devices which has low power and lossy networks (LLNs). These devices have limited resources like memory, energy, processing power, and bandwidth. These features leads the system in challenging environment which may end up in unstable state if proper security measures are not applied. IETF designed a protocol for routing mechanism of these devices known as routing protocol for low power and lossy networks (RPL). RPL is vulnerable to number of routing attacks and does not support mobility of nodes. This paper includes the study of different kind of threats possible on RPL. The proposed method for attack detection by using trust-based intrusion detection system considers trust value of participating nodes.

Keywords Trust · 6LBR · Sink node

1 Introduction

Internet of things (IoT) ia a vast network of worldwide recognizable physical objects or devices which are linked mostly to the Internet and represent themselves in virtual or digital world. It has capability to be revolution in Internet technology. Because of dynamic nature of working background, Internet of things is challenged by significant security threats which extends to other connected systems of it. IoT is combination of different types of sensors nodes or devices with different kind of functionality

A. W. Burange (✉) · V. M. Deshmukh
Department of Computer Science and Engineering, PRMIT&R, Badnera, Maharashtra 444701, India
e-mail: awburange@mitra.ac.in

V. M. Deshmukh
e-mail: vmdeshmukh@mitra.ac.in

J. Tavares et al. (eds.), *Emerging Technologies in Data Mining and Information Security*, Lecture Notes in Networks and Systems 164,
https://doi.org/10.1007/978-981-15-9774-9_16

at individual layer. IPv4 due to its limited address space could not satisfy the large requirements of IoT devices, so IoT uses IPv6 address to hold space in Internet. IoT objects can be deviced with sensors, actuators, RFID's, etc. [1]. These devices are not very good for security. IoT is a combination of diverse networks that brings similar security challenges which are present in sensor networks, mobile ad hoc network and the Internet; in addition to it, IoT also raises issues like network privacy problems, authentication, access control, etc. [2]. Introduction of 6LoWPAN using IPv6 causes more complex routing process and thereby added more security concerns. To maintain uniformity in routing and addressing is big challenge as per the way packets are routed between source and destination. Routing is rising as complex issue in IoT. IoT devices used across the world are of different making, using different physical architectures which create more challenges in routing process [3]. Routing in IoT is more challenging because of constrained devices in network. IETF developed RPL for IoT routing provides the path which enables constrained devices, using a low power and lossy networks, to use the Internet. This paper includes following details: Sect. 2 includes introduction of RPL, Sect. 3 includes different intrusion detection systems to counter various attacks on RPL, in Sect. 4 proposed work, and in Sect. 5, conclusion and future scope.

2 RPL Overview

2.1 Low Power Network

Popular network for low power devices is 6LoWPPAN, which uses IPv6 for addressing. This network is used in many applications and usually designed for low power devices having limited computing and power capabilities. It consist of edge router or border router which is connected to different devices like sensors, actuators, etc., and at other end, it is connected to external Internet. Due to wide range of devices and applications, it is difficult to maintain uniformity and strict security policy in this type of network. IoT hackers have huge opportunity to control non-secured devices and these devices can also be used to attack other networks and devices. Routing and its security are also very important issue for such networks as these networks support multihop feature in which message gets transferred through many on route nodes. Maintaining shorter and safe route, having large number of low power devices is challenging task. Currently, there are some of the routing protocols which works on these type of network like RPL, LEACH, etc. These protocols also having certain limitations and does not fully satisfy the changing demands of user and so as to the network.

RPL is a type of distance vector (DV) protocol developed by IETF, which creates a destination-oriented directed acyclic graph (DODAG). The DODAG structure is generated by considering the node attributes, link costs, and an objective function. Objective function works on grouping of metrics and constraints to build best route.

Rank generation for every node on the DODAG is done by the objective function designed in RPL. RPL sorts the nodes as destination-oriented DAGs (DODAGs), the placement of sinks or DAGs root give a route to the Internet (i.e., gateways). A network may consist of more than one DODAG, which enables RPL instance adapted by unique RPL instance ID. A network can manage a group of RPL instances at the same time. These RPL instances are logically independent. A node is capable to join to several RPL instances, but it must correlate itself to a single DODAG within every instance [4].

RPL-based network topology enables self-organization of nodes based on the parent-to-child relationship. Objective function (OF) within DODAG is used to select and optimize routes as per different metrics. Position of node is decided based on its rank in accordance with sink node. For constructing and keeping the routing, RPL has different types of control messages; they are as follows: DODAG Information Object (DIO) for maintaining and updating the topology; DODAG Destination Object (DAO) to broadcast destination information upwards for route updating progress; and DODAG Information Solicitation (DIS) for a new node which can ask for topology information before joining the network. The routing information of a node such as rank and objective function is included in DIO message so that its neighboring node can refer to it for creation of route. The DAO and DIS messages are used for the initializing a topology set up [5].

2.2 *RPL Attacks*

Following are the different attacks possible on RPL network which extensively impact the network resources and its performance (Table 1).

3 Intrusion Detection System

Intrusion detection system generates the alert when it detects any suspicious activity. Signature based, anomaly based, and hybrid are three main types of intrusion detection systems. Signature-based intrusion detection system mainly detects known attack types, anomaly based intrusion detection systems are generally used to detect unknown attack pattern/ types. In anomaly based IDS, different techniques like fuzzy logic, artificial neural network, etc., can be used for detection or prediction of unknown patterns/ attacks. New IDS called trust based has been recently evolved which detect attacks on trust metric. Relationship between the trustor and trustee is defined as trust. Trustworthiness of any node or device depends on how node actually behaves into the network in accordance to its neighboring nodes. Trust factor of a node can be represented by some index which also depicts node's reputation in the network. Trust also ensures quality of services it renders to its neighboring node. In addition, reward and punishment can be applied to impose trustworthy conformity of

Table 1 Attack types

Attack	Effect on network/topology	Techniques to counter	Remarks
Sybil	A malicious object may use multiple identities by introducing an adverse entity such as node or piece of software	K-mean clustering [6], fine-grained physical channel information [7],	To find broad range of Sybil attacks common strategy is needed for securing large IoT network
Wormhole	Adversary creates a tunnel between two legitimate devices by connecting them through stronger wireless or wired links	Location information and neighbor information [8], fuzzy logic approach [9], merkle tree authentication [10]	Uses minimum resources, high false positive rate, and applicable for static network
Rank	Decrease packet delivery ratio, loop generations in network	Cryptographic techniques, VeRA [11], SVELTE [12]	These techniques detect rank attack but energy usage is more detect for static network
Sinkhole	To be selected as parent node, malicious node deliberately lowers its rank	IDS, INTI- IDS [13], SAD-EIoT [14]	Trust-based methods are used less resources. More false positive values

node [15]. Complex cryptographic and hashing techniques are not generally a part of trust management because of its unsuitability for resource constrained devices like in IoT. Trust management has been evolved as very effective measure to calculate working capacity of IoT device [16] (Table 2).

Z. Khan and P. Herrmann designed trust management scheme to gather the information of near-by devices and their status. They proposed anomaly based intrusion detection is for detecting routing-specific attack like sinkhole attack and version number attack [17].

Prachi Shukla implemented machine learning-based intrusion detection system for RPL network in IoT, which uses unsupervised, supervised, and hybrid which combines k-means and decision tree approach. They identified that hybrid IDS can minimize false positive rate significantly [18].

S. Raza, Linus Wallgren, Thiemo Voigt presented SVELTE in contiki OS. SVELTE detects malicious nodes which initiate sinkhole/selective forwarding attacks. This IDS uses feature of firewall. This technique has small overhead which is suitable for constrained devises which has limited capabilities [12].

A. Mayzaud, R. Badonnel, and I. Chrisment presented a novel system architecture to detect the version number attacks in RPL-based networks [19].

S. Mohammad Hossein and Behrouz Ghahfarokhi proposed hybrid of sinkhole and clone-ID attack called sink-clone attack. They have identified its effect on performance of RPL network and also implemented IDS which they tested on cooja simulator [20].

Table 2 Different intrusion detection systems

IDS	Detection	Architecture	Attack types	Detection performance
Khan et al. [17]	Anomaly based	Distributed	Attacks related with routing: version number attack selective forwarding and sinkhole attack	Detection rate = 50%
Prachi Shukla [18]	Hybrid ML based	Centralized	Wormhole attack	Detection accuracy: 70–80% approx
Raza et al. [12]	Hybrid	Distributed	selective forwarding and sinkhole attacks	approx 80% for 30 nodes on avg
Mayzaud et al. [19]	Specification based	Distributed	Version number attacks	Detection accuracy: 81.97%
Mohammad Hossein [20]	Hybrid	Hybrid	Sinkhole and clone-ID	Simulated their IDS on cooja simulator of contiki OS
Airehrour et al. [21]	Trust based	Distributed	Rank and Sybil attack	Uses trust to make routing decisions and to isolate malicious nodes
Sivaraman, and Boreli [22]	Hybrid	Distributed	Masquareding	Accuracy: 94.25%
Le et al. [23]	Specification	Hybrid	Rank, sinkhole and neighbor attacks	Detection accuracy: 80%
Murali [24]	Bio-inspired analytical model	–	Mobile Sybil attack	Lightweight IDS for mobile Sybil detection

D. Airehrour, J. Gutierrez, Sayan kumar ray proposed and implemented secure trust system for RPL protocol to detect and isolate rank, Sybil attack. They tested the effectiveness of system through extensive simulations and test bed [21].

V. Sivaraman, and R. Boreli focused on smart home and proposed a host-based intrusion detection system using software defined technology (SDN). They also state three prerequisites for an capable IDS for IoT, i.e., scalability, unobtrusive approach, and negligible overheads. They given a choice of detection modules [22].

A. Le, J. Loo, Y. Luo, and A. Lasebae proposed host-based intrusion detection system in contiki OS. The system transmits the data gathered from node level to some centralized system for additional analysis [23].

S. Murali and Abbas Jamalipour proposed a novel artificial bee colony (ABC) Sybil attack modeling and lightweight intrusion detection for Sybil attack. They consider dynamic network by using mobile-aware RPL [24].

4 Proposed Work

6LoWPAN/IoT allows smart objects to join the network; these objects can be monitored and sensed from remote location in certain frameworks. Cyber criminals can gain the control over the communication of these objects or can directly control them to launch other attacks. Attacker can passively monitor messages between two parties. As number of devices in IoT are increasing at very fast pace, it will result in heavy traffic and data within the network. This would cause the intrusion detection system to drop some packets without proper monitoring and security check and in result degrades the effectiveness of intrusion detection system.

(a) External attack: An approach for removing external attacks and detection of malicious node based on analysis of behavior shown by the node can be used. We aimed to use network-oriented intrusion detection system which is able to monitor all traffic and based on that it will create a trust model for the nodes as shown in Fig. 1. This module can be placed on sink nodes as well as on border router as most of the traffic passes through these nodes. Trust value generated by monitoring the traffic will be called as network trust.

(b) Dynamic network: IoT devices can be either mobile or stationary which is attached to the local sink node. We consider the hybrid scenario in which one or more nodes can be static or dynamic (moving); so according to that, parent selection objective function need to be developed based on RSSI value. Previously, some of has been done by different researchers but it has some pros

Fig. 1 Traffic-based trust calculation

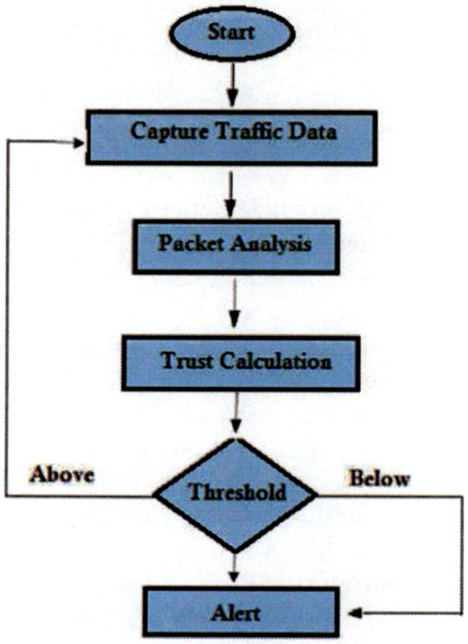

and cons. We aimed to develop optimize objective function for parent selection considering frequent changes.

(c) Internal attack: Trust-based IDS is proposed which will consider the mobility also as trust parameter if any moving node send DIS message using new IPv6 address, node will verify RSSI value of other nodes, etc., trustworthiness of mobile node will be analyzed. RPL does not have any technique to identify the behavior of any node. Trust can be categorized as direct trust and indirect trust. Direct trust is the kind of trust in which one node has surety about genuineness of other whereas indirect trust refers to recommendations by other neighboring nodes. This module will be placed on all nodes to analyze the trust of neighboring nodes. Trust parameter and its location is used for analyzing the best path toward sink node. Sink node maintains history of all nodes and global sink node maintains history of all sink nodes to counter the attacks. This hybrid IDS is used to detect known attacks and this module can be placed on sink nodes.

For removing internal attacks trust based intrusion detection system is used, which will be positioned at sink nodes. These sink nodes are attached to border router. There will be at least one sink node in each DODAG network. Devices which are either mobile or stationary will be connected to these sink nodes. Sink nodes maintain the trust values of other nodes as the data passes through this node. Sink node reduces the number of transmissions and every sink node maintains the DAG upto only limited number of nodes. Data collected by this node is sent to Border router which then forward this data to other sink nodes based on destination addresses. Border router however has traffic analysis module which analyze the each packet received from sink nodes and generate alert or will be able to take some action. The final routing decision will be based on trust obtained from traffic and total trust as shown in Fig. 2

The primary mobility capability for a single node is carried out as an initial stage on cooja simulator of contiki OS as shown in Fig. 3 in which mobility is provided to node 4. The proposed system aimed to build objective function for mobility in RPL as it does not support mobility of nodes. This will help us to study the different anomalies or attacks that carried out in dynamic nature and are yet to be discovered. Dynamic nature is the current need as most low power devices now has moving ability, also RPL has been studied in static nature only. To implement dynamic nature objective function needs to take care of parent selection and route establishment.

Fig. 2 Overall system view

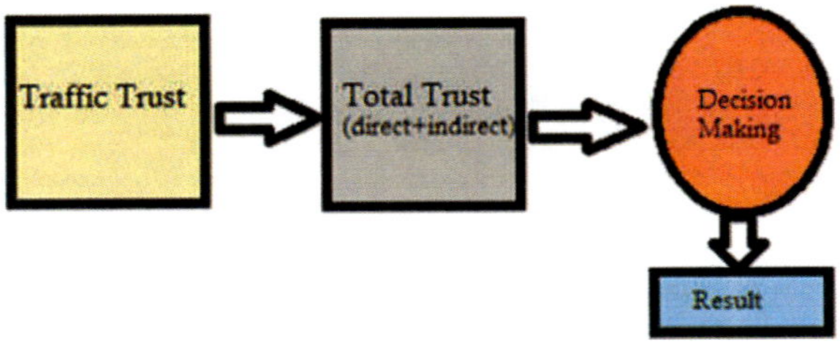

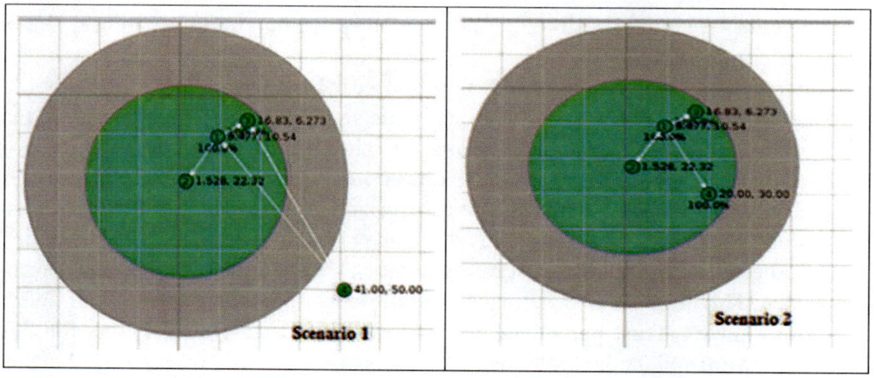

Fig. 3 Mobility of node in RPL

5 Conclusion

6LoWPAN devices are vulnerable to different attacks due to its several limitations. In order to maintain security and normal operation of such devices, different attack detections and mitigation strategies are adopted but they having certain disadvantages as mentioned. Proposed lightweight intrusion detection system is aimed for devices which has less power and computational resources like in Internet of things. Mobility, location information can make system more relevant for detection of attacks like wormhole and Sybil. Traffic monitoring is used for detection of external attacks and malicious nodes. Trust-oriented intrusion detection system is used to minimize internal attacks. This work aimed to counter external and internal attacks using trust-based distributed IDS considering mobility of nodes.

References

1. P. Pongle, G. Chavan, Real time intrusion and wormhole attack detection in internet of things.Int.J.Comput.Appl. **121**(9), 0975–8887
2. K. Zhao, L. Ge, A survey on the internet of things security, in *Proceedings of the 9th International Conference on Computational Intelligence and Security (CIS)*, pp.663–667
3. J. Gubbi, R. Buyya, S. Marusic, M. Palaniswami, Internet of things (IoT): a vision, architectural elements, and future directions. Future Gener. Comput. Syst. **29**, 1645–1660 (2013)
4. A.K. Idrees, A.J.H. Witwit, *A Comprehensive Review for RPL Routing Protocol in Low Power And Lossy Networks*
5. A. Le, J. Loo, Y. Luo, A. Lasebae, The impacts of internal threats towards routing protocol for low power and lossy network performance, in *Presented in IEEE Conference*
6. N. Diwakar, R. Kashyup, Detecting Sysbil attack using hybrid fuzzy k-means algorithm in WSN, in *International Journal of Engineering Development and Research* (2017)
7. C. Wang, L. Zhu, L. Gong, Z. Zhao, L. Yang, Z. Liu, X. Cheng, Accurate Sybil attack detection based on fine-grained physical channel information. Sensors **18**, 878 (2018)

8. S. Eidie, B. Akbari, P. Poshtiban, WANI: Wormhole avoidance using neighbor information, in *IEEE xplore, 7th Conference on Information and Knowledge Technology (IKT)*, May 2015
9. R. Singh, J. Singh, R. Singh, Fuzzy based advanced hybrid intrusion detection system to detect malicious nodes in wireless sensor networks, in *Wireless Communications and Mobile Computing* (2017)
10. R.K. Sundararajan, U. Arumugam, FBMT: fuzzy based Merkle technique for detecting and mitigating malicious nodes in sensor networks. Int. Arab J. Inf. Technol. **16**(6) (2019)
11. A. Dvir, T. Holczer, L. Buttyán, (VeRA)—version number and rank authentication in RPL, in *8th IEEE International Conference on Mobile Ad-hoc and Sensor Systems, MASS* (2011)
12. S. Raza, L. Wallgren, T. Voigt, SVELTE: real-time intrusion detection in the internet of things. Ad Hoc Netw. **11**https://doi.org/10.1016/j.adhoc.2013.04.014
13. R. Stephen, A. Dalvin Vinoth Kumar, L. Arockiam, Deist: dynamic detection of sinkhole attack for internet of things. Int. J. Eng. Comput. Sci. **5**(12), 19358–19362. ISSN: 2319-7242
14. S. Pundir, M. Wazid, D.P. Singh, A.K. Das, J.J.P.C. Rodrigues, Y. Park, Designing efficient sinkhole attack detection mechanism in edge-based IoT deployment. J. Sens. **20**, 1300 (2020)
15. M. Al-Hawawreh, N. Moustafa, E. Sitnikova, Identification of malicious activities in industrial internet of things based on deep learning models. J. Inf. Secur. Appl. **41**https://doi.org/10.1016/j.jisa.2018.05.002
16. W. Meng, Intrusion detection in the era of IoT: building trust via traffic filtering and sampling. IEEE Xplore Comput. **51**, 36–43 (2018). https://doi.org/10.1109/MC.2018.3011034
17. Z. Khan, P. Herrmann, Hive: home automation system for intrusion detection, in *IEEE 31st International Conference on Advanced Information Networking and Applications (AINA)* (2017)
18. P. Shukla, ML-IDS: a machine learning approach to detect wormhole attacks in Internet of Things, in *IEEE xplore, Intelligent Systems Conference (IntelliSys)* (London, 2017), pp. 234–240. https://doi.org/10.1109/IntelliSys.2017.8324298
19. A. Mayzaud, R. Badonnel, I. Chrisment, A distributed monitoring strategy for detecting version number attacks in RPL-based networks. IEEE Trans. Netw. Serv. Manage. **14**(2), 472–486 (2017)
20. S. Mirshahjafari, B. Ghahfarokhi, Sinkhole+CloneID: a hybrid attack on RPL performance and detection method. Inf. Secur. J. Glob. Perspect. 1–13 (2019).https://doi.org/10.1080/19393555.2019.1658829
21. D. Airehrour, S. Ray, SecTrust -RPL: a secure trust-aware RPL routing protocol for internet of things science direct. Future Gener. Comput. Syst. (2018).https://doi.org/10.1016/j.future.2018.03.021
22. M. Nobakht, V. Sivaraman, R. Boreli, A host-based intrusion detection and mitigation framework for smart home IoT using openflow, in*11th International Conference on Availability, Reliability and Security (ARES)* (2016)
23. A. Le, J. Loo, Y. Luo, A. Lasebae, Specification-based ids for securing RPL from topology attacks, in *IFIP Wireless Days (WD)* (2011)
24. S. Murali, A. Jamalipour, A lightweight intrusion detection for sybil attack under mobile RPL in the internet of things. IEEE Internet Things J. 1 (2019). https://doi.org/10.1109/JIOT.2019.2948149

Clinical Decision Support System Using Fuzzy Logic Programming and Data Analysis

Sandip Paul, Kumar Sankar Ray, and Diganta Saha

Abstract Clinical decision support systems (CDSS) are intelligent systems, developed to assist physicians for taking crucial decisions. This work proposes a hybrid architecture for a CDSS using a rule base and a data-driven module. The rule base uses a variant of interval-valued fuzzy answer set programming to represent a physician's knowledge. The data-driven module refines these rules using available patient data. This hybrid structure has reasoning capability of a rule-based module and reliability for using real-world data.

Keywords Expert systems · Clinical decision support system · Rule-based system · Fuzzy logic programming · Uncertainty · Correlation

1 Introduction

CDSSs are developed to assist doctors in medical decision making. General physicians or the doctors at the emergency ward of any hospital do the challenging task of assessing a patient's condition by observing his vital symptoms. Even before the appropriate medical tests are done, these doctors have to diagnose patient's illness and take proper measures. Any mistake in judgment can be proven fatal for the patient. In such a scenario, a CDSS may become useful as it can reduce the risk of error.

S. Paul (✉) · K. S. Ray
Electronics and Communication Sciences Unit, Indian Statistical Institute, 203, B.T.Road, Kolkata 700108, India
e-mail: talking2sandip@gmail.com

K. S. Ray
e-mail: ksray@isical.ac.in

D. Saha
Computer Science and Engineering Department, Jadavpur University, Kolkata 700032, India

J. Tavares et al. (eds.), *Emerging Technologies in Data Mining and Information Security*, Lecture Notes in Networks and Systems 164,
https://doi.org/10.1007/978-981-15-9774-9_17

Approaches toward building a CDSS can be broadly classified as, rule-based approaches [1–3] and machine learning-based data-driven approaches [4–7]. Machine learning-based smart systems are data dependent. They use the huge amount of available medical data to predict diseases from patients' symptoms or test results. However, a solely data-dependent model suffer from several drawbacks [8]: 1. Requirement of huge amount of data and computing power, 2. Re-training required incorporate new cases 3. Cannot be trained for rare diseases, due to lack of sufficient data 4. Incorporating knowledge from hand-written clinical notes in human language is difficult 5. Lacks interpretability as system behaves like a black box and lacks humanly interpretable semantics.

On the other hand, a rule-based system uses symbolic logic to represent knowledge in the form of IF–THEN rules and reasons with them to infer new knowledge. A knowledge engineer can incorporate a doctor's knowledge (about common as well as rare diseases) and information from clinical notes directly in a rule base using suitable symbolic logic. This construction directly reveals the symptom-disease connection without training. Rule-based systems are very intelligible and modular and hence recognition and removal of problematic rules becomes easy [7]. However, without considering the real data, the reliability of rule-based systems remains questionable. The doctor's judgment may be biased in some ways by his personal views. This limitation is called *symbol grounding problem*, which says the elements of the system are not learned from real world data.

As it can be seen, both solely rule-based or completely data-driven systems have certain disadvantages. Instead of that, in this work, we suggest a hybrid architecture, where a rule base is constructed using an interval-valued fuzzy answer set programming paradigm and on top of that there is a data analysis module, which extracts information from patient statistics and modifies the rules appropriately. The logic-based segment provides better representation and the data-driven segment enables learning from data. This paper proposes only a rough skeleton of such a system. Validation with actual medical data is not reported here.

2 Structure of the Proposed Model

Figure 1 shows the main building blocks of the proposed CDSS.

Module 1: The *knowledge base and inference engine* contains the rules that represent one or more physician's knowledge using a suitable logical framework. Any new knowledge, obtained by medical research, is inserted here by the knowledge engineer.

Module 2: The *data analysis module* takes the rules from module 1 and modifies them as per available medical data. Records of the patients are stored here to be used for regular update and validation of the rules.

Module 3: This module contains the rules after they are modified by data. Any rule regarding rare diseases or new research, which do not have sufficient data for validation, directly comes here from module 1. In that way, the knowledge base is richer as compared to solely data driven (i.e., ML or deep learning based) systems.

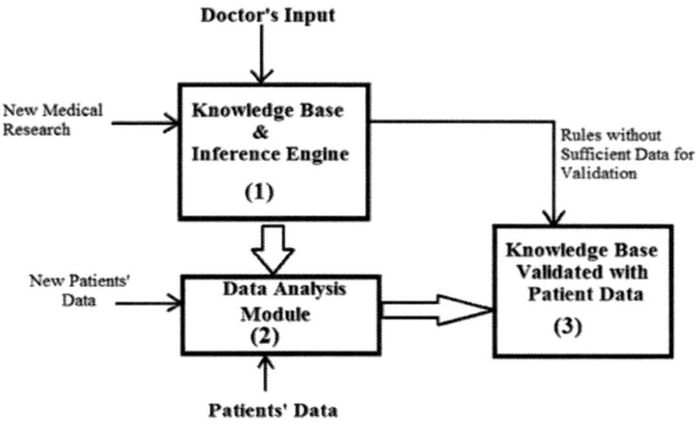

Fig. 1 Architecture of the proposed CDSS

3 Essential Characteristics of Logical Structure for Knowledge Representation

To ensure reliable utility of the CDSS, it is utmost important to choose an appropriate knowledge representation system and inference mechanism, which can deal with the following issues:

1. **Imprecision or Vagueness**: The doctors specify their domain knowledge mostly using natural language. A **Fuzzy Logic**-based system has to be taken to represent vague concepts like "young person", "severe pain", "high blood pressure", etc.
2. **Uncertainty**: Medical decisions have some innate uncertainty. A physician may use terms like "certainly", "rarely", etc., to specify the chance of a certain disease. The system must be capable to capture and reason with the non-stochastic probability or uncertainty of decisions.
3. **Nonmonotonicity or reasoning with rules having exceptions**: Medical decisions are often taken without complete knowledge and later they may have to be withdrawn in light of latest information. *Medical rules may have exceptions*. For instance, a knowledge base concerning breast cancer [9] contains the following two rules: **R1**: If a patient is young, then breast cancer will not reappear after 5 years and **R2**: If more lymph nodes are involved, then breast cancer can reappear in spite of low age. Here, the second rule is an exception to the first.

This requirement of nonmonotonic reasoning makes variants of answer set programming (ASP) more preferable than other monotonic fuzzy [10] or probabilistic fuzzy [1] expert systems. Some approaches, based on argumentation theory, are proposed [9, 11] to deal with nonmonotonicity, but they cannot handle vagueness or uncertainty.

4 Logical Framework for Module 1

The logical framework is based on answer set programming (ASP) [12], which allows nonmonotonic reasoning and explicit representation of rules with exceptions by using a *negation-as-failure* (*not*) operator. For a proposition *p*, *not p* is true if nothing is known about *p*; whereas, for classically negated statement $\wedge p$ is true if *p* is *known to be false.*

In this work, we choose an **interval-valued** (to handle uncertainty) **fuzzy** (to handle vagueness of natural language) **ASP** (to deal with nonmonotonicity) framework. On top of that, some additional mechanisms have to be added in order to incorporate data-dependent modifications.

4.1 System Components

$\mathbf{D} = \{D_1, D_2, \ldots, D_N\}$ is set of N diseases under consideration. $\mathbf{S} = \{S_1, S_2, \ldots, S_M\}$ is the set of symptoms or attributes used to describe the diagnoses. $\mathbf{P} = \{p_1, p_2, \ldots\}$ is the finite set of patients. Each p_i is the unique identification code of the patient. **M** and **Dir** are the set of medicines and directions to be prescribed. Each symptom in **S** is transformed to some number in the interval [0, 1] using a symptom coding scheme. This numeric representation is used for evaluation of the rules.

4.2 Symptom Coding Scheme (SCS)

- If a symptom $s_i \in \mathbf{S}$ is **binary**, for example, present or absent, then the corresponding code $SCS\ (s_i)$ lies in $\{0, 1\}$.
- A symptom may have several categories, for example, color of sputum may be white/yellow/reddish or may have several **discrete** values, for example, rate of pain on a scale 0–10. For such a **multinary** symptom s_i, $SCS\ (s_i)$ gives set of rational numbers taken from [0, 1]. For instance, sputum colors may be coded as 0.25, 0.5, 0.75 and pain can be mapped to the set $\{0, 0.1, \ldots, 0.9, 1\}$.
- Some symptoms attain **continuous** values, for example, body temperature, age, blood pressure, etc. These are often represented by *linguistic variables* and associated *term set*, which is represented by a fuzzy set defined over the whole range of values. The term set for "age" can be {*young, very young, middle aged, old*} and the term set for "blood pressure" can be {*low, very low, normal, high*}. The coding scheme takes a term along with the variable and maps it to a number in [0, 1] using the corresponding fuzzy membership function. Some of the membership functions are shown in Fig. 2a, b.

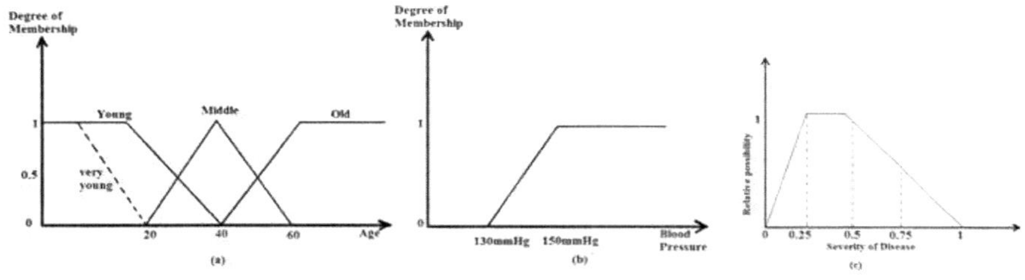

Fig. 2 **a** Membership functions of "age", **b** membership function of high "BP", **c** trapezoid (0, 0.25, 0.4, 1)

4.3 Rules in the Knowledge Base

The rule base consists of symptom-disease rules, disease-directions rules, and ordinary weighted IF–THEN rules of general ASP. The first category is of concern here. These rules take a complex form in order to incorporate vagueness, uncertainty, nonmonotonicity, and provision to be updated based on data. A rule is of the form:

$$r :: f_1(s_1(p_i) : \alpha_1) \wedge \cdots \wedge \sim f_k(s_k(p_i) : \alpha_k) \wedge \cdots \wedge \text{not } f_n(s_n(p_i) \colon \alpha_n)$$
$$\xrightarrow{\beta_r} d_r(p_i) \colon \delta_r$$

4.3.1 Significance of Rule Components

The significance of each component of the rule is as follows:

- r :: is label of the rule; $\wedge$, $\sim$, *not* and $\xrightarrow{\beta_r}$ are conjunction, negation, negation as failure, and weighted implication operator, respectively.
- The left hand side and right hand side of $\xrightarrow{\beta_r}$ are the **body** or **antecedent** and the **head** or **consequent** of the rule, respectively.
- $s_1, \ldots, s_k, \ldots, s_n \in \mathbf{S}$, $p_i \in \mathbf{P}$ and $d_r \in \mathbf{D}$,$s_i(p_i)$ and $d_r(p_i)$ represent symptom s_i and disease d_r of patient p_i, respectively.
- The **annotations** α_j stand for *truth conditions*, i.e., the condition to be met for the symptom s_j to contribute in the diagnosis of disease d_r. $s_j(p_i) \colon \alpha_j$ represents the amount of contribution in the diagnosis. α_j is specified according to the coding scheme for s_j.

For a binary symptom s_j, $\alpha_j \in \{0,1\}$ and $s_j(p_i) \colon \alpha_j = 1$ if *SCS* $(s_j(p_i)) = \alpha_j$, $s_j(p_i) \colon \alpha_j = 0$ otherwise. For instance, *blood_in_sputum* (*p*): 1 is true is *SCS* (*blood_ in_ sputum*(*p*)) = 1, i.e., there is blood in sputum. For a multinary symptom s_j, $\alpha_j \subseteq [0, 1]$ and $s_j(p_i) \colon \alpha_j = 1$ if *SCS* $(s_j(p_i)) \in \alpha_j$, $s_j(p_i) \colon \alpha_j = 0$ otherwise. For a continuous symptom s_j, α_j is a term from the term set of the linguistic variable s_i. In that case, $s_j(p_i)\colon \alpha_j = \mu_{\alpha_j}(s_j(p_i))$. Say, s_j is "age", $s_j(p_i) = 70$ for a 70 year old patient; $\alpha_j = old$ and μ_{α_j} is a membership function for "old" defined by the *SCS*. Then is the degree of membership obtained from the membership function shown in Fig. 2a.

- The **severity** of the disease d_r is denoted by δ_r. If a disease may be only diagnosed to be present or absent δ_r is not used. For diseases with several severity levels like cancer, Alzheimer's δ_r specifies which level is more possible by means of **Trapezoidal fuzzy numbers**. Severity level 3 or "intermediate" is $\delta_r =$ trapezoid$(0.75, 0.75, 0.75, 0.75)$. If stage 1 and 2 cancer are more probable that stage 3 and stage 4 are certainly not the case then $\delta_r =$ trapezoid$(0, 0.25, 0.4, 1)$ (Fig. 2c).
- **Correlation function** f_i specifies the correlation of symptom s_i with disease d_r and the data analysis module evaluates it using available data. If a symptom is indicative to more than one disease then the contribution of the symptom is not same for all those diseases. The different levels of contribution are depicted by f_i.
- The **rule weight** β_r is an interval $[\beta_1, \beta_2] \subseteq [0, 1]$ and it captures the uncertainty, when (1) same set of symptoms may indicate different diseases and (2) when a rule may have exceptions.

4.3.2 Rule Evaluation: Semantics

The components of the rule body are evaluated using the symptom value, annotation, and correlation function. If there is no uncertainty regarding the symptom value, its evaluation gives a number from [0, 1]. If a symptom value is fluctuating within a range or there is some amount of uncertainty then an interval of values is assigned to it. If nothing is known about a symptom value, for example, a test result has not come yet, [0, 1] is assigned. Therefore, **interval-valued** evaluation is used to capture uncertainty and *wider the interval, more is the uncertainty*.

For two intervals $[a_1, a_2], [b_1, b_2] \subseteq [0, 1]$, $[a_1, a_2] \wedge [b_1, b_2] = [a_1 b_1, a_2 b_2]$, i.e., $\wedge$ is the product t-norm; $\sim [a_1, a_2] = [1-a_2, 1-a_1]$ and *not* $[a_1, a_2] = [1-a_1, 1-a_1]$ [13]. If the body of the rule becomes $[a_1, a_2]$ then the consequent d_r is evaluated as $[a_1\beta_1, a_2\beta_2] : \delta_r$ where δ_r stands for severity degree of d_r and the interval $[a_1\beta_1, a_2\beta_2]$ captures the *range of possibilities*.

Example 1 Suppose an old person P (age 70 years) has high blood pressure and he falls frequently. Without ECG or CT scan tests a physician suspects Syncopal attack or heart problem. If along with falling down the patient vomits, it implies stroke, but the patient did not vomit. The associated rules in the rule base are:

1. $f_1(\text{BP(P): high}) \wedge f_2(\text{age}(P)\text{: old}) \wedge f_2(\text{Fall}(P)\text{: Frequent}) \xrightarrow{B_1} \text{Syncope}$
2. $f_2(\text{BP(P): high}) \wedge f_5(\text{age}(P)\text{: old}) \wedge f_6(\text{Fall}(P)\text{: Frequent}) \xrightarrow{\beta_2} \text{Heart\_ Problem}$
3. $f_7(\text{BP}(P)\text{: high}) \wedge f_8(\text{age}(P)\text{: old}) \wedge f_9(\text{Fall}(P)\text{: }[0.1, 1]) \wedge f_{10}(\text{vomiting}) \xrightarrow{[1,1]} \text{Stroke;}$

Since the patient did not vomit, *stroke* is evaluated to [0, 0], i.e., certainly it is not stroke. However, *Syncope* and *Heart_Problem* both are possible. Depending on correlation functions and rule weights (which is assessed by data), the more certain option is chosen.

Example 2: Representing Rules with Exceptions
The case regarding recurrence of breast cancer, as mentioned in Sect. 3, can be represented by the following rules:

$$r1 :: f_1(\text{age}(P)\text{: low}) \wedge \textit{not}\ \text{exception\_}r1 \xrightarrow{[\beta,1]} \text{recurrence}(p)\text{: }0;$$

$$r2 :: \text{node\_involve}(p)\text{: high} \xrightarrow{[1,1]} \text{exception\_}r1;$$

$$r3 :: \text{exception\_}r1 \xrightarrow{[1,1]} \text{recurrence}(p)\text{: }1.$$

Here, rule $r2$ is an exception of rule $r1$ and the presence of negation-as-failure in the premise of rule $r1$ depicts this connection. Expanding the list of exceptions is very straight forward. For instance, just adding a rule r :: new_ exception$\xrightarrow{[1,1]}$exception_$r1$; specifies the attribute *new_exception* as the new exception to rule $r1$.

5 Role of Data Analysis Module

5.1 Determination of Correlation Functions

A particular symptom may have different degrees of correlation with different diseases. The correlation of symptom S_i with disease D_j is measured by the variation in the number of patients with D_j when the symptom S_i becomes more and more prominent. Figure 3a shows almost linear correlation of stroke with blood pressure.

Correlation function $f(x)$: [0, 1] $\rightarrow$ [0, 1] can be can be linear or nonlinear, positive, or negative. Figure 3b shows different linear correlation functions. Linear correlation coefficients 0, 1, -1 are represented by $f(x) = 1, f(x) = x$ and $f(x) =$ 1 - x, respectively. Zero correlation is mapped to the value 1, since 1 is the identity element of $\wedge$. Similarly, nonlinear correlation functions can also be defined using polynomial functions.

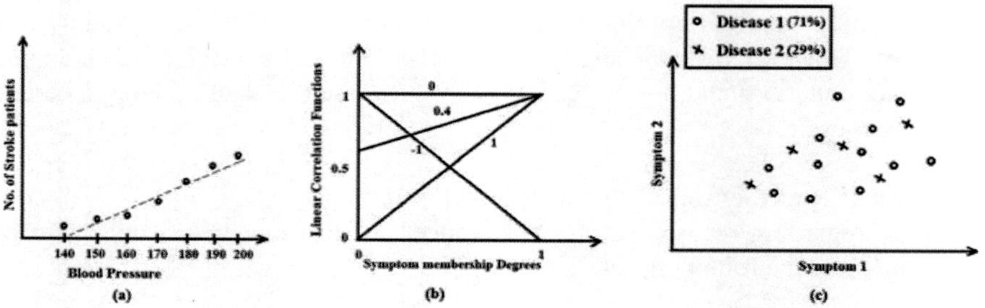

Fig. 3 **a** Correlation of blood pressure with stroke, **b** some linear correlation functions, **c** diseases with same pair of symptoms

5.2 Determination of Rule Weights to Prescribe Uncertainty of the Diagnosis

When same set of symptoms is indicative to more than one disease, there is an innate uncertainty in the decision (as shown in example 1). Now, if d_1, ..., d_k are such diseases with same set of symptoms and $n_{d_1}, \ldots, n_{d_k}$ are, respectively, number of patients having these diseases then their relative chance of occurrence is $n_{d_i}/(n_{d_1} + \cdots + n_{d_k})$. The weight of the for d_i is the interval $[n_{d_i}/(n_{d_1} + \cdots + n_{d_k}), 1]$. Considering Fig. 3c d_1 is 71% likely and d_2 is 29% and the corresponding rule weights will be [0.71, 1] and [0.29, 1], respectively.

5.3 Determining Weights of Rules with Exceptions

In case of rules having exceptions, the rule weight (β in Example 2) is given by enumerating the number of exceptions from the available data. Addition of new exceptions requires re-adjusting the rule weights.

6 Operation of Module 3

Module 3 contains all the refined rules passed from module 1 through module 2 and some rules directly coming from module 1. This is the actual intelligent agent with which the patient interacts. A GUI is developed on top of this block. Patients' data are fed into it to get possible diagnosed diseases and directions for further tests and suggestive medicines.

7 Conclusion

This work presents a skeletal structure of a hybrid CDSS based on a rule base with interval-valued fuzzy answer set programming and a data analysis segment to refine the rule parameters using real-life data. Different sophisticated machine learning tools can be incorporated to make the framework more robust, which is the topic of our future research.

References

1. A.S. Fialho, S.M. Vieira, U. Kaymak, R.J. Almeida, F. Cismondi, S.R. Reti, S.N. Finkelstein, J.M. Sousa, Mortality prediction of septic shock patients using probabilistic fuzzy systems. Appl. Soft Comput. **42**, 194–203 (2016)
2. Y. Jiang, B. Qiu, C. Xu, C. Li, The research of clinical decision support system based on three-layer knowledge base model. J. Healthcare Eng. **2**(3), 1–8 (2017)
3. B. Malmir, M. Amini, S.I. Chang, A medical decision support system for disease diagnosis under uncertainty. Expert Syst. Appl. **88**, 95–108 (2017)
4. S. Anakal, P. Sandhya, Clinical decision support system for chronic obstructive pulmonary disease using machine learning techniques, in *2017 International Conference on Electrical, Electronics, Communication, Computer, and Optimization Techniques (ICEECCOT)* (IEEE, 2017), pp. 1–5
5. M.M. Baig, H.G. Hosseini, M. Lindén, Machine learning-based clinical decision support system for early diagnosis from real-time physiological data, in *2016 IEEE Region 10 Conference (TENCON)* (IEEE, 2016), pp. 2943–2946
6. A. Belard, T. Buchman, J. Forsberg, B.K. Potter, C.J. Dente, A. Kirk, E. Elster, Precision diagnosis: a view of the clinical decision support systems (cdss) landscape through the lens of critical care. J. Clin. Monit. Comput. **31**(2), 261–271 (2017)
7. R. Caruana, Y. Lou, J. Gehrke, P. Koch, M. Sturm, N. Elhadad, Intelligible models for healthcare: Predicting pneumonia risk and hospital 30-day readmission, in *Proceedings of the 21th ACM SIGKDD International Conference on Knowledge Discovery and Data Mining* (ACM, 2015), pp. 1721–1730
8. M. Garnelo, M. Shanahan, Reconciling deep learning with symbolic artificial intelligence: representing objects and relations. Curr. Opin. Behav. Sci. **29**, 17–23 (2019)
9. L. Longo, L. Hederman, Argumentation theory for decision support in health-care: a comparison with machine learning, in *International Conference on Brain and Health Informatics* (Springer, 2013), pp. 168–180
10. R. Safdari, M. Kadivar, M. Langarizadeh, A.F. Nejad, F. Kermani, Developing a fuzzy expert system to predict the risk of neonatal death. Acta Inform. Med. **24**(1), 34 (2016)
11. C. Yan, H. Lindgren, J.C. Nieves, A dialogue-based approach for dealing with uncertain and conflicting information in medical diagnosis. Auton. Agent. Multi-Agent Syst. **32**(6), 861–885 (2018)
12. C. Baral, *Knowledge Representation, Reasoning and Declarative Problem Solving* (Cambridge University Press, 2003)
13. S. Paul, K.S. Ray, D. Saha, *A Unified Framework for Nonmonotonic Reasoning with Vagueness and Uncertainty*. arXiv:1910.06902 (2019)

Analysis and Review on Energy Efficient Cluster-Based Routing Protocol

Kanu Patel and Hardik Modi

Abstract In area of Internet of things, wireless sensors communication is very important part of system. Power consumption is significantly main issue in IoT-assisted wireless sensor networks. Mostly, sensors are battery-operated after some period that sensors are lifeless. In clustering technique, cluster head selection is a key issue. With the cluster-head selection, the cluster formation is one of the biggest challenges. In this paper, we review many clustering-based routings protocols to analysis with three factors that are cluster formation, cluster head selection, and routing protocols. For cluster head selection, we keep considering quality of service impact, energy status, and sensors location which include some six sub-criteria each contain two. These factors can impact the cluster head selection process. Many paper reviews on routing protocols are used to mainly improve the reliability of the data transmission in network.

Keywords Energy efficiency · Clustering · LEACH · Routing · Wireless sensor network · EECRP

1 Introduction

A WSN is a network that collects sensor nodes interconnected by wireless media or wireless channels of communication [1]. The data is collected from the surrounding area by each sensor node which, as shown earlier, actually acts as a small device (Fig. 1).

K. Patel (✉)
Department of Computer Engineering, CHARUSAT University, Changa, Gujarat, India
e-mail: kanu.patel@bvmengineering.ac.in

H. Modi
Department of Electronic and Communication, CHARUSAT University, Changa, Gujarat, India
e-mail: hardikmodi.ec@charusat.ac.in

J. Tavares et al. (eds.), *Emerging Technologies in Data Mining and Information Security*,
Lecture Notes in Networks and Systems 164,
https://doi.org/10.1007/978-981-15-9774-9_18

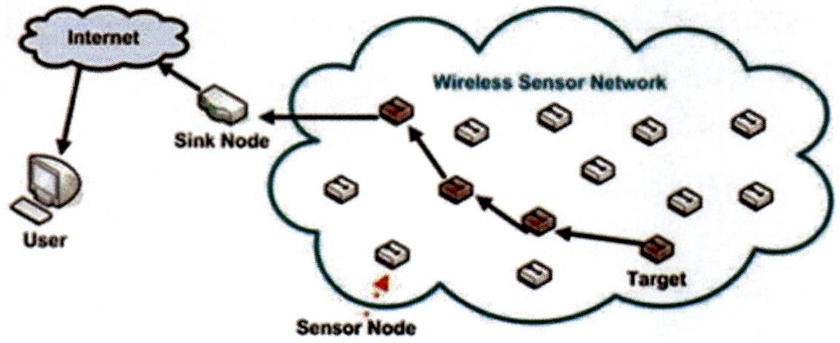

Fig. 1 Wireless sensor network architecture [2]

This also interacts through the use of additional sensor nodes or the BS and also performs basic computations. Typically, the sensor nodes are grouped into a particular county to track and detect environmental events in real time, and are capable of collecting and transmitting information to BS/sink. BS/sink can connect via the Internet or Web to an end-user or source node. The sensor node system structure is affected by many devices, including variation toward internal failure, adaptability, generation costs, working circumstances, sensor arrangement topology, transmission media, and power-use [3].

2 Clustering in WSN

Due to the rare resources into WSN, direct transmission of sensor nodes with BS or multi-hop transmission of sensor nodes toward BS is not accessible like energy consumption is great which outcomes into the quick end of sensor hubs Fig. 2.

Since the WSN cannot increase its entire deal correspondence, direct correspondence or single-level correspondence is not possible for large-scale organizations. Direct correspondence has its drawbacks, for example, great vigor usage, replication of information (sensor hubs that are adjacent to each other), transfer information by exceptionally little variety), & maximum distant hubs kicking the bucket rapidly. To beat these issues, two-level correspondence through various leveled approach is used where hubs are gathered into bunches. Pioneer hub likewise known as CH is in to the charge of collecting the data & when that transfer it toward BS.

2.1 Advantages of Clustering

1. Decrease above aimed at information transmission in clustering NW topologies.
2. Routing time-consuming low memory in the sensor node.
3. It delivers bandwidth (BW) reusability and decreases conflict of channels.

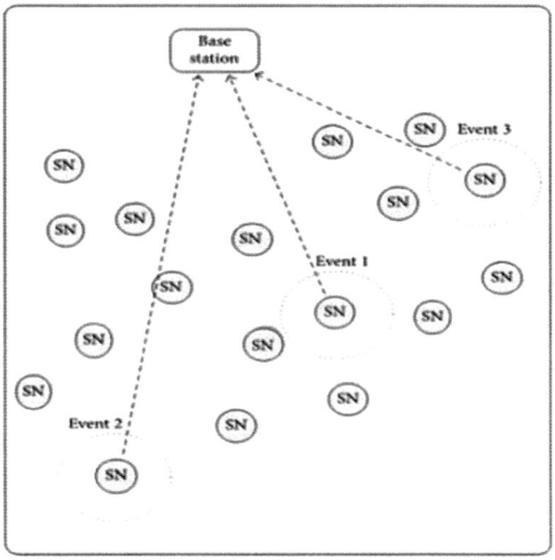

Fig. 2 Direct communication [4]

4. Increases the solidity of the NW because of the low use of power through the cluster.
5. Maintain routing on the cluster head networks of only a few nodes and gateway nodes.
6. Increase network lifetime.

3 Background Study

1. Clustering formation: In WSN, nodes are distributed randomly throughout the monitoring environment. Node and base station locations are fixed, not mobile. The receiver node determines the distance between the two by the transmitter node. Sensor nodes of the same form and data fusion capability will communicate with each other and have unique numbers.
2. Cluster head selection: In wireless sensor network, whole network divide in separate group of clusters using approach of clustering technique like simple clustering or grid clustering. Every cluster contains many members depend on cluster formation and each cluster contain many node and is called cluster member node. Each cluster has one leader called as cluster head; In LEACH protocol, we used the technique called election among the nodes for cluster head selection. The approach for cluster head selection is through highest residual energy as well load consumption among the network. Selected cluster head is responsible for inter- and intra-cluster communications in network. There are two phases in LEACH's protocols name like set-up phase broadly like forming cluster, cluster head selection and steady phase is like data forwarding to base station in network.

$$T(n) = \frac{p}{1 - p^*\left[r * \bmod \frac{1}{p}\right]} \text{ if } n \in G$$

otherwise

where G is remaining node in cluster, r is random number, the leaders of the cluster chose in which cluster they want to be. Cluster leaders collect and submit data from the system to their cluster leader.

3. Cluster node communication: There are two way to communication or data transmission in clustering using intra-cluster and inter-cluster techniques. In clustering member node, collect the data from the environment and communication or data transmitted to base station through above both approaches using data aggregation, compression, and filtering operation. All records are maintained in routing table along with less power consumption of cluster head. There are two approaches in intra-cluster communication one is single-hop communication and multi-hop communication. In single-hop communication, data transmitted from member node to cluster head without intervention of others nodes. In multi-hop communication, data transmitted from multiple node from member node to cluster head and cluster head to base station (Fig. 3),

4. Routing algorithm

The routing protocols proposed for sensor networks can be loosely defined as flat, hierarchical protocols [6]. Hierarchical protocols connect multi-level network nodes. It is usually a method called cluster creation. A cluster consists of a group of geographically proximal sensor nodes; one node is a cluster center.

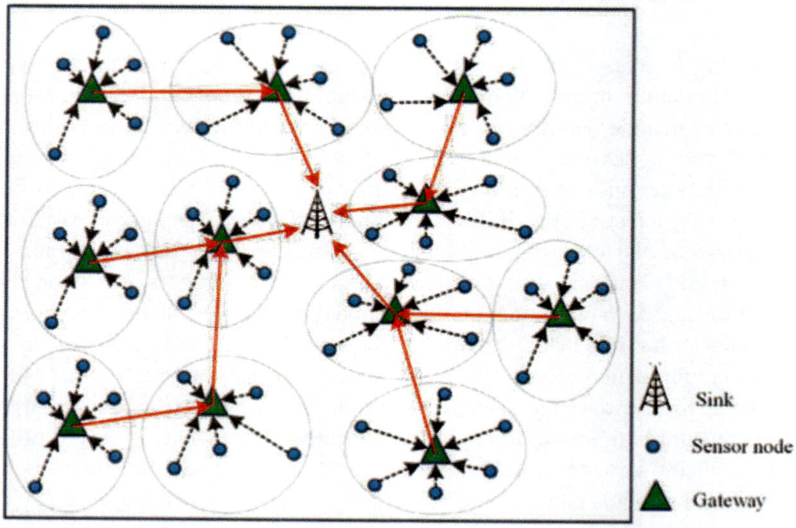

Fig. 3 Wireless sensor network after clustering [5]

4 Related Work

Internet of things (IoT) has emerged as a modern paradigm allowing the development of new smart services and applications through different object interconnections. Key features such as traceability, unique identification, energy efficiency, system heterogeneity, scalability, and ubiquity are important in this sense at the network. It is not capable of providing all these features as a single system, hence the incorporation of wireless sensor networks (WSNs) is an important approach to these solutions.

Some of the recent related works related to energy efficient clustering and routing in IoT-based WSNs are listed below:

Sobral et al. [7] have proposed an appropriate architecture for the provision of the related system structures needed for Internet of things applications collected of two main important part. Among the two parts: one is fuzzy Q-based algorithm, other one fuzzy system-based routing classifier. This paper concludes the fuzzy framework is used for classifying the routing and supporting routing protocols used the application where we used WSN and RFID technology. In which its improves efficiency between data transformation between node by decreasing rate of packet loss, load balancing among the network and energy efficiency. Using the way to select the best routing path for traversing the packet routing. Using this approach reduce the packet loss and this leads indirectly to decreased node energy consumption and improved network load balance.

Shen et al. [8] proposed a novel energy-efficient centroid-based routing protocol (EECRP) for wireless sensor network to assist Internet of thing application for improving network performance. The projected protocol EECRP comprises following main components: a new circulated cluster forming technique that enables local node self-organization, a new set of cluster adaptation, and rotational cluster head (CH) algorithms based on the centroid location to spread the energy load uniformly across all sensor nodes in network, and a novel approach to minimize power consumption for large-scale network communication. In EECRP in particular, the remaining power of nodes is used for finding the position of centroid. Comparison among the EECRP better with LEACH, LEACH-C, and GEEC. Additionally, EECRP is suitable for long-life networks whose base station (BS) is located inside the network.

Bagula et al. [9] revisited the matter of energy consumption in sensor node structure to suggest a clustering approach in which the package transfer of sensor devices is plotted into a model of energy cognizance, it develops a clustering algorithm for system, the clustering algorithm used to identifies service aware clustering (SAC) arrangements in Internet of things situations. The performance evaluation showed the energy consumption in novel SAC protocols compared with similar kind of clustering and routing protocols with respect of power consumption as well battery lifetime of the sensor nodes and the efficiency of their traffic manufacturing in terms of throughput of system and interruption. Those comprise the famous LEACH and improved LEACH-centralized (LEACH-C) protocols, as well as the new protocols like as D.E.C.S.A. [10] and M.O.C.R.N., among others.

The green network-based IoT was accomplished by Rani et al. [11] through a novel implementation scheme designed to balance energy efficiency. A hierarchical architecture model was initially proposed for the IoT implementation. It introduced the function of scalability into the IoT and made it more extensible. After that, based on the proposed structure, an comprise model was considered and this model is energy efficient, which smoothest setup on the road to a green Internet of thing. At last, a least power consumption chain-based cluster coordinator algorithm (ME-CBCCP) was introduced, using cluster-approach node structure and a new broadcast procedure to optimize energy parameters. Compared to conventional solutions for WSNs, the system model has become lighter and more effective MDFSPS [12] and can be easily applied in an energy-intensive IoT.

Muhammad et al. [13] developed an effective pattern change (HABCA-EST) fusion artificial bee colony protocol for the Iot application used WSN technology. There are three key components involved for artificial bee in colony. The components are like employed bees, onlooker bees, and scout bees to find for the outcome collectively. The distinctive value of HABCA-EST is that through effective representation, device development of more serious items, and total use of laying-off info using an EST of smart objects (SDs), it guarantees fast development in the capable worth of each food-item places. The swarm and EST processes in the proposed HABCA-EST combine work for finding the best outcome with fewer fitness evaluations, and fewer SDs schedules. The results of the simulation showed that HABCA was considerably faster than other traditional techniques.

Fouladlou et al. [14] proposed a novel directing system to increase the power consumption of IoT-based wireless sensor network. For the purpose of efficient routing and prolonging network life, the famous inherent procedure was used for clustering application. It processes described involves two phases: the phase of configuration and the phase of stable state. This literature adopts that all activities of the system are carried out in Sink Station (SS). SS had been measured a strong, high-energy procedure centre. Based on the node information obtained, such as battery level and place, SS assigns the sensor-nodes to appropriate clusters. The main idea is to efficiently allocate the cluster head role in the network to decrease overall power used on the network. Experimental tests showed a better performance of the new routing algorithm compared with the IEEE 802.15.4 procedure in standings of end-to-end stay, power consumption, bit rate, and throughput.

Aslam et al. [15] proposed a novel energy-harvesting radio frequency (RF) scheme for Internet of things components joint with the finest conceivable RF source to make best use of the total energy collected by the Internet of things components. For battery balancing in system, the nearby IoT devices were group of node combine. It is rotationally selected IoT devices through the maximum remaining power was chosen as a key important node in cluster (CH). Upon completion of the CH selection, it allocates networks to the IoT application used various devices to account their information using a network sharing structure based on a novel integer linear program (ILP), by satisfying their desired quality of service (QoS). Various parameters were used to perform exhaustive simulations, together with the quantity of IoT devices, the number of collecting base foundations. For the different network structures, the main key thing is gap between radio frequency and the distance highly computed IoT resources. The novel schemes or algorithm shows the better result than traditional approaches through displayed in simulation.

Saptarshi et al. [16] suggested an energy-efficient frequency range-conscious connectivity for device-to-device IoT (SpEED-IoT) communication. They addressed the issues of secondary routing based on dynamic spectrum access (DSA) in a device-to-device (D2D) IoT network, and built power consumption in many network and multiple hop communication networking between IoT devices using a frequency map created by environmental sensing capability (ESC) sensors. A selective flooding technique based on transmission power control is future to blowout the routing requirements in the network without causing overhead for wide network transmission. Using comprehensive simulation-based test bed evaluation, they demonstrated the SpEED-IoT efficiency in guaranteeing IoT network compatibility, system data end to end rate compress, main data security, and fairness of data routing.

John et al. [17] introduced an ESCHS algorithm for optimal cluster head selection process in WSN network that used in Internet of things application. The midpoints of all the individual clusters were obtained in the cluster head election process which emphases on well-organized battery when all the cluster created in network. The size of each node is determined with respect to the midpoints of each cluster. In cluster, the distance is important factor for picking cluster head selection. The minimum distance from sensor node with the mid-point of clustering. The sensor node with preserved more power than the average residual power of cluster. To make an ideal node distribution, both minimum distance between the CH node with sensor node that improve the energy. Using uniform clustering, In this scheme CH become any node from the network who has more battery power, With this concept incres the energy of every sensor node in network. For 2000 rounds the algorithm was simulated. It was completely necessary to increase energy savings in ESCHS by 10.80% compared with LEACH and by 9.14% compared to DLEACH.

Behera et al. [18] suggested a cluster-based routing protocol for the use of hybrid energy for WSN in IoT applications. Not only did the protocol use vector lossy compression to form clusters and active cluster heads, but it also used in shorted path algorithm to find the shortest distance between the sensor node with cluster head name is called Dijkastra. To find the minimum distance from the active cluster to the cluster head and creating shortest path from the cluster head to cluster member. It also showed the minimum distance that is ideally suited for IoT applications between the active cluster heads. As a consequence, it provides a shortest path leading to an energy-efficient communication mode. Considering various metrics, energy consumption in a vast network system such as in IoT has a huge effect on its efficiency. From the simulation test, it was found that the average energy consumption of the network had increased to nearly 60% after the implementation of this protocol.

Shah et al. [19] proposed an power based and interoperable aware routing in clustered Internet of things wireless sensor networks for the optimization of the throughput. The authors intended to increase network throughput, particularly during the routing process, by conserving the energy. The routing technique used proactively by the CLEARMaxT, here the fix level of power pledges the process of route updating which makes it tolerant to fault. In this case, a neighboring table with more than one better alternative path plays a vital role, particularly if any node is dead then it loss of battery as well failed in communication. The decision for a new better transmitting node is not moved to the CH, and the CH rotation decision is not transferred in to the base station's central point. In both respects, CLEARMaxT contributes locally to the current optimum transit node and operational cluster core, respectively. The cluster-based interoperable conscious energy routing maximizes the throughput by reducing end-to-end delay, reducing the packet drop ratio and improving network life.

Preetha et al. [20] proposed a tree-based cluster head selection protocols to ensure power consumption using energy-efficient framework for densely distributed WSNs IoT devices. For the data collection from the sensor node using fuzzy reinforcement learning method called RL-FIS in the closely circulated sensor network based on three parameters: neighbourhood connection, bioactivity key, and nearest neighbour connectivity. In this process, one initial cluster head is responsible for the cluster communication after using tree-based clustering approach for selecting new cluster head in network to achieve power consumption in network. The writers compared the scheme proposed to other schemes. Simulation findings suggested that the proposed scheme offers highly stable data archive, filtering, compressing, picking node, and improves service criteria such as bit error rate, end to end delay, throughput with a decrease in network traffic related to buffer occupancy and offers reduced energy consumption than LEACH, HEED, and MBC through resultant tests.

Sarwesh et al. [21] developed a cross-layer model to optimize the transmission range of nodes in IoT-supported low power wireless networks. In the cross-layer structure that needs the different requirement for achieving the lifelong stability of the network. In the protocol stack, it needs the various network layers like physical, MAC layer, and network layer. In this proposed protocol, a static threshold value is called ETRT value. It is calculated using various parameters like sensor data circulated in network information. MAC created battery protected technique later takes advantage of the ETRT key and allocate to the minimum range for data transmission to each sensor node. The basic concept for the novel cross-layer protocols is to estimate the ETRT key capability of the particular node and using this capability value to find the requirement transmission power of each and every sensor node in network. Therefore, assigning maximum transmission capacity based on ETRT knowledge prolongs the lifespan of the network with increased efficiency and service quality (QoS). It had found from simulation results that the cross-layer model based on ETRT performs twice better than standard mode (Table 1).

5 Conclusion

WSN is the resource-constrained process in an unsecured form environment, so replacing battery recharging is not possible. In this paper, we discussed various protocols in which is energy efficient, data routing, cluster formation and cluster

Table 1 Comparison of various protocols with cluster formation, CH selection, and routing approach

S. No.	Protocol	Cluster formation approach	CH selection approach	Node routing approach	Merits and de merits
1	Sobral et al. [7]	Fuzzy arithmetic logic	Fuzzy arithmetic logic	Multi-node communication routing	• The packet loss rate is decreased • Decreased node energy consumption and improved network load balance
2	Shen et al. [8] (EECRP)	Self-organization	Rotational cluster head	Centroid-based routing	• Improving network performance • Minimize energy consumption for long-distance communication
3	Bagula et al. [9] (SAC)	Non-cluster head become cluster head	Highest residual energy	Single-node communication	• Clustering model where the service delivery of sensor devices is mapped into a model of energy awareness,
4	Rani et al. [11] (MBCCP)	Simple	A distributed method	Single-node communication	• Optimization model was considered and this model is energy efficient, which smoothes progress on the road to a green iot
5	Muhammad et al. [13] (HABCA-EST)	Bio-inspired artificial bee colony	Artificial bee colony	Multi-node communication	• The swarm and EST operations in the proposed HABCA-EST work together to find the optimal solution with fewer fitness evaluations
6	Fouladlou et al. [14]	Genetic algorithm	Gateway	Multi node communication	• The cluster head role in the network to reduce overall energy consumption on the network
7	Aslam et al. [15]	Simple	Highest residual energy	Novel integer linear program	• Novel energy-harvesting radio frequency (RF) scheme for iot devices
8	Saptarshi et al. (2016) (speed-iot)	Non-cluster head becomes cluster head	A distributed method	Dynamic spectrum access, multi-hop	• Addressed the issues of secondary routing
9	John et al. [17]	Simple	Midpoints	Multi-hop routing	• Balances the network load by CH rotation and thus increases energy savings and enhances network life

(continued)

Table 1 (continued)

S. No.	Protocol	Cluster formation approach	CH selection approach	Node routing approach	Merits and de merits
10	Behera et al. [18]	Vector lossy compression	Vector lossy compression	Dijkstra algorithm	• Shortest path resulting in a technique that is energy efficient
11	Shah et al. [19] (clearmaxt)	Non-cluster head becomes cluster head	CH rotation	Multi-hop routing	• Optimization of the throughput • Reducing end-to-end delay, reducing the packet drop ratio, and improving network life
12	Preetha et al. [20] (Rl-FIS)	A distributed method	Tree-based robust cluster head	Multi-hop routing	• Highly stable data gathering node selection and improves service metrics such as throughput, PDR, bit error rate and end-to-end latency
13	Sarwesh et al. [21] (ETRT)	Simple	Highest residual energy	Multi-hop cross layer routing	• Cross-layer model to optimize the transmission range • Lifespan of the network with increased efficiency and service quality (qos)

head selection were key issue in all papers, It was conversed to use properly battery source. Out of the given energy-efficient system of structures, cluster-based schemes are expected to be of comparatively huge interest due to substantial improvements over their lifetime. Different attempts have been made in most open methods to obtain energy efficiency during the hierarchical process of clustering, where the nodes are gathered in the clusters, and information is forwarded to BS via CH.

References

1. V.K. Singh, M. Kumar, A compressed sensing approach to resolve the energy hole problem in large scale WSNs. Wirel. Person Commun. **99**(1), 185–201 (2018)
2. T. Puyalnithi, V.M. Viswanatham, Visualization and statistical analysis of multi-dimensional data of wireless sensor networks using self organising maps. Int. J. Eng. Technol. **8**, 391–395 (2016)
3. S. Harendra, A.N. Jangwan, Energy efficient hierarchical routing protocols for homogeneous and heterogeneous wireless sensor networks: a systematic review. Adv. Wirel. Mobile Commun. **10**(6), 1227–1256. ISSN 0973-6972 (2017)
4. R. Aggarwal1, A. Mittal, R. Kaur, Hierarchical routing techniques for wireless sensor networks: a comprehensive survey. Int. J. Future Gener. Commun. Netw. **9**(7), 101–112 (2016)
5. R. Yarinezhad, S.N. Hashemi, A routing algorithm for wireless sensor networks based on clustering and an FPT-approximation algorithm. J. Syst. Softw. **2019**(155), 145–161 (2019)
6. I.F. Akyildiz, W. Su, Y. Sankarasubramaniam, E. Cayirci, Wireless sensor networks: a survey. Comput. Netw. **38**, 393–422 (2002)
7. J. Sobral, R. Rodrigues, J. Rabelo, N. Lima Filho, H. Sousa, R. Araujo, H. Filho,A framework for enhancing the performance of internet of things applications based on RFID and WSNs. J. Netw. Comput. Appl. **107**, 56–68 (2018)
8. A.W. Shen, C. Wang, P. Hung, C. Lai, An efficient centroid-based routing protocol for energy management in WSN-assisted IoT. IEEE Access **5**, 18469–18479 (2017)
9. A. Bagula, A. Abidoye, G. Zodi, Service-aware clustering: an energy-efficient model for the internet-of-things. Sensors **16**(1), 9 (2015)
10. T. Qiu, Y. Lv, F. Xia, N. Chen, J. Wan, A. Tolba, ERGID: an efficient routing protocol for emergency response internet of things. J. Netw. Comput. Appl. **72**, 104–112 (2016)
11. S. Rani, R. Talwar, J. Malhotra, S. Ahmed, M. Sarkar, H. Song, A novel scheme for an energy efficient internet of things based on wireless sensor networks. Sensors **15**(11), 28603–28626 (2015)
12. S. Kumar, V. Chaurasiya, A multisensor data fusion strategy for path selection in internet-of-things oriented wireless sensor network (WSN). Concurr. Comput. Pract. Exper. **30**(18), 1–14 (2018)
13. Z. Muhammad, N. Saxena, I. Qureshi, C. Ahn, Hybrid artificial bee colony algorithm for an energy efficient internet of things based on wireless sensor network. IETE Techn. Rev. **34**(1), 39–51 (2017)
14. M. Fouladlou, A. Khademzadeh, An energy efficient clustering algorithm for wireless sensor devices in internet of things, in *IEEE Conference on Artificial Intelligence and Robotics*, pp. 39–44 (2017)
15. N.H. Aslam, J. Jang, K. Lee,Optimized energy harvesting, cluster-head selection and channel allocation for IoTs in smart cities.Sensors**16**(12), 1–22 (2016)
16. S. Debroy, P. Samanta, A. Bashir, M. Chatterjee, SpEED-IoT: spectrum aware energy efficient routing for device-to-device iot communication. Future Gener. Comput. Syst. **90**, 1–16 (2018)
17. A. John, A. Rajput, K.V. Babu, Energy saving cluster head selection in wireless sensor networks for internet of things applications, in *IEEE International Conference on Communication and Signal Processing*, pp. 34–38 (2017)

18. T.M. Behera, S.K. Mohapatra, P. Mukherjees, H.K. Sahoo, Work-in-progress: DEEC-VD: a hybrid energy utilization cluster-based routing protocol for WSN for application in IoT, in *IEEE International Conference on Information Technology*, pp. 97–100 (2017)
19. S. Shah, Z. Chen, F. Yin, I. Khan, N. Ahmad, Energy and interoperable aware routing for throughput optimization in clustered IoT-wireless sensor networks. Future Gener. Comput. Syst. **81**, 372–381 (2018)
20. S. Sathya Lakshmi Preetha, R. Dhanalakshmi, R. Kumar,An energy efficient framework for densely distributed WSNs IoT devices based on tree based robust cluster head. Wirel. Pers. Commun. (2018)
21. P. Sarwesh, N. Shet, K. Chandrasekaran,ETRT—cross layer model for optimizing transmission range of nodes in low power wireless networks—an internet of things perspective. Phys. Commun. **29**, 307–318 (2018)

Analysis and Review on Service-Oriented-Based IoT Middleware

Kanu Patel and Hardik Modi

Abstract Internet of things provide the platform to human and machine to learn and communicate using huge numbers of smart sensors, embedded devices, services, and other Internet-connected devices. Currently, IoT applications are used in wide range of sectors like healthcare, industrial, smart cities, etc. This application having different characteristics like real-time, diverse applications, and everything as a service. Middleware is key part or layer of IoT architecture. Middleware is software component that provides communication between devices and application. It provides the interface between heterogeneous devices and IoT application. Actually, Middleware supports many features like heterogeneity, reliability, availability, mobility, scalability, intelligence, privacy, and security. Recently, many proposals for IoT middleware are proposed based on their design types and architecture. The review presented in this paper contains in depth comparison between different service-oriented middleware. In addition of summary, we also explore research challenges, open issues, and research directions in era of service-oriented middleware.

Keywords IoT · Middleware · WSNs · RFID · Service-oriented middleware

1 Introduction

The limitation of intranetworking that overcome by Internet of things. The evolution of Internet of things is being started by increasing number of sensors, embedded devices and actuators are Internet-connected through gateway or any communication channel. The Wireless World Research Forum predicts that by 2020, there will be 50

K. Patel (✉)
CHARUSAT University, Changa, India
e-mail: kanu.patel@bvmengineering.ac.in

H. Modi
Department of Electronics and Communication, CHARUSAT University, Changa, India
e-mail: hardikmodi.ec@charusat.ac.in

J. Tavares et al. (eds.), *Emerging Technologies in Data Mining and Information Security*, Lecture Notes in Networks and Systems 164,
https://doi.org/10.1007/978-981-15-9774-9_19

billion wireless devices serving 7 billion people [1]. The ratio is 6.58 connected wireless devices per person. This large number of devices creates the IoT infrastructure; Internet of things is a core area that takes lead role in next upcoming technologies based on Internet. IoT will use in every sector of our earth like healthcare, automotive, home automation, consumer, industrial, retail, and transportation sector. We also connect this all sector in one place using IoT. For smart cities concept, IoT domain plays major role for making such smart like smart parking, urban noise, traffic problem solution, irrigation, water pollution, waste management, and street light. We can build smart home to provide smart surveillances for security and energy saving devices [2, 3]. In USA, there have been around 620 bridge failures since 1990 [2], With this fact, in all state, the travelling in bridge is dangerous as well big challenges. With the Internet of things technology, they fitted sensors on every bridge to sense the data like materials, condition in bridge, and its vibration. With analysis of this data, they can save many human lives. As per Fig. 1, Internet of things can happen anywhere, anytime, and anything. Internet of things is generally dividing in six levels for connecting device to human or user. First level is device connection; In IoT infrastructure consists of numbers of devices. They all can connect to the system through network. Devices basically consider like different sensors. The role of sensors to sense the data or data collection from their region and send to the system. It considers data sensing or collection level. Communication is third level of Iot. Using this level, data will transfer to the cloud or data storage in system. Internet will use in this level for transferring data from devices to data storage through network and Internet. Later on, all levels are like data analysis, data value, and human value.

Numbers of connectivity protocols are developed for IoT system and it is increasing everyday. Among this protocols MQTT, Constrained application protocol.

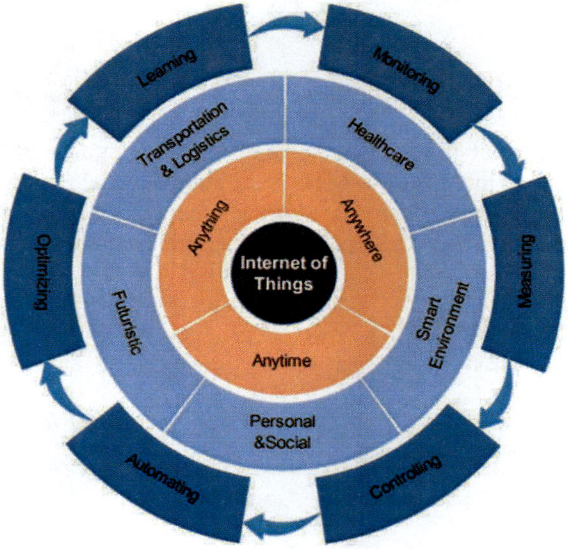

Fig. 1 Internet of things: futures and functionalities [4]

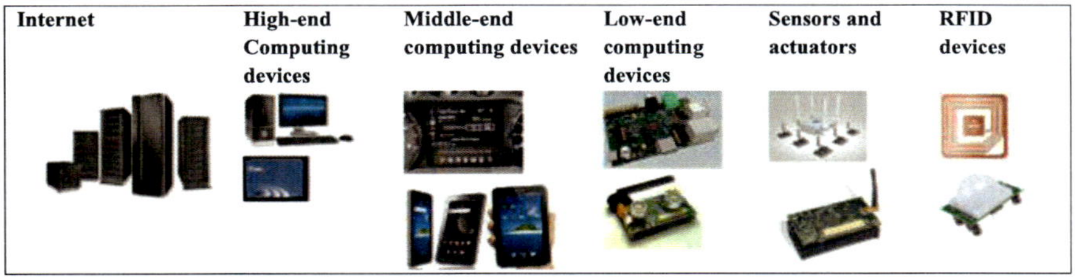

Fig. 2 Heterogeneous devices [6]

Bluetooth is very popular for low-range connectivity protocols designed especially for IoT. As per Fig. 2. In IoT, having four component like wireless sensor networks (WSN), M2M communication, SCADA, and RFID [5]. For supporting diverse application area we must to integrate these four technology. Majority of existing middleware proposals are wireless sensor network centric. Form the paper of Razzque [6], it is evident that no one middleware supports this all requirement of WSN in IoT system. The number of existing middleware supports WSN, RFID, and M2M communication. SCADA is used limited in middleware. Bandyopadhyay [2, 7] has focused the importance of a middleware system in IoT.

The growth of making middleware is rapidly high because it is easy to combine any number of services and others technologies to make new middleware having more capable then previous one. Heterogeneity is one of main challenges of many existing middleware. The main feature of middleware is transparency. Using this feature, we can solve so many problems like architecture mismatch and data loss. Some others features of middleware should be desirable outcome like programming abstraction, flexibility, availability, context awareness, and interoperability. We will detail discussed in a paper.

Considering the important of IoT in various domains, this paper is organized as follows. Section 2 discusses the characteristics and features of middleware, Sect. 3 provides a overview of service oriented middleware, Sect. 4 discusses the review existing research work for service-oriented middleware compare based on features, we discuss and specify the technical few challenges and research issues in Sect. 5, and summary of this paper is in Sect. 6.

2 Middleware

Middleware in IoT is very hot area for research. Many middlewares have been proposed in last few year. No one fully supported all basic and architectural requirement of middleware. These all proposed middleware can be differentiated by their design method and implementation specific and level of abstraction. we grouped middleware based on design approaches and listed like event based, service oriented, virtual machine based, agent based, and database oriented. In one more paper, Ngu

[8] can view middleware based on architecture. It form three class of existing middleware on architectures like service-oriented architecture, cloud based, and actor-based middleware. In this paper, we will discuss service-oriented middleware in details.

Functional requirement of middleware [6], in IoT system, the role of middleware is to identify the resource in IoT infrastructure. It also manages the resource in a system, resource includes heterogeneous devices from different vendors and different specifications. This resource discovery and manage is fully automated in IoT system. A middleware has important service for resource management, and with this service, we can monitor resource, allocation, and provisioned in fair manner. Data management is also important requirement of middleware. This data is key element in IoT applications. This data may refer like sensed data or any information provided by network. Middleware provides the data storage, preprocessing, filtering, data compression, and data aggregation functionality. Many events occur in IoT system so managing event and transform event in meaningful and priority wise events. Managing and deploying code are also challenging tasks. For code allocation and migration process, middleware needs service for it. In code allocation is to select set of node and sensors from infrastructure used in application level. Code migration service is for transferring code from one node or device to another node.

Non-Functional Requirements of middleware [6], Due to the growth of IoT network, service, and device, the middleware should be scalable in IoT system. IPv4 is not scalable solution for IoT network because IPv4 has limited address for network; so for this scalability, we must used IPv6 for large address and its scalable solution. Middleware should be timely or real-time operate. So many applications want real-time data as well real-time service for correctness. If middleware failed to give real-time data, then it is meaningless in IoT system. Availability and reliability are also considered for non-functional requirement. Both works together to protect application with highest fault tolerance. Middleware should be operational over the duration of system whether work or fail is. Middleware reliable means we achieved system reliability. If any failure occurs at any time, then availability and recovery time should be minimum. Security should be considered in all the useful and non-functional squares including the application level. Context awareness may be leakage the personal information of user or any analysis information to the unknown person. IoT middleware is easy to be deployed in system and also easily updated when it required. Without using any additional effort from application or any user, it is deployed in IoT system,

Middleware Features

The middleware having various features and characteristics like scalability, security, etc., that we can see the various combination of system functional or non-functional requirements. In this section, we have seen the various features of middleware.

Scalability: IoT middleware should be scalable to accomplish the growth of IoT system. Any time to add IoT network or services then easily accommodate in system. For the size of IoT network, IPv6 is more preferable for scalable addressability to the services [9]. Virtualization in middleware is great steps for improving scalability especially service level.

Interoperability: In middleware, interoperability is key feature, for heterogeneous devices are placed in network or different sets of application interconnected through networks are communicate or sharing data without any programming efforts by the users. In middleware, interoperability can be viewed from network, syntactic, and semantic [10]. For different devices, services, platforms, protocols, and data need the interoperable features in middleware. Implementing interoperability needs the creating and fulfills the common standards.

Durability: Middleware must remain available to the system during the operation. If middleware is unavailable then the operation considers as failure of system. The middleware level reliability means system level reliability. Even if any failure occurs in system then the recovery time and failure frequency should be minimum. For durability consists two main factor reliability and availability.

Flexibility: Flexibility is one of the features of middleware provide to IoT systems. IoT will integrate different devices like mobile device, wireless devices and resource type's devices. Within a network any new device wants to join the system then its join it and any device is move out of IoT infrastructure rang then its leave from system. There is different kind of flexibility like device level, resource level. Flexibility moves from application level to middleware end at device level and handle different kind of flexibility. The flexibility is more important because more flexibility provide many numbers of connectivity API.

Resource and Data management: Resource and data is key features of middleware. In Resource management, it's started from resource discovery is automated without human involvement. For resource management, application provide with service for manage the resource. This means resources management should be monitored, resource allocation and migration in a fair manner. Service oriented middleware need to facilitated the on the spot service composition. Data may refer sensed data for IoT infrastructure provides to application. Middleware need to provide data management like data storage, processing, filtering and compression.

Real time: Middleware must be providing the real time service otherwise it's meaningless. For correctness operation data must provide to the middleware in real time. IoT system deals many real time application like traffic management, healthcare sector etc. On time delivery information or data is crucial things. Late deliver data or service refers as useless or dangerous to the system.

Intelligence and Context awareness: Context awareness is important features of middleware in building smart system then establish the value from sensed data. It allows users to fix the setting for one application, the same setting will reflect to others application same setting is called context awareness. This feature is called coordination management concepts. IoT middleware is aware about the context of users, services and devices for providing effective services to the user in infrastructure.

Security and privacy: Security can be view in IoT system as inner security of middleware and device security. Security in middleware is big issue because data transferring from device to application through middleware. Using confidentially, privacy, trust and integrity we can achieved system level security. Middleware provide the different security measure like device authentication, secure applications, pervasive infrastructure and access policy. Privacy need to use personal user information and must guarantee to protect the information in middleware.

Distributed: In IoT system use ultra large scale network for application/devices to share the information, This all devices or application are probably located in form of distributed network. Centralized approach is not sufficient for large scale network. A middleware implementation requires to support functions that are distributed across the physical infrastructure.

3 Service-Oriented Middleware

Service oriented architecture is a wide area of research, The service-oriented design approach developer or users to add software or applications in the form of services. It also exploited to point some challenges like service discovery and composition of services in middleware. Service oriented architecture consist many services in middleware part. This services host by some service providers. Service consumer will do service registration and consume using application. Service oriented architecture having three different functionalities. Service discovery, service composition, service access [6]. The role of service discovery is to add new service also find best service for request. Composition handle specific service that cannot discover by service discovery. Service access will provide the communication with the discovered services. In Service oriented middleware facing some challenges like interoperability for services, reliability, security, privacy and trust (Table 1).

For making service oriented middleware platforms for achieving various features like interoperability in heterogeneous devices, network and application scalability as well efficiency for application.

Table 1 Relationship between middleware and requirements [6]

Middleware type	Key functional requirements	Non functional requirements	Architectural
Service-Oriented	Service distributed discovery	Scalability—application level	Programming abstraction supported
	Device distributed discovery	Scalability: network as well WSN level	Network level interoperability
	Resource allocation, Monitor, adaptive composition	Security: confidentially, availability	Context aware, dynamically adaptive
	Data storage, data processing	Soft data real time	Service based, lightweight energy, lightweight memory
	Small scale event management	No reliability at communication and data	Distributed

Hydra [11], presently called as LinkSmart [12]. Basically it is one kind of middleware for ambient intelligence (AmI). It is built on a service oriented architecture and model-driven structure. Its structure consists of a various components of managing services, including a service, event, device, storage, context and security manager. All these modules are clustered into an application, these all components have different layer like semantic, service, network and security layer. This algorithm is responsible for syntactical and semantic level interoperability using semantic web services. Also it maintain self configuration and dynamic reconfiguration. All this managers create hydra smooth by adjusting energy consumption in resource-constrained devices. Disseminated security and social trust parts offer secure and reliable communication within devices. Its security and privacy policy uses virtualization and an implementation of WS-based mechanisms improved by semantic resolution [12]. However, its virtualization can involve security concerns. Also, ontology-based semantic security and interoperability solutions are likely to be unsuitable in IoT because, currently, there are no standard ontology's for ultra large-scale IoT.

GSN [13, 14] is Service based middleware, It's provide common platform for heterogeneous devices, device integration, sharing and deployment on IoT infrastructure. The focal idea is the virtual sensor reflection, which empowers clients/designers to definitively determine XML-based arrangement descriptors to convey a sensor. GSN [13] use same container architecture as in J2EE Where It's host various multiple virtual sensors and its provide life cycle of container. The sensor lifecycle consist persistency, security, notification, resource pooling, and event handling. The contribution to the virtual sensor is at least one information streams which are prepared by the XML particular. These incorporate the examining rate of the information, the sort, and area of the information stream, the persistency of the information, the yield structure of the information, and the SQL preparing rationale for the information stream. Each information stream is related with a wrapper. It's specifies following things. 1. Which arrange convention to be utilized to the interface, interact, and communication with the physical -sensor when initially introduced. 2. How to read data from sensors. 3. How to deal with data that received from sensors. GSN [14] use SQL based data storage of all sensors data as permanent storage using storage attribute set true in XML specification.

Rimware and Huo [15] introduce Rimware service based middleware. The motto behind this middleware is the position of this middleware on rim that cover the cloud data storage inside and network management on the outside. This middleware is in between devices and IoT applications. In this layer it's include gateways, data storage cloud and common component that use in both side. This middleware provides various key requirements like scalability, interoperability, privacy, security, and saving power via gateway adapter and knowledge-based and access controller in cloud. Its provide interoperability at device level so like different vendor made same product with different configurations. Interoperability at device level use profile based protocols. For communication it's used Bluetooth 4.0 connectivity between device and middleware. Using BLE, we can reduce the power consumption in system. For Initial steps, the different sensors authenticate with basic technique and connection through a trusted gateway to the cloud via an adapter on the gateway which can substitute the role of application on smart phones. Then, the cloud side operation is to mapping among device profiles, web APIs and various cloud operations.

TinySOA [16] is a SOM that proposes a complex concept of the structure for the growth of WSN applications. It offers a simple SOA based API, using it application designers may use WSN resources. It controls WSN device and conversation level heterogeneity, and provides simple combination of Internet applications with WSNs which will allow them to gather data from the devices or sensors. This algorithm has modest and deterministic policies for WSN resource. It provides only a few primary functional requirements like abstraction, resource discovery and management.

DropLock and Vinh [17] introduce a middleware architecture to integrate various devices, different sensors, and cloud computing. Using various cloud service it's provide large data storage, data filtering also power saving management along with scalability. Aim of this middleware to mapping between MCC (mobile cloud computing) and IoT services. Primary concerns of this middleware is to provide security at middleware layer. The authors presented a framework of CoT Middleware along with network management, Device management, Data management with security and privacy. For security concerns, before communicating between different devices, it's mandatory check the basic authentication steps. For resource constraint devices, authentication could be performed by any information security algorithms like encryption-decryption method. Data must be sent in encrypted form to the cloud. Droplock middleware [17] provide the privacy using their privacy policy and privacy management services. The authors demonstrate Droplock middleware architecture in smart city scenario [17].

CHOReOS [18, 19] CHOReOS is set of software component that provide to large scale web service implementation and execution. It's converge the Internet of things, Cloud computing and Internet service [20, 21] empowers enormous compositions of adjustable, QoS-aware, and heterogeneous services in IoT. It addresses scalability, interoperability, mobility, and adaptability concerns in through approaches like scalable probabilistic thing-based service archives and discoveries.[18, 19]. CHOReOS is mixture of four components: 1. eXecutable Service Composition (XSC) is to responsible for coordinate the composition of services and things, 2. eXtensible Service Access (XSA) to retrieving and selecting services and things, 3. eXtensible Service Discovery (XSD) to manage service level protocols and processes for discovery of services and things, and the Cloud and Grid middleware to handle computational resources and drives the deployment of choreographies. Grid Middlewares have three components: (1) Service for storage that install a database server on cloud and define security and privacy policies. (2) Grid Service, which provides resource and application with high configuration. (3) The Hybrid Engine is the main layer of the Cloud Grid and Middleware and is a composition of the CHOReOS middleware. MobIoT, a key component of CHOReOS [18, 19], is a thing-based Service oriented middleware. We discuss further about MobIoT in next m,iddleware. For the same reason, reliability in these middleware could be an open issue and challenge ontology-based semantic support will be very challenging in heterogeneous IoT environments.

MobIoT [22] is basically for Mobile Internet of things. It's also service oriented middleware which aim to deals with large scale network, device heterogeneity and dynamic networking environments. MobIoT [22] provide the service discovery and composition using probabilistic ways. This middleware provide two level of abstraction first one is things as a service and things actions as a service. We consider service provider is things as a service and service consumer is things action as service. MobIoT [22] provides various key requirements or features like its overcome of heterogeneity of the network things, using probabilistic approaches for service registration and composition, scalability of network infrastructure, This middleware implemented using Java and android platform. Its consist two components mobile middleware and MobIoT web services. Middleware is deployed in mobile device like smart-phone, tablets, etc. MobIoT provide the device level heterogeneity or we can sys that interoperability at device level.

Xively [23] is a Platform as a Service which offers middleware services to build products and solutions for IoT. Public cloud-based Xively provide designers a standards-based directory, data, and business services. Directory services support to discover suitable objects with appropriate authorization. Data management services, using a high performance and time-series database, store and retrieve data reliably. Its online tools simplify data, regulate and other application difficulties of IoT development. Xively's device lifecycle administration and actual message bus provisions significant and actual placements in IoT. Significantly, it provides provision for end-to-end safety over the whole platform to confirm IoT solutions' integrity. The absence of storage safety [24] can be an problem in various IoT applications. It provisions several data formats, though, it cannot normalize the received data so data handling must be done separately for each source or it requires an earlier plotting process to normalize it. This makes an overburden in the system. Also, it provides a list of software and hardware mixtures required to design and develop IoT applications, but its support for interoperability is inadequate.

CarrIoTs [25] is a cloud-based service-oriented middleware for IoT, particularly for Mobile to mobile conversation, and concentrations on: cost effective Mobile to Mobile application development, scalability, and simplicity of usage. The main benefit of CarrIoTs is that it cares scalability at network level. One can put trigger on various stages of the data handling cycle to push data to an exterior system. Like Xively, CarrIoTs does not normalize the entering data. It also does not assurance for storage safety, and provides finite support for interoperability [24]. Though, CarrIoTs [25] and Xively [23] provide advantage from dynamic online groups, and are popular among designers and innovative adopters.

AUSOM [26] (Autonomic Service-Oriented Middleware) is autonomic service based middleware to support MAPE-K [27] loop (M: Monitor, A: Analyze, P: Plan, E: Act Using K: Knowledge) to provide autonomic computing, adaption using information available at IoT bottom device layer. Monitor component is responsible for register services, detect any event generated. Analyze component is responsible for generation of new workflows, deletion of unnecessary workflows, modification of existing workflows. Planning module do the implementation of workflow via Artificial intelligence. Execution component is responsible for execute selected or modified workflow by planning module. When application call the middleware then particular workflow executed by component.

SmartCityWare [28] is used for smart city or smart building application, In this paper they discussed about for intelligent traffic light control application in smart city. It's also used for forecast traffic patterns and regular traffic light flow. It's developed to use of Cloud of Things and Fog computing for use in smart city as services. All function of smart city seen as a services. Service can be classified in two different categories like Core services and Environmental services. Core services are like service registration, invocation, security and location aware. Main function of SmartCIty-Ware is to enable and integrate all the services to perform the operation for smart city application, It's provide the more flexibility to add more new component as well new services accommodate in middleware as per application requirement (Table 2).

4 Open Research Challenges

Deep analysis on all existing service oriented middleware of IoT system is suggested that service oriented middleware is live research areas. Our surveys of service oriented middleware give the following open challenges and issues.

- Service discovery: The main challenge of service oriented middleware is discover the service in range of IoT system. Provide the environment to the application communicate to the evolve services continuously and new services come online at any time in environment and existing service might be unavailable. It's also important to discover service Compatible to the environments at right time as well right place.
- Heterogeneity: Many middleware support one or two types of resources but it's not sufficient for IoT heterogeneous resources. For application level heterogeneity is limited because of manual process of service registration, composition and discovery. Above all middleware are limited to resource constrained and application heterogeneity also its limited for small scale network only. New middleware proposed for mange the resource and autonomous application level heterogeneity for ultra large scale network.
- Interoperable: Interoperability challenges for middleware started form initial level like heterogeneous devices having different specification are work together and share the information in system. Network interoperability supported by all above existing middleware but semantic and syntactic middleware is not fully supported by all middleware. Semantic interoperability is big challenge in IoT because heterogeneity of devices and their different standards. Service oriented middleware is limited for syntactic middleware.
- Self adaptive: Middleware should be self adaptive. Middleware consider the network conditions and persistence links availability is still a challenge in IoT.
- Soft Data: The applications need real time data for accurate analysis. This aspects of IoT application have not been addressed in above mention middleware.
- Intelligence: Intelligence is one of the most challenges for middleware. Above discuss middleware are intelligent for resource and service discovery. Middleware

Table 2 Relationship between middleware and it's features

	Scalability	Durability	Flexibility	Interoperability	Resource and data management	Real-time	Intelligence and context awareness	Security and privacy	Distributed
Hydra [11]	AL, NL	No	DA	NW, SI, SE	Yes	SRT	Yes	S, NP	Yes
Rimware [15]	AL, NL	No	DA	NW	Yes	NO	NO	NO	Yes
GSN [13, 14]	AL, NL	Yes	DA	NO	NO	SRT	NO	S, P	Yes
TinySOA [16]	AL, NL	Yes	NI	NW	Yes (partially)	SRT	NO	NO	Yes
DropLock	AL, NL	Yes	DA	NO	No	NO	NO	S, P	Yes
CHOReOS [18, 19]	AL, NL	Yes	DA	NW, SE	Yes	NO	Yes	NO	Yes
MobIoT [22]	AL	Yes	NI	NW	Yes	SRT	Yes	S, P	Yes
Xively [23]	AL, NL	Yes	DA	NW	Yes (partially)	HRT	Yes	S	Yes
CarrIoTs [25]	AL, NL	Yes	DA	NW	Yes (partially)	HRT	No	S.NI	Yes

Legend *AL* application level, *NL* network level, *DA* dynamic, *NW* network, *SI* syntactic, *SE* semantic, *SRT* soft real time, *HRT* hard real time, *S* security, *P* privacy, *NI* no information

should be intelligent to detect the failures occurs in middleware or it's an isolate mode. Intelligent challenges in terms of failures and isolate of middleware.

- Security and Privacy: Security, trust and privacy is not fully resolved in RFID, WSNs, M2M Communication technologies. For IoT Application having basic authentication solutions but it's not sufficient in middleware. So Security of middleware architecture is open challenge, Along with security data privacy is to be consider as challenges of middleware.

5 Conclusion

Internet of things will play key role in near futures. With platform of IoT we can connect peoples with billions of devices. In this paper we show six different types of devices from low power consumption to high power consumption. Middleware is necessary component for developing application and services in Internet of things. The middleware cab be categorize depends on design approach. In this paper we discussed heavy weight service oriented middleware, it's requirements and features of it. We discussed nine different service oriented middleware architecture with their characteristics. We also present the summary of all existing middleware with key features. We outline the key challenges and research issues like heterogeneity, service discovery, interoperability and real-time data of service oriented middleware. Some features and related issues remain unexplored like platform interoperability, scalability, security and privacy in middleware. There is significant scope for future work in area of service oriented middleware.

References

1. K. Paridel, E. Bainomugisha, Y. Vanrompay, Y. Berbers, W.D. Meuter, Middleware for the internet of things, design goals and challenges. Electron. Commun. EASST**28** (2010)
2. S. Bandyopadhyay, M. Sengupta, S. Maiti, S. Dutta, A survey of middleware for internet of things, vol. 162, in *Recent Trends in Wireless and Mobile Networks* (Springer, New York, NY, USA, 2011) pp. 288–296
3. J. Gubbi, R. Buyya, S. Marusic, M. Palaniswami, Internet of things: a vision, architectural elements, and future directions. Future Gener. Comput. Syst.**29**(7), 1645–1660 (2013)
4. A. Farahzadi, P. Shams, J. Rezazadeh, R. Farahbakhsh, Middleware technologies for cloud of things-a survey. Dig. Commun. Netw. ISSN 2352 8648 (2017)
5. C. Perera, A.B. Zaslavsky, P. Christen, D. Georgakopoulos, Context aware computing for the internet of things: a survey. IEEE Commun. Surveys Tuts. **16**(1), 414–454 (2013)
6. M.A. Razzaque, M. Milojevic-Jevric, A. Palade, S. Clarke, Middleware for internet of things: a survey. IEEE Internet Things J. **3**(1), 70–95 (2016)
7. S. Bandyopadhyay, M. Sengupta, S. Maiti, S. Dutta, Role of middleware for internet of things: a study. Int. J. Comput. Sci. Eng. Survey **2**(3), 94–105 (2011)
8. A.H. Ngu, M. Gutierrez, V. Metsis, S. Nepal, Q.Z. Sheng, IoT middleware: a survey on issues and enabling technologies. IEEE Internet Things J. **4**(1), 1–20 (2017)
9. L. Atzori, A. Iera, G. Morabito, The internet of things: a survey. Comput. Netw. **54**(15), 2787–2805 (2010)

10. A. Katasonov, O. Kaykova, O. Khriyenko, S. Nikitin, V.Y. Terziyan, Smart semantic middleware for the Internet of Things, in *Proceedings of ICINCO-ICSO*, pp. 169–178 (2008)
11. M. Eisenhauer, P. Rosengren, P. Antolin, Hydra: a development platform for integrating wireless devices and sensors into ambient intelligence systems, in *The Internet of Things* (Springer, New York, NY, USA, 2010), pp. 367–373
12. Linksmart Middleware Portal [Online]. Available https://linksmart.eu
13. K. Aberer, M. Hauswirth, A. Salehi, A middleware for fast and flexible sensor network deployment, in *Proceedings of 32nd International Conference on Very Large Data Bases* (Seoul, South Korea, 2006), pp. 1199–1202 [Online]. Available https://dl.acm.org/citation.cfm?id=1182635.1164243
14. Global Sensor Networks [Online]. Available: https://lsir.epfl.ch/research/current/gsn/ (2004)
15. C. Huo, T.-C. Chien, P.H. Chou, Middleware for IoT-cloud integration across application domains. IEEE Des. Test **31**(3)
16. E. Avils-Lpez, J. Garca-Macas, TinySOA: a service-oriented architecture for wireless sensor networks. Serv. Oriented Comput. Appl. **3**(2), 99–108 (2009)
17. T.L. Vinh, S. Bouzefrane, J.-M. Farinone, A. Attar, B.P. Kennedy, Middleware to integrate mobile devices, sensors and cloud computing. Proc. Comput. Sci. **52**, 234–243(2015)
18. T. Teixeira, S. Hachem, V. Issarny, N. Georgantas, Service oriented middleware for the internet of things: a perspective, in *Proceedings of 4th European Conference on Towards Service Based Internet*, pp. 220–229 (2011)
19. S. Hachem, Service-oriented middleware for the large-scale mobile Internet of Things. M.S. theses, Université de Versailles-Saint Quentin en Yvelines, Versailles, France, Feb 2014
20. A. Ben Hamida et al., *The Future Internet*, pp. 81–92 (2012)
21. A. Ben Hamida et al., *Integrated CHOReOs Middleware—Enabling Large-Scale, QoS-Aware Adaptive Choreographies*, Sept 2013
22. MobIoT [Online]. https://projects.ow2.org/mobiot/
23. Xively [Online]. Available https://xively.com/
24. S. Satyadevan, B. Kalarickal, M. Jinesh, Security, trust and implementation limitations of prominent IoT platforms, vol. 328, in *Proceedings of Frontiers Intelligent Computing Theory Application (FICTA'14)*, pp. 85–95 (2015)
25. Carriots [Online]. Available https://www.carriots.com/
26. U. Bellur, N.C. Narendra, S.K. Mohalik,AUSOM: autonomic service-oriented middleware for IoT-based systems, in *2017 IEEE World Congress on Services (SERVICES)* (Honolulu, HI, 2017), pp. 102–105
27. V. Issarny et al., Service-oriented middleware for the future internet: state of the art and research directions. J. Internet Serv. Appl. **2**(1), 23–45 (2011)
28. N. Mohamed, J. Al-Jaroodi, I. Jawhar, S. Lazarova-Molnar, S. Mahmoud, SmartCityWare: a service-oriented middleware for cloud and fog enabled smart city services. IEEE Access **5**, 17576–17588 (2017)

Colon Cancer Classification Using Binary Particle Swarm Optimization and Logistic Regression

Nimrita Koul and Sunilkumar S. Manvi

Abstract Cancer is a genetic disease characterized by various types of changes in the activity of genes at cell level. The molecules produced by activity of genes are captured by the expression values of genes as obtained by genome sequencing technologies like microarray technology. This gene expression data can be computationally analyzed to detect abnormalities in one or more genes of an organism. Thus, diseases like cancer can be successfully diagnosed and classified into subtypes by analyzing gene expression datasets. Since this data is very high dimensional and very few genes are associated with incidence of a type of cancer. Therefore, many methods have been proposed for selecting subset of most relevant genes for classification task. In this paper, we have proposed a method for feature selection from cancer gene expression data based on binary particle swarm optimization and logistic regression. The ensemble consists of binary particle swarm optimization and logistic regression classifier. The proposed method was compared to existing feature selection-classification approaches and the results show the proposed method to perform better with respect to classification accuracy and learning rate.

Keywords Cancer classification · Binary particle swarm optimization · Logistic regression · Feature selection · Classification accuracy

N. Koul (✉) · S. S. Manvi
REVA University, Bangalore, India
e-mail: emailnk1@gmail.com

S. S. Manvi
e-mail: ssmanvi@reva.edu.in

J. Tavares et al. (eds.), *Emerging Technologies in Data Mining and Information Security*, Lecture Notes in Networks and Systems 164,
https://doi.org/10.1007/978-981-15-9774-9_20

1 Introduction

1.1 DNA Microarray Technology

Central dogma of systems biology postulates the process of protein synthesis to consist of two steps. First, the process of transcription manufactures mRNA from DNA and then mRNA is translated into amino acids which are building blocks of proteins. These mRNA levels manufactured within a cell at a particular time are measured by gene expression values of a genome at that moment. A DNA microarray [12] is a glass or silicone slide containing thousands of single strands of cDNA or oligonucleotides. It helps to measure the expression levels of all the genes present in that sequence. It reveals the information about which genes are activated and which ones are deactivated under the diseased condition of the sample. Though DNA microarray provides the expression values of thousands of genes together, yet the number of experiments or the number of samples for which we get this data is small as compared to the number of genes. It can been established from clinical studies that very few genes actually influence development of most of the cancers in humans. Therefore, in computational research, as well, gene subset selection is an important task as a prelude to further tasks of classification, clustering, or drug target discovery. In this paper, we have presented an evolutionary computation approach for selection of small subset of most relevant genes which influence the incidence of colon cancer. We have used colon cancer gene expression dataset downloaded from [11]. The dataset contains expression values of 2000 genes and 62 samples. There are two classes in the data: tumorous and normal.

1.2 Feature Selection

Feature selection [1] is an important computational step that involves eliminating the irrelevant features and retaining the subset of most relevant features which maximize an objective function. For machine learning tasks like classification, the objective function generally is the classification accuracy of a classifier module. The methods for feature selection can consider one feature at a time to observe the influence of the feature on the objective function, or they can consider groups of features simultaneously. The former method is called as univariate feature selection while the later method is called as multivariate feature selection. A drawback of univariate methods is that they ignore the interrelations and interdependencies among features; however, univariate methods are simpler and more intuitive. Prominent methods for univariate feature selection include signal-to-noise ratio, information gain, mutual information, and correlation coefficient. Multivariate feature selection which analyzes groups of features at a time is divided into three categories: filter methods, wrapper methods, and embedded methods. Filter feature selection methods do not depend on machine learning algorithm, they rank the features according to their influence on the target variable as measured by information theoretic or statistical measures like correlation, ANOVA, chi-square, linear discriminant analysis, variance, *z*-score [2], etc. Wrapper methods have a machine learning algorithm tightly coupled with them. Based on the accuracy of this model with current set of features, the features are dropped or retained. For each set of input features, the classification accuracy is noted and the subset that gives maximum classification accuracy is returned as the optimal feature subset.

1.3 Particle Swarm Optimization Algorithm

Particle swarm optimization (PSO) [3] is an evolutionary stochastic optimization algorithm inspired by nature. The algorithm mimics the collective intelligent behavior of groups of natural organisms like birds, bees, fish, etc., known as swarms. The swarms exhibit a cooperative and intelligent behavior in activities like search for food sources. Every member of a swarm shares its knowledge with others and receives knowledge from others to reach a common optimum goal. The algorithm searches the solution space for the optimum value of the objective function. There have been multiple variants of particle swarm optimization created by researchers to suit the different types of computational problems. Original PSO was meant to handle optimization problems with continuous variables, limited to small number of dimensions and was prone to locally optimum values. Five important criteria in the computational adaptation of swarm theory are: proximity, quality, diverse responses, stability, and adaptability. Proximity means that the members of swarm should be able to perform computations with respect to time and space. Quality mean that members of the swarm should be able to measure and respond to changes in the search space, by diverse response we mean that the swarm members should be able to search in separate areas of search space and respond differently, stability and adaptability are trade-offs where the members should maintain a steady state but still be able to change behavior if necessary. Particle swarm optimization algorithm therefore is a heuristic search based on the behavior of the swarms. Each member of the swarm is known as a particle and represents an individual in the swarms of organisms. Each particle is a potential solution to the search problem and has the ability to memorize its velocity, optimal position of itself and of the swarm. In binary PSO [4] applied for feature selection from gene expression data, the '*n*' genes in the dataset are represented by a binary bit string of length '*n*', Xi is a positional vector that represents the ith subset of genes. If value of a bit string is 1, the corresponding gene is selected else not.

Rest of the paper is organized as follows—Section on literature survey covers the related work in the field of feature selection in gene expression data, section on 'Proposed Approach' covers the details of the approach proposed by us for feature selection and classification from colon cancer gene expression data. Section called 'Results and Discussion' covers the experimental results and their interpretation, and last section concludes the paper.

2 Literature Survey

Many variants of PSO and many other methods of feature selection have been proposed in literature. In this section, we present a brief survey of the methods proposed in related works. In [1], authors have proposed a nature-inspired method called particle swarm methodology for optimization of nonlinear functions. They have presented two paradigms, discussed their performance gains, and benchmark testing. In [2], authors have proposed a modified particle swarm algorithm for cancer gene selection. They used sigmoid function to increase probability of a bit being zero. The testing was performed on ten gene expression datasets. In [3], authors have proposed an improved algorithm over the binary particle swarm optimization; they have introduced a change in the method of changing velocity of the particles. In [4], authors have developed a hybrid of tabu search and particle swarm optimization such that tabu search acts as local search procedure to overcome local optima. They tested the method on three datasets. In [5], authors have presented an algorithm which is a hybrid of support vector machine and discrete particle swarm optimization for classification of cancers. In [6], authors have presented a two-stage hybrid algorithm that combines correlation coefficient and binary particle swarm optimization to perform dimensionality reduction on the cancer gene expression dataset for the purpose of cancer classification in to two or more classes. They have used Naive-Bayes classifier to prevent early convergence at the locally optimal solutions. They tested their solution on 11 datasets. In [7], authors have proposed an algorithm that is a hybrid of imperialist competition algorithm for global search and tabu search at local level for gene selection from cancer gene expression data sets with 12,600 genes. In [8], authors have proposed a modified binary particle swarm optimization procedure; the K-NN algorithm is used as evaluator to evaluate fitness of each gene for the cancer classification problem. They obtained good classification accuracy. In [9], authors have used a hybrid algorithm combining correlation-based approach to feature selection with Taguchi genetic algorithm to gene selection from eleven cancer gene expression data sets. In [10], there is the detail of logistic regression algorithm for classification tasks in sklearn package of Python.

3 Proposed Method

In this section, we present our approach for gene selection from cancer gene expression datasets. The method has been tested on colon cancer dataset containing 2000 genes and 62 samples. The dataset was downloaded for [11]. Our method is an ensemble of binary particle swarm optimization and logistic regression. The implementation was carried out on a Windows 10 machine using Python 3.7. We used train–test split of 80–20% for training the logistic regression classifier. The prediction accuracy of each feature was taken as its fitness function. It is this fitness function that was optimized by the classifier. The experiment was repeated 20 times the average

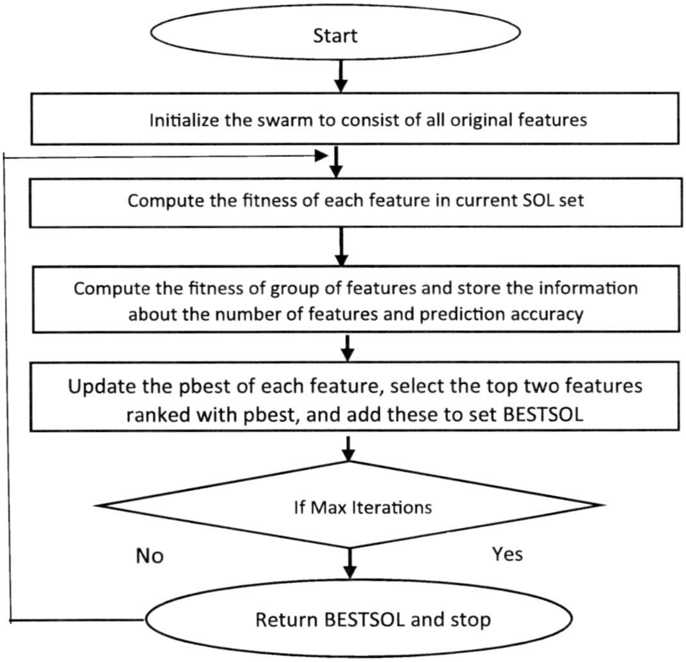

Fig. 1 Schematic representation of the particle swarm optimization for feature subset selection

values are reported for the performance parameters. The parameters are classification accuracy and learning rate. The features obtained from our method were also evaluated using two classification algorithms viz. Naïve Bayes and SVM with RBF kernel and the comparative performance was observed to be above 94%. The detailed schematic of the feature selection part using binary particle swarm optimization is shown in Fig. 1.

4 Results and Discussion

In this section, we present the results of the experiment and their analysis. The experiment was carried out on the colon cancer dataset with 80:20 train–test split of the data set with fivefold cross validation. The logistic regression classifier was used to evaluate the performance of the genes selected by BPSO. The results show that our methods gives very good accuracy on the chosen dataset. The features obtained from our method were also evaluated using two classification algorithms viz. Naïve Bayes and SVM with RBF kernel and the comparative performance was observed to be above 94%. Figures below shows the performance parameters for the results of the experiment.

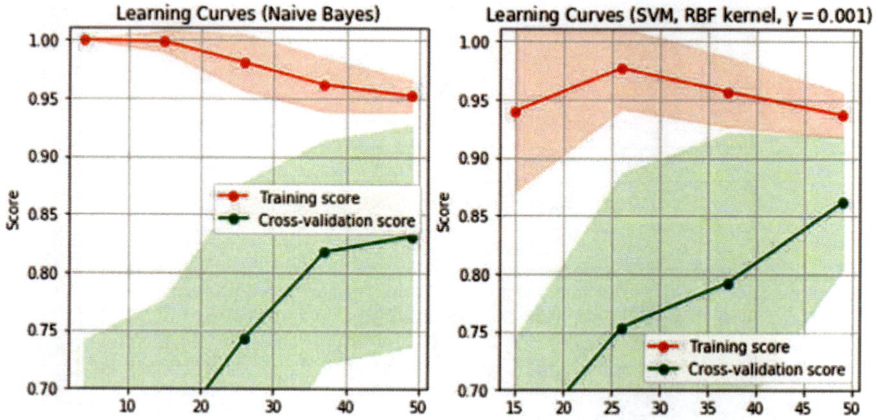

Fig. 2 Learning curve for the feature selected using BPSO with Naïve Bayes and SBM classifiers

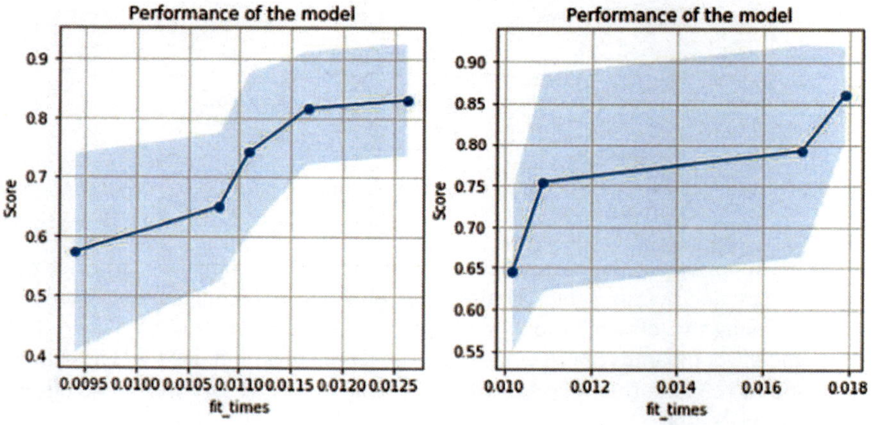

Fig. 3 Comparison of model performance for the features selected by BPSO with NB and SVM

Figures 2 and 3 show the comparison of learning rate and classification accuracy obtained when the features selected by our algorithm were used to train the classifiers Naïve Bayes and SVM. The results show a classification accuracy of more than 95% with SVM.

5 Conclusion

Evolutionary algorithms like particle swarm optimization and its variants like binary particle swarm optimization, discrete particle swarm optimization are metaheuristic global search algorithms which are used to search for optimum solution to a problem. In this paper, we used logistic regression as evaluator for binary particle swarm optimization algorithm to use it as feature selection procedure for gene selection from cancer gene expression data sets. The results showed a classification of above 90% and a good learning rate with average training time of 0.09 ms. The method is a promising approach to the problem of feature selection for cancer classification. In the future, the method can be applied on more number of datasets.

Acknowledgements Authors would like to thank Department of Science and Technology (DST), Government of India, for supporting this research work under the grant DST-ICPS 2019.

References

1. R.C. Eberhart, J. Kennedy, A new optimizer using particle swarm theory, in *Proceedings of the 6th International Symposium on Micro Machine and Human Science* (Nagoya, Japan, Mar 13–16, 1995), pp. 39–43
2. M.S. Mohamad, S. Omatu, S. Deris, M. Yoshioka, A. Abdullah, Z. Ibrahim, An enhancement of binary particle swarm optimization for gene selection in classifying cancer classes. Algorithms Mole. Biol. **8**(15) (2013). https://www.almob.org/content/8/1/15
3. M.S. Mohamad, S. Omatu, S. Deris et al., Particle swarm optimization for gene selection in classifying cancer classes. Artif. Life Robotics **14**, 16–19 (2009). https://doi.org/10.1007/s10015-009-0712-z
4. Q. Shen, W.M. Shi, W. Kong, Hybrid particle swarm optimization and tabu search approach for selecting genes for tumor classification using gene expression data. Comput. Biol. Chem. **32**(5), 3–60 (2008)
5. Q. Shen, W.M. Shi, W. Kong, B.X. Ye, A combination of modified particle swarm optimization algorithm and support vector machine for gene selection and tumor classification. Talanta **71**(4), 1679–1683 (2007). https://doi.org/10.1016/j.talanta.2006.07.047
6. I. Jain, V.K. Jain, R. Jain, Correlation feature selection based improved-binary particle swarm optimization for gene selection and cancer classification. Appl. Soft Comput. **62**, 203–215 (2018). https://doi.org/10.1016/j.asoc.2017.09.038
7. S.A. Wang, W. Kong, W. Zeng, X. Hong, Hybrid binary imperialist competition algorithm and tabu search approach for feature selection using gene expression data. Biomed. Res. Int. 9721713 (2016). https://doi.org/10.1155/2016/9721713
8. L.Y. Chuang, H.W. Chang, C.J. Tu et al., Improved binary PSO for feature selection using gene expression data. Comput. Biol. Chem. **32**, 29–38 (2008)
9. L.Y. Chuang, C.H. Yang, K.C. Wu, C.H. Yang, A hybrid feature selection method for DNA microarray data. Comput. Biol. Med. **41**(4), 228–237 (2011). https://doi.org/10.1016/j.compbiomed.2011.02.004
10. F. Pedregosa et al., Logistic regression. J. Mach. Learn. Res. **12**, 2825–2830 (2011)
11. https://datam.i2r.astar.edu.sg/datastes/krbd/ColonTumor/ColonTumor.html
12. T.R. Golub, D.K. Slonim, P. Tamayo et al., Molecular classification of cancer: class discovery and class prediction by gene expression monitoring. Science **286**(5439), 531–537 (1999). https://doi.org/10.1126/science.286.5439.531

Accident Detection and Avoidance Using Spot Sensor Technology

Prathamesh R. Mahadik, Praj Shete, Shrenik Kabade, Shlok Gaikwad, and N. S. Shirsat

Abstract In this paper, we propose a smart accident detection system depending upon present environmental condition. With the increase in number of accidents daily, human lives have been claimed. Due to which we have come up with a solution that may help the government organizations to apply some brakes on these accidents. Hence, our project will focus on present environmental parameters and will analyze the data generated regarding these parameters to get some useful information so that give us the real-time data some actions may be taken in each situation. To capture this data at that particular spot sensor technology will be used, this will give us the real-time data. Our proposed project includes GSM, GPS, smartphones mobile applications and vehicular ad-hoc network technologies. Our system gives timely information of emergency services regarding location of accident which helps to save valuable lives. When an accident has occurred, these sensor data will be sent to nearest hospital, police station also. The data that is continuously being captured will also help the other citizens to be aware of and ignore to go through those conditions which are terrible and take some another route. With this, the occurrence of an accident will not completely stopped, but some amount of possibility may be avoided as the data which has been mined will be useful for some further analysis and different measures can be taken by the government to take some necessary actions in the situations where the number of accidents have crossed the line. Reports will help the government organization to take further actions.

P. R. Mahadik (✉) · P. Shete · S. Kabade · S. Gaikwad · N. S. Shirsat
Pune Vidyarti Gruha College of Engineering, Pune, India
e-mail: prathamesh98mahadik@gmail.com

P. Shete
e-mail: prajshete17@gmail.com

S. Kabade
e-mail: shrenikk98@gmail.com

S. Gaikwad
e-mail: shubhamgaikwad204@gmail.com

J. Tavares et al. (eds.), *Emerging Technologies in Data Mining and Information Security*, Lecture Notes in Networks and Systems 164,
https://doi.org/10.1007/978-981-15-9774-9_21

Keywords Spot sensor technology · Data mining · Smart accident detection · Real-time data generation

1 Introduction

In recent years, road accidents are nowadays turning out to be fatal for one's life, their evaluation is considered on the basis of fate, and the actual fact or parameters due to which the accident has taken place are often ignored. A set of variables which are discrete in nature are considered for a road accident. To obtain various goals, we will be using data mining techniques. Here, classification of raw data can categorize the accident into different categories.

With the increase in number of accidents daily, according to 2018 reports, there have been over 6 lakh accidents occurred out of which around 2 lakh are major, and around 80,000 human lives have been claimed [1, 2]. Due to which, we have come up with a solution that may help the government organizations to apply some brakes on these accidents. Some important and productive results are obtained by using data mining [3] as the technique which analyzes the road accidents. To measure the intensity and severity of the accident feature selection, clustering and classification techniques provide productive results. A challenge waits in front of the local emergency departments when a certain accident takes place [4]. As far as many injured could lose their lives and since no on-site medical assistance has been provided promptly as a result of: (1) delay in reporting the accident, (2) no proper geographic location and (3) lack of number of casualties, hence the need for an intelligent and automated system to face this severe issue is mandatory. In some cases, it was observed that no major harm was done to the casualty, but due to late arrival of the rescue team, the casualty had turned severe [5–7]. People often lose their lives when in transit, and it was observed that it happened only when the ambulance delayed its arrival.

In our Country, the number of deaths could have been reduced if the rescue team would have been on time. The rescue team should not be blamed since the location of accident was not provided to them on time [8, 9]. In the event of a road accident, hence through real-time traffic accident detection system (RTTADS), proper authorities can be informed intelligently using wireless interface technology so that no delay in the entire system could take place [10, 11].

So, our proposed work is to capture the real-time road accident data and evaluate it using data mining algorithms to predict patterns in accidents and extract information useful for assistance of road accident.

In this paper, we present a novel accident detection and avoidance using spot sensor technology, and proposed and existing systems are discussed in Sect. 2. The analysis and design of proposed system is presented in Sect. 3. However, results of proposed system are discussed in Sect. 4, and Sect. 5 discusses the conclusion of this system.

2 Proposed and Existing System

2.1 Existing System

Auto accident App, developed by Platinum Peak LLC, is a mobile application that provides assistance to the casualties when an accident occurs. The main drawback of this system is that the system is not automated and provided manual reporting of the mishap that took place. In some cases the casualty may not be in healthy condition to manually enter the details.

Auto Accident App, developed by the Murphy Battista, is an application that features time saving forms that allows the users to collect the information but again the system not being automated is the major drawback.

2.2 Proposed System

In our system, we will be collaborating the smartphone, mobile application the sensor (IoT) technology to capture this real-time data. A small car model will be placed in which different sensors namely photo diode (light sensor), humidity ensor (climatic conditions), shock sensor (to detect collision), speed sensor, GPS unit to track the location will be placed from which the data will be collected.

As soon as the collision is detected, the sensors which will be providing values continuously, the data corresponding to the highest value of the impact sensor will be notified to the different emergency services such as hospitals, police stations and highway authorities.

3 Analysis and Design of Proposed System

3.1 Analysis

3.1.1 Capturing the Data Using Spot Sensor Technology

In our system, we will be collaborate the Smartphone, Mobile Application the Sensor (IoT) technology to capture this real time data. A small car model will be placed in which different sensors namely photo diode (light sensor), humidity sensor (climatic conditions), shock sensor (to detect collision), speed sensor, GPS unit to track the location will be placed from which the data will be collected. As soon as the collision is detected, the sensors which will be providing values continuously to our cloud, the data corresponding to the highest value of the impact sensor will be compared with threshold value of the impact sensor, if it satisfies, then accident will be detected.

3.1.2 Notifying the Data to Emergency Departments

As soon as the collision is detected, the sensors which will be providing values continuously, the data corresponding to the highest value of the impact sensor will be notified to the different emergency services such as hospitals, police stations and highway authorities. There will be an Android app for different emergency departments where the admin will login himself/herself and then the notification of accident (if any) will be displayed with the location of the accident.

3.1.3 Gathering the big data for further analysis

(A) Clustering Algorithms
There are various types of clustering algorithms are available in the research papers. The clustering algorithm divides the available data into different groups or clusters, in such a manner that the objects within a cluster are similar to each other, while objects from other clusters are different from each other. K means, latent class clustering (LCC) and hierarchical clustering technique (e.g., Ward method, single linkage, complete linkage, etc.), can be used in road accident detection analysis. An enhanced version of K means algorithm is K-modes clustering. To calculate number of clusters, LCC clustering technique is preferable. As LCC algorithm is consistently used for the purpose of road accident data analysis. But as the data consists of variety of categorical attributes, so LCC has practically some limitations, and hence, it can be minimized by using K-modes algorithm. As K-mode algorithm is good for the selection of K value for the analysis purpose. LCC and K-modes have same efficiency in retrieving the required cluster structure, are mentioned in the earlier Monte Carlo simulations. For generation of locally minimal clustering result, K-modes are much faster and efficient than that of LCC algorithm.

(B) Association Rules
The extraction of hidden relations between different attributes in a large data set is given by association rule. Association rule mining gives rise to the set of rules that interprets the underlying patterns in the respective data set. The associativity of any two characteristics of accident is decided by the repetition of their occurrence together in the respective data set. The rule A associate with B ($A \rightarrow B$) indicates that if A occurs then B will also occur.

3.2 *Design*

3.2.1 Architecture Diagram

Here, Fig. 1 shows the architecture diagram of project. According to architecture diagram, system consists of three main components:

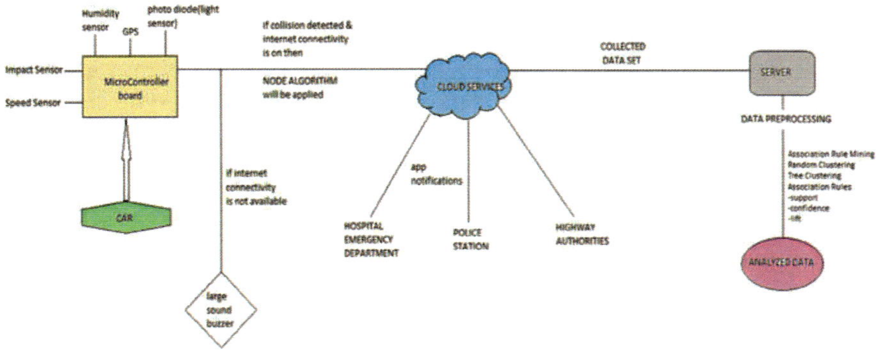

Fig. 1 Architecture diagram

1. Microcontroller or Node-MCU: It is a small minicomputer which can handle operations like a normal computer. To this microcontroller unit, we have attached different sensors including the DHT, GPS unit, impact sensor and the buzzer which will provide us the real-time data with duration of 5 s.
2. Cloud Server/Firebase: The collected data will directly be uploaded to the cloud. Here, we are using the Firebase as our cloud server which is free, user friendly and even provides security.
3. Database Server/MySQL: The data uploaded by the microcontroller and the sensors to the Firebase cloud will be downloaded to the MySQL server data management system where a query will be fired against this data and the accident can be identified by applying parameter checks on the impact sensor.

4 Proposed System Results

4.1 Testing Strategy

We followed the following testing strategy for the project.

1. Unit Testing for every feature—Functional testing to verify if functional requirements are satisfied.
2. Integration Testing—Integrating every module together and testing their cohesive behavior for pitfalls and errors.

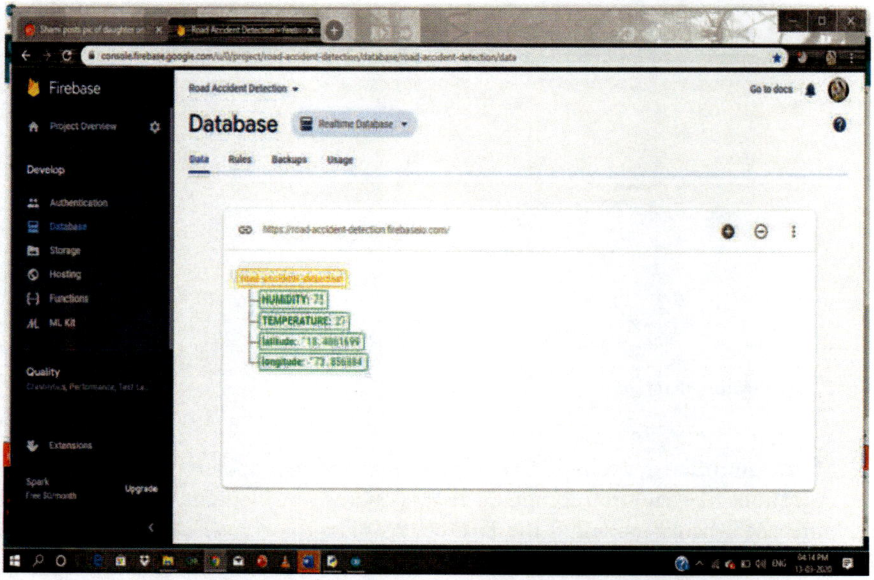

Fig. 2 Capturing uploading over the Firebase cloud

4.2 Results/Evaluation/Testing Description Per Module

4.2.1 Capturing the Data Using Spot Sensor Technology

1. The microcontroller used is Node-MCU. We selected Node-MCU over Raspberry Pi for its size, feasibility, overcome of heating problem and wireless Wi-Fi antenna.
2. We tested whether the data uploaded to the cloud by the sensors is precise and is getting uploaded in timely manner.

The details regarding the data being uploaded to the Firebase cloud are shown in Fig. 2. The attributes are temperature, humidity and GPS attributes to trace the location.

4.2.2 Notifying the Data to Emergency Departments

1. We checked whether the data uploaded to the cloud server is being downloaded to our local sever Xampp in a timely manner.
2. We checked the working of our impact sensor, whether the sensor is turning 'HIGH' on any impact or not.
3. We checked whether that particular accident record is getting separated from the rest of the data.

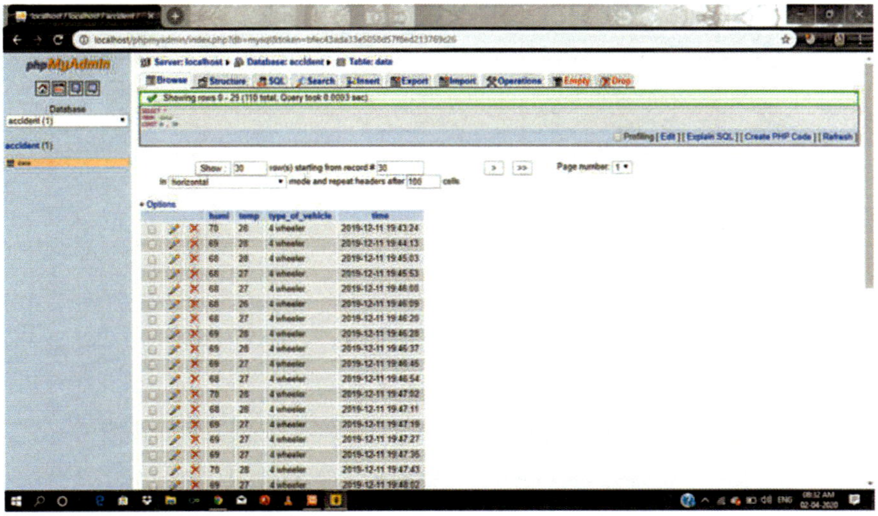

Fig. 3 Data captured in MySQL

4. We checked whether this data is being notified to the application or not.

Figure 3 speaks for the detailed representation of the extracted data. It shows the descriptive tables made in MySQL XAMPP server which gives the information about the attributes and the extracted data contained by these attributes. There is huge amount of data being generated which is being extracted and saved into XAMPP MySQL server, separated into tables. Figure 3 represents the detail representation of data being extracted.

Figure 4 gives the précised schematics of mobile application that represents the data being read by the sensors. This application can be provided to the end user to connect to the server through the corresponding mobile interface and get the details regarding the environment. Same details will be forwarded to the cloud, and as soon as the accident occurs, the details will be sent to emergency departments.

As shown in Fig. 5, the application will send an emergency push notification containing the location received by the GPS module in terms of geographical coordinates (longitude, latitude). This notification will be sent to the registered close relatives of the user and the emergency departments.

5 Conclusion

Our system has presented a technique which detects the accidents through sensors, records and analyzes it automatically using data mining techniques. Mainly, in our system, we used some hardware components such as built-in sensors like GPS

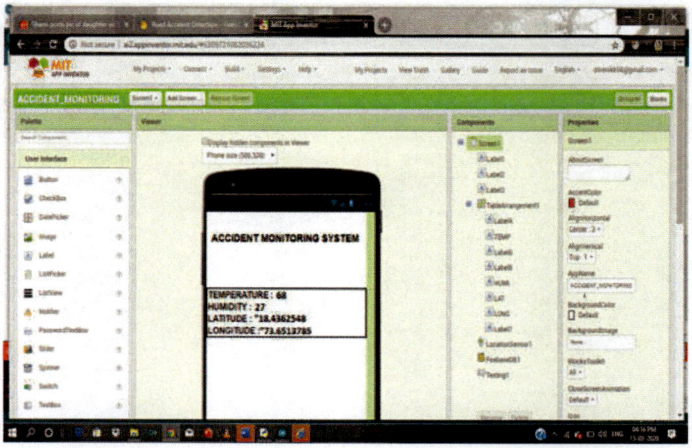

Fig. 4 Notification in mobile application

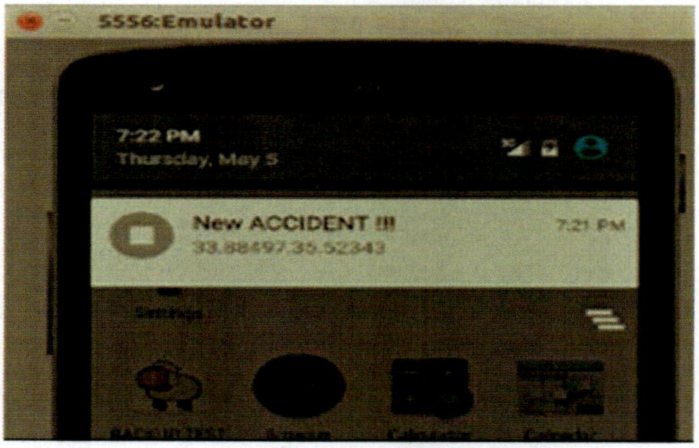

Fig. 5 Notification as a text message

module, DHT sensor, impact sensor which are attached to the vehicle and connected to the Arduino, so that we could record the constraints such as conditions of weather, lighting and vehicle type. These constraints are the base on which we classified the preprocessed data and distributed that data into labels. This classification resulted into creation of the datasets, which are then clustered using random clustering technique. Apriori algorithm is used for the application of association rule mining on the clustered datasets. This algorithm is used to gather semantics of those accident patterns so that it can predict the necessary constraints. The amount of the information which can be revealed is totally dependent upon the amount of the data provided.

Acknowledgements Our special thanks to PVG, Savitribai Phule Pune University (SPPU) Pune, for providing all facilities and software.

References

1. S. Priya, R. Agalya, Association rule mining approach to analyze road accident data, in *Proceeding of 2018 IEEE International Conference on Current Trends toward Converging Technologies* (Coimbatore, India, 2018)
2. H.M. Sherif, M. Amer Shedid Samah, A. Senbel, Real time traffic accident detection system using wireless sensor network, in *International Conference of Soft Computing and Pattern Recognition*, 978-1-4799-5934-1/14/$31.00©2014. IEEE
3. N. Arun Prasath, M. Punithavalli, A review on road accident detection using data mining techniques. Int. J. Adv. Res. Comput. Sci. **9**(2) (2018)
4. E. Nasr, E. Kfoury, D. Khoury, An IoT approach to vehicle accident detection, reporting, and navigation, in *2016 IEEE International Multidisciplinary Conference on Engineering Technology (IMCET)*
5. U. Khalil, T. Javid, A. Nasir, Automatic road accident detection techniques: a brief survey, in *Conference Paper* November 2017. IEEE. https://doi.org/10.1109/ISWSN.2017.8250025. 978-1-5386-1556-0/17/$31.00©2017
6. A. Ajith Kumar, V. Jaganivasan, T. Sathish, S. Mohanram, Accident detection and alerting system using GPS amp; GSM. Int. J. Pure Appl. Math. **119**(15), 885–891 (2018)
7. K. Nandaniya, N.V. Choksi, A. Patel, M.B. Potdar, Automatic accident alert and safety system using embedded GSM interface. Int. J. Comput. Appl. **85**(6), 0975–8887 (2014)
8. H.S. Ahire, M.B. Kamble, P.K. Khade, R.A. Ghare, B.V. Jadhav, Vehicle accident detection and alerting system. Int. J. Res. Appl. Sci. Eng. Technol. (IJRASET) **6**(I), 67 (2018)
9. Prabha C, Sunitha R, Anith R, Automatic vehicle accident detection and messaging system using GSM and GPS modem. Int. J. Adv. Res. Electr. Electron. Instrument. Eng. **3**(7) (2014)
10. M. Nirbhavane, S. Prabha, Accident monitoring system using wireless application. Int. J. Adv. Res. Comput. Eng. Technol. (IJARCET) **3**(4) (2014)
11. F.B. Basheer, J.J. Alias, C. Mohammed Favas, V. Navas, N.K. Farhan, C.V. Raghu, Design of accident detection and alert system for motor cycles

Classification of Focal and Non-focal EEG Signal for Epileptic Seizure Detection with Entropy Features Using KNN Classifier

N. Samreen Fatima, M. K. Mariam Bee, Abhishek Bhattacharya, and Soumi Dutta

Abstract Electroencephalography is the recording of the brain activity. It is used to diagnose the diverse disease conditions in the brain. In case of epilepsy, a specific part of the brain is affected. The EEG recorded from the affected part of the brain is called as focal EEG (FEEG), and the EEG recorded from the other portion is called as non-focal EEG (NFEEG). In this paper, an automatic method to classify the EEG signal has been proposed. Bern Barcelona database has been used in this method. Entropy features like approximate entropy (ApEn), sample entropy (SampEn), fuzzy entropy (FuzzyEn), and Reyni entropy (ReEn) features are analyzed. In this method, KNN classifier has been used to get the highest accuracy.

Keywords EEG · FEEG · NFEEG · ApEn · SampEn · ReEn · FuzzyEn

1 Introduction

Epilepsy is the persistent brain disorder which can affect anyone at any age [1]. It is defined by the onset of seizures in the brain. Seizures have been caused due to the electrical discharge of the brain activity. Epilepsy is a neurological disorder which can lead to involuntary convolution in the patients muscle and can lead to loss of consciousness.

N. Samreen Fatima (✉) · M. K. Mariam Bee
Saveetha School of Engineering SIMATS, Chennai, India
e-mail: fathimasamreen25@gmail.com

M. K. Mariam Bee
e-mail: mariamresearch09@gmail.com

A. Bhattacharya · S. Dutta
Institute of Engineering and Management, Saltlake, India
e-mail: abhishek.bhattacharya@iemcal.com

S. Dutta
e-mail: soumi.it@gmail.com

J. Tavares et al. (eds.), *Emerging Technologies in Data Mining and Information Security*, Lecture Notes in Networks and Systems 164,
https://doi.org/10.1007/978-981-15-9774-9_22

In this world, about 40 million people are affected with epilepsy. Electroencephalography (EEG) is widely used to analyze the brain activity. Epilepsy has been clinically analyzed by the EEG. Some people affected with epilepsy become resistant to drugs thus for such patient's removal of affected part of the brain has been suggested in order to get rid of the disease. That portion which is affected with epileptic seizures is called as epileptogenic foci. Nowadays, surgery has become common in the society. Hence, an automatic system to identify the FEEG and NFEEG signal will assist doctor to identify the regions affected and will help in the surgical evaluation of the brain portion. The Bern Barcelona database has been considered in this work and used toward building algorithm for signal classification.

Further analysis can enable doctors to classify the EEG characteristic and to diagnose the disease condition. MRI is used as other modality to diagnose the epilepsy. EEG is considered as a source to diagnose epilepsy due to its low cost. A seizure can be defined as a clinical manifestation which can lead to excessive discharge of neurons in brain cells.

Consequently it leads to irregular brain pattern generation within the brain cells. It has been reported that 20% of the people are affected with generalized epilepsy which is affected to the entire brain and about 60% of the people have been affected by focal (partial) epilepsy which is affected to a very smaller portion of the brain [2], which cannot be treated only with medication. In some cases, it requires isolation of epileptic zones. Some of the treatment clinically requires the removal of the affected portion of the brain.

According to the international system of signal classification, the signal is classified as the partial seizures and generalized seizures which are more dominant. A damaged neuron can result in the indication of the seizures in the brain network. About 75% of the generalized seizures spread randomly across the brain cells. One must understand the implication and generalization of the focal seizure.

Classification of EEG signals is always carried out in two phases, in which the features are extracted in phase 1 which are same fed to machine learning algorithm in phase 2. In the phase 2, the features are extracted in the higher-order moments [3]. The EEG signal is decomposed in both time and frequency domain using wavelet transform.

The aim of this paper is to classify the EEG signals as focal and non-focal EEG signal. This classification of the EEG signals can help the doctors to predict the occurrence of the seizures. The identification of focal and non-focal regions in the brain can clinically help people to find out the regions affected in the brain cells and can also lead to removal of the affected portion depending on the clinical situation of the person's health (Fig. 1).

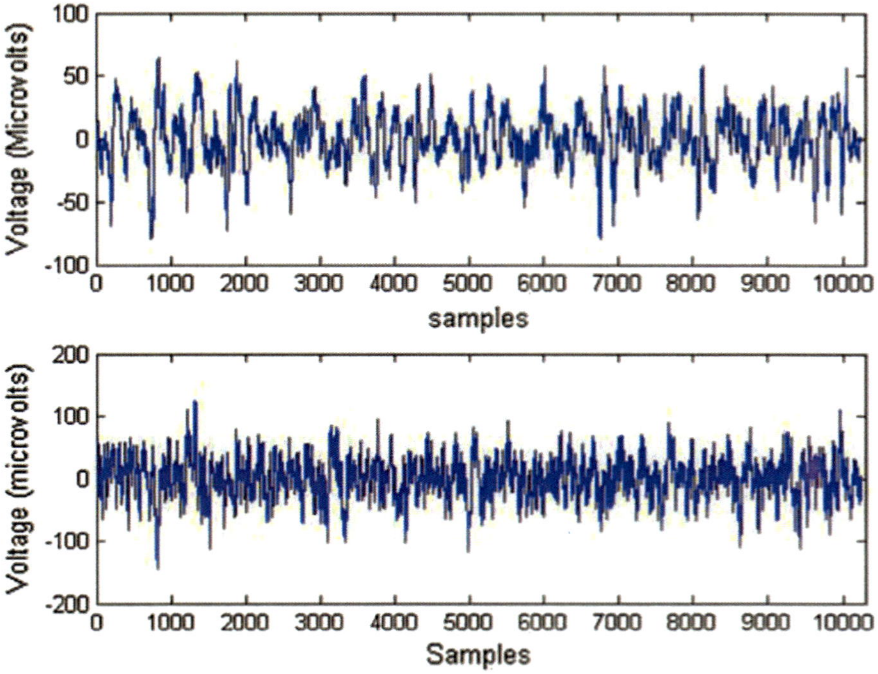

Fig. 1 Represents the sample signal of focal and non-focal EEG signal, respectively

2 Literature Survey:

Acharya et al. [4]: In this study, a set of 23 nonlinear features was extracted in this work. The classifier used in this work is LS-SVM classifier. This classifier has resulted in an accuracy of 87.93% and sensitivity of 89.97% and specificity of 85.89%.

Raghu et al. [5]: In this study, SVM classifier has been utilized in this work. The experimental results occurred are sensitivity 97.6%, specificity 94.4%, accuracy 96.1%, positive predictive rate 92.9%, negative predictive rate 98.8%, and the area under the curve is 0.96.

Rahman et al. [6]: In this study, stand-alone classifier has been utilized in this work. The accuracy obtained is 91.3%. He has also utilized stacked classifier which resulted in the accuracy of 95.2% in his work for the same dataset considered.

Gupta et al. [7]: In this study, the flexible analytic wavelet transform has been considered in this work. In this (FAWT), the signal is subjected to 15 levels of sub-band decomposition. This feature resulted in the specificity of 96% and sensitivity of 93%.

Bhattacharrya et al. [8]: In this study, Tunable Q wavelet transform has been utilized in this work and which achieved an accuracy of 85%.

Bhattacharrya et al. [9]: In this study, empirical mode decomposition wavelet transform has been utilized in this work which yielded sensitivity of 88% and specificity of 92%.

Sriraam et al. [10]: In this study, nonlinear features were extracted based on frequency decomposition which resulted in the accuracy of about 92%.

Kannathal et al. [11]: In this study, ANFIS classifier has been utilized in this work. In this work, different entropy estimators were utilized to yield a promising entropy accuracy of 90%.

Sharma et al. [12]: In this study, EMD technique is used to decompose the EEG signals in narrow band of various IMFs functions of a signal. LS-SVM classifier has been utilized in this work which has resulted in the accuracy of 85%.

Zhu et al. [13]: In this work, it has been noted that the proposed methodology has resulted in the accuracy which is based on the DPE index = 21, which has 18% higher than that of the PE index. This resulted in the classification accuracy of 84%.

3 Methods and Materials

This section describes about the method designed to automatically classify the EEG signal as FEEG and NFEEG signal.

Data: The EEG database of Bern Barcelona database has been utilized in this work. The details about this database can be further referred from [4]. The EEG information is from five patients with fleeting projection epilepsy. Initially we have utilized 50 sets of data of information for assessing our calculation, with the testing rate of procurement of about 512 Hz. We have utilized 3750 sets of FEEG and NFEEG signal dataset of 10,540 examples sets. We took 50 sets flags and frame it into a group of 100 flags to be specific focal and non-central EEG gathering. The idea of the entropy was first started in the year 1803 by mathematician Lazare Carnot who found that the vitality is due to loss of dissemination and grinding [14, 15]. This thermodynamic entropy was then brought into the field of data hypothesis with the name data entropy by Shannon in the year 1948 [16]. In this work, approximate entropy (ApEn), sample entropy (SampEn), fuzzy entropy (FuzzyEn), and Reyni entropy (ReEn) are the features extracted from the given EEG signal.

3.1 Preprocessing

EEG signals are subjected to the preprocessing in order to remove the noise or artifact prior the signal processing. Wavelet transform has been used to preprocess the EEG signal. These signals are subjected to the differencing operation to obtain EEG signals in x–y series before the features are extracted [17].

3.2 Features

3.2.1 Approximate Entropy (ApEn)

Approximate entropy (ApEn) has been widely used in the literature from a long time. It was first proposed by Pincus [18]. Approximate entropy is used to measure the regularity of the signal. This approximate entropy has been widely used by more number of researchers [19–22]. The details of this entropy can be further referred by [23, 24].

$$\text{ApEn} = \sum -\log(1 + A_i)/(1 + B_i)$$

3.2.2 Sample Entropy (SampEn)

Sample entropy is the modified form of the approximate entropy. Sample entropy is used to measure the complexity of the signal. The advantage of using sample entropy is that it does not depend on the length of the data as well it improves the relative consistency. Sample entropy has been widely used in literature, and it is used toward the diagnosis of epilepsy [25, 26].

$$\text{SampEn} = -\log A/B$$

3.2.3 Reyni Entropy (ReEn)

Reyni entropy was first discovered by Alfred Reyni [17] in the year 1961. The Reyni entropy is used to estimate the spectral complexity of the signal for event of f. It has been discussed in [27].

$$\text{Re} = -\log\left(\sum_f pf^2\right)$$

The complexity of the signal has been widely used in the literature to detect the epileptic seizures [7].

3.2.4 Fuzzy Entropy (FuzzyEn)

Fuzzy entropy has been literature from a long time. It is used to measure the complexity of a varying signal. The fuzzy entropy is usually measured to calculate the varying entropy. A detailed about the fuzzy entropy can be referred from [10].

$$\text{FuzzyEn} = \sum x - p(x)\log_2 \pi(x)$$

3.3 *Classifier*

In this proposed system, K-nearest neighbor classifier (KNN) has been used in order to get highest accuracy. The KNN algorithm is a simple easy to implement algorithm. It is one of the machine learning algorithms which can solve both classification and regression problems. It is one of the supervised machine learning algorithms. This algorithm hinges the closest value of proximity or nearing distance. It calculates the nearest distance in the graph and estimates the proximity value. The KNN classifier assumes values which are true enough. The KNN captures the idea of similarity, closeness, distance, and proximity calculating the distance between the points of the graph (Fig. 2).

Fig. 2 KNN algorithm

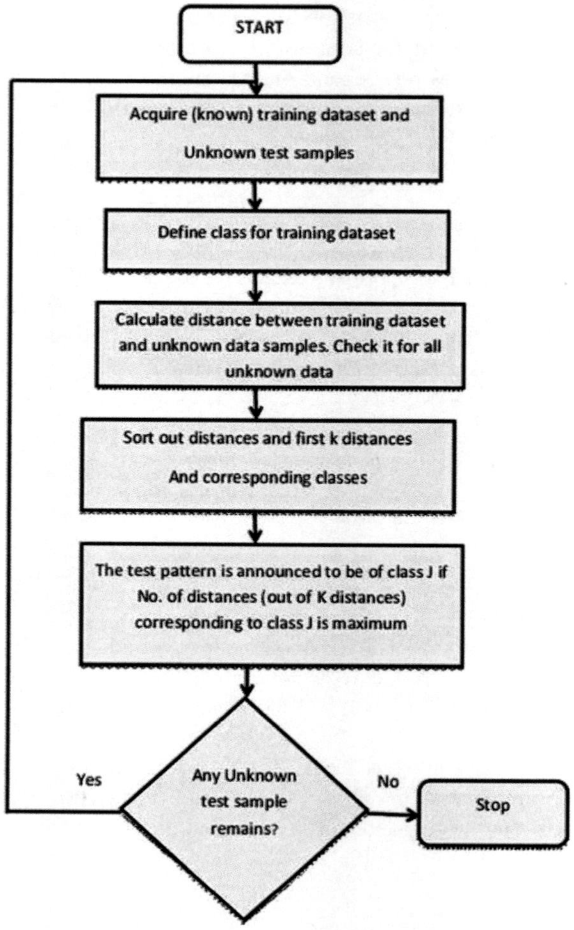

The KNN Algorithm

1. The dataset is loaded.
2. The K value is initialized to a chosen number of neighbors.
3. The distance and the indexed are added to ordered collection of the dataset.
4. The ordered collection of the dataset is sorted according the indices of smallest to the largest distance.
5. The first K entry is picked up from the sorted dataset.
6. The K labels are selected for regression and classification.
7. If want to know the regression labels of K, mean value is returned.
8. If we want to know the classification labels of K, mode value is returned.

Choosing the Right Value for K

To choose the K value directly for the dataset, we can run the KNN algorithm by estimating a few of the K values. We choose the value which gives the less number of errors. This shows the precision of the chosen KNN algorithm. It makes the precise prediction for every dataset.

1. As we decrease the estimation of the K value to 1, the prediction becomes less stable. Imagine $K = 1$ we have an encompassed to few reds and green strands. The question point at this session is supposed to be red but when $K = 1$, it incorrectly predicts the value to be green.
2. Inversely as we increase the K value, the estimation becomes more progressively stable due to majority voting and averaging which is used to make more accurate predictions.
3. In the start, we realize that we are getting more number of errors. It means that we have driven the K value very far.
4. In such situations, we prefer to keep K labels as odd numbers

Advantages

1. This is a very simple and an easy to implement.
2. There is no other reason, just tune a few parameters or make extra suppositions.
3. It is one of the versatile algorithms, which is used for both classification and regression problems.

Disadvantages

The calculation gets fundamentally slow as the number of independent models increase.

3.4 Proposed System Flow Diagram

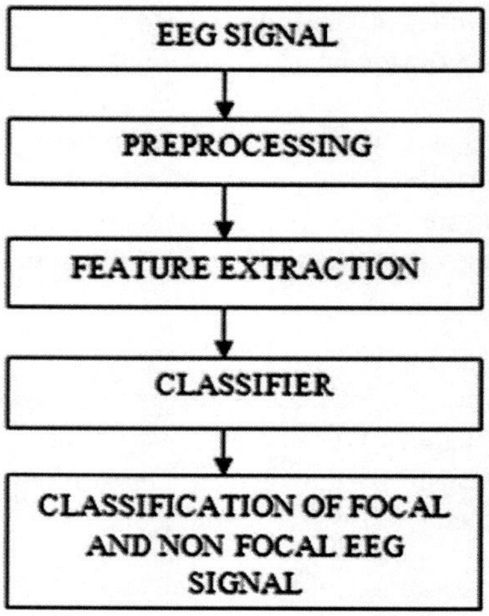

4 Results

In this work, the features considered are ApEn, SampEn, FuzzyEn, and ReEn. These features are applied to the EEG database with the full length of sample of 10,240. These features considered are given as input to the KNN classifier. This helps to classify the EEG signals. Table 1 shows the performance measure of this classifier. This proposed methodology achieved the possible accuracy of about 99% using the KNN classifier.

Table 1 Comparison with the same database with different sets of features and classifiers

Classifier	Accuracy (%)
SVM	87.93
Stacked	92.5
ANFIS	90
Our work	99

5 Discussion

The entropy features considered in this work have been applied to the some databases that have been used by other researchers [28–31], and they have achieved the highest possible accuracies of about 98 and 99.7% [28–31]. In this work, a different combination of entropy features is considered which is applied to the Bern Barcelona database and applied toward the signal classification of EEG signals and achieved a highest accuracy of 99%. This shows that the features considered are good with the relation of signal classification.

The works considered in the literature made use of the EEG database that used intrinsic mode function (IMF) [11, 12]. The entropy features considered for this method worked on the IMFs value of the EEG signal of the database. We have used about 10,240 samples in our work but they have used only about 4096 samples in their work [11]. They have used six different entropy features which were applied to the IMFs function of the EEG database. Our method has resulted in the highest accuracy of 99% which was better than the accuracy of 87% which was achieved in their work [13], and they have used delayed permutation entropy features with SVM classifier which had resulted in the accuracy of 84%. The work resulted in [12] has achieved the highest accuracy with others when compared.

Our method has resulted in the highest possible accuracy of about 99% in the signal classification. This shows that a good improvement of the features considered related to the signal classification. The KNN classifier shows the highest classification accuracy when compared to other classifiers.

6 Conclusion

Epilepsy has become one of the common issues nowadays. Many individuals undergo surgical procedures in order to remove the portion affected with epilepsy. Finding the epileptogenic center is one of the difficult parts which can enable specialist to diagnose and get completely rid of epilepsy disorder. In this work, we are utilizing the basic entropy features to classify the signals as FEEG and NFEEG signals which have achieved the highest possible accuracy of about 99%. Likewise the highlights considered in this work have fundamentally less calculation time. It ought to likewise be noticed that the calculation time of our highlights is just 1.14 s which is imperative in connection to ongoing preparing of the EEG information. In addition, unprocessed EEG information as accessible in the database is connected. In this way, there is no preprocessing done but in the prior works they had connected these information to the IMFs on these information and determining the entropy esteems. In our work, there is no need of feeding the information to the IMFs and then determining the entropy esteems. The entropy features are directly calculated, and the EEG signal is classified as FEEG and NFEEG signal.

References

1. R.S. Fisher, W.V.E. Boas, W. Blume, C. Elger, P. Genton, P. Lee and J.J. Engel, Epileptic seizures and epilepsy: definitions proposed by the international league against epilepsy (ILAE) and the international bureau for epilepsy (IBE). Epilepsia **46**(4), 470–472 (2005)
2. S. Pati, A.V. Alexopoulos, Pharmacoresistant epilepsy: from pathogenesis to current and emerging therapies. Clevel. Clin. J. Med. **77**(7), 457–467 (2010)
3. S.M.S. Alam, M.I.H. Bhuiyan, Detection of seizure and epilepsy using higher order statistics in the EMD domain. IEEE J. Biomed. Health Inform. **17**(2), 312–318 (2013)
4. U.R. Acharya, Y. Hagiwara, S.N. Deshpande, S. Suren, J.E.W. Koh , S.L. Oh, N. Arunkumar, E.J. Ciaccio, C.M. Lim, Characterization of focal EEG signals: a review (2019)
5. S. Raghu, N. Sriraam, Classification of focal and non-focal EEG signals using neighborhood component analysis and machine learning algorithms (2018)
6. M. Rahman, M.I. Hassan, A. Bhuiyan, D. Das, Classification of focal and non-focal EEG signals in VMD-DWT domain using ensemble stacking (2019)
7. V. Gupta, T. Priya, A.K. Yadav, R.B. Pachori, U.R. Acharya, Automated detection of focal EEG signals using features extracted from flexible analytic wavelet transform. Pattern Recogn. Lett. **94**, 180–188 (2017)
8. A. Bhattacharyya, R.B. Pachori, U.R. Acharya, Tunable-Q wavelet transform based multivariate sub-band fuzzy entropy with application to focal EEG signal analysis. Entropy **19**(3), 99 (2017)
9. A. Bhattacharyya, M. Sharma, R.B. Pachori, P. Sircar, U.R. Acharya, A novel approach for automated detection of focal EEG signals using empirical wavelet transform. Neural Comput. Appl. **29**(8), 47–57 (2018)
10. N. Sriraam, S. Raghu, Classification of focal and non focal epileptic seizures using multifeatures and SVM classifier. J. Med. Syst. **41**, 160 (14 p.) (2017)
11. N. Kannathal, P.K. Sadasivan, Entropies for detection of epilepsy in EEG **80**(3), 187–194 (2005)
12. R. Sharma, R.B. Pachori, *Empirical Mode Decomposition based Classification of Focal and Non-focal EEG Signals* (2014)
13. G. Zhu, Y. Li, P. Paul Wen, S. Wang, M. Xi, *Epileptogenic Focus Detection in Intracranial EEG Based on Delay Permutation Entropy* (2013)
14. D. Gajic, Z. Djurovic, S.D. Gennaro, F. Gustafsson, Classification of EEG signals for detection of epileptic seizures based on wavelets and statistical pattern recognition. Biomed. Eng. Appl. Basis Commun. **26**(02), 1450021 (2014)
15. A.S. Zandi, M. Javidan, G.A. Dumont, R. Tafreshi, Automated real-time epilepticseizure detection in scalp EEG recordings using an algorithm based onwavelet packet transform. IEEE Trans. Biomed. Eng. **57**(7), 1639–1651 (2010)
16. V. Bajaj, R.B. Pachori, Classification of seizure and nonseizure EEG signalsusing empirical mode decomposition. IEEE Trans. Inf. Technol. Biomed. **16**(6), 1135–1142 (2012)
17. R.J. Hyndman, G. Athanasopoulos, 8.1 stationarity and differencing, in *Forecasting: Principles and Practices* (Melbourne, Australia, OTexts, 2013)
18. A.B. Das, M.I.H. Bhuiyan, Discrimination and classification of focal andnon-focal EEG signals using entropy-based features in the EMD-DWT domain. Biomed. Signal Process. Control **29**, 11–21 (2016)
19. W. Yang, Z. Peng, K. Wei, P. Shi, W. Tian, Superiorities of variational modedecomposition over empirical mode decomposition particularly intime-frequency feature extraction and wind turbine condition monitoring. IET Renew. Power Gener. **11**(4), 443–452 (2017)
20. T. Zhang, W. Chen, M. Li, AR based quadratic feature extraction in the VMDdomain for the automated seizure detection of EEG using random forest classifier. Biomed. Signal Process. Control **31**, 550–559 (2017)
21. N.E. Huang, Z. Shen, S.R. Long, M.C. Wu, H.H. Shih, Q. Zheng, N.-C. Yen, C.C. Tung, H.H. Liu, The empirical mode decomposition and the Hilbert spectrumfor nonlinear and nonstationary time series analysis. Proc. R. Soc. Lond. A Math. Phys. Eng. Sci. **454**(1971), 903–995 (1998)

22. Z. Wu, N.E. Huang, X. Chen, The multi-dimensional ensemble empirical modedecomposition method. Adv. Adapt. Data Anal. **01**(03), 339–372 (2009)
23. R. Sharma, R. Pachori, U. Acharya, An integrated index for the identification of focal electroencephalogram signals using discrete wavelet transform and entropy measures. Entropy **17**(12), 5218–5240 (2015)
24. K. Dragomiretskiy, D. Zosso, Variational mode decomposition. IEEE Trans. Process. **62**(3), 531–544 (2014)
25. Z. Wu, N.E. Huang, Ensemble empirical mode decomposition: a noise-assisteddata analysis method. Adv. Adapt. Data Anal. **01**(01), 1–41 (2009)
26. J.-R. Yeh, J.-S. Shieh, N.E. Huang, Complementary ensemble empirical modedecomposition: a novel noise enhanced data analysis method. Adv. Adapt. Data Anal. **02**(02), 135–156 (2010)
27. K. Fu, J. Qu, Y. Chai, T. Zou, Hilbert marginal spectrum analysis for automaticseizure detection in EEG signals. Biomed. Signal Process. Control **18**, 179–185 (2015)
28. U.R. Acharya, F. Molinari, S.V. Chattopadhyayd, K.-H. Nge, J.S. Suri, Automated diagnosis of epileptic EEG using entropies. Biomed. Signal Process. Control, 401–408 (2012)
29. N. Nicolaou, J. Georgiou, Detection of epileptic electroencephalogram based on permutation entropy and support vector machines. Expert Syst. Appl. 202–209 (2012)
30. U.R. Acharya, S. Bhat, H. Adeli, A. Adeli, Computer-aided diagnosis of alcoholism-related EEG signals. Epilepsy Behav. **41**, 257–263 (2014)
31. U. Melia, M. Guaita, M. Vallverdú, C. Embid, I. Vilaseca, M. Salamero, J. Santamaria, Mutual information measures applied to EEG signals for sleepiness characterization. Med. Eng. Phys. (2015)
32. N. Arunkumar , K. Ram Kumar , V. Venkataraman, Entropy features for focal EEG and non focal EEG

Prediction of Future Career Path Using Different Machine Learning Models

Subhabrata Sengupta, Anish Banerjee, and Satyajit Chakrabarti

Abstract The main purpose of this paper is to examine the strength and weaknesses of a student based on their performance in different exams. Students are classified using the K-means classification algorithm and decision tree. The proposed model will help teachers to comprehend their students well and will also assist the students to get their most serviceable job. The data mining technique capable of analyzing relevant results is used over the students' information to produce relevant correlations and produce different aspects to understand more about the students. The paper proposes a model based on a classification approach in finding an enhanced evaluation method for students and predict the placement prospects.

Keywords Machine leaning · Classification · Data mining

1 Introduction

With the technology of personalized recommendation, the problem of information overload on recruiting Web sites can be solved by the job recommendation system. It can provide the appropriate fields for the applicants and can even recommend a suitable employee to the recruiting body based on their requirement analysis. Thus, the advanced recommendation system comprises of a job seeker subsystem that is

S. Sengupta · A. Banerjee (✉)
Department of Information Technology, Institute of Engineering and Management (IEM), Kolkata 700091, India
e-mail: anish.banerjee1998@gmail.com

S. Sengupta
e-mail: subhabrata.sengupta@iemcal.com

S. Chakrabarti
Department of Computer Science and Engineering, Institute of Engineering and Management (IEM), Kolkata 700091, India
e-mail: satyajit.chakrabarti@iemcal.com

J. Tavares et al. (eds.), *Emerging Technologies in Data Mining and Information Security*, Lecture Notes in Networks and Systems 164,
https://doi.org/10.1007/978-981-15-9774-9_23

designed for job applicants and an e-recruiting subsystem, used by recruiters. As a result of the comparability in the suggestion standards of two subsystems, the activity searcher subsystem is taken for its wide scope of utilization in reality for portrayal. The traditional job applicant subsystem takes some personal information and work intention of a job aspirant for his or her user profile and uses them to generate results by employing particular recommendation algorithms.

2 Objectives

The main objective of the thesis is to build a highly personalized job recommendation system which suggests specific job from a set of job postings to the user based on their skill sets, and it has a brief introduction to the application and the background is detailed. Besides, the approach and methodologies used are also explained. Later it includes the evaluation dataset description and associated results. The concluding chapter summarizes the proposed work.

The paper addresses the following issues:

A. Finding the eligibility of a student for placement.
B. The dependency of a student getting placed or not based on the number of hours of study.
C. Feature selection for a more accurate result.

The paper focuses on one of the important applications of building a relationship between student eligibility for placement: application of data mining in case of placement analysis that has played an important role to predict the possibility of a student will get placed or not. Support vector machine played an important role in describing the accuracy result of the whole prediction. Placement prediction system (PPS) using logistic regression algorithm helped the model to get more accuracy while predicting the result based on their marks and previous performance.

3 Background Study

The placement prediction system would predict the likelihood of a student understudy getting set in an IT organization by applying the AI model of K-nearest neighbors' characterization. The qualification measures of understudies in different organizations are increasingly significant and this can be acknowledged by the model. For this, we will consider the scholastic history of the candidate, programming aptitudes, relational abilities, diagnostic abilities, and cooperation, which are tried by employing organizations during the enrolment procedure. Decision tree C4.5 algorithms are applied to company's earlier year information and the current necessity to create the model, and this model can be utilized to anticipate the understudies' qualifications in different organizations (Fig. 1).

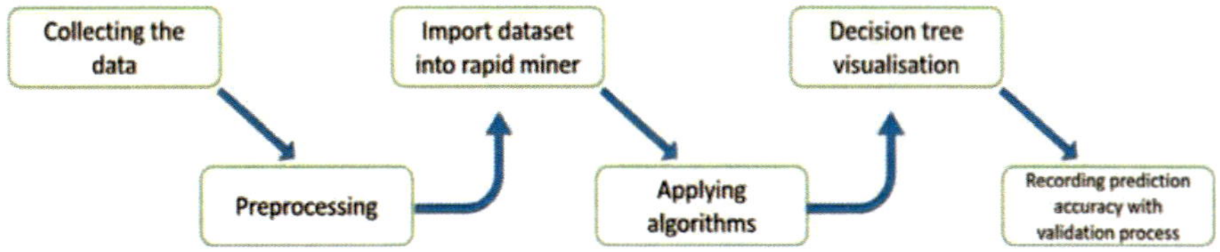

Fig. 1 Ways to collect data

3.1 Eligibility Filtration

Eligibility filtration, on the other hand, does not need anything else except students' historical performance on different phases. Because it is based on historical data, the core assumption here is that the students who have academic performance in the past tend to also have performance in the future. Higher studies' marks are the marks student got in their higher studies that are class 12th. Minimum aggregate in college semesters suggests student's eligibility directly, such as marks in an operating system, data structure, computer networks, algorithm, and DBMS, whether they got the minimum eligibility marks or not. Based on marks in different semester and divisions and clubbed with other credentials, the filtration process was carried.

3.2 Prediction Computation

The most significant advancement in a collaborative filtering system is to produce the yield interface regarding expectation. When we separate the arrangement of most comparative things based on the likeness, the following stage is to investigate the objective clients' evaluations and utilize a strategy to acquire forecasts. Here, we think about two such strategies.

Weighted sum: As the name suggests, this technique processes the expectation on an object I for a client u by registering the aggregate of the ratings given by the client on the things like I. The weighted aggregate is scaled by the entirety of the comparability terms to ensure the expectation is inside the predefined range.

Regression: This methodology is like the weighted whole strategy however rather than legitimately utilizing the appraisals of comparative things it utilizes an estimate of the evaluations dependent on the regression model. All things considered, utilizing the crude appraisals of the so-called comparable thing may bring about the poor forecasts. The fundamental thought is to utilize a similar equation as the weighted entirety procedure, yet as opposed to utilizing the comparable thing N's "crude" appraisals value Ru, N's, this model uses their approximated qualities R's, N dependent on a linear regression model.

3.3 Literature Review

El-Seoud et al. [1] focused main two points in their paper: An online application which is customized to one's educational preferences and norms.

The application using self-learning mechanisms using its previous experiences is based on semantic analysis. Academic progress and performance of students provide educational authorities with valuable information. This offers diverse opportunities for decision making for the said authorities and assisting the students to achieve a great progress in their academics [2].

Annabel Latham et al. discussed a new dimension in the field of comprehensive tutoring systems in learning style prediction in the paper "A conversational intelligent tutoring system to automatically predict learning styles." Authors [3] proposed the idea of dynamic learning of the system based on various previous experiences and real students have enabled the prediction mechanism in a real-time environment.

ElGamal directed a study for predicting a students' performance in programming courses [4]. The investigation draws in factors like the candidate's scientific foundation, programming fitness, critical thinking abilities, sexual orientation, related knowledge, secondary school science grades, region, past coding experience, and e-learning utilization.

Sayali Rajesh Sural et al. in the paper "Quality Improvisation of Student Performance Using Data Mining Techniques" [5] identified students' performance by applying the classification rule.

Shradha Shet in the paper "Approach for Predicting Student Performance Using Ensemble Model Method" [6] discussed the knowledge retrieval process from the educational field.

ID3 algorithm is used to construct the decision tree by using a top-down, greedy search with different datasets. The same types of data can be there inside the dataset. Abeer Badr El-Din Ahmed et al. proposed some salient features to predict the students' performances based on several types of data in their paper [7].

Fred Niederman communicated that the inspiration driving thinking about what is to come is not to foresee with exactness of what will happen, be that as it may, the explanation behind existing is to hone us toward picking pathways and taking exercises that extend the probability of the possibilities we might want and decay the probability of future states we should keep up a vital good ways from [8].

This investigation approved the model by the corresponding utilization of AI calculations nearby an old-style SEM-based methodology based on information gathered from 308 college students. The SEM results demonstrated that the understudies' data management (for example, recover, store, share, and apply) rehearses were essentially connected with their perspectives, which were altogether connected with the social aims [9].

García-Peñalvo introduced a contextual analysis that includes information from the Spanish Observatory for Employability and Employment (OEEU). Utilizing information from this venture (data around 3000 understudies), prescient models have been fabricated that characterize how these understudies find a new line of work in the wake of concluding their degrees [10].

Andrea De Mauro expressed that the fast development of big data analytics is constraining organizations to reexamine their human resource (HR) needs. In any case, simultaneously, it is muddled which kinds of employment jobs and abilities establish this zone [11].

Prescient demonstrating utilized in anticipating understudy execution is identified with a few learning tasks, for example, classification, regression, and clustering. Different strategies have been proposed by past analysts to build up the best execution model utilizing an assortment of understudies' information, methods, calculations, and instruments. Al assessed flow look into exercises identified with scholastic examination concentrating on foreseeing understudy scholarly execution [12].

Prescient demonstrating utilized in anticipating understudy execution is identified with a few learning tasks, for example, classification, regression, and clustering. The examination applied a choice tree calculation to arrange understudy execution with a forecast aftereffect of 90.91% for competent, 66.67% for capable, and 100% for challenged. This outcome will give the organization to evaluate the readiness of the understudies in the workplace [13].

The encoding incorporates getting to discrete substances containing setting data for the client's profile, developing a majority of implanting vectors from the setting data and producing a setting vector from the majority of inserting vectors. The deciphering incorporates applying a long short-term memory (LSTM) model to the setting vector to create play out the expectation of the client's next organization and next title for introduction in a UI [14].

As saw by Hilbert [15], throughout the most recent two decades, we have seen an exponential increment of data fellow, stock, and computational force. The properties of information are in like manner advancing quickly: data is directly progressively "individual," suggesting that its prevailing part is respected to be made and eaten up clearly by individuals, as they partner with others and machines through their own devices [15].

To reflect the activity of the insightful model at an institutional level, an academic assessment is essential parts that encased all activities in higher education [16].

Predicting understudy execution is an irksome task considering the way that a huge segment of the envisioning procedure does not deal with a dynamic environment [17].

The specific desire for understudy execution will be valuable in order to provide guidance in the learning strategy [18].

Most of the experts used understudy total assessment point ordinary (CGPA) as a parameter since it has impressive and important characteristics for future guidance similarly as expert adaptability [19].

It is likewise been a benchmark whether an understudy will graduate on time without returning in the all-encompassing semester and handle professional openings at the opportune time [17].

This recommendation will ensure that students are not misguided in choosing the field that are equivalent to their result and their interest [20].

4 Proposed Strategy

4.1 Naive Bayes Classifier

A Naive Bayes classifier is a straightforward probabilistic classifier dependent on applying Bayes' hypothesis (from Bayesian insights) with solid independent suppositions. A progressively descriptive term for the model would be the "autonomous component model." Based on the preciseness of the probability model, Naive Bayes classifiers can be trained effectively in supervised learning atmosphere. In numerous pragmatic applications, parameter estimation for Naive Bayes models utilizes the technique for greatest probability.

4.2 Multilayer Perceptron

A multilayer recognition (MLP) is a feed-forward artificial neural system model that maps sets of input information onto a lot of fitting yields. MLP uses a supervised learning strategy called backpropagation for preparing the network. MLP is an alteration of the standard linear perceptron.

4.3 Decision Tree

Data mining can be utilized in the instructive field to upgrade the comprehension of the learning procedure to concentrate on finding, separating, and approving factors identified with the understudy learning process.

Knowing the elements for the situation of understudies can support the instructors and heads to take important activities with the goal that the achievement level of placements can be improved. The expectation of models incorporates all close to home, social, and mental, and different factors are required for the viable forecast of the position of the understudy.

An instructive foundation contains understudy records which are flourishing of data however is unreasonably enormous for one individual to comprehend completely.

4.4 Data Mining and Preprocessing Process

Knowing the components for placement of candidates can support the educators and heads to take vital activities with the goal that the achievement level of the situation can be improved.

5 Data Preparations

This includes the collection and preprocessing the received data which includes data cleaning and data wrangling. We have acquired the raw data from universities and colleges and the dataset consists of the continuous academic performance of the student including the internals and their semester marks. Apart from the academic details, the personal information of the student including the economic class and language proficiency was the important required fields.

6 Data Preprocessing

This is the most important process that includes data cleaning and wrangling to make it suitable for use and designing the model and proper training and testing. The unimportant field like contact details and address, etc., was dropped and the overview of the attributes were taken. In this kind of data, there are chances of outliers to be present which are not exceptions or wrong entries but very natural, so they are not removed. Rather, NAN values of the fields are replaced by the median to avoid the influence of outliers on them.

7 Data Selection and Transformation

In this step, the relevance of fields and new features was generated on the basis of various analysis of fields and inference drawn from numerous plots. Heat-map and correlation plots like violin plots and pair plots reveal the degree of correlation and also priority influence of each field toward the classification was calculated and on all these results new features were generated with the clubbing of existing features and irrelevant features were dropped, e.g., instead of having separate marks of each internal, an average was calculated that increased the accuracy and also reduces the number of fields.

8 Implementation of Mining Model

The classify panel empowers the client to apply regression and classification to the subsequent dataset, to assess the precision of the subsequent prescient model, and to picture wrong forecasts, or the model itself. Since there is no different assessment informational index, this is important to get a sensible thought of the precision of the produced model. This prescient model gives an approach to foresee whether another student will be placed or not in any esteemed organization.

8.1 Results

To more readily comprehend the significance of the input factors, it is standard to examine the effect of input variables during students' placement achievement, in which the effect of certain input variables of the model on the yield variable has been analyzed. Tests were led utilizing four tests for the evaluation of information factors on the following tests: chi square, info gain, and gain ratio.

8.2 Process Flow

See Fig. 2.

9 Data Analysis and Feature Engineering

Plots and analysis of data fields reveal a lot of informative insights into the real-world scenario. The plots reveal that the percentage of eligible students is way more (Fig. 3).

Plotting the percentage logarithmic student count versus the hours of input they gave into placement related activities held by the organization shows a more or less uniform distribution (Fig. 4).

Performance plot of students with respect to their density reveals some unusual results that portray that even with same hours of input there is not a drastic difference between the eligible count and not eligible count which accounts for some external

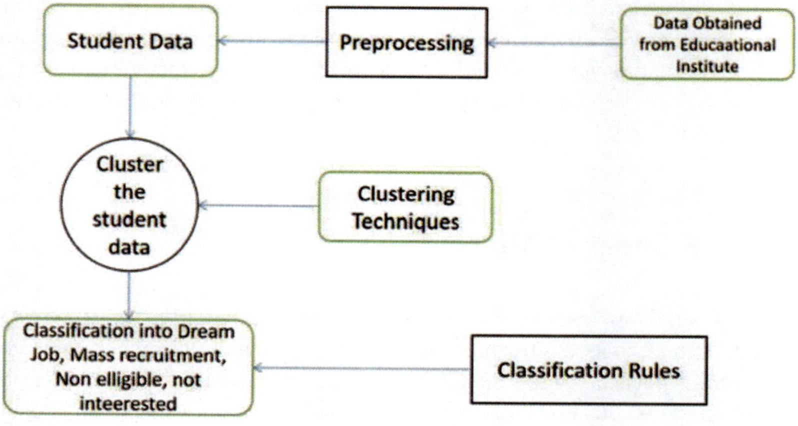

Fig. 2 Activity diagram

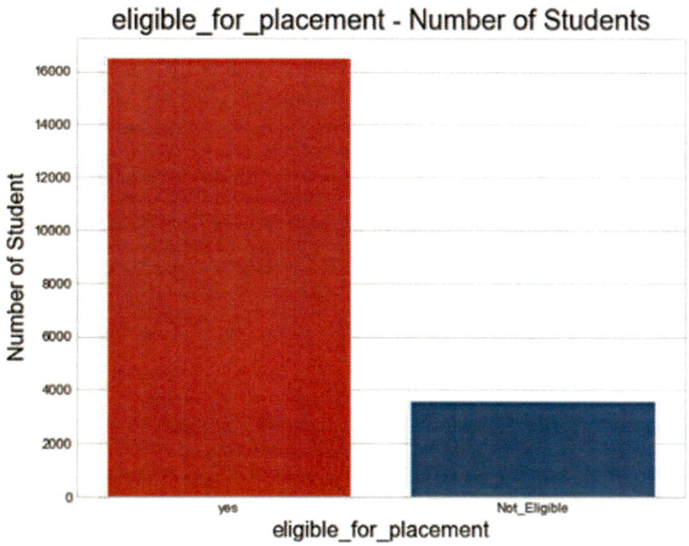

Fig. 3 Filtration of students' data

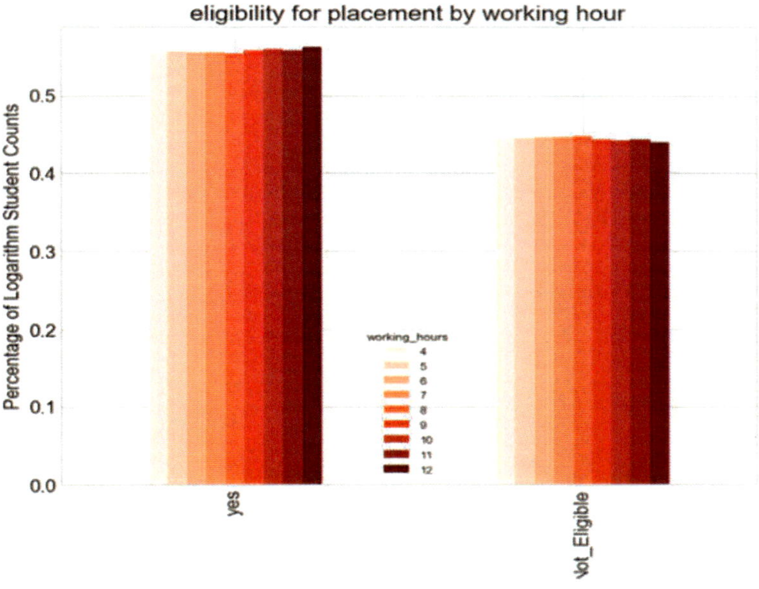

Fig. 4 Performance comparison on working hour

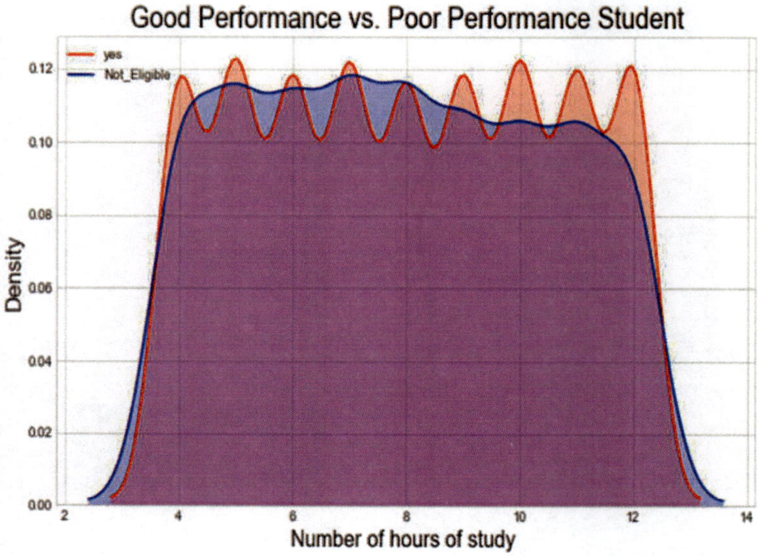

Fig. 5 Performance comparison on study

influence like merit and IQ or might be better understanding or prior knowledge of the subject. But the gradual spread of "Not_Eligible" toward the end also depicts that with higher hours of study as well, eligibility criteria cannot be met (Fig. 5).

10 Results on Practical Implementation

Relative accuracy analysis of different models

10.1 Chi-square Model Performance

Accuracy Score: 0.40173(approximately)—40.173%.

10.2 Random Forest Classifier Result After Data Cleaning and Processing

Accuracy score: 0.998822(99.9%).

10.3 Proposed Model Performance Based on SVC and Cross Validation

SVC model score: 97.457%,
Cross-validation score: 82.03%

10.4 Prediction Accuracy Graph

See Fig. 6.

11 Conclusion and Future Scope

This paper outlines how well unique characterization strategies are utilized as prescient apparatuses in the information mining area and in the wake of contrasting their exhibitions. From the outcomes, it is demonstrated that the logistic regression calculation is generally suitable for anticipating understudy execution. The calculated regression calculation gives 82% forecast which is generally higher than different calculations. This investigation is an endeavor to utilize grouping calculations for anticipating understudy execution and looking at the exhibition of Naive Bayes simple, decision tree, random forest, SVC, and logistic regression calculations.

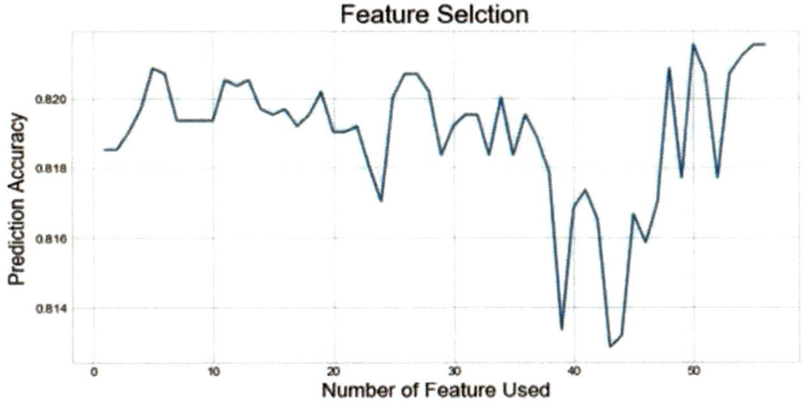

Fig. 6 Prediction results on feature selection

11.1 Future Scope

- We are trying to improve our project by designing the app to predict the most appropriate job one should opt.
- The app may predict the most desired way to improve their marks in other semester.
- Add more features to the app such as predicting the regions to work on in worst case for not eligible students.

References

1. S.A. El-Seoud, H.F. El-Sofany, O.H. Karam, The semantic web architecture and its impact on e-learning systems development. Int. J. Eng. Technol. **10**(5) (2015)
2. A. Sharabiani, F. Karim, A. Sharabiani, M. Atanasov, H. Darabi, IEEE Glob. Eng. Educ. Conf. 832–837 (2014)
3. A. Latham, K. Crockett, D. Mclean, B. Edmonds, A conversational intelligent tutoring system to automatically predict learning styles **59**(1), 95–109 (2012)
4. A.F. Elgamal, An educational data mining model for predicting student performance in programming course. Int. J. Comput. Appl. **70**(17), 0975–8887 (2013)
5. S.R. Suyal, M.M. Mohod, Quality improvisation of student performance using data mining techniques. Int. J. Sci. Res. Publ. **4**(4) (2014)
6. G. Shradha Shet, Approach for predicting student performance, using ensemble model method. Int. J. Innov. Res. Comput. Commun. Eng. **2**(5) (2014)
7. A.B.E.D. Ahmed, I.S. Elaraby, Data mining: a prediction for student's performance using classification method. J. Comput. Appl. Technol. **2**(2), 43–47 (2014).https://doi.org/10.13189/wjcat.2014.020203
8. F. Niederman,*The Future of IT Work: Computers and People*
9. I. Arpaci,*A Hybrid Modeling Approach for Predicting the Educational Use of Mobile Cloud Computing Services in Higher Education*
10. F.J. García-Peñalvo, *Proposing a Machine Learning Approach to Analyze and Predict Employment and its Factors*
11. A. De Mauro, M. Greco, Human resources for big data professions: a systematic classification of job roles and required skill sets. Inf. Process. Manage. (2017)
12. A. Mat, *A Review on Predictive Modeling Technique for Student Academic Performance Monitoring*
13. A.L.A. Ramos, *Prediction Model for Internship Deployment Using Decision Tree*
14. Yang et al., *Next Career Move Prediction with Contextual Long Short-Term Memory Networks* (2019)
15. M. Hilbert, Big data for development: a review of promises and challenges. Develop. Policy Rev. **34**, 135–174 (2016)
16. B. Tulasi, Int. J. Comput. Appl. **68**(14), 21–23 (2013)
17. P. Kavipriya, Int. J. Adv. Res. Comput. Sci. Softw. Eng. **6**(12), 101–105
18. M. Ramaswami, Int. J. Comput. Intell. Inform. **1**(4), 231–235 (2012)
19. A. Mohamed Shahiri, W. Husain, N. Abdul Rashid, Proc. Comput. Sci. **72**, 414–422 (2015)
20. K. Sin, L. Muthu, ICTACT J. Soft Comput. **05**(04), 1035–1049 (2015)

Holographic Renormalization with Machine Learning

Eric Howard

Abstract At low energies, the microscopic characteristics and changes of physical systems as viewed at different distance scales are described by universal scale invariant properties investigated by the Renormalization Group (RG) apparatus, an efficient tool used to deal with scaling problems in effective field theories. We employ an information-theoretic approach in a deep learning setup by introducing an artificial neural network algorithm to map and identify new physical degrees of freedom. Using deep learning methods mapped to an effective field theory, we develop a mechanism capable to identify relevant degrees of freedom and induce scale invariance without prior knowledge about a system. We show that deep learning algorithms that use an RG-like scheme to learn relevant features from data could help to understand the nature of the holographic entanglement entropy and the holographic principle in context of the AdS/CFT correspondence.

Keywords Renormalization group · AdS/CFT correspondence · Restricted Boltzmann machines

1 Introduction

Deep learning methods such as deep neural networks, deep belief networks [1–3] and recurrent neural networks use learning data representations to process, predict, optimize and discover new features and patterns in big data [4]. Recent progress in the field of machine learning [5] has shown significant promise in applying machine learning tools like classification or pattern recognition to identify phases of matter [6]. Our paper provides a connection between the properties of many-body quantum states and neural network based machine leaning. Restricted Boltzmann machines (RBM) attracted great interest from researchers in the field of quantum many-body

E. Howard (✉)
Department of Physics and Astronomy, Macquarie University, Sydney, NSW, Australia

Centre for Quantum Dynamics, Griffith University, Brisbane, Australia
e-mail: eric.howard@mq.edu.au

J. Tavares et al. (eds.), *Emerging Technologies in Data Mining and Information Security*, Lecture Notes in Networks and Systems 164,
https://doi.org/10.1007/978-981-15-9774-9_24

physics. Carleo and Troyer [7] trained an RBM like wavefunction to study quantum many-body systems away from equilibrium. Deng [8] developed an exact RBM model for multiple topological states. Huang and Wang [9, 10] developed a recommender system RBM-based in order to boost Monte Carlo modelling of quantum many-body systems. Torlai and Melko [11] used an RBM for modelling the thermodynamics of statistical physics models. A low-energy effective theory with universal characteristics can be deducted from finding and identifying de degrees of freedom that are physically relevant to the theory. Since de discovery of critical systems [12], asymptotic freedom in quantum chromodynamics and Kosterlitz-Thouless phase transition, the Renormalization Group [13–15] methods model the effective behaviour of systems depending on the scale at which they are observed. Quantum many-body problems assume an exponential scaling of the Hilbert space dimension with the system size therefore requiring a large amount of information computationally daunting and intractable. The construction of efficient representations for physical states that access only partially the Hilbert space makes possible the use of much less numerical resources. A low-energy effective theory with universal characteristics can be deducted from finding and identifying the degrees of freedom that are physically relevant to the theory. We find similarities between the conceptual framework of RG and deep learning, where depth and scale play similar roles. RG predicts the behaviour of a system as dependent on a scale or energy parameters while here, the same parameter is treated as depth. The predictive success of the scaling theory supports the idea that near criticality the correlation length represents the only important length scale, while the microscopic lengths become irrelevant. The critical behaviour will be generated by the fluctuations that are statistically self-similar up to that scale.

We employ RG in momentum space, developed by Wilson [16] to study systems displaying scale invariant behaviour. We find connections between RBMs and an effective field theory by mapping a fully connected graph to a field theory described by a Hamiltonian density. RG in momentum space traces a path in the coupling space for the Hamiltonian and its couplings. The differential equations of the couplings become constraints for the connection weights of the system. The assumption of self-similar patterns eliminates any correlated degrees of freedom at specific length scales up to the point where the system contains simple uncorrelated degrees of freedom at a renormalised length scale.

We find that a holographic approach of the effective theory may be connected to deep learning processes associated with the partition function of a quantum field theory. According to the holographic principle all information in the bulk of the spacetime is encoded at its boundary. In this sense the AdS/CFT correspondence [17, 18] is a realization of the holographic principle. The partition function in AdS gives us the correlation functions of the boundary CFT. AdS/CFT correspondence has been employed to compute the two point functions at the boundary and the RG transformations of the CFT are identified with the AdS geometry. Our goal is to provide an alternative way to understand the entanglement structure of the spacetime [19, 20] as encoded in the graph of a deep neural network and show that deep neural networks can explain the holographic nature of the spacetime. We already know that the entanglement structure of the spacetime has a Ryu–Takayanagi [21] form in AdS/CFT but we find that the entanglement structure of a deep neural network also has a Ryu–Takayanagi form. Spacetime can be understood as an emergent process from the deep neural network of quantum states. The boundary theory retains information on the gauge degrees of freedom of the bulk theory. We find that the emergence of a holographic gravitational theory is connected to a deep learning architecture encoded in the quantum field theory. Deep learning works here because it is based on a computational mechanism similar to the mathematical formulation of the Holographic Principle [22, 23].

2 Deep Learning Mapping Model of Renormalization Group

The first step of the RG method is the decreasing the resolution by changing the minimum length scale from a smaller scale to a larger one. This is done by integrating out fluctuations of the fields that occur on length scales finer than the renormalization scale. The result is a renormalisation of the Hamiltonian which leads to an effective Hamiltonian expressed in terms of a coarse-grained field. As the Hamiltonian H depends on fields and couplings, the system is described by the partition function, a scale dependent effective action functional encoded in the functional integral over the Hamiltonian. The RBM is composed by layers of hidden neurons and visible neurons represented by spins. If the RBM is trained, the hidden units learn to extract useful features from training data. The RBM distribution is defined through a function that assigns a number to each possible state of the visible and hidden variables. The RBM contains an ensemble of N spins $\{v_i\}$. The index i with $(i = 1, 2, \ldots, N)$ gives the position of spins v_i in the lattice. The data is given by the probability distribution $P(\{v_i\})$ of $\{v_i\}$ spins with $i = 1, \ldots, N$. The probability distribution is also called a Gibbs distribution or partition function. The visible neurons are represented by the physical spin variables $\{v_i\}$. The hidden neurons of the neural network contain further new spin variables $\{h_j\}$. The RG transformation generates a deep learning like architecture as the spins on each layer are hidden spins during the RG transformation on the previous layer. The hidden spins on the next layer are given by the new Hamiltonian using couplings between the neighbour spins. The deep learning architecture is described by spins coupled between consecutive layers. Correlations between the spins $\{v_i\}$ generate correlations between the coarse-grained spins $\{h_j\}$. Each new top layer in the RBM learns exponentially higher level features from the input data. The new hidden spin variables $\{h_j\}$ $(j = 1, \ldots, M)$ are coupled to the visible spins and the interactions between them are described by the energy function

$$\mathbf{E}(\{v_i\}, \{h_j\}) = \sum_i b_j h_j + \sum_{ij} v_i w_{ij} h_j + \sum_i c_i v_i, \tag{1}$$

with $\lambda = \{b_j, w_{ij}, c_i\}$ are variational parameters of the model.

The probability of observing the full system composed of both hidden and visible spins is

$$p_\lambda(\{v_i\}, \{h_j\}) = \frac{e^{-\mathbf{E}(\{v_i\},\{h_j\})}}{\mathscr{Z}}. \tag{2}$$

The visible spins are described by the distribution

$$p_\lambda(\{v_i\}) = \sum_{\{h_j\}} p_\lambda(\{v_i\}, \{h_j\}) = \mathrm{T}r_{h_j} p_\lambda(\{v_i\}, \{h_j\}) \tag{3}$$

and the hidden spins by:

$$p_\lambda(\{h_j\}) = \sum_{\{v_j\}} p_\lambda(\{v_i\}, \{h_j\}) = \mathrm{T}r_{v_i} p_\lambda(\{v_i\}, \{h_j\}). \tag{4}$$

At thermal equilibrium, the probability of a spin configuration is calculated using the Boltzmann distribution

$$P(\{v_i\}) = \frac{e^{-\mathbf{H}(\{v_i\})}}{Z} = \frac{e^{\sum_i K_i v_i + \sum_{ij} K_{ij} v_i v_j + \sum_{ijk} K_{ijk} v_i v_j v_k + \cdots}}{Z}, \tag{5}$$

where $H(\{v_i\})$ is the Hamiltonian, and Z is the partition function

$$Z = \mathrm{T}r_{v_i} e^{-\mathbf{H}(\{v_i\})} \equiv \sum_{v_1,\ldots,v_N=\pm 1} e^{-\mathbf{H}(\{v_i\})}. \tag{6}$$

where $\mathbf{K} = \{K_s\}$ is the coupling constant. The Hamiltonian depends on the set of couplings $\mathbf{K} = \{K_s\}$, describing the set of all possible Hamiltonians. The spins $\{h_j\}$ are coarse-grained. The coupling constant is renormalized as $\tilde{\mathbf{K}} = \{\tilde{K}_s\}$. The auxiliary spins $\{h_j\}$ are coupled to physical spins $\{v_i\}$ and the visible spins have to be averaged out in order to give a coarse-grained picture of the physical spin system. The free energy of the spin system is given by

$$F^v = -\log Z = -\log\left(\mathrm{T}r_{v_i} e^{-\mathbf{H}(\{v_i\})}\right). \tag{7}$$

while the free energy of the coarse grained spin system is

$$F^h_\lambda = -\log\left(\mathrm{T}r_{h_i} e^{-\mathbf{H}^{\mathrm{RG}}{}_\lambda(\{h_i\})}\right). \tag{8}$$

RG is generated by repeating the computational step of solving the effective Hamiltonian in order to obtain multiple effective spins. This step becomes depth in deep learning. The coarse-grained Hamiltonian is

$$\mathbf{H}^{\mathrm{RG}}(\{h_j\}) = -\sum_i \tilde{K}_i h_i - \sum_{ij} \tilde{K}_{ij} h_i h_j - \sum_{ijk} \tilde{K}_{ijk} h_i h_j h_k + \cdots, \tag{9}$$

Each step of the Renormalization group represents a layer in the deep neural network. The adjustable function $\mathbf{T}_\lambda(\{v_i\}, \{h_j\})$ has the parameters $\{\lambda\}$. The Hamiltonian $\mathbf{H}^{RG}_\lambda[\{h_j\}]$ of the coarse-grained RG degrees of freedom is also the Hamiltonian for the hidden spins in a deep neural network.

$$\mathbf{H}^{\mathrm{RG}}_\lambda[\{h_j\}] = \mathbf{H}^{\mathrm{RBM}}_\lambda[\{h_j\}]. \tag{10}$$

The RG transformation provides the approximate solution, dependent on the function $\mathbf{T}_\lambda(\{v_i\}, \{h_j\})$

$$e^{-\mathbf{H}^{\mathrm{RG}}(\{h_j\})} \equiv Tr_{v_i} e^{\mathbf{T}_\lambda(\{v_i\},\{h_j\}) - \mathbf{H}(\{v_i\})}, \tag{11}$$

The parameter λ minimizes the free energy difference $\Delta F = F^h_\lambda - F^v$ between physical and coarse grained spins and the long-distance physical observables remain invariant.

$$\Delta F = 0 \iff Tr_{h_j} e^{\mathbf{T}_\lambda(\{v_i\},\{h_j\})} = 1 \tag{12}$$

The parameters $\lambda = \{b_j, w_{ij}, c_i\}$ are adjustable. The correlations between the hidden spins are given by $\{\tilde{K}\}$. The probability of a configuration of hidden neurons is

$$p_\lambda(\{h_j\}) = \sum_{\{v_i\}} \frac{e^{-\mathbf{E}(\{v_i\},\{h_j\})}}{\mathcal{Z}} = \sum_{\{v_i\}} \frac{e^{-\sum_j b_j h_j - \sum_{ij} v_i w_{ij} h_j - \sum_i c_i v_i}}{\mathcal{Z}}, \tag{13}$$

The couplings are mapped as $\{K\} \to \{\tilde{K}\}$. The Hamiltonian for hidden neurons in the RBM algorithm is

$$p_\lambda(\{h_j\}) = \frac{e^{-\mathbf{H}^{\mathrm{RBM}}_\lambda(\{h_j\})}}{\mathcal{Z}}. \tag{14}$$

The conditional probability distribution $p_\lambda(\{h_j\}|\{v_i\})$ can map the layer of visible spins to the hidden spins layer. Once trained, the network uses the hidden layer as a feedback response to a visible new layer as input data for learning the next hidden spins layer. The Renormalization group and the deep neural network are directly mapped

$$\mathbf{T}_\lambda(\{v_i\}, \{h_j\}) = -\mathbf{E}(\{v_i\}, \{h_j\}) + \mathbf{H}(\{v_i\}), \tag{15}$$

$$\mathbf{H}^{RG}_\lambda(\{h_j\}) = \mathbf{H}^{RBM}_\lambda(\{h_j\}). \tag{16}$$

The $\mathbf{T}_\lambda(\{v_i\}, \{h_j\})$ is dependent on the set of parameters $\{\lambda\}$ and describes the statistical correlations interactions between the physical and coarse-grained degrees of freedom. The function $\mathbf{T}_\lambda(\{v_i\}, \{h_j\})$ is an approximation of the conditional probability of hidden spins based on visible spins. The couplings between hidden and visible spins are given by $\mathbf{T}_\lambda(\{v_i\}, \{h_j\})$. The energy function $\mathbf{E}(\{v_i\}, \{h_j\})$ plays the role of the RG coupling between hidden and visible spins. For each RG transformation we have

$$Tr_{h_j} e^{\mathbf{T}_\lambda(\{v_i\}, \{h_j\})} = 1 \tag{17}$$

For a number $\tilde{N}$ of nodes given by physical spins in a sub-region A of the network, the number of steps is $\tau = \log_2 \tilde{N}$. Every layer has a link that must be removed, in order to separate A from the rest of the network. As A is coarse-grained to one node, the minimum number of links to be removed is $n \propto \tau$. If we treat the neural network as a tensor network, the entanglement entropy S_A of A is proportional to the number of removed links. The scaling takes the form of an area law, stating that the entanglement entropy of a sub-region A grows with the area of the boundary of the sub-region rather than its volume. We therefore have

$$S_A \propto cn \propto c\tau = c \log_2 \tilde{N}, \tag{18}$$

that has the same form with the logarithm law for one dimensional systems. Here c is a proportional constant related to the logarithm of the dimensions of the Hilbert space of every link. The measure of entanglement $S(A)$ is the mutual information $I(A, B)$ between two disjoint adjacent subsystems A and B. In the holographic bulk the mutual information has a geometric interpretation where the entanglement entropy $S(A)$ of the region A is proportional to the area of the minimal surface γ_A. In AdS_3, this area is mapped to the geodesic line between two boundary points in the region A. The entanglement entropy will be given by the single systems

$$S(A \cup B) = S(A) + S(B) - I(A, B) \tag{19}$$

If $I(A, B)$ is small, the entropy is mainly dominated by the additive part $S(A \cup B) \simeq S(A) + S(B)$. The mutual information of the two subsystems becomes

$$I(A, B) = S(A) + S(B) - S(AB) = \frac{1}{4G_N}(|\gamma_A| + |\gamma_B| - |\gamma_{AB}|) \tag{20}$$

and captures the total amount of correlations between the two systems. The mutual information also offers an upper bound for the connected two-point functions between the operators associated with the systems A and B. But so far, the system did not contain any matter. Let's suppose we introduce in the system a fermion of mass m. The mass of the fermion provides a bound for the IR region at the scale $z_{\mathrm{IR}} \sim \ln \xi$. The correlation length ξ of the fermion is finite and is given by

$$\xi^{-1} = \frac{1}{2} \ln \frac{1 + |m|}{1 - |m|} \tag{21}$$

At the same scale, the entanglement between adjacent regions is described by a high energy theory, in UV. The fermion mass will stop the network to grow deeper at a boundary $z_{\mathrm{IR}} = \log_2 \xi$ where the holographic space terminates. The AdS_3 geometry emerges as the mass decreased and moves the boundary toward the CFT_2. As the fermion mass increases, the scale z_{IR} fades away from large z in IR towards small z in UV. The partition function of the boundary CFT field $\{v_i\}$ has the deep neural representation

$$\mathscr{Z} = \sum_{\{v_i\},\{h_j\}} e^{-\mathbf{E}(\{v_i\},\{h_j\})} = \sum_{\{v_i\},\{h_j\}} e^{-\sum_j b_j h_j - \sum_{ij} v_i w_{ij} h_j - \sum_i c_i v_i}. \tag{22}$$

where $\{h_j\}$ is the bulk field. Also $\sum_{ij} v_i w_{ij} h_j$ can be seen as $\int \mathrm{d}^d x \phi(x) O(x)$ term in GKP-W, where $\phi(x)$ is bulk field and $O(x)$ is boundary field. The task here is to identify new $M < N$ coarse-grained binary spins $\{h_j\}$ after all short distance fluctuations on smaller scale are averaged out. The length scale of the spin lattice system is changed, the new degree of freedom creating more space available for each h_j spin. In a one dimensional spin system with translational invariance, the nodes are formed by spin variables with the first line as the physical spin $\{v_i\}$, and the rest being coarse-grained spins $\{h_j\}$. The network s defined in coordinates (τ, x). A coarse-grained spin is in the same equal-x line, and spins in every step of RG are in the equal-τ line. We consider a discrete d-dimensional space described by a d-dimensional lattice viewed as a fixed graph Γ within $\mathbb{R}^d$. Every vertex v is associated with a site random variable X_v taking any n possible value. The lattice's state is given by the joint probability distribution for all variables. At thermal equilibrium the state of the system can be encoded in the Hamiltonian, proportional to the logarithm of the probability distribution. The array of new variables predicted is described by a new renormalized Hamiltonian. The relevant features at large length scales are extracted by integrating out the short distance degrees of freedom. The hidden spins generate effective interactions among the visible ones, while there is no direct connection between the visible spins. If the number of hidden spins is increased, more complex functions of the visible units can be processed. We divide the visible units into two different sets $v = \{v_{i\in A}, v_{i\in B}\}$. Given a distribution Ψ_{RBM} of the visible variables obtained by integrating out the hidden units, the reduced density matrix becomes

$$\rho = \sum_{\{v_{i\in B}\}} \Psi^*_{\mathrm{RBM}}\left(\{v'_{i\in A}, v_{i\in B}\}\right) \Psi_{\mathrm{RBM}}\left(\{v_{i\in A}, v_{i\in B}\}\right). \tag{23}$$

By integrating out the hidden variables, the RBM distribution of the visible spins takes the form

$$\Psi_{\mathrm{RBM}}(v) = \sum_h e^{-E(v,h)} = \prod_{i,j} e^{a_i v_i}\left(1 + e^{b_j + \sum_i v_i W_{ij}}\right), \tag{24}$$

The information content of the distribution Ψ_{RBM} is given by the entanglement entropy of the system, which filters the correlations between any two neighbouring regions or partitions. The von Neumann entanglement entropy is $S = -\mathrm{Tr}(\rho \ln \rho)$.

The spins that remain uncorrelated show no entanglement entropy for any bipartition, while the long range correlations will increase the entanglement entropy. The maximum entanglement entropy of an RBM can be quantified using the bond dimension of the network. Larger bond dimension means more entanglement. In order to model a highly entangled system, the bond dimension (and therefore the memory and computation time) will grow exponentially with the entropy. The maximal entanglement entropy bound is represented by the upper bond dimension $\ln D$, for witch all eigenvalues of the reduced density matrix become $1/D$. Instead of layers of visible and hidden units, we can think of regions or partitions where the hidden units (spins) are been traced out while the sites (regions) are put in contact with each other. The network is cut in a bipartition of visible units with two regions A and $B = B_1 \cup B_2$, where B_1 describes the visible spins that can be reached from A, and B_2 describes the rest. The product state of maximally entangled pairs connecting two neighbouring regions.

$$\Psi_{\mathrm{RBM}}(v) = \psi\left(\{v_{i\in A}, v_{i\in B_1}\}\right)\phi\left(\{v_{i\in B_1}, v_{i\in B_2}\}\right). \tag{25}$$

If m is the number of spins at the interface region of the bipartition, the bond dimension is

$$D = 2^m \tag{26}$$

The bond dimension gives an (upper) bound for the maximal entanglement entropy at the bipartition

$$S_{\max} = m \ln 2. \tag{27}$$

The entanglement entropy bound is dependent on the number of connections in the RBM. If each visible spin has m effective connections to the neighbour visible spins, we need 2^m parameters to describe the system.

3 Conclusion

Tensor networks have been used for studying the holographic duality because their entropy is bounded by the area law in agreement with the Ryu–Takayanagi holographic entanglement formula. RBM directly mapped to tensor networks provide a strong framework to study holographic principle as they obey the entanglement area law. The area law provides only an upper bound for the entanglement entropy, corresponding to a maximum saturation of the network. In the case of deep learning, the entropy is used in the learning phase of a neural network. As the learning process progresses, more hidden nodes are saturated, increasing the impact of relevant units while dampening the less important ones. The less relevant units can be then eliminated to reduce the memory requirements and computational cost for training the neural network. Such a deep learning approach can provide a well-suited framework for exploiting and studying the holographic principle, with potential life-long application in AdS/CFT correspondence. Properties of deep learning networks can be extrapolated and extended to assist with a new information theoretic scenario of holographic duality for a better understanding of locality and RG. Entanglement entropy between a bulk region and the boundary in a given tensor network can be better understood under the new umbrella of deep learning/renormalization group direct mapping, consistent with the expectations of AdS/CFT.

References

1. D.H. Ackley, G.E. Hinton, T.J. Sejnowski, Connectionist models and their implications: readings from cognitive science, *A Learning Algorithm for Boltzmann Machines* (Ablex Publishing Corp., Norwood, NJ, USA, 1988), pp. 285–307
2. K. He, X. Zhang, S. Ren, J. Sun, Deep residual learning for image recognition. CoRR, abs/1512.03385, 2015
3. G.E. Hinton, S. Osindero, Y.-W. Teh, A fast learning algorithm for deep belief nets. Neural Comput. **18**(7), 1527–1554 (2006)
4. E.M. Howard, Astron. Data Anal. Softw. Syst. XXV **512**, 245 (2017)
5. R.M. Neal, Connectionist learning of belief networks. Artif. Intell. **56**(1), 71–113 (1992)
6. A. Tanaka, A. Tomiya, Detection of phase transition via convolutional neural network. J. Phys. Soc. Jpn. **86**(6), 063001 (2017)
7. G. Carleo, M. Troyer, Solving the quantum many-body problem with artificial neural networks. Science **355**, 602 (2017)
8. D.-L. Deng, X. Li, S.D. Sarma, Quantum entanglement in neural network states. Phys. Rev. X **7**, 021021 (2017)
9. Y. Huang, J.E. Moore, Neural network representation of tensor network and chiral states
10. L. Wang, Discovering phase transitions with unsupervised learning. Phys. Rev. B **94**, 195105 (2016)
11. G. Torlai, R.G. Melko, Learning thermodynamics with Boltzmann machines. Phys. Rev. B **94**, 165134 (2016)
12. A. Morningstar, R.G. Melko, Deep learning the Ising model near criticality
13. C. Bény, Deep learning and the renormalization group
14. P. Mehta, D.J. Schwab, An exact mapping between the variational renormalization group and deep learning
15. A. Paul, S. Venkatasubramanian, Why does deep learning work?—A perspective from group theory
16. K.G. Wilson, Renormalization group and critical phenomena. 1. Renormalization group and the Kadanoff scaling picture. Phys. Rev. B **4**, 3174 (1971). K.G. Wilson, Renormalization group and critical phenomena. 2. Phase space cell analysis of critical behavior. Phys. Rev. B **4**, 3184 (1971)
17. J.M. Maldacena, The large-N limit of superconformal field theories and supergravity. Adv. Theor. Math. Phys. **2**, 231 (1998) [Int. J. Theor. Phys. **38**, 1113 (1999)]
18. G. 't Hooft, Dimensional Reduction in Quantum Gravity, THU-93/26
19. G. Vidal, Entanglement renormalization. Phys. Rev. Lett. **99**, 220405 (2007) [cond-mat/0512165]. G. Vidal, Class of quantum many-body states that can be efficiently simulated. Phys. Rev. Lett. **101**, 110501 (2008)
20. B. Swingle, Entanglement renormalization and holography. Phys. Rev. D **86**, 065007 (2012)
21. S. Ryu, T. Takayanagi, Holographic derivation of entanglement entropy from the anti-de Sitter space/conformal field theory correspondence. Phys. Rev. Lett. **96**, 181602 (2006)
22. E. Witten, Anti-de Sitter space and holography. Adv. Theor. Math. Phys. **2**, 253 (1998)
23. L. Susskind, The World as a hologram. J. Math. Phys. **36**, 6377 (1995)

Optimization of Energy Consumption and Throughput in Cognitive Radio Network Using Swarm Intelligence Techniques

Ashim Jyoti Gogoi, Romen Gogoi, Anirban Gogoi, and Deepjyoti Talukdar

Abstract The design of energy and throughput efficient cognitive radio system holds great importance in the present scenario of wireless communications. The distinctive point of the paper is that it proposes a scheme that optimizes both energy consumption and throughput simultaneously in a cognitive radio network. To do the optimization, an optimal ratio is formulated considering both consumed energy and throughput in the cognitive radio network. Moreover, to make the system more energy efficient, energy harvesting is done. The optimization of throughput and energy consumption in the cognitive radio system is done using swarm intelligence techniques like particle swarm optimization (PSO), human behavior-based particle swarm optimization (HPSO), and particle swarm optimization with aging leader and challengers (ALCPSO). Finally, it is shown that the proposed optimization scheme can improve the energy consumption and throughput in the cognitive radio network as compared to that of the conventional schemes.

Keywords Cognitive radio network · Spectral efficiency · Throughput · Particle swarm optimization · Energy harvesting

A. J. Gogoi (✉) · R. Gogoi · A. Gogoi
Department of Electronics and Communication Engineering, Kaziranga University, Assam, India
e-mail: ashimgogoi209@gmail.com

R. Gogoi
e-mail: romengogoi5496@gmail.com

A. Gogoi
e-mail: anirbangogoi23@gmail.com

D. Talukdar
Department of Electrical Engineering, Kaziranga University, Assam, India
e-mail: deepjyoti.talukdar@kazirangauniversity.in

J. Tavares et al. (eds.), *Emerging Technologies in Data Mining and Information Security*, Lecture Notes in Networks and Systems 164,
https://doi.org/10.1007/978-981-15-9774-9_25

1 Introduction

With the advancement of wireless communications, the demand of using radio spectrum is rapidly increasing. In a study done by the Federal Communication Commission (FCC), it is revealed that under static spectrum allocation, only 5–15% of the spectrum is utilized generally [1].

To solve the spectrum under-utilization problem, the regulatory bodies show interest to permit opportunistically access to the unused licensed primary bands (referred to as white spaces) by unlicensed users. To overcome the scarcity, cognitive radio [2] is introduced, which can dynamically allocate the spectrum unused by licensed or primary users to unlicensed or secondary users. The cognitive radio performs spectrum sensing to find the spectrum band unused by primary users, so that the unused band can be utilized by secondary users. Cognitive radio network enables secondary users to opportunistically access the licensed frequency bands [3–6].

Cognitive radio network has received substantial attention as a promising approach to increase spectrum efficiency. Previous works on cognitive radio network mainly focused on sensing accuracy, spectral efficiency, and throughput [7–14]. The optimization of transmission energy in the relay-based cognitive radio network is done by Chatterjee et al. [7]. Banerjee et al. [8] formulate a residual energy maximization problem in which the cognitive radio network opportunistically harvests energy from the signals of primary users. Optimization of throughput in multiuser cognitive radio network using swarm intelligence-based techniques is done in [9]. In [10], the trade-off between sensing time and throughput is addressed using a particle swarm optimization (PSO). The limitation of the previous works [7–14] is that they do not optimize the energy consumption and throughput simultaneously in a cognitive radio network. Therefore, the main aim of the paper is to optimize the throughput and energy consumption simultaneously in the network. To achieve the goal, energy harvesting and different types of swarm intelligence techniques are used. The contributions of the paper are:

Firstly, the optimization of energy consumption and throughput is done in cognitive radio network simultaneously, which is done by formulating an optimal ratio.

Secondly, to make the network more efficient in terms of energy consumption, an energy harvesting scheme is deployed.

Thirdly, optimization of energy consumption and throughput is done simultaneously using swarm intelligence-based techniques such as PSO, HPSO, and ALCPSO.

The remainder of the paper is arranged as follows: the system model is elaborated in Sect. 2, whereas Sect. 3 describes the energy harvesting process and Sect. 4 formulates the optimization problem. Section 5 presents the results, and finally, Sect. 6 concludes the paper.

2 System Model

The system model is shown in Fig. 1, which consists of a primary network and a secondary network. The primary network contains a primary user (PU_s) and a primary destination (PU_d). The secondary network consists of a cognitive source (SU_s), an amplify-and-forward (AF) relay, and a cognitive destination (SU_d). In addition there exists an radio frequency (RF) energy harvesting unit, which harvests energy from primary users. In the primary network, the PU_s transmits data to PU_d through primary base station. In the secondary network, SU_s transmits data to SU_d through the AF relay. With the help of the harvester circuit, the SU_s harvests energy from the signals of PU_s.

In the cognitive radio network, a channel is allocated to SU_s for time frame of duration T_f, which has been split into two slots for spectrum sensing (T_s) and data transmission (T_d), respectively, as shown in Fig. 2. During sensing time T_s, the channel is sensed by SU_s to monitor the existence of primary users. In the sensing process, if none of primary users found active in the sensed channel then SU_s will use remaining duration of the time frame i.e., ($T_f - T_s$) for the purpose of data transmission. If SU_s do not detect idle licensed spectrum, it will not transmit data

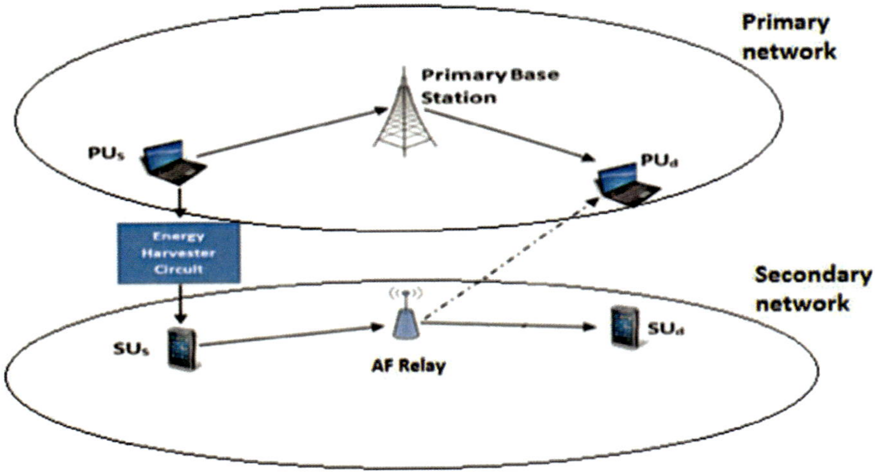

Fig. 1 System model

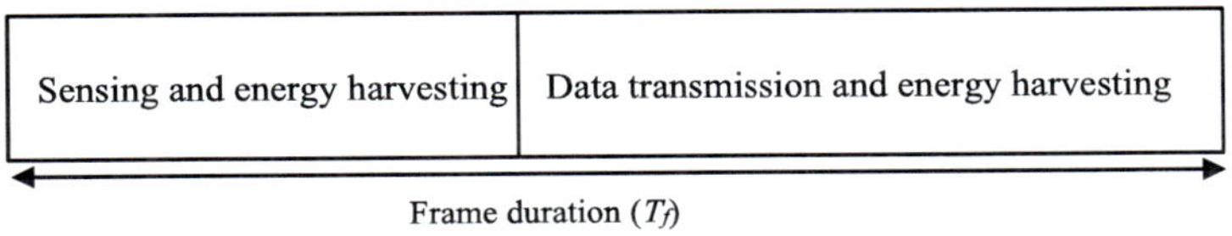

Fig. 2 Time frame of cognitive radio network

during the time frame. As shown in Fig. 2, the energy harvesting process also takes place simultaneously with the spectrum sensing and data transmission process. The energy harvesting process is briefly described in Sect. 3.

2.1 Throughput

The cognitive radio network performs spectrum sensing to detect the existence of primary users in the allotted channel. Once cognitive network detects the absence of primary user in the spectrum, the SU_s would start data transmission. The data transmission process is categorized into two stages: (1) the cognitive source (SU_s) sends data to AF relay and (2) AF relay forward data to cognitive destination (SU_d). The overall throughput of the network is given as [10].

$$R = P(H_0)(1 - P_f)\frac{T_d}{T_f}C_0 \tag{1}$$

Here, P_f is the probability of false alarm, $P(H0)$ is the probability that primary user to be inactive, C_0 is the capacity of cognitive relay network, which can be represented as

$$C_0 = \frac{1}{2}\log_2\left(1 + \frac{P_s h_{sd}^2}{\sigma_n^2} + \frac{\beta^2 P_s h_{sr}^2 h_{rd}^2}{1 + \beta^2 h_{rd}^2 \sigma_n^2}\right) \tag{2}$$

Here, h_{sr}, h_{sd}, h_{rd} signify the fading coefficients between SU_s and AF relay, SU_s and SU_d, AF relay and SU_d, respectively. To protect the primary user, the value probability of detection is taken with the constraint $\overline{P}_d \geq 0.9$ [10]. In (2), β is the amplifying gain of the AF relay and is given as [9]

$$\beta = \sqrt{\frac{P_r}{P_s h_{sr}^2 + \sigma_n^2}} \tag{3}$$

where P_s and P_r are transmit power of the secondary user and the relay, respectively,σ_n^2 is variance of Gaussian noise.

2.2 Energy Consumption

In cognitive radio network, mainly three kinds of energy consumption take place, which are [7] circuit power consumption, spectrum sensing energy consumption, and data transmission energy consumption. The total energy consumption in the network can be expressed as

$$E = T_s P_\tau + P_c T_f + (T_f - T_s)\beta^2 P_s h_{sr}^2 h_{rd}^2 \left[P(H_0)\left(1 - P_f\right) + P(H_1)(1 - P_d)\right] \quad (4)$$

where P_τ is the sensing power and P_c is the circuit power in the network.

3 Energy Harvesting

The cognitive radio network harvests energy from primary users with the help of RF energy harvester circuit. The harvesting process helps to make the system more energy efficient. The RF energy harvester circuit contains a power splitting device, harvester unit, and a receiver unit. Firstly, the receiver unit accumulates the collected RF signal of the primary user and/or noise. The power of received signal is splitted into different proportions for energy harvesting, spectrum sensing, and data transmission with the help of splitter device. The RF energy is harvested by harvester unit with a conversion efficiency η ($\eta > 0$) during both spectrum sensing and data transmission duration of a time frame as shown in Fig. 2. In this paper, the value of η is considered as 0.5 (=50%), because the value is suitable for the given secondary users' power of -10 dBm [15]. The harvested energy during the sensing time T_s is given as

$$E_{h1} = \eta(1 - \rho)T_s\left[\mathrm{P}(H_1)\left(\sigma_n^2 P_p + \sigma_n^2\right)\right] \quad (5)$$

where ρ is power splitting factor, P_p is transmitted power of PU_s, and σ_n^2 is the noise variance. The amount of energy harvested during transmission time can be expressed as

$$E_{h2} = \eta(T_f - T_s)\left[P_d \mathrm{P}(H_1)\left(\sigma_n^2 P_p + \sigma_n^2\right)\right] \quad (6)$$

where P_d is the probability of detection and $P(H1)$ is the probability of primary users to be active. Total harvested can be expressed as

$$E_h = E_{h1} + E_{h2} \quad (7)$$

Putting the value of E_{h1} and E_{h2} from (5) and (6), respectively, in (7),

$$E_h = \eta(1 - \rho)T_s\left[\mathrm{P}(H_1)\left(\sigma_n^2 P_p + \sigma_n^2\right)\right] + \eta(T_f - T_s)\left[P_d \mathrm{P}(H_1)\left(\sigma_n^2 P_p + \sigma_n^2\right)\right] \quad (8)$$

The harvested energy can be utilized for data transmission and spectrum sensing purposes. So, the harvested energy reduces overall energy requirement in cognitive radio network. Therefore,

Effective energy consumption (E_{eff}) = total energy consumption-total harvested energy.

or,

$$E_{\mathrm{eff}} = E - E_h \tag{9}$$

Putting the value of E and E_h from (4) and (8), respectively, in (9),

$$\begin{aligned} E_{\mathrm{eff}} = {} & T_{\mathrm{s}} P_{\tau} + P_{\mathrm{c}} T_{\mathrm{f}} + (T_{\mathrm{f}} - T_{\mathrm{s}})\beta^2 P_{\mathrm{s}} h_{\mathrm{sr}}^2 h_{\mathrm{rd}}^2 [P(H_0)(1 - P_{\mathrm{f}}) + P(H_1)(1 - P_{\mathrm{d}})] \\ & - \eta(1 - \rho) T_{\mathrm{s}} \big[\mathrm{P}(H_1)\big(\sigma_n^2 P_{\mathrm{p}} + \sigma_n^2\big)\big] \\ & - \eta(T_{\mathrm{f}} - T_{\mathrm{s}}) \big[P_{\mathrm{d}} \mathrm{P}(H_1)\big(\sigma_n^2 P_{\mathrm{p}} + \sigma_n^2\big)\big] \end{aligned} \tag{10}$$

4 Problem Formulation

The aim of the paper is to optimize throughput and energy consumption simultaneously in the network. To achieve the objective, a ratio of throughput to the energy consumption in the cognitive radio network is considered. The ratio is taken in such a way that as the ratio increases, the energy consumption decreases and throughput increases in the network. The ratio is referred as optimal ratio in this paper, which is given as,

$$\alpha = \frac{R}{E_{\mathrm{eff}}} \tag{11}$$

Putting the value of R and E_{eff} from (1) and (8), respectively, in (9),

$$\alpha = \frac{P(H_0)(1 - P_{\mathrm{f}}) \frac{T_{\mathrm{d}}}{T_{\mathrm{f}}} C_0}{\begin{aligned} & T_{\mathrm{s}} P_{\tau} + P_{\mathrm{c}} T_{\mathrm{f}} + (T_{\mathrm{f}} - T_{\mathrm{s}})\beta^2 P_{\mathrm{s}} h_{\mathrm{sr}}^2 h_{\mathrm{rd}}^2 [P(H_0)(1 - P_{\mathrm{f}}) + P(H_1)(1 - P_{\mathrm{d}})] \\ & -\eta(1 - \rho) T_{\mathrm{s}} \big[P(H_1)\big(\sigma_n^2 P_{\mathrm{p}} + \sigma_n^2\big)\big] - \eta(T_{\mathrm{f}} - T_{\mathrm{s}}) \big[P_{\mathrm{d}} P(H_1)\big(\sigma_n^2 P_{\mathrm{p}} + \sigma_n^2\big)\big] \end{aligned}} \tag{12}$$

The optimization problem is formulated considering the threshold values of interference $\big(\overline{I}\big)$ and outage probability $\big(\overline{P_{\mathrm{out}}}\big)$ of cognitive radio network. For this, an optimal limit for transmitted power (P_{s}) considering both $\overline{I}$ and $\overline{P_{\mathrm{out}}}$ is formulated, which is given as,

$$\frac{\ln\big(1 - \overline{P}_{\mathrm{out}}\big)^{-1} h_{\mathrm{sr}}^2 h_{\mathrm{rd}}^2}{R_{\mathrm{th}} \sigma_n^2} \le P_{\mathrm{s}} \le \frac{\overline{I}}{h_{\mathrm{sr}}^2 T_{\mathrm{d}}} \tag{13}$$

where R_{th} is the threshold value of throughput in the network.

$$\begin{aligned} & \max \alpha = \frac{R}{E_{\mathrm{eff}}} \\ & \text{s.t.} P_{\mathrm{d}} \ge \overline{P_{\mathrm{d}}} \end{aligned}$$

Table 1 Values of parameters for optimization

Parameter	Value	Parameter	Value	Parameter	Value
T_d	0.22425 ms	$P(H0)$	0.95	σ_n^2	1
T_s	0.012 ms	$P(H1)$	0.7	h_{sr}^2	0.31
β	2.5	$\overline{I}$	−20 dBW	h_{rd}^2	0.031
η	0.5	$\overline{P}_{out}$	0.1	h_{rd}^2	0.31
$\overline{P}_d$	90%	$\overline{P}_f$	10%	R_{th}	0.2 bits/s/Hz

$$P_{\mathrm{f}} \leq \overline{P_{\mathrm{f}}}$$

$$\frac{\ln\left(1-\overline{P}_{\mathrm{out}}\right)^{-1} h_{\mathrm{sr}}^2 h_{\mathrm{rd}}^2}{R_{\mathrm{th}}\sigma_n^2} \leq P_{\mathrm{s}} \leq \frac{\overline{I}}{h_{\mathrm{sr}}^2 T_{\mathrm{d}}}$$

where $\overline{P}_{\mathrm{d}}$ and $\overline{P}_{\mathrm{f}}$ are target probability of detection and target probability of false alarm, respectively. The values of and $\overline{P}_{\mathrm{f}}$ are considered as per the guidelines of FCC [10]. In this paper, a particular scenario of channel condition is considered, and the values of channel coefficients, noise variance, etc., are chosen accordingly. The values of the different parameters considered for the optimization process are given in Table 1.

5 Results and Discussions

The optimization process has been done with the help of standard swarm intelligence techniques such as PSO [16], HPSO [17], and ALCPSO [18]. The reasons to choose the swarm intelligence-based optimization algorithms are that these techniques have been found to be computationally efficient in solving optimization problems in terms of good computational time complexity, fast convergence, and are derivative-free [16–18]. The convergence plots of PSO, HPSO, and ALCPSO are given in Figs. 3, 4 and 5, respectively. The convergence plots shows the values of the optimal ratio for different iterations. From the converge plots, it is noticed that the plot of ALCPSO converges at higher value as compared to that of the other two algorithms, which signifies that the ALCPSO gives greater value of the optimal ratio. The results of the optimization process are given in Table 2. Moreover, the achieved optimal values of energy consumption and throughput in the network are given in Table 3. From Table 3, it is revealed that the system optimized with ALCPSO algorithm consumes less energy than that of the other two algorithms, although throughput achieved is almost same as that of HPSO algorithm. So, it can be inferred that the system optimized with ALCPSO algorithm is more efficient in terms of throughput and energy consumption simultaneously in the cognitive radio network.

In Fig. 6, a comparison of achieved throughput with the proposed scheme and that

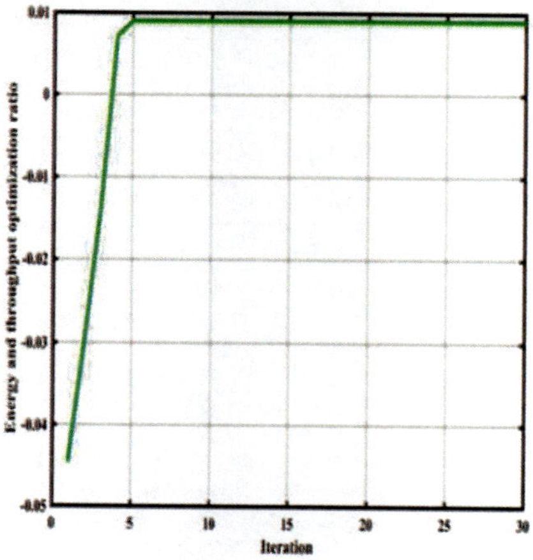

Fig. 3 Convergence plot of PSO

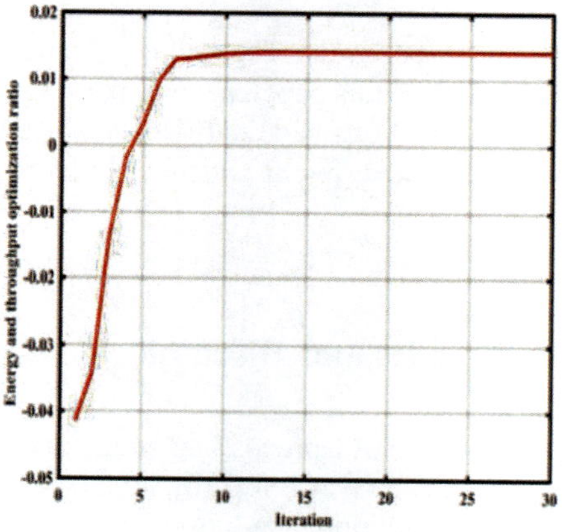

Fig. 4 Convergence plot of HPSO

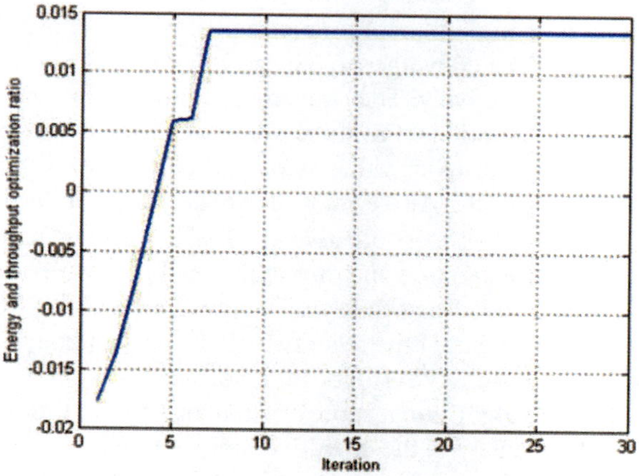

Fig. 5 Convergence plot of ALCPSO

Table 2 Results of the optimization process

Parameters	PSO-based optimum values	PSO-based optimum ratio	HPSO-based optimum values	HPSO-based optimum ratio	ALCPSO-based optimum values	ALCPSO-based optimum values
P_s	0.3980	0.0091	0.2910	0.0127	0.3980	0.0135
P_f	0.0200		0.0500		0.0300	
P_d	0.9000		0.9000		0.9100	

Table 3 Optimized values of energy consumption and throughput

Optimization technique	Energy consumption (mJ)	Throughput (bit/sec/Hz)
PSO	1.197	0.011
HPSO	0.801	0.010
ALCPSO	0.7152	0.010

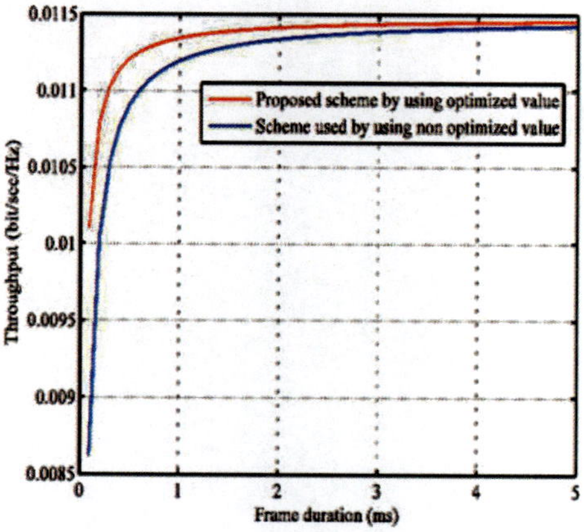

Fig. 6 Comparison of throughput obtained with non-optimized value throughput

of the scheme with non-optimized scheme is shown, which shows the graphs between throughput versus frame duration. From Fig. 6, it is revealed that the proposed scheme achieves more throughput than the other scheme in the cognitive radio network. A comparison of energy consumption in the network using proposed scheme with that of the scheme used in [3] is shown in Fig. 7, which shows the variation of energy consumption with respect to received signal-to-noise ratio (SNR) in SU_d. Figure 7

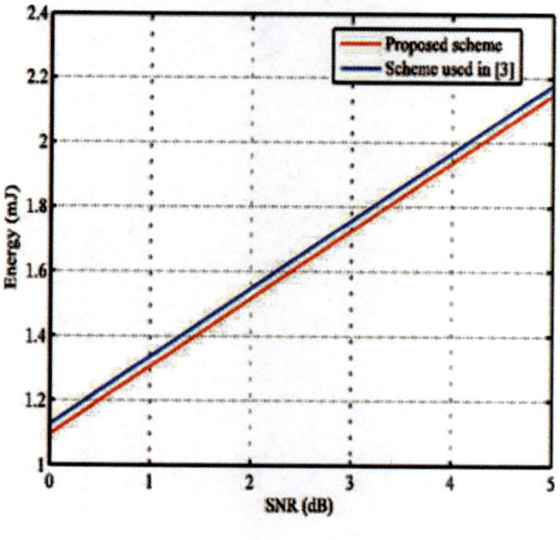

Fig. 7 Comparison of energy consumption

signifies that the proposed scheme makes the network more energy efficient than that of the other. Hence, the comparisons of the obtained results with other existing results, signify the improvement in the throughput and energy consumption. The optimal ratio that is formulated with the combination of throughput and energy consumption in the network helps to optimize both parameters simultaneously. The deployment of energy harvesting scheme is helpful to make the network energy efficient. Moreover, the application of swarm intelligence-based techniques contributes to optimize the energy consumption and throughput simultaneously and more efficiently than the conventional schemes.

From the above results, it is clear that the proposed scheme efficiently optimizes both energy consumption and throughput simultaneously. But in conventional schemes only one of the parameters, i.e., either energy consumption or throughput is optimized at a time [3–10]. Moreover, it can be noticed from Figs. 6 and 7, that the proposed scheme is more energy and throughput efficient than the aforementioned schemes.

6 Conclusions

In this paper, a scheme to optimize energy consumption and throughput simultaneously in cognitive radio network has been proposed. In the proposed model, due to deployment of the energy harvesting scheme, the network becomes energy efficient. To optimize energy consumption and throughput simultaneously, an optimal ratio is formulated. In the optimization process, the optimal limits of system parameters like probability of false alarm, probability of detection, and transmitted power of cognitive source are considered to make the system reliable. Comparative results show that the proposed scheme makes the system more efficient in terms of energy consumption and throughput simultaneously than the existing systems. To increase the efficiency and reliability of cognitive radio network, more parameters of the network like routing time, interference temperature, etc., will be optimized using multiobjective optimization techniques. For better spectral efficiency in cognitive radio network, prediction-based spectrum sensing techniques will be implemented.

References

1. Spectrum Efficiency Working Group, *Report of the Spectrum Efficiency Working Group," Federal Communications Commission* (Tech, Rep, 2002).
2. J. Mitola, G.Q. Maguire, Cognitive radio: making software radios more personal. IEEE Pers. Commun. **6**, 13–18 (1999)
3. A. J. Gogoi, H. A. Choudhury and K. L. Baishnab,*Swarm Intelligence Based Optimization of Energy Consumption in Cognitive Radio Network*, vol. 36, no. 3, pp. 2399–2407 (2019)

4. Y.Pei,A.T. Hoang, Y.C Liang, Sensing-Throughput tradeoff in cognitive radio networks: how frequently should spectrum sensing be carried out? in*The 18th Annual IEEE International Symposium on Personal, Indoor and Mobile Radio Communications* (2007)
5. L. Li, X. Zhou, H. Xu, G. Li., D. Wang, and A. Soong, Energy-Efficient transmission in cognitive radio networks, in*7th IEEE Consumer Communications and Networking Conference*, pp. 1–5 (2010)
6. S. Huang, H. Chen, Y. Zhang, Optimal power allocation for spectrum sensing and data transmission in cognitive relay networks. IEEE Wirel. Commun. Lett. **1**(1), 26–29 (2012)
7. S. Chatterjee, S.P. Maity, T. Acharya, Energy efficient cognitive radio system for joint spectrum sensing and data transmission. IEEE J. Emerging Sel. Topics Circ. Syst. **4**(3), 292–300 (2014)
8. A. Banerjee, S.P. Maity, S. Roy, On residual energy maximization in energy harvesting cognitive radio network, in*IEEE International Conference on Wireless Communications and Networking Conference (WCNC)*, pp. 1–6 (2017)
9. A.J. Gogoi, N.M Laskar, C.L Singh, K.L. Baishnab, *Throughput Optimization of Cognitive Radio Network Using Swarm Intelligence Techniques*, vol. 14, no. 11, p. 443 (2016)
10. R.A. Rashid et al., Efficient in-band spectrum sensing using swarm intelligence for cognitive radio network. Can. J. Electr. Comput. Eng. **38**(2), 106–115 (2015)
11. P. Verma, B. Singh, Threshold optimization in energy detection scheme for maximizing the spectrum utilization. Proc. Comput. Sci. **33**, 191–198 (2016)
12. A.O. Ercan, M.O. Sunay, Energy sensing strategy optimization for opportunistic spectrum access. IEEE Commun. Lett. **16**(6), 828–830 (2012)
13. P. Yiyang, L. Ying-Chang, Energy-Efficient design of sequential channel sensing in cognitive radio networks: optimal sensing strategy, power allocation, and sensing order. IEEE J. Sel. Areas Commun. IEEE J. **29**(8), 1648–1659 (2001)
14. Y. Wu, D.H.K. Tsang, Energy-efficient spectrum and transmission for cognitive radio system. IEEE Commun. Lett. **15**(5), 545–547 (2011)
15. M. Abdallah, J. Costantine, A.H. Ramadan, C.G. Christodoulou, K.Y. Kabalan, Wide power range RF energy harvesting circuit, in *IEEE International Symposium on Antennas and Propagation & USNC/URSI National Radio Science Meeting*, 2011, pp. 1296–1297
16. J. Kennedy, Particle swarm optimization, in *Encyclopedia of Machine Learning* (Springer, US, 2011), pp. 760–766
17. L. Hao, Human behavior-based particle swarm optimization. Sci. World J. (2014)
18. W.-N. Chen, Particle swarm optimization with an aging leader and challengers. IEEE Trans. Evol. Comput. **17**(2), 241–258 (2013)

Estimation of Various Parameters for AES, DES, and RSA

Jyoti Agarwal, Manoj Kumar, and Atul Kumar Srivastava

Abstract Securing the data from unauthorized access is turn out to be a major challenge nowadays. Large volume of data is available in every field for which the security is most crucial. To secure the data, the security experts in last few decades implement various techniques. The main domain, which provide data security, is known as cryptography. This provide security to data against the unauthorized access. Encryption and decryption are two basic terms in cryptography. With the evolution of techniques and intelligence, the chances of getting unauthorized access on data are increased with multifold scenario. To overcome with this, the researchers use various security algorithms. In this research paper, a practical analysis between AES, DES, and RSA is implemented to assess their performance parameters in respect of their space complexity, encryption time, and throughput. Experiments are carried out on various type of data files so that the performance of three can be analyzed.

Keywords Cryptography · Data security · Encryption · DES · AES · RSA

1 Introduction

In today's scenario, securing of the information from unauthorized access is a major concern. Organizations are working for providing data security. Cryptography provides a way of securing the data information (plain text) from unauthorized access

J. Agarwal · A. K. Srivastava
Amity University Tashkent, Tashkent, Uzbekistan
e-mail: jagarwal1@amity.edu

A. K. Srivastava
e-mail: aksrivastava1@amity.edu

M. Kumar (✉)
School of Computer Science, University of Petroleum and Energy Studies (UPES), Dehradun, India
e-mail: wss.manojkumar@gmail.com

J. Tavares et al. (eds.), *Emerging Technologies in Data Mining and Information Security*, Lecture Notes in Networks and Systems 164,
https://doi.org/10.1007/978-981-15-9774-9_27

by encrypting the data (cipher text) in such a way that it can be decrypt only be the authorized receiver. Cryptography is an indication of loading and spreading data in a way that only the approved users are able to interpret the data [1]. Encryption can provide the implications, that are required in day-to-day security scenarios [2]. Confidentiality, authentication, data integrity, non-repudiation, and access control are five main concerns for any security algorithm [3]. Cryptography means to convert plain data into cipher text before sending it to the receiver and to get back the plain text by the receiver by decrypting the cipher text. This is achieved by keys, which may be same or different for sender and receiver [4]. If same keys are used by sender and receiver, then it will be called as symmetric key but if keys are not similar, then it will be called as asymmetric key [5, 6]. Asymmetric key cryptography is more secured compared to symmetric key cryptography, but more processing is required for this [7]. As there is increase in the ciphering unit, the overheads get slightly high in public key cryptography. There are various security algorithms using symmetric and asymmetric key cryptography. All the algorithms have their own advantages and disadvantages. It is required to study about these security algorithms to identify their strength and weakness, so that one can select the correct algorithm according to the need of their research work. The main aim of this work is to study and understand various cryptographic algorithms. The algorithms, which are part of this work, are AES, DES, and RSA. A performance comparison is discussed to select the appropriate algorithms as per the research requirement.

The paper is classified into five sections. Literature review is described in Sect. 2. Section 3 deals with studied algorithms followed by the experimental work in Sect. 4. Conclusion and the future scope of the research are summarized in Sect. 5.

2 Literature Review

This section discusses various past studies related to security algorithms. Cryptography may be symmetric key or asymmetric key cryptography [8]. The difference between both methods lies in terms of number of keys. Same key is used in symmetric key cryptography for encryption and decryption [9, 10]. Blowfish, Twofish, Rijndeal, Data Encryption Standard (DES), Triple DES, Advanced Encryption Standard (AES) are example of symmetric key cryptography algorithms. In asymmetric key cryptography, public and private key pair are used for encryption and decryption [9, 10]. Cipher text is generated by applying the sender's public key but private key of the receiver is used for decrypting the cipher text back into the plain text. Diffie–Hellman (DH), Rivest–Shamir–Adleman (RSA), Secure Shell (SH) are examples of asymmetric key cryptography algorithms. Figures 1 and 2 show the basic working of symmetric and asymmetric cryptography.

These algorithms have their own pros and cons. In this paper, focus has been given on three most popular security algorithms, which are data encryption standard (DES) algorithm, advanced encryption standard (AES) algorithm, and Rivest–Shamir–Adleman (RSA) algorithms.

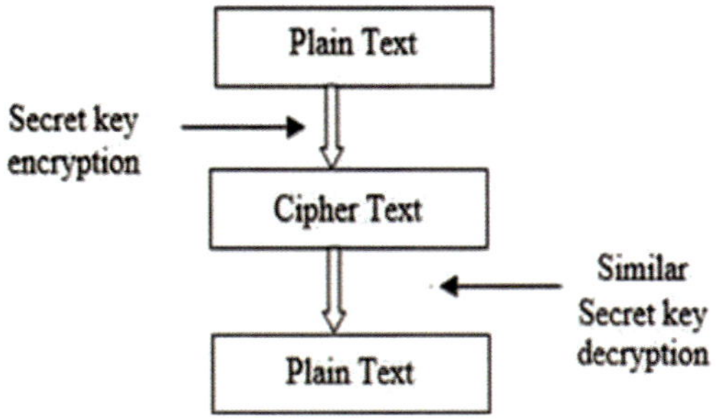

Fig. 1 Symmetric key cryptography

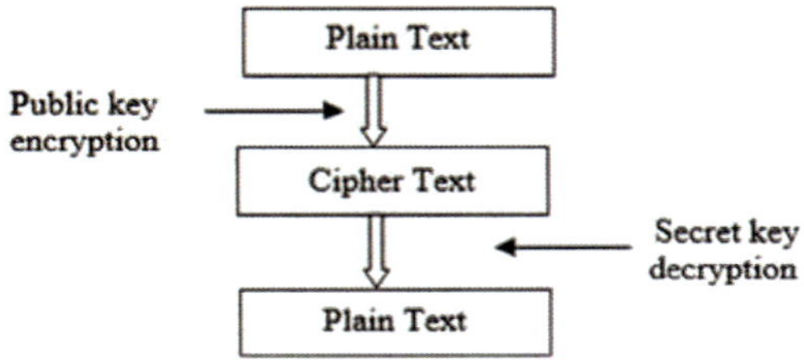

Fig. 2 Asymmetric key cryptography

DES has been applied on the images to enhance the transmission security on web. Experiment was performed on different size of images [11]. A hybrid approach of AES and RSA algorithm was used to provide data security in Bluetooth transmission [12]. AES algorithm was also used to provide at a security in android phones. Results were identified for different size of files to secure the files from unauthorized access [13]. Research was done to improve the Kerberos protocol using fast RSA algorithm [14]. Wang and Zang [15] applied RSA algorithm to secure the personal information of the customers. AES, RSA, and DES algorithms were also used to provide security to cloud data [16]. Performance of AES and DES algorithms was compared, and it was analyzed that AES algorithm performs better than DES algorithm in terms of throughput and execution time [17]. Singh and Supriya [18] have performed a comparative study between RSA, DES, 3-DES, and AES algorithms.

3 Studied of Assessed Cryptographic Algorithms

The paper implements three cryptographic algorithms, respectively, DES, AES, and RSA. The explanation of these algorithms is discussed in this section.

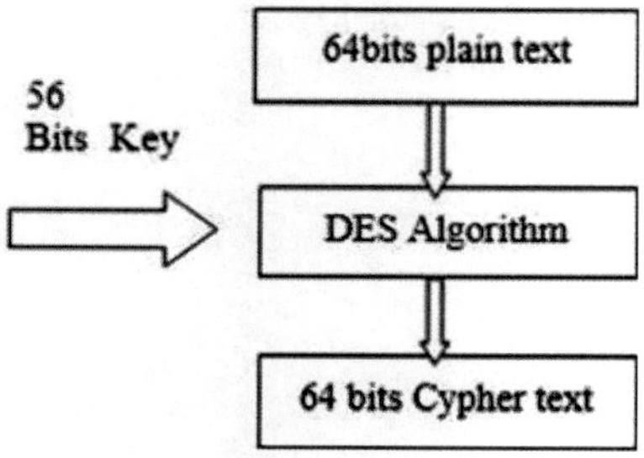

Fig. 3 Working of DES algorithm

3.1 *Data Encryption Standard Algorithm (DSA)*

IBM gave this algorithm in the year 1970. The key size for this algorithm is of 56 bits. It works on the data in-group of 64 bits. The algorithm works on two different data blocks (left and right part of the block). Each block is of 32 bits. DES is symmetric metric key cryptography algorithm. This algorithm takes total 16 steps to complete the task. The general framework of the algorithm is shown in Fig. 3.

3.2 *Advanced Encryption Standard Algorithm (AES)*

Joan Daemen identified this algorithm in 1998 [19]. It works on 128 bits of data and follow the concept of symmetric key cryptography. The key size in this algorithm is not fixed, it can vary from 128 to 256 bits. Based on the key size, total number of steps in these versions of AES algorithm are 10, 12, and 14, respectively. The algorithm is well described by Mahajan and Sachdeva [20].

3.3 *Rivest, Shamir, and Adleman Algorithm (RSA)*

This algorithm was given by Rivest, Shamir, and Adleman in 1977 [19]. The size of data block and key can vary in this algorithm. Prime numbers are used to generate the public and private key between size of 102-4096 bits. It follows the concept of asymmetric key cryptography. A comparison study between above mentioned algorithms is done, and after analysis, it was observed that the results identified by AES algorithm were better in terms of time, speed, and throughput [21]. DES, 3DES, AES, RSA, and Blowfish algorithms were also compared against the prevention of security attacks [19]. They identified that blowfish algorithm is better in terms of time and memory while DES algorithm is better if confidentiality and integrity are the major concerns. Performance of AES, DES, and RSA algorithm is also checked by applying the cryptography on audio, video, text, and image file [6]. A hybrid algorithm (AES-DES-RSA) was used for securing the data, and the results were

Table 1 Comparison of security algorithms

	Algorithms		
Security factors	AES	DES	RSA
Encryption time	Very low	Moderate	High
Memory Used	Low	Moderate	High
Throughput	High	Low	Very low

Table 2 Difference between security algorithms

	Algorithm		
Factors	AES	DES	RSA
Proposed in (year)	2000	1977	1978
Key size	128, 192, 256 bits	56 bit	>1024 bits
Block size	128 bits block size	64 bits block size	Minimum 512 bits block size
Type of key	Symmetric	Asymmetric	Asymmetric
Scalability	Not scalable	scalable	Not scalable
Encryption time	Fast	Moderate	Slow
Decryption time	Fast	Moderate	Slow
Security	Highly secured	Secured	Least secured
Types of attacks	Brute force	Brute force, linear, differential cryptanalysis	Brute Forced, Oracle
Rounds	10/12/14	16	1

also compared against the RSA, DES, and AES algorithm and it was analyzed that hybrid algorithm provide better results by adopting the advantages of all the three algorithms [22]. Differentiation between these algorithms is given in Table 1 based in three parameters, i.e., encryption time, memory, and throughput. Various security factors are also used to compare the performance of these algorithms and described in Table 2.

4 Experimental Work

In this section, comparison between these algorithms is performed based on three factors, i.e., encryption time, memory used, and throughput.

4.1 Time Required for Encryption

The amount of time, required for the encryption depends on the size ok key, block size of original message and the conversion mode [23]. Encryption time influences the performance of the system. Less encryption time results in the better performance of the system in terms of response time [24].

4.2 Memory Used

Any encryption algorithm required a variable memory size for its implementation. This amount of memory will be based on size of the key, number and type of actions performed by the security algorithm. System cost is also affected by the memory requirement. Cost of the system increases with high memory requirement [20].

4.3 Throughput

Throughput means rate of production. Throughput is the rate at which information or message is getting encrypted [25]. High encryption throughput will be commendable and will lead to safe exchange and securing of information through Internet or web. Encryption time, memory used, and throughput for all the three algorithms are analyzed for the data and presented in Table 3. Graphical representation of this analysis is shown in Figs. 4, 5 and 6, respectively.

In this research paper, the best algorithm is identified that provides better results using various quality parameters on data. The estimated results state that AES provides better results as it takes minimum amount of time and less memory space. Highest throughput was also achieved by AES algorithm. Thus, the result shows that AES algorithm provides better results for text data files.

Table 3 Analysis based on encryption time, memory, and throughput

	Algorithm		
Factors	AES	DES	RSA
Encryption time (ms)	0.562	3.651	5.897
Memory used (kb)	14.2	18.9	31.7
Throughput (bits/s)	890.5487	482.1548	46.2888

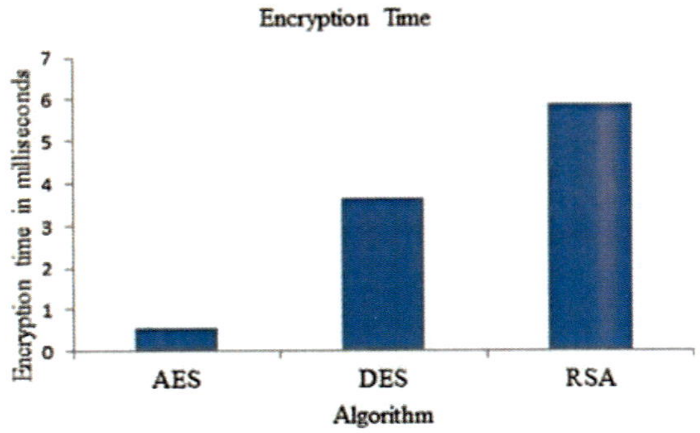

Fig. 4 Encryption time estimation

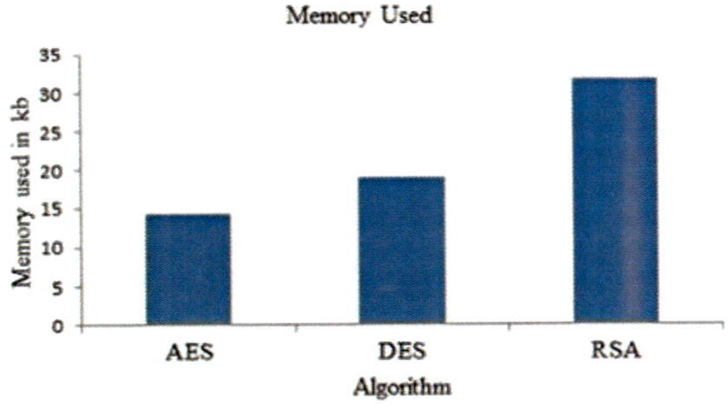

Fig. 5 Memory estimation

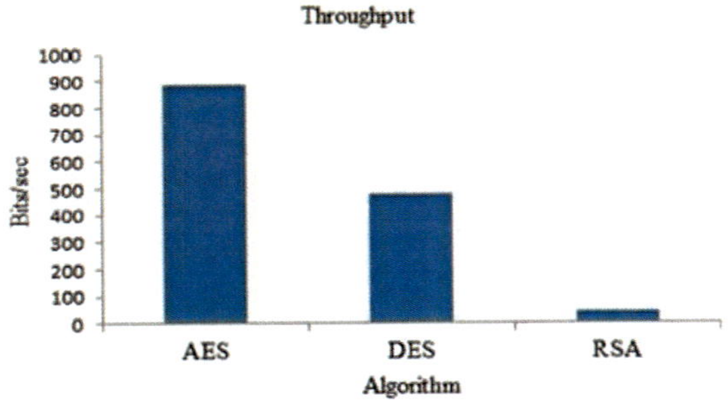

Fig. 6 Throughput estimation

5 Conclusion and Future Scope

This research paper is based on three most popular security algorithms, i.e., DES, AES, and RSA. Experiments are carried out on a text message to identify the encryption time, memorys used, and throughput for all the three algorithms. From the results, it is demonstrated that AES algorithm gives better results for text messages as it takes least amount of time, least memory and provide highest throughput. In this work, key size is 128 bits for AES algorithm, 56 bits for DES algorithm, and 2056 bits for RSA algorithm is used. In future, different key size can be used for AES algorithm

(192 or 256 bits) and RSA algorithm and analysis can also be done on a large data message. New hybrid cryptographic algorithms can also be developed by applying the advantages of these algorithms.

References

1. S. Kumari, A research paper on cryptography encryption and compression techniques. Int. J. Eng. Comput. Sci. **6**(4), 20915–20919 (2017)
2. N. Arora, P. Goel, A. Sehgal, S. Bhatia, Cryptography and related techniques. Int. J. Sci. Eng. Sci. **1**(12), 58–60 (2017)
3. G. Singh, A study of encryption algorithms (RSA, DES, 3DES and AES) for information security. Int. J. Comput. Appl. **67**(19), 33–38 (2013)
4. K.E. Prasetyo, T.W. Purboyo, R.E. Saputra, A survey on data compression and cryptographic algorithms. Int. J. Appl. Eng. Res. **12**(33), 13589–13595 (2017)
5. M.A. Hossain, M.B. Hossain, M.S. Uddin, S.M. Imtiaz, Performance analysis of different cryptography algorithms. Int. J. Adv. Res. Comput. Sci. Softw. Eng. **6**(3), 659–665 (2016)
6. N. Arora, K. Metha, K. Anjali, M. Bhatt, K. Aradhana, K.C. Mishra, M. Martolia, Performance analysis of AES, DES and RSA cryptographic encryption algorithms: a novel study. Int. J. Electr. Electron. Comput. Sci. Eng. **3**(3), 5–13 (2016)
7. J. Thakur, N. Kumar, DES, AES and Blowfish: symmetric key cryptography algorithms simulation-based performance analysis. Int. J. Emerg. Technol. Adv. Eng. **1**(2), 6–12 (2011)
8. Y. Kumar, R. Munjal, H. Sharma, Comparison of symmetric and asymmetric cryptography with existing vulnerabilities and countermeasures. Int. J. Comput. Sci. Manage. Stud. **11**(3), 60–63 (2011)
9. R. Tripathi, S. Agarwal, Comparative study of symmetric and asymmetric cryptography techniques. Int. J. Adv. Found. Res. Comput. **1**(6), 68–76 (2014)
10. D.S.A. Elminaam, H.M.A. Kader, M.M. Hadhoud, Performance evaluation of symmetric encryption algorithms. Int. J. Comput. Sci. Netw. Secur. **8**(2), 280–286 (2008)
11. P.P. Dang, P.M. Chau, Image encryption for secure internet multimedia applications. IEEE Trans. Consum. Electron. **46**(3), 395–403 (2000)
12. K. Rege, N. Goenka, P. Bhutada, S. Mane, Bluetooth communication using hybrid encryption algorithm based on AES and RSA. Int. J. Comput. Appl. **71**(2), 10–13 (2013)
13. S. Tayde, S. Siledar, File encryption, decryption using AES algorithm in android phone. Int. J. Adv. Res. Comput. Sci. Softw. Eng. **5**(5), 550–554 (2015)
14. D. Hu, Z. Du, An improved Kerberos protocol based on fast RSA algorithm, in *Proceedings of IEEE International Conference on Information Theory and Information Security*, pp. 274–278 (2010)
15. L. Wang, Y. Zhang, A new personal information protection approach based on RSA cryptography, in *Proceedings of IEEE International Symposium on IT in Medicine and Education*, vol. 1, pp. 91–593 (2011)
16. R. Arora, A. Parashar, Secure user data in cloud computing using encryption algorithms. Int. J. Eng. Res. Appl. **3**(4), 1922–1926 (2013)
17. S. Pavithra, Performance evaluation of symmetric algorithms. J. Glob. Res. Comput. Sci. **3**(8), 43–45 (2012)
18. G. Singh, A. Supriya, A study of encryption algorithms (RSA, DES, 3DES and AES) for information security. Int. J. Comput. Appl. **67**(19), 33–38 (2013)
19. M.N.A. Wahid, A. Ali, B. Esparham, M. Marwan, A comparison of cryptographic algorithms: DES, 3DES, AES, RSA and Blowfish for Guessing Attacks Prevention. J. Comput. Sci. Appl. Inf. Technol. 1–7 (2018)
20. P. Mahajan, A. Sachdeva, A study of encryption algorithms AES, DES and RSA for security. Glob. J. Comput. Sci. Technol. **1**(2), 1–6 (2013)

21. G. Singh, A. Singla, K.S. Sandha, Cryptography algorithm comparison for security enhancement in wireless intrusion detection system. Int. J. Multidisc. Res. **1**(4), 143–151
22. S.N. Ghosh, Performance analysis of AES, DES, RSA and AES-DES-RSA hybrid algorithm for data security. Int. J. Innov. Emerg. Res. Eng. **2**(5), 83–88 (2015)
23. O.G. Abood, S.K. Guirguis, A survey on cryptography algorithms. Int. J. Sci. Res. Publ. **8**(7), 2250–3153 (2018)
24. S. Pavithra, E. Ramadevi, Study and performance analysis of cryptography algorithms. Int. J. Adv. Res. Comput. Eng. Technol. **1**(5), 82–86 (2012)
25. N. Kaur, S. Sodhi, *Data Encryption Standard Algorithm (DES) for Secure Data Transmission International Conference on Advances in Emerging Technology (ICAET)*, pp. 31–37 (2016)

Semi-automated Ontology Creation and Upgradation for Rail-Road Incidents: A Case of a Steel Plant in India

Anima Pramanik, Sobhan Sarkar, V. Sai Siddharth, J. Maiti, and Pabitra Mitra

Abstract Most of the earlier studies for ontology creation and upgradation are carried out using structured data. Very few of them focused on unstructured texts for this purpose, which are restricted due to huge human involvement. In order to address this issue, in this study, we have proposed a methodology for ontology creation and upgradation using both unstructured (text) and structured data of rail-road incidents obtained from a steel plant. Features are predicted from the texts using machine learning (ML) techniques and added to structured data, which is finally used for ontology upgradation. This is performed in a semi-automated way, thereby substantially reducing level of human efforts. Ontology is created using Web Ontology Language (OWL) language in Protégé software. Two ML techniques, Random Forest and Support Vector Machine are used to see how accurately the logical relationships are established between classes/sub-classes used in this ontology. Hence, the study proves the effectiveness of the methodology for ontology creation and upgradation.

Keywords Text classification · Semi-automated ontology · Protégé · OWL language · Rail-road incidents

A. Pramanik · J. Maiti
Department of Industrial & Systems Engineering, IIT Kharagpur, Kharagpur, India
e-mail: apramanik17@gmail.com

J. Maiti
e-mail: jhareswar.maiti@gmail.com

S. Sarkar (✉)
Management Science, Business School, The University of Edinburgh, Edinburgh, UK
e-mail: sobhan.sarkar@gmail.com; sobhan.sarkar@ed.ac.uk

V. Sai Siddharth
Department of Ocean Engineering and Naval Architecture, IIT Kharagpur, Kharagpur, India
e-mail: sidvud98@gmail.com

P. Mitra
Department of Computer Science & Engineering, IIT Kharagpur, Kharagpur, India
e-mail: pabitra.mitra@gmail.com

J. Tavares et al. (eds.), *Emerging Technologies in Data Mining and Information Security*, Lecture Notes in Networks and Systems 164,
https://doi.org/10.1007/978-981-15-9774-9_28

1 Introduction

The occurrence of incidents is a staggering issue for any industry like steel manufacturing industry. These incidents have a high impact on human life, safety, profits, and environment [1–4]. Due to the advancements in technology, information or knowledge regarding these incidents are stored for future use in the form of reports. An ample amount of priority is rendered by industries to analyse and extract useful information from these reports so that after analysing the reports [11], the number of occurrences can be mitigated by adopting proactive measures. This can be possible if the ontological relationships between the classes correspond to each incident is established. Therefore, ontology creation and utilization is very essential for analyzing any incident in future. By definition, an ontology is a set of concepts and categories in a subject area or domain that shows their properties and the relations between them [5]. It is used to decrease the complexity of the problem and organize data based on classes and their properties. It offers a better format of relationships and visualization of the same. Creating an ontology from the available reports helps enormously to organize the data. Along with ML techniques, it also assists in analysis, decision-making, and knowledge management tasks. Several studies have been carried out for ontology creation and upgradation [6, 7]. Majority of them are based on structured data. Very few focused on unstructured texts due to manual involvement [5, 8–10]. Here, features are extracted from texts by using either manual checking or text pre-processing. Text classification is a method of categorizing the texts or documents into disjoint classes. Using machine learning (ML), this task can be automated, fast, and efficient. With natural language processing (NLP) as a powerful ML tool, each document or text is assigned to a set of categories that makes it easier to manage, sort, and analyse the data. These categories extracted from texts can be applied in ontology creation and upgradation in such a manner so that human efforts can be minimized. With ever-increasing volume of data, it is necessary to deploy a powerful approach for information retrieval from those data. Though many of the approaches are proposed for web information retrieval [6], semantic analysis [7], many of them cannot be applied on unstructured data. Data containing ample amount of noise obstruct the extraction of useful information. Under the circumstances, ontology can be extremely useful in structuring the data in a straightforward and understandable format. Additionally, it enables analysts to extract the relationships among different classes and consequently, many useful information can be retrieved by simple queries.

To exemplify, Sanchez-Pi et al. [8] proposed information extraction technique to contend with tremendous volume of information collected from oil industry, whose task of analysis was previously daunting. In 2014, Sanchez-Pi et al. [10] proposed in their paper to automate the process of producing structured information from unstructured text. This was endeavored to get impeccable and concise bundle of results rather than thousands of abstracts. Another interesting work by Chi et al. [12] explores a strategic way of utilizing pre-existing safety equipment and infrastructure in construction industry to reduce the requisite human efforts while maintaining or even increasing the effectiveness of job hazard analysis. Sanchez-Pi and Martí [13] explored an information retrieval approach for occupational health and security application. It is premised on previous work of ontology-based heuristic algorithm. In 2017, Mujtaba et al. [14] aimed to develop an automatic multi-class classification system for predicting incident-related causes of death from free text autopsy reports by using expert-driven feature selection. Among the recent works, Hughes et al. [5] proposed an ontology-based approach to extract meaning from multi-lingual safety incident reports with overall true positive rate of 98.5%.

Based on the above-mentioned discussion, though in brief, the following two issues can be identified:

(i) Most of the incident reports from industries remain unstructured and unorganized in nature that raises the level of difficulty in sorting and analysing the data.
(ii) Very limited works are done so far, which use text processing with machine learning techniques in order to create and update the ontology.

Based on these research issues, the present study contributes in the following ways:

(i) The paper proposes a way for creation and upgradation of an ontology using text processing with ML techniques. Features are predicted from the unstructured text and added with structured data to upgrade the ontology. For the advent of any new report, if predicted features are already present in the existing ontology, then new relationship between classes is established automatically, otherwise, human interaction is required to make the relationship. This results in semi-automated ontology upgradation. It creates a simple and efficient way of information retrieval using structured query language (SQL) queries that can be used by any user.
(ii) The proposed methodology has been applied to build ontology for rail-road incidents in order to improve rail-road safety of a steel plant.

The remainder of this paper is organized in the following ways: in Sect. 2, the proposed methodology has been explained in details. A case study of a steel plant is presented in Sect. 3. The experimental results are reported in Sect. 4 and finally, the conclusions including scope for future works are discussed in Sect. 5.

2 Proposed Methodology

Our main aim is to create an ontology for rail-road incidents in a steel plant and upgradation of ontology using text classification. Our investigation consists of two phases (refer to Fig. 1): (i) data pre-processing, and (ii) ontology creation and upgradation. A short description of each phase is given in the following sections.

2.1 Phase 1: Data Pre-processing

In this phase, brief description (unstructured texts) of each rail-road incident is used in ontology creation. Text pre-processing is required to convert unstructured texts into structured form. Some basic text pre-processing techniques [15], including stemming, part-of-speech (POS) tagging, and bag-of-words are used in this study. After data pre-processing, key information (words) are retrieved from each text. Thereafter, two new attributes, namely "hazardous element" and "initiating mechanism" are created using the retrieved information for each descriptive text. Hazardous element here refers to the source that can cause the incident and initiating mechanism denotes the triggering element for hazardous element to cause the incident. Other categorical attributes-value (class) correspond to each rail-road incident are extracted from the data. Finally, the classes of existing attributes are added with the classes of hazardous element and initiating mechanisms are added and used in ontology creation, which is described in the following section.

2.2 Phase 2: Ontology Creation and Upgradation

It has two basic parts of operations: (i) ontology creation, and (ii) upgradation of ontology, which are explained below.

2.2.1 Ontology Creation

Ontology is the process by which a logical relationship is established between features/classes. Features are nothing but attributes present in the data, and classes indicate the categories of each attribute. Features can be divided into a set of classes based on some characteristics, such as sets, collections, concepts, etc. Classes can be divided into a set of sub-classes based on same properties. Class creation involves the aforesaid two steps. After class creation, a logical relationship is established between classes/sub-classes using some properties, which is shown in the phase/block (refer to Fig. 1), named 'object property creation' in Fig. 1. Data type property is used for each class/subclass to make an ontological relationship between them. After creating the object property, individuals of each class/sub-class are created (refer to 'Individual creation' in Fig. 1). Individuals are the instances or objects of each class that satisfy the object properties with their corresponding domain and range which correspond to class and a set of its related classes, respectively.

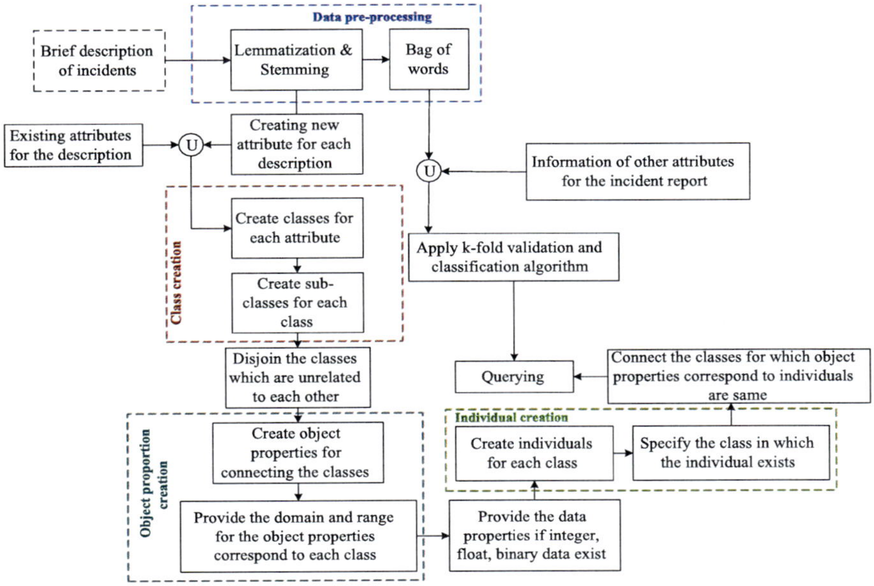

Fig. 1 Flowchart of the proposed methodology

2.2.2 Ontology Upgradation

Upgradation of ontology is essential when an information of new incident comes. The upgradation has two tasks: (i) feature-vale prediction using ML-based classification, and (ii) querying using predicted feature-values. As said earlier, the classes of two features, namely 'hazardous element' and 'initiating mechanism' are retrieved from the texts. Therefore, these two feature-values are predicted using ML techniques. We have trained two ML models, namely Support Vector Machine (SVM) [16, 17] and Random Forest (RF) [15, 18, 19] using the existing ontology-values to retrieve the actual information for the aforesaid two features used for ontology upgradation. To upgrade/update the ontology, the predicted individuals (feature-values) are fed into the ontology through a query in SPARQL. If the ontology provides information about the features that are not present in the report, then a new relation will be created by human involvement to upgrade the ontology. Otherwise, all the predicted features are treated as individuals present in the existing ontology. A new relationship between all the classes is established automatically, using the aforesaid individuals. As the data become structured after the generation of predicted features using ML-techniques and querying by ontology, it can provide good classification accuracy. We have applied this methodology to prepare the ontology for rail-road incidents in a steel plant, as described in the next section.

3 Case Study

In this research, we have conducted a case study of a steel plant in India. During this study, we collected rail-road incident data from this plant. Whenever any incident is occurred in plant rail-road, it is logged into the IT system, and a dataset is generated called preliminary investigation report (PIR). A total of 188 such incident records are collected and used in this study for ontology creation. The attributes in the dataset used for ontology creation are given as follows: (i) *Division*: It implies the division where an incident takes place. It has fourteen classes; (ii) *Department*: It implies the department where an incident takes place. Under each division, there are more than twenty departments present; (iii) *Brief description*: It is a text data that contains the cause and consequence of any incident taken place; (iv) *Impact*: It indicates whether any person, property, or environment is affected due to the incident; (v) *Potential*: It represents the outcome of the incident, such as injury, property damage, or environment damage; (vi) *Behavior*: It implies human behavior. It has three classes, such as unsafe act, unsafe condition, and unsafe act and unsafe condition; (vii) *Primary cause*: It is the root cause of the incident. It has 12 categories; (viii) *Hazardous element*: This is source that leads to any incident. It has seven classes: wagon/loco, track line, materials, scotch block, operator, operation and maintenance crew; (ix) *Initiating mechanism*: This is the triggering element that helps hazardous element occur the incident. Initiating mechanism is reported to have a set of forty six classes; and (x) *Injury status*: This implies whether incident occurred with or without injury. In addition, another attribute, say 'Frequency' is generated after the creation of ontology, which is used to show the number of incident occurs at a particular location with a given category under each attribute.

4 Results and Discussions

The performance of the proposed methodology is demonstrated by making and updating of an ontology for rail-road incidents from a steel plant. This section has three main parts: (i) experimental setup, (ii) ontology creation using Protégé software,[1] and (iii) performance evaluation of the models. All these are explained in the following sections.

4.1 Experimental Setup

The computer used for the programming and observations was an I5 fifth generation laptop with 4 GB DDR3 RAM. It has a clock speed 2.2 GHz. R language is used for the programming of Natural Language Processing algorithms [17, 20]. Protégé 4.3, a

[1] https://protege.stanford.edu/download/protege/4.3/installanywhere/Web_Installers/.

Java-based open source OWL ontology creator and editor by Stanford University, is used to build ontological relationship between the features and extracting information using SPARQL queries.

4.2 Ontology Creation Using Protege

To develop an ontology using Protégé, 'Classes', 'Object Properties' and 'Individuals' need to be defined. In the railroad dataset, all the features are considered as disjoint classes with no sub-classes. Here, 'Classes' and 'Subclasses' are created in ontology using two commands 'ClassOf' and 'SubclassOf'. 'ObjectProperty' is used to present the property of an object. 'DataTypeProperty' is used to present the data saving type of an object. An object property is defined for each relation, as shown in Table 1. Each class has its specific set of individual present in the dataset, as defined in Table 2. The algorithms used are programmed manually in Protégé 4.3. The relationships are retrieved by programming a query in SPARQL window. For example, Protege-generated Ontology relationship between a central event and corresponding causes is shown in Fig. 2.

4.3 Performance Evaluation of the Models

We have used two models/classifiers, such as SVM and RF in order to get how accurate (i) the features are predicted from text, and (ii) the relationship is created between the classes/sub-classes using the ontology. For training of ontology upgradation, we have used k-fold cross validation with $k = 10$ and RF with number of trees $= 100$ and SVM with linear kernel algorithm. RF results 87% accuracy, 84% precision, and 92% recall. Whereas, SVM ensures 83% accuracy, 82% precision, and 87% recall. From these results, it is evident that ontology creation and upgradation using the

Table 1 Object properties connecting the classes

Object property	Domain	Range
AtLocation	Central event	Location
elementCaused	Hazardous element	Primary cause
hasBehaviour	Cause	Behaviour
hasElement	Location	Hazardous element
hasMechanism	Location	Initiating mechanism
hasPotential	Cause	Potential
mechanismCaused	Initiating mechanism	Cause

Table 2 Individuals of each class

Central event	Behavior	Location	Potential	Hazardous element	Cause
Near miss	Unsafe act	Div1	Injury	Loco operator	High speed
Property damage/injury	Unsafe condition	Div2	Property damage	Maintenance crew	Poor track condition
	Unsafe act and condition	Div3	Environmental damage	Operation	Poor wagon condition
		Div4		Track	Two way formation
		Div5		Track and operation	
		Div6		Wagon	
		Div7			

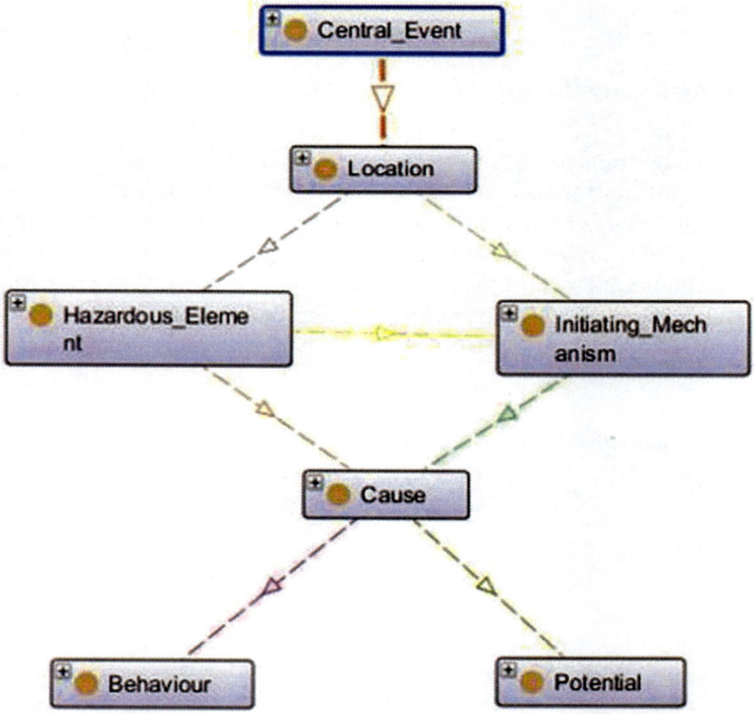

Fig. 2 Example generated through Protégé 4.3

proposed methodology holds accurate relationship between the classes/sub-classes. Hence, its effectiveness is proved.

5 Conclusions

Although several researches have been carried out for ontology creation and upgradation using text, they are still limited due to human involvement. Addressing the issue, we have proposed a semi-automated ontology creation and upgradation approach using text processing and machine learning techniques. As a case study, we have applied our proposed methodology over rail-road incident data collected from a steel plant in India. Two machine learning techniques, such as support vector machine (SVM) and random forest (RF) are applied over the upgraded ontology to see how accurate the logical relationship is established between classes/sub-classes corresponding to any incident. RF and SVM show 87% and 83% accuracy, respectively of the upgraded ontology. Our method is generic. One can use it in making ontology for any kind of domain. As a limitation, fuzziness in text is not handled in this research. In future, ontology upgradation can be done using fuzzy text clustering-based approach [21] embedded in a decision support system [22] can be developed.

Acknowledgements The work is funded through UAY project, GOI (Project Code: IITKGP-022). Authors acknowledge Safety Analytics and Virtual Reality Laboratory (www.savr.iitkgp.ac.in), Department of Industrial and Systems Engineering, IIT Kharagpur for research facilities for this work.

References

1. A. Verma, S. Chatterjee, S. Sarkar, J. Maiti, Data-driven mapping between proactive and reactive measures of occupational safety performance, in *Industrial Safety Management* (Springer, Singapore,2018), pp. 53–63
2. K. Singh, N. Raj, S. Sahu, R. Behera, S. Sarkar, J. Maiti, Modelling safety of gantry crane operations using petri nets. Int. J. Inj. Control Saf. Promot. **24**(1), 32–43 (2017)
3. S. Sarkar, A. Verma, J. Maiti, Prediction of occupational incidents using proactive and reactive data: a data mining approach, in *Industrial Safety Management* (Springer, Singapore, 2018), pp. 65–79
4. S. Sarkar, S. Baidya, J. Maiti, Application of rough set theory in accident analysis at work: a case study, in *2017 Third International Conference on Research in Computational Intelligence and Communication Networks (ICRCICN)* (IEEE, 2017), pp. 245–250
5. P. Hughes, R. Robinson, M. Figueres-Esteban, C. Van Gulijk, Extracting safety information from multi-lingual accident reports using an ontology-based approach. Saf. Sci. **118**, 288–297 (2019)
6. R. Subhashini, V.J.S. Kumar, A framework for efficient information retrieval using NLP techniques, in *International Conference on Advances in Communication, Network, and Computing* (Springer, 2011), pp. 391–393
7. M.V.G. Aziz, A.S. Prihatmanto, D. Henriyan, R. Wijaya, Design and implementation of natural language processing with syntax and semantic analysis for extract traffic conditions from social

media data, in *2015 5th IEEE International Conference on System Engineering and Technology (ICSET)* (IEEE, 2015), pp. 43–48
8. N. Sanchez-Pi, L.M. Orosa, A.B. Garcia, Information extraction techniques for health, safety and environment applications in oil industry, in *International Conference Intelligent Systems and Agents* (2013), pp. 115–117
9. S. Sarkar, N. Ejaz, J. Maiti, Application of hybrid clustering technique for pattern extraction of accident at work: a case study of a steel industry, in *2018 4th International Conference on Recent Advances in Information Technology (RAIT)* (IEEE, 2018), pp. 1–6
10. N. Sanchez-Pi, L. Martí, A.C.B. Garcia, Text classification techniques in oil industry applications, in *International Joint Conference SOCO'13-CISIS'13-ICEUTE'13* (Springer, 2014), pp. 211–220
11. S. Sarkar, A. Kumar, S.K. Mohanpuria, J. Maiti, Application of Bayesian network model in explaining occupational accidents in a steel industry, in *2017 Third International Conference on Research in Computational Intelligence and Communication Networks (ICRCICN)* (IEEE, 2017), pp. 337–392
12. N.W. Chi, K.Y. Lin, S.H. Hsieh, Using ontology-based text classification to assist job hazard analysis. Adv. Eng. Inform. **28**(4), 381–394 (2014)
13. N. Sanchez-Pi, L. Martí, A.C.B. Garcia, Improving ontology-based text classification: an occupational health and security application. J. Appl. Logic **17**, 48–58 (2016)
14. G. Mujtaba, L. Shuib, R.G. Raj, R. Rajandram, K. Shaikh, M.A. Al-Garadi, Automatic ICD-10 multi-class classification of cause of death from plaintext autopsy reports through expert-driven feature selection. PLoS ONE **12**(2) (2017)
15. S. Sarkar, A. Pramanik, J. Maiti, G. Reniers, Predicting and analyzing injury severity: a machine learning-based approach using class-imbalanced proactive and reactive data. Saf. Sci. **125**, 104616 (2020)
16. S. Sarkar, S. Vinay, R. Raj, J. Maiti, P. Mitra, Application of optimized machine learning techniques for prediction of occupational accidents. Comput. Oper. Res. **106**, 210–224 (2019)
17. S. Sarkar, V. Pateshwari, J. Maiti, Predictive model for incident occurrences in steel plant in India, in *2017 8th International Conference on Computing, Communication and Networking Technologies (ICCCNT)* (IEEE, 2017), pp. 1–5
18. S. Sarkar, A. Patel, S. Madaan, J. Maiti, Prediction of occupational accidents using decision tree approach, in *2016 IEEE Annual India Conference (INDICON)* (IEEE, 2016), pp. 1–6
19. S. Sarkar, R. Raj, S. Vinay, J. Maiti, D.K. Pratihar, An optimization-based decision tree approach for predicting slip-trip-fall accidents at work. Saf. Sci. **118**, 57–69 (2019)
20. S. Sarkar, A. Pramanik, N. Khatedi, J. Maiti, An investigation of the effects of missing data handling using 'r'-packages, in *Data Engineering and Communication Technology* (Springer, 2020), pp. 275–284
21. S. Sarkar, V. Lodhi, J. Maiti, Text-clustering based deep neural network for prediction of occupational accident risk: a case study, in *2018 International Joint Symposium on Artificial Intelligence and Natural Language Processing (iSAI-NLP)* (IEEE, 2018), pp. 1–6
22. S. Sarkar, M. Chain, S. Nayak, J. Maiti, Decision support system for prediction of occupational accident: a case study from a steel plant, in *Emerging Technologies in Data Mining and Information Security* (Springer, 2019), pp. 787–796

Blockchain and Machine Learning: Background, Integration Challenges and Application Areas

Abhishek Goel, Bharat Bhushan, Bhavish Tyagi, Harshita Garg, and Siddharth Gautam

Abstract The blockchain is a new era technology that gives secure mechanical effects on different trades. It thoroughly rotates over the disseminated database. That system assists with disposing of the disappointment at any single point that may influence in sparing time and for the decrease of cost. Till now, a large portion of the work in regards to blockchain is restricted to developing advancements that are identified with the scrambled type of cash. With the progression of different shrewd gadgets and the field of artificial intelligence and machine learning, interdisciplinary cooperation with blockchain technology might be extremely useful for future inquiries about. In this paper, we present a decentralized structure and immense measure of information is to be broken down in the machine learning zone, there are gigantic investigating capacities in the joint effort of the innovation. Right now, parts of the use of the blockchain innovation structure with regards to machine learning are examined.

Keywords Blockchain · Machine learning · Smart applications · Integration · Challenges

A. Goel (✉) · B. Tyagi (✉) · H. Garg · S. Gautam
HMRITM, Delhi, India
e-mail: goelabhishek206@gmail.com

B. Tyagi
e-mail: bhavishtyagi@gmail.com

H. Garg
e-mail: harshigarg2000@gmail.com

S. Gautam
e-mail: siddharthinfo92@gmail.com

B. Bhushan
Department of Computer Science and Engineering, Sharda University, Greater Noida, India
e-mail: Bharat_bhushan1989@yahoo.com

J. Tavares et al. (eds.), *Emerging Technologies in Data Mining and Information Security*,
Lecture Notes in Networks and Systems 164,
https://doi.org/10.1007/978-981-15-9774-9_29

1 Introduction

Data is the king of new era and acts as a very essential source which is mostly used in pre-processing techniques merely improving research areas. Its primary sources include questionnaires, interviews, studies and surveys [1]. The most available data is generated electronically from Internet. There is a lot of data generated from various sources all over the world every day. Not all the data generated is useful and can be used for analysis. It is proper quality and quantity makes it useful for classification and primarily prediction in machine learning. Starting from marketing and health care to transportation and e-commerce, almost all the areas need machine learning models for its working which indeed is dependent on data generated. Blockchain technology and machine learning are related to each in terms of data manipulation and extraction [2, 3]. Blockchain technology gains appreciation through cryptocurrency. It was established by Satoshi Nakamoto in 2008 and was primarily used as a form of digitized money in cryptocurrency [4].

But now, BT has a variety of applications in numerous fields other than Bitcoin like agriculture, forecasting, Ethereum and automation [5]. It has some characteristics due to which it continues to gain fame: decentralized, persistency, audibility and anonymity. Blockchain basically describes the transaction in a P2P network. This network is concentrated between nodes and the transaction is validated from nodes itself. After verification, the verified transaction forms a block which is abiding and unaltered in every condition. Similarly, different blocks containing varied information continue to add which grows the chain continuously [6]. Figure 1 represents the structure of blockchain.

On the other side, machine learning is closely related to artificial intelligence. ML is known to perform various types of tasks such as acting as a robot controller, working in planning and assimilation of the process, predicting the right information and diagnosing the element. ML is not fully detached from the concept of BT. Earlier; it was used for forecasting the prices of Bitcoin cryptocurrency [7]. Its architecture is

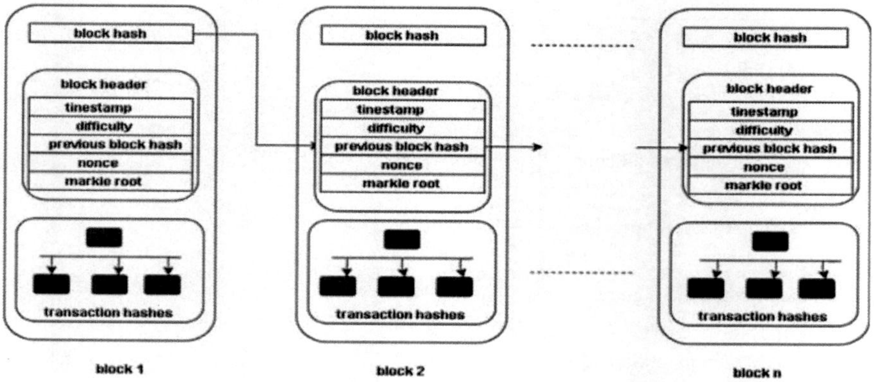

Fig. 1 Structure of blockchain

fully based on the competency of AI agents to grasp the environment's mobility [8]. This paper presents the state-of-the-art review on integration of ML and blockchain technology, challenges faced in this integration and a case study on energy trading system.

The rest of the paper is organized as follows. Section 2 presents the integration of blockchain with machine learning. Section 3 presents the background of blockchain technology. In Sect. 4, various types of research challenges with ML are discussed. To understand more about the integration of blockchain with ML, a recent case study in Sect. 5 is explained and finally Sect. 6 concludes our work on blockchain and machine learning.

2 Integration of Blockchain with ML

ML includes teaching a computer without making a program for it in an explicit manner. Machine learning includes the involvement of a judicious amount of data to make more accurate and rigorous decisions. This is primarily achieved through data sharing and data reliability. Machine learning combined with blockchain technology gives immense results and ML applications such as SVM, bagging, clustering and DL algorithms like CNN and LSTM can be fundamentally accessed to analyze and prevent attacks on any network which is primarily based on blockchain [9]. ML itself is a broad concept as it revolves around many algorithms such as that of data mining, non-deterministic algorithms for human face recognition and algorithms related to adaptable programming. On the other side, blockchains are a cryptographic set of records which is almost like an accounting ledger. Decentralized structure in blockchain nodes creates the difference between the two.

Blockchain consists of a chain of blocks that are linked by the method where the first block contains the hash of the last block. A replica of blockchain is made applicable with every individual block and these blocks are permanent and unaltered [10, 11]. Blockchain can be made even smarter and easily accessible with the use of ML. ML can be used to improve the data sharing routes and decentralized system of BT. ML and BT are linked together with the process of data collection. Data acquired from devices like IoT, sensors and smart devices are processed through blockchain smart applications. Further, ML is applied to this process in the form of real-time and data analytics. Its main function is to do proper analysis and production of the data collected with the help of datasets. One of the main advantages of using ML in blockchain-based applications is that ML is known to reduce the errors in the data collected like that of duplicacy of data, missing figures and noise. Table 1 describes the benefits of blockchain-based applications when integrated with ML.

Table 1 Pillars of blockchain-ML integration

S. No.	Pillars	Specifications
1.	Ledger	Users can be easily identified through ML algorithms when performing a request to join any blockchain network or perform further transactions [3]
2.	Smart contracts	BT ensures that the previous terms and conditions provided are well followed and sustained. Smart contracts basically are digitized terms of contract which can be generated by the person who is treating the datasets [4]
3.	ML model improvement	BT reassures that its users are contributing data [1]. It helps in impeccable implementation of the incentive-based system, thus improving machine learning model performance [2]
4.	Health care	Patients can also get appropriate guidance without consulting a doctor through the trained model and natural language processing

3 Blockchain Background

BT grew up through Bitcoin cryptocurrency where the transactions can be done from any end, i.e., from any node without any permission [12]. Permissioned blockchain came into existence when business enterprises started to commute information on a common platform in an assured and personalized manner [13]. Need of permissioned blockchain raised due to the fact that there are no restrictions in Bitcoin and any information can be used by anyone in any way, be it positive or negative. There are some of the components of permissioned blockchain that makes it stronger and easier to use. These components are detailed in the subsections below.

3.1 Shared Ledger

Shared ledger is also known as distributed ledger. It is known to contain the replicated, synchronized and shared digital data that is spread across countries and institutions. It is a type of allocated system where all the peers in the network have the track of all the transactions going on in the network. This reduces the frauds that may happen in a simple Bitcoin system due to lack of security. Shared ledger spreads across several nodes or devices in a peer-to-peer network and the work of each node is to replicate and save an identical copy of the ledger to get updated independently.

3.2 Peer Network

Peer-to-peer network is one where the nodes are connected to each other directly without any external or central server. No intermediaries like banks and clearing houses work in the process of transaction. This network provides a safer environment to work in as the transactions validated by the peers add to the block. No invalid transaction is made in a block as transactions added in a block are unaltered and irreplaceable. This makes the data distribution model a more defining characteristic of blockchain technology where the centralized authorities are not a part of the communication and all the transaction is done between the nodes directly.

3.3 Certificate Authority

The certificate or the membership authority possesses the role of providing certificate users to the blockchain. The main services of a certificate authority include the transactions invoked on a blockchain, the enrollment user and TLS-secured connections between various users of the blockchain. Not all the users get the opportunity to enter the blockchain transactions. Every user needs to ask for permission from the membership authority. After proper authentication, they give the users right to access the transactions and get included in the blockchain.

3.4 Smart Contract

Smart contracts are the lines of code that are stored in a blockchain and perfectly executed when the terms and conditions that are already defined get fulfilled. This basically defines the software running on the shared as well as personal ledger [14]. This creates the learning model using blockchain technology, machine learning models and artificial intelligence. It is a digitized form of contract which monetizes skill-sets of machine learning algorithms. The benefits of the smart contracts are mostly apparent in some of the business collaborations in which all the participants are enforced in a discussion or agreement and the outcome is determined without any involvement of any intermediary. Machine learning and blockchain technology work together mostly for the prediction of certain models which helps to analyze them correctly and make the right decision. Machine learning merely predicts the Bitcoin prices through Bayesian regression. The factors which work are total Bitcoins, trade volume, day high, number of transactions and block size. The dataset is usually trained using box-cox normalization, z-score and log techniques. The practical administration of these theoretical principles has been done on numerous cryptocurrencies like Litecoin, Ethereum, Ripple, Dash and of course Bitcoin.

4 Challenges Faced in ML

There are many challenges faced in machine learning such as manipulation, privacy threat, ambiguity and hackable datasets. Blockchain is prone to safety issues and threats. The most important concern is due to attacks, so gathering efficiency in a less number of blocks would handle the blocks altered in the blockchain [15]. Which specified query exists in private-blockchains. Blockchain private ones are not at all affected by these attacks as it has identification of each and every node present. Blockchain is an answer if non-trusted information is accessed, and several entities are more in different environments. If peak performance is the requirement, so a complex database would be a better solution. Thus, a system of a blockchain has to be functionable prior to the usage in any program or software. Data imposed in a blockchain is visible to the whole node. It may lead to a safety issue of information that is to be kept secret. This type of issue can be explained with the help of controlled blockchain and private entities. Thus, the machine learning projects adopt this restricted information or data imposes a stoppage of analytics.

Blockchain hardwares and network architectures are improving the peak performance of maximum blockchain applications [16]. This involves administration networks, mining capacity, centralized space or storage and communication of networks. The products altered to be used in blockchains are under construction or development. Major-scale blockchain network may require a huge amount of agreements. Execution in such blockchains introduces major problems; for example, if there is a huge requirement of network bandwidth, that is, tough to decrease, proceedings will become loaded for this type of network in that time frame. Amount of blockchain will keep on increasing as fresh boxes are introduced [17]. The whole chain must be fixed in nodes, which will create a specific memory constraint on systems. Here, the peak performance is also reflected as there is an increase in size of chains. Apart from that, the storage of unwanted data is also wasting the resources of computation. So, the management of storage is a concern faced by maximum executions of blockchain networks [18]. Figure 2 gives the diagrammatic representation of research challenges.

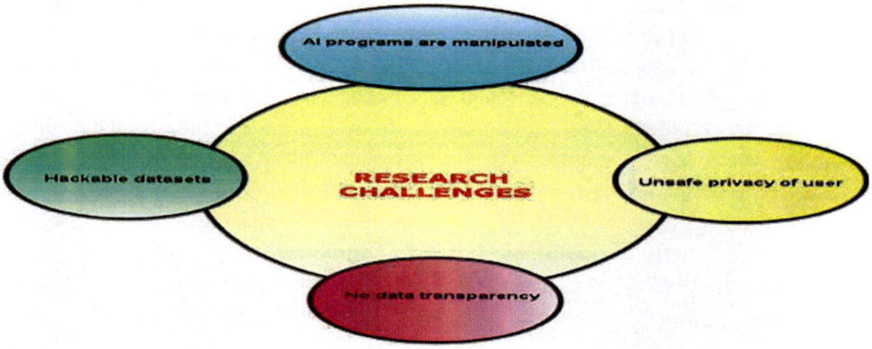

Fig. 2 Research challenges associated with ML

5 Case Study

To explain the proposal of architecture for machine learning, the approach of blockchain softwares, the team prepared a case study, represented in Fig. 6. In the case study, the team explains blockchain-related Sur-Gard (SG) system of resources of energy transfer with the (Bitcoin) cryptocurrency. The energy is transferred with the completion of deep learning approach [19]. Here, the peer-to-peer (P2P) energy system is reliable. Byzantine fault tolerance algorithm plays a vital role in producing a high system output [20]. A short signature and hash function help in generating blocks [17]. Figure 3 represents the energy trading platform a case study to understand the integration of blockchain and ML better.

Intrusion detection system [21] detects fraud transactions and attacks on networks on the basis of recurrent networks. This approach has five parts: (a) block creation phase, (b) change view phase, (c) consensus-creation phase, (d) agreement phase and (e) setup phase [22]. The performance of the intrusion detection system is described in different energy datasets: (i) energy retailer, (ii) energy purchasers, (iii) blockchains and (iv) IDs [23]. In a peer-to-peer (P2P) energy trade system, energy buyers might look at the displayed price and energy. The energy purchaser keeps a check on their account if they have the maximum cryptocurrency (Bitcoin) for buying some energy resources. So, the source of energy required for energy purchasers and the price quoted defines their needs at this point of time, he purchases the energy from energy sellers through the network of blockchain [24]. After validity from purchasers on the networking of blockchain who deals with the energy to the purchaser, and then accepts the cryptocurrency or Bitcoin. The peer-to-peer (P2P) power trade system

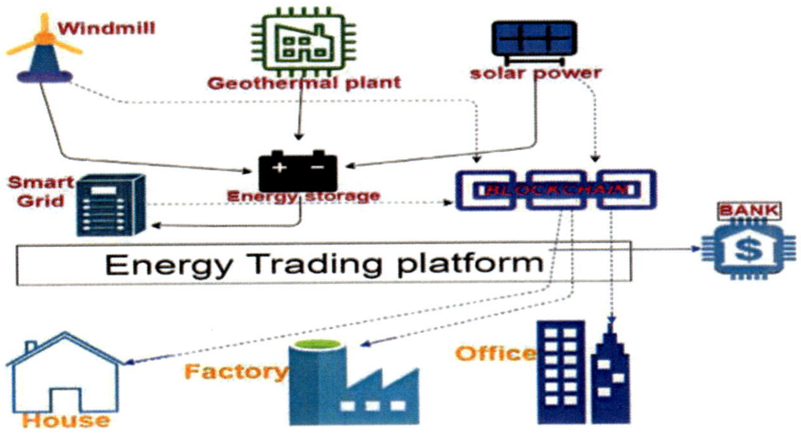

Fig. 3 Energy trading system

Table 2 Description of IDS in energy trading systems

S. No.	Components	Description
1.	Energy purchasers	An energy purchaser has a sufficient amount of cryptocurrency to buy the minimum assets required for selling the energy, and an energy buyer is a commercial, industry or company. Energy purchasers can be located in the house area networks (HAN), neighbor area networks (NAN) and building area networks (BAN)
2.	Energy retailers	An energy retailer has a sufficient amount of energy for sale. The retailer is a neighbor utility provider and neighborhood energy resource. Energy sellers trade with energy buyers [25]
3.	Blockchain	It helps in allowing all use of energy stored in a Sur-Gard system. It works as an intermediate between the energy seller and the energy buyer
4.	IDS	IDS is used to help the useful undertakings and threats on networks for marketing energy [26] The report displays the peak performance of the software system compared to that of stat-arch approach

requires data to be analyzed to point the regular purchaser and dealer for detecting the fake deals. Table 2 provides the brief description of various energy datasets.

6 Conclusion

The ongoing headways in blockchain and ML have made them way breaking advancements. The appropriated record has the likelihood to function as the foundation of different savvy applications, for example, shrewd urban areas, UAV, SG and information exchanging. Right now, we have introduced nitty gritty data on BT and ML, alongside their utilizations in brilliant applications and proposed ML-BT-based engineering. This engineering can be utilized to plan and send a ML-BT-based information investigation framework. A conversation and correlation of different existing overviews are introduced. At that point, we introduced the ML-BT arrangement scientific categorization, concentrating on objective situated, layer arranged, countermeasures and shrewd application measurements. A relative investigation of accessible philosophies and approaches is introduced in each measurement. We additionally underlined various research prospects, for example, foundation accessibility, quantum versatility and security or safety that can fill in as future research bearings right now. At that point, we introduced a contextual research on the vitality exchanging framework to confirm the viability of the proposed design and finished up the paper finally.

References

1. Y. Yuan, F. Wang, Blockchain and cryptocurrencies: model, techniques, and applications. IEEE Trans. Syst. Man Cybern. Syst. **48**(9), 1421–1428 (2018). https://doi.org/10.1109/tsmc.2018.2854904
2. H. Saini, B. Bhushan, A. Arora, A. Kaur, Security vulnerabilities in Information communication technology: blockchain to the rescue (A survey on Blockchain Technology), in *2019 2nd International Conference on Intelligent Computing, Instrumentation and Control Technologies (ICICICT)*. https://doi.org/10.1109/icicict46008.2019.8993229 (2019)
3. S. Tanwar, Q. Bhatia, P. Patel, A. Kumari, P.K. Singh, W.-C. Hong, Machine learning adoption in blockchain-based smart applications: the challenges, and a way forward. IEEE Access **8**, 474–488 (2020). https://doi.org/10.1109/access.2019.2961372
4. S. Nakamoto et al., Bitcoin: a peer-to-peer electronic cash system (2008)
5. T. Varshney, N. Sharma, I. Kaushik, B. Bhushan, Authentication and encryption based security services in blockchain technology. in *2019 International Conference on Computing, Communication, and Intelligent Systems (ICCCIS)*. https://doi.org/10.1109/icccis48478.2019.8974500 (2019)
6. T. Sharma, S. Satija, B. Bhushan, Unifying blockchian and IoT: security requirements, challenges, applications and future trends, in *2019 International Conference on Computing, Communication, and Intelligent Systems (ICCCIS)*. https://doi.org/10.1109/icccis48478.2019.8974552 (2019)
7. M. Rahouti, K. Xiong, N. Ghani, Bitcoin concepts, threats, and machine-learning security solutions. IEEE Access **6**, 67189–67205 (2018). https://doi.org/10.1109/access.2018.2874539
8. A. Malik, S. Gautam, S. Abidin, B. Bhushan, Blockchain technology-future OfIoT: including structure, limitations and various possible attacks, in *2019 2nd International Conference on Intelligent Computing, Instrumentation and Control Technologies (ICICICT)*. https://doi.org/10.1109/icicict46008.2019.8993144 (2019)
9. W. Xiong, L. Xiong, Smart contract based data trading mode using blockchain and machine learning. IEEE Access **7**, 102331–102344 (2019). https://doi.org/10.1109/access.2019.2928325
10. P. Louridas, C. Ebert, Machine learning. IEEE Softw. **33**(5), 110–115 (2016). https://doi.org/10.1109/ms.2016.114
11. D. Arora, S. Gautham, H. Gupta, B. Bhushan, Blockchain-based security solutions to preserve data privacy and integrity, in *2019 International Conference on Computing, Communication, and Intelligent Systems (ICCCIS)*. https://doi.org/10.1109/icccis48478.2019.8974503 (2019)
12. S. Jadon, A. Choudhary, H. Saini, U. Dua, N. Sharma, I. Kaushik, Comfy smart home using IoT. SSRN Electron. J. (2020). https://doi.org/10.2139/ssrn.3565908
13. F. Adoma, Big data, machine learning and the blockchain technology: an overview. Int. J. Comput. Appl. **180**(28), 1–4 (2018). https://doi.org/10.5120/ijca2018916674
14. M. Jindal, J. Gupta, B. Bhushan, Machine learning methods for IoT and their future applications, in *2019 International Conference on Computing, Communication, and Intelligent Systems (ICCCIS)*. https://doi.org/10.1109/icccis48478.2019.8974551 (2019)
15. Y. Zhang, J. Wen, The IoT electric business model: Using blockchain technology for the internet of things. Peer-to-Peer Networking and Applications **10**(4), 983–994 (2016). https://doi.org/10.1007/s12083-016-0456-1
16. S. Soni, B. Bhushan, A Comprehensive survey on blockchain: working, security analysis, privacy threats and potential applications, in *2019 2nd International Conference on Intelligent Computing, Instrumentation and Control Technologies (ICICICT)*. https://doi.org/10.1109/icicict46008.2019.8993210 (2019)
17. V.K. Chaurasia, A. Yunus, M. Singh, Blockchain technologies blockchain technology for smart cities, in *An Overview of Smart City: Observation, Technologies, Challenges and Blockchain Applications* (2020), pp. 133–154. https://doi.org/10.1007/978-981-15-2205-5_7

18. T. Varshney, N. Sharma, I. Kaushik, B. Bhushan, Architectural model of security threats and their countermeasures in IoT, in *2019 International Conference on Computing, Communication, and Intelligent Systems (ICCCIS)*. https://doi.org/10.1109/icccis48478.2019.8974544 (2019)
19. M.A. Alam, S. Jain, Blockchain implementation using smart grid-based smart city, in *Handbook of Research on Blockchain Technology*, pp. 133–169. https://doi.org/10.1016/b978-0-12-819816-2.00006-x (2020)
20. N. Sharma, I. Kaushik, R. Rathi, S. Kumar, Evaluation of accidental death records using hybrid genetic algorithm. SSRN Electron. J. (2020). https://doi.org/10.2139/ssrn.3563084
21. I. Kaushik, N. Sharma, N. Singh, Intrusion detection and security system for blackhole attack, in *2019 2nd International Conference on Signal Processing and Communication (ICSPC)*. https://doi.org/10.1109/icspc46172.2019.8976797 (2019)
22. X.A. Inbaraj, T. Rama Chaitanya, Need to know about combined technologies of blockchain and machine learning, in *Handbook of Research on Blockchain Technology* (2020), pp. 417–432. https://doi.org/10.1016/b978-0-12-819816-2.00017-4
23. B. Hu, C. Zhou, Y. Tian, Y. Qin, X. Junping, A collaborative intrusion detection approach using blockchain for multimicrogrid systems. IEEE Trans. Syst. Man Cybern. Syst. **49**(8), 1720–1730 (2019). https://doi.org/10.1109/tsmc.2019.2911548
24. S. Gupta, S. Sinha, B. Bhushan, Emergence of blockchain technology: fundamentals, working and its various implementations. SSRN Electron. J. https://doi.org/10.2139/ssrn.3569577 (2020)
25. S. Gautam, A. Malik, N. Singh, S. Kumar, Recent advances and countermeasures against various attacks in IoT environment, in *2019 2nd International Conference on Signal Processing and Communication (ICSPC)* (2019), pp. 315–319
26. A. Khamparia, B. Pandey, Association of learning styles with different e-learning problems: a systematic review and classification, in *Education and Information Technologies*. (Springer, 2019). https://doi.org/10.1007/s10639-019-10028-y

Various Types of Cybercrime and Its Affected Area

Harshit Yadav, Siddharth Gautam, Aniket Rana, Jatin Bhardwaj, and Nitin Tyagi

Abstract In today's world, cybercrime is a computer-related crime which is an offense in the cyber world. Nowadays, we are totally dependent on technology which will grow the cybercrime threats. Protect our personal data and cyber security is an authoritatively mandating task for all web users. It is very difficult to find out the various types of cybercrime. We utilize computer technology to pay our electricity bill, for banking purpose, for buying several types of tickets, and so many other activities of our day-to-day life. These numbers of online transactions by users are additionally increasing the crimes which are called cybercrimes or computer-related crime. In this paper, we present the types of cybercrime along with cyberspace and areas affected by cybercrimes have been discussed to develop the more understanding about cybercrimes. We also provide a set of recommendations on how to approach these challenges in order to ameliorate currently subsisting cybercrime reporting processes. In the end, we condensed a research direction and recommendations to ascend up user's cognizance and arm them well to the battle with cybercrimes.

Keywords Cybercrimes · Cyber terrorism · Hacking · Malware · Ransomware

H. Yadav · S. Gautam · A. Rana · J. Bhardwaj · N. Tyagi (✉)
Branch of Information Technology, HMRITM, Delhi, India
e-mail: nitintyagi5631@gmail.com

H. Yadav
e-mail: yharshit648@gmail.com

S. Gautam
e-mail: siddharthinfo92@gmail.com

A. Rana
e-mail: ishurana289@gmail.com

J. Bhardwaj
e-mail: Jatin200.mps@gmail.com

J. Tavares et al. (eds.), *Emerging Technologies in Data Mining and Information Security*, Lecture Notes in Networks and Systems 164,
https://doi.org/10.1007/978-981-15-9774-9_30

1 Introduction

Nowadays, with the advancement in technology and the growth rate of computer owners is increasing rapidly. With the increase of computer owners, the risk of cybercrime is also increasing simultaneously. Cybercrime is a relatively new concern for the officials of law enforcement [1]. India has been the sole prey of cybercrime in few years [2]. Hence, new models should be developed in order to prevent the user from various types of vulnerability. Earlier there was no technology for preventing cybercrime, but in recent years cybercrime is widespread globally which leads to the loss of huge amounts of money. So, new technologies are inventing to stop the cybercrimes. Almost all the sectors are affected from the cybercrime globally but in India there are main four sectors that are affected by cybercrimes, namely healthcare, manufacturing, finance, IT sector, and telecoms [3].

Each cyber attack has different precautions and every cyber attack target different parts of the system but all cyber attacks have only one motive that is to leak information and harm the data of the user. In recent years, there are millions of cases about cyber fraud globally and the only way to stop is to spread awareness and to develop a good firewall security setup so that no one can enter your system unauthorised [4]. Term cyberspace is used for a medium or a virtual environment created by interconnected computers and computer networks on the Internet. This medium is a conceptual electronic space which is not bounded by distance or other physical drawbacks. Cyberspace indicates a dynamic and universal domain defined by electromagnetic spectrum and electronics, whose aim to share, modify, inform, use, eliminate, exchange, eliminate, and create physical assets.

The rest of the paper is organized as follows. In Sect. 2, we present the virtual environment which is also known as cyberspace. Section 3 presents the areas affected by cybercrimes. In Sect. 4, various types of cybercrimes are discussed. To understand more about the cybercrimes, a recent case study in Sect. 5 which is based mainly on cyberstalker is explained and finally Sect. 6 concludes our work on cybercrime.

2 Cyberspace

Cyberspace has mainly two goals which are primary and secondary goals. The target of primary goal is to make a contact between visualization and data acquisition. Visualization displays data using a three-dimensional model, emerges the old two-dimensional model which do not allow any involvement in the Visualization Universe [5]. The study part is component in order to figure out problems related to visualization data acquisition analyze the quality of data, work on a given network, data transport, network itself. Secondary goal starts with the end of application. The view point will be distinct due to a particular application is something that allows supplementary considerations. For example, an user sitting in front of a flat screen we start with a query [1]. (a) origin from these results, what can be enhanced to build exact grained applications? (b) What are the upcoming metrics to add to the system to allow a good communication? (c) The growth of the system among many possible goes in which directions? Cyberspace is an addition of visualization and networking.

Data visualization is a part of science which aims to allow data classification. Computer networking abstracts the communication between two or more members who need to interact with each other through networking. Many countries have their own social networks like Facebook, instagram, Twitter, hike, and some countries using other countries' networking. In this case, chances of cybercrime increased due to snooping. Countries used this in the name of security alert, where it is a technique in which criminals or enemy countries get unauthorized data. It should be any type of data (person's data, Hospital databank data, etc.,) Experts suggest that it can be reduced only when our country used self-made network so, people cannot share any personal information to any other countries. As a social experience, individuals can interact, share information, exchange ideas, provide social support, conduct business, direct actions, create artistic media, play games, engage in political discussion, and so on using this universal network. In all these activities sometimes, we share some personal information with hackers. Experts say that it can be reduced only when our country uses self built networks, so people cannot share any personal information to any other countries [1].

3 Areas Affected by CyberCrime

Cyber security is one of the most concerning issues nowadays. As cybercrime is rapidly increasing all over the world, all the people around the world are concerned about their privacy due to the increase in cybercrime. After the conduction of a survey, a report is created by OGL computers which shows that technology is advancing and how cyber securities are used to increase the growth rate of businesses and different industries [6]. But with the advancement in technologies, there are various sectors which are prone to cybercrime. Figure 1 shows the four sectors are represented that are most vulnerable to cyber attacks:

3.1 Healthcare Sector

On the one hand where the healthcare sector saves the lives of many people and takes care of every patient in spite of any issues. On the other hand, healthcare departments are fighting against the cyber threat or cybercrime [7]. Healthcare industry is plagued by cyber issues. These issues consist of malware activities. Hacker cracks the firewall and enters the system of the health care and leaks the confidential information of the patient. Cyber attacks also result in the financial loss of the healthcare sector. According to a report, 75% of healthcare departments suffer from cyber attacks.

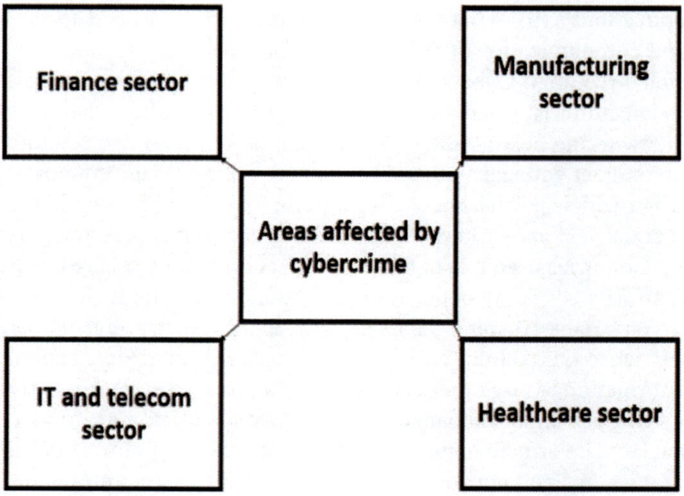

Fig. 1 Areas affected by cybercrime

3.2 IT and Telecoms Sector

The IT and telecom sectors store the confidential information of customers and help to keep the whole world connected. But these sectors are no longer safe, and the only reason is cyber attacks. Many IT companies all over the world stores the large amount of customer data in the form of cloud storage [8]. In spite of many technologies, the data of these companies got hacked, and finally, customers have to suffer. Telecommunication companies that keep the whole world connected also face the same problem as they also got hacked and personal information of customers like phone numbers, email addresses leaks. According to a report, 77% of IT and telecom companies suffer cyber attacks.

3.3 Finance Sector

Finance the word explains everything. A sector in which involves money, Finance sector consists of banks, insurance companies, managing companies, etc. [8]. The most common financial attack is money laundering. Finance sector deals with the exchange of money so these sectors are one of the most prone sectors to cyber threat. Generally, a high profile criminal targets the finance sector. Once the finance sector gets cyber attacked, there will be a huge amount of loss of money. According to reports, only 50% of the finance sector got cyber attacks [9].

3.4 Manufacturing Sector

The manufacturing sector consists of automotive industries, electronic industries, and pharmaceutical companies. Manufacturing sector is at high risk due to the intellectual property having high value [10]. Manufacturing industry has a high risk of cyber attack because their database consists of basic information of building a product. Also, manufacturing sector deals with the huge amount of exchange of money to buy raw material. So these are the reasons, the manufacturing sector got a cyber attack. According to the reports, 57% manufacturing industries got cyber attacks.

4 Types of CyberCrime

As different types of crimes related to computers are occurred nowadays, on the basis of which different types of cyber attacks are defined. In Fig. 2, main 15 types of cyber attacks are described.

Due to the advancement in technology, workloads on people are decreasing with the help of the Internet. But due to the excessive use of the Internet, the risks of cybercrimes or computer-related crimes are increasing. Table 1 describes the different types of cybercrimes.

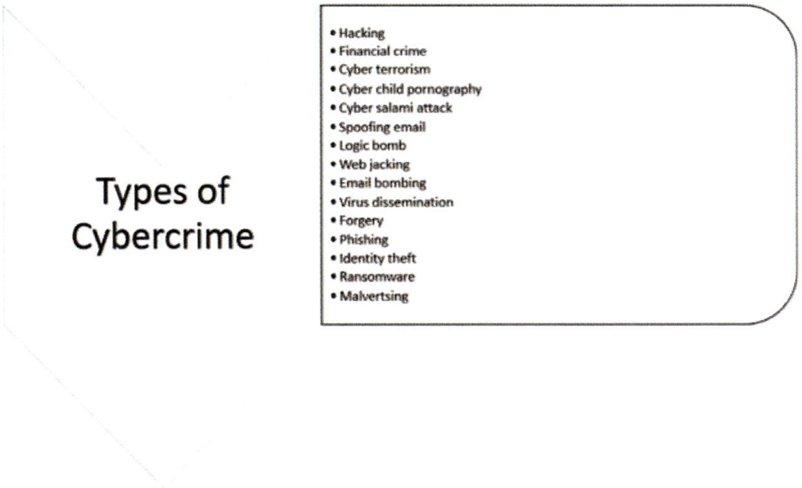

Fig. 2 Types of cybercrime

Table 1 Types of cybercrimes

Types	Description	Example
Hacking	• used to find the weakness of system • It is done to access someone's system • Misuse their private information without the owner's knowledge	• Recently, in may 2019 'GnosticPlayers' hacked the Australian graphic design website named Canva and leaked the username, addresses and emails of 137 million users [11]
Financial crime	• Financial crime is a crime that involves money • One of the most prevalent financial crimes is money laundering • It involves the change of ownership of properties without the owner's knowledge by cracking the firewall	• One of the most high profile cyber attacks took place in Bangladesh when a person stole $81,000,000 from the central bank of Bangladesh by cracking their firewall security. It is one of the largest bank robberies till now
Cyber terrorism	• Use of cyber terrorism to promote violent acts which further results in the terror activities such as bomb blast • It is done with the help of public networks and internet • Such types of activities are done by narrow minded people who want to take revenge	• In 2020, Indian government will implement two citizenship acts CAA and NRC, which takes the citizenship of persons who are illegally residing in the country. Due to CAA and NRC, riots and crimes took place in different parts of the country. These riots were spread with the help of a public network and internet
Cyberchild pornography	• Using cyberspace to make, display and distributes the pornographic content to below 18 children • It is used to engage children in any kind of sexual act to promote the sexual crimes	• One of the biggest pornography crime took place, when FBI catches and jailed a British citizen Thomas Reddy for 1,335 years for promoting the pornographic content among below 18 children
Cybersalami attack	• It takes place in small amounts for a very long time and then further turns into a major attack • It cannot be detected easily by anyone • Hackers use the bank information of a customer and start taking very small amounts of money from many accounts for a particular period of time	• Recently, an employee of a bank in the USA is dismissed as he plant a logic bomb in bank system which deduct 10 cents from all the accounts in the bank and transfer it into a account named Ziegler. Bank came to know about it when a person named ziegler opened an account in that bank and found large amount of money in his account

(continued)

Table 1 (continued)

Types	Description	Example
Spoofing email	• An email is sent to the victim who likes to appear from one source which is false as email originates from another source or culprit source [12] • It is done to give monetary damage to the victim	• On July 21, 2017, Inc. V. Federal Insurance Company registered a case for loss of $4.8 million due to email spoofing in court
Logic bomb	• It is a special type of malware and dependent program • Depend on some special type of malware to produce malfunctioning	• Recently, an employee of a bank in the USA was dismissed as he planted a logic bomb in the bank system which deducts 10 cents from all the accounts in the bank
Web jacking	• A cybercrime related to the malicious website • Hacker forcefully takes control over a website by cracking its firewall system and breaking the security barrier • Jacker changes the password of the website so that no one can access it	• Recently, in the USA a hacker took control over a website known as hobby. The jacker demanded a ransom of $1 million but the owner does not take the threat seriously. After 3 days, she received calls from all over the country that the website is being hacked
Email bombing	• Bomber bombs the email of the victim • Bomber sends large number of emails to the victim, from sending mails in large amount • The server of the victim got crashed and the victim is unable to receive any mail until the server is being established again	• In 2018 a case of email bombing was registered. When a foreigner is residing in India from 30 years, wants to get a scheme of buying a house at very low rate, when the foreigner mails to the department about it. She receives a mail that this scheme is for Indians only than she crashed the server of the department by email bombing
Virus dissemination	• Viruses are the special types of malware programs • Virus attaches themselves to the file in a system • Virus replicates themselves and moves from one file to another in no time	• In 2010 a virus named love bug was invented by a filipino undergraduate. Love bugs have become the world's most widespread computer virus. Love bug is present in every 1 of the 5 computer systems. Love bug results in the loss of $10 billion

(continued)

Table 1 (continued)

Types	Description	Example
Forgery	• Unauthorized currency notes, revenue stamps, marksheets can be forged with the help of advanced computers, scanners and the printers • Mostly unauthorized and fake marksheets, certificates, and degrees of colleges are produced through forgery [13]. Later these degrees are sold to the students illegally in thousands of rupees	• On 16 May 2018, Mrs. Doris Victor filed a case in the court about forgery in which her agent impersonated Mrs. Doris Victor and created a Power of Attorney in which ownership of the property is transferred to her agent after her death. Mrs. Doris Victor gets the information about forgery, after the aforesaid transactions from her account [15]
Phishing	• One of the rare cyber-attacks that take place • Hackers target a particular victim • Hackers contacted the victim by sending a text message, telephone or by an email in the name of any legislative institution [14] • It is done to get the personal information, banking details, and passwords of the victim [16]	• In December 2015, Ukranian power grid got cyber attacked by the process of email phishing. In this cyber attack, many emails are sent by the Ukranian power grid to the users for their personal information but this fraud was further detected by the Russian cyber intelligence operators
Identity theft	• Hackers break through the firewall system and steal the personal information of the victim [18] • Personal data consists of passwords, bank statements, credit and debit cards details	According to a report in 2019, 3 million cases will be filed in 2018 from all over the world of identity theft and fraud. Out of these 3 million cases, 25% cases consist of money laundering in it [17]
Ransomware	• Most destructive cyber attacks till now. Ransomware is the malware based cyber attack [19] • Spreads through email with invoice attachment	• In May 2017, 4 lakh computers were affected by a Ransomware named WannaCry. WannaCry causes the loss of about $5 billion worldwide [20]
Malvertising	• A method in which the whole website is covered from the malicious advertisement [21] • These viruses alter the data and delete the important information from the system	• In recent years, there are many cases filed against malvertising. As it leads to the loss of confidential data [22]

5 Case Study

Cybercrime includes criminal acts carried out with computers, any electronic device with the help of the Internet. Cybercrime is fatal and has many facets or types such as cyber terrorism, frauds, and assaults due to threatening [23], child pornography, cyberstalking, and much more. Figure 3 shows a case of cyberstalking in detail.

Thousands of cases of cyberstalking are reported every year, and this includes 90% women victims. One such famous case was evident in the USA where a person named Jason White, the owner of a Temecula art gallery in the city of Los Angeles, approached his employers, association members and other working professionals through Internet. He created vituperative websites in the name of art publishers and threatens her by sending manipulated pictures of her child with abusing comments. He targeted different publishers one by one adopting a variety of techniques to bully her/him and threaten to kill their child, harm their prestige, and cause trouble with inappropriate messages. In this way, the stalker demanded a large sum of money and was successful to extort as much as $3,000,000 from art world professionals. After huge efforts, this person was identified and arrested by police, gone under prosecution and sentenced to 60 months punishment behind the bar. Since cyberstalkers use fake identity, thus threats are ominous to the victim which encourages cyberstalkers to continue his or her wrong deeds. Usually, cyberstalkers find their targets from sources like chat rooms, online forms, discussion boards, and various social sites.

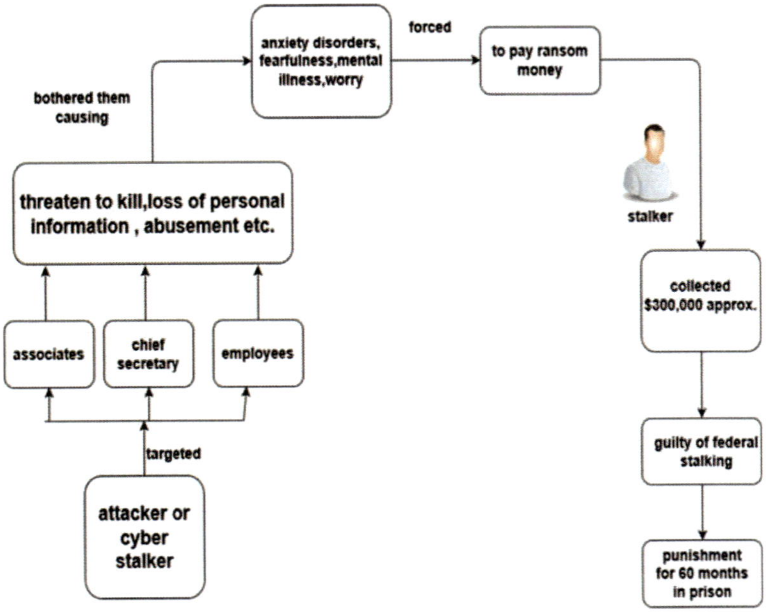

Fig. 3 Schematic diagram of case study

6 Conclusion

After looking at the various statistics from many sources, it is confirmed that there has been a sudden rise in the number of cybercrime attacks over a few years. Cybercrime has become a topic of deep concern as it has shown tremendous growth in this digital era. The cybercrime suspects are generally of age between 18 and 45 years, and the general case of cybercrimes deals with money laundering. Earlier, there were many incidents of cybercrime such as unauthorized access to a network, leakage of personal information like passwords and bank details but these were restricted by ethical hackers, operators of the intelligence department. In this paper, we have discussed various aspects of cybercrime with the case studies in content of the same. From this paper, it is concluded that cybercrime is much more destructive and disastrous than any other crime. As these cybercrimes not only leads to the theft of money, but also for the loss of credibility of people. It may be apprehended that the reported cybercrimes may increase exponentially with the incrementation in the technological pace, if the due attention is not paid.

References

1. G. Tsakalidis, K. Vergidis, A systematic approach toward description and classification of cybercrime incidents. IEEE Trans. Syst. Man Cybern. Syst. **49**(4), 710–729 (2019). https://doi.org/10.1109/tsmc.2017.2700495
2. P. Sinha, A.K. Rai, B. Bhushan, Information security threats and attacks with conceivable counteraction, in *2019 2nd International Conference on Intelligent Computing, Instrumentation and Control Technologies (ICICICT)*. https://doi.org/10.1109/icicict46008.2019.8993384 (2019)
3. G. Xian, Cyber intrusion prevention for large-scale semi-supervised deep learning based on local and non-local regularization. IEEE Access **8**, 55526–55539 (2020). https://doi.org/10.1109/access.2020.2981162
4. S. Soni, B. Bhushan, Use of machine learning algorithms for designing efficient cyber security solutions, in *2019 2nd International Conference on Intelligent Computing, Instrumentation and Control Technologies (ICICICT)*. https://doi.org/10.1109/icicict46008.2019.8993253 (2019)
5. Cybercrime at a glance. IEEE Spectrum **43**(4), 17 (2016)
6. M. Husak, J. Komarkova, E. Bou-Harb, P. Celeda, Survey of attack projection, prediction, and forecasting in cyber security. IEEE Commun. Surv. Tutorials **21**(1), 640–660 (2019). https://doi.org/10.1109/comst.2018.2871866
7. B. Bhushan, G. Sahoo, Requirements, protocols, and security challenges in wireless sensor networks: an industrial perspective, in *Handbook of Computer Networks and Cyber Security* (2020), pp. 683–713. https://doi.org/10.1007/978-3-030-22277-2_27
8. R. Pillai, R. Sarngapani, H. Thukral, Indian manual for cyber security in power systems. CIRED Open Access Proc. J. **2017**(1), 2787–2790 (2017). https://doi.org/10.1049/oap-cired.2017.0523
9. B. Bhushan, G. Sahoo, Recent advances in attacks, technical challenges, vulnerabilities and their countermeasures in wireless sensor networks. Wirel. Pers. Commun. **98**(2), 2037–2077 (2017). https://doi.org/10.1007/s11277-017-4962-0
10. J.-Y. Yu, E. Lee, S.-R. Oh, Y.-D. Seo, Y.-G. Kim, A survey on security requirements for WSNs: focusing on the characteristics related to security. IEEE Access **8**, 45304–45324 (2020). https://doi.org/10.1109/access.2020.2977778

11. I. Kaushik, N. Sharma, N. Singh, Intrusion detection and security system for blackhole attack, in *2019 2nd International Conference on Signal Processing and Communication (ICSPC)*. https://doi.org/10.1109/icspc46172.2019.8976797 (2019)
12. M.C. Sanchez, J.M.C.D. Gea, J.L. Fernandez-Aleman, J. Garceran, A. Toval, Software vulnerabilities overview: a descriptive study. Tsinghua Sci. Technol. **25**(2), 270–280 (2020). https://doi.org/10.26599/tst.2019.9010003
13. D. Mitropoulos, P. Louridas, M. Polychronakis, A.D. Keromytis, Defending against web application attacks: approaches, challenges and implications. IEEE Trans. Dependable Sec. Comput. **16**(2), 188–203 (2019). https://doi.org/10.1109/tdsc.2017.2665620
14. M. Sharma, A. Tandon, S. Narayan, B. Bhushan, Classification and analysis of security attacks in WSNs and IEEE 802.15.4 standards: a survey, in *2017 3rd International Conference on Advances in Computing, Communication and Automation (ICACCA) (Fall)*. https://doi.org/10.1109/icaccaf.2017.8344727 (2017)
15. BabitaPandey AdityaKhamparia, A novel method of case representation and retrieval in CBR for e learning. Edu. Inform. Technol. **22**(1), 337–354 (2015). https://doi.org/10.1007/s10639-015-9447-8
16. P. Sinha, V.K. Jha, A.K. Rai, B. Bhushan, Security vulnerabilities, attacks and countermeasures in wireless sensor networks at various layers of OSI reference model: a survey, in *2017 International Conference on Signal Processing and Communication (ICSPC)*. https://doi.org/10.1109/cspc.2017.8305855 (2017)
17. M.A. Qbeitah, M. Aldwairi, Dynamic malware analysis of phishing emails, in *2018 9th International Conference on Information and Communication Systems (ICICS)*. https://doi.org/10.1109/iacs.2018.8355435 (2018)
18. J.-L. Zhong, C.-M. Pun, An end-to-end dense-inception net for image copy-move forgery detection. IEEE Trans. Inf. Forensics Secur. **15**, 2134–2146 (2020). https://doi.org/10.1109/tifs.2019.2957693
19. A. Sharma, A. Singh, N. Sharma, I. Kaushik, B. Bhushan, Security countermeasures in web based application, in *2019 2nd International Conference on Intelligent Computing, Instrumentation and Control Technologies (ICICICT)*. https://doi.org/10.1109/icicict46008.2019.8993141 (2019)
20. A.K. Sood, S. Zeadally, R. Bansal, Exploiting trust: stealthy attacks through socioware and insider threats. IEEE Syst. J. **11**(2), 415–426 (2017). https://doi.org/10.1109/jsyst.2015.2388707
21. A. Singh, A. Sharma, N. Sharma, I. Kaushik, B. Bhushan, Taxonomy of attacks on web based applications, in *2019 2nd International Conference on Intelligent Computing, Instrumentation and Control Technologies (ICICICT)*. https://doi.org/10.1109/icicict46008.2019.8993264 (2019)
22. A.K. Luhach, A. Khamparia, R. Sihag, R. Kumar, Honey Bee optimization based sink mobility aware heterogeneous protocol for wireless sensor network. Scalable Comput. Prac. Exp. **20**(4), 591–598 (2020)
23. T. Varshney, N. Sharma, I. Kaushik, B. Bhushan, Architectural model of security threats and their countermeasures in IoT, in *2019 International Conference on Computing, Communication, and Intelligent Systems (ICCCIS)*. https://doi.org/10.1109/icccis48478.2019.8974544 (2019)

A Mathematical Recommendation Model to Rank Reviewers Based on Weighted Score for Online Review System

Runa Ganguli, Priyanka Banerjee, Sanjib Halder, and Soumya Sen

Abstract Recommendation systems play a very important role in business from several aspects. New systems, concepts are being evolved to enrich the business from different perspectives. Online reviews provide valuable information about products and services to consumers. Generally, online reviews are done on products to understand the usefulness or popularity of a product. However, it is being found that often fake reviews are given to increase the popularity of own product or to defame competitors' products. This imposes research challenge to validate the reviews or Trustworthiness of reviewers. A recommendation model in online review system aims to filter the authenticated reviewers and then rank the top reviewers or emphasize on more impactful reviews. In this paper, a mathematical model is presented to rank the reviewers by assigning weighted scores based on certain parameters. A pre-processing technique is used before applying mathematical model to filter out the quality reviews. The pre-processing technique and data analysis are experimented on a real-life dataset to show the effectiveness of the proposed model.

Keywords Recommendation systems · Services to consumers · Data analysis

R. Ganguli (✉) · P. Banerjee · S. Halder
Department of Computer Science, The Bhawanipur Education Society College, Kolkata, India
e-mail: runa.ganguli@gmail.com

P. Banerjee
e-mail: priyankabanerjee333@gmail.com

S. Halder
e-mail: sanjibdothalder@gmail.com

S. Sen
A.K. Choudhury School of I.T, University of Calcutta, Kolkata, India
e-mail: iamsoumyasen@gmail.com

J. Tavares et al. (eds.), *Emerging Technologies in Data Mining and Information Security*, Lecture Notes in Networks and Systems 164,
https://doi.org/10.1007/978-981-15-9774-9_31

1 Introduction

Modern life has become easier due to the widespread use of Internet technology for communication and information sharing. Corporate and business houses try to use the benefits of Web medium at the fullest to explore their business to a new scale. Internet marketing is one such way to increase the volume of business by reaching the consumers directly in a customized way. Electronic Market [1] can present a range of offerings available in a market segment so that purchasers can compare the prices of the offerings and make wise purchase decisions. Customers can also get 24×7 support, place orders anytime, anywhere from any location and get quicker delivery. Consumers before making any online purchase seek out for online reviews whether it is for a product or a service. This has become a standard practice of the buying process. Also, after purchasing products or a service, a customer can put review comments about it to provide feedback and recommendations [2, 3] from a customer's perspective which are visible to other online buyers. In fact, a huge number of customers are more likely to purchase from a site if it has product ratings and reviews.

Basically, the review system can be of two types. In an open review system, literally anyone can leave a feedback against products and services. There is no real verification process before the review is written and posted, to check if the reviewer is an actual customer or not. This leaves review system open to manipulation by anyone, which includes both actual customers as well as fraud customers. The motive of these kinds of users may be uplifting or downgrading any product or service by unfair means. Even there is a growing tendency of merchants to hire professionals to write fake reviews and promote their products or defame the products of their competitors. A closed review system ensures that only verified customers leave reviews. But it is found that all reviews against products or services are not authentic every time. These spam reviews [11] can mislead the consumers about the products. Therefore, it is necessary to identify the authenticity, trustworthiness of the reviewers. Henceforth, it is an interesting research work to rank the reviewers based on their review.

In this paper, we present an appropriate method to detect the Trustworthiness of the reviewers by calculating a reviewer helpfulness score and assign a rank to the reviewers on the basis of the score. Before the calculation process, we pre-process the collected data so that we could filter those reviewers who have written certain number of reviews as well as we choose the products for which certain numbers of reviews are available. We apply our proposed method on real life data for experimental purpose. In this paper, we have used Amazon Gift Cards dataset (Ratings Only) (collected from Github [13]).

The paper is organized into the following sections. A brief survey work on reviewer trustworthiness is discussed in Sect. 2. The proposed methodology of our research work is presented in Sect. 3, which has two sub-sections. The first sub-section deals with data pre-processing and the next sub-section discusses about the data analysis steps. Section 4 demonstrates the experimental results and finally the paper is concluded in Sect. 5 along with the future scope of work.

2 Related Work

Graph theoretic approaches are useful for analyzing data for recommendation systems. A concept of a review graph [4] is proposed to identify the relationships between reviewers, reviews, and stores for which the reviewers have reviewed and tried to detect fake reviewer. The researchers also proposed an iterative computation framework [4] to calculate the trustworthiness of reviewers, the honesty of reviews and the reliability of stores within the range (−1,1). In another research work, a graph theory-based algorithm [5] has been proposed to identify the trustworthiness of the reviewers and the honesty of the reviews and tested their model on Amazon dataset. In order to categorize spam reviewers, they evaluated the faith of reviewers and their association with usefulness. In another interesting analysis by Banerjee et al. [6], reviewer's honesty and its impact on business is studied. The researchers proposed few hypotheses in their work based on both business characteristics and reviewer characteristics including business sponsorship, online reputation of business, number of reviewers engaged in reviewing, reviewer's positivity, experience, reputation, competence, and sociability. Moreover, the research work [6] could classify reviewers into two categories depending on high and low trustworthiness using linear and robust logistic regression method. Chua and Banerjee [7] examined review helpfulness as a function of reviewer reputation, review rating, and review depth. Review depth moderates the relationship between reviewer reputation and review helpfulness. They used a dataset collected from Amazon.com. The authors preferred Tobit estimation [7] over ordinary least square estimation. The result of their examination supported all their hypotheses successfully except that the review rating had no curvilinear relationship with review helpfulness. Yun Wan et al. [8] examined the reliability of online review helpfulness in their research paper. Along with chi-square tests and Monte Carlo simulation [8], they also used binomial test to compare the ratings. Mathematical models are suitable for building recommendation system and can be applied in different applications. Methods such as linear regression [9], nonlinear regression [10] have been used extensively to rank/recommend data/attributes in different business domains.

3 Proposed Methodology

The entire data analysis of this research work is performed on Amazon dataset of Gift Cards (Ratings Only) collected from Github [13]. This dataset consists of 147,194 records having three main fields: Product ID, Reviewer ID, and Rating Score. The Rating Score is marked on a scale of 0–5. The primary objective of the work is to calculate a reviewer helpfulness score and rank them on the basis of the score. For this analysis, we have filtered reviewers having a high review count from the large dataset. This section is divided into two sub-section: data pre-processing followed by data analysis.

3.1 Data Preprocessing

1. Sorted the entire data by Reviewer's ID.
2. Applied pivot table for getting count of reviews per reviewer and average reviewer rating. Pivot table is used for selective data filtering.
3. Sorted this record in descending order of review count.
4. Selected first ten reviewers from the sorted list for analysis.
5. Again, the original dataset is sorted separately by Product's ID.
6. Applied pivot table for getting count of reviews per product and average rating per product.

3.2 Data Analysis

The ranking of product or reviewer based on review count has got its significance. A reviewer which shares similar review rating with a greater number of other reviewers should be given higher priority compared to one whose rating is in agreement to a smaller number of reviewers. For this particular purpose, the calculated rank acts as the weight.

The objective of the work is to rank the reviewer based on two factors.

(i) The difference between the individual reviewer rating with the average rating of the product (taken as modulus). Low difference gets higher priority.
(ii) If one product has higher number of reviews than another product, the review difference calculated in previous step gets more weightage.

We first propose a mechanism to assign weightage to products based on their review count. This Product Rank Based on Review Count (Prod_Rank) is calculated by assigning rank beginning from 1 in ascending order based on product's review count in descending order. Next, the Weighted Rating Difference (WD) for each product is calculated which gives the difference in rating opinion of a reviewer with respect to other reviewers, weighted with product rank. Weighted Rating Difference (WD) against each Product ID is defined by (1).

$$\mathrm{WD}_i = (\mathrm{Prod\_Rank_i}) * |\mathrm{Reviewer\,Rating} - \mathrm{Average\,Product\,Rating}| \quad (1)$$

For each reviewer, we have calculated a Reviewer Score (R_Score) which gives the average difference in rating opinion defined by (2).

$$\mathrm{R\_Score} = \mathrm{Average\,of\,all\,WD}_i \mathrm{\,for\,Reviewer\,R}_i \quad (2)$$

As, we have computed Product Rank, similarly we also compute Reviewer Rank Based on Review Count (Rev_Rank) by assigning rank beginning from 1 in ascending order based on reviewer's review count in descending order. Finally, each of these Reviewer scores (R_Score) is multiplied by corresponding reviewer rank (Rev_Rank) to get Weighted Reviewer Score (WR_Score) which gives priority to more active reviewers over others. A reviewer having a low value of WR_Score is considered to be more helpful and ranked accordingly. Final Reviewer Rank is obtained by assigning rank beginning from 1 in ascending order based on Weighted Reviewer Score in ascending order. The workflow of the data analysis approach is depicted in the Fig. 1.

The **algorithm** of the proposed model is as follows:

Step 1: For each reviewer R_i in the selected reviewer list:

1.1 Get all the product IDs P_j for $jj = 1$ to $k[k =$ total review count of reviewer $R_i]$
1.2 Based on the number of reviews per product, each product P_j is ranked in descending order of review count.
1.3 For each product P_j, a **weighted rating difference** is calculated as:

$$\text{WD}_j = \text{ProdRank}_j * |\text{Reviewer Rating} - \text{Average Product Rating}|$$

1.4 A **Reviewer score** is calculated as:
R_Score$_i$= Average (WD$_j$) for $j = 1$ to k [$k =$ total review count of reviewer R_i].

Step 2: Based on the review count of each reviewer, R_i in the selected reviewer list is ranked in descending order of review count.
Step 3: For each reviewer R_i, a **Weighted Reviewer Score** is calculated as:

$$\text{WR\_Score}_i = (\text{Rev\_Rank}_i) * (\text{R\_Score}_i)$$

Step 4: **Final reviewer rank** is computed by ranking in ascending order of Weighted Reviewer Score.

4 Experimental Results

The experiment is conducted on Amazon dataset of gift cards (Ratings Only) collected from Github [13]. This dataset consists of 147,194 records having three main fields: Product ID, Reviewer ID, and Rating Score. During the data pre-processing stage of our work, the entire dataset was filtered based of review count of reviewer. Using a threshold value of ten for review count, we obtained 62 reviewers. A final shortlisting is done from which top ten reviewers were selected for our case study. During the data analysis phase, ten computational tables for each selected reviewer were calculated. Using these tables, we computed the final reviewer ranking score table where reviewers were ranked based on Weighted Reviewer Score or in other words difference in rating opinion. One of the computational tables for Reviewer 1 is shown in Fig. 2. It is clearly seen that how all the product IDs for a particular reviewer is listed and ranked based on review count. These are the products

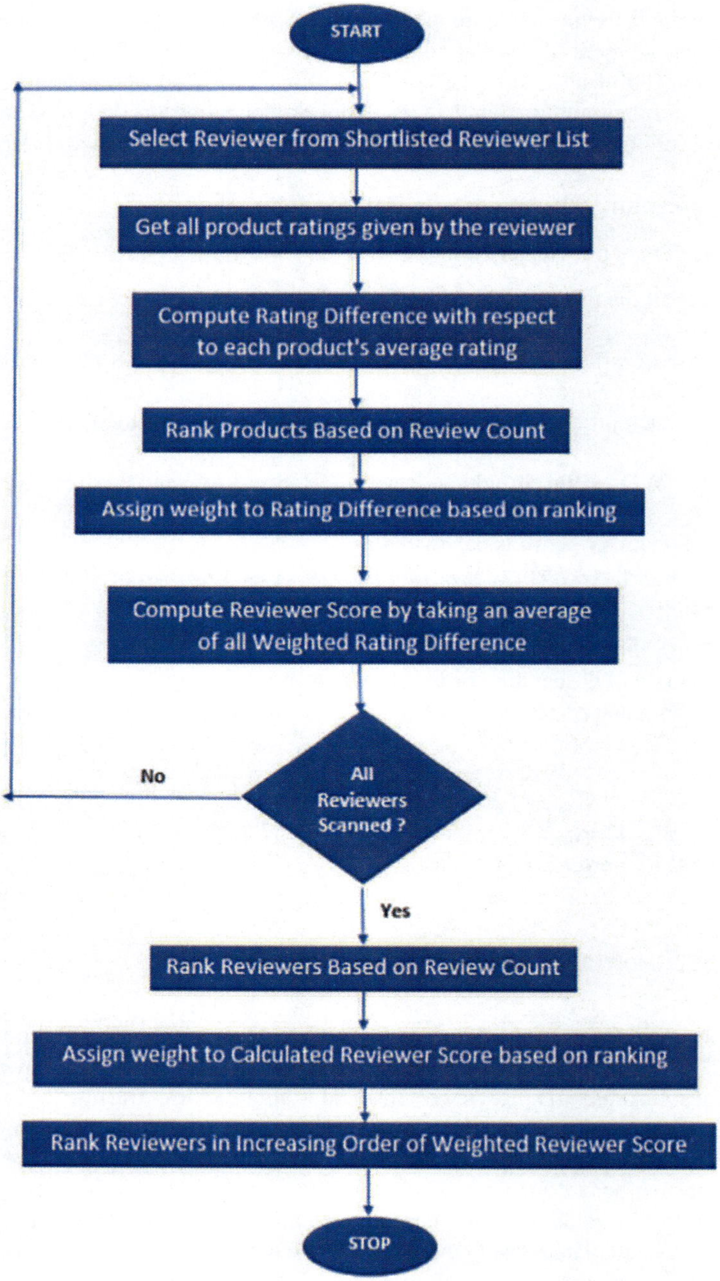

Fig. 1 Sample workflow of reviewer rating analysis

Product ID	Reviewer ID	Review Count per product	Product rank based on review count	Reviewer Rating	Average Product Rating	Difference in Rating	Weighted Difference
B00AR51Y5I	A13H0YP0J8PM6V	1687	1	5	4.706579727	0.293420273	0.293420273
B00MV9H6VY	A13H0YP0J8PM6V	560	2	5	4.476785714	0.523214286	1.046428571
B00PG85020	A13H0YP0J8PM6V	544	3	5	4.661764706	0.338235294	1.014705882
B00GU2ZZ8Y	A13H0YP0J8PM6V	352	4	5	4.534090909	0.465909091	1.863636364
B00MV9P8MS	A13H0YP0J8PM6V	318	5	5	4.534591195	0.465408805	2.327044025
B0078EPRPE	A13H0YP0J8PM6V	300	6	5	4.616666667	0.383333333	2.3
B01H5PPJT4	A13H0YP0J8PM6V	295	7	5	4.552542373	0.447457627	3.13220339
B015WY0DOQ	A13H0YP0J8PM6V	278	8	5	4.758992806	0.241007194	1.928057554
B00GRLUECA	A13H0YP0J8PM6V	273	9	5	4.663003663	0.336996337	3.032967033
B00BXLSZPM	A13H0YP0J8PM6V	261	10	5	4.689655172	0.310344828	3.103448276
B00MV9GCYQ	A13H0YP0J8PM6V	258	11	5	4.565891473	0.434108527	4.775193798
B00F2RZMEA	A13H0YP0J8PM6V	257	12	5	4.447470817	0.552529183	6.630350195
B00OX9OKO6	A13H0YP0J8PM6V	234	13	5	4.585470085	0.414529915	5.388888889
B0188LO3UK	A13H0YP0J8PM6V	209	14	5	4.751196172	0.248803828	3.483253589
B00MV9N316	A13H0YP0J8PM6V	207	15	5	4.70531401	0.29468599	4.420289855
B01CT5PHJ2	A13H0YP0J8PM6V	197	16	5	4.52284264	0.47715736	7.634517766
B00MV9M0UQ	A13H0YP0J8PM6V	175	17	5	4.434285714	0.565714286	9.617142857
B00MV9LSHC	A13H0YP0J8PM6V	173	18	5	4.421965318	0.578034682	10.40462428
B00MV9FQXO	A13H0YP0J8PM6V	160	19	5	4.7375	0.2625	4.9875
B01BLV4R8M	A13H0YP0J8PM6V	152	20	5	4.532894737	0.467105263	9.342105263
B00KFLZVQM	A13H0YP0J8PM6V	146	21	5	3.931506849	1.068493151	22.43835616
B00BXLURUI	A13H0YP0J8PM6V	142	22	5	4.788732394	0.211267606	4.647887324
B00MV9OQYO	A13H0YP0J8PM6V	126	23	5	4.53968254	0.46031746	10.58730159
B00MV9FPLM	A13H0YP0J8PM6V	118	24	5	4.593220339	0.406779661	9.762711864
B00JJ59UA8	A13H0YP0J8PM6V	88	25	5	4.340909091	0.659090909	16.47727273
B0188LOK7G	A13H0YP0J8PM6V	76	26	5	4.828947368	0.171052632	4.447368421
B00BXLVD8I	A13H0YP0J8PM6V	75	27	5	4.746666667	0.253333333	6.84
B00MV9OHUM	A13H0YP0J8PM6V	62	28	5	4.741935484	0.258064516	7.225806452
B01571YV42	A13H0YP0J8PM6V	58	29	5	4.672413793	0.327586207	9.5
B00FRL7G7M	A13H0YP0J8PM6V	54	30	5	4.833333333	0.166666667	5
B00NU8FTK0	A13H0YP0J8PM6V	50	31	5	4.36	0.64	19.84
B00MV9H2B8	A13H0YP0J8PM6V	44	32	5	3.409090909	1.590909091	50.90909091
B00GRLGF6O	A13H0YP0J8PM6V	43	33	5	4.581395349	0.418604651	13.81395349
B01BN45ICQ	A13H0YP0J8PM6V	33	34	5	4.666666667	0.333333333	11.33333333
B00MV9MON4	A13H0YP0J8PM6V	26	35	5	4.076923077	0.923076923	32.30769231
B00OGPXG9W	A13H0YP0J8PM6V	16	36	5	4.375	0.625	22.5
B00MV9MEO8	A13H0YP0J8PM6V	6	37	5	4.833333333	0.166666667	6.166666667
B01GOQIKRI	A13H0YP0J8PM6V	4	38	5	5	0	0
B00OD4ELZ4	A13H0YP0J8PM6V	2	39	5	5	0	0
						Final Score	8.731364592

Fig. 2 Computational table for Reviewer 1

for which the reviewer has given rating. Now, each one of these products are also rated by other reviewers and hence have average rating. This average product rating is desired to be as close as possible to the particular reviewer's rating. The computed difference in rating is weighted with product rank just to give emphasis on the more reviewed products. The calculated final score at the bottom of Fig. 2 is the reviewer score.

A reviewer whose opinion differs with the average opinion by a high margin could be treated as outlier. This often happens when a particular reviewer tries to spam a rating. Such rating deviations need to be filtered out whenever any average case analysis is performed on these types of datasets. Often these outliers mislead a certain interpretation and hence results in erroneous data analysis. In Fig. 3, we see the top ten selected reviewers are each assigned a Weighted Reviewer Score. The weight here is the reviewer rank based on review count, i.e., highlighting the reviewers whose review writing is more frequent than others. The Weighted Reviewer

Reviewer Serial No.	Reviewer ID	Count of Reviews	Reviewer rank based on review count	Calculated Reviewer Score	Weighted Reviewer Score	Final Reviewer Rank
1	A13H0YP0J8PM6V	39	1	8.731364592	8.731364592	1
2	A1U1G73EI5IRZF	32	2	10.75947679	21.51895359	4
3	A3OHGWD8LIDZ8K	26	3	4.932649623	14.79794887	2
4	A1F2NKB1ZKMO2V	23	4	7.917395466	31.66958187	8
5	A2RTTRR421J9KG	22	5	3.367360083	16.83680041	3
6	A1SVYJFIASQ46Z	20	6	7.18387044	43.10322264	10
7	A2QKY7ZG920FT7	19	7	4.387081404	30.70956983	7
8	ALQUMN89FI5HD	19	8	5.130870341	41.04696273	9
9	A1L2NJYQ9O10TP	17	9	3.269056764	29.42151087	6
10	A33SGA7BQ1X7XG	16	10	2.80865161	28.0865161	5

Fig. 3 Final reviewer ranking score table

Score can be treated as the helpfulness attribute often found in Amazon or similar review datasets. The final reviewer ranking is done on the basis of this score. A low score is desired more over a high score and hence ranked accordingly.

5 Conclusion and Future Scope

In this paper, we discussed and implemented an algorithm to assign trustworthiness score and rank users reviewing items in an online shopping website. The proposed framework is generic in nature and can be customized for any other online system to rate the reviewers. In the work, we considered several issues related to the system. First, we have considered the reviewers and products over a certain threshold limit to avoid the possible fake reviews. Secondly, we found that a particular reviewer often gives same review for many products. In order to address this, we employ a weighted score scheme based on many parameters such as deviation from average score, assigning more weight to the reviews of a product for which higher numbers of reviews are available. The proposed model performs the computation on review score which is given in numerical format. The work could be extended by considering other significant attributes like review content, helpfulness votes, reviewer characteristics, etc. Reviews are mostly given in text format; therefore, NLP methods [12] will be suitable to understand the exact sentiment of the customer. It is often found that there is a gap between the review score given and the text message written by the reviewers. Sentiment analysis is another dimension to extend the work. Even for certain systems, we can view our problem as a typical binary classification problem to identify trusted and untrusted reviewers. Different mathematical models and machine learning techniques could be explored on heterogeneous dataset and several business domains for more effective business values.

Acknowledgements The research has been conducted under a research project named as "**Review Spam Detection and Product Recommendation System using Machine Learning Techniques**" sponsored by The Bhawanipur Education Society College, Kolkata. Authors are thankful to the Research and Publication Cell of the College for the Research Grant which provided the computational and other facilities.

References

1. G. Maji, S. Mandal, N.C. Debnath, S. Sen, A novel delivery model to expand e-commerce customers based on telecom data mining, in *Proceedings of 34th International Confer*ence, vol. 58 (2019), pp. 336–344
2. R. Ganguli, A. Mehta, N.C. Debnath, S. Aljahdali, S. Sen, An integrated framework for friend recommender system using graph theoretic approach, in *Proceedings of 35th International Confer*ence, vol. 69 (2020), pp. 242–255
3. S. Ghosh, S. Roy, S. Sen, An efficient recommendation system on e-learning platform by query lattice optimization, in *Springer 4th International Conference on Data Management, Analytics and Innovation (ICDMAI)*, New Delhi, India (17–19 Jan, 2020)
4. G. Wang, S. Xie, B. Liu, S.Y. Philip, Review graph based online store review spammer detection, in *2011 IEEE 11th International Conference on Data Mining*. (IEEE, 2011), pp. 1242–1247
5. D. Shinzaki, K. Stuckman, R. Yates, Trust and helpfulness in amazon reviews (2013)
6. S. Banerjee, S. Bhattacharyya, I. Bose, Whose online reviews to trust? Understanding reviewer trustworthiness and its impact on business. Decision Supp. Syst. **96**, 17–26 (2017)
7. A.Y. Chua, S. Banerjee, Understanding review helpfulness as a function of reviewer reputation, review rating, and review depth. J. Assoc. Inform. Sci. Technol. **66**(2), 354–362 (2015)
8. Y. Wan, M. Nakayama, The reliability of online review helpfulness. J. Electron. Commer. Res. **15**(3), 179 (2014)
9. P. Ghosh, S. Sen, N. Chaki, Materialized view construction using linear regression on attributes, in *2012 Third International Conference on Emerging Applications of Information Technology*. (IEEE, 2012), pp. 214–219
10. S. Roy, R. Ghosh, S. Sen, Materialized view construction based on clustering technique, in *IFIP International Conference on Computer Information Systems and Industrial Management*. (Springer, Berlin, Heidelberg, September 2015), pp. 254–265
11. M. Crawford, T.M. Khoshgoftaar, J.D. Prusa, A.N. Richter, H. Al Najada, Survey of review spam detection using machine learning techniques. J. Big Data **2**(1), 23 (2015)
12. L. Dutta, G. Maji, S. Sen, A study on spatiotemporal topical analysis of Twitter data, in *Emerging Technology in Modelling and Graphics*. (Springer, Singapore, 2020), pp. 699–711
13. Github Weblink, https://nijianmo.github.io/amazon/index.html

Virtual Numeric Authentication System Using Contour Detection of Color-Banded Fingertips

Aman Gurung, Siddharth Gautam, Tushar Garg, Yash Bhardwaj, and Harshul Gupta

Abstract In this paper, simplistic and much-awaited solution to the enigma of a secure user authentication system is proposed. For the same, we designed a virtually controllable numpad authentication system. To provide an effortless, immersive, and robust virtual experience which is color recognized, we employed a web camera along with the applied algorithm. The proposed modal intends to accomplish particular objectives, i.e., detection, recognition of an object, specialized graphical user interface, and authentication. Proposed methodology makes use of preprocessing tasks to convert a sequence of frames to detect and recognize a specified color. Numeric keys are placed at random positions in a definitive structure to provide a security feature. As a result, using our system, users can control a virtual numeric keypad by the movements of their color-banded fingertips. The webcam captures the consecutive frames and compares them to detect a difference in the contour, perceiving it as a keypress.

Keywords User authentication · Color-banded fingertips · Contour detection · Thresholding · Masking

A. Gurung (✉) · S. Gautam · T. Garg · Y. Bhardwaj · H. Gupta
Department of Information Technology, HMRITM, GGSIPU, New Delhi, India
e-mail: cradmercer@gmail.com

S. Gautam
e-mail: siddharthinfo92@gmail.com

T. Garg
e-mail: i.tushar.dev@gmail.com

Y. Bhardwaj
e-mail: yashb6625@gmail.com

H. Gupta
e-mail: gharshul98@gmail.com

J. Tavares et al. (eds.), *Emerging Technologies in Data Mining and Information Security*,
Lecture Notes in Networks and Systems 164,
https://doi.org/10.1007/978-981-15-9774-9_32

1 Introduction

With the advancement of cameras and computer vision algorithms, several strides have been made to interpret and execute user commands. All the attempts have faced certain challenges such as to create a secure and robust implementation to gather input or minimize use of physical space for compact style. Similarly, in this paper, those challenges were faced and defeated. The proposed authentication system strives to produce a hassle-free and impregnable system to gather user input and authenticate it. Previous approaches include the use of a physical device whose keystrokes could be tracked, jeopardizing security and privacy. An on-screen interface seems more reliable than a physical device but still not entirely secure as key coordinates visible on the screen can be detected. To overcome the above shortcomings, this paper acquaints the use of random positions for all keys but in a definitive structure, so that security concerns can be averted.

This paper intends to accomplish particular objectives that contribute to its completeness and application. Following are the objectives achieved:

- Detection of an object being the first objective helped in differentiating between multiple objects in front of a camera [1].
- Recognition, the second objective was achieved by the identification of color bands through HSV (hue–saturation–value) ranges of different colors [2, 3].
- To provide meaningful input, a specialized graphical user interface (GUI) was designed which eventually fulfilled the third objective [4].
- The user input was authenticated ultimately completing the main objective.

Firstly, an introduction is given describing in brief about what the proposed paper intends to achieve and the challenges faced in doing so. Next, literature survey gives an account of the previous studies and solutions found in the field of taking user input by different means and authenticating it [1, 2, 5–11]. In the methodology section, under preprocessing subsection, certain processes like changing color space, masking, noise reduction, and thresholding have been well defined which are also depicted graphically with the help of a flowchart. The preprocessed output frames are then used in live contour detection and recognition of the predefined colored object. Next section of result and analysis gives an account of the security feature added to the GUI paving way to an impregnable authentication system using a virtual numeric keypad [4].

2 Literature Study

This section briefly describes the groundwork laid and previous work done in accepting user input securely through the use of various systems with their merits and demerits. Most commonly used input mechanism is a full-length keyboard that is projected in front of user but this is not as secure as conventional ways of gathering

Table Literature study

S. No.	Title	Author	Ref. no.	Year
1	Study on object detection using OpenCV–Python	Gupta B	[1]	2017
2	Finger recognition and gesture-based augmented keyboard	Jagannathan MJ	[2]	2018
3	Real-time finger tracking and contour detection for gesture recognition using OpenCV	Gurav RM	[5]	2015
4	Bigkey: A virtual keyboard for mobile devices	Al Faraj K	[6]	2009
5	A virtual keyboard implementation based on finger recognition	Zhang Z	[7]	2017
6	Webcam-based virtual keyboard using shadow analysis	Adajania Y	[8]	2010
7	User-defined custom virtual keyboard	Kumar R	[9]	2016
8	Virtual keyboard: A human–computer interaction device based on laser and image processing	Su X	[10]	2015
9	Feature extraction of edge detected images	Wankhade MR	[11]	2017

input such as physical keyboard. The reference literature is given in Table 1. Gupta et al. [1] gave an account of all the possible object detection techniques including color-based detection which led this paper to achieve its first objective of detection of an object using a webcam. Jagannathan et al. [2] paved the way toward the second objective of the recognition of an object through the study of fingertip gesture but recognition through color bands seemed more economical and viable alternative for this paper. Gurav et al. [5] helped realizing that a gesture can be much easily recognized using color-banded fingertips instead of hand gestures. Al Faraj et al. [6] suggested the use of virtual projected keyboard to enable user to input data in case of small screens but failed to account for user's convenience. Zhang et al. [7] used a layout of keyboard scripted on paper to allow webcam to monitor it and take input by interpreting user interaction with the paper keyboard but failed to account for availability of paper every time an input is required. The idea of on-screen number pad was coined after carefully contemplating all the mentioned papers and their shortcomings such as lack of physical space to project a full-length keyboard or even use a screen to render it [6–10].

3 Proposed Methodology

In this section, the use of live contour detection around the selected colored object to feed as an input to the virtual numeric system. This system gathers input if the center of the contour lies in between the region of the key. The key label then supplies the value of a key and is implemented on the virtual numeric keypad. In this method, we use a random position generator function for all numeric values every time the system is accessed. This changes the classical approach of the numeric keypad and secures the system like never before.

The user is granted access if the entered numeric value matches the stored password and is prompted to re-enter the password a maximum of five times for an incorrect attempt, after which the system is terminated.

3.1 Preprocessing

Preprocessing the input frame sequence will result in the color detection and recognition through which the final contour retrieval will take place making use of the bounding rectangle. The approach used is discussed in detail in further subsections, and those steps are graphically represented in the flowchart through Fig. 1.

Change color space

A color space also known as color system is simply the range of colors as tuples of numbers. It is important to change color space of an image from (say) blue-green-red (BGR) to HSV color space because rather than using a vast infinity of numbers to represent colors, a scale in range of 0...1 is used to simplify detection.

$$V \leftarrow \max(R, G, B) \tag{1}$$

$$S \leftarrow \begin{cases} \frac{V - \min(R,G,B)}{V} & \text{if } V \neq 0 \\ 0 & \text{otherwise} \end{cases} \tag{2}$$

$$H \leftarrow \begin{cases} 60(G - B)/(V - \min(R, G, B)) & \text{if } V = R \\ 120 + 60(B - R)/(V - \min(R, G, B)) & \text{if } V = G \\ 240 + 60(R - G)/(V - \min(R, G, B)) & \text{if } V = B \end{cases} \tag{3}$$

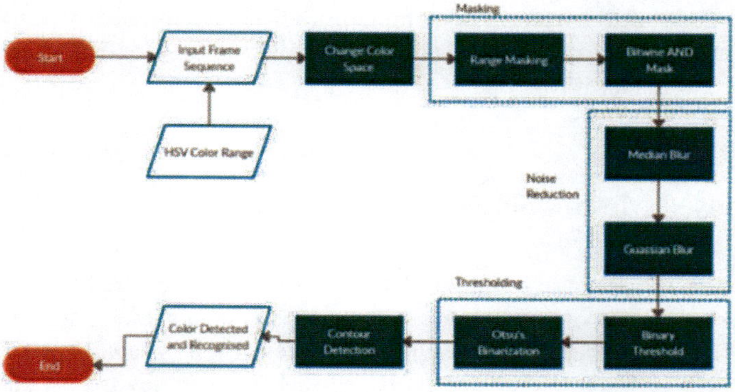

Fig. 1 Preprocessing workflow

Table 2 HSV values for destination data types

Image size	*V* (Value)	*S* (Saturation)	*H* (Hue)
8	255 V	255S	H/2
16	−65535 V	−65535S	−H
32	V	S	H

If $H > 0$ then $H \leftarrow H + 360$. On output $0 \leq V \leq 1.0 \leq S \leq 1.0 \leq H \leq 360$.

Here, Eq. (1) depicts that *V* (value) is maximum of red-green-blue (RGB) values, subsequently Eq. (2) shows that saturation (*S*) is equal to zero if *V* is zero, and lastly, hue (*H*) is also dependent upon *V*. Equations (1), (2), and (3) are referred from [12].

The values are then converted to the destination data type as shown in Table 2.

Masking

Masking in image processing is a process of hiding needless portions of an image or frame and revealing meaningful portion from it.

Range Masking

The lower and upper HSV bound output from HSV color range is fed as an input to the range masking process to create a sieve, allowing the values that lie in between the bounds to pass. Figure 2 depicts the working of masking. In this manner, the masking is done. Equation (4) shows single-channel input array for every element:

$$\mathrm{dst}(I) = \mathrm{lower}b(I)_0 \leq \mathrm{src}(I)_0 \leq \mathrm{upper}b(I)_0 \tag{4}$$

Equation (4) is referred from [13]. Here, dst(*I*) is set as 255 (all 1 bit) if src(*I*) is specified as 1D, 2D, 3D, … box or 0 otherwise. "$(I)_0$" is a depiction of single-channel input.

Fig. 2 Range masking

Fig. 3 Bitwise AND mask

Bitwise AND mask

Bitwise AND masking performs bitwise logical conjunction per element between arrays or an array and scalar.

When the two arrays src1 and src2 have the same size:

$$\mathrm{dst}(I) = \mathrm{src1}(I) \wedge \mathrm{src2}(I) \text{ if } \mathrm{mask}(I) \neq 0 \tag{5}$$

An array and a scalar when src2 is constructed from scalar or have the same number of elements as src1.channels(), Eq. (5) and (6) are referred from [16]:

$$\mathrm{dst}(I) = \mathrm{src1}(I) \wedge \mathrm{src2} \text{ if } \mathrm{mask}(I) \neq 0 \tag{6}$$

Here, Eqs. (5) and (6) depict conjunction between source arrays either scaler or not but only if the masking is not equal to zero. Figure 3 shows live bitwise masking of the image.

Noise reduction

Noise reduction helps to suppress the noise from the image which are caused due to various factors such as environment. So, to gain meaningful frames or image we need such techniques that help us in image restoration, image registration, and image segmentation. The final processed image is shown in Fig. 4.

Median blur

It replaces the central pixel in a kernel window of defined positive odd integer size with the median value of all the pixels present inside the kernel window. This is performed as to remove salt-and-pepper noise from the image. The kernel window size should be greater than 1. Median blur effectively reduces the noise.

Fig. 4 Median and Gaussian blur

Gaussian Blur

The resultant median blurred frame is then transformed using Gaussian blur. This image blurring filter applies to each pixel of the image reducing details and image noise. This noise reduction technique is a smooth blur that resembles an image through a translucent screen.

Formula of gaussian function in two dimensions:

$$G(x, y) = \frac{1}{2\pi\sigma^2} e^{-\frac{x^2+y^2}{2\sigma^2}} \tag{7}$$

For one dimension:

$$G(x) = \frac{1}{\sqrt{2\pi^{\sigma^2}}} e^{-x^2/2\sigma^2} \tag{8}$$

Here, for calculating the transformation to apply to each pixel in the image in one dimension and two dimensions, Eqs. (7) and (8) are used, respectively. Equations are referred from [14].

Thresholding

Thresholding is an image segmentation process, that retrieves the required value from the image from its background. Further processing after thresholding of image is given in Fig. 5.

Binary Thresholding

After the noise reduction techniques, the threshold is applied to the frame. If the pixel value is greater than the provided threshold, then the assigned value is 1 usually white else the value is 0 or black. The input frame in this case must be in grayscale as converted in masking. The final frame is then passed for Otsu's binarization.

Fig. 5 Thresholding

$$d(x, y) = \begin{cases} \text{max val if src}(x, y) > \text{thresh} \\ 0 \qquad \text{otherwise} \end{cases} \tag{9}$$

Here, if source pixel has a value greater than threshold value, it is changed to maximum value and to zero if less than or equal to threshold value. Equation (9) is referenced in [15].

Otsu's binarization

It calculates the threshold value from an image histogram for a bimodal image (image whose histogram has two peaks). It then finds the optimal threshold value for the image and returns the value.

This algorithm finds a threshold value(t) which minimizes the weighted within class variance. The relation is given as:

$$\sigma_\omega^2(t) = q_1\sigma_1^2(t) + q_2\sigma_2^2(t) \tag{10}$$

Equation (10) is retrieved from [17], where

$$q_1(t) = \sum_{i-1}^{t} P(i) \quad \text{and} \quad q_2(t) = \sum_{i=t+1}^{I} P(i)$$

$$\mu_1(t) = \sum_{i-1}^{t} \frac{iP(i)}{q_1(t)} \quad \text{and} \quad \mu_2(t) = \sum_{i=t+1}^{I} \frac{iP(i)}{q_2(t)}$$

$$\sigma_1^2(t) = \sum_{i-1}^{t} [i - \mu_1(t)]^2 \frac{P(i)}{q_1(t)} \quad \text{and} \quad \sigma_2^2(t) = \sum_{i=t+1}^{I} [i - \mu_2(t)]^2 \frac{P(i)}{q_2(t)}$$

Hence, a value of t which lies between two peaks is found such that variance to both classes as minimum.

3.2 Contour Detection

Contour is a curve that joins all the continuous points along the boundary of the shape that have same color or intensity. We use binary images for color detection. It is usually used for shape analysis and object detection. The steps for contour detection are represented below with the help of a flowchart in Fig. 6.

Contour Approximation Method

Contours approximation method sets the boundary of a shape with same intensity. It stores the coordinates of the boundary depending on the method used. In our case, we store the boundary points of a line. This removes all the redundant points that are not required and compresses the contour, hence saving memory.

Contour Retrieval

The contour is then retrieved using the retrieval tree method using border following algorithm [18]. Binary image is required for the retrieval of the contour. In this technique, all the contour is retrieved and reconstructed as full hierarchy of nested contours.

Bounding rectangle and center retrieval

Use of moments helps in extracting shape features such as area and centroid. An image moment is basically a certain weighted average (moment) intensities of the image's pixels, usually with some attractive property or interpretation.

Centroid is given by $C_x = \frac{M_{10}}{M_{00}}$, $C_y = \frac{M_{01}}{M_{00}}$ and area is given by M['m00'].

The bounding rectangle returns rotated rectangle, as it gives minimum area of the rectangle. Figure 7 is the result of contour detection.

4 Result and Analysis

This paper has minimalist needs in terms of hardware and software requirements which boasts of its economical and simple-to-use feature. For hardware, webcam either computer integrated or external can be used to capture the sequence of frames which will eventually be processed with the help of software's including programming language Python with OpenCV module [19]. Several Python libraries were

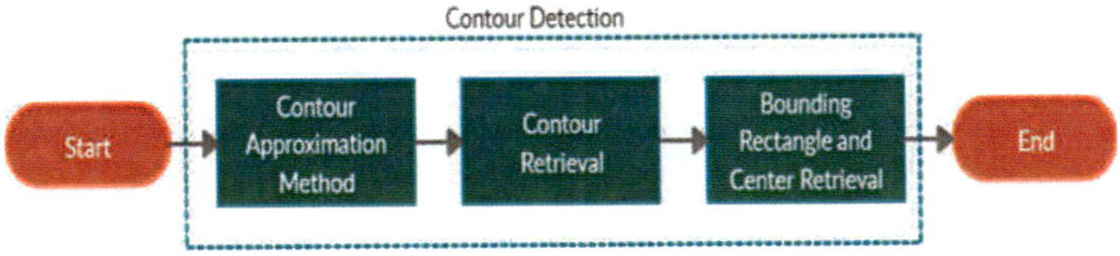

Fig. 6 Contour detection workflow

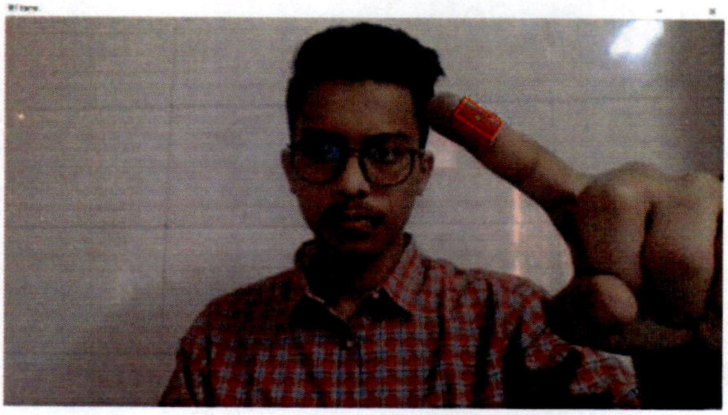

Fig. 7 Live contour detection

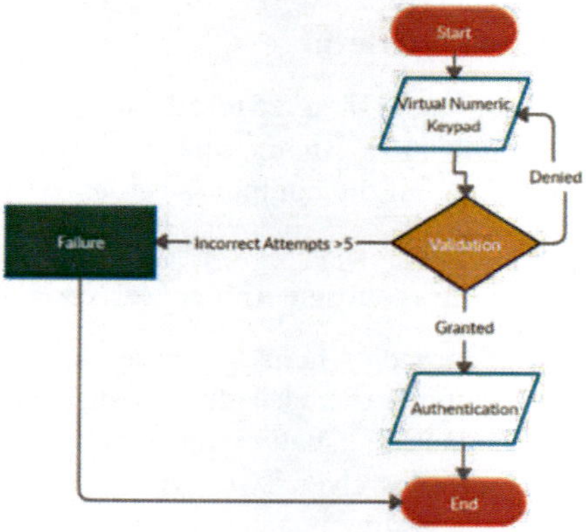

Fig. 8 System flowchart

also used such as NumPy which helps in working with n-dimensional arrays [20]. Figure 8 provides suitable flowchart for the system.

4.1 Implementation

A specialized graphical user interface (GUI) was implemented through the use of random position generator function, which arbitrary places the key labels across the numeric keypad [4–21]. This random generator shuffles the keys and hence always picks a random pattern for placing the key values. This ensures the system remains impregnable. The resultant unique numeric GUI is show in Fig. 9.

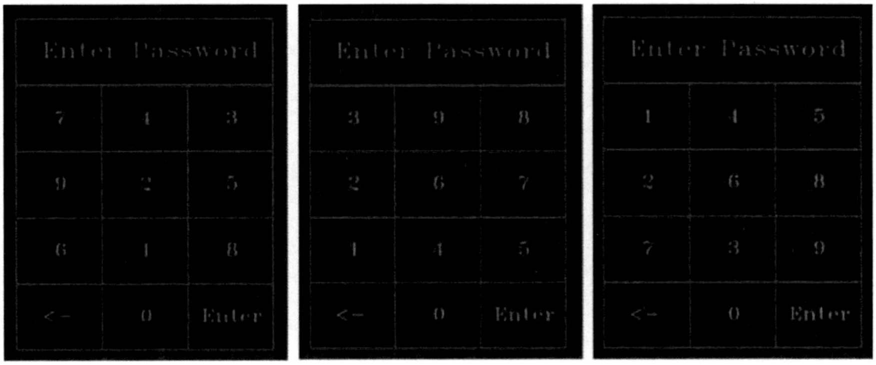

Fig. 9 Unique numeric GUI

4.2 Results

A virtual numeric keypad is generated after the preprocessing which is ready for input and authentication. Access is granted if the input data matches the stored password. A prompt of incorrect password appears a maximum of 5 times before the system is forced to terminate due to excess of invalid attempts. All the process snippets have been shown through Fig. 10.

5 Conclusion

This paper proposed a much-secure way of accepting user input and then authenticating it. A sequence of frames is preprocessed with processes, i.e., changing color space, masking, noise reduction, and thresholding to give the output frame sequence which undergoes contour formation around the specified color-banded fingertips. A keypress is perceived by the system only when the center of the formed contour lies in between the boundary of the key. If by some means a person records the user input pattern, they will not be able to authenticate themselves as the user interface will be randomized, stopping the illegitimate attack. In the future, we would like to improve the algorithm to recognize not only color but gestures as well, also to enable a two-step authentication with first being face recognition and second being user password.

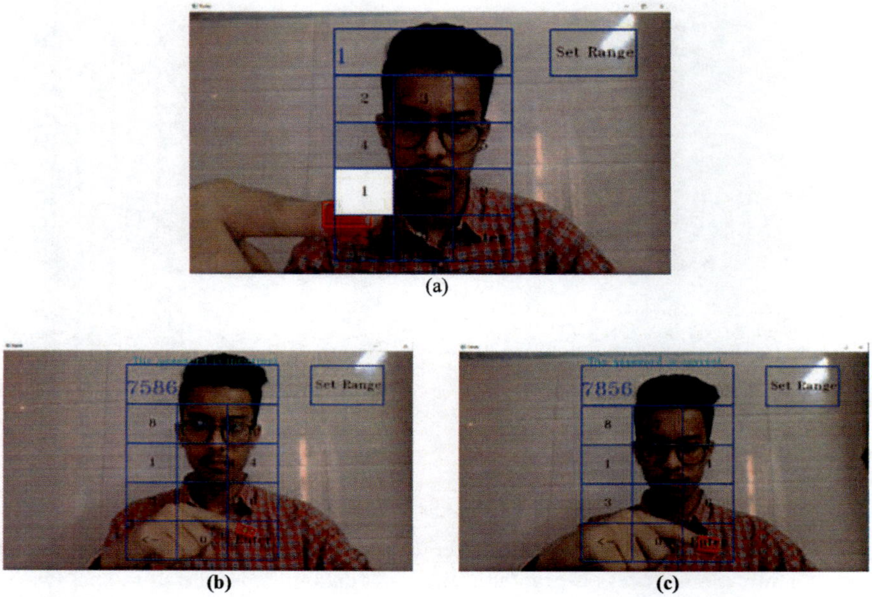

Fig. 10 Virtual keypad and authentication. **a** Giving inputs virtually, **b** incorrect password, **c** correct password

6 Copyrights

The actor performing tests and implementing the proposed methodology is one of the co-authors of the paper and the snapshots in the paper should not be used for any purpose without consent.

References

1. B. Gupta, A. Chaube, A. Negi, U. Goel, Study on object detection using open CV-Python. Int. J. Comput. Appl. **162**(8), 17–21 (2017)
2. M.J. Jagannathan, M. Surya, B.T. Aqib Muhammed, R.J. Poovaraghavan, Finger recognition and gesture based augmented keyboard (2018)
3. HSL and HSV, https://en.wikipedia.org/wiki/HSL_and_HSV. Last accessed 2019/11/30
4. Graphical User Interface, https://en.wikipedia.org/wiki/Graphical_user_interface. Last accessed 2019/11/30
5. R.M. Gurav, P.K., Kadbe, Real time finger tracking and contour detection for gesture recognition using OpenCV, in *2015 International Conference on Industrial Instrumentation and Control (ICIC).* (IEEE, 2015), pp. 974–977
6. K. Al Faraj, M. Mojahid, N. Vigouroux, Bigkey: a virtual keyboard for mobile devices, in *International Conference on Human-Computer Interaction.* (Springer, Berlin, Heidelberg, 2009), pp. 3–10

7. Y. Zhang, W. Yan, A. Narayanan, A virtual keyboard implementation based on finger recognition, in *2017 International Conference on Image and Vision Computing New Zealand (IVCNZ)*. (IEEE, 2017), pp. 1–6
8. Y. Adajania, J. Gosalia, A. Kanade, H. Mehta, N. Shekokar, Virtual keyboard using shadow analysis, in *2010 3rd International Conference on Emerging Trends in Engineering and Technology*. (IEEE, 2010), pp. 163–165
9. R. Kumar, P. Chaudhary, User defined custom virtual keyboard, in *2016 International Conference on Information Science (ICIS)*. (IEEE, 2016), pp. 18–22
10. X. Su, Y. Zhang, Q. Zhao, L. Gao, Virtual keyboard: a human-computer interaction device based on laser and image processing, in *2015 IEEE International Conference on Cyber Technology in Automation, Control, and Intelligent Systems (CYBER)*. (IEEE, 2015), pp. 321–325
11. M.R. Wankhade, N.M. Wagdarikar, Feature extraction of edge detected images. Int. J. Comput. Sci. Mob. Comput. (IJCMSC) **6**, 336–345 (2017)
12. Miscellaneous Image Transformations—OpenCV 2.4.13.7 documentation, https://docs.opencv.org/2.4/modules/imgproc/doc/miscellaneous_transformations.html. Last accessed 2019/11/30
13. Operations on Arrays—OpenCV 2.4.13.7 documentation, https://docs.opencv.org/2.4/modules/core/doc/operations_on_arrays.html. Last accessed 2019/11/30
14. Gaussian Blur, https://en.wikipedia.org/wiki/Gaussian_blur.html. Last accessed 2019/11/30
15. Binary Thresholding—OpenCV 2.4.13.7 documentation, https://docs.opencv.org/2.4/modules/imgproc/doc/miscellaneous_transformations.html. Last accessed 2019/11/30
16. A. Mordvintsev, K. Abid, Opencv-python tutorials documentation. Obtenido de https://media.readthedocs.org/pdf/opencv-python-tutroals/latest/opencv-python-tutroals.pdf
17. Otsu's Binarization—OpenCV 2.4.13.7 documentation, https://opencv-python-tutroals.readthedocs.io/en/latest/py_tutorials/py_imgproc/py_thresholding/py_thresholding.html#otsus-binarization. Last accessed 2019/11/30
18. S. Suzuki, Topological structural analysis of digitized binary images by border following. Comput. Vis. Graph. Image Process. **30**(1), 32–46 (1985)
19. OpenCV, https://pypi.org/project/opencv-python. Last accessed 2019/11/30
20. NumPy, https://numpy.org/. Last accessed 2019/11/30
21. Random, https://docs.python.org/3/library/random.html. Last accessed 2019/11/30

Efficient Authentication Technique in Medical Application by Using Homomorphic Encryption on Finger Print and RFID Tags

Sonali S. Lunawat, Hitesh Bhandari, Madhuri Badole, and Ompriya Kale

On the contrary, in [1], suggested Hierarchical Quadrature Amplitude Modulation (HQAM) for better protection of high-priority data during image transmission. Nevertheless, no channel model such as AWGN is considered in this work.

Abstract Nowadays, Internet of things had made easier to connect any devices and bring everything online with enormous amounts of new data which will make our life easier and smarter. Researchers estimated that 50 billion new devices will be connected together online in the next 5 years. For such networks, security is main concern as devices retrieve data from sensors and provides valuable information for many applications. In Internet of things, devices have high capability of computing and storing information in those with different constraints as resources, memory, bandwidth, and battery by reliable transportation over networks. In order to secure, many different techniques are present on of which is providing high security is radio frequency identification (RFID) tags security of sender node (i.e., patient) and receiver node (i.e., doctor) is very important. Using RFID with some Encryption will provide high security. In this paper, we proposed a system which will be highly secure in medical applications for patient to securely pass there inputs and report to doctors without man-in-the-middle attack. In this paper, we have identified an approach of combining homomorphic encryption on biometric authentication and using RFID tag identification using Internet of things. To implement this concept,

S. S. Lunawat (✉) · M. Badole · O. Kale
Department of Computer Engineering, Pimpri Chinchwad College of Engineering and Research, Ravet, Pune, India
e-mail: sonali.lunawat@pccoer.in

M. Badole
e-mail: madhuri.badole@pccoer.in

O. Kale
e-mail: ompriya.kale@pccoer.in

H. Bhandari
Tata Consultancy Services, Hinjewadi, Pune, India
e-mail: hitesh.bhandari@gmail.com

J. Tavares et al. (eds.), *Emerging Technologies in Data Mining and Information Security*, Lecture Notes in Networks and Systems 164,
https://doi.org/10.1007/978-981-15-9774-9_33

RFID send tag information of patient which will be encrypted data with it finger print by applying homomorphic encryption is sent and receiver will able to decrypt that by following set of rules. A server will perform homomorphic encryption and using feature extraction for authentication of a Patient.

Keywords *Cryptography* · Encryption · Homomorphic encryption · Biometric · RFID

1 Introduction

The Internet of things in combination of technology and techniques of device connection with sensors for communication and information providing to node using Internet. Each object who provides data are known as SMART in the Internet of things. The term "Internet of Things" was defined by Kevin Ashton in the year 1982 [2]. Internet of things is used for radio frequency identification ("RFID") tags that can be placed on patient's card. Where each tags provide information related to user authentication. Research tell that this information may be hacked. The "things" in the Internet of things as in Fig. 1 does not means Desktop, laptop, computing device, smart phones, and but any object which can be any physical devices which do not have any computation power or nor works on electricity.

Fig. 1 Internet of things

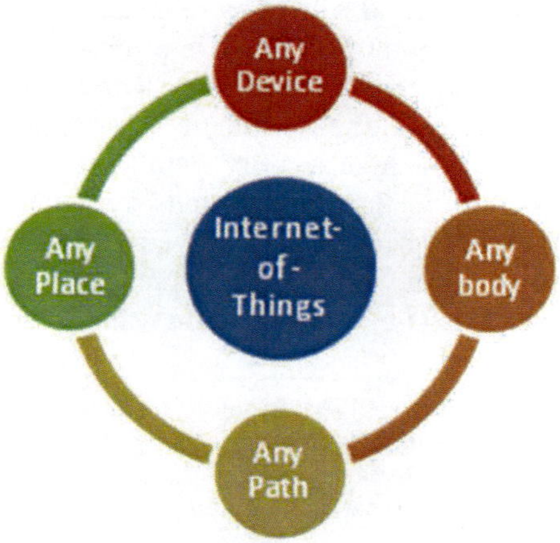

2 Theoritical Background

2.1 *Various Components of Internet of Things*

The information exchanged between two "things" without any trigger to provide input is what called as Internet of things. This process of automated information exchange between two physical devices or "things" through some specific communication or network [3].

2.1.1 Wireless Sensor Networks (WSN)

Wireless sensor networks techniques are where connection between heterogeneous nodes whose wireless communication takes place over allocated frequency and bandwidth. The wireless sensor node consists of the following part consist of different types of sensors, processing capacity, and low memory. Power supply is provided to devices. The data transfer is carried using trans receiver. If communication is facing problem, the multi-hop technology is used.

2.1.2 Radio Frequency Identification (RFID)

RFID technology is mainly uses tags for interacting with each other automatically. RFID tags use radio frequency waves for exchanging information between one another without physical connect. RFID use two components as follows.

RFID Tags

In a RFID tag, a microchip is embedded with an antenna. The RFID tag has a unique identification code to authenticate tag holder known as electronic product code (EPC). The function of the EPC in each tag follows standards that are universal identification. RFID tags have two types as

(1) **Active tag**: This type of tag has an internal battery and uses a unique EPC with its surrounding EPCs remotely from a limited distance.
(2) **Passive tag**: There is no internal battery in the passive tags electromagnetic signal emitted by a tag through inductive coupling as a source of energy [2].

RFID Reader

The RFID reader does identification of each detected tag by its interaction with the EPC of the scan tag [2].

3 Cryptography

Cryptography is a technique which converts the plain text into cipher text using different algorithms and is classified as symmetric and asymmetric cryptography. Symmetric cryptography uses one key for decrypting and encrypting data. Following algorithm uses symmetric cryptography

1. DES
2. 3DES and
3. AES.

In asymmetric cryptography uses pair of keys which are public key and private key for encryption and decryption. Following algorithm uses asymmetric cryptography RSA.

4 Biometric Authentication

Reliable user authentication is becoming needed in this Web-enabled world [4]. The issues of an insecure authentication system in a corporate or enterprise environment can include loss of confidential information, denial of service (DoS), and loss of data integrity and confidentiality. The value of reliable user authentication is major concern nowadays. In day-to-day life, many applications in everyday life also require user authentication. Security is major concern in some application like banking, medical, etc. [4]. The techniques used as passwords and PINs or OTP (personal identification numbers) can be cracked easily as it consist of number or some character with limited length.

Passwords and PINs can be acquired by simple algorithm and many crack software or tools available. If hacker can crack password then will have full access to your data. Fortunately, biometrics, and fingerprint technology in particular, can provide a much more accurate and reliable user authentication method. Biometrics is growing rapidly in the field have less chances for hackers. Examples of biometrics include fingerprint, face, iris, and speech recognition.

5 Homomorphic Encryption

Homomorphic encryption (HE) was proposed by Rivest in 1978 with the concept of "privacy homomorphism" [5]. This encryption technique can directly manipulate cipher text. While modern encryption algorithms are virtually unbreakable as they requires processing power and the process of breaking, it is too costly and require more time. Homomorphic encryption helps to protect the integrity of your data by allowing others to manipulate its encrypted form while no one can access its decrypted values (Fig. 2 and Table 1).

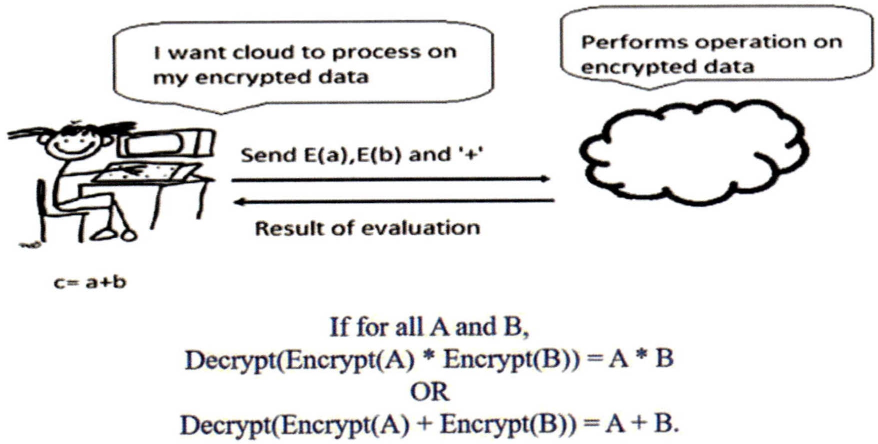

Fig. 2 Homomorphic encryption

Table 1 Comparison among different authentication techniques [6]

S. No.	Authentication technique	Problems/benefits
1	Password and PIN	Can easily be cracked
2	Biometric authentication	Can have less changes to modify. Best way of authentication
3	Cryptography	Require much processing and utilize more time
4	Homomorphic encryption	Highly secured and less processing and utilize less time

The paper is organized as follows. Section 1 gives an introduction, Sects. 2, 3 and 4 refers to the literature Survey on different techniques like cryptography, biometric, and homomorphic encryption, and Sect. 4 covers need of security, and Sect. 5 covers proposed system and conclusion.

6 Proposed System

The system aims that client which is called as patient and server is doctor. The architecture of the system is as shown (Fig. 3).

From the above literature survey and analysis, we proposed a system which be highly secure in medical applications for patient to securely pass their inputs and report to doctors without man-in-the-middle attack.

Client: Client is acting as patient having RFID tags and finger print with applied homomorphic encryption

Fig. 3 Architecture of proposed system

Internet: Communication between client and Sserver remotely using wireless sensor network.
Server: Will be acting as authorize server and process the comparison of given input with database data. Processing and computation of all happen on server.

Our aim is to

(1) To prevent authentication of patient using RFID tags.
(2) To ensure integrity and confidentiality of the RFID tag by using homomorphic encryption.
(3) To get information of patient.
(4) To provide flexibility for patient and doctor to choose a privacy policies. Our system aim to integrate biometric authentication process with RFID card.

Our scenario involves a doctor acting as server having central authority. Intruder is a person who wants to identify a patients report for harming him/her by leaked information. As shown in Fig. 4, more focus is on protecting the identity of homomorphic encryption and RFID tags. A patient is an RFID which store keys to encrypt tag identifiers and has the ability to write the tag identifiers also can send finger

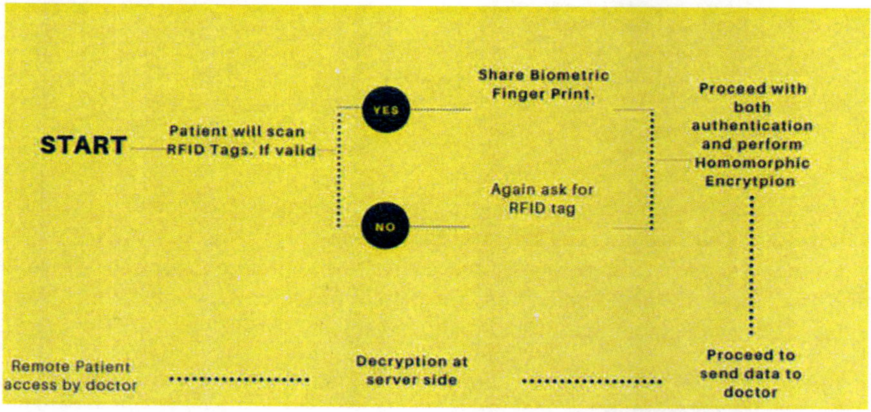

Fig. 4 Flowgraph of the proposed system

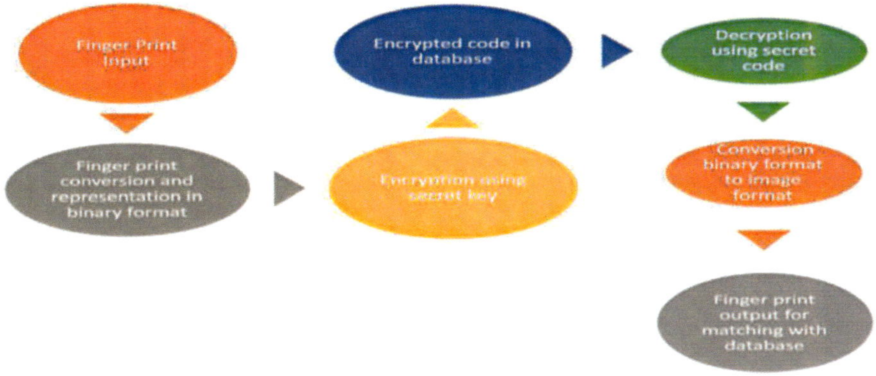

Fig. 5 Finger print and homomorphic encryption

print by applying authentication algorithm to the server. The server is an authority to coordinate with client to implement homomorphic encryption. Server has details of RFID tags, including their original identifiers having finger print in encrypted form.

Where authentication will be done by checking the input with the data stored in database.

The server system has these data passed with identity of authorize person for interacting with each other over a secure channel.

The intruder gets identity of the RFID tags and also manipulate the biometric authentication but will not understand the anything as due to encrypted data. With this doctor and patient information will never be hacked and also it can save patient as finger prints of one person cannot be same with other as well as homomorphic encryption applied. Also with homomorphic encryption, it will provide high security. Finger print will be encrypted using homomorphic encryption as (Fig. 5).

7 Benefits of Proposed System

The key benefits of

- Lowered cost.
- Improved security.
- Remote disease management.
- Improved quality of life.

8 Conclusion

In this paper, we have surveyed all the security existing techniques for user authentication. Healthcare monitoring and treatment require strong security. Researchers across the world are struggling to improve the security protocol and healthcare application which cannot be compromised. In conclusion, we would like to suggest that the proposed system will provide high security and applying homomorphic encryption on finger print will be highly secured.

References

1. N. Vartak, A. Patwardhan, A. Joshi, T. Finin, P. Nagy, Protecting the privacy of passive RFID tags. Department of Computer Science and Electrical Engineering University of Maryland, Baltimore County
2. I.R. Adeyemi, N.Bt. Ithnin, Users Authentication and Privacy control of RFID Card. University of Teknologi Malaysia
3. Inside Internet-of-Things White paper by Deloitte University Press
4. H. Wimberly, L.M. Liebrock, Using fingerprint authentication to reduce system security: an empirical study, in *2011 IEEE Symposium on Security and Privacy*
5. N.W. Wasankar, A.V. Deorankar, A study paper on homomorphic encryption in cloud computing. Int. Res. J. Eng. Technol. (IRJET) **5** (2018)
6. T. Borgohain, U. Kumar, S. Sanyal, Survey of security and privacy issues of Internet-of-Things
7. R. Mary Lourde, D. Khosla, Fingerprint identification in biometric security systems. Int. J. Comput. Electr. Eng. **2**(5) (2010)
8. Internet of things research study a report by Hewlett Packard Enterprise
9. G. Shen, B. Liu, The visions, technologies, applications and security issues of Internet of Things, in *2011 International conference on E-Business and E-Government (ICEE)*. (IEEE, 2011)
10. A. Juels, RFID security and privacy: a research survey. IEEE J. Sel. Aread Commun. **24**(2), 381–394 (2006)
11. S. Lunawat, A. Patankar, Efficient architecture for secure outsourcing of data and computation in hybrid cloud, in *2014 International Conference on Reliability Optimization and Information Technology (ICROIT)*

Automated Waste Monitoring and Alert System Using IoT

Meegan Scariah, Kavya Ann Salbi, Rintu Ponnachan, Sruthi Anna Joseph, and Er. Jinu Thomas

Abstract The uncollected waste materials when the waste bin is full are a common problem nowadays. To build a healthy and hygienic environment, efficient waste disposal and management system is a necessity. Through this paper, we introduce an automated waste monitoring and alert system using the emerging technology called the Internet of Things (IoT). The system is electronically engineered using various sensors like an ultrasonic sensor which is connected to Raspberry Pi to monitor the level of waste inside the bin and the tracked information, which is the level of waste, is sent to the assigned phone number using GSM module. In addition to this, the GPS which is connected inside the bin finds the location of the bin, and using the Google Map APIs, the shortest distance is also provided. Hence, the person can come and collect the waste more systematically through the shortest path.

1 Introduction

Recent studies have shown that nearly 62 billion tonnes of waste are produced every year in India. There is an average annual growth of 4%. The amount of waste emitted continues to rise due to the increasing population, and only less than 5% of the waste is

M. Scariah · K. A. Salbi · R. Ponnachan · S. A. Joseph (✉) · Er. Jinu Thomas
Department of Computer Science and Engineering, Saintgits College of Engineering, Kottayam 686536, Kerala, India
e-mail: sruthi.anna1620@gmail.com

M. Scariah
e-mail: meeganscariah1997@gmail.com

K. A. Salbi
e-mail: kavyanns@gmail.com

R. Ponnachan
e-mail: rintup1998@gmail.com

Er. Jinu Thomas
e-mail: jinu.thomas@saintgits.org

J. Tavares et al. (eds.), *Emerging Technologies in Data Mining and Information Security*, Lecture Notes in Networks and Systems 164,
https://doi.org/10.1007/978-981-15-9774-9_34

being recycled. Therefore, when there is no proper management for waste collection, the country will face a problem. An effective waste material collection is essential to prevent waste from affecting human health and polluting the environment. If it is not properly managed, these large quantities of uncollected waste material can affect society's health and ecological system. As the population increase, waste too increases [1].

Hence, one of the problems to solve the uncollected waste problem is by knowing when the waste bin is full and ready for waste collection. This information can help the municipalities to properly schedule the garbage truck for waste collection within the managed area. This will enhance the efficiency of the waste collection system.

Waste management is the most important aspect that needs to be considered in terms of making the ecosystem in urban areas safer. It has become necessary and challenging to manage the waste with rapid urbanization and increased population growth which in turn increases the cause of contagious diseases. A better waste management solution can help to improve the wellbeing of a community and built up a better and cleaner neighborhood. Municipalities that want to achieve cleaner and healthier environments can implement IoT based solutions [2].

An IoT based system that can monitor everyday garbage enables waste bin monitoring, active scheduling, and also provides routes for garbage collector trucks in smart cities [3]. Among the existing IoT based solutions in smart cities, proper waste management solution brings together the state-of-the-art solution for example in terms of self-powered solutions [4].

IoT is a platform that has become very handy for people in this world. It has become the core of revolutionary developing engines. It is also efficient due to sufficient power supply and good Internet connectivity [5]. It is commonly used to describe a framework where sensors are connected to objects and help these objects to share their 'digital voice' with the external world over an Internet connection. This paper introduces an automated waste monitoring and alert system in Raspberry Pi, GPS, and GSM modules, using IoT technology. Section 3 describes the system development and the overall project strategy. The experimental result of the developed system is also discussed. Finally, Sect. 4 provides the concluding remarks, and it points out the ideas for future extension of this work.

2 Related Works

In [1] paper, a smart garbage clearance alert system is proposed where the municipal internet server is given an alert signal for instant dustbin clearance, also verification is done to whether the garbage has been filled or not. It can read, compile, and exchange data over the Internet. These data can be used to dynamically control waste collection mechanism when placed into a spatial-temporal context and can be processed with optimized algorithms. A framework with IoT and WSN is proposed in [6], and here, the waste bins are fitted with sensors, and then, it is networked using WSN. The sensors that are embedded within the bins collect the information at any given interval. Once the edge is reached, it gives a signal to the garbage collector agent (GCA) and so that person will handle the signals of all the filled tricks and then will communicate using IoT framework. In [7], an automatic waste collection system is proposed by considering the waste level, the smell level, truck loading, and the user's reservation.

3 Methodology

The project development and methodology will be explained briefly in this section. This project proposes an automated system that prevents the overflow of waste material from the bin, and warning is sent to clear the waste. The waste collection status can be monitored effectively by using the anticipated system. The overflow of the bin can be avoided as the waste level is monitored at regular intervals. The ultrasonic sensor is used to sense the level of the waste in the bin and is also used to monitor the original height of the waste bin. The programming done in the Raspberry Pi 3 is in Python language. The code is written in such a way that when the sensor senses a particular level, a message will be sent to the concerned authority for clearing the waste from the waste bin.

Figure 1 shows the block diagram of the developed system. Referring to the block diagram, the anticipated system consists of (1) ultrasonic sensor (2) SIM800A GSM module (3) GPS module (Neo 6 m v2). These components are connected to the Raspberry pi.

The developed system consists of the hardware development phase. In this phase, the sensor is used for the detection of waste levels in the bin, the GSM module is used to send the message to the user, and the GPS module is used to obtain the coordinates of the waste bin. Now, we will discuss each part of the system in detail.

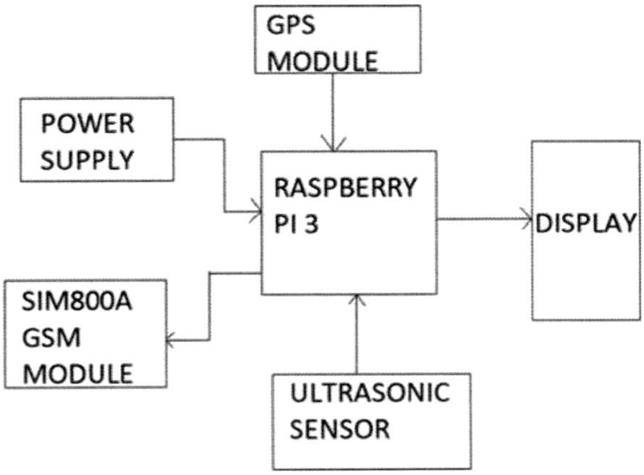

Fig. 1 Block diagram of automated waste management system

The first part is the ultrasonic sensor. Digital pins are used instead of analog pins. The object range is sensed using ultrasound reflection which calculates the time for the ultrasonic waves to reflect from the object. Ultrasound is used because it is not audible to the human ear, and it accurately measures short distances. There are one or more transmitters, a receiver, and a control circuit in a basic ultrasonic sensor. High-frequency ultrasonic sounds are emitted from a transmitter. These sounds bounce off from nearby objects. Some of these sounds are reflected and are detected by the receiver. The control circuit which receives the returned signal calculates the time difference between the transmission and receiving of the signal. The distance between the sensor and the object can be determined using the obtained time.

There are four pins in the ultrasonic sensor to connect with the Raspberry Pi3. These pins are ground (GND), echo pulse output (ECHO), 5 V supply (Vcc), trigger pulse input (TRIG). The 5 V supply is used to power up the module. An ultrasonic pulse will be sent by the sensor, which is triggered when TRIG receives a signal from Raspberry pi. Pulse signals bounce off any object nearby, and some of these signals are reflected in the sensor. The received signals are determined by the sensor. Calculations are done to find out the time difference between the trigger and the returned signal. A 5 V signal is then sent to the ECHO pin. Initially, the ECHO will be 0 V (low). It will change once the sensor is triggered. It will be set to high (5 V) when the returned pulse is located. The time difference between the sensor sending an ultrasonic pulse and the detection of the returned pulse is called pulse duration. The Python code written will measure the pulse duration and use it to determine the distance.

The ECHO signal is marked as 5 V on the ultrasonic sensor. But the input pin is labeled as 3.3 V on the Raspberry Pi. When a 5 V signal is sent into the 3.3 V input port, the GPIO pins get damaged. To overcome this problem, a voltage divider circuit is used which consists of two resisters. This helps to lower the voltage from the sensor to a voltage the Raspberry pi can handle.

Figure 2 shows the schematic diagram for the connection between the Raspberry Pi and the ultrasonic sensor. From the figure, the VCC pin of the sensor is plugged into the positive line of the breadboard, and the ground pin is connected to the negative line of the breadboard. The GPIO 5 V (Pin 2) is connected to the positive line and the GPIO GND (Pin 6) into the negative line. Plug TRIG of the ultrasonic sensor into a blank line. Plug the ECHO pin into a blank line; link another blank line with a 1 KΩ resistor. Link the 1 KΩ resistor line with the GND line with a 2 KΩ resistor. Space is left between the two resistors. The GPIO 24 (Pin 18) is connected to the line with the 1 KΩ resistor. This GPIO pin is placed between the two resistors.

We write a Python code to calculate the distance using the following formula:

$$\text{Speed} = \text{Distance}/\text{Time} \tag{1}$$

The speed/velocity of sound changes according to the medium in which it is traveling. It is taken as 343 m/s at sea level. The time which is measured needs to be

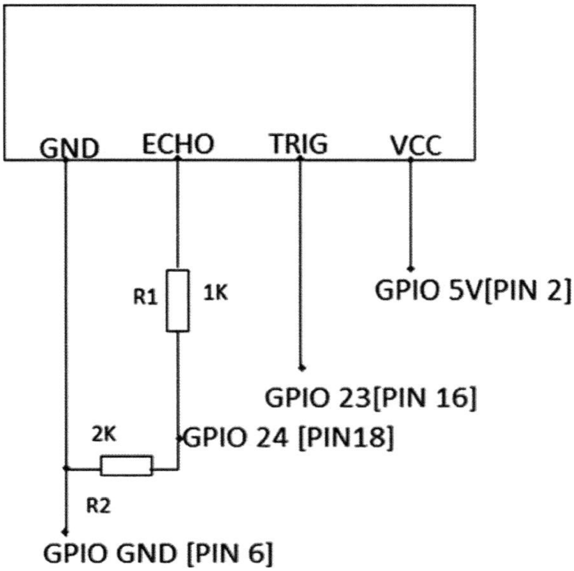

Fig. 2 Circuit diagram showing the connection between the Raspberry pi and ultrasonic sensor

divided by two as the calculated time is the time taken for the pulse to travel from the sensor to the object and back. The actual distance that is to be measured is the distance between the sensor and object. Substituting the values:

$$34{,}300 = \text{Distance}/(\text{Time}/2) \tag{2}$$

$$17{,}150 = \text{Distance}/\text{Time} \tag{3}$$

$$\text{Distance} = \text{Time} \times 17{,}150 \tag{4}$$

We use this formula to calculate the actual distance between the sensor and the object in the bin.

The output of the Python program to measure the distance is shown in Fig. 3. First, we measure the distance from the sensor to the bottom of the bin to get the height of the bin. Then, the distance of the waste from the sensor is measured. Then, we calculate the percentage of the bin that is filled by using the formula:

$$\text{Percentage} = \text{Distance}/(\text{height of bin}/100) \tag{5}$$

The ultrasonic sensor will be placed under the lid of the bin in such a way that the transmitter and the receiver of the sensor are facing the bottom of the bin when closing the lid. This is done, so that the ultrasonic pulses when emitted by the transmitter hit the objects in the bin and reflect it to the receiver.

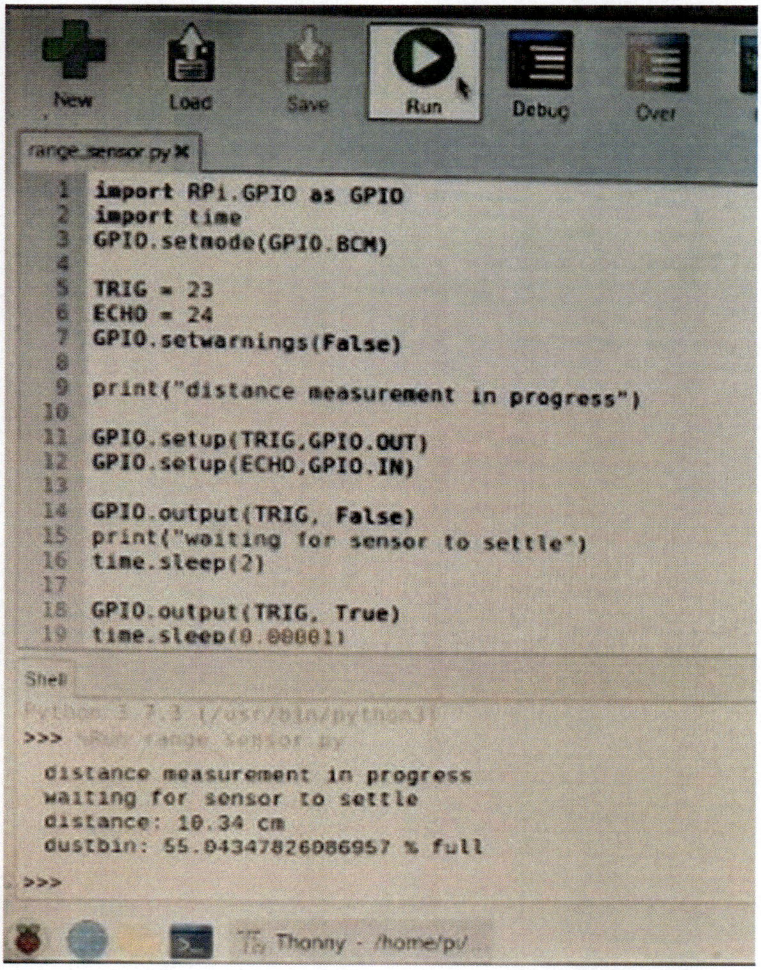

Fig. 3 Result of Python code to calculate the distance

Next, we will discuss the GSM module. GSM module is implemented to rectify the dial-up modem's disadvantage based on sim card acceptance. It is identical to a mobile phone network because it runs over a mobile operator subscription. From the viewpoint of mobile operators, a GSM module looks like a cell phone. The module will receive and transmit messages using the transmission and reception pins, and the PC or a microcontroller is used to interface with it. This property allows the module to reside on embedded applications at a relevant location. Raspberry Pi is a high-end embedded computer, and a GSM module interface can lead to the development of an intelligent network.

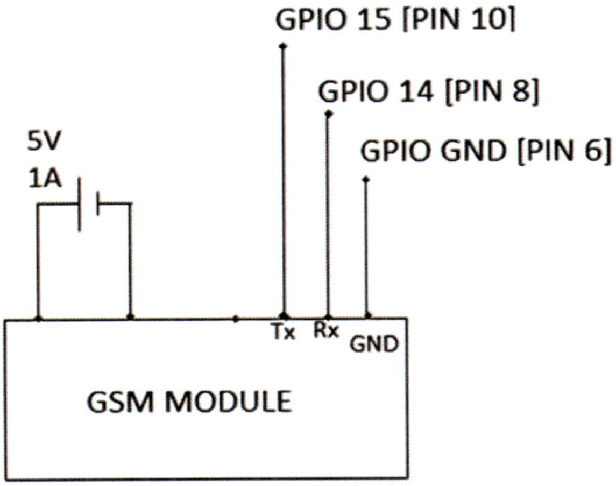

Fig. 4 Circuit diagram showing the connection between the GSM module and Raspberry pi

The GSM module is inserted with a sim card, and Fig. 4 shows the connections that are made. The pins used for transmission and reception should be connected in the reverse order that is Tx pin of the GSM module is connected to the Rx pin of Raspberry pi. In the same way, the Tx pin of Raspberry pi is connected to the Rx pin of the GSM module. The pins used for ground are shorted. A point to note is that the sim which is inserted into the module should have enough balance to receive and send messages. The module should be placed in an area where there is an adequate range for that particular network.

The GSM module is powered up and sim initialization takes a few seconds. A Python script is written to send a message through this modem. The message will be sent to the assigned mobile number. It will be sent only when the percentage value is greater than or equal to 80. The output of the Python code is shown in Fig. 5.

Next, we will discuss the GPS module. It is used to find the location details of any place on the earth. The satellite coordinates are obtained for every second, with the time and date, respectively. This module sends the data of the location that is to be tracked in real time, and it also sends large amounts of details in NMEA format. There are several sentences in NMEA format, in which we only need one sentence which starts from the term $GPGGA and contains other important values like the coordinates, time, and other useful information.

The smart real-time navigation system can be developed when Raspberry pi is interfaced with this module. There are four pins in the GPS module. They are GND, Tx, Vcc, and Rx. The input power of 5 V 100 mA is required, which a suitable supply source can provide. The Rx pin of Raspberry pi is connected to the Tx pin of the module. The ground pins of both the Raspberry pi and the module should be properly grounded. The module is powered up when the connection is all properly made. It should be placed where it has a clear sky view for reliable signal integrity.

The first thing we need to do to get this project started is to initialize the Raspberry pi, so that it can communicate with the module via UART. Neo 6 m v2 GPS module is

Fig. 5 Alert SMS

used here. First, we have to update the Raspberry Pi and then set up the UART in the Raspberry Pi. Next, we have to disable the Raspberry Pi serial getty service. Then, we have to activate the AMAO which symbolizes the module. The GPS module sends serial data which is obtained by writing suitable Python code. From this serial data, we require only the latitude and the longitude. Thus, the user can find the location of the bin by applying obtained values in the options provided in Google Maps.

4 Conclusion

We see many cities and towns filled with waste materials. The heap of waste is a breeding ground to many pathogens like house flies, mosquitoes, etc. So, through this paper, we introduce an IoT based waste monitoring and alert system technology. The main objective is to deal with waste bins safely and technically. The waste bins are electronically engineered to detect the level of waste present inside, and using the WiFi module present inside the bin, the message is sent to the authorized people, so that they can come and collect the waste. GPS module provides the location of the bin, and with Googlemap APIs the application so produces the shortest route to collect wastes from bins which have a critical level. Thus, it reduces the number of times the person has to come and check the bin. Many of the similar systems lack the data streaming technology, because of that when the bin is not filled up to the desired level, the authorities will not clear it so that the waste material started emitting foul smell and become the breeding ground for many pathogens. But, in our system, the level data is continuously streamed,and after a preset number of days, the personnel collects the waste. The system can be adopted for hospitals, educational institutions, or large organizations, etc. Thus, the system helps in keeping the environment clean and healthy.

References

1. Z.H.C. Soh, M.A. Al-Hami Husa, S.A.C. Abdullah, M.A. Shafie, Smart waste collection monitoring and alert system via IoT, in *2019 IEEE 9th Symposium on Computer Applications and Industrial Electronics (ISCAIE)*, Malaysia, vol. 2019 (2019), pp. 50–54
2. S.S. Chaudhari, V.Y. Bhole, Solid waste collection as a service using IoT-solution for smart cities, in 2018 *International Conference on Smart City and Emerging Technology (ICSCET)*, Mumbai, vol. 2018 (2018), pp. 1–5
3. W. Chen, Y. Wang, P. Huang, Y. Huang, M. Tsai, A smart IoT system for waste management, in *2018 1st International Cognitive Cities Conference (IC3)*, Okinawa, vol. 2018 (2018), pp. 202–203
4. E. Al-Masri, I. Diabate, R. Jain, M.H. Lam, S. Reddy Nathala, Recycle.io: an IoT-enabled framework for urban waste management, in *2018 IEEE International Conference on Big Data (Big Data)*, Seattle, WA, USA, vol. 2018 (2018), pp. 5285–5287
5. S. Paul, S. Banerjee, S. Biswas, Smart garbage monitoring using IoT, in *2018 IEEE 9th Annual Information Technology, Electronics and Mobile Communication Conference (IEMCON)*, Vancouver, BC, vol. 2018 (2018), pp. 1181–1185
6. N.S. Kumar, B. Vuayalakshmi, R.J. Prarthana, A. Shankar, IOT based smart garbage alert system using Arduino UNO, in *2016 IEEE Region 10 Conference (TENCON)*, Singapore, vol. 2016 (2016), pp. 1028–1034
7. G.K. Shyam, S. Kumar, S. Manvi, P. Bharti, Smart waste management using Internet-of-Things (IoT), in *2017 2nd International Conference on Computing and Communications Technologies (ICCCT)*, Chennai, India, vol. 2017 (2017), pp. 199–203
8. T. Anagnostopoulos, A. Zaslavsy, A. Medvedev, S. Khoruzhnicov.: Top–k query based dynamic scheduling for IoT-enabled smart city waste collection, in *2015 16th IEEE International Conference on Mobile Data Management*, Pittsburgh, PA, vol. 2015 (2015), pp. 50–55
9. P. Chowdhury, R. Sen, D. Ray, P. Roy, S. Sarkar, Garbage monitoring and disposal system for smart city using IoT, in *2018 Second International Conference on Green Computing and Internet of Things (ICGCIoT)*, Bangalore, India, vol. 2018 (2018), pp. 455–460
10. G.S. Rohit, M.B. Chandra, S. Saha, D. Das, Smart dual dustbin model for waste management in smart cities, in *2018 3rd International Conference for Convergence in Technology (I2CT)*, Pune, vol. 2018 (2018), pp. 1–5

English to Bengali Neural Machine Translation Using Global Attention Mechanism

Sheikh Abujar, Abu Kaisar Mohammad Masum, Abhishek Bhattacharya, Soumi Dutta, and Syed Akhter Hossain

Abstract Neural Machine Translation (NMT), increased the accuracy of machine translation and recently become more popular in machine translation research community. An attention mechanism has been added with NMT and the result of MT is improving, respectively. Using different attention mechanism in encoder-decoder, along with NMT models are achieving good results considering mentionable improvements in source sentences during the translation. Translating from English to another language is a widely used machine translation problem. For this regard, English-Bengali translation is a researchable area in Bengali machine translation. Bengali is one of the usable languages in the world. Therefore, conversion of Bengali to other language is very essential for people. The translator is a way to understand the meaning of different language. But sometimes the translator does not provide the accurate Bengali translation for the corresponding English text. So, generating the correct English to Bengali translation requires more research. In this research, we build an attention-based NMT model for English to Bengali machine translation instead of the normal NMT model and that achieves 22.3 BELU for model evaluation. Also, we build a seq2seq model without attention for checking the machine

S. Abujar (✉)
Department of Computer Science and Engineering, Independent University, Dhaka, Bangladesh
e-mail: sheikh.csesets@iub.edu.bd

A. K. M. Masum
Department of Computer Science and Engineering, Daffodil International University, Dhaka, Bangladesh
e-mail: mohammad15-6759@diu.edu.bd

A. Bhattacharya · S. Dutta
Department of Computer Science, School of Information Technology, Institute of Engineering and Management, Kolkata, India
e-mail: abhishek.bhattacharya@iemcal.com

S. Dutta
e-mail: soumi.dutta@iemcal.com

S. A. Hossain
Department of Computer Science and Engineering, University of Asia Pacific, Dhaka, Bangladesh
e-mail: akhter.hossain@ulab.edu.bd

J. Tavares et al. (eds.), *Emerging Technologies in Data Mining and Information Security*, Lecture Notes in Networks and Systems 164,
https://doi.org/10.1007/978-981-15-9774-9_35

performance on translation. That achieves 15.3 BELU for the applied dataset. The analysis shows that the attention-based seq2seq model performs better than seq2seq learning for English to Bengali translation.

Keywords Neural machine translation · Encoder · Decoder · Seq2Seq · Attention

1 Introduction

The noticeable advancement of Machine Translation (MT) one language to another has experienced with the uses of sequence-to-sequence (seq2seq) [1] models. Despite this, the early studies include rule-based systems started back in 1950s [2]. Since then, researchers are trying to solve MT problems with different approaches. However, neural network approaches has been introduced in MT fields in the late 20th century [3]. Statistical Machine Translation (SMT) has been showed prominent performances in the last decade in the MT community [4]. However, SMT doesn't perform well and act more complex by using more additional features. Researchers focused on deep learning technologies due to this limitation of SMT.

The study of basic Neural Machine Translation (NMT) research consist of encoder-decoder, in which encoder takes the vector form of words as input to process and decoder uses the vector to predict the most likely output. To get the maximum performance of the translation output, NMT model jointly learns the parameters [5–7] with minimum domain knowledge. However, NMT has limitation to deal with the long sentences [8]. However, compare to resource-rich languages, the development of MT systems is very sparse for the Bangla-English language pair. The literatures also report that the development of large-scale parallel corpus for Bangla-English is very scarce.

Although NMT model needs a large amount of parallel corpus to maximize the performance, we used a small amount of data to train our NMT model. In this study, we mainly aim to explore both Bangla-to-English and English-to-Bangla NMT. Our achievements include, (i) running experiments using seq2seq and attention-based NMT models, (ii) evaluating AmaderCAT corpus using the approaches that we used in our study, and (iii) present an analysis about the results.

The rest of this paper is as follows. Section 2, gives a brief overview about the literature review of MT systems for Bangla-English language pair. In Sect. 3, we discuss about the dataset and the technique we used in our study. The results of our study using the neural network approach has been discussed in Sect. 4. Finally, we conclude our paper in Sect. 5.

2 Literature Review

NMT is a system of producing a neural network where the model produces conditional probability of the target translation $(y_1, y_2, \ldots, y_n)$ based on the given source as input $(x_1, x_2, \ldots, x_n)$. Encoder and Decoder are the two basic components of NMT, where encoder handles all the input sources sentences and the decoder generate single words at a time [9]. Most of the recent works are mostly similar except some differences in their encoder and decoder—RNN architecture. Many used single RNN layer for their encoder but some of the work also used bidirectional RNN and achieved good performance [6, 10]. Sutskever et al. and Luong et al. [11] both developed Deep RNN and LSTM for both encoder and decoder.

Kalchbrenner et al. used RNN for decoder and Convolutional Neural Network (CNN) for encoder [7]. And some other works achieved a good result on a combined architecture of RNN and Long Short-Term Memory (LSTM) in both for encoder and decoder [8, 10]. Jinsong Su, et al. [9] introduced different decoders—"*neural backward decoder*" and "*neural forward decoder*" over the conventional attention encoder-decoder NMT model, and achieved significant improvements over the conventional RNN. Most of the recent works used different neural network for encoding and decoding, and [7] used RNN for decoding and CNN for encoding.

There are some other facts with the sentence length if the vector dimension is fixed, while encoding. Bahdanau et al. shows those facts on why degrading quality of translation?—Because of sentence length and the vector spaces. If the vector dimension is fixed for encoder and then sentence is longer, so the encoder has to encode the whole source of sentences in a single vector and this is creating a problem for sentence with long variable length. And solved this problem through attention mechanism by keeping track on the hidden states of source memory and then commencing those to the relevant ones [8] using context vector.

3 Methodology

Seq2Seq learning approach is widely used for machine translation problem. It gives a good result for any machine translation problems. But attention added better performance of seq2seq learning approaches for NMT. Attention-based translation system performs well [12] than any seq2seq model. In this section, we discussed our attention method based on global attention mechanism [5] for English to Bengali sentence translation.

3.1 Dataset Properties

For making an accurate system large dataset is required. There is no Standard English to Bengali translation dataset is available. Here we used "AmaderCAT" dataset [13] to implement this model. This dataset is built for Bengali and English parallel machine translation. The dataset contains Bengali and English sentence combination. The dataset has a total 1781 data. And total 9600 is total unique token of both language text. The dataset is built for the glossary suggestion with Bengali and English sentence combination.

3.2 Machine Translation

All machine translation problems depend on conditional probability [5, 6, 8]. The probability of output directly depends on the input text. Let, is English source text $x_1, \ldots, x_n$ and $y, \ldots, y_m$ is Bengali output text then $p(y|x)$ is the conditional probability (Fig. 1).

Every translation model has encoder and decoder. Encoder contains source input text and decoder contains the target output text. Now the final conditional probability will,

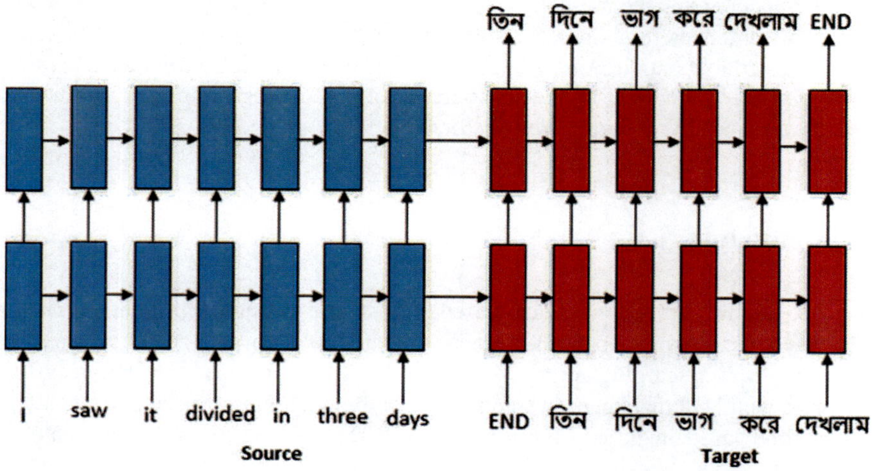

Fig. 1 Seq2Seq model for English to Bengali machine translation

$$P(y_1, \ldots, y_j|x_1, \ldots, x_i) = \prod_{t-1}^{j} P(y_j|S, y_1, \ldots, y_{t-1}) \tag{1}$$

Seq2Seq learning is a major part of today's machine translation problem. All sequence are memorized by the RNN [7]. LSTM is widely used for any seq2seq learning-based model. Encoder and Decoder both used LSTM for maintain the text sequence [5, 10]. Encoder pass the input sequence with LSTM and Decoder receive and provide the output sequence using decoder LSTM. Output is generated by the probability of decoder with output layer the activation function.

$$P(y_j|S, y_1, \ldots, y_{t-1}) = \text{softmax}(W_o.h_{t+1} + b) \tag{2}$$

y_j is the output of each translated word of the machine. h_{t+1} is hidden unit with the bias b of activation function. S is used a source hidden unit. That are maintain the all translation process in whole model.

3.3 Global Attention

Attention-based translation model is used to extract information from the sequence context. Predicting content from all source position is main working process of a global attention mechanism [11]. Using all hidden state of the encoder is the main idea of global attention from context vector c_t where t is each time step. a_t is alignment vector. The size of the alignment vector is equal to the time step of the source text. All are derive by hidden state h_t of the target data with hidden state $\bar{h}_s$ of the source data. a_t is calculated total weights of the source, also connected with the context vector. Decoder predict the output from the context vector c_t (Fig. 2).

All score are count by three function. Firstly, hidden source state is dot with the hidden state of the target sate. Then, weight is multiply with hidden source. Finally, activation function is used for concatenate the layer.

3.4 Model Architecture

Our model is implemented based on the seq2seq with attention. For creating a neural network we used Keras API. Passing the source text we used encoder and decoder for the target text. LSTM is used in both for maintain the sequence of input and output (Fig. 3).

The model architecture shows that encoder inputs and decoder inputs shapes are dots at the beginning of attention layer. In the attention layer, the combination of input and output shapes have crossed and entered into the second dot layer. The

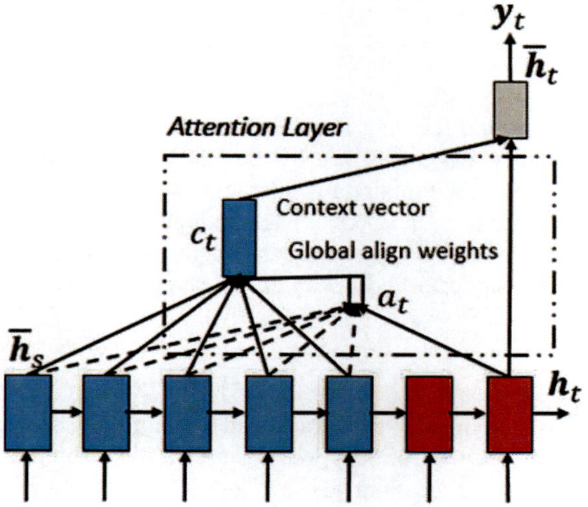

Fig. 2 Luong [11] global attention model for machine translation

activation function is used in the attention layer. Decoder input shape is added in the second dot layer. Then concatenate the second dot layer with added the encoder input shape. Finally, the dual time distribution layer is added with the final shapes.

Algorithm 1: Model algorithm for global attention

1 Encoder = Input()
2 Decoder = Input()
3 Dot_1 = (Encoder, Decoder)
4 Attention = (Activation(Dot_1))
5 Dot_2 = (Attention, Decoder)
6 Concatenate = (Dot_2, Encoder)
7 Time_Distribution_1 = (Concatenate)
8 Time_ Distribution_2 = (Time_Distribution_1)

4 Experiment Result

NMT models are evaluated by the value of the BELU score. High BELU score refers to more accurate machine. This section, we discuss the experiment result and finding of our applied model. For the research purpose, we used Seq2Seq model with and without attention. Both models are performed well for translation. And we check the model performance English and Bengali in bi-directionally.

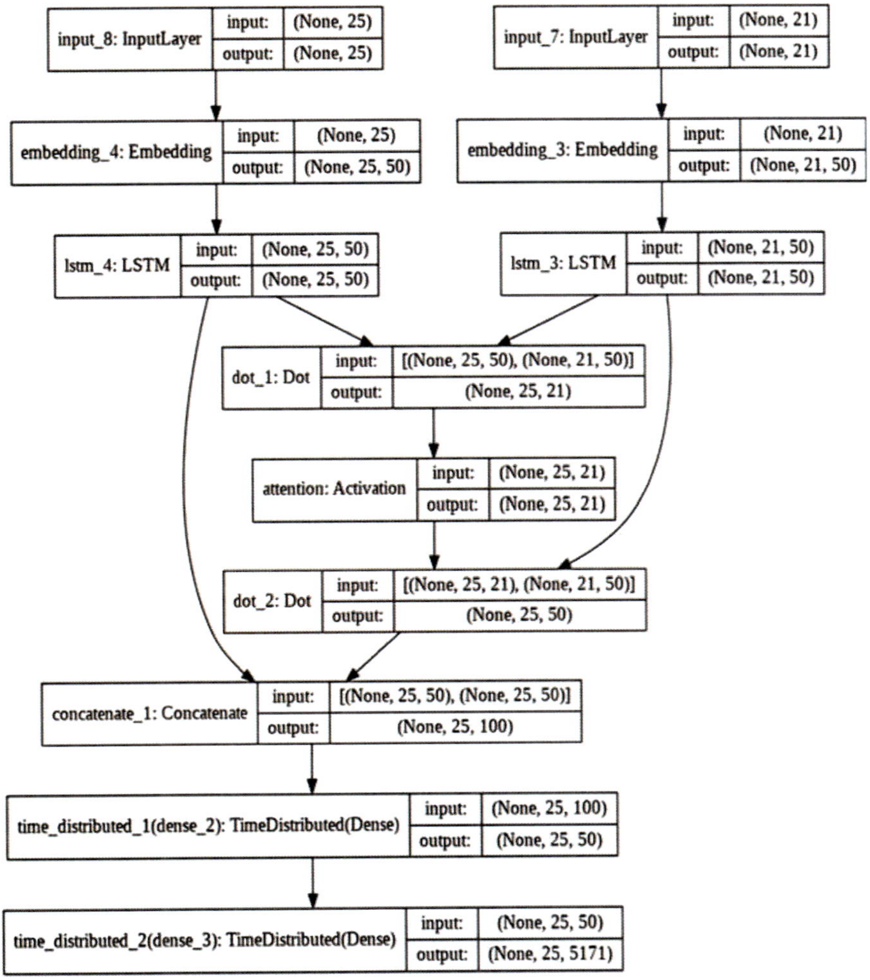

Fig. 3 Model architecture for English to Bengali machine translation

4.1 Parameters Setup

The dataset has two parameters one is the source and another is target text. For the first experiment, we used source language as English and Bengali is the target language. We used to START and END token for the define start and end for the target sequence. Then, we count the unique word of target and source language.

In other experiments, we used Bengali is the source language and English is the target language. Both word input details are given in Tables 1 and 2 (Table 3).

Table 1 Input parameter for English-Bengali NMT experiment

Input parameter for English-Bengali	
Unique source token	4024
Unique target token	5171
Max length of source	21
Max length of target	23

Table 2 Input parameter for English-Bengali NMT experiment

Input parameter for Bengali-English	
Unique source token	5169
Unique target token	4430
Max length of source	23
Max length of target	23

Table 3 Data distribution for NMT model

Data distribution	
Training data	65%
Test data	20%
Validation data	15%

Before training, data need to distribute properly. We have a total of 1781 data. For the testing set, we set 20% of the total dataset. And for the validation, 15% of data is defined. Rest of the 65% data is available for the training.

4.2 Training Details

We use LSTM in encoder and decoder and set 50 embedding size for both models. In seq2seq learning, we used softmax in the decoder as activation function. Also, in the attention model softmax is used in the attention layer and tanh and softmax are used in decoder layer. We fit the both model using 64 batch size with 30 epochs for seq2seq model and 50 epochs for attention model. For optimizing the model we used rmsprop for optimization function. Calculating the loss we used the categorical cross entropy.

Table 4 English-Bengali BELU for the AmaderCAT dataset

Model	BELU
Seq2Seq	15.3
Seq2Seq + global attention	22.3

4.3 English-Bengali Results

After the train our both models we find a good BLUE score for English to Bengali NMTs. For the seq2seq model, we achieve 15.3 BELU for data. After using the global attention in the seq2seq model we achieve 22.3 BELU (Table 4).

High BELU score shows that seq2seq with attention gives the better performance for the English to Bengali machine translation.

Table 5 shows the translation sample of our model. Here, source is the English sentence and a Bengali sentence is the target. The model translate the sentence of more similarly with the target sentence.

4.4 Bengali-English Results

We achieve BELU 12 for Bengali to English machine translation using seq2seq and BELU 2.7 for attention based seq2seq model (Table 6).

After analysis BELU score this dataset is not provided good result for Bengali to English translation. But seq2seq model gives an accurate NMT than the attention-based model for Bengali to English translation.

5 Conclusion

BLEU score helps to the evaluation of NMT model. Our attention based seq2seq model gives high BELU, also seq2seq model gives a good BELU for English to Bengali transition. Here we used global attention which is used for looking all source at position wise.

Structure of Bengali language is different from other languages. For this conversion from another language to the Bengali language is more challenging. An automatic system can help language conversion like google translator. But predicting accuate result an automatic system need large data. Our Test dataset is not sufficient. In future making an accuate an automatic conversion system form another language to Bnegali need a large dataset.

Table 5 English-Bengali BELU for the Amader CAT dataset

Sample response of English-Bengali MT	
Source Sentence	Elo Khushir Eid a show for children starts after that
Target Sentence	এরপর শুরু হয় ছোটদের অনুষ্ঠান এল খুশির ঈদ
Response Sentence	এরপর হয় ছোটদের অনুষ্ঠান

Table 6 Bengali-English BELU for the Amader CAT dataset

Model	BELU
Seq2Seq	12.2
Seq2Seq + global attention	2.7

Acknowledgements We gratefully acknowledge support from DIU NLP and Machine Learning Research LAB for providing GPUs support. We thank, Dept. of CSE, Daffodil International University for providing necessary supports. And also thanks to the anonymous reviewers for their valuable comments and feedback.

References

1. O. Kuchaiev, B. Ginsburg, I. Gitman, V. Lavrukhin, J. Li, H. Nguyen, C. Case, P. Micikevicius, Mixed-precision training for NLP and speech recognition with open seq2seq (2018)
2. T. Poibeau, *Machine Translation.* (MIT Press, 2017)
3. A. Waibel, A. N. Jain, A. E. McNair, H. Saito, A. Hauptmann, J. Tebelskis, Janus: a speech-to-speech translation system using connectionist and symbolic processing strategies, in *Proceedings of the ICASSP* (1991), pp. 793–796
4. M.A. Hasan, F. Alam, S.A. Chowdhury, N. Khan, Neural machine translation for the Bangla-English language pair, in *2019 22nd International Conference on Computer and Information Technology (ICCIT).* (IEEE, 2019), pp. 1–6
5. K. Cho, B. Van Merriënboer, D. Bahdanau, Y. Bengio, On the properties of neural machine translation: encoder-decoder approaches. arXiv preprint arXiv:1409.1259 (2014)
6. I. Sutskever, O. Vinyals, Q.V. Le, Sequence to sequence learning with neural networks, in *Advances in Neural Information Processing Systems* (2014), pp. 3104–3112
7. N. Kalchbrenner, P. Blunsom, Recurrent continuous translation models, in *Proceedings of the 2013 Conference on Empirical Methods in Natural Language Processing* (2013), pp. 1700–1709
8. D. Bahdanau, K. Cho, Y. Bengio, Neural machine translation by jointly learning to align and translate. arXiv preprint arXiv:1409.0473 (2014)
9. J. Su, X. Zhang, Q. Lin et al., Exploiting reverse target-side contexts for neural machine translation via asynchronous bidirectional decoding. Artif. Intell. **277** (2019)
10. K. Cho, B. Van Merriënboer, C. Gulcehre, D. Bahdanau, F. Bougares, H. Schwenk, Y. Bengio, Learning phrase representations using RNN encoder-decoder for statistical machine translation. arXiv preprint arXiv:1406.1078 (2014)
11. M.T. Luong, H. Pham, C.D. Manning, Effective approaches to attention-based neural machine translation. arXiv preprint arXiv:1508.04025 (2015)
12. Z. Tu et al., Modeling coverage for neural machine translation. arXiv reprint arXiv:1601.04811 (2016)
13. M.A. Hasan, F. Alam, S.R.H. Noori, A collaborative platform to collect data for developing machine translation systems, in *Proceedings of International Joint Conference on Computational Intelligence. Algorithms for Intelligent Systems* edited by M. Uddin, J. Bansal. (Springer, Singapore, 2020)

Taxonomy of Cyber Security in Medical Science

Jatin Bhardwaj, Siddharth Gautam, Harshit Yadav, Nitin Tyagi, and Shafiqul Abidin

Abstract Recently, an incrementing number of cyber-attacks in the medical field have resulted in great losses. In the health care industry, since medical information plays an essential role in human health. To prevent from these types of cyber attacks, there are various steps are taken by Hospital Institute Management Systems (HIMS) regarding to its security which is also called cyber security as it is related to computer technology. In response, we have been discussing this misconduct and trying to increase the awareness of researchers, but it seems that there is a lack of research that presents guidelines for editors to help them protect themselves against these threats In this paper, we understand more about cyber security in medical field and a brief overview of cyber securities is presented along with building of cyber securities is explained for the better understanding. We then explain the details involved in each type of attack.

Keywords Cyber security · Malware · Ransomware · Clouds · Phishing · Cyber crimes · Cyber-attacks

J. Bhardwaj · S. Gautam · H. Yadav (✉) · N. Tyagi · S. Abidin
HMRITM, Delhi, India
e-mail: yharshit648@gmail.com

J. Bhardwaj
e-mail: Jatin200.mps@gmail.com

S. Gautam
e-mail: siddharthinfo92@gmail.com

N. Tyagi
e-mail: nitintyagi5631@gmail.com

S. Abidin
e-mail: shafiqulabidin@yahoo.co.in

J. Tavares et al. (eds.), *Emerging Technologies in Data Mining and Information Security*, Lecture Notes in Networks and Systems 164,
https://doi.org/10.1007/978-981-15-9774-9_36

1 Introduction

Medical device network and other health facilities in today's world are extremely useful. They have so much potential in medical field that they can change whole scenario and they are very useful for curing patients. But they can lead to the leak of information about the patients and the doctors or health care providers by breaching of the cyber security wall in the networked medical devices. They have a high risk of being hacked or attacked by some viruses or malwares by some malicious applications. They are more prone or we can say that they are more vulnerable to the unauthorized access by the malicious users. Patients safety issues, their information related to their injury or death are more vulnerable. Therefore, the cyber security of medical sciences is a major risk factor and the security should be improved in the concerned field [1]. Breaching of data costs the health industry about 5.6 billion dollars every year, according to the Becker's hospital review. Therefore, security has become a critical condition in medical networked devices. Verizon's 2016 data breach investigations reports revealed that most of the cyber vulnerabilities and the breaches are about the money and information the attacker wants to know.

Attackers usually take the easiest routes to know the information they want to know. Some of the easiest routes are—phishing attacks, malware, ransomware, cloud threats, misleading websites and employees' error. So, the medical field one should take proper preventive measures and implant proper security measures to counter the attackers. Some of the strategies that can be taken to prevent the attacks are establishing a security culture, protected mobile phones, using a firewall, maintaining good computer habits, install and maintain the antivirus software, plan for the unexpected, control access to protected health information, use strong passwords and change them regularly and control physical access and also limit network access [2]. To protect the information in health care facilities the organization must hire some informatics professionals that can not only collect and manage data but also protect the data. In the further sections we will discuss more about the cyber security in medical science and the challenges in the cyber security and the strategies we can take against the challenges [3].

The rest of the paper organised as follows: Sect. 2 illustrated the three main building blocks of cyber security in medical science. In Sect. 3 cyber security in medical science is explained. In Sect. 4 we have discussed some important strategies for improving the cyber securities in medical science. And In the last, conclusion of the paper is presented in Sect. 5.

2 Building Blocks of Cyber Security in Medical Science

There are three main building blocks in cyber security in medical science on which cyber security depends, that are: Integrity checking, Authentication [4] and encryption [5]. Figure 1 shows the building blocks of cyber security in medical science.

2.1 Integrity Checking

It means that in cyber security it checks the digital messages whether they are altered or not during the transmission period [6]. One of the most common approaches in integrity checking is using one mathematical algorithm known as cryptographic hashing function [7]. We can use a function called 'hash' for integrity checking. The algorithm can be applied to any message of some random size which gives output of fixed size [8]. The output is generally of smaller size than the input size making it easy for the hash function for integrity checking. Then the hash function and the message are sent together to our desired location or we can say that the hash function [9] and the message are both stored together wherever they are needed to be stored. Now to check whether the message is modified or not, the same hashing algorithm is applied to the new location of the message where it is stored resulting in generating a new hash function with the message. For security reasons the cryptographic hashing function [10] is generally applied to the message containing a secret key. This key is only known to the sender of the message and the receiver of the message, making it difficult for the malicious user or we can say attacker to modify or alter the message between the sender party and the receiver party [11]. The use of a secret key with the

Fig. 1 Building blocks of cyber security in medical sciences

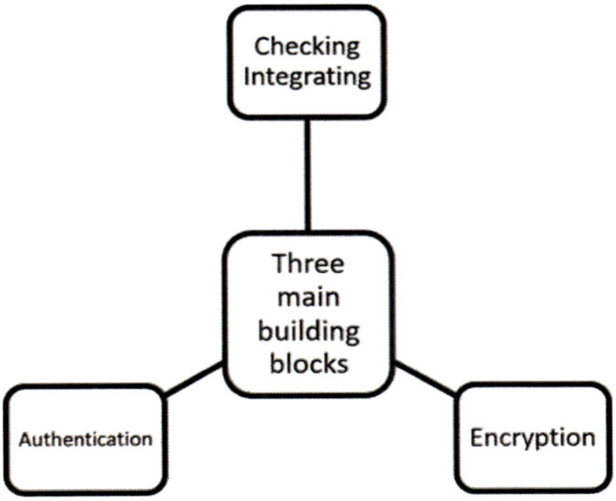

message will also generate a new hashing function making the hash function with a property that it will not leak any information about the secret key. This property is a sign of a good cryptographic Hashing function [2].

2.2 *Authentication*

Authentication means that whenever any new node or message is entering into the network then with the help of procedure the node will confirm that the node or message is the same message or node whatever it claims to be [12]. Authentication checking takes place by a well-known approach that is asking the node (i.e. it can be any person or any message) something that only the node knows about it (any secret code or message that can be used to prove their originality) [13]. If they give the right answer then they are confirmed to be original otherwise authentication fails [14]. Authentication basically relies on the other cryptographic functions like hashing and encryption. For example, self-authentication [15] is one of the authentication approaches that rely on hashing as well as encryption [16].

2.3 *Encryption*

Basically speaking, encryption means that converting the digital message into the form of codes that only the verified users can see it by decrypting the message [17]. Those users who are not authorized to access it cannot see the message what's in it. Authorized users have a key that they use it for decrypting the message. Encryption is one of the best ways of cyber security in today's world. Many algorithms for encryption are available in the market like AES [18]. AES stands for advanced encryption standard which is a well-known algorithm for encrypting the messages. AES is an example of symmetric encryption technique [19]. In AES encryption the parties must have a shared key among them who want to communicate with each other. As AES is an example of symmetric encryption function, a well-known example of asymmetric encryption function is RSA public key encryption system. In this both the encryption key and decryption key are different from each other. Decryption is the opposite of encryption. In encryption there are two inputs that are the message to be encrypted and an encryption key and the output is scrambled (encrypted) data. While in decryption there are two input keys that are the message to be decrypted and a decryption key and the output in unscrambled (decrypted) data.

3 Cyber Security Challenges in Medical Science

Nowadays, there is no falsehood in the statement that for cybercriminals, the healthcare industry is a top target. Every year billions of cases are reported in which industries of healthcare are going through data breaches by some cybercriminals and it is found that data breach in one health data affects millions of people life daily due to the leak of the patient personal information globally [20]. In today's time, whole data and most of the record related to health are dependent on this technology, which also increases the risks of the cyber-attacks on the medical sciences from the cybercriminals. To improve administration and patient care, healthcare uses everywhere lots of software as well as hardware. This modernization in devices of healthcare system leads to development of a variety of cyber security challenges which are described below. In Table 1 various types of challenges in healthcare has been discussed.

Table 1 Some challenges of cyber security in healthcare

Challenges	Description
Cloud threats	• Special type of software present in the computer system. It stores the data of the user in it • Accumulative quantity of secured haleness knowledge is essentially gathered on the cloud • Create certain loopholes that can pose a threat for certainty of well-being concerning institutions
Malware and ransomware	• Most frequently used in cyber-attacks by hackers • Cyber attackers handle malware and ransomware to close down particular devices, assistants or regular absolute associates • Hackers demand a ransom to uplift the encryption which is blocking the working of the organisation
Ambiguous websites	• Ambiguous websites are the special type of malicious website that is created by the hacker • Highly experienced cyber criminals make the same websites as that of the original site of the healthcare department • For example, .gov is replaced by .com. In these cases the user thought that the user entered the original website [11]
Phishing attack	• Hacker targets the particular person randomly • Stated approach sends away bulk chunks of emails in distinction to achieve the trust of users • Delicate data from them which is now considered as one of the biggest challenges for cyber security in healthcare
Employee error	• Agents can abandon health caution associations amenable to assail • Employees can leave healthcare organizations susceptible to attack through weak passwords
Encryption irrational spot	• Important data of the user or the owner is converted into a special type of code • Code is only understood by the user or the owner • Done to prevent unauthorized access to the system from unknown persons • Create blind spots where hackers cannot detect the data breaches

4 Strategies for Improving Cyber Security in Medical Science

The well-being informatics along with different authority are acting as crucial characters that aesculapian federation abide confined [21]. Health IT Government says that, particular well-being anxiety associations may enhance cyber assurance by applying the subsequent conventions [22]. Figure 2 shows the Strategies for improving cyber security in medical science.

4.1 Defend Mobile Device

Mobile devices must be protected from cyber attacks as these types of devices are mostly prone to cyber attacks. As these devices store the crucial information of the owner in it such as passwords, banking details, photographs etc. Due to the accumulative amount of well-being anxiety donors are applying mobile gadgets at work. Symbols and different defensive assessments are crucial to assure that each instruction on the following device is assured and protected from every way as possible. Therefore a proper and strong firewall system must be made for these types of devices as well to prevent the cyber attacks [23].

4.2 Check Password Regularly

According to a report more than 50% of chances of data breaching takes place due to the password. And this data theft or data breaching is done due to the use of weak passwords, stolen passwords and default passwords generated by the system. To prevent this data breaching in the cyber security of medical sciences strong passwords

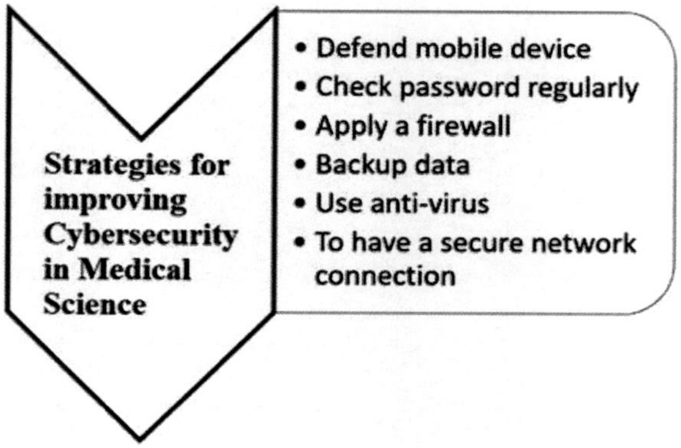

Fig. 2 Strategies for improving cyber security in medical science

must be created or generated [24]. One more way to stop the data theft is to change the passwords of the biomedical system regularly after a particular period of time. In this way biomedical data can be stored in the system and protected from cyber-attacks [25].

4.3 Apply a Firewall

Firewall is the special type of the wall used in computer networks. Firewall is used to protect the personal data of the person from any malicious acts such as hacking, cyber threats [26], cyber terrorism etc. Firewall is very much necessary as anything connected to the internet should have a firewall system. Firewall acts as a barrier between the data and the hacker. For any theft done, a hacker must have to crack the firewall system of the network or the computer, to access the data. This data can only be secured by creating a good firewall system.

4.4 Backup Data

Due to the frequent increase in the cyber-attacks on the medical sciences, risk of data loss also increases. Loss of this crucial data also risks the life of the patient [27]. So there must be more than one system in the Hospital Database Management System which stores the information of the patient. The files present in these database systems must be backed up on a regular basis and the restoration of the data also takes place regularly. If possible, authority should store this data away from the main system. These are some of the steps which prevent cyber-attacks.

4.5 Use Anti-virus

Nowadays, due to the increase in the number of cyber-attacks risk of data loss also increases; and normally installing anti-virus in the system or computer network is not sufficient as due to the increase in technology risk of data theft also increases. Therefore, constant changes are necessary for assured well-being of the computer systems. Therefore, computer anti-virus must be continuously updated with the particular interval of time. These are some of the points which must be remembered and followed to reduce the risk of cyber-attacks in medical sciences [28].

4.6 To Have a Secure Network Connection

All application software also alternative access to the unknown network or unknown site through the actual system which contains important data of patients may be risky. As the access to these sites may increase the risk of the cyber-attacks to the system [29]. Therefore, the user must not go to these kinds of malicious networks or websites for any work as these malicious websites allow the virus to enter our system and damage the data of the system [30].

5 Conclusion and Future Scope

Due to the presence of different types of vulnerabilities in medical sciences there is always a risk of loss of data. Even after doing lots of work in the field of cyber security there is still more space for research and development in medical cyber security and there are many challenges, pros and cons regarding cyber security in medical sciences which we have already discussed in detail. These strategies are not only useful in medical sciences but also in different—different fields all over the world. These strategies are very much helpful to nullify the problems like cybercrimes etc. There is lots of scope in future for cyber security in medical sciences, in recent years the market of cyber security reached to few billion dollar for medical sciences only and there is many other fields in which cyber security is required so simply we can conclude that how vast and how fast will cyber security will expand is beyond our thoughts and to predict it is also beyond our scope. The main reason for this is the sudden and very great increment in the salaries of the cyber security operators and this increment is very fast in medical sciences than any other field in the world. So we can say that there will be a sudden fall in the rate of cyber attacks in a few years because of awareness about cyber crimes and also due to the advancement in technology which makes our firewall system strong and makes it untraceable one day.

References

1. C. Peng, M.M. Hossain, Cyber-physical security for ongoing smart grid initiatives: a survey. IET Cyber Phys. Syst. Theor. Appl. https://doi.org/10.1049/iet-cps.2019.0039 (2020)
2. Security for industrial cyber-physical systems, in *2018 IEEE Industrial Cyber-Physical Systems (ICPS)*. https://doi.org/10.1109/icphys.2018.8390806 (2018)
3. J. Chen, Q. Zhu, Security as a service for cloud-enabled internet of controlled things under advanced persistent threats: a contract design approach. IEEE Trans. Inf. Forensics Secur. **12**(11), 2736–2750 (2017). https://doi.org/10.1109/tifs.2017.2718489
4. F. Li, Y. Shi, A. Shinde, J. Ye, W. Song, Enhanced cyber-physical security in internet of things through energy auditing. IEEE Internet Things J. **6**(3), 5224–5231 (2019). https://doi.org/10.1109/jiot.2019.2899492

5. M. Span, L.O. Mailloux, R.F. Mills, W. Young, Conceptual systems security requirements analysis: aerial refueling case study. IEEE Access **6**, 46668–46682 (2018). https://doi.org/10.1109/access.2018.2865736
6. H. Orojloo, M.A. Azgomi, Modelling and evaluation of the security of cyber-physical systems using stochastic petri nets. IET Cyber Phys. Syst. Theor. Appl. **4**(1), 50–57 (2019). https://doi.org/10.1049/iet-cps.2018.0008
7. S. Kim, Y. Won, I.-H. Park, Y. Eun, K.-J. Park, Cyber-physical vulnerability analysis of communication-based train control. IEEE Internet Things J. **6**(4), 6353–6362 (2019). https://doi.org/10.1109/jiot.2019.2919066
8. Y. Zhang, Y. Xiang, X. Huang, A cross-layer key establishment model for wireless devices in cyber-physical systems, in *Proceedings of the 3rd ACM Workshop on Cyber-Physical System Security—CPSS 17*. https://doi.org/10.1145/3055186.3055187 (2017)
9. T. Varshney, N. Sharma, I. Kaushik, B. Bhushan, Architectural model of security threats and their countermeasures in IoT, in *2019 International Conference on Computing, Communication, and Intelligent Systems (ICCCIS)*. https://doi.org/10.1109/icccis48478.2019.8974544 (2019)
10. V. Venkataramanan, A. Hahn, A. Srivastava, CP-SAM: cyber-physical security assessment metric for monitoring microgrid resiliency. IEEE Trans. Smart Grid **11**(2), 1055–1065 (2020). https://doi.org/10.1109/tsg.2019.2930241
11. X. Li, K.W. Hedman, Enhancing power system cyber-security with systematic two-stage detection strategy. IEEE Trans. Power Syst. **35**(2), 1549–1561 (2020). https://doi.org/10.1109/tpwrs.2019.2942333
12. I. Kaushik, N. Sharma, N. Singh, Intrusion detection and security system for blackhole attack, in *2019 2nd International Conference on Signal Processing and Communication (ICSPC)*. https://doi.org/10.1109/icspc46172.2019.8976797 (2019)
13. S. Mili, N. Nguyen, R. Chelouah, Transformation-based approach to security verification for cyber-physical systems. IEEE Syst. J. **13**(4), 3989–4000 (2019). https://doi.org/10.1109/jsyst.2019.2923818
14. A. Sharma, A. Singh, N. Sharma, I. Kaushik, B. Bhushan, Security countermeasures in web based application, in *2019 2nd International Conference on Intelligent Computing, Instrumentation and Control Technologies (ICICICT)*. https://doi.org/10.1109/icicict46008.2019.8993141 (2019)
15. A. Ahmed, V.V.G. Krishnan, S.A. Foroutan, M. Touhiduzzaman, A. Srivastava, Y. Wu, S. Sindhu et al. Cyber physical security analytics for anomalies in transmission protection systems, in *2018 IEEE Industry Applications Society Annual Meeting (IAS)*. https://doi.org/10.1109/ias.2018.8544672 (2018)
16. P. Sinha, A.K. Rai, B. Bhushan, Information security threats and attacks with conceivable counteraction, in *2019 2nd International Conference on Intelligent Computing, Instrumentation and Control Technologies (ICICICT)*. https://doi.org/10.1109/icicict46008.2019.8993384 (2019)
17. A. Ahmed, V.V.G. Krishnan, S.A. Foroutan, M. Touhiduzzaman, C. Rublein, A. Srivastava, S. Suresh et al., Cyber physical security analytics for anomalies in transmission protection systems. IEEE Trans. Ind. Appl. **55**(6), 6313–6323 (2019). https://doi.org/10.1109/tia.2019.292850
18. S. Soni, B. Bhushan, Use of machine learning algorithms for designing efficient cyber security solutions, in *2019 2nd International Conference on Intelligent Computing, Instrumentation and Control Technologies (ICICICT)*. https://doi.org/10.1109/icicict46008.2019.8993253 (2019)
19. Comparative analysis of 3-D password using various techniques. Int. J. Emerg. Technol. Innov. Res. **6**(6), 711–718. ISSN 2349-5162, www.jetir.org (2019)
20. B. Bhushan, G. Sahoo, Requirements, protocols, and security challenges in wireless sensor networks: an industrial perspective. in *Handbook of Computer Networks and Cyber Security*, pp. 683–713. https://doi.org/10.1007/978-3-030-22277-2_27 (2020)
21. A. Carelli, A. Vallero, S.D. Carlo, Performance monitor counters: interplay between safety and security in complex cyber-physical systems. IEEE Trans. Device Mater. Reliab. **19**(1), 73–83 (2019). https://doi.org/10.1109/tdmr.2019.2898882

22. B. Bhushan, G. Sahoo, Recent advances in attacks, technical challenges, vulnerabilities and their countermeasures in wireless sensor networks. Wirel. Pers. Commun. **98**(2), 2037–2077 (2017). https://doi.org/10.1007/s11277-017-4962-0
23. M. Evans, Y. He, C. Luo, I. Yevseyeva, H. Janicke, E. Zamani, L.A. Maglaras, Real-time information security incident management: a case study using the IS-CHEC technique. IEEE Access **7**, 142147–142175 (2019). https://doi.org/10.1109/access.2019.2944615
24. Aditya Khamparia, Babita Pandey, A novel method of case representation and retrieval in CBR for e-learning. Edu. Inform. Technol. **22**(1), 337–354 (2015). https://doi.org/10.1007/s10639-015-9447-8
25. A. Nourian, S. Madnick, A systems theoretic approach to the security threats in cyber physical systems applied to Stuxnet. IEEE Trans. Dependable Sec. Comput. **15**(1), 2–13 (2018). https://doi.org/10.1109/tdsc.2015.2509994
26. P. Sinha, V.K. Jha, A.K. Rai, B. Bhushan, Security vulnerabilities, attacks and countermeasures in wireless sensor networks at various layers of OSI reference model: a survey, in *2017 International Conference on Signal Processing and Communication (ICSPC)*. https://doi.org/10.1109/cspc.2017.8305855 (2017)
27. Y. Cui, F. Bai, Y. Liu, Y. Liu, A measurement source authentication methodology for power system cyber security enhancement. IEEE Trans. Smart Grid **9**(4), 3914–3916 (2018). https://doi.org/10.1109/tsg.2018.2826444
28. Y. Liu, A. Liu, X. Liu, M. Ma, A trust-based active detection for cyber-physical security in industrial environments. IEEE Trans. Ind. Inf. **15**(12), 6593–6603 (2019). https://doi.org/10.1109/tii.2019.2931394
29. A. Singh, A. Sharma, N. Sharma, I. Kaushik, B. Bhushan, Taxonomy of attacks on web based applications, in *2019 2nd International Conference on Intelligent Computing, Instrumentation and Control Technologies (ICICICT)*. https://doi.org/10.1109/icicict46008.2019.8993264 (2019)
30. A.K.. Luhach, A. Khamparia, R. Sihag, R. Kumar, Honey bee optimization based sink mobility aware heterogeneous protocol for wireless sensor network. Scalable Comput. Prac. Exp. **20**(4), 591–598 (2020)

Tendulkar's Cat and Schrodinger's Bat—Knowledge-Enhanced Real-Word Error Correction

Shailesh Sridhar, E Saahitya, Satyam Shivam Sundaram, and Bhaskarjyoti Das

Abstract Real-word errors are those errors in which a word is lexically correct or syntactically correct in the sentence it belongs to, but is contextually incorrect. As a result, they are difficult to identify and arguably even more difficult to correct. In this work, we attempt to create a model based on total word similarity which is able to correct real-word errors in sentences based on the simple assumption that a real-word error will not be semantically cohesive with respect to its context, and that the actual word which was intended, will be. The main aspects of our approach are the usage of semantic context, knowledge context, and syntactic context, all together at once. To this approach, we also retain the concept of edit penalty, which is an integral component of most spelling checkers. We modified the Microsoft Research Sentence Completion task of 2011 for measuring the performance of real-word error correction models and achieve 80% accuracy.

Keywords Real-word spelling error · Word embedding · Syntactic context · Semantic context · Knowledge context · Edit penalty · Knowledge graph · ConceptNet

1 Introduction

Take the example of the title of this paper "Tendulkar's Cat and Schrodinger's Bat." A cricket fan who is familiar with physics or vice versa will be able to identify that the words Bat and Cat are probably interchanged. However, a person without the

S. Sridhar · E. Saahitya (✉) · S. S. Sundaram · B. Das
PES University, Bengaluru, India
e-mail: saahitya.e@gmail.com

S. Sridhar
e-mail: shailesh.sridhar@gmail.com

S. S. Sundaram
e-mail: satyspy007@gmail.com

B. Das
e-mail: bhaskarjyoti01@gmail.com

J. Tavares et al. (eds.), *Emerging Technologies in Data Mining and Information Security*, Lecture Notes in Networks and Systems 164,
https://doi.org/10.1007/978-981-15-9774-9_37

knowledge of Tendulkar or Schrodinger may not be able to do so. It is possible that a person who has simply encountered the terms "Schrodinger's Cat" and "Tendulkar's Bat" is able to detect the errors.

The examples cited here are not typical spelling error corrections. A spell-checker used in a word processing software or a mobile phone is not likely to catch these errors. Typical error correction depends on variants of edit distance algorithm and an implementation of a standard lexicon. Such error correction methods primarily focus on spelling errors corrections and only work on correcting errors in words that do not exist in the standard lexicon. Hence, such methods will not spot the real-word errors mentioned above as the words are in the standard lexicon but they are contextually wrong. However, correcting real-word errors is an important problem and has serious applications in OCR, video subtitle generation, etc.

An important hypothesis regarding real-word error correction is that a real-word error does not fit in with its surrounding context. In a typical natural language processing pipeline, the goal is to find the meaning of a piece of text and the key steps in meaning disambiguation are resolutions of syntax and semantics. Hence, both syntax and semantics can be good candidates in the word error resolution process. Knowledge is usually not part of the meaning resolution process in natural language processing but can really contribute a lot in resolution of meaning as the usage of words bears lot of relevance to the domain context in which they are used. Hence, knowledge context should also be a candidate in the word error correction problem. The work described in this paper uses all three kinds of context while keeping the traditional edit distance approach. Since all three kinds of context can be used in the algorithm, a suitable weight of each should be decided in any such implementation. Hence, it is possible to have candidate words for correcting a real-word error by using all three different contexts and the edit distance of each candidate. One can then score each candidate and the candidate with the highest score can determine the correct real-word error correction. Distributional semantics is an alternative to lexical semantics as any lexicon tends to be limited. Distributional semantics comes up with an embedding for each word in a language vocabulary based on usage. The key assumption here is that the embedding of two words will be adjacent in the embedding space if these two words tend to appear in a small window in real usages as observed in a very large corpus of text. Hence, word embedding can be used to have a vector representation of the context in which the candidate word can be used in a sentence.

Knowledge engineering and machine learning so far have continued on two different axis. While knowledge engineering depends on specifics and hierarchy as captured by the "subject, predicate, object" triple, machine learning depends on as much data as possible. Machine learning, in spite of lot of training data, may still deliver doubtful results. Knowledge engineering, on the other hand, may present a challenge as knowledge graph can keep growing for ever and may never be considered completely done. However, both taken together can provide an interesting approach that harnesses the strength of both. But searching in knowledge graph for relevant context is an approach that is hard to scale. Hence, researchers nowadays have come up with approaches that capture the facts of a knowledge graph into a suitable embedding scheme. An embedding approach of knowledge graph also makes it amenable to modern techniques of natural language processing that already uses word embedding. In real-word error correction problem as well such an approach makes sense to capture knowledge context. In this work, that knowledge embedding approach has been adopted as well.

2 Literature Survey

2.1 Spelling Error Correction and Real-Word Errors

Singh et al. [23] discussed error correction and detection methods in text. The authors divide typographic errors into two classes, i.e., spelling real-word errors. Both these types of errors are not limited to text alone. Henry Liberman et al. [13] explained the frustrating problem of homophone in speech dictation systems. Hirst et al. [9] refer to real-word errors as "malapropisms." Their work is based on the basic assumption that real-word errors in a document do not fit into the context of the rest of the document. This hypothesis is a very important one and is the basis of our approach as well. The authors use a set of rules to generate possible spelling variations of the identified malapropism. Each possible correction is then evaluated with its neighboring words for semantic similarity, and the distance metric used is Jiang–Conrath distance based on WordNet. While a lot of the intuition in this paper is valid, the techniques utilized are quite outdated. In the last decade, word embeddings have emerged and cosine similarity between word vectors has been shown to be a much more reliable metric of word similarity. Furthermore, using a word-to-word distance does not capture the context of the entire system.

The CoNLL-2014 shared task [18] was focused on grammatical error correction and most attempts were language model based. Mays et al. [16] came up with a statistical approach to address contextual error correction. Gupta et al. [8] use a conjunction of trigram approach and Bayesian model trained on Brown corpus for real-word error correction. These classical techniques use either n-grams [22] or a probabilistic noisy channel model [10]. These methods perform reasonably well but have their own limitations. N-grams typically are unable to represent long-term context. Probabilistic methods can be highly powerful but depend heavily upon the corpus used for training.

In recent times, with the advent of deep learning and its applications, contextual spelling error correction has been approached with the latest of neural network models. This is a parallel to the effort that has been happening in speech recognition domain. Guo et al. [6] investigated attention-based sequence to sequence models in speech recognition. Ge et al. [4] investigated the sequence-to-sequence models in automatic error correction. Recently Lu et al. [15] used word2vec-based embedding for health-related questions. Similarly, Yuan et al. [28] used neural machine translation for grammatical error correction. Etoori et al. [2] built a RNN-based approach for automatic error correction for low resource Indian languages. Xie et al. [27] use a character-based attention model toward automatic error correction. Ahmadi [1] models error correction as a monolingual machine translation model. In [11], a nested RNN model is used. Ghosh et al. [5] propose an architecture which consists of a combination of character level CNN and gated recurrent unit (GRU) encoder along with a word-level gated recurrent unit (GRU) attention decoder. They even use a relatively smaller corpus (in comparison to the training corpora utilized by most deep learning architectures) of 12 million words.

Comparatively, very few research attempts were made to use knowledge for automatic error correction. Peters et al. [20] integrate WordNet and a subset of Wikipedia into BERT and come up with knowBERT as a knowledge-enhanced embedding. However, no attempt has been made to use this in word error correction. Ling et al. [14] use knowledge-enhanced named entity recognition in biomedical text.

2.2 Sentence Completion

Given a sentence with a blank in it and several options, "sentence Completion" is the task of predicting the appropriate word to fill that blank with respect to the surrounding context.

The Microsoft Research Sentence Completion Challenge [29] of 2011 had the aim of providing a common metric for evaluation of language models on the spelling completion task. Figure 1 shows an example of such a task. Further information on the dataset and the nature of the task can be found in the dataset section. The authors themselves suggest an n-gram method and a latent semantic analysis-based approach for the task. Zweig et al. [30] analyze various other approaches, predominantly modeled around the use of vectors generated using LSA. One of these methods is total word similarity, which is the basis of our model in this work. Total word similarity between a candidate word and the target sentence is defined as sum of cosine similarity between the candidate word and each word of the target sentence. However, most of these methods rely on training on similar data. The Microsoft Dataset consists of sentences taken from Project Gutenberg texts, and the LSA vectors are generated by training on a similar dataset of texts. However, considering the nature of word errors

Fig. 1 An example of the sentence completion task

1. I have seen it on him , and could _____ to it.
   (a) write
   (b) migrate
   (c) climb
   (d) swear
   (e) contribute

being investigated, one cannot depend on this sort of training to generate word vectors. We hence resort to pretrained word embeddings, as described in the following section.

The current state of the art for this task can be attributed to Aubrie Woods [26]. The author uses a "pointwise mutual information" method to generate word vectors for the words. They then perform total word similarity. However, instead of weighting each word evenly as is done with normal standard total word similarity, the authors assign different weights to each word according to the importance it has, on the basis of POS tags and features derived from dependency parsing. We attempt to use this sort of weighting in our model.

2.3 Knowledge Graphs and Knowledge Enhancement

Knowledge graphs are graphical implementation of knowledge bases. These are large networks of entities (nodes) and their semantic relationships (edges). This kind of graph consists of triples, i.e., subject, predicate, object which are arranged in a hierarchy. They serve as a powerful tool that changes the way we do data integration, search analytics, and context-sensitive recommendation. Knowledge enhancement is the concept of incorporating the information from knowledge bases and knowledge graphs into NLP models. Several works have highlighted how knowledge enhancement can improve performance in various ways. Peters et al. [19] carry our knowledge enhancement on Bidirectional Encoder Representations from Transformers (BERT) to improve its perplexity and fact-recalling capabilities, among other things. Li et al. [12] have done several experiments on the importance of knowledge enhancement for the task of natural language inference and conclude that it indeed leads to improved performance in several aspects.

ConceptNet Numberbatch [24] is a system of word embeddings which incorporate information from a knowledge graph called ConceptNet into Glove and Google Word2Vec word embeddings. As described in [25], the authors have shown superior performance of ConceptNet Numberbatch in several tasks compared to standard embeddings. They are easy to use as they can simply be used as a replacement for the original set of word embeddings. In this work, we utilize ConceptNet Numberbatch embeddings.

3 Methodology

In our approach, we use the hypothesis that real-word errors are not semantically cohesive [9] with respect to their context. We also connect this hypothesis with the task of sentence completion, where the contextually correct word from a set of options is to be selected to fill a blank in a sentence.

The model we propose here is inspired by the one proposed by Fivez et al. [3], which utilizes a word embeddings-based approach to represent context and predict the correct spelling by identifying the spelling variation which is most similar to the context vector. This approach is a variant of total word similarity, even though the name of this algorithm is not mentioned explicitly in the work. The word embeddings used are trained on medical text using a skip gram model. The model performs well on the task of performing spelling correction on clinical-free text. In our work, instead of the medical domain-specific word embeddings generated by training, we propose using pretrained word embeddings which have a generic purpose. We also add the concept of weighting of the context words so as to produce a more accurate representation of the context, by applying more and less weight to more and less important words. To increase the performance, we consider the knowledge context as well taking advantage of ConceptNet [24].

Our approach is based on four main components. The first three components are ways of representing the context.

- Semantic context: Word vector using distributional semantics
- Weighted context: Application of appropriate weights to context words
- Knowledge context: Knowledge-enhanced word vector
- Penalty based on edit distance: This is not a way of improving the generation of the context vector but the main building block in existing spelling error correction in software.

Figure 2 provides an example of how choice of candidate words can be done solely based on semantic context and edit penalty.

3.1 Semantic Context and the Total Word Similarity Algorithm

Word embeddings represent words as dense unit vectors of real numbers, where vectors that are close together are semantically related. This approach to machine learning about semantics is sometimes referred to as distributional semantics as words which are used in similar context and in similar fashion end up being close together in embedding space. By adding up all the vectors (embeddings) that form the context of a word, we can arrive at a vector which is a reasonable representation of the context. By finding the cosine similarity between the word vector and the context vector, we obtain a score of how well a given word fits in its context. By calculating the scores for all possible candidates for replacing this word, we can identify which word best fits in the context. We utilize this method for determining the appropriate correction for the error word.

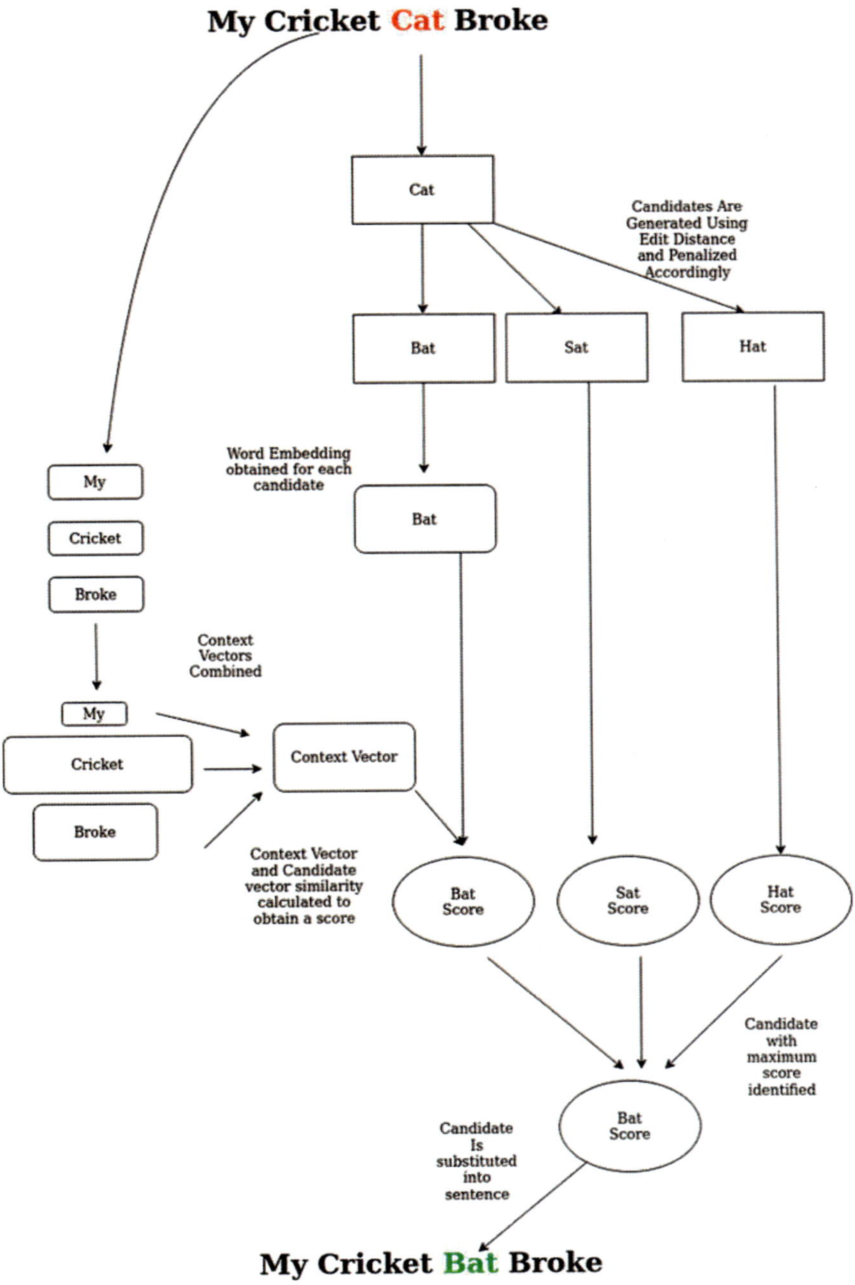

Fig. 2 Choosing candidate word based on semantic context and edit penalty

3.2 Weighted Context

While simply summing up the context vectors results in a good representation of the context, it is possible to obtain a better representation by finding the weighted sum of the context vectors, where the weight of each word is proportional to its importance in the context w.r.t to the error word. We use two types of weighting in this work.

- Reciprocal Weighting Similar to the weighting used in [3], applies a weight of 1/d to the context word in question, where d represents the distance to the error word in number of words. It is based on the simple assumption that words which are farther away from the error word are likely to be less important in representing the context with respect to the error word.
- Syntactic Context Based on the algorithm used in [26] for applying weights to PMI values, we weight context word vectors based on their belonging to the following feature sets:

1. **Reduced context**: This feature set consists of words in the sentence after removing determiners, coordinating conjunctions, pronouns, and proper nouns. These words are identified in [26] as not important.
2. **Dependencies**: Sentence words that share a semantic dependency with the candidate word(s) are included in this set. Absent from the set of dependencies are words removed during the first phase. Figure 3 shows such a dependency graph produced by a dependency parser.
3. **Keywords**: Providing the model with a collection of salient tokens effectively increases the tokens' associated weights. Using common nouns is an effective method to do this.

The weights applied to each context word are then simply the number of the above feature sets related to it. Stanford NLP Group's Stanza toolkit [21] is used to carry out all POS tagging and dependency parsing.

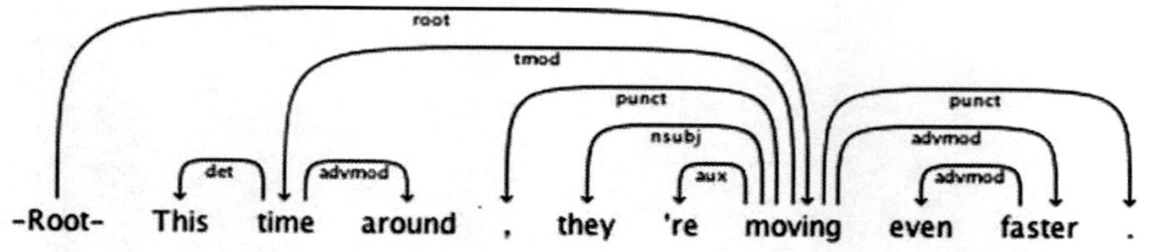

Fig. 3 A sample dependency parse showing the dependencies words in a sentence may have on each other

3.3 Knowledge Context

In some cases, simply utilizing information obtained from distributional semantics is not enough to adequately represent words in vector space. Recent research [25] has shown that enhancing word embeddings with knowledge helps to improve the performance of many natural language processing task. Instead of calculating path distance between different concepts in the knowledge graph, ConceptNet Numberbatch provides a ready to use word embeddings that incorporate knowledge. It uses Glove and Word2Vec embeddings fortified with knowledge from ConceptNet and are shown to outperform several word embeddings. We use these embeddings here in the same manner as we use normal word embeddings.

3.4 Edit Penalty

Edit penalty refers to the penalizing of every candidate by dividing the final scores by their edit difference from the original error word. This is an important building block in all spelling checkers and operates based on the assumption that a human is more likely to make a smaller number of errors than a larger number with respect to the word he/she intended to type.

We incorporate all these features and compare and contrast to obtain the best results. Figure 4 shows the high-level flow of the model though it does not explicitly mention the syntactic and knowledge context. Figure 5 shows the overall algorithm.

4 Dataset

Our test dataset consists of 825 sentences with 4 candidates, exactly one being correct. The original MSR Sentence Completion Challenge Data [29], consists of 1040 sentences, and each sentence consists of one ground truth and 4 imposter sentences where a specific word has been replaced with a syntactically correct but semantically incorrect impostor word. We modified the original dataset to align with our task of real-word error correction. Table 1 lists the source, size, and description of the dataset used.

Our dataset consists of sentences from the MSR dataset, replacing specific manipulated words with blanks, each with 4 options, among which only one option is correct. We generated real-word candidates for each of the correct options within 2 Damerau–Levenshtein distance from the correct option. We used WordNet synset as the dictionary and a list of all the words within Damerau–Levenshtein distance of 2 from the correct option were generated from the dictionary. We randomly choose three options from the generated list, discarding lists with less than three options, along with the corresponding sentence. For example,

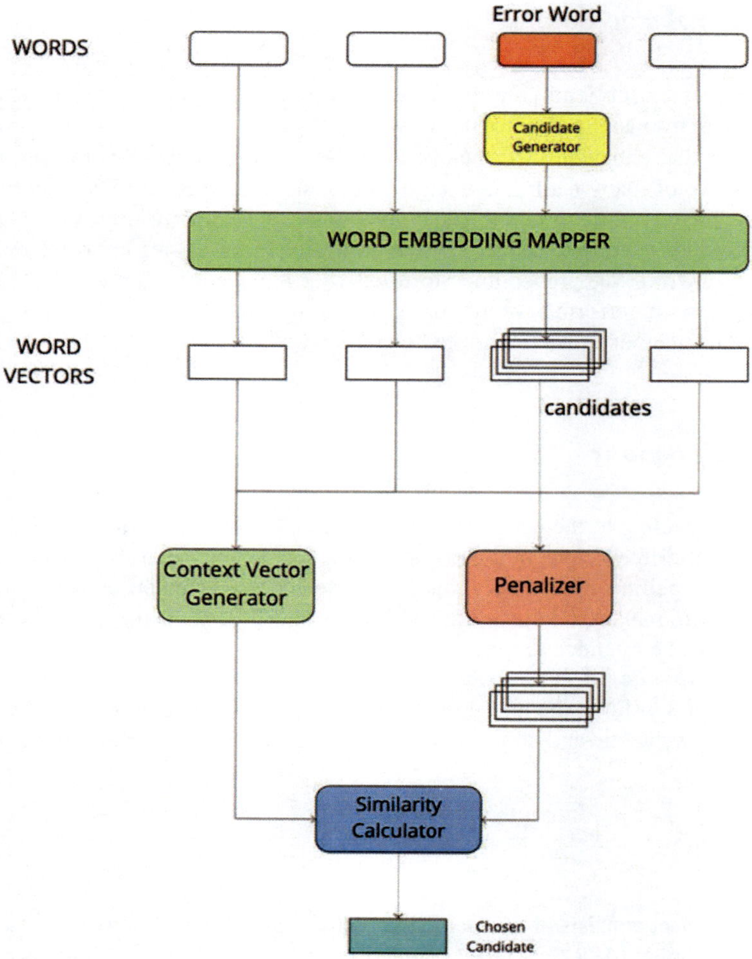

Fig. 4 High-level flow of our model

- Generated dataset options: citation, titration, saturation, actuation, situation
- Original dataset options: situation, fence, food, instrument, building.

As mentioned in the methodology section, all the information our model required to be able to make predictions was found within the pretrained word embeddings. No additional training was done on any corpus.

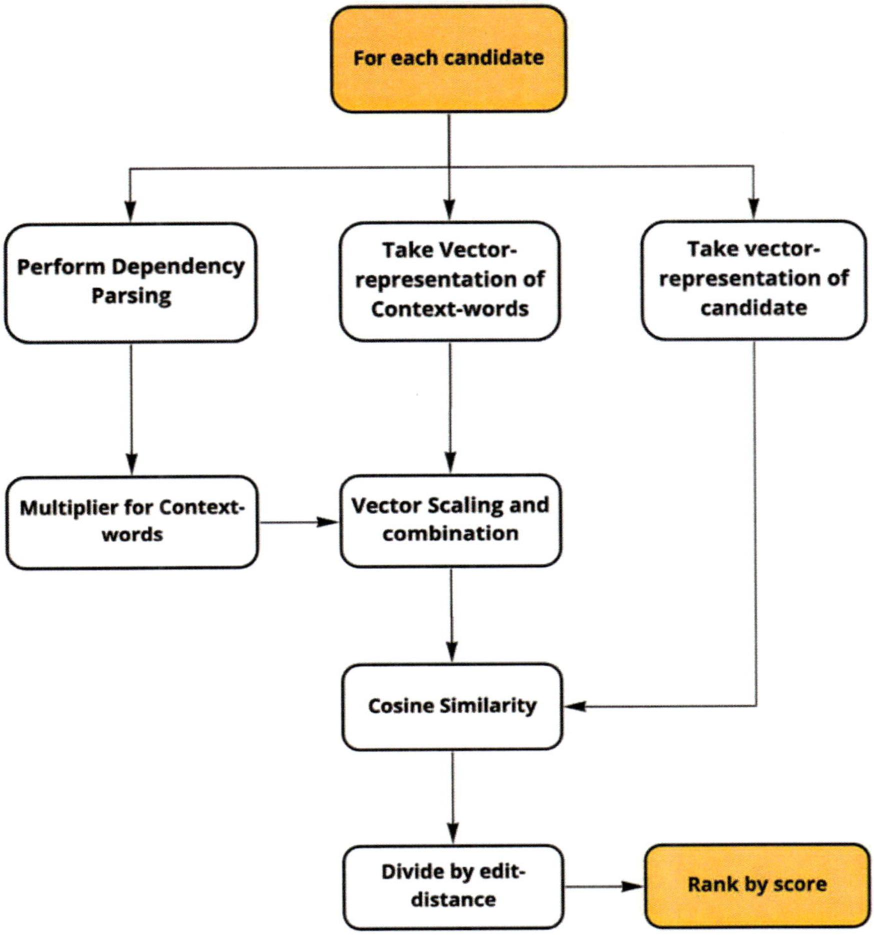

Fig. 5 Algorithm flow of our model

Table 1 Source, size, and description of the dataset

Source	Size	Description
Microsoft research sentence completion dataset	825 sentences	Real-word candidate generation, edit-distance, randomized selection

5 Results

Tables 2 and 3 list the results obtained. Embedding like Glove is used for ascertaining the semantic context. Knowledge context is incorporated by ConceptNet Numberbatch embedding. Syntactic context is used as one of the weight options in our algorithm. Edit penalty is a building block that is used as an option as well.

After running the experiments on the different models, we find that we achieve that the best results with the model that uses the ConceptNet embedding, syntactic context weighting and edit distance penalty to successively correct the maximum number of real-word errors in the dataset chosen.

While both word embeddings perform very similarly, it is the ConceptNet Numberbatch embeddings which were responsible for our best result. Our hypothesis is that the additional knowledge possessed by the ConceptNet Numberbatch embeddings (due to its use of the ConceptNet semantic graph) made a difference and allowed better representation of the words in the sentence and in turn the context of the sentence.

Reciprocal weighting impacts performance very slightly, but positively, showing that while applying weights to different words that are inversely proportional to the distance from the real-word error definitely matters, the representation of context using reciprocal weighting is not significantly better than that using uniform weighting.

However, syntactic context-based weighting identifies the most important words in the sentence and weights the important words more significantly, and this in conjunction with the knowledge advantage possessed by ConceptNet results in models with higher accuracy.

Table 2 Percentage accuracy for GloVe-enhanced real-word error correction

GloVe%	Uniform weighting	Reciprocal weighting	Syntactic context weighting
Without edit penalty	67.1 (baseline)	66.7	63.7
With edit penalty	77.2	78.0	75.3

Table 3 Percentage accuracy for ConceptNet-enhanced real-word error correction

ConceptNet (knowledge enhanced)%	Uniform weighting	Reciprocal weighting	Syntactic context weighting
Without edit penalty	65.1	66.5	69.9
With edit penalty	76.5	77.3	**80.0**

Applying an edit penalty improves the performance of the model by a very significant amount. This is due to the fact that people making real-word errors tend to make errors that are slightly different from the actual word and hence candidates generated from the model with smaller edit distances are more likely to be the actual word than other candidates with greater edit distances.

6 Conclusion

The outcome of this work is a successful model based on the total word similarity algorithm which corrects real-word errors, and those errors in which a word is lexically correct or syntactically correct in the sentence it belongs to, but is contextually incorrect.

We enhance the model far beyond its baseline of 60% using the vanilla total word similarity algorithm, by incorporating the semantic context, syntactic context, and knowledge context of a sentence into its functioning. Knowledge context is brought by using the knowledge-enhanced ConceptNet Numberbatch as the primary word embeddings, which have been shown to be performed better compared to other more conventional choices of word embeddings. Dependency parsing and resultant feature set-based techniques, along with reciprocal-weighting are incorporated for providing syntactic context and allowing the model to make use of the structure of the sentence around the error to improve its prediction. Edit penalty, a time-tested basic building block of standard spelling checkers, is also incorporated in order to further enhance the model's performance.

We test the model on a version of the Microsoft Research Sentence Completion task, which has been appropriately modified for performing real-word error correction tasks. We combine the techniques mentioned above, which encompass several ways of representing the context of a sentence, and provide our model with the best information possible, to achieve our best accuracy of 80%.

As a next step, we plan to extend the work by creating a larger dataset and investigating the effectiveness in speech to text translation that shows lot of errors due to homophones and can immensely benefit from this work [7, 17].

References

1. S. Ahmadi, Attention-based encoder–decoder networks for spelling and grammatical error correction. arXiv preprint arXiv:1810.00660 (2018)
2. P. Etoori, M. Chinnakotla, R. Mamidi, Automatic spelling correction for resource-scarce languages using deep learning, in *Proceedings of ACL 2018, Student Research Workshop* (2018), pp. 146–152
3. P. Fivez, S. Šuster, W. Daelemans, Unsupervised context-sensitive spelling correction of clinical free-text with word and character n-gram embedding, in *16th Workshop on Biomedical Natural Language Processing of the Association for Computational Linguistics* (2017), pp. 143–148

4. T. Ge, F. Wei, M. Zhou, Reaching human-level performance in automatic grammatical error correction: an empirical study. arXiv preprint arXiv:1807.01270 (2018)
5. S. Ghosh, P.O. Kristensson, Neural networks for text correction and completion in keyboard decoding. arXiv preprint arXiv:1709.06429 (2017)
6. J. Guo, T.N. Sainath, R.J. Weiss, A spelling correction model for end-to-end speech recognition, in *ICASSP 2019—2019 IEEE International Conference on Acoustics, Speech and Signal Processing (ICASSP)* (IEEE, 2019), pp. 5651–5655
7. A. Gupta, P. Goyal, S. Sarkar, M. Gattu, Fully contextualized biomedical NER, in *Advances in Information Retrieval—41st European Conference on IR Research, ECIR 2019*, Cologne, Germany, 14–18 April 2019, Proceedings, Part II, ed. by L. Azzopardi, B. Stein, N. Fuhr, P. Mayr, C. Hauff, D. Hiemstra. Lecture Notes in Computer Science, vol. 11438 (Springer, 2019), pp. 117–124. https://doi.org/10.1007/978-3-030-15719-7_15
8. S. Gupta et al., A correction model for real-word errors. Procedia Comput. Sci. **70**, 99–106 (2015)
9. G. Hirst, A. Budanitsky, Correcting real-word spelling errors by restoring lexical cohesion. Nat. Lang. Eng. **11**(1), 87–111 (2005)
10. M.D. Kernighan, K.W. Church, W.A. Gale, A spelling correction program based on a noisy channel model, in *Proceedings of the 13th conference on Computational linguistics*, vol. 2 (Association for Computational Linguistics, 1990), pp. 205–210
11. H. Li, Y. Wang, X. Liu, Z. Sheng, S. Wei, Spelling error correction using a nested RNN model and pseudo training data. arXiv preprint arXiv:1811.00238 (2018)
12. T. Li, X. Zhu, Q. Liu, Q. Chen, Z. Chen, S. Wei, Several experiments on investigating pretraining and knowledge-enhanced models for natural language inference. CoRR **abs/1904.12104** (2019). http://arxiv.org/abs/1904.12104
13. H. Lieberman, A. Faaborg, W. Daher, J. Espinosa, How to wreck a nice beach you sing calm incense, in *Proceedings of the 10th International Conference on Intelligent User Interfaces* (2005), pp. 278–280
14. Y. Ling, S.A. Hasan, O. Farri, Z. Chen, R. van Ommering, C. Yee, N. Dimitrova, A domain knowledge-enhanced LSTM-CRF model for disease named entity recognition, in *AMIA Summits on Translational Science Proceedings 2019* (2019), p. 761
15. C.J. Lu, A.R. Aronson, S.E. Shooshan, D. Demner-Fushman, Spell checker for consumer language (CSPELL). J. Am. Med. Inform. Assoc. **26**(3), 211–218 (2019)
16. E. Mays, F.J. Damerau, R.L. Mercer, Context based spelling correction. Inf. Process. Manag. **27**(5), 517–522 (1991)
17. T. Murphy, T. McIntosh, J.R. Curran, Named entity recognition for astronomy literature, in *Proceedings of the Australasian Language Technology Workshop, ALTA 2006*, Sydney, Australia, 30 Nov–1 Dec 2006, ed. by L. Cavedon, I. Zukerman (Australasian Language Technology Association, 2006), pp. 59–66. https://www.aclweb.org/anthology/U06-1010/
18. H.T. Ng, S.M. Wu, T. Briscoe, C. Hadiwinoto, R.H. Susanto, C. Bryant, The CoNLL-2014 shared task on grammatical error correction, in *Proceedings of the Eighteenth Conference on Computational Natural Language Learning: Shared Task* (2014), pp. 1–14
19. M.E. Peters, M. Neumann, R.L. Logan IV, R. Schwartz, V. Joshi, S. Singh, N.A. Smith, Knowledge enhanced contextual word representations, in *Proceedings of the 2019 Conference on Empirical Methods in Natural Language Processing and the 9th International Joint Conference on Natural Language Processing, EMNLP-IJCNLP 2019*, Hong Kong, China, 3–7 Nov 2019, ed. by K. Inui, J. Jiang, V. Ng, X. Wan (Association for Computational Linguistics, 2019) pp. 43–54. https://doi.org/10.18653/v1/D19-1005
20. M.E. Peters, M. Neumann, I. Logan, L. Robert, R. Schwartz, V. Joshi, S. Singh, N.A. Smith, Knowledge enhanced contextual word representations. arXiv preprint arXiv:1909.04164 (2019)
21. P. Qi, Y. Zhang, Y. Zhang, J. Bolton, C.D. Manning, Stanza: a python natural language processing toolkit for many human languages. CoRR **abs/2003.07082** (2020). https://arxiv.org/abs/2003.07082

22. P. Samanta, B.B. Chaudhuri, A simple real-word error detection and correction using local word bigram and trigram, in *Proceedings of the 25th Conference on Computational Linguistics and Speech Processing (ROCLING 2013)* (2013), pp. 211–220
23. S. Singh, S. Singh, Review of real-word error detection and correction methods in text documents, in *2018 Second International Conference on Electronics, Communication and Aerospace Technology (ICECA)* (IEEE, 2018), pp. 1076–1081
24. R. Speer, J. Chin, C. Havasi, ConceptNet 5.5: an open multilingual graph of general knowledge (2017), pp. 4444–4451. http://aaai.org/ocs/index.php/AAAI/AAAI17/paper/view/14972
25. R. Speer, J. Lowry-Duda, ConceptNet at SemEval-2017 task 2: extending word embeddings with multilingual relational knowledge. CoRR **abs/1704.03560** (2017). http://arxiv.org/abs/1704.03560
26. A. Woods, Exploiting linguistic features for sentence completion, in *Proceedings of the 54th Annual Meeting of the Association for Computational Linguistics, vol. 2: Short Papers* (Association for Computational Linguistics, Berlin, Germany, 2016), pp. 438–442. https://doi.org/10.18653/v1/P16-2071
27. Z. Xie, A. Avati, N. Arivazhagan, D. Jurafsky, A.Y. Ng, Neural language correction with character-based attention. arXiv preprint arXiv:1603.09727 (2016)
28. Z. Yuan, T. Briscoe, Grammatical error correction using neural machine translation, in *Proceedings of the 2016 Conference of the North American Chapter of the Association for Computational Linguistics: Human Language Technologies* (2016), pp. 380–386
29. G. Zweig, C.J. Burges, The Microsoft research sentence completion challenge. Tech. Rep. MSR-TR-2011-129 (2011). https://www.microsoft.com/en-us/research/publication/the-microsoft-research-sentence-completion-challenge/
30. G. Zweig, J.C. Platt, C. Meek, C.J.C. Burges, A. Yessenalina, Q. Liu, Computational approaches to sentence completion, in *The 50th Annual Meeting of the Association for Computational Linguistics, Proceedings of the Conference, 8–14 July 2012, Jeju Island, Korea, vol. 1: Long Papers* (The Association for Computer Linguistics, 2012), pp. 601–610. https://www.aclweb.org/anthology/P12-1063/

Multi Task Learning for Financial Forecasting

Deepali Prabhu, Arnav Chhabra, Saksham Goyal, and Bhaskarjyoti Das

Abstract Financial forecasting plays a crucial role in investment decision making and portfolio diversification. A given stock's future value can be predicted based on its history on the assumption that there will be a repetition of trends. However, stocks are not only affected by their history but are also affected by various other factors. Usually, such environmental factors are taken as feature inputs to the stock prediction model. In the approach followed in this work, a multitask learning model is used that simultaneously tries to predict the future values of stocks whilst sharing domain specific knowledge among them. It is assumed that sufficiently correlated stocks from a particular domain capture enough environmental influences and a model that considers such correlated stocks may be as useful as a sophisticated model that considers diverse extraneous factors. However, consideration of all stocks at once is impractical and introduces noise in the network. So a stock selector that finds the dynamic correlation between the stocks and selects the most correlated stocks as input to our multitask learning model is used. It was observed that the multitask learning model had an improvement of around 3–4% of relative accuracy when compared to the baseline models that operate with single stock in similar settings.

Keywords Multi task learning · Financial forecasting · Stock prediction · Soft parameter sharing · Dynamic time warping

D. Prabhu · A. Chhabra (✉) · S. Goyal · B. Das
PES University, Bengaluru, India
e-mail: arnav.chh@gmail.com

D. Prabhu
e-mail: deepali846@gmail.com

S. Goyal
e-mail: sakshamgoyal808@gmail.com

B. Das
e-mail: bhaskarjyoti01@gmail.com

J. Tavares et al. (eds.), *Emerging Technologies in Data Mining and Information Security*, Lecture Notes in Networks and Systems 164,
https://doi.org/10.1007/978-981-15-9774-9_38

1 Introduction

In current day and age, the stock markets are said to impact the economy and development of the country in various ways. However stocks are affected by various factors such as inflation, demand, supply, macroeconomic factors, human behaviour, political or geopolitical events etc. Given the significance of financial markets, forecasting stocks occupies a paramount position in investment decision making and portfolio diversification. It is almost impossible to accurately predict the prices of stocks accurately due to the ingredients of low signal-noise ratio and heavy tailed distribution.

The stock prices are determined by the short term effect of the beliefs of thousands of investors around the globe which are in turn affected by hundreds of variables. Some investors believe in investing when the markets are going up and refrain in investment activities when the market is falling. The systematic investment plans (SIPs) is a step to address this challenge. Nevertheless, the history of the stock continues to play a crucial role in it's forecasting.

There are two streams of methods to predict stock prices i.e. fundamental analysis and technical analysis. Fundamental analysis involves predicting the stocks by studying the markets, companies and trends whereas technical analysis involves using historical data of securities based on the assumption that trends repeat itself given few circumstances. Even though technical analysis is a challenging task, classical statistical models such as ARIMA [1], VARMA [2] etc. and machine learning models such as neural networks [3] and recurrent network [4] have shown promising results.

In this work, machine learning solutions to financial forecasting using historical data of various stocks is used. Recurrent neural networks with LSTM [5] or GRU [6] cells have proved superior to other machine learning methods [7] due to its ability to "remember" when dealing with time series data. They contain memory cells which help en-capture the input sequence and deliver high forecasting capacity. Using a standalone recurrent neural network has the disadvantage that it only en-captures history of one stock and does not consider the effects of other stocks on it. In that sense, it is a single-dimensional approach and it puts a limit on the capability of the network to accurately predict the movement of stocks. However, considering the effect of all stocks on the given stock may introduce higher computational complexity and a large amount of noise. Hence, first we need to extract highly correlated stocks and only consider their effects in stock prediction.

In order to consider the effects of stocks on each other and then predict the future value, we use multitask learning [8]. Multitask learning is a method of machine learning which enables us to carry out multiple tasks in parallel with information being shared among them. Multitask learning helps unlock further generalization by taking advantage of the domain specific knowledge encompassed among related tasks, in our case prediction of individual stocks. Multi task learning introduces a form of inductive bias which helps in the prioritizing of some parameters over the other. It not only helps learning networks share important information but also help reduce over fitting. Our work focuses on building a multitask learning model to simultaneously predict the closing prices of multiple correlated stocks and compare its performance to a baseline network under similar settings. A subset of stocks selected based on dynamic correlation are used to train and test both the models.

2 Literature Survey

2.1 Current Stock Prediction Technologies

The two main types of methods used for stock price prediction are statistical models and neural network models. The statistical models include models like the Time Series linear model, Holt winters model, ARIMA model etc. One issue with these models is the performance drop once the amount of data increases. This happens as statistical models do not capture the non-linearity of the data whereas stock data is non-linear. On the other hand, neural networks like ANN and RNNs tend to perform better than the statistical model. They consider only the history of the stock in its prediction and not the other external factors which affect the values of the stock price. As we all know that the stock price of a company is not only affected by its history but also is affected by multiple external factors. The models which consider only the history of the stock price are missing out on key additional data which can help improve the accuracy of the prediction. There have been significant research to capture the effect of such external events on stock. Mittal et al. [9] did stock prediction with the help of twitter data. Chen et al. [10] have proposed a multi-task framework of fine-grained events from financial news for stock prediction. Zhang et al. [11] built their model by fusing heterogeneous information. The approach adopted in this project relies on gathering these external cues from related stocks instead of miscellaneous sources. The multi-task learning architecture can capture these external factors by considering a basket of related stocks and can possibly make a more accurate prediction.

2.2 Multi-task Learning Techniques

Luong et al. [12] looked at positive effect of multi-task learning in a text sequence to sequence model for the three specific scenarios of one to many, many to many and many to one scenarios. Liu et al. [13] showed that in a text classification scenario, multi-class learning can improve performance. Sebastian Ruder et al. explains that the decision making of a multi-task learning architecture is very similar to a human mind as it considers all the different inputs and after considering all these inputs the model makes its final decision. The two most common techniques used for multi-task learning are hard parameter sharing and soft parameter sharing. Hard parameter sharing is implemented by the hidden layers of the model being shared for all the tasks while keeping multiple task-specific output layers. The drawback of this technique is that the architecture needs to be hard coded and thus reduces the flexibility of the model. In soft parameter sharing, each task has its model and parameters between the tasks are regularized in such a manner that the distance between them is minimized. Meyerson et al. [14] have shown that instead of parallel ordering, a permuted ordering of layers between parallel stack can enable better learning. The current work done in soft parameter sharing is mostly implemented using neural networks and not LSTMs for stock prediction [15]. However, multi-task learning does not help improve performance in all scenarios [16]. The multi-task learning model helps in improving accuracy only when the tasks are related to each other.

2.3 *Multi View Learning*

Multi-view learning is a machine learning model that considers the effect of other external factors for the prediction and can be used to improve the normal machine learning models. Nguyen et al. [17] have done some research about using external factors to improve the performance of machine learning models. Their prediction task was for time series data and they used a novel approach to decompose the data which can then be passed to the machine learning models. Instead of adding additional feature sets for data they propose a self-boosted approach in which they divide the time series data decomposed into multiple time series. They considered a combination of multi-view learning and multi-task learning in the prediction task. This technique despite consider external factors is quite different from that of multi-task learning. This method considers the external factors as additional features to the parameter we are predicting While multi-task learning shares the domain knowledge across these different features which help in predicting multiple features simultaneously.

2.4 *Stock Correlation*

Kim et al. [18] constructed a pattern matching trading system (PMTS) used to efficiently extract the optimized pattern of the proposed representative pattern in time series data to find the optimal training exit point. They proposed an approach that could match the time-series pattern of the index futures data with the representative pattern using the naive dynamic time warping (DTW) algorithm. Using the dynamic time warping algorithm it can interpret the liquidity, volatility and trend changes of the market to give decisions regarding the trading system. Dynamic time warping can be used to get the most correlated stocks.

3 Overall Approach

The work described in this paper is split into four high level modules as shown in Fig. 1.

- **Data set extraction** considers various histories of stocks across industries spread across the globe. The stock data for over 40 stocks is considered for a time period of ten years on with closing prices collected on a day to day basis. The future closing stock price value is predicted using the closing price value history of the stocks only.
- **Stock analysis and selector** performs dynamic time warping among the histories of the graphs and selects two most correlated stocks among the given stocks that can benefit from each other's histories. Dynamic time warping between the values ensure that each value on a given time step of one series is optimally matched with a proximal value of the other series. This enables us to match trends rather than a straight forward point to point match.
- **Stock prediction networks** are of two types. The data set of the two most correlated stocks are used to train and test the above networks
  - **Baseline networks** are individual stock prediction recurrent neural network with GRU cells.
  - **Multitask learning network** Multi task learning model for joint learning with underlying recurrent neural networks with GRU cells. We use the soft parameter sharing method of multitask learning. The multitask learning model has the same underlying recurrent neural network as the baseline model for both the stocks

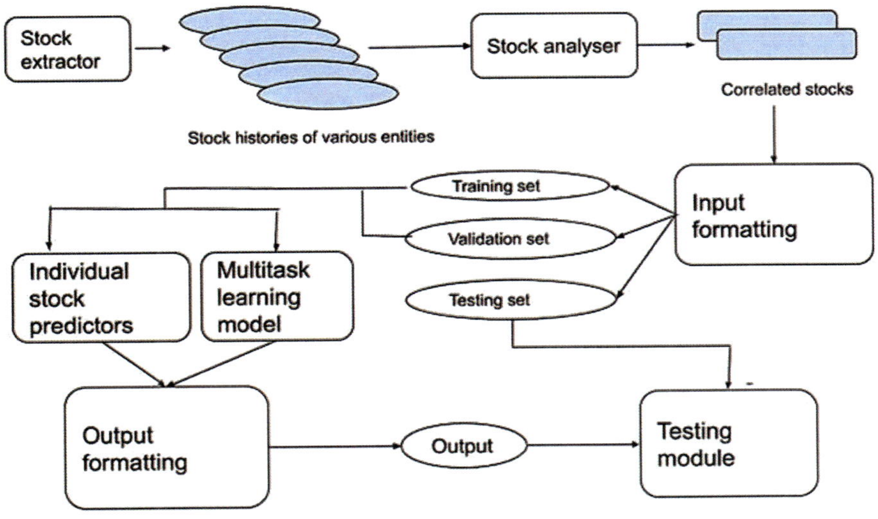

Fig. 1 High level architecture depicting the dependencies among the modules

but the parallel layers among the networks are constrained using a combined loss function that enables the overall network to share domain specific knowledge.

- **Testing and performance analysis** compares the performance of individual stock prediction GRU networks against the performance of our multitask learning model for both the stocks.

4 Implementation

4.1 Data Set

The data set contains stock histories of five companies i.e. Apple, Google, General motors, Toyota and Tesla. The stock data is collected by querying Yahoo finance API. The stock history of each company consists of the date and closing price on that date over a period of ten years collected on a day to day basis. This stock data needs to be transformed into sequences of equal sequence length. Sequence length is taken to be 20 where n to $n + 19$th value is considered as input sequence and the 20th value is considered as the target value. Hence, the input values are transformed in the shape (number of samples $\times$ number of time-steps $\times$ input features) which in our case is $n \times 19 \times 1$. The values also need to be normalized as to be able to train our stock prediction models. The formatted data is split into training, testing and validation set in the ratio 7:2:1.

4.2 Stock Analysis and Selection

Using the stock price data obtained from the data extraction module now we perform data analysis on this data. We do this to extract the stocks which are most related to each other and which can be used for our multi-task learning model. To perform the analysis we decided to use dynamic time warping which is used to calculate the similarity between any 2 given time series. The fastdtw library available in python is used to implement the dynamic time warping. Function Call: distance, path = fastdtw(array1, array2)

- distance: the distance between the two companies data lists
- path: the indices matched between the two lists
- Array1: Stock price data for company one
- Array2: Stock price data for company two.

Since the bucket of stock chosen had five stocks, we passed these five different company's data to the analysis and selector model. The selector model then compared each company's data with each other using the fastdtw function. The model then returns the two most related stocks from the five i.e. the companies whose data has the least distance between them.

4.3 *Stock Prediction Networks*

For the machine learning model, we decided to use recurrent neural networks with GRU cells as they are apt for dealing with time series data.

4.3.1 Baseline Models

We constructed two identical baseline networks as shown in Fig. 2, one for each stock derived from the stock analyzer and selector. Each network had one GRU layer followed by a dense layer to provide the output and have the following hyper parameters.

- Loss function: Mean square error

$$\text{MSE} = \frac{1}{n} * \sum_{n=1}^{n} (\text{expected} - \text{predicted})^2$$

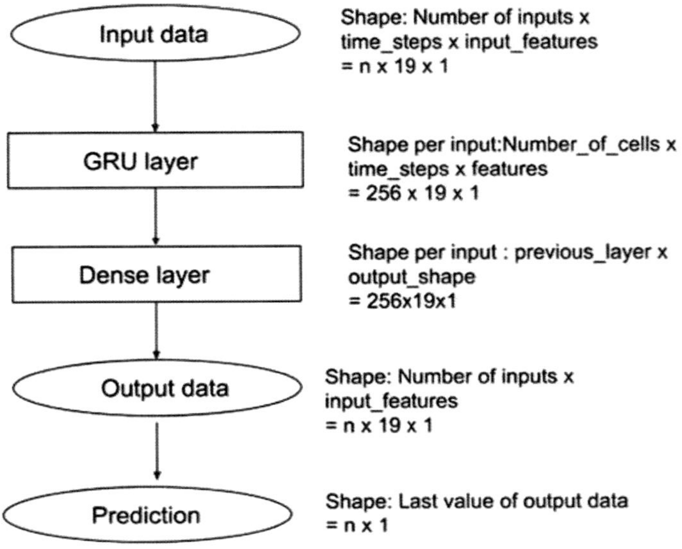

Fig. 2 Architecture of the individual baseline recurrent neural networks with GRU cells

- Optimization function: Ada grad optimizer
- Learning rate: 0.01
- Batch size: 185
- Training epochs: 1000.

4.3.2 Multitask Learning Network

The multitask learning network attempts to predict the future values of both the stocks simultaneously whilst sharing domain specific knowledge among the stocks. Each of the stocks have their own individual network same as the baseline model but the layers are constrained using the loss function where the networks are optimized to find the optima of weights with respect to both the stocks. The setting for multitask learning model is identical as the baseline model except the modified loss function (Fig. 3).

- Modified loss: loss of network A + loss of network B
- Optimization: AdagradOptimizer().optimize(Modified loss).

Hence, the two networks are simultaneously trained keeping in mind each other's performance, thereby constraining the layers and reducing the distance between the two networks.

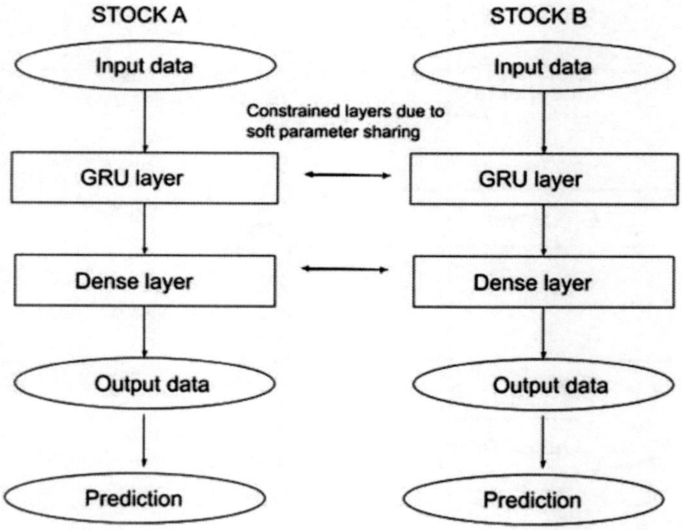

Fig. 3 Multitask learning model with soft parameter sharing

4.4 Evaluation

Testing data set is fed to the trained network to derive the output. The output from the network is denormalized to get the actual predicted value. Since the predicted and expected outputs are continuous values, conventional metrics such as f1_score, precision, recall cannot be applied. The output and expected output are compared in several ways to derive various metrics. The metrics on which the models are compared on:

- Mean squared error
- Root means squared error
- Mean average error
- Relative accuracy percentage.

The relative accuracy percentage is given by:

$$\text{RelativeAccuracy} = 1 - \frac{1}{n} * \sum_{n=1}^{n} \frac{(\text{expected} - \text{predicted})}{\text{expected}}$$

Here, for each sample in the testing data set, the amount of match between the expected and predicted value is taken. Each of these values lie between 0 and 1 with one being a perfect match. The mean over all samples is taken to form the relative accuracy metric.

5 Results

5.1 Stock Correlation Visualization

The prediction begins with finalising a bucket of stocks that are correlated with each other. This requires performing analysis on the stocks extracted from stock extractor. DTW distance is used as a parameter to check for the relatedness between stocks. Additionally, using the normalised prices of these stocks, a visualisation of their trends is created using a line graph.

5.1.1 DTW Distance Matrix of Top 5 Correlated Stocks

As we can observe from Table 1, the distance between Tesla and Toyota is 39.8840. This distance is much more when compared to other stocks. The same can be seen visually in Fig. 4 where the trend between the 2 stocks does not correlate at all.

On the other hand, when General Motors and Toyota are considered, the distance between them is the least out of the 5 companies, i.e. 22.1651. The same can be

Table 1 Distance matrix between top 5 stocks

	TSLA	Apple	GM	Toyota	Google
TSLA	0	35.05	43.35	39.88	54.23
Apple	35.05	0	43.33	38.27	55.94
GM	43.35	43.33	0	22.16	33.32
Toyota	39.88	38.27	22.16	0	31.15
Google	54.23	55.94	33.32	31.15	0

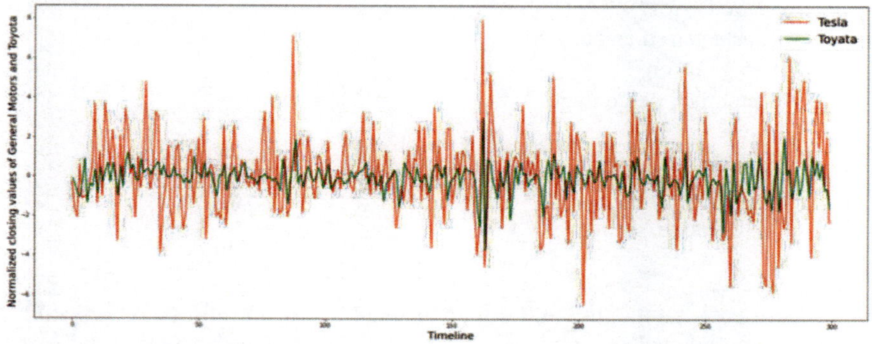

Fig. 4 Plot of the normalized history of TESLA versus Toyota

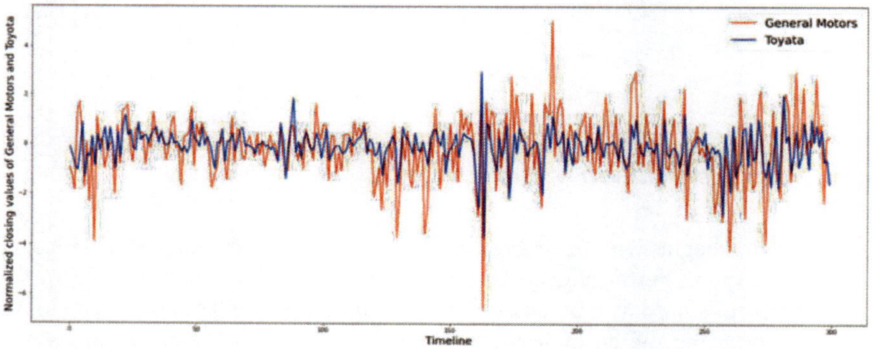

Fig. 5 Plot of the normalized history of GM versus Toyota

observed visually in Fig. 5. The change between normalised closing prices of both the companies follow identical trends.

5.2 *Comparison Between Baseline and Multitask Learning Model*

General Motors and Toyota were the two most correlated stocks, finalised after the stock selector and analyzer. The closing price of stocks of these 2 companies were formatted and split into training and testing sets. The baseline and the multitask learning model was then trained on data of these 2 companies.

5.2.1 Performance Comparison

For accurate comparison, we fed both the networks with same amount of training data and trained them for same number of epochs. The underlying architecture i.e. recurrent neural networks have exactly the same architecture in both the models. It should be noted that predicting exact value of stock prices is a really difficult task. This is because there are various other factors influencing the stock which have not been considered. However, we can see in Table 2 that the gap between the expected output and predicted output during training, is lesser in case of multitask learning as compared to the baseline models.

Some important metrics like mean squared error, root mean squared error and mean absolute error have been calculated to compare the prediction by baseline and multi task learning method. Results are shown in Table 3 and it can be observed that there is major improvement in our model compared to baseline model.

Table 2 Comparison between the mean square errors during training of networks

Model	MSE
Baseline: Toyota	0.00136
Baseline: GM	0.00116
MultiTask learning model: Toyota	0.00100
MultiTask learning model: GM	

Table 3 Comparison between the various performance metrics for testing data

Model	Stock	MSE	RMSE	MAE	Accuracy
Baseline	GM	8.46	2.90	2.40	91.89
Multi-task	GM	2.08	1.44	1.12	96.15
Baseline	Toyota	244.30	15.63	13.27	88.44
Multi-task	Toyota	46.17	6.79	6.12	94.78

5.2.2 Performance Visualisation

5.2.3 Stock Prediction for Toyota

Figure 6 shows the performance of baseline model on Toyota's closing prices. We can see that though there is a match in the trend for some time periods, there are certain time periods where the gap is large. This happens due to the misleading individual history of the stocks. Since the network only depends on one dimension of a single stock i.e. the history of the stocks, the past may sometimes mislead the target value and it does not have any domain specific knowledge. Hence only some identical or repeating trends can be predicted.

On the other hand, Fig. 7 shows the performance of multitask model on Toyota's closing prices. On comparison with baseline model's performance graph, it is seen that there is a lower gap between the expected and predicted values. We can also observe that the trend of the predicted price values closely follows the trend of expected values with a slight gap.

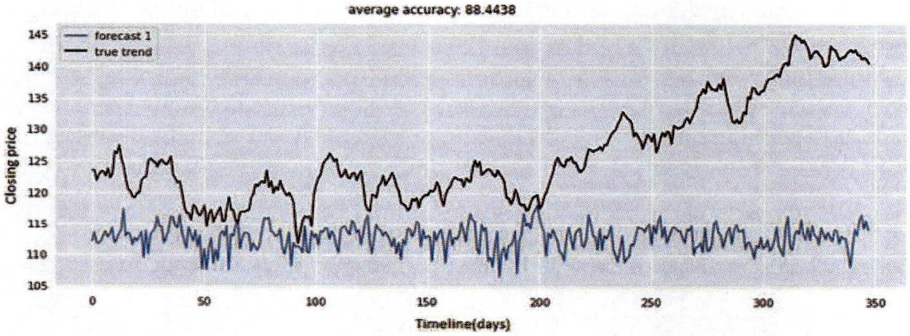

Fig. 6 Graph of predicted price versus expected price by baseline model for Toyota

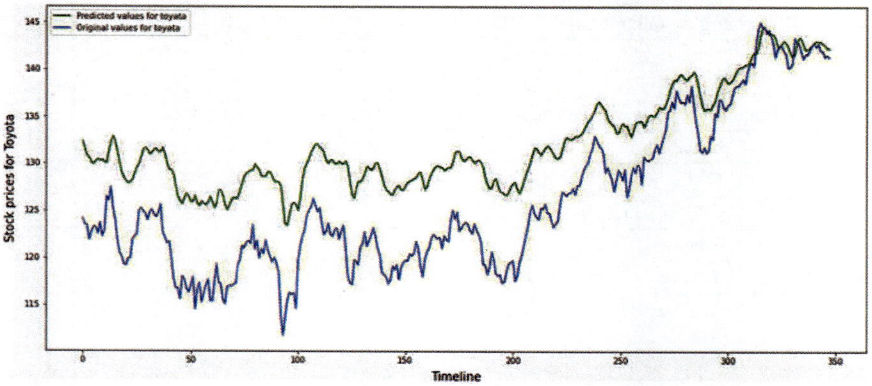

Fig. 7 Graph of predicted price versus expected price by multitask model for Toyota

5.2.4 Stock Prediction for General Motors

Figure 8 shows the performance of baseline model on General Motor's closing prices. We can see that though there is a match in trends for some time periods, there are certain time periods where the gap is either large or very low. The baseline model is not able to learn the long range trends and is unable to pull up the output with change in input sequence. This is because of the rapid unpredicted behaviour history of the stock.

On the other hand, Fig. 9 shows the performance of multitask model on General Motor's closing prices. It can be observed that the multitask learning model is able to catch the trends of the stock and reduce the gap between the trend of expected price value and predicted price value.

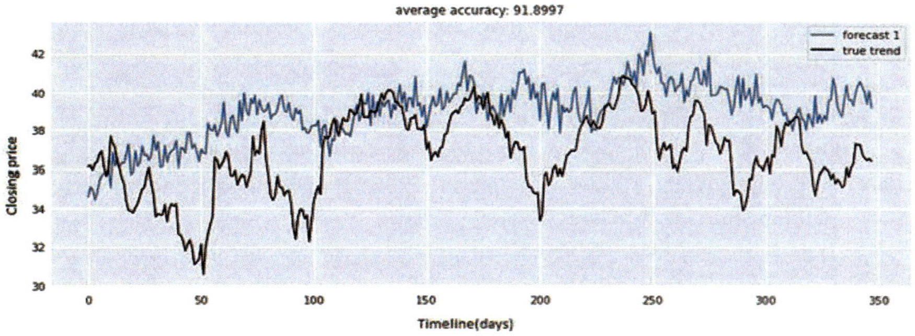

Fig. 8 Graph of predicted price versus expected price by baseline model for General Motors

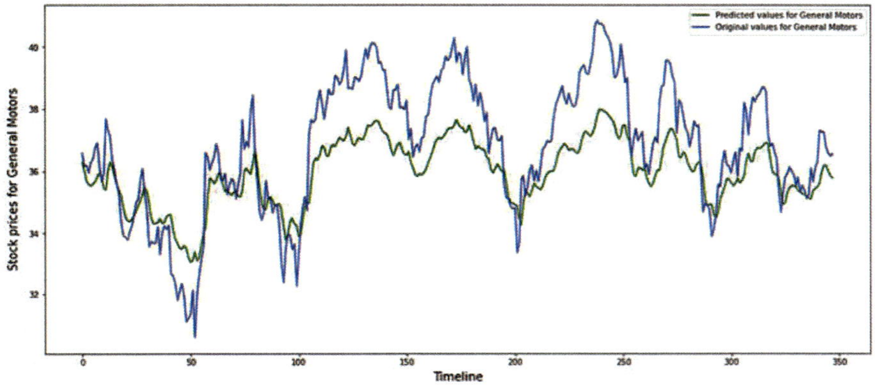

Fig. 9 Graph of predicted price versus expected price by multitask model for General Motors

6 Conclusion

Stock prices are very volatile in nature and are affected by various events. Considering only one dimension of the stock i.e. the history of the stock puts a cap to the performance any machine learning model can provide. Stocks in general are affected by competitor companies and general market trends. Inclusion of this domain specific data can help improve the performance of stock prediction models. The multi task learning approach proposed in this work, takes into consideration multiple correlated stocks for simultaneous prediction of their closing prices and has shown a significant improvement in performance compared to the baseline method of considering only the history of a single stock. This approach works as it captures the effect of various external factors that influence the stock and is an effective alternative method to model these external factors. It does this by considering the basket of correlated stock rather than depending only on the history of the stock in question.

Given that the data set of stock histories collected on a daily basis is a comparatively small data set, the underlying machine learning model must not contain too many stacked layers as it may lead to over-fitting and be unable to regress values. Though inclusion of all stocks helps share domain specific knowledge, if the stocks are unrelated it may introduce unnecessary noise to mislead the target. Hence it is important to analyze the stocks beforehand and select only a few correlated stocks for this kind of multi task learning.

As a next step, the same multi-task learning architecture can be extended to predict multiple stocks instead of one single stock. Such a set up will facilitate in prediction of all stocks in a particular segment. Also, with a larger set of computing resources, the same experiment can be extended to a larger set of stock as compared to the current experimental set-up of five stocks.

References

1. A.A. Ariyo, A.O. Adewumi, C.K. Ayo, Stock price prediction using the ARIMA model, in *2014 UKSim-AMSS 16th International Conference on Computer Modelling and Simulation* (IEEE, 2014), pp. 106–112
2. E. Isufi, A. Loukas, N. Perraudin, G. Leus, Forecasting time series with VARMA recursions on graphs. IEEE Trans. Signal Process. **67**(18), 4870–4885 (2019)
3. M. Billah, S. Waheed, A. Hanifa, Stock market prediction using an improved training algorithm of neural network, in *2016 2nd International Conference on Electrical, Computer & Telecommunication Engineering (ICECTE)* (IEEE, 2016), pp. 1–4
4. D.M. Nelson, A.C. Pereira, R.A. de Oliveira, Stock market's price movement prediction with LSTM neural networks, in *2017 International Joint Conference on Neural Networks (IJCNN)* (IEEE, 2017), pp. 1419–1426
5. S. Hochreiter, J. Schmidhuber, Long short-term memory. Neural Comput. **9**(8), 1735–1780 (1997)
6. J. Chung, C. Gulcehre, K. Cho, Y. Bengio, Empirical evaluation of gated recurrent neural networks on sequence modeling. arXiv preprint arXiv:1412.3555 (2014)

7. E.W. Saad, D.V. Prokhorov, D.C. Wunsch, Comparative study of stock trend prediction using time delay, recurrent and probabilistic neural networks. IEEE Trans. Neural Netw. **9**(6), 1456–1470 (1998)
8. S. Ruder, An overview of multi-task learning in deep neural networks. arXiv preprint arXiv:1706.05098 (2017)
9. A. Mittal, A. Goel, Stock prediction using twitter sentiment analysis. Stanford University, CS229 (2011, http://cs229.stanford.edu/proj2011/GoelMittal-StockMarketPredictionUsingTwitterSentimentAnalysis.pdf) 15 (2012)
10. D. Chen, Y. Zou, K. Harimoto, R. Bao, X. Ren, X. Sun, Incorporating fine-grained events in stock movement prediction. arXiv preprint arXiv:1910.05078 (2019)
11. X. Zhang, Y. Zhang, S. Wang, Y. Yao, B. Fang, S.Y. Philip, Improving stock market prediction via heterogeneous information fusion. Knowl.-Based Syst. **143**, 236–247 (2018)
12. M.T. Luong, Q.V. Le, I. Sutskever, O. Vinyals, L. Kaiser, Multi-task sequence to sequence learning. arXiv preprint arXiv:1511.06114 (2015)
13. P. Liu, X. Qiu, X. Huang, Recurrent neural network for text classification with multi-task learning. arXiv preprint arXiv:1605.05101 (2016)
14. E. Meyerson, R. Miikkulainen, Beyond shared hierarchies: deep multitask learning through soft layer ordering. arXiv preprint arXiv:1711.00108 (2017)
15. J. Ghosn, Y. Bengio, Multi-task learning for stock selection, in *Advances in Neural Information Processing Systems* (1997), pp. 946–952
16. S. Ruder, J. Bingel, I. Augenstein, A. Søgaard, Latent multi-task architecture learning., in: Proceedings of the AAAI Conference on Artificial Intelligence, vol. 33 (2019), pp. 4822–4829
17. L.H. Nguyen, Z. Pan, O. Openiyi, H. Abu-gellban, M. Moghadasi, F. Jin, Self-boosted time-series forecasting with multi-task and multi-view learning. arXiv preprint arXiv:1909.08181 (2019)
18. S.H. Kim, H.S. Lee, H.J. Ko, S.H. Jeong, H.W. Byun, K.J. Oh, Pattern matching trading system based on the dynamic time warping algorithm. Sustainability **10**(12), 4641 (2018)

Embedding-Assisted Auto-Tuned Community Detection

Akshara Subramaniasivam, Kaushik Ravichandran, Aishwarya Poomuttam Sreedas, and Bhaskarjyoti Das

Abstract State-of-the-art techniques for community detection in graphs either use feature-based clustering algorithms or structure-driven spectral clustering approaches. The former takes into consideration only the semantic relations based on node features, and the latter uses structural information for an analysis in eigenspace. Hence, neither of these two approaches can provide a very accurate detection of communities. The existing node2vec algorithm translates structural information into an embedding space. We propose a novel approach that combines all of these and obtains encouraging results in large graphs. The proposed approach in this work brings in few innovations in the form of an auto-tuning framework to address the existing shortcomings of these methods. It stabilizes an important hyper-parameter of spectral clustering and brings in a step-heuristic along with a proposed BinKMeans (binary search-driven k-means) algorithm to better detect a distinct drop in eigenvalue while using spectral clustering. Additionally, the pipeline implements NodeFeat2Vec to bring in feature information in node2Vec. An in-depth evaluation with real-world graph data showed that this proposed approach obtained superior results when evaluated using appropriate metrics on communities in labelled real-world graphs.

Keywords Community detection (CD) · Node2Vec with features (NodeFeat2Vec) · Node embeddings · Clustering · Binary search driven K means (BinKMeans) · GraphSAGE (GS) · Node2Vec (N2V) · Graph auto encoders (GAE) · Auto-tuned spectral clustering

A. Subramaniasivam · K. Ravichandran · A. Poomuttam Sreedas (✉) · B. Das
Computer Science and Engineering Department, PES University, Bengaluru, India
e-mail: aishwarya.ps31@gmail.com

A. Subramaniasivam
e-mail: akshara.sivam@gmail.com

K. Ravichandran
e-mail: kaushikrchandran@gmail.com

B. Das
e-mail: bhaskarjyoti01@gmail.com

J. Tavares et al. (eds.), *Emerging Technologies in Data Mining and Information Security*, Lecture Notes in Networks and Systems 164,
https://doi.org/10.1007/978-981-15-9774-9_39

1 Introduction

The modern science of networks has brought significant advances to our understanding of complex systems with graph representations. A social graph is one that represents social relations between entities and is often referred to as "the global mapping of everybody and how they're related". Predicting this relation is commonly called link prediction problem. Once the links or edges are established, community structures are determined. A set of densely connected vertices is grouped into communities which are loosely coupled with the rest of the graph. Detecting communities in graphs may have applications such as detecting link farms in web graphs that enhance PageRank computation, improve the performance of World Wide Web services by assigning dedicated mirror servers, making recommendation systems more efficient, handle navigational queries, analyse co-authorship networks [8], discover functionally related objects [8, 19, 22], identifying protein structures in protein interaction networks [1, 14], infer missing attribute values [2, 5], and predict unobserved connections [4].

A huge percent of contributions in the field of graphs research caters to the exponential growth rate of dynamic graphs and how to exploit the information they contain effectively. Community detection in graphs [7] has had many innovations in the past, ranging from statistics or mathematical-based solutions [17, 23]. We propose one such innovation which addresses the shortcomings in community detection in attributed graphs. We evaluate our performance on well-known clustering metrics to show that our approach performs better than completely structural-based and completely feature-driven approaches for community detection.

Spectral clustering performs exceedingly well in detecting highly nonlinear, elongated and multi-dimensional clusters [3] and has gained considerable attention in the recent past due to its well-defined mathematical framework [20] and the fact that it does not make strong assumptions on the form of the clusters. Spectral clustering is accomplished by solving a relaxation of the normalized min-cut problem on the similarity matrix of a graph [18]. As opposed to k-means, where the resulting clusters form convex sets, spectral clustering can solve problems such as the intertwined spirals and be implemented efficiently even for large datasets, as long as we ensure that the similarity graph is sparse [20]. The vanilla spectral clustering has shown significant improvements when used along with other techniques such as weighted kernel k-means [6], Jacobi splitting [12], nonparametric Bayesian models [16] and also using ratios of eigenvectors [13].

In this paper, we propose novel approaches to address the problems in algorithms used for community detection in large graphs. Our key contributions are listed below:

- We propose a stabilized hyper-parameter for the commonly used spectral clustering algorithm which is less susceptible to variation in the input graph
- We propose a step heuristic that can be used in accordance with the same hyper-parameter in the spectral clustering algorithm to detect a jump in eigenvalues better

- We propose a new embedding technique, NodeFeat2Vec which incorporates the use of node features in the vanilla node2vec algorithm
- We propose a new clustering algorithm, BinKMeans (binary search-driven k-means) that can be used in any variation of spectral clustering
- We propose a novel pipeline that can be used for the task of community detection.

The rest of this paper is structured as follows. In the following section, we highlight important research work that has been carried out for this task. We then give a brief description of an approach that is the current state of the art for detecting communities in real-world graphs. We highlight major shortcomings of this approach in Sect. 2, and community detection in general and talk about how we address them in Sects. 3–6. We present the results we obtained in Sect. 7 followed by a through discussion of the same in the same section. We conclude this paper by highlighting directions for improvement in Sect. 8.

2 Shortcomings

The core idea of the spectral clustering algorithm is used throughout the paper. There are two parts to the algorithm, i.e. to obtain the affinity matrix from node embeddings provided and to obtain communities using spectral clustering from the adjacency or affinity matrix. Embeddings are vectors of fixed dimensions that are provided for each node. The embeddings are converted to an affinity matrix using the formula

$$A_{ij} = e^{\frac{-d^2(s_i, s_j)}{\sigma^2}} \tag{1}$$

A distance function such as euclidean distance to find distance between embedding vectors s_i and s_j and σ is a hyper-parameter provided. This matrix is constructed by setting $A_{ii} = 0$ and A_{ij} obtained using the above formula. This affinity matrix can then be used in place of the adjacency matrix for spectral clustering. Spectral clustering involves finding the eigendecomposition of the standard normalized Laplacian of the input matrix using,

$$L = D^{\frac{-1}{2}} A D^{\frac{-1}{2}} \tag{2}$$

where D is the diagonal matrix of input graph matrix A. The expected number of communities, k is estimated by finding the largest difference between two consecutive eigenvalues in their sorted list. The k eigenvectors corresponding to the k largest eigenvalues are then given as input to a clustering algorithm to obtain the communities in the graph.

Most algorithms perform clustering on the adjacency matrix of the graph which contains only edge information. The implementation by Hu et al. in [11] uses embeddings generated by node2vec. However, node2vec does not account for node features either. Thus, communities that are generated by all these algorithms are not using the full information available for the graph.

The hyper-parameter in Eq. 1 causes significant variation in the affinity matrix obtained by change in its value. The algorithm provided by Zelnik-Manor et al. in [24] obtains the affinity matrix by localizing the hyper-parameter σ. Here, σ for a particular node of the graph is taken as the distance of the ith embedding vector from the embedding vector of the node under consideration. The affinity matrix still does vary by changing the value of i for some graphs more than others.

The nature of the eigenvalues obtained from eigendecomposition theoretically has values between 0 and 1. A value of 1 indicates a completely isolated component in the graph. Thus, the eigenvalues for a graph with k communities will have k values close to 1 after which the values will drastically drop. Using this theory, current methods attempt to find the exact point at which the eigenvalues encounter maximum drop by comparing consecutive eigenvalues. However, Zelnik-Manor et al. in [24] have shown that the nature of eigenvalues does not have a strict pattern like the theory suggests. The eigenvalues do not necessarily have a sharp drop over consecutive values. Thus, the techniques that depend on this sharp drop will work for a few graphs, but for majority of graphs it will return a value that is not indicative of the actual number of communities in the graph.

Node2vec is an algorithmic framework for representational learning on graphs. Using node2vec in spectral clustering was proposed by Aditya Grover et al. [9]. This improves the quality of the affinity matrix but does not address the problem of using attribute information for the same. Node features are often encoded as node embeddings to map high-dimensional sparse feature vectors into a lower-dimensional space while preserving information. However, the traditionally used node2vec algorithm cannot be used for this task since it only considers locality while embedding the nodes and must be modified to take into account node features.

The clustering technique most commonly used in the last step of vanilla spectral clustering is k-means clustering. However, this can be replaced by any other clustering algorithm. The disadvantage of k-means clustering is that it requires the number of clusters, k as input. Thus, if we do not already know the communities, we should either provide k from a highly accurate algorithm that detects the number of clusters or we should land on the right number of clusters by trial and error. Graphs are dynamic and ever growing, making it almost impossible to know the exact number of communities.

3 Spectral Clustering

We address shortcomings in existing algorithms used for spectral clustering [11, 24]. We have demonstrated two improvements for spectral clustering in this paper. Both of these attempt to make the algorithm more robust to changes in hyper-parameter values due to the varying nature of the input graph.

3.1 Stabilised Alpha

The shortcoming in the hyper-parameter alpha that is required in computation of affinity matrix from the node embeddings is discussed in Sect. 2. The latest approach of localising alpha by taking the distance of the ith makes the affinity matrix less prone to change in the hyper-parameter. However, intuitively, it is clear that if we take the average of the distance of the i closest neighbours rather than just the distance from the ith embedding, the model will be more robust to change. This is derived from the fact that on changing from i to $i + 1$, the change in distance from the ith neighbour will be more than the change in the average of the distance from the i closest neighbours.

3.2 Step Heuristic

This is an improvement to spectral clustering by mitigating the shortcoming observed which was covered in Sect. 2, where we show that the nature of eigenvalues in reality does not always follow the theory. Although there is no single, sudden drop in the eigenvalues, there exists a range of eigenvalues over which the drop will occur. The step heuristic attempts to locate this range rather than the futile search for the drop over consecutive eigenvalues. Thus, instead of finding the largest difference between consecutive eigenvalues, we aim to find a range over which there is maximum drop in eigenvalues. Throughout the development of the model, the range or step was set as $\min(n/5, 5)$ where n in the number of nodes in the graph. This formula produced good results during our experimental analysis. While this significantly reduces abnormalities caused by the nature of eigenvalues, there is still a dependency, but to a much lesser extent. We further propose a novelty covered in later sections that removes this dependency altogether.

4 Using Node Attributes in NodeFeat2Vec

Since node2vec [9] cannot capture feature information, it needs to be modified to incorporate this. The random walks in node2vec are parameterized on number of walks, length of each walk, and the unnormalized transition probability between nodes. This unnormalized probability is calculated using the edge weight between the two nodes it connects. By appropriately modifying this edge weight, we can ensure that the node attributes are incorporated while calculating the transition probabilities. A high similarity between two nodes should translate to a high edge weight as it would imply a high transition probability between the two nodes. We use similarity metrics to compute similarity between two nodes, which is used as the edge weight.

The similarity metrics that were experimented on are Pearson's coefficient, cosine similarity, and Euclidean distance.

Computing similarity between nodes cannot be done directly on the feature vector, as the input dimensions are too large to be reduced to a single number in one shot. First, dimensionality reduction techniques were used to reduce these sparse feature vectors into a more condensed vector space. The reasoning for trying various dimensionality reduction and normalization techniques is given below.

Standard Scaler is a normalization technique which standardizes the data by scaling to unit variance. This is done feature wise and helps standardize data with negative values. Principle component analysis (PCA) [21] is a dimensionality reduction technique that is aimed at increasing interpretability and reducing information loss. This is done by constructing a co-variance matrix of the data and later truncating the less important features. Truncated singular valued decomposition was tried as it does not centre the data before reducing its dimensions. It uses tf-idf vectorizers for this purpose and is also known as latent semantic analysis.

Results obtained with various combinations of similarity metrics and dimensionality reduction techniques are quoted and discussed in Table 7.

5 Binary Search-Driven *k*-Means

We propose BinKMeans, a novel clustering technique, to perform two operations at once. It does not apply *k*-means to provide clusters but also detect the number of communities in a way that does not depend on eigenvalues. This is essential in building a robust model since the nature of eigenvalues is unpredictable. At a high level, it can be thought of as having a protective layer around *k*-means. It runs *k*-means multiple times to get the highest *k* for which we obtain a substantial number of nodes per cluster. The motivation behind this can be seen in Table 12 where spectral clustering for Cora gives 99.3% of nodes in one cluster. For example, even obtaining a prediction for number of clusters as 7 is not representative of the results of the algorithm.

BinKMeans performs binary search over a range of k to obtain the highest number of communities such that there is a minimum configurable percentage of nodes in each community. The range can be automatically set as $[0, 2 * k]$ where k is the number of communities predicted by the step heuristic function. Thus, BinKMean can also be thought of as an extension to step heuristic that makes the results obtained significantly more accurate. Step heuristic removes dependency on nature of eigenvalues to an extent which can be further eliminated by providing it as input to BinKMeans. Even if the *k* provided by step heuristic is incorrect, BinKMeans will do additional computation to provide an accurate prediction for the number of communities.

6 Pipeline

Our novelty lies in the following pipeline which was constructed ensuring that the best information is preserved at every step and also avoiding over-fitting. Figure 1 shows the pipeline we designed for this task.

The input graph data is first pre-processed. We serialize the node IDs and convert the graph into an undirected one. In this process, we also represent the graph in the input format we require. The dimensions of the node features are reduced optionally depending on the embedding technique used and the size/sparsity of the input features. These node features along with the adjacency matrix are provided as input to the node embedding algorithms. The node embedding algorithms that were tried are NodeFeat2Vec, GraphSAGE [10] and Graph Auto Encoders [15].

Once these node embeddings are generated, we use them to construct the affinity matrix which is used for spectral clustering. We then perform eigendecomposition on this affinity matrix to obtain the number of clusters by observing the nature of the eigenvalues. Conventional clustering techniques such as agglomerative hierarchical clustering and k-means (the default clustering technique used by spectral clustering) are performed to obtain the communities. We also validate the performance of BinKMeans in this step. Novelties in constructing the affinity matrix, estimating the number of communities, and the BinKMeans algorithm have been covered earlier and are hence avoided here.

7 Evaluation and Results

In this section, we evaluate the performance of embedding-assisted auto-tuned spectral clustering against the vanilla spectral clustering algorithm. All experiments were performed on real-world networks with ground-truth communities. It should be

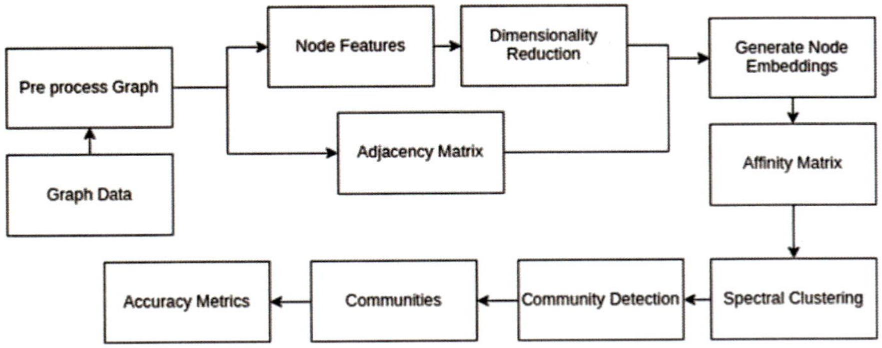

Fig. 1 Process pipeline

noted that results generated for the Karate dataset with embedding algorithm set as node2vec use the implementation by [9] which is without accounting for node features since the Karate club network is not an attributed graph. All experiments were carried out on Python 3.6 with a CPU8-core Intel(R) Core(TM) i9-9980HK CPU@2.4GHz processor, 16 GB DDR4 RAM. The metrics that were used to compare the algorithms which are Coverage, Performance, Calinski–Harabasz Index, Homogeneity, Completeness, and V Measure.

7.1 Datasets

The pipeline along with each of the novelties proposed in this paper was tested on three datasets: The karate club network, the Cora citation network and the Facebook social graph. Metadata for the datasets is given in Table 1. The number of communities for Karate and Cora is provided as part of the dataset, while Facebook provides egonets from which the communities are derived. Thus, unlike Cora, Facebook's ground-truth communities are based on structural data alone. Additionally, Facebook provides a fixed but different set of features for each egonet. Pre-processing was performed on this data to get the total set of features per node obtained by combining the features of all egonets since most embedding algorithms require a fixed number of features for all nodes.

7.2 Results

In this section, we first validate the communities obtained for correctness and quality. We then compare the performance of our individual contributions listed through Sects. 5–8. We finally show how the proposed approach outperforms the current state of the art in community detection techniques by providing a comparison on the datasets and metrics explained above.

Table 1 Metadata of datasets

	Nodes	Edges	Features per node	Communities
Karate	34	78	0	2
Cora	2708	5429	1433	7
Facebook	4039	88,234	1283	10

7.2.1 Community Validation

Number of nodes obtained per community is reported in Tables 2, 3 and 4 for all datasets mentioned in Sect. 7.1. It can be seen that embedding-assisted spectral clustering algorithms do not perform well on small, featureless datasets such as the Karate club network. However, for the Cora citation network, spectral clustering provides skewed results by classifying almost all the nodes into a single cluster. Embedding-assisted spectral clustering outperforms the baseline significantly with GraphSAGE performing the best among the three. When it comes to the Facebook web graph, our proposed approach performs as well as spectral clustering and also gives a good distribution of nodes in clusters. Overall, embedding-assisted spectral clustering gives better distributions with minimum skew.

7.2.2 Contributions Analysis

Given below is a comparison and discussion of performance of our proposed approach with the baseline when evaluated on the mentioned accuracy metrics. The value used for step was 5 for the larger graphs and 3 for Karate. The evaluation of the pipeline itself is covered in the section below with the metrics. Table 5 shows the impact of the step heuristic which either makes the detected number of communities (k) better or as in the case of Karate, remains the same as when compared with step. Cora shows the disadvantage of the existing as well as proposed approach since the detected value of k is because of a small but sharp drop after 2646 values from the first eigenvalue.

Table 2 Number of nodes per community for Karate

Karate	Spectral clustering	Node2Vec	GraphSAGE	GAE
0	19	31	33	33
1	15	3	1	1

Table 3 Number of nodes per community for Cora

Cora	Spectral clustering	NodeFeat2Vec	GraphSAGE	GAE
0	2691	2510	625	1846
1	5	50	366	155
2	2	16	436	153
3	2	89	289	100
4	4	2	338	207
5	2	8	357	128
6	2	33	297	119

Table 4 Number of nodes per community for Facebook

Facebook	Spectral clustering	NodeFeat2Vec	GraphSAGE	GAE
0	718	90	208	345
1	518	312	242	105
2	511	203	267	162
3	318	112	242	172
4	26	2216	1838	2926
5	198	95	230	200
6	53	157	247	42
7	294	272	252	83
8	183	270	268	1
9	1220	312	245	3

Table 5 Performance of step heuristic

k, spectral clustering	With step	Without step
Karate	4	4
Cora	86	2646
Facebook	11	52

This is not indicative of the number of clusters and is an anomaly in behaviour. The step heuristic we propose performs much better than the baseline by 2646 by giving 86 and can be considered significantly more reliable. With Facebook, we can see the improvement clearly by adding step. In addition to that, the step heuristic detects 11 which is an extremely good prediction of the number of clusters for the Facebook graph (10 clusters).

The existing localized alpha performs well but is unstable and causes too much variation in results with a small change. The numbers quoted are the difference in k while changing the local scale parameter from 15 to 20 for the bigger graphs and 7 to 12 for Karate. There was no difference in k when using stabilized alpha as opposed to the original alpha in which the difference in k increases with increase in size of the graph (Table 6).

Table 6 Difference in k with and without stability

Diff(k), NodeFeat2Vec	With alpha	With stabilized alpha
Karate	3	0
Cora	15	0
Facebook	50	0

Table 7 NodeFeat2Vec with dimensionality reduction on Cora

k (coverage, performance)	Pearson's coefficient	Euclidean	Cosine
Standard scaler (SS)	2500+	2500+	2500+
PCA (size = 270)	10 (84, 15)	17 (90, 8)	3 (57, 43)
SS + PCA (size = 270)	10 (93, 7)	17 (60, 40)	3 (99, 0.3)
TSVD (size = 270)	3 (99, 0.3)	17 (60, 40)	3 (90, 9.8)
SS + TSVD (size = 270)	3 (99, 0.3)	10 (55, 44)	3 (83, 16)
PCA (size = 677)	7 (91, 8.5)	10 (55, 44)	3 (99, 0.3)
SS + PCA (size = 677)	3 (99, 0.3)	2500+	10 (78, 20)
TSVD (size = 677)	3 (93, 8.4)	2500+	3 (99, 0.3)
SS + TSVD (size = 677)	3 (99, 0.3)	2500+	2500+

Table 7 shows the results we obtained on varying the dimensionality reduction technique, number of reduced dimensions and similarity metric used for the NodeFeat2Vec algorithm. Standard Scaler in general was expected to perform well when used with a dimensionality reduction technique, as it normalizes the data feature wise. This seemed to always overestimate the number of communities, as there were few features which dominated the entire feature vector for that node. Truncated SVD was tried as it is known to work well with sparse data. This didn't turn out to be beneficial as the performance depended on other factors such as similarity metric used and the reduced number of dimensions as well. We discarded this technique as it showed a huge variation in predicted number of clusters, with a change in input graphs. PCA as expected performed the best as it was the best technique for preserving the most important information. By experimentation, we prove that truncating the feature vector to maintain only the right number of features boosts its performance by a huge amount and gives an accurate exact count of clusters even with a variation in input graphs.

Tables 8, 9 and 10 show the performance of various clustering algorithms that were used to cluster the k largest eigenvectors obtained be decomposing the Laplacian matrix of the input graph. K means clustering is what is usually used and will be used after completion of identifying k using step. Here, we introduce two other clustering techniques, viz. agglomerative clustering and a newly proposed binary search-driven K means clustering. Both of these techniques have the property to fix the number of clusters in a way that it is not dependent on the nature of eigenvalues as well. Thus, the performance is evaluated based on the k identified by these clustering techniques. Agglomerative clustering will not put two nodes in the same cluster if the distance between their embeddings is above a certain threshold. The threshold used is the maximum distance between any two eigenvectors obtained for the graph. However, the results are not as accurate as BinKMeans which predicts k accurately for all combinations of embeddings and datasets apart from Cora with NodeFeat2Vec. This can be explained by the fact that a large percentage of the nodes fall in one class, as shown in Table 3.

Table 8 k for different clustering techniques with NodeFeat2Vec

k, NodeFeat2Vec	Step	Agglomerative	BinKMeans
Karate	4	2	2
Cora	7	1	2
Facebook	4032	1	11

Table 9 k for different clustering techniques with GraphSAGE

k, GraphSAGE	Step	Agglomerative	BinKMeans
Karate	2	1	1
Cora	2	23	10
Facebook	2	5	7

Table 10 k for different clustering techniques with GAE

k, GAE	Step	Agglomerative	BinKMeans
Karate	1	1	2
Cora	2	1	8
Facebook	27	1	8

7.2.3 Performance Metrics

Finally, we assess the performance of the pipeline for each dataset and with all three embeddings against spectral clustering. It is critical to note that for the Facebook graph, the ground-truth communities is obtained from egonets which are networks with nodes having similar structural information and is only an approximate estimate of the actual communities. This explains why spectral clustering performs as well as feature rich embeddings for the Facebook graph. However, despite the scale tipped against embeddings, GraphSAGE produces values which are comparable to the ground truth, with false classifications from lack of attributes in the ground-truth information. With Cora, embedding-assisted spectral clustering with our pipeline outperforms other techniques. Majority of the better results are using GraphSAGE, followed by GAE and then NodeFeat2Vec as the embedding technique. The value of 0.99 as coverage for spectral clustering is not indicative of its performance as the high value was due to the fact that almost all the nodes were in clustered into one community. Karate is too small (and unattributed) to efficiently describe the performance with good results split across the baseline and our approach. All results are documented in Tables 11, 12 and 13.

Table 11 *k* for different clustering techniques with NodeFeat2Vec

Facebook	NodeFeat2Vec	GraphSAGE	GAE	Spectral clustering
Coverage	0.346	0.226	0.548	0.464
Performance	0.669	0.754	0.462	0.827
Calinski–Harabasz Index	1.46	3.59	0.887	7.73
Homogeneity	0.568	0.63	0	0.643
Completeness	0.684	0.659	0.675	0.646
V measure	0.621	0.644	0.479	0.645

Table 12 *k* for different clustering techniques with GraphSAGE

Cora	NodeFeat2Vec	GraphSAGE	GAE	Spectral clustering
Coverage	0.98	0.15	0.505	0.99
Performance	0.022	0.845	0.517	0.014
Calinski–Harabasz Index	0.79	0.99	1	0.834
Homogeneity	0.003	0.03	0.011	0.004
Completeness	0.1	0.029	0.017	0.15
V measure	0.007	0.03	0.013	0.008

Table 13 *k* for different clustering techniques with GAE

Karate	NodeFeat2Vec	GraphSAGE	GAE	Spectral clustering
Coverage	0.883	0.961	0.922	0.423
Performance	0.27	0.178	0.174	0.704

8 Conclusion and Future Work

In this paper, we address the shortcomings in state-of-the-art techniques used to detect communities in large, attributed real-world graphs. We propose a novel approach that can be used for detecting communities that are accurate in representation of the real communities of the graph. The pipeline we propose was selectively refined to optimize the algorithm for correctness and efficiency. The results we quoted superseded the baseline when compared on the metrics mentioned. This can be attributed to the innovations in auto-tuned detection of number of clusters, stabilized parameters for spectral clustering, NodeFeat2Vec, a technique for embedding node features, BinKMeans, a clustering algorithm customized for community detection and a pipeline that uses node attributes for the task of community detection.

Experimentation with a large quantity of possibilities led us to discard approaches that did not work well. All results are still quoted in this paper so that they can be inferred from and reasoned or improved further. This would require experimentation on differently varying input graphs. Synthetic graphs with synthetically generated features could be used to understand the working of this model better. The auto-tuning heuristics proposed are just a first look into possible ideas. This could be analysed further to introduce a single hyper-parameter which is stable and accurate in determining the number of communities.

References

1. Y.Y. Ahn, J.P. Bagrow, S. Lehmann, Link communities reveal multiscale complexity in networks. Nature **466**(7307), 761–764 (2010)
2. R. Balasubramanyan, W.W. Cohen, Block-lda: jointly modeling entity-annotated text and entity-entity links, in *Proceedings of the 2011 SIAM International Conference on Data Mining* (SIAM, 2011), pp. 450–461
3. H. Chang, D.Y. Yeung, Robust path-based spectral clustering. Pattern Recogn. **41**(1), 191–203 (2008)
4. J. Chang, D. Blei, Relational topic models for document networks, in *Artificial Intelligence and Statistics* (2009), pp. 81–88
5. M. Coscia, G. Rossetti, F. Giannotti, D. Pedreschi, Demon: a local-first discovery method for overlapping communities, in *Proceedings of the 18th ACM SIGKDD International Conference on Knowledge Discovery and Data Mining* (2012), pp. 615–623
6. I.S. Dhillon, Y. Guan, B. Kulis, Kernel k-means: spectral clustering and normalized cuts, in *Proceedings of the tenth ACM SIGKDD International Conference on Knowledge Discovery and Data Mining* (2004), pp. 551–556
7. S. Fortunato, Community detection in graphs. Phys. Rep. **486**(3–5), 75–174 (2010)
8. M. Girvan, M.E. Newman, Community structure in social and biological networks. Proc. Natl. Acad. Sci. **99**(12), 7821–7826 (2002)
9. A. Grover, J. Leskovec, node2vec: scalable feature learning for networks, in *Proceedings of the 22nd ACM SIGKDD International Conference on Knowledge Discovery and Data Mining* (2016), pp. 855–864
10. W. Hamilton, Z. Ying, J. Leskovec, Inductive representation learning on large graphs, in *Advances in Neural Information Processing Systems* (2017), pp. 1024–1034
11. F. Hu, J. Liu, L. Li, J. Liang, Community detection in complex networks using node2vec with spectral clustering. Phys. A: Stat. Mech. Appl. 123633 (2019)
12. F. Hu, M. Wang, Y. Wang, Z. Hong, Y. Zhu, An algorithm J-SC of detecting communities in complex networks. Phys. Lett. A **381**(42), 3604–3612 (2017)
13. J. Jin et al., Fast community detection by score. Ann. Stat. **43**(1), 57–89 (2015)
14. P. Kapadia, S. Khare, P. Priyadarshini, B. Das, Predicting protein–protein interaction in multi-layer blood cell PPI networks, in *International Conference on Advanced Informatics for Computing Research* (Springer, 2019), pp. 240–251
15. T.N. Kipf, M. Welling, Variational graph auto-encoders. arXiv preprint arXiv:1611.07308 (2016)
16. D. Niu, J. Dy, Z. Ghahramani, A nonparametric bayesian model for multiple clustering with overlapping feature views, in *Artificial Intelligence and Statistics* (2012), pp. 814–822
17. J. Reichardt, S. Bornholdt, Statistical mechanics of community detection. Phys. Rev. E **74**(1), 016110 (2006)
18. J. Shi, J. Malik, Normalized cuts and image segmentation. IEEE Trans. Pattern Anal. Mach. Intell. **22**(8), 888–905 (2000)
19. A.P. Streich, M. Frank, D. Basin, J.M. Buhmann, Multi-assignment clustering for boolean data, in *Proceedings of the 26th Annual International Conference on Machine Learning* (2009), pp. 969–976

20. U. Von Luxburg, A tutorial on spectral clustering. Stat. Comput. **17**(4), 395–416 (2007)
21. S. Wold, K. Esbensen, P. Geladi, Principal component analysis. Chemom. Intell. Lab. Syst. **2**(1–3), 37–52 (1987)
22. J. Yang, J. Leskovec, Overlapping community detection at scale: a nonnegative matrix factorization approach, in *Proceedings of the Sixth ACM International Conference on Web Search and Data Mining* (2013), pp. 587–596
23. J. Yang, J. McAuley, J. Leskovec, Community detection in networks with node attributes, in *2013 IEEE 13th International Conference on Data Mining* (IEEE, 2013), pp. 1151–1156
24. L. Zelnik-Manor, P. Perona, Self-tuning spectral clustering, in *Advances in Neural Information Processing Systems* (2005), pp. 1601–1608

Optimization of Coverage Hole Identification in 5G SON Using Data Mining

Biswadeep Deb Nandy and Kartick Chandra Mondal

Abstract Recently, Self Organizing Networks (SONs), have been immensely researched. SON solutions must become the brains of the 5G Network. 5G has been envisaged to deploy in above 6 GHz mostly in 28 GHz or above. Beam-forming will be used to mitigate coverage loss due to high propagation loss. Line of sight communication is a must for above 28 GHz operations thereby presenting severe challenges in maintaining UE connection. Optimizing coverage and capacity are the major requirements of operators. The main hindrance in achieving optimized coverage and capacity is the complexity of adjusting all the configuration parameters which are affecting them. Most of the existing solutions are based on 4G and currently, very few solutions have been proposed for 5G considering line of sight operations above 6 GHz. Here, firstly we identify the challenges which hamper the current SON paradigm from meeting the requirements of 5G in terms of providing coverage. We then propose a data mining technique based approach to optimize the working in terms of performance and capacity for coverage hole identification in 5G SON network.

Keywords 5G · Self-organizing networks · Coverage hole · Beam-forming · Data mining

1 Introduction

As we rapidly move towards the next generation of telecom networks [5G] with our ever-increasing requirement for speed and reliability, 5G promises a lot more beyond such improvements. It is expected to deliver us with three main services: enhanced

B. D. Nandy
Samsung R&D Institute, Bengaluru, India
e-mail: biswadeep1997@gmail.com

K. C. Mondal (✉)
Department of Information Technology, Jadavpur University, Kolkata, India
e-mail: kartickjgec@gmail.com

J. Tavares et al. (eds.), *Emerging Technologies in Data Mining and Information Security*, Lecture Notes in Networks and Systems 164,
https://doi.org/10.1007/978-981-15-9774-9_40

mobile broadband (eMBB), massive machine-type communications (mMTCs), and ultra-reliable low-latency communications (URLLCs). 5G is expected to bring into existence a massive IoT (Internet of Things) ecosystem with the right trade-offs between cost, speed, and latency. Operators worldwide and the GSM Association (GSMA) expect 5G connections to reach 1.8 billion by 2025 with developed Asia and the US leading the way. It is around 23% of the total mobile connections. It also forecasts overall operator revenues to grow at a Compound Annual Growth Rate (CAGR) of 2.5%, to reach USD 1.3 trillion.

Minimization of life-cycle cost at the time of deployment done by eliminating manual configuration of network elements for running SON mobile network [1]. Present-day SON results usually assume that some temporal information is fully or partly available for SON based compensation related problem. For example, the position of coverage holes, handover ping-pong zones, or bottleneck spots are assumed to be known by the SON engine. In 4G, this is typically based on PHY layer measurements conveyed by base station and User Equipment (UE) under that base station [2]. As we are moving towards 5G, the concept of Massive MIMO (Multiple Input Multiple Output) to increase capacity and beam-forming to improve coverage is getting much more attention.

In this paper, we propose a solution for the joint optimization of coverage hole detection and its accurate region identification thereby trying to solve a very common problem with a new technique, with the help of data mining. We provide this result to SON for better network throughput in terms of fewer coverage holes and network distortions in any region. Following on, identifying the challenges impede the current SON paradigm from satisfying the 5G requirements.

In Sect. 2, we discuss the problem statement and its current understanding in depth. Followed by in Sect. 3, we showcase the implementation of our solution based on the data mining approach in a step-wise fashion for a better understanding of the optimization technique. Finally, concluding remarks are present in Sect. 4.

2 Problem Definition

Coverage holes are areas having terrible wireless connectivity from macro base stations [3]. So we will look into an optimization technique for the identification of such coverage holes. However, cell densification, coverage resolution, and indoor-outdoor path-loss variation can increase the complexity of system optimization. Capacity of the network and coverage optimization are the classical task of SON operation.

To minimize human intervention, automation of the above tasks through base station is needed. Intelligent mining algorithms optimizes the parameters automatically to solve this problem. We consider the case where a 5G network is deployed and thick building walls creates is a coverage hole or a dense population and cars of downtown area creates more data demand. The proposed framework reliably detects and compensate detected coverage holes in an autonomous fashion.

Say you are in a bunch of buildings and you are trying to make a phone call. Your signal is reverberating off of the buildings and crisscrossing with other signals of users in the area. These signals are received by massive MIMO base station along with timing and direction of arrival [4]. Before going forward with our implemented approach, a brief understanding of localization and distance calculation of detected UE's is required. As we will be taking into account such methods for accurate calculation of UE distance from the base station or the gNB (Next Generation Node B). Some of these methods are put forward below.

A few well-known techniques are available which are used for estimating the location of UE. in a wireless system, UE location tracing involves the location information from signals between UE, gNB and a collection of reference UEs. Positioning techniques determines the location or distance of a UE from the base station [4] by using angle of arrival (AOA), received signal strength (RSS), or time of arrival (TOA) information.

Means of antenna arrays used to measure angles of the target UE seen by reference UE which is used in AOA based positioning techniques. AOA based technique for localization has been widely used in wireless sensor networks. By processing the received signals, relevant signal metrics can be extracted for localization. It is a metric characterizing the arriving direction of the signal at the agent, and it can be obtained using an array of antennas and spatial filtering, then the angle measurements translate to location information by triangulation [5]. In a 2-D space, location of a node can be determine by the value of the angles of the straight lines that connect the UE and two references UE's [6, 7].

The RSS technique relies on a path-loss model, the distance between the base station and UE can be calculated by measuring the energy of the received signal at the UE. This distance-based technique requires at least three reference UE's to determine the 2-D location of a given UE, using the well-known triangulation approach [7]. We have seen that in indoor environments signal strength localization has some limitation. Signal strength approaches are widely applicable to wireless sensor networks without additional localization hardware [8].

Travel time of signals between base station and UE is used in time-based positioning techniques. If both have a common clock, the UE receiving the signal can determine the time of arrival (TOA) or the time-difference-of-arrival (TDOA) of the incoming signal that is time-stamped by the base station, this also is a widely used technique for localization in wireless sensor networks [7]. TOA is a time-based metric obtained via measuring the signal propagation time between the anchor and agent; then the distance measurements translate to location information by trilateration [5].

3 Proposed Solution and Observation

Data Mining includes finding anomalies, correlations, patterns, and trends to predict outcomes which can be applied to results obtained from beamforming. Data mining involves almost always working with large datasets to extract usable data or unforeseen information.

The Apriori Algorithm is a classic and influential algorithm for mining frequent itemsets. It falls under the category of Association Rule Mining algorithms. This class of algorithms helps to determine relations among variables in a large database [9]. An association rule is represented in the form of Antecedent ≥ Consequent [support].

The principle of the algorithm on the basis of which we will be using it is that, if an item set is infrequent, then all its subsets must also be infrequent. The idea of using Apriori is to prune the increasing search space using anti-monotonicity. It is devised to work on databases containing a lot of transactions, specifically a dense database as more number of transactions helps in generating more rules, relations, and patterns for analysis and robust data filtering [10]. Following on our optimization approach for coverage hole detection using beamforming is explained stepwise.

3.1 Step 1—Data Gathering from Beamforming

According to Fig. 1, as a beam (say B1) travels for a distance '*d*', it takes into consideration the distance of each UE as it connects with the beam. Localization techniques used are AOA and TOA which are the most popular and widely used methods as mentioned above in the paper and hence thereby help in finding the accurate distance of the UE. The Transaction table is populated in time T1 with the distances of the connected UE's as shown in Table 1. Similarly, it populates the database with each UE distance value respective to their connection timestamp T11, T12, T13 and so on for Beam B1 (refer Table 1). Transaction Id T11 corresponds to Beam 1 at time T1 of the target Base Station and 7, 14, 18 are the distances from the Base Station of UE111, UE112, UE113 respectively served with Beam 1. Transaction

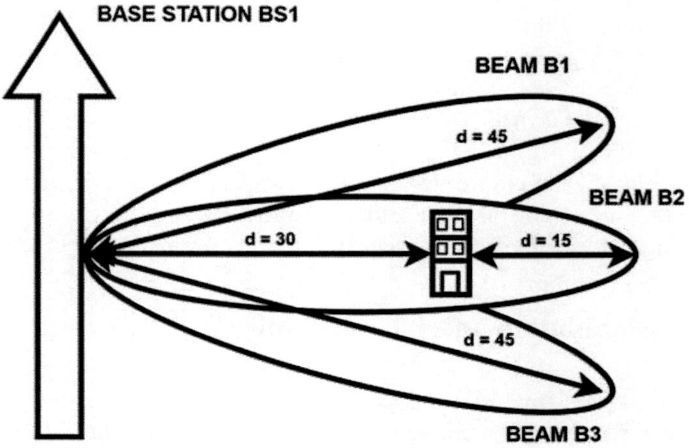

Fig. 1 Beam forming

Table 1 Beam-wise UE distance

Transaction	UE distance	Transaction	UE distance	Transaction	UE distance
T11	7, 14, 18	T21	1, 13, 24	T31	5, 19, 21
T12	9, 14, 40	T22	5, 13, 15	T32	11, 19, 42
T13	7, 9, 14, 40	T23	1, 5, 13, 15	T33	5, 11, 19, 42
T14	9, 40	T24	5, 15	T34	11, 42
T15	7, 9, 14, 40	T25	1, 5, 13, 14, 15	T35	5, 11, 19, 42
Beam 1		Beam 2		Beam 3	

Id T12 corresponds to Beam 1 at time T2 of the target Base Station and 9, 14, 40 are the distances from the Base Station of UE121, UE122, UE123 respectively served with Beam 1 and so on as represented in Table 1. Similarly, the same procedure is repeated for every beam and we generate a table for every individual beam.

3.2 *Step 2—Data Filtering*

Here we use a data mining technique to filter out the distance value for those UE's, which are inconsistent as compared to other UE's served by the same Beam. The Apriori algorithm is used for this purpose. Calculate the minimum support for every single item-set. Omitted values as shown in Table 2 and Filtered values as shown in Table 3 for B1, B2 and B3 and based on the given threshold (20%) are selected from Table 1 for Beams B1, B2 and B3, respectively. All the above UE distance values served by individual beams from the base station are integrated together in the KDD. Let us take a min-support of 20% (1 transaction).

3.3 *Step 3—Hole Identification*

Here we first compute the average distance traveled by each beam (refer Table 1 for Beam B1). The base station BS1 is set with the mean beam probing of maximum distance. When we find that a particular beam has an average distance less than the

Table 2 Beam-wise omitted distance

Item set	Support	Item set	Support	Item set	Support
18	1	14	1	21	1
		24	1		
Beam 1		Beam 2		Beam 3	

Table 3 Beam-wise filtered values

Item set	Support	Item set	Support	Item set	Support
7	3	5	4	5	3
14	4	13	4	19	4
7, 14	3	15	4	5, 19	3
9	4	1	3	11	4
40	4	1, 13	3	42	4
7, 9	2	1, 5	2	5, 11	2
7, 40	2	1, 15	2	5, 42	2
7, 9, 14	2	1, 5, 13	2	5, 11, 19	2
7, 9, 40	2	1, 5, 15	2	5, 11, 42	2
7, 14, 40	2	1, 13, 15	2	5, 19, 42	2
7, 9, 14, 40	2	1, 5, 13, 15	2	5, 11, 19, 42	2
Beam 1		Beam 2		Beam 3	

probing distance, we may conclude that this particular beam has a blockage in its path as shown in Fig. 1, which is preventing it from covering its designated region. The coverage hole thus lies after the average distance computed for this Beam as shown for Beam B2 in Fig. 1.

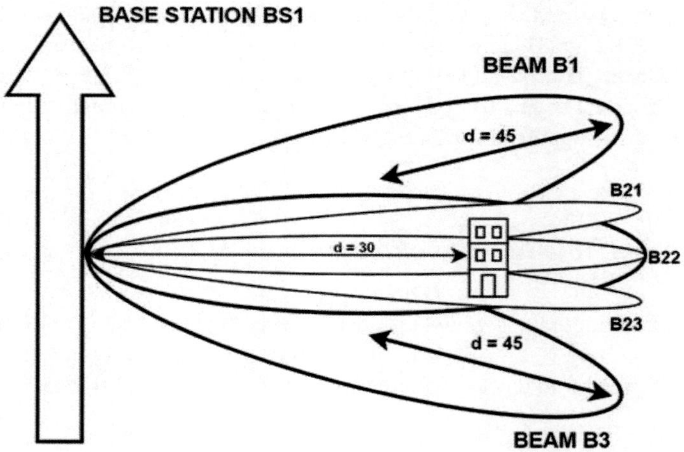

Fig. 2 Affected beam steering

3.4 Step 4—Beam Steering

Now the detection problem has been reduced to a particular beam which is Beam 2 in Fig. 1. We apply the beam steering technique to make n sub-beams (3 sub-beams in Fig. 2 in order to narrow down a particular region at which the coverage hole has occurred. This can be done by repeating the above steps 1 and 2. Figure 2 shows a detailed view of affected beam steering which again further helps to identify the exact sub-beam B22 of the original beam B2, by calculating the average distance probed by a beam from the actual mean probing distance set for all beams at the base station BS1. The acquired distance data of each connected UE is taken down to analyze the average distance probed by each beam as shown here. Thereby this approach further narrows down the affected region which will help us in better diagnosis and recovery.

3.5 Step 5

As mentioned above for the beam steering mechanism we need to follow the above steps 1 and 2 over again. So again we collect the detected UE distances for Beam B21, B22 and B23. The following Beam 2 with its beam steering, UE distance data table is shown in Table 4.

3.6 Step 6

In the previous step in Step 2, the data filtering technique is applied to this data as well. The Apriori algorithm for data mining is used to remove the inconsistent UE's then the minimum support for each item-set is calculated. Omitted values as shown in Table 5 and Filtered values as shown in Table 6 for B21, B22 and B23 and based on the given threshold (20%) are selected from Table 4 containing the Beam 2 sub-beams B21, B22, B23 steering values, respectively for Beams B21, B22 and B23.

Table 4 Beam-wise UE distance after beam 2 steering

Transaction	UE distance	Transaction	UE distance	Transaction	UE distance
T211	8, 22, 34	T221	1, 23, 30	T231	9, 28, 37
T212	15, 22, 43	T222	7, 14, 23	T232	21, 28, 44
T213	8, 15 ,22, 43	T223	1, 7, 14, 23	T233	9, 21, 28, 44
T214	15, 43	T224	7, 14	T234	21, 44
T215	8, 15, 22, 43	T225	1, 7, 19, 14, 23	T235	9, 21, 28, 44
Beam 21		Beam 22		Beam 23	

Table 5 Beam 2 steered sub beam-wise omitted distance

Item set	Support	Item set	Support	Item set	Support
34	1	19	1	37	1
		30	1		
Beam 21		Beam 22		Beam 23	

Table 6 Beam 2 steered sub beam-wise filtered values

Item set	Support	Item set	Support	Item set	Support
8	3	7	4	9	3
22	4	23	4	28	4
8, 22	3	14	4	9, 28	3
15	4	1	3	21	4
43	4	1, 23	3	44	4
8, 15	2	1, 7	2	9, 21	2
8, 43	2	1, 14	2	9, 44	2
8, 15, 22	2	1, 7, 23	2	9, 21, 28	2
8, 15, 43	2	1, 7, 14	2	9, 21, 44	2
8, 22, 43	2	1, 23, 14	2	9, 28, 44	2
8, 15, 22, 43	2	1, 7, 23, 14	2	9, 21, 28, 44	2
Beam 21		Beam 22		Beam 23	

3.7 Step 7—Providing Input to SON for Coverage Optimization:

Affected beam service area which can be termed a hole, is provided to Self-Organizing Network (SON) which then takes care of the diagnosis and optimization of the network.

Results obtained above helps data cleaning by filtering out wrong and inconsistent measured distances of UE's for each beam. These distances are now inputted to KDD to identify problematic beams whose coverage distance is less than the total average of distance served by all beams of this target base station. Similarly, as we worked with a min-support of 20% (1 transaction), if we increased the min-support to 40% (2 transactions) for even better optimization we could remove every item-set in the KDD with a support count of 2 transactions. Thus, feeding this much smaller input dataset to SON for accurate coverage hole detection.

4 Conclusion

In this paper, an approach using a data mining algorithm has been presented for the optimization of coverage hole identification in 5G networks. Data mining approaches have long been used for finding anomalies and frequent itemsets in large databases. So the problem of working with such algorithms on a large amount of data can be ruled out. Moreover, this proposed technique paves a new path for the application of data mining in such anomaly detection making scenarios in 5G networks or SONs.

From the simulation results, we can conclude that our approach is able to acquire a robust identification policy for different complex scenarios. It also provide a significantly better output in terms of both accuracy and performance for coverage hole identification. The process of beam steering in the proposed approach narrows down the search space to the last possible extent providing a more accurate distance measurement of the effected region. In densely populated regions, the Apriori algorithm would provide very few infrequent itemsets. Fewer the number of itemsets imply fewer the number of beams to be steered. Hence an effective way of saving time and complexity overhead in the network.

It is evident that the affected beam coverage is much lower where a coverage hole scenario exits. The proposed approach efficiently improves the coverage hole identification, capacity and performance in a dynamic environment.

References

1. K. Chowdhury, K. Porika, R. Ramankutty, R. Koodli, *System and Method for a Self Organizing Network*. U.S. Patent, US20100299419A1 (2010)
2. A. Zoha, A. Iran, A. Abu-Dayya, Challenges in 5g: how to empower son with big data for enabling 5g. IEEE Network **28**(6), 27–33 (2014)
3. B. Rong R. Q. Hu F. Qi, S. Sun, Y. Qian, Cognitive radio based adaptive son for lte-a heterogeneous network, in *IEEE Global Communications Conference*, pp. 4412–4417 (2014)
4. A. Iqbal Khan, J.A. Sheikh, F. Mustafa, A new responsibility domain based architecture for 5g networks, in *International Conference on Recent Developments in Science, Humanities & Management* (2018)
5. X. Zhang, M.Z. Win, Y. Han, Y. Shen, H. Meng, Performance limits and geometric properties of array localization. IEEE Trans. Inf. Theor. **62**(2), 1054–1075 (2016)
6. R. Jäntti, K. Ruttik, M. Costa, E.Y. Menta, N. Malm, K. Leppänen, On the performance of aoa-based localization in 5g ultra-dense networks. IEEE Access **7**, 33870–33880 (2019)
7. S. Gezici, Z. Tian, G.B. Giannakis, H. Kobayashi, A.F. Molisch, H.V. Poor, Z. Sahinoglu, Localization via ultra-wideband radios: a look at positioning aspects for future sensor networks. IEEE Signal Process. Mag. **22**(4), 70–84 (2005)
8. E. Elnahraway, X. Li, R.P. Martin, The limits of localization using rss, in *Proceedings of the 2nd International Conference on Embedded Networked Sensor Systems, SenSys'04*, pp. 283–284 (Association for Computing Machinery, New York, NY, USA, 2004)
9. K.C. Mondal, B.D. Nandy, A. Baidya, A factual analysis of python implementation for apriori algorithm, in *Computational Intelligence Communications and Business Analytics (CICBA-2018)* (Springer, 2018)
10. J. Han, M. Kamber, *Data Mining: Concepts and Techniques* (Morgan Kauffman, San Francisco, 2000)

A Machine Learning Based Heuristic to Predict the Efficacy of Online Sale

Aditya Vikram Singhania, Saronyo Lal Mukherjee, Ritajit Majumdar, Akash Mehta, Priyanka Banerjee, and Debasmita Bhoumik

Abstract It is difficult to decide upon the efficacy of an online sale simply from the discount offered on commodities. Different features have different influence on the price of a product which must be taken into consideration when determining the significance of a discount. In this paper we have proposed a machine learning based heuristic to quantify the "*significance*" of the discount offered on any commodity. Our proposed technique can quantify the significance of the discount based on features and the original price, and hence can guide a buyer during a sale season by predicting the efficacy of the sale. We have applied this technique on the Flipkart Summer Sale dataset using Support Vector Machine, which predicts the efficacy of the sale with an accuracy of 91.11%. Our result shows that very few mobile phones have a significant discount during the Flipkart Summer Sale.

Keywords Mobile price prediction · Significance of discount · Machine learning · Support vector machine

A. V. Singhania · S. L. Mukherjee · A. Mehta (✉) · P. Banerjee
Department of Computer Science, The Bhawanipur Education Society College, Kolkata, India
e-mail: akash93mehta@gmail.com

A. V. Singhania
e-mail: mailingadityavs@gmail.com

S. L. Mukherjee
e-mail: saronyolal@gmail.com

P. Banerjee
e-mail: priyankabanerjee333@gmail.com

R. Majumdar · D. Bhoumik
Advanced Computing & Microelectronics Unit, Indian Statistical Institute, Kolkata, India
e-mail: majumdar.ritajit@gmail.com

D. Bhoumik
e-mail: debasmita.ria21@gmail.com

J. Tavares et al. (eds.), *Emerging Technologies in Data Mining and Information Security*, Lecture Notes in Networks and Systems 164,
https://doi.org/10.1007/978-981-15-9774-9_41

1 Introduction

Online retailers such as Amazon, Flipkart etc. provide lucrative sale offers in almost every season. Discount offered for various products often lures the customers into buying items which may not be of actual or immediate requirement to them. However, a question which can hardly be answered from these sale offers is whether the discount offered for a particular product is *significant* for its features. For example, a laptop with a nominal SSD storage can be priced higher than another laptop with a higher HDD storage. The same discount for the former is arguably better than that for the later. Therefore, it is necessary to take into consideration the influence of individual features towards the price of the commodity, and then decide on the significance of the discount provided. A machine learning based technique is proposed in this paper to decide on the efficacy of online sale which can guide a buyer by predicting the "*significance*" of the discount offered on any commodity.

Machine learning (ML) is the technique of training a model to perform a specific task without using explicit instructions [1]. ML algorithms can either be initially trained with some data so that it can perform predictions over similar, but previously unseen dataset; or it can be expected to extract patterns from the dataset all by itself. The former is termed as Supervised Learning [2], while the later is the Unsupervised Learning [3]. Apart from these, other learning algorithms such as Reinforcement learning [4], Semi-supervised learning [5] etc. also have a wide range of application.

Fields such as air fare prediction [6], stock market analysis [7], real-estate price prediction [8] etc. have seen a plenty of applications of ML. In this paper, we have proposed a ML based heuristic to predict the efficacy of the sale price for any commodity. Our technique partitions the commodities into *discount classes* which essentially determines the *amount of significance* of a discount. In other words, we quantify the "*significance*" of the discount offered for a commodity, as a function of the standard deviation of its price class, in order to guide the buyer to decide whether he/she should buy it or not.

The proposed technique is then applied on the *Flipkart summer sale* to decide the "*significance*" of the discount offered for mobile phones during this sale. To obtain the results for the said sale, we have (i) created our own sale and non-sale dataset through web crawling; (ii) heuristically classified the set of mobile phones into four price classes and have used Support Vector Machine to train the model with the data of non-sale season; (iii) applied the trained model on the sale dataset to predict the price class of each mobile phone based on their features with 91.11% accuracy; (iv) determined the "*significance*" of the discount for each mobile phone. Our results show that Flipkart offered a *significant* discount for very few mobile phones. Although this result is largely dependent on the dataset and the heuristic grouping, our proposed technique is universal and can be applied on any dataset with any number of groups created through any heuristic.

The rest of the paper is organized as follows: In Sect. 2 we discuss the proposed mathematical technique for quantifying "*significance*" of a discount price. Section 3 elaborates the Web Crawling technique used to create the dataset on which we have applied the proposed technique. Sections 4 and 5 respectively presents the training of ML algorithm for the created dataset and the resultant efficacy as obtained by our technique. We conclude in Sect. 6.

2 Quantifying the Efficacy of Online Sale

Consider an E-commerce website which sells N commodities (of the same type, e.g. mobile phones or laptops), where the original (non-sale) and discounted (sale) price of the ith commodity are n_i and s_i respectively. The set of N commodities are first partitioned into g price classes according to their price range. Let G_j, $1 \leq j \leq g$ denote the j price class. Each price class G_j is associated with a price range $[p_j^{\text{start}}, p_j^{\text{end}}]$, and a commodity i is assigned to G_j only if its *non-sale* price $n_i \in [p_j^{\text{start}}, p_j^{\text{end}}]$. Therefore, $\sum_{j=1}^{g} |G_j| = N$, and $G_k \cap G_l = \phi$, $\forall\ 1 \leq k \neq l \leq g$. For each price class G_j, we also determine the mean price $\bar{G_j}$ and the standard deviation (S.D.) σ_{G_j}. Indeed, σ_{G_j} governs the variation of price for the commodities in the price class G_j. Therefore, we have considered the *level of significance* of a discount as the number of standard deviations which the discount price is away from the original price.

For each commodity, a set of features $\mathbf{f} = \{f_1, \ldots, f_m\}$ are selected, and an ML algorithm is trained, where each training data is a tuple $(\mathbf{f}_l, G_l)$, with $\mathbf{f}_l$ as the feature vector for the lth commodity and G_l as its assigned price class.

In the sale dataset, for each commodity i, we run the ML algorithm to predict its assigned price class G_i. Furthermore, let G_{d_i} be the price class for the discount price for the same commodity. Our heuristic is that:

1. If G_{d_i} is lower than G_i, i.e., the discount takes the commodity to a lower price class, then we define the discount to be '∞-**fold significant**'.
2. If $G_{d_i} = G_i$, then the discount is defined to be '**n-fold significant**' if $(n+1)k\sigma_{G_i} \leq s_i - d_i < nk\sigma_{G_i}$, for some fixed $k \in \mathbb{R}$ and $0 < k \leq 1$; $n \in \mathbb{Z}^+$.

This quantification of "*significance*", as presented above, allows a lot of flexibility. In our original definition of '∞-**fold significant**' we did not consider how much lower G_{d_i} is from G_i. However, if necessary, one can define '∞_x-**fold significant**' if G_{d_i} is x classes lower than G_i. Obviously '∞_x-**fold significant**' is better than '∞_{x-h}-**fold significant**' $\forall\ h \geq 1$. Furthermore, since $k \in \mathbb{R}$ and $n \in \mathbb{Z}^+$, it is possible to make k arbitrarily close to 0, and n arbitrarily large. However, making k very small will lead to a large number of *'folds'* with very small range for each and hence most of such ranges will most likely remain empty. Making n large will soon result in moving from one price class to another, and then it falls in the realm of '∞-**fold significant**'. One interesting scope of study, which we have not looked at in this paper, is how best to define the significance class for a given dataset.

The flexibility in quantifying "*significance*" and the use of ML to determine the influence of different features on the price makes our technique usable in a wide range of application. Apart from the prediction of sale efficacy, this technique can as well be used for more complex tasks, such as, one can learn the choice of the folds from prior knowledge of the buying pattern of a person, and use our technique as a recommender system for that buyer during a sale.

In this paper, we have applied this technique for determining the efficacy of the discount provided in *Flipkart Summer Sale* for mobile phones.

3 Creation of the Dataset

There are quite a few datasets available for mobile price prediction in the internet. However, we could not find any dataset in which the price of phones, both during a sale and the original price, were available. Therefore we created our own dataset via web crawling.

A web crawler is an internet bot that systematically browses the internet for the purpose of web indexing. It starts with a list of URLs to visit, called the seeds. As the crawler visits each URL, it identifies all the hyperlinks in that page and adds them to the list of URLs to visit henceforth. The web crawler can be used to copy and save the information of each page that it visits. A similar technique, called web data extraction, is deployed to use HTML tags in order to identify important information from the visited URL, and the extracted data, according to those tags, is saved in a local database. For our dataset creation, we have used two Python library modules, namely (i) *request*, which sends HTTP requests to the URL and returns a Response Object with all the response data (content, encoding, status, etc), and (ii) *beautifulsoup4* [9], which is used for data extraction from HTML, XML and other markup languages.

For the purpose of this research, we have used data extraction on the Flipkart website only for mobile phones. The first sale datatset was created via web data extraction on 19th March 2020, when the *Flipkart Summer Sale* was active, and the second non-sale dataset was created on 24th March, 2020 after the termination of the summer sale. The features which we have considered in this paper are (i) RAM, (ii) Storage, (iii) Megapixel of Front Camera, (iv) Megapixel of Back Camera, (v) Battery Capacity, and (vi) Internet connectivity. Our crawling resulted in a dataset with 1193 entries. However, we have excluded all those basic phones which do not have internet connection, and have henceforth worked with the first five features only. This is because those basic phones had a significant deviation in the other five features from the smart phones, which could lead to a biased training of the ML algorithm. Furthermore, hardly any people buy those phones now, and it can be a safe assumption that not many will be interested to buy the basic phones at discounted rates.

The crawled dataset resulted in some entries where one or more of the features had null values. After dropping all such data, and also dropping duplicate data (same mobile set sold in different colors), we were left with 733 entries, which constitutes

Table 1 Heuristic grouping of the price classes

Price range	Assigned class	# phones	μ	σ
(0, 5000)	"LOW"	426	1350.92	1051.63
[5000, 15,000)	"BUDGET"	204	9476.41	3007.14
[15,000, 30,000)	"MID RANGE"	76	19,483.81	3914.78
>30,000	"PREMIUM"	27	51,372.44	19,097.43

our final dataset. The non-sale and sale price of each mobile phone, along with the features, were stored separately for the training and testing purposes respectively.

3.1 Heuristic Grouping of the Price Classes

For the prediction of sale efficacy with this dataset, we partitioned the crawled dataset into four price classes as depicted in Table 1. For this particular choice of price classes, we also calculate the mean (μ) and standard deviation (σ) of each price class.

We note from Table 1 that the grouping into classes is not symmetric, and the distribution of the number of phones is skewed towards the lower price classes. This heuristic for the grouping was developed via an informal survey carried out by the authors amongst their friends and relatives, where they were asked to define the price range for which they will mark a mobile phone as 'low cost', 'budget phone', 'mid range' or 'premium'. From that survey, this grouping seemed to be the most obvious one. However, one can readily use the original technique (from Sect. 2) to form a different set of price classes.

4 Training the Model with Support Vector Machine

We now use the proposed technique from Sect. 2 on our crawled dataset to predict the efficacy of discount on mobile phones during *Flipkart summer sale* with four price classes. For the quantification of "*significance*", we have considered the fixed value of $k = \frac{1}{2}$. Furthermore, we have truncated the number of '**n-fold significant**' discount classes to three based on the result obtained from our dataset. Increasing the number of such classes would have only resulted in multiple classes with zero or very few entries. The quantification of the '**n-fold significant**' classes, and their assigned names are depicted in Table 2.

In other words, the '∞-**fold significant**' class has been renamed as "EXCELLENT", and by definition, it is not possible to have "EXCELLENT" deals for "LOW" price class.

Table 2 Quantification and nomenclature of "*significance*" classes

Range of discount	Assigned "*significance*" class name
$0 \leq s_i - d_i < \frac{\sigma}{2}$	"POOR"
$\frac{\sigma}{2} \leq s_i - d_i < \sigma$	"ACCEPTABLE"
$s_i - d_i > \sigma$	"GOOD"
Different price class	"EXCELLENT"

Machine Learning is now used to train the model with the non-sale dataset to predict the price class of any mobile phone based on the features mentioned earlier. We have used Support Vector Machine (SVM) Algorithm for the training purpose.

4.1 SVM Algorithm and Training with Non-sale Data

Classification [10] is a subclass of Supervised Learning. A k-class classification problem, where $C_1, \ldots, C_k$ are the class labels, consists of m training samples and n testing samples. Each training sample is a tuple (s_l, C_{s_l}) where the later is the designated class of s_l. After the training phase, for each test data t_l, the algorithm predicts a class C_i such that

$$\text{Prob}(t_l \in C_i) > \text{Prob}(t_l \in C_j), \forall j \neq i.$$

Support Vector Machine (SVM) [11] is one of the most extensively used supervised machine learning algorithm for classification problem [12, 13]. In SVM, the input is a set of labeled training data and the output is an optimal hyper-plane. The m training sets are of the form $(x_1, y_1), (x_2, y_2), \ldots, (x_m, y_m)$, where $x_i \in \mathbb{R}^d$ is a feature, and the class label is $y_i \in \{+1, -1\}, i = 1, \ldots, m$. SVM creates an optimal separating hyper-plane based on a mathematical function known as the kernel function. A kernel function is defined as $k(X_i, X_j) = \phi(X_i^T)\phi(X_j)$, where X_i, X_j are feature vectors and ϕ is the kernel function. Figure 1 shows a the hyper-plane created by SVM for a linearly separable dataset using linear kernel function.

However, every dataset does not contain vectors which are linearly separable. In such cases, we require a nonlinear SVM classifier which uses other forms of kernel function such as polynomial kernel, RBF kernel and sigmoid kernel.

In the heuristic grouping, elaborated in Sect. 3, there are four price classes, and hence this is a multi-class classification problem. There are two techniques [14] for multi-class SVM: (i) One-against-one which combines several binary classifiers, and (ii) One-against-all which considers all data at once.

The entire non-sale dataset was divided into the training and testing part where 80% of the dataset was used for training purpose and the remaining 20% for testing purpose. Training was done using the SVM algorithm with the RBF kernel and

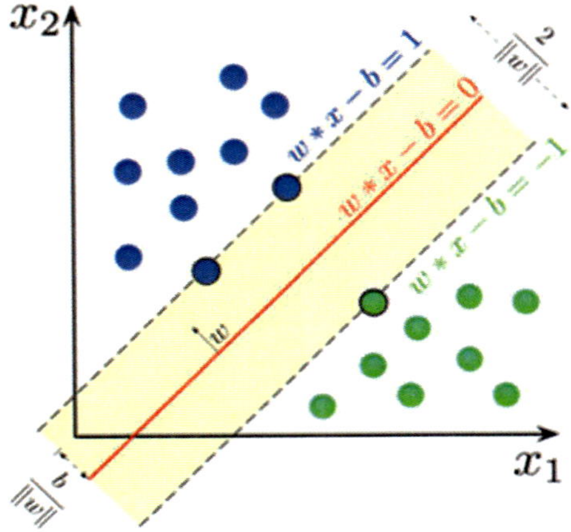

Fig. 1 Linear support vector machine

the One-against-all technique, as they were predicted best for our data via the cross-validation [15] method. The training and testing accuracy thus obtained were 98.66% and 91.11% respectively.

5 "*Significance*" of discount in Flipkart Summer Sale

We have applied the trained SVM model on the Flipkart Summer Sale mobile phone dataset. The price classes and the "*significance*" class are as depicted in Tables 1 and 2 respectively. The SVM algorithm predicts an overall 710 "POOR", 7 "ACCEPTABLE", 3 "GOOD", and 13 "EXCELLENT" discounts. We have selected two mobile phones from our dataset to show the detailed calculation of the "*significance*" of discount for them, where one is an "ACCEPTABLE" discount and the other one is an "EXCELLENT" discount. The summary of our result is depicted in Table 3.

The feature vector **f** for each of the two phones follow the ordering: (RAM, Storage, Front camera, Back camera, Battery capacity), and the price vector **p** follows the ordering: (original price, sale price). **o** and **s** denotes the price class for original and sale price respectively, and $\sigma_\mathbf{o}$ is the S.D. of the original price class.

1. **Company**: iBall, **f**: (2, 16, 0, 5, 2800), **p**: (4399, 3749), **o**: Low, **s**: Low, $\sigma_\mathbf{o}$: 1051.63. Since $\frac{\sigma_o}{2} \leq s_i - d_i = 650 < \sigma_o$, this discount is "ACCEPTABLE".
2. **Company**: LG, **f**: (4, 128, 5, 16, 3300), **p**: (60,000, 14,999), **o**: Premium, **s**: Budget. Since the discount price falls in a lower price class, this discount is "EXCELLENT".

From Table 3 it can be readily inferred that Flipkart does not provide significant discount on most of the mobile phones during its Summer Sale season.

Table 3 "*Significance*" of discount in flipkart summer sale

Discount class name	"*Significance*" of discount				
	Excellent	Good	Acceptable	Poor	Total
Low	0	1	1	424	426
Budget	4	1	3	196	204
Mid-range	5	1	2	68	76
Premium	4	0	1	22	27
Total	13	3	7	710	733

6 Conclusions and Future Work

In this paper, we have proposed a ML based technique to quantify the "*significance*" of the discount offered by online retailers for a commodity during sale season. Our proposed technique quantifies the "*significance*" of discount based on the features, original price and the discount class of the product. The application of this technique on Flipkart Summer Sale shows with an accuracy of 91.11% that significant discount is offered on very few mobile phones during this sale. We intend to extend the application part of this paper for other commodities, other sale seasons, and other online retailers as well. Such a cross-platform comparative study can be used as a guiding system for a buyer during online sale.

References

1. E. Alpaydin, *Introduction to Machine Learning* (MIT press, 2020)
2. R. Caruana, A. Niculescu-Mizil, An empirical comparison of supervised learning algorithms, in *Proceedings of the 23rd International Conference on Machine Learning*, pp. 161–168 (2006)
3. Z. Ghahramani, Unsupervised learning, in *Summer School on Machine Learning*, pp. 72–112 (Springer, 2003)
4. L.P. Kaelbling, M.L. Littman, A.W. Moore, Reinforcement learning: a survey. J. Artif. Intell. Res. **4**, 237–285 (1996)
5. X. Zhu, A.B. Goldberg, Introduction to semi-supervised learning. Synth. Lect. Artif. Intell. Mach. Learn. **3**(1), 1–130 (2009)
6. K. Tziridis, T. Kalampokas, G.A. Papakostas, K.I. Diamantaras, Airfare prices prediction using machine learning techniques, in *2017 25th European Signal Processing Conference (EUSIPCO)*, pp. 1036–1039 (IEEE, 2017)
7. R.T.F. Nazário, J.L. e Silva, V.A. Sobreiro, H. Kimura, A literature review of technical analysis on stock markets. Q. Rev. Econ. Finance **66**, 115–126 (2017)
8. Byeonghwa Park and Jae Kwon Bae, Using machine learning algorithms for housing price prediction: the case of fairfax county, virginia housing data. Expert Syst. Appl. **42**(6), 2928–2934 (2015)
9. L. Richardson, *Beautiful Soup Documentation* (2007)
10. S.B. Kotsiantis, I. Zaharakis, P. Pintelas, Supervised machine learning: a review of classification techniques. Emerg. Artif. Intell. Appl. Comput. Eng. **160**, 3–24 (2007)

11. C. Cortes, V. Vapnik, Support-vector networks. Mach. Learn. **20**(3), 273–297 (1995)
12. K. Nurhanim, I. Elamvazuthi, L. Izhar, T. Ganesan, Classification of human activity based on smartphone inertial sensor using support vector machine, in *2017 IEEE 3rd International Symposium in Robotics and Manufacturing Automation (ROMA)*, pp. 1–5 (IEEE, 2017)
13. A. Dey, D. Bhoumik, K.N. Dey, Automatic multi-class classification of beetle pest using statistical feature extraction and support vector machine, in *Emerging Technologies in Data Mining and Information Security*, pp. 533–544 (Springer, 2019)
14. C.M. Bishop, *Pattern Recognition and Machine Learning* (Springer, 2006)
15. T. Hastie, R. Tibshirani, J. Friedman, *The Elements of Statistical Learning: Data Mining, Inference, and Prediction* (Springer Science & Business Media, 2009)

Evaluation Metrics of Spatial and Spatiotemporal Data Mining Techniques

S. Sharmiladevi and S. Siva Sathya

Abstract In this era of big data, copious amount of spatial and spatiotemporal data is being generated in various domains like transportation, climate science, earth science, neuroscience, public safety, ecology, social science, etc. Since, inferring knowledge from these data is difficult, automation of knowledge discovery (KDD) process is necessary. Mining useful knowledge from spatial or spatiotemporal data is referred as spatial data mining (SDM) or spatiotemporal data mining (STDM), respectively. There are various techniques or algorithms available for knowledge discovery, wherein the quality and correctness of each technique is evaluated by some metrics or measures. In this article, several measures used for analysing SDM and STDM techniques are listed and classified based on the type of patterns obtained: co-location pattern, incremental co-location patterns, co-occurrence pattern and sequential patterns.

Keywords Evaluation metrics · Spatial co-location · Spatiotemporal co-occurrence · Data mining

1 Introduction

Spatial and spatiotemporal data analysis is a tedious task because of the underlying complex relation between the objects. Many techniques have been proposed for deducing useful knowledge from this information-rich data. The various spatial and spatiotemporal data mining techniques belong to one of the following pattern families: spatial and spatiotemporal outliers, spatial and spatiotemporal association and tele-connection, spatial and spatiotemporal predictions, spatial and spatiotemporal

S. Sharmiladevi (✉) · S. Siva Sathya
Department of Computer Science, Pondicherry University, Puducherry, India
e-mail: sharmiladevi94@gmail.com

S. Siva Sathya
e-mail: ssivasathya@gmail.com

J. Tavares et al. (eds.), *Emerging Technologies in Data Mining and Information Security*, Lecture Notes in Networks and Systems 164,
https://doi.org/10.1007/978-981-15-9774-9_42

clustering, spatial and spatiotemporal change detection [1–3]. In this article, we focus on spatial and spatiotemporal association pattern mining.

Mining association patterns in spatial data is called as spatial co-location pattern (SCOP) mining. SCOP mining is the process of discovering subsets of spatial features that are located together or found in nearby neighbourhood. SCOP can be obtained from all the three spatial model, viz. object model, field model and spatial network. The object model consists of vector data such as point, line, polygon. The field model consists of raster data which can be regular or irregular space. A spatial network is a graph where each node is a spatial point and the edges represent the link between nodes. The SCOPs can be mined using a join, joinless or partial join approach.

Discovering frequent association patterns from spatiotemporal data is called as spatiotemporal co-occurrence pattern (STCOP) mining. The spatiotemporal data can be grouped based on the manner in which temporal information is added to the spatial data. The temporal models are generally classified as temporal snapshot model, temporal change model, temporal event model. The ST data are event data, point reference data, Trajectory, raster data, evolving data. The STCOPs are discovered using the divide and conquer or coupling approach.

In this article, various metrics used to evaluate the SCOP and STCOP techniques are listed. These measures are classified based on the types of patterns mined by the SCOP or STCOP algorithm. The various types of patterns are co-location patterns, incremental co-location patterns, high utility patterns, totally ordered co-occurrence patterns, partially ordered patterns, unordered co-occurrence patterns. Henceforth, in this paper, the terms spatial data mining or spatiotemporal data mining refers to the spatial or spatiotemporal association mining, respectively.

The rest of the paper is organized as follows. Section 2 explains the different SCOP and STCOP data mining techniques available. Section 3 gives an idea about the SCOP and STCOP evaluation metrics. Section 4 concludes the paper.

2 Spatial and Spatiotemporal Data Mining Techniques

The various spatial and spatiotemporal pattern mining techniques or approaches are discussed in this section.

Spatial co-location is the process of identifying features that are often found in a neighbourhood. The neighbourhood distance is fixed by the user. Mining spatial co-location patterns is different from general association because there are no transactions in spatial data. This property also makes co-location mining a unique and tedious task. The spatial co-locations can be found by using a join, joinless, partial join approach. In a join-based approach, the candidate patterns are generated by performing a computationally expensive join operation between the underlying features. Then, pruning strategies are applied to find prevalent SCOP [4]. In partial join, the number of join operations is reduced by forming transactions using the neighbourhood relation [5]. A joinless approach uses the star neighbourhood or clique neighbourhood instead of the expensive join operation to find neighbour relationship [6]. These approaches or algorithm can be implemented in a CPU or GPU or MapReduce framework [7, 8].

Spatiotemporal data as the name suggests has both space- and time-related information; hence, finding patterns from it is difficult compared to spatial data mining. Spatiotemporal data mining can be defined as finding features/events in close geographic and temporal proximity. STCOP mining is an extension of SCOP mining algorithm. Based on the manner in which time is added, STCOP can be classified into two types: divide and conquer and coupling method. Divide and conquer is a timeslot-based approach, where the temporal data is divided into a set of time slots. SCOP in each time slot is mined, and then, its prevalence across all the time slots is computed to get STCOP [9, 10]. In coupling method, time is added as an additional dimension. Events which are close in space and time and mined directly using coupling method [11–13]. STCOP mining algorithms can be grouped into three types based on the temporal ordering as: totally ordered (sequence), partially ordered (cascading), unordered (co-occurrence).

3 Spatial and Spatiotemporal Data Mining Evaluation Metrics

There are numerous spatial and spatiotemporal data mining techniques available. The metrics used for evaluating these techniques or algorithm differ based on their problem statement. For example, in some spatial data mining technique, the main aim is to mine efficient co-location patterns, so processing– time related metrics are used for evaluating these approaches [7, 14, 15]. Table 1 lists out all the measures used in spatial and spatiotemporal data mining. These measures are classified as co-location, ordered STCOP and unordered STCOP based on the types of pattern for which they are used. Partially ordered patterns are considered as ordered patterns in this article because they are a subset of totally ordered patterns. Table 1 presents the various evaluation measures and the category under which it is used. Some of the basic definition related to the metrics and parameters of study are given in Sect. 3.1 and 3.2.

3.1 Metrics

Processing Time The processing time is the total time taken for executing an algorithm. While evaluating an algorithm's performance with respect to processing time, the algorithm which has least processing time is the best. In [14] the processing time taken for join based, joinless, PCPM_OC is compared by varying the distance threshold.

Spatial Participation Index/Prevalence Threshold It is used to find the prevalence of a co-location pattern 'p'. The participation index (PI) is the minimum of participation ratio (PR). The participation index satisfies the anti-monotonicity property. For small prevalence threshold, the running time and condensation power will be more [16].

$$\text{PI}(p) = \min_{f_j \in p}\{\text{PR}(p, f_j)\}, \text{ where } f_j \text{ is the feature}$$

Participation ratio (PR) of a feature is defined as given below

Table 1 Evaluation metrics of SCOP and STCOP

	Metrics	Variables	Inference	Papers
1	Processing time	Neighbourhood Distance/distance threshold	The distance threshold is varied to find its impact on processing time. For small distance threshold, the running time is less because very few instances will be able to form patterns. As the distance value increases, many instances are included in pattern generation so more running time	SCOP [8, 14, 16, 18, 19] Unordered STCOP [27–29]
2		Spatial participation index	For smaller spatial prevalence, the running time will be more because more patterns will be generated the vice versa happens for large threshold	SCOP [8, 14, 16] Unordered STCOP [10, 30]
3		No:of spatial events/features/objects	The number of spatial events present in the dataset has a huge impact on the processing time. As the events increase, the number of candidate patterns and its instances increase; therefore, the processing time increases along with it	SCOP [14, 16, 18, 19, 31] Unordered STCOP-[10, 27, 32] Ordered STCOP- [24]

(continued)

Table 1 (continued)

	Metrics	Variables	Inference	Papers
4		No:of spatial instances/dataset size	An instance is an occurrence of an event/feature. As the dataset size increases, the number of instances for each candidate pattern increases thus rise in execution time. When pruning methods or filtering strategies are used, the dataset size has minimum impact on execution time	SCOP [14, 16, 18] Ordered STCOP- [33]
5		Processing nodes	The processing nodes are the worker nodes in a Hadoop environment. When the worker nodes are increased, the running time is reduced because the co-location mining job are given to more nodes for execution	SCOP [14]
6		Incremental ratio/changed ratio	As the changed ratio increases, the execution time increases for naïve algorithms but not for utility update co-location algorithms. Because utility algorithm will just find the changed patterns	SCOP [18, 19] Unordered STCOP- [17]
7		Utility threshold	When utility threshold increases, the running time decreases, because very few co-location patterns will satisfy this increased threshold value	SCOP [18]
8		No. of time slot	The execution time increases as the number of time slots increases, but for algorithms employing pruning strategy the execution time is less comparatively	Unordered STCOP [10, 27, 32]

(continued)

Table 1 (continued)

	Metrics	Variables	Inference	Papers
9		Time prevalence threshold	A ST pattern is said to be time prevalent if it is already spatial prevalent, and it satisfies the time prevalence threshold. As the threshold value increases, the running time decreases because very few patterns will satisfy the large threshold value	Unordered STCOP [27]
10		Pattern size	When a filtering/pruning strategy used, the change in pattern size does not have a great impact on execution time	Unordered STCOP [21] Ordered STCOP [24]
11		Co-occurrence co-efficient	The execution time for J measure is more than OMAX because J requires union geometries which is an expensive task	Unordered STCOP [23]
12		Lifetime of objects	Synthetic spatiotemporal dataset generators are used to create ST objects with different lifetime. This dataset containing objects of different lifetime is used to find its effect on execution time. If a pruning strategy is used, the rise in lifetime has no impact on execution time; otherwise, the running time increases	Unordered STCOP [27]

(continued)

$$\mathrm{PR}(p, f_j) = \frac{\text{No:of unique instances of } f_j \text{ in } p}{\text{Total No:of instances of } f_j}$$

Temporal Prevalence Time prevalence is used to find the persistence of a co-located pattern across various time slots. It is calculated as the ratio of time slots where the pattern occurs to the total number of time slots. Similar to spatial prevalence for time prevalence, also the processing time is high for small threshold value [10].

Table 1 (continued)

	Metrics	Variables	Inference	Papers
13		Temporal constraint and spatial constraint	Temporal constraint is given in terms of hours. Spatial constraint is given in terms of kilometre. When the temporal/spatial value increases, the execution time also increases. But if downward closure strategy is used, the execution time remains unchanged	Ordered STCOP [25]
14		Filtering process	The filtering process is the pruning strategy used to mine non-prevalent patterns in the early stage. If a pruning strategy is used, the execution time is less compared to the algorithms that are not using pruning	SCOP [20] Unordered SCTOP [10] Ordered STCOP [25, 33]
15		Slicing size	The optimal slicing size has to be chosen to achieve better execution time. The execution time is high if the slicing size is too small, or too large. This can be attributed to the partition overhead or busy page swap, respectively	Ordered STCOP [24]
16		Sequence index threshold	The sequence index threshold has no impact on the running time for in memory experiments	Ordered STCOP [24]

(continued)

Incremental Ratio or Changed Ratio It is the ratio of incremental dataset over the original dataset. The incremental dataset consists of both added points and disappeared points compared to the original data. The added points are the newly updated points with respect to the original dataset. The disappeared points are the existing points in the original dataset that has vanished in the updated dataset or incremental dataset. This ratio is varied to find its effect on processing time [17–19].

Table 1 (continued)

	Metrics	Variables	Inference	Papers
17		Generation of 2 sequence	The processing time required for generation of size 2 sequence or candidate patterns increases as the head and tail interval value increases	Ordered STCOP [34, 35]
18	Memory Usage	co-occurrence co-efficient	The memory usage for J measure is more than OMAX because J requires union geometries which consumes more memory comparatively	Unordered STCOP [21, 23]
19		Pattern size	The memory usage decreases as the pattern size increases because very few patterns of large size will be present in a dataset	Unordered STCOP [21, 23]
20	Speedup	Spatial prevalence	When the spatial prevalence is increased, the speedup slowly decreases. Because for large prevalence value, the number of candidate patterns generated is small thus leading to worse use of GPU	SCOP [20]
21		Neighbourhood distance	The speedup increases as the neighbourhood distances increases. Because more patterns will be identified as the neighbourhood is expanded	SCOP [20]
22	Condensation power	Spatial participation index/prevalence	If the condensation power is large then the performance is better. But as the spatial prevalence value increases, the condensation power decreases, because very few prevalence patterns have to be mined now	SCOP [16]

(continued)

Table 1 (continued)

	Metrics	Variables	Inference	Papers
23		Neighbourhood distance	As distance increases, the condensation power increases because more prevalent patterns will be mined now	SCOP [16]
24	Precision	Temporal constraint and spatial constraint	Variation in temporal/spatial constraint affects the precision results, but no general trend can be told as it behaves differently for each dataset	Ordered STCOP [25]
25	RMSE	Temporal constraint and spatial constraint	Variation in temporal/spatial constraint does not show any clear trend in RMSE value	Ordered STCOP [25]

(continued)

Utility Threshold In utility pattern mining, for each feature f_j, an external utility (e_u) value is assigned. The internal utility (i_u) value of a feature f_j is the number of unique instances of it in a pattern *p*. The product of external and internal utility gives the utility value of f_j in 'p', $u(f_j, p) = e_u \times i_u$. By summing up the utility values of all feature in *p*, the utility value for a pattern *p* is obtained $U(p) = \sum_{f_j \in p} u(f_j, p)$. The total utility value for the database is computed by summing up the utilities of all features in it. The utility threshold value ϑ is defined by the user. A pattern *p* is called as high utility pattern if its utility value $U(p)$ is greater than ϑ. The effect of utility threshold on processing time is compared in [18].

Speedup The speedup is defined as the ratio of the serial runtime (T_s) of the best sequential algorithm for solving a problem to the time taken by the parallel algorithm (T_p) to solve the same problem on *p* processors. The effect of spatial prevalence, distance, number. of instances on speedup is compared in [8, 20].

$$S = T_s / T_p$$

Condensation Power Condensation power is calculated to find the percentage of redundancy reduced by an algorithm. In [16], a redundancy reduction algorithm is proposed, and to check its performance, the condensation power (%) is used.

$$\text{Condensation power} = (|S_{\text{PI-closed}} - S_{\text{SPI-closed}}|)/|S_{\text{PI-closed}}|$$

where $S_{\text{PI-closed}}$ is the set of PI closed co-location and $S_{\text{SPI-closed}}$ is the set of SPI-closed co-location. The larger the condensation power, the better is the performance.

Table 1 (continued)

	Metrics	Variables	Inference	Papers
26	Instance count	Number of time slot	When the number of time slots in a ST dataset is less, the generated instances is also less, but as the time slots increases the instance count also rises. The use of filter strategy reduces the instances count because of early pruning	Unordered STCOP [10]
27		Time prevalence threshold	The number of instances generated gradually decreases as time prevalence threshold increases, if and only if the algorithm employs filtering strategy. Otherwise, the instances generated are same for all the threshold values	Unordered STCOP [10]
28		Spatial prevalence threshold	The number of instances generated gradually decreases as time prevalence threshold decreases, if and only if the algorithm employs filtering strategy	Unordered STCOP [10, 23, 27]
29	Protopattern	Participation index threshold	When the participation index increases, the number of proto-patterns that needs to be refined is reduced because the filter procedure removes many patterns that could not satisfy the threshold	Unordered STCOP [22]

(continued)

Memory Usage Efficiency of SCOP and STCOP algorithms can be compared by memory consumed for finding patterns. For example, in [21], the memory usage of FastSTCOP miner and naïve STCOP algorithm was compared by varying the pattern size for J and OMAX measure. It is observed that J measure consumes more memory than Omax because J measure requires union geometries of the pattern instances [21].

Table 1 (continued)

	Metrics	Variables	Inference	Papers
30	Number of Jaccard operations	Participation index threshold	When the participation index increases, the number of proto-patterns that needs to be refined is reduced because the filter procedure removes many patterns that could not satisfy the threshold, thus resulting in reduced Jaccard operations	Unordered STCOP [22]

Proto-patterns Proto-patterns consist of events which participate in at least one pattern instance [22]. Computing proto-patterns is very cost efficient, and it also reduces the search space from which STCOP patterns have to be mined.

Co-occurrence Co-efficient Co-occurrence co-efficient (CCE) is a measure for finding the strength of a STCOP [23]. It is used in mining STCOPs from evolving ST data. Modified version of Jaccard (J) and OMAX measure was used to find whether the STCOP had strong overlap relation or not. Jaccard (J) is defined as the ratio of intersection volume to union volume of instances in a pattern, whereas OMAX is the ratio of intersection volume to the maximum volume of the instances. Computing J is an expensive task, and it takes more memory and processing time. But using J measure reduces the number of patterns generated.

$$J = \frac{V(i_1 \cap i_2 \ldots i_{k-1} \cap i_k)}{V(i_1 \cup i_2 \ldots i_{k-1} \cup i_k)}$$

$$\text{OMAX} = \frac{V(i_1 \cap i_2 \ldots i_{k-1} \cap i_k)}{\max(V(i_1), V(i_2) \ldots V(i_k))}$$

Sequence Index The sequence index is used to determine whether an event E' follows another event E. A threshold value 'θ' is fixed for this index to find significant spatiotemporal sequence. The sequence index threshold value is varied to compare processing time of the algorithms. All the other threshold values are fixed to a constant. When the θ value is closer to 0, the performance of the algorithms is similar [24].

Precision Precision is the fraction of relevant patterns in the retrieved pattern. In [25], precision is used to find the quality of a framework. Here, the relevant patterns (p_t) are the ground truth ST patterns generated by the baseline method, and retrieved patterns (p_g) are the ones obtained by the top k pattern generation method. The impact of different spatial and temporal constraints on precision is also analysed in [25].

$$\text{Precison} = \frac{|p_g \cap p_t|}{p_g}$$

Root Mean Squared Error (RMSE) RMSE is used to measure the differences between the values predicted by a model.

$$\text{RMSE} = \sqrt{\frac{1}{k}\sum_{i=1}^{k}(rg_i - rt_i)^2}$$

In [25], RMSE is used for evaluating the rank values (rg) generated by the top k ST patterns to the rank values (rt) generated by the ground truth ST pattern method. Smaller RMSE value means the accuracy is better.

F measure For measuring the accuracy of a model, F measure is used. It considers both precision and recall. In [26], precision is the faction of true positive patterns (TP) to the total co-occurrence patterns retrieved (P), and recall is the fraction of true positive pattern to the total dataset size (S).

$$\text{F-measure} = 2 \times \frac{\text{Precision} \times \text{Recall}}{\text{Precision} + \text{Recall}}$$

3.2 *Parameters*

Though several metrics like the ones mentioned are used to evaluate the performance of an algorithm, several parameters play a key role in determining the performance of the algorithm, a knowledge of which is essential in corroborating the results obtained. Hence, the parameters that influence the metrics are also reviewed here in this section.

Spatial Features/Events The number of spatial features in a spatial data or event in a spatiotemporal data has strong impact on processing time, number of patterns and instances generated. An increase in spatial events means the pattern size increases and all instances of this pattern have to be generated, thereby causing a rise in processing time [19]. The processing time increases when the features increases because of the increase in number of join operations [14].

Spatial Instances or Dataset Size A spatial instance is an occurrence of a feature or event. For example, if hospital is a feature, then Jipmer Hospital is an instance of it. Varying the number of instances is same as varying the dataset size. The dataset size is varied to check its effect on processing time, speedup [20].

Nodes The MapReduce framework has worker nodes which perform the computation. By increasing the number of worker nodes, the running time decreases [14].

Slicing Size When ST dataset is large, mining sequence pattern is costly because of the numerous database scans, and hence, partitioning or slicing is necessary. Compared to spatial partitioning, temporal-based partitioning or slicing is preferred because of its unidirectional property. In [24], the effect of slicing size on processing time is analysed. Small slicing size reduces the processing time because of the overhead in handling interior and exterior sets. If the slicing size is too large, then the slice could not fit into the memory resulting in busy page swaps and poor performance.

Distance or Neighbourhood Threshold Distance or spatial distance is used to determine the spatial closeness between features or events. As distance increases, the condensation power decreases, processing time increases and number of pattern instances mined increases [16, 18]. An increase in distance threshold means more instances could become neighbours; this causes the rise in processing time and number of pattern instances mined.

Timeslot In a divide and conquer STCOP mining approach, the time attribute is divided into time slots. Frequent co-location patterns are found for each time slot, and then, using temporal prevalence index, STCOP pattern are discovered. Increase in time slots increases processing time. By increasing the number of time slots, the processing time of different algorithms is compared. In [10] Fast MDCOP requires less processing time compared to other algorithms because of early pruning/filtering of time non-prevalent patterns. Time slots effect on number of instances generated for each pattern size is also compared.

Pattern Size The size of a patterns is measured by the number of features found in it. The size of the patterns to be mined is varied to analyse the performance of algorithms. It can also be called as length of the pattern.

Life Time of Objects For a spatiotemporal dataset having different object types and a set of time slots $\mathrm{TS} = [T_0, \ldots, T_{n-1}]$, the lifetime of an object can be defined as a total number of time slots where the instance of that object/feature appears [27].

In Table 1, the various SCOP and STCOP evaluation metrics are listed along with the type of patterns for which it is used with reference to the literature.

4 Conclusion

Spatial and spatiotemporal data mining is used in various domains for finding useful information. Finding patterns in spatial and spatiotemporal data is a challenging task because the information is unevenly distributed. Therefore, discovering hidden patterns from spatial or spatiotemporal data is of great interest in recent times. Efficiency of the methods used for mining these patterns has to be checked along with the usefulness of the pattern mined. In this article, an overview of various SCOP and STCOP evaluation measures is given. The evaluation measures are grouped into three categories, namely co-location patterns, unordered STCOP pattern, ordered STCOP pattern. The parameters that influence the metrics are also discussed.

Acknowledgements This work was supported by Council of Scientific & Industrial Research, India, under the scheme Direct-SRF with grant File No: 09/559(0141)/19-EMR-I.

References

1. Z. Jiang, S. Shekhar, Spatial and spatiotemporal big data science, in *Spatial Big Data Science* (Springer, Cham, 2017), pp. 15–44
2. G. Atluri, A. Karpatne, V. Kumar, Spatio-temporal data mining: a survey of problems and methods. ACM Comput. Surv. **51**(4), 1–37 (2018)
3. S. Shekhar et al., Spatiotemporal data mining: a computational perspective. ISPRS Int. J. Geo-Inform. **4**(4), 2306–2338 (2015)
4. Y. Huang, S. Shekhar, H. Xiong, Discovering co-location patterns from spatial datasets: a general approach. IEEE Trans. Knowl. Data 612–626 (2002)
5. J.S. Yoo, S. Shekhar, A partial join approach for mining co-location patterns, in *GIS: Proceedings of the ACM International Symposium on Advances in Geographic Information Systems* (2004), pp. 241–249
6. J.S. Yoo, S. Shekhar, A joinless approach for mining spatial colocation patterns. IEEE Trans. Knowl. Data Eng. **18**(10), 1323–1337 (2006)
7. J. Yoo, D. Boulware, D. Kimmey, Incremental and parallel association mining for evolving spatial data: a less iterative approach on MapReduce (2015)
8. W. Andrzejewski, P. Boinski, Parallel approach to incremental co-location pattern mining. Inf. Sci. **496**, 485–505 (2019)
9. F. Qian, L. Yin, Q. He, J. He, Mining spatio-temporal co-location patterns with weighted sliding window, in *Proceedings—2009 IEEE International Conference on Intelligent Computing and Intelligent Systems, ICIS 2009*, vol. 3 (2009), pp. 181–185
10. M. Celik, S. Member, S. Shekhar, J.P. Rogers, J.A. Shine, Mixed-drove spatiotemporal co-occurrence pattern mining. IEEE Trans. Knowl. Data Eng. **20**(10), 1322–1335 (2008)
11. B. Aydin, R.A. Angryk, Spatiotemporal event sequence (STES) mining. Springer Briefs Comput. Sci. 71–96 (2018)
12. R. Campisano et al., Discovering tight space-time sequences, in *Lecture Notes in Computer Science (including subseries Lecture Notes in Artificial Intelligence and Lecture Notes in Bioinformatics)*, vol. 11031 (LNCS, 2018), pp. 247–257
13. Y. Huang, L. Zhang, P. Zhang, A framework for mining sequential patterns from spatio-temporal event data sets. IEEE Trans. Knowl. Data Eng. **20**(4), 433–448 (2008)
14. W.X. Yang Peizhong, W. Lizhen, A parallel spatial co-location pattern mining approach based on ordered clique growth, in *Database Systems for Advanced Applications. DASFAA 2018. Lecture Notes in Computer Science*, vol. 10827 (2018), pp. 734–742
15. J.S. Yoo, D. Boulware, D. Kimmey, A parallel spatial co-location mining algorithm based on MapReduce, in *Proceedings—2014 IEEE International Congress on Big Data, BigData Congress 2014* (2014), pp. 25–31
16. L. Wang, X. Bao, H. Chen, L. Cao, Effective lossless condensed representation and discovery of spatial co-location patterns. Inf. Sci. **436–437**, 197–213 (2018)
17. Y.C. Chen, C.Y. Wu, S.Y. Lee, Incremental maintenance of topological patterns in spatial-temporal database, in *Proceedings—IEEE International Conference on Data Mining, ICDM*, December 2011, pp. 853–860
18. W. Xiaoxuan, L. Wang, Incremental mining of high utility co-locations from spatial database, in *2017 IEEE International Conference on Big Data and Smart Computing, BigComp 2017* (2017), pp. 215–222

19. J. Lu, L. Wang, Y. Fang, X. Bao, A novel method on incremental mining of spatial co-locations, in *2016 International Conference on Big Data and Smart Computing, BigComp 2016* (2016), pp. 69–76
20. W. Andrzejewski, P. Boinski, Efficient spatial co-location pattern mining on multiple GPUs. Exp. Syst. Appl. **93**, 465–483 (2018)
21. K.G. Pillai, R.A. Angryk, B. Aydin, A filter-and-refine approach to mine spatiotemporal co-occurrences, in *GIS: Proceedings of the ACM International Symposium on Advances in Geographic Information Systems* (2013), pp. 104–113
22. S.M. Hamdi, B. Aydin, R.A. Angryk, A pattern growth-based approach for mining spatiotemporal co-occurrence patterns, in *IEEE International Conference on Data Mining Workshops, ICDMW*, vol. 0 (2016), pp. 1125–1132
23. K.G. Pillai, R.A. Angryk, J.M. Banda, D. Kempton, B. Aydin, P.C. Martens, Mining at most top-k% spatiotemporal co-occurrence patterns in datasets with extended spatial representations. ACM Trans. Spat. Algorithms Syst. **2**(3) (2016)
24. Y. Huang, L. Zhang, P. Zhang, A framework for mining sequential patterns from spatio-temporal event data sets. **20**(4), 433–448 (2008)
25. B.H. Chen, S.Y. Teng, K.T. Chuang, Mining spatio-temporal chaining patterns in non-identity event databases. Intell. Data Anal. **21**(S1), S71–S102 (2017)
26. H. Tran-The, K. Zettsu, Finding spatiotemporal co-occurrence patterns of heterogeneous events for prediction, in *Proceedings of the 3rd ACM SIGSPATIAL International Workshop on the Use of GIS in Emergency Management, EM-GIS 2017* (2017)
27. M. Celik, Partial spatio-temporal co-occurrence pattern mining. Knowl. Inform. Syst. 27–49 (2015)
28. M. Akbari, F. Samadzadegan, R. Weibel, A generic regional spatio-temporal co-occurrence pattern mining model: a case study for air pollution. J. Geogr. Syst. **17**(3), 249–274 (2015)
29. M. Akbari, F. Samadzadegan, Identification of air pollution patterns using a modified fuzzy co-occurrence pattern mining method. Int. J. Environ. Sci. Technol. **12**(11), 3551–3562 (2015)
30. M. Celik, N. Azginoglu, R. Terzi, Mining periodic spatio-temporal co-occurrence patterns: a summary of results, in *INISTA 2012—International Symposium on INnovations in Intelligent SysTems and Applications* (2012)
31. S. Maiti, R.B.V. Subramanyam, Mining co-location patterns from distributed spatial data. J. King Saud Univ. Comput. Inform. Sci. (2018)
32. Z. Wang, X. Peng, C. Gu, B. Huang, Mining at most top-K% mixed-drove spatiotemporal co-occurrence patterns, in *2013 9th Asian Control Conference, ASCC 2013* (2013)
33. P. Mohan, S. Shekhar, J.A. Shine, J.P. Rogers, Cascading spatio-temporal pattern discovery. **24**(11), 1977–1992 (2012)
34. B. Aydin, R.A. Angryk, A graph-based approach to spatiotemporal event sequence mining, in *IEEE International Conference on Data Mining Workshops, ICDMW*, (1) (2017), pp. 1090–1097
35. B. Aydin, R. Angryk, Discovering spatiotemporal event sequences, in *Proceedings of the 5th ACM SIGSPATIAL International Workshop on Mobile Geographic Information Systems, MobiGIS 2016*, (2) (2016), pp. 46–55

ECG Assess Heartbeat rate, Classifying using BPNN while Watching Movie and send Movie Rating through Telegram

Sima Das and Aakashjit Bhattacharya

Abstract The paper has been designed for heartbeat rate classification while watching movies and sending movie ratings using Telegram Bot. The heartbeat of a movie viewer can increase abruptly at any point in the movie, which can harm the health of the person adversely with serious problems like cardiac arrest. To record heart activity in terms of heartbeat rate electrocardiogram is used. The Electrocardiogram is the technique to capture heart-signal provoked by the electric field and captured by a sensor. In this paper, the heartbeat rate is measured while watching a movie and is getting classified using Backpropagation Neural Network, that whether watching the movie will result in a slow, normal, or fast heartbeat rate. Observations are collected and analyzed and from next time when a person starts watching that movie, an alert message is sent using Telegram to notify the type of a movie. So from next time, before watching that movie, a person can get alert about the type of that particular movie based on the previous ratings and can watch or avoid watching the movie based on their discretion as it may affect their heart. The system is designed for weak-hearted, teenagers and be used as a parental guide.

Keywords Electrocardiogram · Machine Learning · Internet of Things · Backpropagation-Neural-Network · Bandpass filter · Telegram Bot

1 Introduction

Movies are used as a means of entertainment all over the world for a long time. It creates an attachment with people massively. In today's world movie is one of the best ways for entertainment and different kinds of people like different genres of movies which impacts the heart and it has the potential to let overcome the human

---

S. Das (✉) · A. Bhattacharya
Department of Computer Science and Engineering, Maulana Abul Kalam Azad University of Technology, Kolkata, West Bengal, India
e-mail: cstsima3@gmail.com

A. Bhattacharya
e-mail: aakashjit4mtech@gmail.com

J. Tavares et al. (eds.), *Emerging Technologies in Data Mining and Information Security*, Lecture Notes in Networks and Systems 164,
https://doi.org/10.1007/978-981-15-9774-9_43

boringness, as well as it connects us to real-life situations that are often ignored by us. Although movies have many advantages, it has many disadvantages too. Sometimes it hurts people's minds. Some movies or some portion of movies have some sexual content and violent scenes that impact our society and at the same time can harm a person's heart as it has happened previously a lot of time in various other movies. We know that emotional regulations and memory formations are related to human sensory functions and involve the olfactory system [1]. Many teenagers are attracted by movies and that adversely affects their studies and changed their behavior also. In the age group of 13–17 years, schoolboys' attitudes have changed and they have become violent after watching violent movies [2]. Some Bollywood movies have changed the attitude of the viewers, with some movies having a negative impact on their life [3]. Normal people's pulse rate increases while watching horror movies [4].

Heartbeat rate is central to the process of heart and its functions are also important because oxygen and blood are circulating to the whole body through heart. For normal people, the heartbeat is slower when they are sleeping or resting, the heartbeat is normal when they doing normal work and heart-beat is faster when they have more pressure like workload, angry, scare, tension, etc. Slow heart rate is good for mammals as it helps to have better longevity [5]. Heart attack can be caused by high blood pressure [6]. Electrocardiogram (ECG) is the process to measure electrical signals through electrodes placed on ventricles of heart, and measured with respect to time that how long it takes to pass through the heart and changes in cardiac muscle. According to electrical signal passes by heart and muscle, it can measure the condition of heart-beat are normal or abnormal. Electrocardiogram signals were captured while watching Fantasia movie during 120 min and R-R interval time series has been used for detrended fluctuation analysis which is collected from subject's fractal scaling related to age [7, 8].

Researchers have invented many techniques for measuring and classifying heart-beats. In [9] Nayak et al, they have shown horror movies to a group of people of the age group of 21–23 years, and have collected ECG data while watching movies, studied by heart rate variability feature, and classified data using the autonomic nervous system (ANS). In [10] Tsai et al. used a wearable ECG sensor to measure tension using Mark Exciting Segment algorithm while watching movies and adventure riding. From [11] Hsu et al., using an ECG sensor for automatic emotion recognition system while playing music. Here these features are extracted by generalized Discriminant Analysis and classifying using least squares support vector machine. In [12] Pandey et al., used a deep convolutional neural network that automatically classifies as a true positive, false positive, false negative, true negative of the electrocardiogram heart-beat on arrhythmia patient's accordingly ANSI-AAMI standards synthetic minority oversampling technique (SMOTE) method used for generating new sample without duplicating.

Internet of Things (IoT) being quite challenging [13] and innovative [14] is evolving at a quite high pace and is expanding into many aspects of human society [15]. It has been said in [16], that IoT has been defined as the next logical stage of the internet along with its extension in the physical world. It is playing a crucial role in many aspects of our daily life [17] and this trend will increase in the upcoming years

[18]. Over the years IoT has evolved and has got its application in the domain of health and fitness monitoring [19], and in combination with Machine Learning/Deep Learning, it is used to create and implement intelligent systems [20]. So, we have used it to implement our proposed system. IoT has played an integral part in our proposed system because, without the concept IoT, the monitoring and alerting system could not be implemented in such a user-friendly manner in our work. Telegram is a Cloud-based service and we have used its Bot APIs for the messaging and alerting service. It is a quite popular app and ranks 8th in the list of instant messaging apps as of July 2019 and supports end-to-end encryption [21]. In various countries where there is a restriction of the social media platforms, Telegram has become a leading social networking platform in those countries.

In this paper, we have discussed the measurement of heartbeat rate during watching movies. The proposed system designed helps guardians to choose which movies are suitable for their children and how long they can watch the movies without a negative impact on their hearts. The proposed system is designed for teenagers as well as old persons and weak-hearted people to show the rating and know before viewing the movie. We have proposed an Internet of Things based alert system using Telegram Bot that is alerting people about the rating of the movie, as a result, they have received an alerting message using Telegram Bot. This is because some parents are busy with their work at the office. So, to observe their children from office who have logged into see the movies using parent's mobile numbers, these techniques used to be a parietal guide for movie rating. So in this paper, we measure heartbeat rate and classified as slow, normal, and fast rate of heartbeat and also rating movie about pressure on the heart to alert people which type movie it is that's why they make their decision for watching a movie.

The rest of the paper organized as the following order—Sect. 2 introduced proposed method for the current paper, Sect. 3 shows the experimental result, and Sect. 4 is the last section which is represented in the conclusion and future work.

2 Proposed Method

Our proposed system (from Fig. 1) collects data from the heart using ECG sensor, and clean data using filtering using bandpass filter and artifact remove process using PAN Tompkins algorithm, after preprocessing, feature selection and extraction are done based on age, R-R interval and heartbeat rate of the subject and then it classifies heartbeat rate as slow, normal or fast using back-propagation-neural-network (BPNN) after classified result store in the cloud and warns other viewers who are watching a movie later and The possible heartbeat rate from previous observations will be sent to the registered Telegram Account.

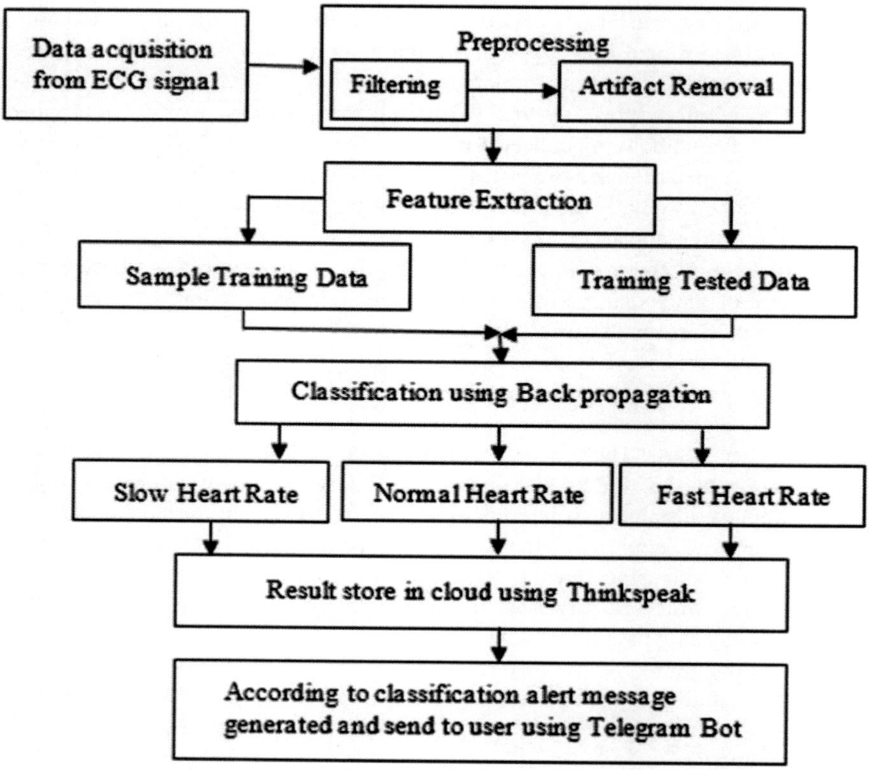

Fig. 1 Heart rate monitoring, classifying using BPNN and send movie rating through Telegram Bot while watching movie and storing result in ThingSpeak Cloud Platform

2.1 *Preprocessing*

Raw dataset is not feasible for analysis that's why there is a need to clean data and that process is called pre-processing. In machine learning, for accurate results data needs to pre-processed properly. It is the first step of machine learning to process data in a formatted way that is suitable for the machine. Filtering and artifact fact removing are the types of data preprocessing to eliminate electrical interference, Electromyography (EMG) noise.

1. Filtering—Filtering (from Fig. 2) is the first step of pre-processing. Different kinds of noises like muscle noise and power line interference are removed by a bandpass filter with a range of 5–15 Hz.
2. Artifact Removal—After applying a bandpass filter PAN-Tompkin algorithm is applied for identified Artifact from the ECG database as shown in Fig. 3. In this algorithm we have implemented the following steps in the following order—at first derivation, then squaring, after that average and threshold, the last step is

Fig. 2 Filtered electrical noise from raw ECG dataset

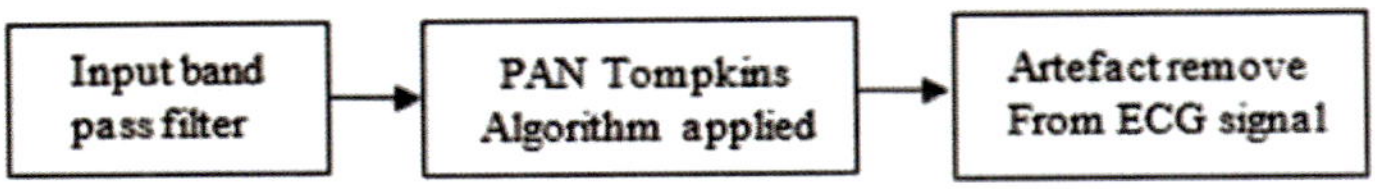

Fig. 3 Artifact remove from ECG signal

peak detection and QRS detection. By applying these algorithm artifacts like a power line, baseline wondering are removed and datasets are more cleaned and formatted.

2.2 *Feature Extraction*

Feature selection and extraction plays a vital role in machine learning. In this work, features are selected for classification based on:

1. Age—For this experiment 20 participates are present at the age between 21–34 and 68–85 age group.
2. R-R interval—R-R interval is discussed below. To feature selection, we used Discrete Wavelet Transform that selected *Q*, *R*, *S* all features of heart-beat. In *QRS* there have time delay denoted as *RR* and others are RR_0 is the interval between previous beat and predecessor, RR_1 is the interval between current and previous beat and RR_2 is the interval between current and subsequent beat

$$R1 = \frac{RR_0}{RR_2} \tag{1}$$

$$R2 = \frac{RR_2}{RR_1} \tag{2}$$

$$R3 = \frac{RR_n}{RR_1} \tag{3}$$

where, RRn = mean (RR_0, RR_1, RR_2) and $R1$ used for Ratio1 (as shown in Eq. 1), $R2$ used for Ratio2 (as shown in Eq. 2), and $R3$ denotes Ratio3 (as shown in Eq. 3).

3. Heart Beat Rate—Heartbeat rate is also an important feature. Given formula in Eq. 4, calculates the heartbeat rate.

$$\mathrm{HBR} = \frac{60}{R - R\ \mathrm{Interval}} \tag{4}$$

2.3 *Classification*

After feature extraction, selected features are used for heartbeat rate classification. Figure 4 describe the method of classification, it is done by backpropagation neural network (BPNN). BPNN is a statistical technique used in machine learning and many biological neural networks. It is iterative gradient descent algorithm that input cleaned and formatted data to the hidden layer and finds propagation error then weight updated, when the error of propagation is in the backward direction it changes it layer by updating weight value. This algorithm designed for minimizing mean square error which occurs on the desired output and actual output.

$$\mathrm{CE} = \frac{1}{2}\sum_{J=1}^{M}(D_J - A_J)^2 \tag{5}$$

where CE is used for detecting the cumulative error, A_J is the desired output and D_J is the actual output of the J Eq. 5.

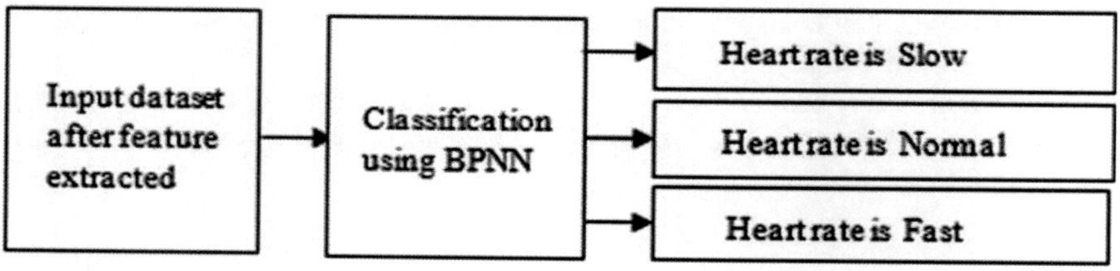

Fig. 4 Classification using backpropagation neural network

Table 1 Training data of R-R interval and heartbeat rate

Classification	R-R Interval	Conditions of heart beat rate (BPM)
Slow	1 for male 0 for female	Heart rate <60
Normal	1 for male 0 for female	Heart rate ≥60 && heart rate ≤100
Fast	1 for male 0 for female	Heart rate >100

2.4 Rating

In this experiment, we have measured the heartbeat rates while watching a movie Fantasia and that is classified into three groups slow, normal, and fast heartbeat rate that depends upon movie scenes. Concerning time movie is separated into different frames and we are measuring the heartbeat of participants of each separated frame then classifying this frame and mark as any one type of heartbeat rate among three. Then apply the formula which is given below in Eq. 6.

$$\text{Movie\_Rating} = \frac{F_1 \times \text{slow} + F_2 \times \text{normal} + F_3 \times \text{fast}}{F_1 + F_2 + F_3} \quad (6)$$

F_1 denotes how many times the heartbeat rate is slow, F_2 denotes how many times the heartbeat rate is normal and F_3 denotes how many times the heartbeat rate is fast while watching Fantasia movie.

IF Movie_Rating >= 0 && Movie_Rating <= 1
THEN "As per our past studies, Heart Beat Rate Slow for this movie"
ELSE IF Movie_Rating >= 1.1 && Movie_Rating <= 2
THEN "As per our past studies, Heart Beat Rate Normal for this movie"
ELSE IF Movie_Rating >= 2.1
THEN "As per our past studies, Heart Beat Rate Fast for this movie"

From the above condition, we can get the result of movie ratings according to the heartbeat rate, and the rating is sent through Telegram to the user and is used as parental controls for the user before watching a particular movie.

2.5 Telegram Bot

In our experiment, we have sent the message using Telegram Bot and the Movie Rating calculated is transmitted to the cloud to help future viewers of that particular movie and send them the collective average rating of the movie from different past viewers. Other messaging apps can also be used if required and if there is the support of APIs for those messaging apps. We have used the Telegram Messenger because the APIs are free of cost and this provides a cross-platform and cross-device support.

3 Experimental Results

3.1 ECG Recording and Data Collection

The ECG signals were collected underwent 120 min continue resting and watching movie "Fantasia" and the blood pressure signals are digitalized at 250 Hz. From

	YOUNG SUBJECTS					OLD SUBJECTS				
	Y1	Y2	Y3	Y4	Y5	O1	O2	O3	O4	O5
HEART BEAT ECG READINGS	0.736	1.152	0.9	1.284	0.916	0.992	0.964	0.8	1.236	1.016
	0.744	0.952	0.864	1.2	0.944	0.98	0.936	0.828	1.24	0.996
	0.752	1.012	0.82	1.152	0.976	0.988	0.96	0.832	1.244	0.996
	0.756	1.016	0.816	1.232	0.996	1	0.944	0.844	1.264	1.004
	0.72	0.904	0.84	1.044	0.988	0.972	0.968	0.868	1.264	1.016
	0.756	0.856	0.852	1.316	0.972	0.964	0.968	0.88	1.296	1.02
	0.78	0.888	0.888	1.188	0.96	0.98	0.964	0.884	1.256	1.028
	0.768	0.928	0.996	1.116	0.956	0.984	0.976	0.892	1.276	1.04
	0.712	0.896	1.02	1.4	0.948	1	0.98	0.884	1.256	1.044
	0.724	0.872	0.952	1.12	0.952	1.004	0.976	0.864	1.252	1.04
	0.748	0.828	0.944	1.128	0.952	0.992	0.98	0.864	1.26	1.044
	0.76	0.856	0.964	1.356	0.956	0.98	0.984	0.864	1.236	1.04
	0.728	0.892	0.952	1.1	0.912	0.98	1	0.884	1.208	1.048
	0.756	0.828	0.912	1.264	0.888	1.004	0.992	0.908	1.204	1.044
	0.764	0.772	0.916	1.304	0.836	1.016	0.992	0.948	1.208	1.044
	0.74	0.792	0.952	1.3	0.892	0.98	1	0.988	1.228	1.048
	0.748	0.888	0.928	1.196	0.98	0.972	0.996	1.008	1.232	1.056
	0.78	0.812	0.88	1.32	1.004	0.976	0.992	1.028	1.236	1.052
	0.816	0.904	0.876	1.376	1.004	0.936	1.008	1.064	1.26	1.056
	0.748	1.096	0.916	1.296	1.008	0.916	1.012	1.04	1.252	1.056

Fig. 5 Dataset collection using ECG while watching movie

Fig. 5, we can see a sample of ECG dataset during watching movie Y1, Y2, Y3, Y4, Y5 are young and O1, O2, O3, O4, O5 are old individual's heartbeat rate. ECG dataset has been collected for this experiment while watching "Fantasia" movie of 20 participants at the age of 21–34 years and 68–85 years old healthy subjects from a free website "https://physionet.org/content/fantasia/1.0.0/".

3.2 *Measuring Heartbeat Rate*

The duration of data collection for the "Fantasia" movie is 120 min and we used 100 data per frame and the movie has several epochs. By measuring each frame and using BPNN we can be classifying the data as slow, normal, and fast heartbeat rate. By applying Eq. 6, we got the answer of movie rating is 1.5 that is in between the range of 1.1–2 that's mean the movie is a normal rating movie according to heartbeat rate which is given in Table 2.

4 Conclusion and Future Work

Fantasia movie from Walt Disney is a quite innovative and revolutionary animated classic, that combines the masterpiece of Western classical music along with colorful imaginative visuals, with Leopold Stokowski and the Philadelphia Orchestra. It is quite impressive, free-flowing, abstract, and often with surrealistic pieces. In "The

Table 2 Screenshot of sending movie rating using Telegram Bot

Screenshot of the Fantasia Movie	**Screenshot of sending movie rating using Telegram Bot**
	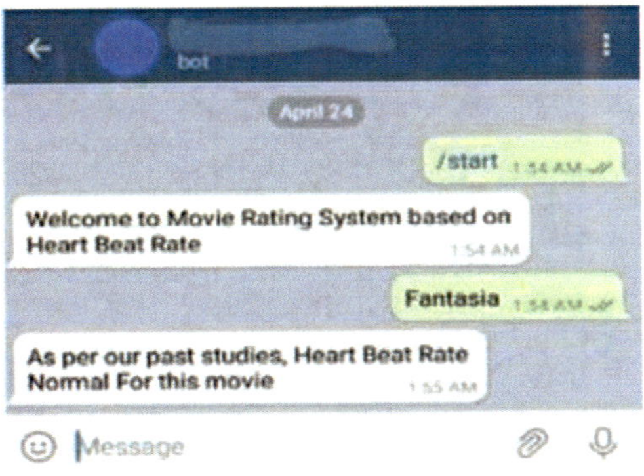

Sorcerer's Apprentice" we find Mickey Mouse has been shown an aspiring magician who has overstepped all his limits. The movie is quite significant, critical, and is having commercial successes. Through this movie a lot of types of emotions have been reflected from the viewer that's why we have chosen this movie for the experiment to measure heartbeat rate of the viewer using ECG to measure heartbeat rate of the participates while watching the movie and classifying heartbeat rate using back-propagation neural network to find which type of heartbeat occurs in which scene and give a message using Telegram to inform heartbeat is which category- slow, normal or fast. Movie rating technique used for rating the movie which is a parental guide for teenagers and a weakhearted person.

In the future, we are planning to extend this work with brain cognitive load detection while watching the movie.

Acknowledgements We gratefully acknowledge the Maulana Abul Kalam Azad University of Technology, West Bengal, which has allowed us to do this wonderful project and also thanks to "physionet", an open-source website for sharing ECG based movie's dataset.

References

1. M. Nardelli, G. Valenza, A.Greco, A. Lanatá, E.P. Scilingo, R. Bailón, Quantifying the lagged Poincaré plot geometry of ultrashort heart rate variability series: automatic recognition of odor hedonic tone. Med. Biol. Eng. Comput. 1–14
2. M.S. Hassan, M.N. Osman, Z.S. Azarian, Effects of watching violence movies on the attitudes concerning aggression among middle schoolboys (13–17 years old) at international schools in Kuala Lumpur, Malaysia. Eur. J. Sci. Res. **38**(1), 141–156 (2009)

3. M.H. Itoo, K. Nagar, Impact of negative portrayal of a destination in bollywood movies on viewers' attitude towards the destination, intention to visit and destination image. Pac. Bus. Rev. Int. **10**(5), 71–82 (2017)
4. M.I. Qadir, M. Asif, Does normal pulse rate correlate with watching horror movies. J. Cardiol. Curr. Res. **12**(2), 67–68 (2019)
5. J.M. Arnold, D.H. Fitchett, J.G. Howlett, E.M. Lonn, J.C. Tardif, Resting heart rate: a modifiable prognostic indicator of cardiovascular risk and outcomes? Can. J. Cardiol. **24**, 3A–15A (2008)
6. World Health Organization, Avoiding heart attacks and strokes: don't be a victim, in *Avoiding Heart Attacks and Strokes: Don't be a Victim*, 2005 (pp. 44–44)
7. N. Iyengar, C.K. Peng, R. Morin, A.L. Goldberger, L.A. Lipsitz, Age-related alterations in the fractal scaling of cardiac interbeat interval dynamics. Am. J. Physiol. Regul. Integr. Comp. Physiol. **271**(4), R1078–R1084 (1996)
8. A.L. Goldberger, L.A.N. Amaral, L. Glass, J.M. Hausdorff, P.C. Ivanov, R.G. Mark, J.E. Mietus, G.B. Moody, C.K. Peng, H.E. Stanley, PhysioBank, PhysioToolkit, and PhysioNet: components of a new research resource for complex physiologic signals. Circulation **101**(23):e215-e220 (2003)
9. S. Nayak, S. Sahoo, B. Champaty, K. Pal, Effect of horror clips on the physiology of ANS & heart using ECG signal classification. in *2014 International Conference on Control, Instrumentation, Communication and Computational Technologies (ICCICCT)* (IEEE, 2014), pp. 480–485
10. H.H. Tsai, C.W. Yi, Wearable ECG for tension assessment in movie watching and adventure riding, in *2015 11th International Conference on Mobile Ad-hoc and Sensor Networks (MSN)* (IEEE, 2017), pp. 23–30
11. Y.L. Hsu, J.S. Wang, W.C. Chiang, C.H. Hung, Automatic ecg-based emotion recognition in music listening. *IEEE Transactions on Affective Computing* (2017)
12. S.K. Pandey, R.R. Janghel, Automatic detection of arrhythmia from imbalanced ECG database using CNN model with SMOTE. Australas. Phys. Eng. Sci. Med. **42**(4), 1129–1139 (2019)
13. M. Al-Emran, S.I. Malik, M.N. Al-Kabi, A survey of internet of things (IoT) in education: opportunities and challenges, in *Toward Social Internet of Things (SIoT): Enabling Technologies, Architectures and Applications* (Springer, Cham, 2020), pp. 197–209
14. T. Srivastava, S. Singh, Impact of internet of things on societal applications, in *Intelligent Communication, Control and Devices* (Springer, Singapore, 2020), pp. 579–587
15. J.J. Kang, Systematic analysis of security implementation for internet of health things in mobile health networks, in *Data Science in Cybersecurity and Cyberthreat Intelligence* (Springer, Cham, 2020), pp. 87–113
16. F. Firouzi, B. Farahani, M. Weinberger, G. DePace, F.S. Aliee, IoT fundamentals: definitions, architectures, challenges, and promises, in *Intelligent Internet of Things* (Springer, Cham, 2020), pp. 3–50
17. C.S. Kolli, C.K. Reddy, M. Kavitha, A critical review on internet of things to empower the living style of physically challenged people, in *Emerging Research in Data Engineering Systems and Computer Communications* (Springer, Singapore, 2020), pp. 603–614
18. S. Alam, S.T. Siddiqui, A. Ahmad, R. Ahmad, M. Shuaib, Internet of things (IoT) enabling technologies, requirements, and security challenges, in *Advances in Data and Information Sciences* (Springer, Singapore, 2020), pp. 119–126
19. V. Saranya, M.C.M. Belinda, G.R. Kanagachidambaresan, An evolution of innovations protocols and recent technology in industrial IoT, in *Internet of Things for Industry 4.0* (Springer, Cham, 2020), pp. 161–175
20. M.A. Al-Garadi, A. Mohamed, A. Al-Ali, X. Du, I. Ali, M. Guizani, M., A survey of machine and deep learning methods for internet of things (IoT) security. IEEE Commun. Surv. Tutorials (2020)
21. A. Jalilvand, M. Neshati, Channel retrieval: finding relevant broadcasters on telegram. Soc. Netw. Anal. Min. **10**, 1–16 (2020)

Software-Defined Networking with Multiple Routing Utilizing DFS to Improving QoS

F. M. Javed Mehedi Shamrat, Pronab Ghosh, Md. Dipu Sultan, Anup Majumder, and Imran Mahmud

Abstract Massive advancement of network traffic has been observed lately since the significant usage of online application and cloud administration. Thus, various routing issues need to be managed without increasing the latency to ensure even traffic. Software-defined networking (SDN) is developed to address the above-mentioned issue. SDN allows the network's programmable features and responses gradually. The suggested system addresses the concerns raised due to multipath steering through SDN. The application of OpenFlow convention and Ryu controller is done to implement SDN with the help of the Mininet-WiFi emulator. The SDN network is designed using a graphical UI program, and the module is implemented with Python programming along with OpenFlow based-Ryu controller. The breadth-first search (BFS) and depth-first search (DFS) schemes are portrayed to calculate the briefest path in the system. Finally, a round trip time (RTT) is observed to ensure the most suitable search algorithm among them to be used in the module resulting in DFS being the preferred algorithm.

Keywords Software-defined networking · Depth-first search · Breadth-first search

F. M. Javed Mehedi Shamrat (✉)
Department of Software Engineering, Daffodil International University, Dhaka, Bangladesh
e-mail: javedmehedicom@gmail.com

P. Ghosh · Md. Dipu Sultan
Department of Computer Science and Engineering, Daffodil International University, Dhaka, Bangladesh
e-mail: pronab1712@gmail.com

Md. Dipu Sultan
e-mail: dipusultan090@gmail.com

A. Majumder
Department of Computer Science and Engineering, Jahangirnagar University, Dhaka, Bangladesh
e-mail: anupmajumder@juniv.edu

I. Mahmud
Department of Information Technology and Management, Daffodil International University, Dhaka, Bangladesh
e-mail: imranmahmud@daffodilvarsity.edu.bd

J. Tavares et al. (eds.), *Emerging Technologies in Data Mining and Information Security*, Lecture Notes in Networks and Systems 164,
https://doi.org/10.1007/978-981-15-9774-9_44

1 Introduction

Logically, the existing system deteriorates the extensive choice hooked on stream switch which decides the absolute broadcast percentage depending on the base way cost, directing repair in what way to part the stream between accessible ways [1]. Software-defined networking is another sort of system design that sending and control are isolated and can be legitimately constrained by encoding [2]. In the rest of the system, system handle and sending are all pursue loyally on the network. Software-defined networking isolates switch work starting the system gear and acknowledges brought together governor of system and programmability of uses by institutionalization. Hence, the software-defined networking altogether varies the customary system construction [3].

System operative necessity enhanced approaches to arrange and regulate the systems. Inappropriately, the present systems usually include joining and interconnection of numerous proprietorships, on a level horizontally incorporated gadgets [4]. This perpendicular reconciliation creates it particularly hard for propellers to indicate elevated parity system-wide approaches utilizing the existing advancements. Modernization in Web organization has appropriately been confined to stop-gap methods and procedures, for example, devices that assess lower strategies of action to identify mistakes or generally answer to system measures [5]. Selective programming and shut improvement in system gadgets by a bunch of vendors enlarge it very hard to present and convey new conventions [6]. Gradual modernizes to design strategies and directions are commonly managed independently by merchants. Incidentally, administrators' necessities for greater usefulness and progressively multipart network methodologies keep on expanding. In software-defined networking, network gadgets become straightforward bundle sending devices (information plane), while the cerebrum or control method of reasoning is applied in the controller [7].

A few briefest way algorithms similar Dijkstra's, Bellman Ford, A* search, Floyd–Warshall, BFS and DFS are actualized on behalf of recognizing the straight route in the conventional network structure. Dijkstra's is executed in the SDN [8]. Achieving the most concise way algorithm in a software-defined network, the quality of service (QoS) is improved as depicted in [9]. Nonetheless, the depth-first search is not executed at this point in a software-defined network to discover the best proper shortest way. In this manner, we completed the depth-first search (DFS) algorithm to improve the parcel transmission rate.

At the beginning of our research, we have worked with some traditional network topologies, such as mesh topology and ring topology. And we have used the DFS algorithm because we can easily optimize the multipath routing problem in the software-defined networking. After implementing the DFS algorithm in both mesh and ring topology, we found the minimum cost and distance to reach the destination. So, we used a DFS algorithm for multipath routing problems.

This paper is shown as pursues. Section 2 delineated the research methodology. Execution is portrayed in Sect. 3. Section 4 evaluates the exhibition to show that outcomes get together quality of service. Section 5 concludes the paper with future work.

2 Methodology

Mininet-WiFi is a kind of framework reenactment arrangement that can make virtual switches, controllers and system affiliations. With frivolous demonstration development, it makes the system practically identical to the genuine. Consequently, we can make a simulated system setting on personal computer terminals for study and progress. Mininet-WiFi maintains the OpenFlow convention 1.3. Programs executing in Mininet-WiFi can use Python programming. So, it is attainable to make a simulated system condition of the software-defined network, not long after, forward to the programs in a virtualization stage to the network system. Likewise, Mininet-WiFi underpins multifaceted and client characterized topology and gives Python API. It is additionally advantageous for different individuals to grow cooperatively.

3 Implementation

The segment defines the employment of the multipath direction-finding tricky expansion in the SDN. The OpenFlow system virtualization called Mininet and the OpenFlow regulator called Ryu structure have been utilized at the time of execution.

3.1 Environmental Setup

An AMD Ryzen 73,800x (8 core with 32 Mb cache memory), among other things 32 GB RAM established work, is selected on behalf of performing the system arrangement. Ubuntu 19.10 LTS 64-bit OS is utilized on behalf of executing the local area network (LAN) building in the software-defined network.

The usage and reenactment have been implemented utilizing Python 2.7 by introducing the Anaconda bundle. At that point, a fast prototyping emulator called Mininet-WiFi has been introduced for making the virtual SDN condition. To build and envision the virtual system, a graphical user interface-based editorial manager specifically MiniEdit has been added. At last, empowering the OpenFlow controller, Ryu structure has been introduced and played out the reproduction. In Fig. 1, we demonstrate the environmental set up of Mininet for calculating the minimum cost selection.

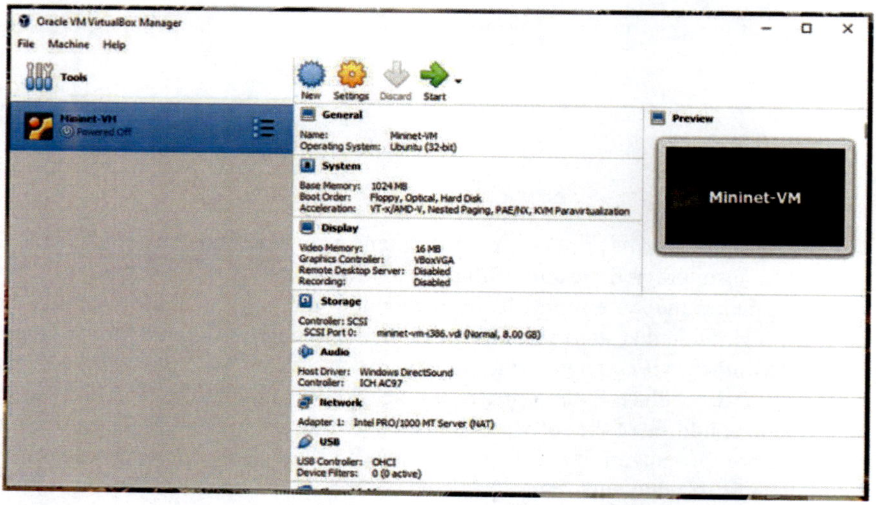

Fig. 1 Environmental setup

3.2 *Network Configuration*

A virtual LAN topology (Fig. 2) is made utilizing MiniEdit for software-defined system organization. The said topology holds six hosts, ten switches and one controller. This controller is chosen to be the remote controller (Fig. 3). In the same TCP convention, OpenFlow upheld OpenV switch is chosen (Fig. 4). For cost estimation, edge weight is doled out as 1.

The controller is characterized as cc1, with switches and hosts which have been defined as sj and hj greetings individually, where j is the whole number variable characterized as the number beginning from 0. Hosts are doled out the IP address which appointed 10.0.0.1 for h0. Sequentially 10.0.0.2 is for h1 and so on. The controller allotted the IP 127.0.0.1 with the port location 6633.

3.3 *DFS Implementation*

To locate the shortest way choice algorithm, depth-first search is executed to help with the Ryu system. Utilizing the iterative technique, DFS can identify the most profound vertex in a diagram using a stack. For multipath steering, DFC stores the profundity in the stack using dropping requests. From that stack, the briefest way is set in the last position. Due to this unique element, software-defined networking with multiple routings utilizes DFS to improving QoS.

DFS has been utilized for finding the briefest way by investigating all conceivable vertices with the least expense. The DFS algorithm is actualized in [10] as follows.

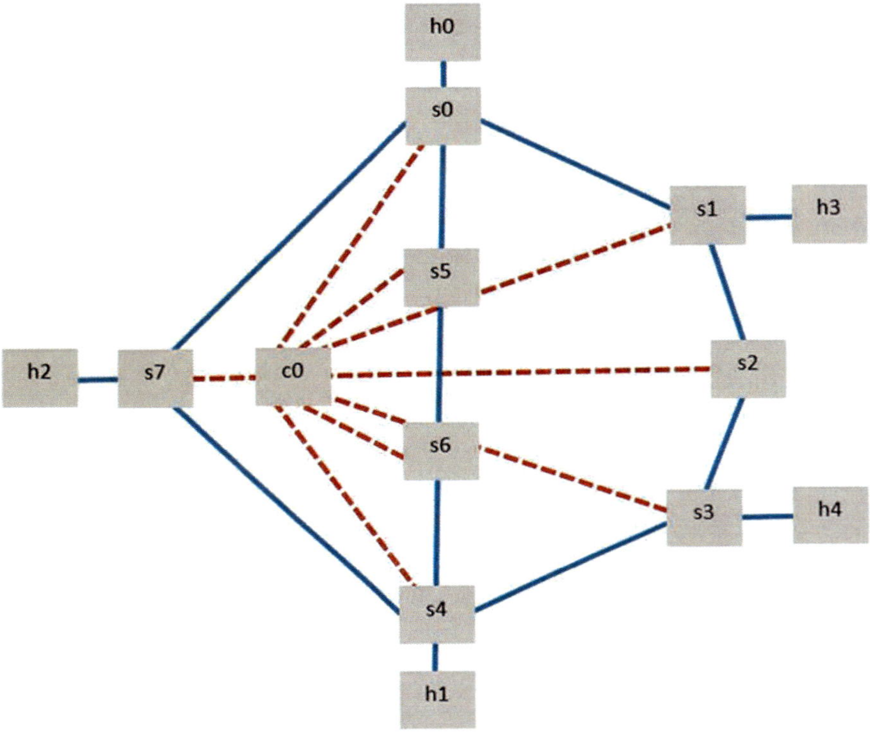

Fig. 2 LAN architecture of software-defined network

Fig. 3 Ryu controller architecture

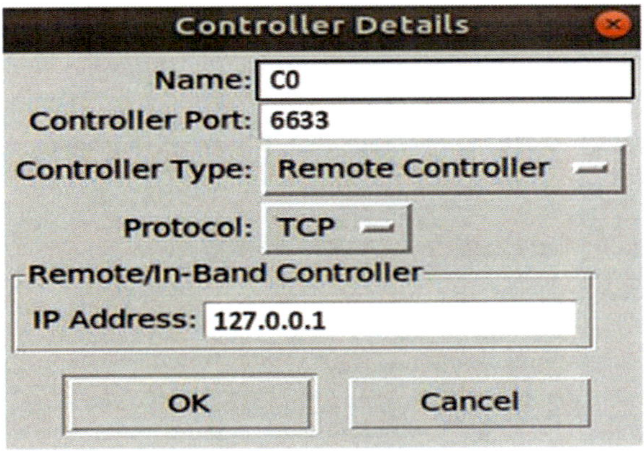

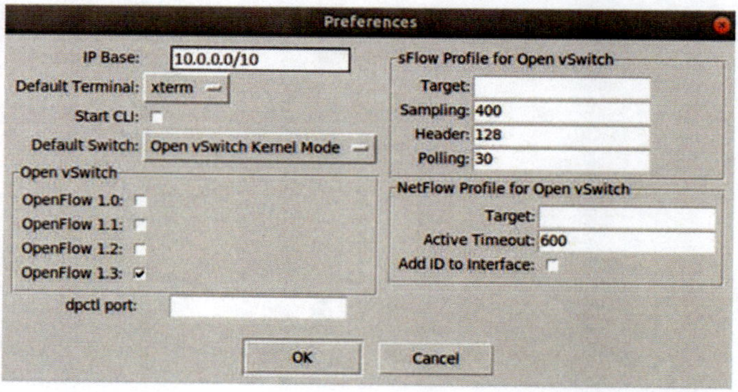

Fig. 4 Version selection

Algorithm

```
Step 1: Function DFS(G , V)
        Input: Graph = G and Initial Vertex = V
Step 2: Stack St := {};   -> (Begins with empty stack)
Step 3: for each vertex uV
Step 4:   |        set visited[uV] := false;
Step 5:   | push St, V;
Step 6:   | while (St is not empty) do
Step 7:   |  |      uV := pop St;
Step 8:   |  |    if (not visited[uV]) then
Step 9:   |  |    |        visited[uV] := true;
Step 10:  |  |    |   for each unvisited neighbor wV of uV
Step 11:  |  |    |   | push St, wV;
Step 12:  |  |    |   end for
Step 13:  |  |    end if
Step 14:  | end while
Step 15: end for
Step 16: END DFS()
```

3.4 Simulation

To get the outcome from the reproduced workstation, we have made ping demands inside every single imaginable host. Natural ways are appeared in the terminal from each source to goal pair and the other way around. Recreation is run a few times to watch the conduct for result examination.

```
Available paths from 1 to 8 : [[1,8], [1,6,7,5,8], [1,2,3,4,5,8]]
[1,8] cost = 1
Shortest path = [[1,8]] Cost = 1
Path Installation finished in 0.00124568264543
Available paths from 1 to 2: [[1,2], [1,6,7,5,4,3,2], [1,8,5,4,3,2]]
[1,2] cost = 1
Shortest path = [[1, 2] Cost = 1
Path Installation finished in 0.00124522164411
Available paths from 1 to 4 : [[1,6,7,5,4], [1,2,3,4], [1,8,5,4]]
[1,2,3,4] cost = 1
Shortest path = [[1,2,3,4]] Cost = 3
Path Installation finished in 0.00144911074101
```

Fig. 5 Path selection between h0, h1

4 Result Analysis

This section inspects the reenacted outcomes and assesses the advancement of the DFS calculation in SDN-empowered LAN to arrange. Point-by-point investigation is depicted in the accompanying subsections.

4.1 Path Selection

In light of the DFS calculation, the way has been pushed on the stack. At that point, the techniques have appeared in the terminal. While pinging from h0 to h1 for a cross-check approval, the rundown of all conceivable accessible ways has been performed as shown in Fig. 5. We have determined the can weight and cost from h0 to h1 and the other way around. We have discovered that, the same approach has been chosen in both cases; however, in switch requests the host h0 begins with switch number 0, while h1 is connected with switch number 4.

4.2 Cost Calculation

The DFS algorithm restores the rundown of ways that are unweighted. Along these lines, for ascertaining the expense of an idea, a container load recipe has been utilized. This condition has been depicted in [11]. Open the briefest way, the first (OSPF) cost estimation system is using to appraise the connection cost. The count of the total expense of all the associated ways is appeared in Eq. (1).

Table 1 Shortest routing path and minimum cost selection

Source (host)	Destination (host)	All-possible path (switches)	Shortest-path (switches)	Cost
1	3	[1, 8], [1, 5–8], [1–5, 8]	[1, 8]	1
1	4	[1, 2], [1–7], [1–5, 8]	[1, 2]	1
1	5	[1, 4–7], [1–4], [1, 4, 5, 8]	[1–4]	3
2	3	[5, 8], [1, 5–8], [1–5, 8]	[5, 8]	1
2	4	[2–5], [1, 2, 6, 6, 7], [1, 2, 5, 8]	[2–5]	3
2	5	[4, 5], [1–7], [1–5, 8]	[4, 5]	1
3	4	[1, 2, 5–8], [2–5, 8], [1, 2, 8], [1–8]	[1, 2, 8]	2
3	5	[4, 5, 8], [1–8], [1, 4–8], [1–4, 8]	[4, 5, 8]	2

$$\mathrm{WB}(p) = \left(1 - \left(\frac{\mathrm{WP}}{\sum \mathrm{WP}(p)[N]}\right)\right) \times 10 \tag{1}$$

where for a path p:

WB = Weight of the bucket, $0 \leq \mathrm{WB}(p) < 10$
WP = Weight of the path (*OSPF cost style*),
N = Total number of available paths.

4.3 *Cost Evaluation*

To assess the briefest path to each couple of source and goal, the recreation has been executed a few times. Because of the reproduction results, the point-by-point navigating data, including the expense, is classified in Table 1. Throughput, speed, deferral and cost are the essential parameter of the nature of administration (QoS). Amid all the QoS parameters, the executed algorithm has accomplished least expense. Fig. 6 represents the graphical interface of Table 1 data.

From the above investigation, it has been seen that among all the conceivable accessible ways, the briefest way is in every case accurately related to the least cost. In this manner, one of the QoS parameters has been upgraded.

5 Conclusion

In this study, we have executed the DFS algorithm in the product characterized system to upgrade the expense in multipath routing. Because of the shortest way, the bundle

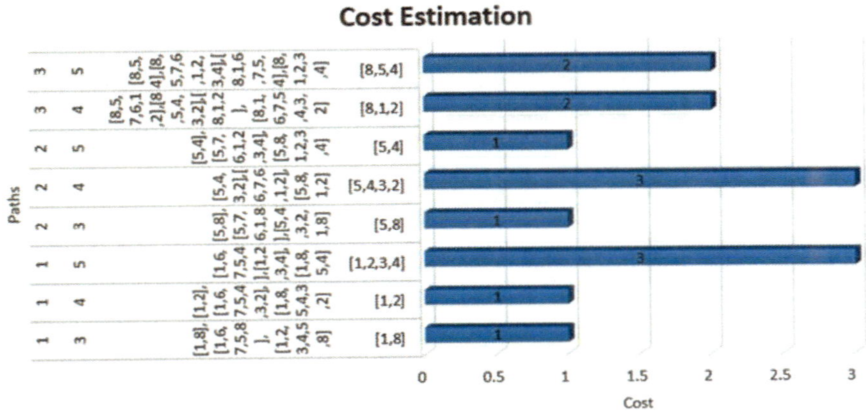

Fig. 6 Graph of minimum cost selection

transmitted starting with one hub then onto the next. Along these lines, the transmission time, deferral and cost were limited. Our mimicked outcomes demonstrated that the QoS is upgraded for the SDN development for the neighborhood. Later on, it will be extremely intriguing to execute this calculation alongside BFS in certifiable use and explore the QoS.

References

1. S. Parnin, M.M. Rahman, Human location detection system using micro electromechanical sensor for intelligent fan, in *International Conference on Mechanical, Automotive and Aerospace Engineering* (2016). https://doi.org/10.1088/1757-899x/184/1/012042
2. K.L. Ku, J.S. Liaw, M.Y. Tsai, T.S. Liu, Automatic control system for thermal comfort based on predicted mean vote and energy saving, in *IEEE Transactions on Automation Science and Engineering*, vol. 12, no. 1 (2015)
3. M. Aftab, C. Chau, P. Armstrong, Smart air-conditioning control by wireless sensors: an online optimization approach, in *International conference on Future energy systems* (2013). https://doi.org/10.1145/2487166.2487192
4. D.V. Babu, D.K. Yadav, N.K. Yadav, V.K. Verma, MEMS based smart and secure home automation system with multi-way control and monitoring facility using smart phone. J. Chem. Pharm. Sci. **9**, 357–359 (2015)
5. A. Aswani, N. Master, J. Taneja, D. Culler, C. Tomlin, Reducing transient and steady state electricity consumption in hvac using learning-based model-predictive control. Proc. IEEE **100**(1), 240–253 (2012). https://doi.org/10.1109/JPROC.2011.2161242
6. J.Y.H. Wang, Y.K. Chuah, S.W. Chou, T.H. Lo, Non-invasive Zigbee wireless controller for air conditioner energy saving, in *7th International Conference on Wireless Communications, Networking and Mobile Computing*, Wuhan, pp. 1–4 (2011). https://doi.org/10.1109/wicom.2011.6040311
7. Thermal camera.: https://learn.adafruit.com/adafruit-amg8833-8x8-thermal-camera-sensor/overview, 28.09.2018

8. Raspberry pi zero.: https://learn.adafruit.com/introducing-the-raspberry-pi-zero?view=all, 28.09.2018
9. IR sensor.: https://learn.adafruit.com/ir-sensor?view=all, 28.09.2018
10. N. McKeown, T. Anderson, H. Balakrishnan, G. Parulkar, L. Peterson, J. Rexford, S. Shenker, J. Turner, OpenFlow: enabling innovation in campus networks. ACM SIGCOMM Comput. Commun. Rev. **38**(2), 69–74 (2008)
11. M. Al-Fares, A. Loukissas, A. Vahdat, A scalable, commodity data center network architecture. ACM Sigcomm Comput. Commun. Rev. **38**(4), 63–74 (2008)

Direct Farmer and Customer Commerce Using Data Mining

Siddharth Singh, Aditya Kumar, Aditya Chopra, and Mitchell D'silva

Abstract Farmers in India are finding it more and more difficult to sustain their livelihoods. There are several causes that are responsible for the prevalent agrarian distress in the country, but the predominant reason for this is the substandard prices at which farmers are compelled to sell their produce. Farmers lack the means to transport their produce and sell it to the customers directly. Instead, the produce passes through a long chain of merchants, traders, wholesalers etc. before they reach the customers. This in turn leads to a monumental surge in prices. The only long term, sustainable solution that can benefit both farmers and customers is to empower them to buy and sell directly from each other. But there are 2 major challenges that obstruct direct interaction between farmers and customers. At first, it is not feasible for farmers to cater to individual customers as an individual is likely to buy not more than 10–15 kg at a time while farmers produce in tens of thousands of kilograms. Secondly there is no existing Supply Chain Management or Freight Management solution that can execute deliveries from farm to customers directly. This paper deals with the common challenges faced by the farmer's and a detailed investigation is carried out to find the possible solutions for the problem. We try to explore a possible solution to aid the dystopian situation of farmers through the system. The system tries to empower and enable farmers to engage in direct interaction and commerce with the consumers by developing a digital platform and integrating all the stakeholders in that platform.

S. Singh · A. Kumar (✉) · A. Chopra · M. D'silva
Department of Information Technology, Dwarkadas J. Sanghvi College of Engineering, Mumbai, India
e-mail: aditya2722@gmail.com

S. Singh
e-mail: sidds236@gmail.com

A. Chopra
e-mail: chopraaditya82@gmail.com

M. D'silva
e-mail: Mitchell.dsilva@djsce.ac.in

J. Tavares et al. (eds.), *Emerging Technologies in Data Mining and Information Security*, Lecture Notes in Networks and Systems 164,
https://doi.org/10.1007/978-981-15-9774-9_45

Keywords Android app · E-grocery · Logistics · Demand-manipulation · Supply chain · Dijkstra's algorithm · Ant colony algorithm and distribution · Knapsack algorithm · Travelling salesman algorithm · HDBSCAN algorithm

1 Introduction

More than 50% of India's workforce is involved in agricultural and farming activities. Despite such a large-scale involvement of people, the contribution of agriculture in India's GDP is just 15% [1]. The farm income growth rate recorded in the year 2018–19 was just 3%. This has resulted in serious distress to the farmers. Most farmers are not equipped to sell their produce to the consumers and markets in the cities. Instead they are forced to sell their produce to a nefarious nexus of middlemen. The farmers are in many cases coerced to sell their produce at prices set by the middlemen and their agents which is one of the major causes of low farm income growth rate. Additionally, due to the commission charged by this chain of middlemen, distributers, wholesalers and retailers, the price of food products increases multifold which results in an acute hike in retail inflation across the country. To provide a fair price to farmers, the government purchases several crops from farmers at a guaranteed Minimum Support Price. These crops are then stored in the government godowns and storage facilities. But the procurement, supply chain process and storage system are riddled with inefficiencies and malfunctions in India. As per the Food and Agricultural Organization of United Nations, approximately US$14 billion (more than ₹1 lakh crore) worth of food is wasted in India every year [2]. More than 40% of the produced food gets spoiled even before it reaches the consumers.

To solve some of the inefficiencies of the current system and improve the incomes of farmers in India, it is imperative to reduce their dependency on middlemen and the government and provide a direct access to consumer markets to sell their produce. The farmer needs to be empowered and be independent to decide the selling price for his produce. In the last decade or so, various e-commerce portals have come up claiming that their produce comes directly from the farmers. These portals also end up charging exorbitant rates at par with the retailers in many cases. While, these portals reduce the number of middlemen involved, they do not necessarily result in a substantial economic benefit for the customers. The price at which farmers sell their produce to these e-commerce portals is also not known.

There is a serious need to find ways to increase the farmer's income in the coming years by enabling the farmers to sell their produce directly in the markets and reach the customers without any interference from the middlemen. The proposed solution aims to tackle the above-mentioned problems faced by the farmers and the customers by enabling direct buying and selling among customers and farmers. The system aims to empower farmers by allowing them to set their own selling price for their produce. Moreover, the system will also provide customers with a real time comparison of prices with other systems in order to make a more informed choice. Food products will become more affordable for the consumers with systematic and optimized pickup and delivery of agricultural produce thereby reducing wastage.

2 Literature Survey

Farmers face various problems while selling their produce in the market. On most occasions they aren't able to make a profit. Despite selling their produce at low prices, the amount that the customer has to pay for the vegetables can be exorbitant. This occurs by a phenomenon known as demand manipulation wherein a nexus of middlemen hoards the produce for a long time to create an artificial sense of shortage in the market and manipulate the prices. Hence it has become imperative to eliminate this nexus of middlemen to ensure greater profitability for farmers as well as lower prices for customers. In [3] an Android application was suggested to connect the farmer and the retailer on a common portal.

Selling products through digital platforms has gained tremendous popularity and expanded monumentally over the years in all domains. The expansion of e-commerce has culminated in the severe disruption in the business of traditional bricks and mortar retail outlets. Numerous retail outlets have either been forced to a complete shutdown or have been made to operate at dramatically lower sales than before in turn resulting in lower profitability. But the rise of e-commerce has not impacted all industries and products with same severity. One such example is Groceries like fruits, vegetables and cereals. Groceries are still a niche in online trading but are expected to grow fast. Study in [4] compare the rise of e-grocery in France and Germany and try to establish the advent and increasing popularity of e-grocery. They also find considerable evidence to conclude that e-grocery will become a trend worldwide in the years to come [4].

For successfully integrating farmers, customers and other stakeholders in a common digital platform, the willingness and acceptability of such a platform must be established among all the stake holders. The feasibility of one such platform was assessed in Thailand in study [5] in accordance with the Thailand 4.0 economic model. The paper tries to suggest methodologies to tackle the supply chain issues faced by the farmers. After the crop production is completed, ensuring timely distribution of products into the cities, overcoming logistical challenges and fulfilling the marketing needs is a major hurdle faced by farmers all throughout the world. Without sufficient means and right channels of distribution, farmers are forced to sell their produce for dismally low prices. To address this issue the study in [5] proposes the development of a "one-stop service supply chain mobile application for the purpose of marketing, product distribution and location-based logistics for elderly farmers and consumers". This paper further tries to come up with ways to select efficient delivery routes using the Dijkstra's and Ant Colony Algorithms. The application attempts to tackle the hurdles related to product distribution and supply chain in a value-based economy with accuracy rates as high as 97%. The resourcefulness of the mobile application was observed to be high among aspects like user experience and understanding. The mobile application could be effectively used as a "tool to support farmers to distribute their agricultural products in the one-stop service supply chain" with respect to Thailand 4.0.

Similar to the Thailand Economic Model, study in [6] suggests the development of an Android Application to connect customers and farmers and eliminate the middlemen. While the aim of the study was to provide a fair price to farmers, it fails to address the transportation issues and does not incorporate a delivery system which is imperative for the success of such an integrated model.

In order to implement an efficient e-grocery delivery system, it is essential to analyse and compare the various scenarios related to e-grocery deliveries. Any integration of Farmers with Customers requires the setup of a robust delivery network. A study in [7] discusses scenarios related to "e-grocery distribution developments, in order to identify and analyse the impacts of new forms of proximity deliveries on households". The research elaborates upon the existing logistical models namely that are being used by businesses namely in-warehouse picking and in-store picking. It then classifies the three main proximity delivery approaches in e-grocery. Furthermore, the system analyses the development trends and patterns of e-grocery delivery network across the cities to understand the prevailing deficiencies and potential areas of improvement. The relationship between the development of e-grocery and the flow of goods within the city is analysed as well through the system. Finally, it studies the impact of these systems on consumer's purchasing trips and uses an empirical simulation approach in order to make a comparison of these three scenarios. The results of this study can be extrapolated to examine the feasibility and the acceptability of an e-grocery delivery model. In essence, the study tries to establish the possibility of an Uber like platform for trucks.

The development of e-grocery has empowered customers to purchase food online and benefit from the perks of having goods delivered at their doorstep. The advent of online delivery has given birth to a new challenge i.e. a high rate of failed deliveries. A high delivery failure rate due to reasons like customer's issues with the order or system error leads to serious revenue loss for the company further resulting in degradation of logistical efficiency and inventory management issues especially for perishable food. Hence, it is necessary to create safeguards and minimize failed deliveries due to absence of customers. Study [8] addresses the problem pertaining to failed deliveries and proposes an innovative approach to use customer-related data to optimize e-grocery home delivery. The approach attempts to calculate the absence probability of a customer based past electricity consumption data. The approach proposes the use of various data mining techniques to estimate the absence probability of a customer with an aim to improve the success rate of delivery and optimize transportation.

With the advent of new e-commerce portals and the rising popularity of online orders, a need to develop an efficient delivery mechanism has become more important than ever. Hyper local delivery model while popular ad convenient for many has turned out to be extremely costly in many situations. There is a need to devise more cost-effective methods for delivery without compromising on convenience. A study in [9] discusses the development of one such method using the IOT platform with the creation of Unmanned Aerial Vehicles (UAVs). UAVs have enormous prospects when it comes to deliveries in highly clustered cities. The study emphasizes the need to develop a truck-drone parcel delivery system ad proposes the creation of a "novel routing and scheduling algorithm". This algorithm groups the customers using DBSCAN clustering technique and assigns a drone to truck-drone pair to each cluster for delivery. The algorithm is similar to hyper local delivery model but it also ensures a minimization of costs.

2.1 *Existing Systems 1 and 2*

They are among the most popular e-commerce portals that are used in India. They use hyper local delivery model and inventory model. Their methods increase the prices exponentially, sometimes even more than Retail Prices. They operate on huge databases and hence make use of big data. They do not always deal directly with farmers and involve middlemen while procuring various items, hence increasing the cost for customers. These systems make use of various Predictive analysis techniques and BI tools to give recommendations to customers based on their past orders and current trends. Goods are purchases in advance in these e-commerce platforms and is then delivered to customers based on individual orders placed.

2.2 *Existing System 3*

It is an initiative to provide non-perishable food products to people at more affordable prices and a superior quality. They deal with various wholesalers and mill owners. They do not deal with farmers at all. Their operations are very slow as customers are able to place order only once a month and hence are forced to buy in bulk. The prices offered to customers are approximately 20% cheaper than Retail. They make use of Clustering concept by making groups of 12–15 customers in a given geographical area. But their clusters are not built dynamically with the use of computer technology, instead they register customers physically by going to their building and their registration involves paperwork. Orders can be placed via a website on a fixed day. Customers don't have the flexibility to order whenever they want. Due to the long delivery durations and the limitedness of the products offered, Existing System 3 hasn't gained popularity as a portal.

2.3 *Existing Systems 4 and 5*

These are Mobile Applications that attempt to connect the customers directly with farmers. The application provides customers with basic details pertaining to the farmers like their contact details, location, products sold etc. There is no online delivery mechanism provided by these portals. The customers are required to call the farmers to place an order.

3 Proposed Architecture

Farm Direct is an integrated web application that combines three different user modules i.e. Customers, Farmers and Truckers under one portal. Unlike other E-commerce websites that make use of Hyper local delivery model, Farm direct makes use of Data Mining algorithms to determine a common delivery point for customers and then completes the delivery with trucks. The system follows a 3-tier architecture as shown in Fig. 1.

3.1 System Architecture

1. Presentation Layer: This layer consists of the user interface with which any user depending upon requirements can interact. There are 3 types of users that use the application. A customer who can shop for available products, a farmer looking to sell his products online can upload his yield on the portal and a Freight Management company looking to expand its business can assign its trucks on the portal.

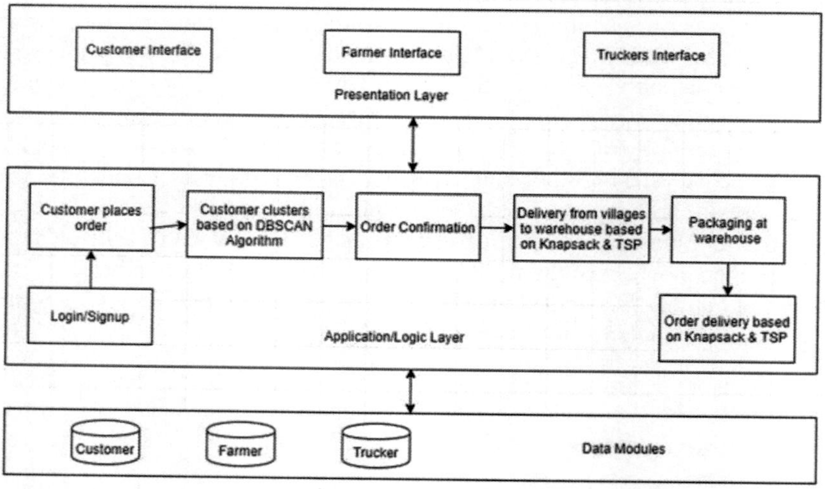

Fig. 1 System architecture

2. Application/Logic Layer: This layer provides the business logic and describes the way the system functions. The backend processing and the functionalities of the system are implemented here. The happens in the following manner:
   (a) Login/Signup: The first step to using the application for any user is to register on the portal. The user has to specify whether he/she is a customer or farmer or a trucker.
   (b) In the 2nd step, the customer places the order based on availability.
   (c) Customer Clusters: After the customers in a given region have placed their orders, a Hierarchical Density Based Spatial Clustering of Algorithms with Noise (HDBSCAN) is run on the geolocations of the customers and clusters are formed based on the density distribution of the orders in the entire city.
   (d) Once the orders are merged to form clusters, a confirmation is sent to the customers and a meeting point is assigned to them within that cluster. Details such as date and time of arrival, truck license plate number, etc. are also sent.
   (e) Order Pickup from villages: Once the orders are placed by the customers, the orders are grouped based on villages and trucks are assigned to these villages. The allotment of trucks is done by a combination of Knapsack 0–1 and Travelling Salesman Algorithm. The orders are picked from the villages and brought to the warehouse.
   (f) Processing at Warehouse: Once the goods arrive at the warehouse, they are separated and packaged based on the customer requirements.
   (g) Delivery to Customers: Once the separation and packaging is done, the orders are again grouped on the basis of the customer clusters and trucks are assigned to deliver these orders by using a combination of Knapsack 0–1 and Travelling Salesman algorithm.
3. Data Tier: The data tier or the database server consists of the customer, farmer and the trucker's database.

3.2 Application Flow

1. Customer: After registering and logging in, the customer places his order on the portal. Once all the orders are placed for the day, the HDBSCAN algorithm is run on all the orders placed in the system and the customers are allotted their respective groups (clusters) (Fig. 2).
2. Farmers: The farmers who are registered on the portal can input the different products that they wish to sell and the respective order quantities. Once the HDBSCAN algorithm is run on the customer orders, a combined total order is sent to the farmers with details about the pickup time and truck details.
3. Transport: The transport module provides some of the core functionalities of the application. It makes use of a combination of Knapsack and Travelling Salesman Algorithm to perform its functions. The transport module consists of following steps:

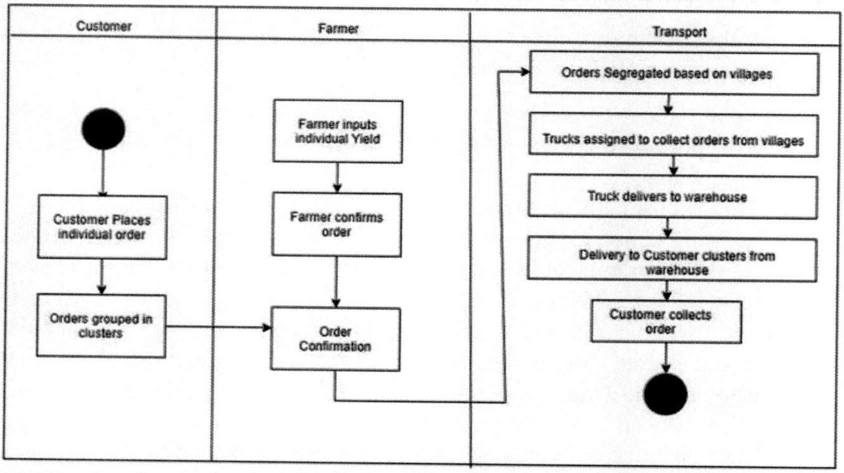

Fig. 2 Application flow

(a) Orders are first grouped based on the source or village of origin. The orders of individual villages are added to form a combined total village order (sack). These sacks are then given as input to the knapsack algorithm.
(b) The knapsack algorithm is run and trucks are assigned to the various villages based on the order quantity of the sack. The trucks are assigned the optimum delivery route from the villages to the warehouse using Traveling salesman algorithm.
(c) The truck picks up orders from villages and delivers to the warehouse.
(d) After the processing and packaging of order is done in the warehouse, a similar combination of Knapsack and Traveling salesman algorithm is used for delivering the products to the customers.
(e) The customer picks up order from the cluster meeting point.

There are various delivery scenarios that Farm Direct has to deal with. These scenarios are broadly divided based on the order quantity of the clusters formed. All customers are served at Farm Direct and no order is rejected. The scenarios pertaining to the quantity of orders are listed as follows:

Scenario 1: Total order in a day exceeds the Total truck capacity
After the customers are done placing their orders, the weight of the total order placed in the city is computed. If the total order placed exceeds the total capacity of the trucks, then orders are selected based on First Come First Serve (FCFS) and the orders beyond the FCFS threshold are kept on hold and their delivery is delayed by a day.

Scenario 2: Total order of a Cluster exceeds the capacity of the delivery truck
If the total order of any cluster exceeds the capacity of a single truck, then multiple trucks need to be allocated to that cluster for delivery. The orders in that cluster are listed based on timestamp and orders are allocated to the required number of trucks based on Knapsack Algorithm. Some of the trucks that are allocated in this case are filled with orders from just 1 cluster and hence they carry out delivery at just 1 location. The delivery route is sent to the trucks using Travelling Salesman Problem.

Scenario 3: Total order of a Cluster is less than the capacity of the delivery truck
If the total order of any cluster is less than the total capacity of any truck, then the cluster order is allocated the required space in a truck using the Knapsack Algorithm. The same truck performs delivery to multiple locations in this case. The optimum delivery route is sent to the trucks using Travelling Salesman Problem.

4 Results/Performance Comparison

There are several techniques especially for clustering that could have been used in our proposed system to implement the given functionalities apart from the ones that are mentioned in this paper. The criteria for the selection of any tool or method to solve a problem in our system is performance. In this section we will try to analyse the parameters based on which we have selected the algorithms for the implementation of our proposed system.

Route Determination: Travelling Salesman (TSP) Algorithm has been used to find the optimum pickup and delivery route in our system. TSP implements Google Maps API to compute the shortest and most efficient path. This is a standard method to determine route which is used by all freight aggregators, 3 PL companies and taxi apps like Uber and Ola.

Allocation of Load: For any logistics and dispatch management system, efficient allocation of loads is key to avoiding overheads and bottlenecks. There are 2 criterions based on which we need to allocate the trucks for pickup and delivery of produce i.e. space utilization and value of produce. Space utilization means allocating the given demand efficiently in as few trucks as possible in order to save transport cost.

In our system the value of produce is determined by the weight/distance metric. This means that cluster orders that have a higher weight by distance value will be given preference for delivery.

Based on these requirements, the most efficient dynamic programming technique for allocation of load in trucks is Knapsack 0–1 algorithm. Knapsack 0–1 uses a combination of weight and value to allocate items into a sack.

Clustering Technique: There are several clustering techniques that can be used to form geographical clusters in our system like K-means, DBSCAN, HDBSCAN and Hierarchical. The analysis of each method is given below:

Table 1 Comparison between DBSCAN and K-means

	DBSCAN	K-means
Hyperparameters	min_samples = 1, eps for 2 km radius	K = 11
No of clusters found	10	11
No of outliers found	0	0
Silhouette score: ignoring outliers	0.3109	0.4967
Silhouette score: outliers as singletons	0.3109	0.4967
Silhouette score: adjusting outliers in other clusters	–	–

K-means: This method is more beneficial for mathematical one-dimensional data as it computes clusters by finding averages and grouping close values together. Geographical data involves 2-dimensions (longitude, latitude) and hence this method becomes ineffective. Moreover, K-means forms K clusters regardless of the number of input points. This makes it difficult to treat data with varying densities.

Hierarchical: As the name suggests, this technique makes a hierarchy of clusters by gradually grouping close points together. While this is a good method to analyse the obtained data, it fails to provide good meeting points or cluster centres that can used for delivery.

DBSCAN: This technique can deal with varying densities and form clusters of varying shapes. It is also efficient in handling outliers and finding cluster centres for delivery. The only problem with this technique is the creation of excessive outliers. In our proposed system, the delivery is happening to every individual in the city and hence outlier management is an important performance criterion for any algorithm that creates the geographical clusters.

HDBSCAN: This method is an improvement on the DBSCAN clustering. It handles data with varying density better than DBSCAN, which makes it more useful for real-world scenarios. The performance of HDBSCAN in terms of outlier management and cluster creation is significantly better. The cluster creation performance is measured by an indicator known as the silhouette score.

The given table displays the performance of the given algorithms based on a sample data of 32 points in Mumbai (Tables 1, 2 and 3).

5 Conclusion

The system that has been developed between customers and farmers meets the required objectives in the paper. The portal empowers both farmers and customers by ensuring greater profitability for the farmers as well as reasonable prices for the customers.

Table 2 Comparison between DBSCAN and DBSCAN with KNN

	DBSCAN	DBSCAN with KNN
Hyperparameters	min_samples = 2, eps for 2 km radius	min_samples = 2, eps for 2 km radius
No of clusters found	4	4
No of outliers found	7	0
Silhouette score: ignoring outliers	0.3358	–
Silhouette score: outliers as singletons	0.3109	–
Silhouette score: adjusting outliers in other clusters	–	0.3281

Table 3 Comparison between HDBSCAN and HDBSCAN with KNN

	HDBSCAN	HDBSCAN with KNN
Hyperparameters	min_cluster_size = 2, min_samples = 1, eps for 2 km radius	min_cluster_size = 2, min_samples = 1, eps for 2 km radius
No of clusters found	5	5
No of outliers found	1	0
Silhouette score: ignoring outliers	0.4846	–
Silhouette score: outliers as singletons	0.4846	–
Silhouette score: adjusting outliers in other clusters	–	0.457

The system successfully performs the tasks that it is required to do. Farmers, Customers and Truckers can all access the system and buy and sell the products they want. The system successfully performs the following operations:

(1) The system can be successfully used to obtain and compare prices from e-grocery vendors like BigBasket and Grofers with the prices quoted by the farmers in real time for customers to make real-time comparisons.
(2) Clusters based on DBSCAN algorithm is created on the backend based on the locations of the order placed.
(3) The pickup point, pickup date, order status and other such details are displayed to the customers after placing the order.
(4) Farmers, Customers and Truckers are able to register and use the portal for their intended purpose.
(5) The farmers are able to upload their yield and access all the details regarding any orders received or details about the truck assigned for the pickup of the order.

(6) Truck drivers get the allotted places to carry out the pickup and delivery of products.

The system empowers farmers and customers to buy and sell farm products directly from each other. The system gives farmers greater control and independence over the pricing of their products as compared to selling middlemen and wholesalers in the APMC markets thereby allowing farmers to increase the profitability of their products. As the products come directly from the farm eliminating a chain of middlemen on the way, the system ensures products at more reasonable prices for customers. Moreover, the proposed system creates demand in a new segment for freight management companies and truckers thereby ensuring greater business opportunities and fruits of digitization for the truckers.

References

1. Ministry of statistics and programme implementation (2018–2019) Sector-wise contribution of GDP of India http://statisticstimes.com/economy/sectorwise-gdp-contribution-of-india.php Accessed 10 Mar 2019
2. H. Thacker, Food wastage in India (2018). https://thecsrjournal.in/food-wastage-india/ Accessed 10 Mar 2019
3. P.J. Gupta, H.R. Mhatre, A.P. Chaudhari, R. Pathari, Smart livestock trading, in *2018 International Conference on Smart City and Emerging Technology (ICSCET)*, Mumbai, 2018, pp. 1–7. https://doi.org/10.1109/icscet.2018.8537359
4. S. Saskia, N. Mareï, C. Blanquart, Innovations in e-grocery and logistics solutions for cities. Transp. Res. Procedia **12**, 825–835 (2016)
5. S. Nuanmeesri, Mobile application for the purpose of marketing, product distribution and location-based logistics for elderly farmers. Appl. Comput. Inf. (2019)
6. K. Khodaskar, Virtual fruits market—an application for farmer,in *2015 Fifth International Conference on Communication Systems and Network Technologies*, Gwalior, 2015, pp. 1369–1371. https://doi.org/10.1109/csnt.2015.235
7. B. Durand, J. Gonzalez-Feliu, Urban logistics and e-grocery: have proximity delivery services a positive impact on shopping trips? Procedia-Soc. Behav. Sci. **39**, 510–520 (2012)
8. S. Pan, V. Giannikas, Y. Han, E. Grover-Silva, B. Qiao, Using customer-related data to enhance e-grocery home delivery. Ind. Manag. Data Syst. **117**(9), 1917–1933 (2017)
9. W. Desheng, et al., Routing and scheduling for hybrid truck-drone collaborative parcel delivery with independent and truck-carried drones. IEEE Internet Things J. **6.6** (2019), 10483–10495

Automatic Question and Answer Generation from Text Using Neural Networks

Sonam Soni, Praveen Kumar, and Amal Saha

Abstract Generating Questions is a major action in educational learning. Questions provide various stages of complication in the educational learning procedure. Knowledge building is costly for computer Assisted evaluation when setting exercise questions, teachers use test producers to build question Banks. The paper introduces a system that produces questions with answers automatically applying natural language processing and openNMT. Previous work question generation automatically, creates questions from sentences applying syntax and semantic parser. The paper presents a new approach for generating question and answer.

Keywords Recurrent neural network · Natural language processing · Automatic question generation · Long short-term memory

1 Introduction

In This Humans nature have eager to know or learn something. Everyone ask question to grow additional information and Knowledge, and seek to connect them with other ideas that we know. Our day-to-day lives involve raising questions in discussions. Student questions compete a crucial position in their understanding and learning procedure and encourage them to learn additional from their educators. To review their knowledge teachers questions performs a crucial role. Questions play a leading resources from regular discussions to verbal tutorings and evaluations. The Majority of learners are not great at raising questions. In 1998 Dunlosky and Graesser states

S. Soni (✉) · P. Kumar
Amity University Uttar Pradesh, Noida, India
e-mail: sonamsoni.nmims@gmail.com

P. Kumar
e-mail: pkumar3@amity.edu

A. Saha
NetWest Markets, Gurugram, India
e-mail: amal.k.saha@gmail.com

J. Tavares et al. (eds.), *Emerging Technologies in Data Mining and Information Security*, Lecture Notes in Networks and Systems 164,
https://doi.org/10.1007/978-981-15-9774-9_46

that learners have a problem with finding their own understanding deficits. The Automation of question generation systems can facilitate learners to obtain their own knowledge differences by assisting them to achieve their significant inquiries. Learners are not just one who encounter limitations in their question generation skills.

Such as IBM Watson, Automation of question production systems support automated question answering systems to complete self-teaching.

As a replacement depending on human experts to manually classify ground truth answers from the questions, automatic question generation systems automatize this procedure. Smart tutoring systems can additionally get advantage from that. Instead of relying on human professionals to manually acquire questions resources (study material), each user can identify its own training system automatically from the study information. After All, question generation systems support the advancement of annotated datasets for question answering and studying comprehension.

On the contrary, in [1], suggested Hierarchical Quadrature Amplitude Modulation (HQAM) for better protection of high-priority data during image transmission. Nevertheless, no channel model such as AWGN is considered in this work.

Here, only salt and pepper noise is taken into account. A raised cosine filter is used in the AWGN channel is introduced in [2]. Here, authors evaluate the performance of the communication channel through the bit error rate (BER). In [3], M array QAM is used in AWGN.

2 Related Work

The latest technique Neural network signify the unique approach in in automatic generating question (GQ) from corpus. Du et al. [4] recommended a sequence to sequence (Seq-to-Seq) structural design for GQ which is encouraged by machine translation using neural networks In our prior work, we (2018) aimed to enhance each word with linguistic highlights and encrypt the highly appropriate important answer to the corpus while generating questions. On the other hand, [5] convert ground-truth answers, apply the copy process and as well employ context matching to obtain collaborations among the answer and its framework within the passage. In our work [1] we proposed a transformer based building to automatically generate complex multihop questions from knowledge graphs. In [2] we proposed a cross lingual training technique for automatically generating questions from text in low resource languages. Very newly DRL (Deep reinforcement learning) has been effectively employed to natural language generation process like summarization of a document and make a short abstract and dialogue generation. Content summarization is method to produces and paraphrases sentences that depict applicable points of the corpus. Generating questions in addition includes deciding different types of questions such as 'who', 'when', 'what' etc., the system is precise on which keywords to extract the question, exiting lingering keywords for the answer. The process involves the advancement of a precise probabilistic procreative template. In

[6] recommended generative adversarial network (GAN) framework with improved discri minator to predict question type. Newly [7] proposed a bidiscriminator structure for visual question generation. They devise the job of question generation visually as a language generation chore with some grammatical as well as subject matter certain characteristics (Fig. 1).


```
{
   "paragraphs":[
     {
       "context": "Super Bowl 50 was an American football game to
       determine the champion of the National Football League (NFL)
       for the 2015 season. The American Football Conference (AFC)
       champion Denver Broncos defeated the National Football
       Conference (NFC) champion Carolina Panthers 24\u201310 to
       earn their third Super Bowl title. The game was played on
       February 7, 2016, at Levi's Stadium in the San Francisco
       Bay Area at Santa Clara, California. As this was the 50th
       Super Bowl, the league emphasized the \"golden anniversary\"
       with various gold-themed initiatives, as well as temporarily
       suspending the tradition of naming each Super Bowl game with
       Roman numerals (under which the game would have been known
       as \"Super Bowl L\"), so that the logo could prominently
       feature the Arabic numerals 50.",
       "qas": [
         {
           "answers": [
             {
               "answer_start": 177,
               "text": "Denver Broncos"
             }
           ],
           "id": "56be4db0acb8001400a502ec",
           "question": "Which NFL team represented the AFC at
           Super Bowl 50?"
         },
         {
           "answers": [
             {
               "answer_start": 403,
               "text": "\"golden anniversary\""
             },
             {
               "answer_start": 521,
               "text": "gold-themed"
             },
             {
               "answer_start": 521,
               "text": "gold"
             }
           ],
           "id": "56bea9923aeaaa14008c91b9",
           "question": "What was the theme of Super Bowl 50?"
         }
       ]
     }
   ]
}
```


Fig. 1 Example data from squad

3 Proposed Method

This Rajpurkar et al. published the dataset in 2016 for Question Answering the Stanford Dataset which is also knows as SQuAD. SQuAD comprises of 100,000+ questions on 500+ articles. SQuAD is a reading comprehension dataset.

The datasets present provides Questions which are asked by group employees on a programmed of Wikipedia articles. The dataset overcomes with the issues like size and semi-synthetic for every paragraph, there are separate human generated questions. For each human created question, answers are provided along with their answer start index with respect in the paragraph that the question is generated.

3.1 Approach

The model uses sequence_to_sequence and LSTM (Long Short-Term Memory). The model shows one of the exceedingly powerful models in deep learning. The model started off in translation but has since moved on to question answering systems (Siri, Cortana etc.), audio transcribing etc. the model is useful for converting one sequence to another sequence.

It includes RNN as encoder as well as RNN as decoder which is the major concept of the model. The input sequence 'Understand' and the decodes that is output decoder in thought vector and generates a output a sequence. The brand-new approach for machine translation is Neural machine translation. The approach commanded a notable enhancement, especially in terms of human assessment, related to rule-based and statistical machine translation (SMT) systems. The approach at first established using seq-to-seq models and enhanced upon using attention-based variants. For performing various NLP tasks like parsing, translation, summarization etc. NMT is one of the grown methods and extensively applied approach for machine translation [8]. The approach in NMT are highly standard, as it develops extra importance for the machine translation. The NMT community developed accessible implementations for scientists as well as researchers to learns from and expand the approach. Just as the SMT community benefited substantially from toolkits like Moses for phrase-based SMT and CDec for syntax-based SMT, NMT toolkits can provide a foundation to build upon. The aim is to deliver a schematic image from a toolkit of machine translation. The sentences first converted into word vectors. As the converted vectors are than given as an input to a recurrent neural network. The prediction $p(wt|w1{:}\ t-1, x)$ of the next word is produce at each time step, after applying the attention is on the Recurrent Neural Network. After that the final output will be merged with the existing hidden state for prediction.

3.2 LSTM Seq-to-Seq

Generating questions Automatically from corpus can be formulated as a learning problem in a sequence-to-sequence form. Provided a text A, series of words, the objective of the series is to produce a question Q, which is syntactically and semantically accurate, significant, appropriate and natural. Additionaly explicitly, the major purpose is to find out the original conditional probability distribution $P\theta(Q|A)$ parameterized by θ. The aim is to understand a model θ through training applying text-question pairs such that the probability $P\theta(Q|A)$ is expanded across the given training dataset.

More correctly let us represent the text A as a sequence of words (a_1, a_2, a_3,..., a_K) of length K, and a question Q as a sequence of words (b_1, b_2, b_3,..., b_P) of length P. Scientifically, the model is meant to generate $Q*$ such that $Q* =$ argmax $QP\theta(Q|A)$ where $P\theta(Q|A)$ is defined as:

$$P\theta(Q|A) = P\prod \mathrm{i} = 1\mathrm{P}\theta(b_i|b_1, \ldots, b_{i-1}; a_1, \ldots, a_K) \quad (1)$$

Usually, such a prototype is trained by enhancing the cross-entropy loss. As motivated earlier, such a loss is difficult to decoding-time errors.

The agenda for question generation comprises of a generator and an evaluator. We refer to the generator as the agent and to the generation of the next word as an action. The possibility of translating a word $P\theta$ gives nonrandom policy. An evaluator allocates a reward for each and every single token that is sequence of output. On the basis of the rewards given by the evaluator, generator enhances and revise its current policy.

Let us imply the reward at time step t by rt. The aggregate reward computed at the end of the sequence generated is represented by $R = \sum Tt = 0rt$. The goal in RL-based question generation is to determine a generator (policy) that maximizes the expected return:

$$\text{Loss } RL(\theta) = EP\theta(y0 : T|X)T \sum t = 0rt(yt; X, Y0 : t - 1) \quad (2)$$

X is the existing input, $Y0$: $t - 1$ is the expected sequence until time $t - 1$.

4 Simulation and Experimental Results

The associated perplexity constantly varies on the actual goal data when accessible (training, validation and gold data). Else it is the perplexity of the model's own prediction. It is calculated by:

$$\exp(\text{loss}/\text{no. of goal words})$$

The loss will be in a negative likelihood of the target data. Such that is will convey, how the presented model will be deliver theses targets.

PS (score_pred) is a score of generated sequences by log likelihood

PAS (score_avg_pred) is the score of log possibility per generated word

PPPL (PPL_pred) is the perplexity of the model's own predictions (exp(-PAS))

In the above model, we use $d = e$ (Where e is exponential, and it can be $d = 2$).

M given here is a total counts of words (number of samples), we are working with unigram. The lowest rate is when $q(xi) = 1$, the perplexity will be exp(0) = 1 as it be predict the samples using highest confidence. The highest rate is when $q(xi) = 0$, the perplexity will be exp(+inf) = +inf as it be predict the samples using lowest confidence. But in our case we have used only pred_score we don't need avg of pred score nd pred perplexity.

$$\text{Pred\_Score} = d - 1/N \sum N1 \log b\, q(xi)$$

From Table 1 the text has been taken as an input, after applying the algorithm the set of questions as well as answers are generated using the model and Squad dataset. The generated Questions with answers from the system is evaluated in the above table on the basis of prediction score.

From Fig. 2 the bar graph represents the prediction score of generated question on the basis of perplexity score. There are more ways to find the accuracy of questions like BLEU, ROUGE-L, METEOR.

Table 1 Generated prediction score of questions

Text	Question and answer generated	Prediction score
The army has been involved in tour wars with neighboring Pakistan and one with China	What military army has been involved in four wars? Indian Air	−2.67
	How many Wars? Four	−3.100
Other major operations undertaken by the army includes operations Vijay, operations meghdoot and operation cactus	What country has a major component of national power? China	−6.28
	What is the name of operation ten-go? Operation cacms	−10.10
Apart Horn conflicts, the army has conducted large peace time exercises like operation brasstacks and exercise shooveer, and it has also been an active participant in nuineriberia, Mozambique, South Sudau and Somalia	Along with Libya, what cities did Namibia get active? South Sudan	−11.20

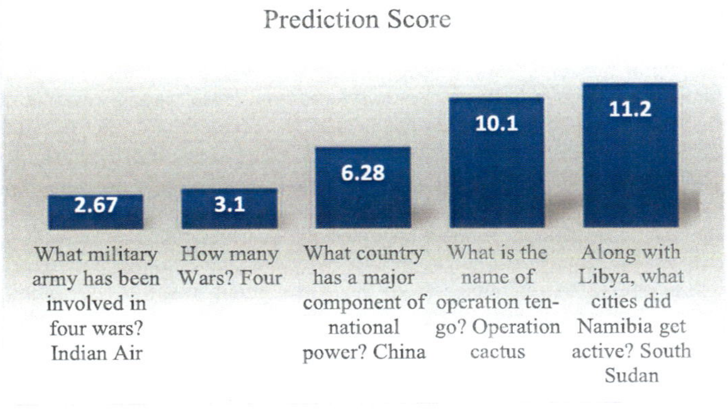

Fig. 2 Bar graph of generated prediction score

5 Conclusion

The paper introduced a unique, complete and identical concept of question generation (QG) using a LSTM sequence to sequence. The system presented in the paper basically factoring in question syntax and semantics, and finds the most crucial answers, identifies contextually crucial words and prevents meaningless replications. As we can see from Table 1 the quality of questions is not up to the mark. The sentences which contains a country as a NER, the question should be generated like which country. Instead of 'which' the system is generating a 'what' type question. The answer prediction is much better than the previous algorithm. As future work we need to focus more on quality of questions as compared to this system.

References

1. V. Kumar et al., Difficulty-controllable multi-hop question generation from knowledge graphs, in *International Semantic Web Conference* (Springer, Cham, 2019)
2. V. Kumar et al., Cross-lingual training for automatic question generation. arXiv preprint arXiv:1906.02525 (2019)
3. B. McCann, N.S. Keskar, C. Xiong, R. Socher, The natural language decathlon: multitask learning as question answering. arXiv preprint arXiv:1806.08730 (2018)
4. X. Du, J. Shao, C. Cardie, Learning to ask: neural question generation for reading comprehension. arXiv preprint arXiv:1705.00106 (2017)
5. L. Song, Z. Wang, W. Hamza, Y. Zhang, D. Gildea, Leveraging context information for natural question generation, in *Proceedings of the 2018 Conference of the North American Chapter of the Association for Computational Linguistics: Human Language Technologies*, vol. 2, (Short Papers, 2018), pp. 569–574
6. K. Yao, L. Zhang, T. Luo, L. Tao, Y. Wu, Teaching machines to ask questions,. in *IJCAI*, pp. 4546–4552 (2018)

7. Z. Fan et al., A reinforcement learning framework for natural question generation using bi-discriminators. in *Proceedings of the 27th International Conference on Computational Linguistics* (2018)
8. A. Chaturvedi, P. Kumar, S. Rawat, Proposed noval security system based on passive infrared sensor, in *2016 International Conference on Information Technology (InCITe)-The Next Generation IT Summit on the Theme-Internet of Things: Connect your Worlds*. (IEEE, 2016), pp. 44–47
9. A. Celikyilmaz, A. Bosselut, X. He, Y. Choi, Deep communicating agents for abstractive summarization. arXiv preprint arXiv:1803.10357 (2018)
10. V. Kumar et al., Automating reading comprehension by generating question and answer pairs, in *Pacific-Asia Conference on Knowledge Discovery and Data Mining*. Springer, Cham (2018)
11. A. See, P.J. Liu, C.D. Manning, Get to the point: summarization with pointer-generator networks. arXiv preprint arXiv:1704.04368 (2017)
12. Y. Zhao, X. Ni, Y. Ding, Q. Ke, Paragraph-level neural question generation with maxout pointer and gated self-attention networks, in *Proceedings of the 2018 Conference on Empirical Methods in Natural Language Processing*, pp. 3901–3910 (2018)
13. N. Zaidi, H. Kaushik, D. Jain, R. Bansal, P. Kumar, Analytical comparison of concurrency control techniques, in *Information Systems Design and Intelligent Applications*. (Springer, Singapore, 2018), pp. 1012–1021
14. A. Tharad, A. Bhatt, S. Srivastava, P, Kumar, Analysis and impact of current captcha approaches and its significance, in *2018 International Conference on Computational Techniques, Electronics and Mechanical Systems (CTEMS)*. (IEEE 2018), pp. 487–493
15. V. Kamra, P. Kumar, M. Mohammadian, Formulation of an elegant diagnostic approach for an intelligent disease recommendation system, in *2019 9th International Conference on Cloud Computing, Data Science & Engineering (Confluence)*. (IEEE 2019), pp. 278–281

A Multi-Criteria Prioritization Based Slot Allocation Scheme in IEEE 802.15.4 for Wireless Body Area Networks

Sathi Roy and Chandreyee Chowdhury

Abstract Wireless Body Area Network (WBAN) is a network of wearable and/or implantable sensor nodes that transmits body vitals to the sink for remote health monitoring of patients. Data and sensor node prioritization in WBAN is a challenging problem as loss of important data may prove to be fatal. In this paper, a multi-criteria environment is considered to allocate Guaranteed Time Slots (GTSs) to the sensors in IEEE 802.15.4 MAC protocol standard. We have focused on allocating GTSs to the sensors based on their need. For that the sensors are prioritized that is hardly considered by the existing works. Fuzzy Analytical Hierarchy Process is applied here to prioritize the sensors for allocating them the GTSs based on the following criteria—Remaining energy, Event Occurrence, Buffer Index, Data Slot, Link Quality and Recently Used Index. The results indicate that each sensor eventually gets a chance to transmit thus preventing data loss and starvation.

Keywords WBAN · FAHP · MAC · Priority · Starvation · Remote health

1 Introduction

Nowadays, there is a huge demand for remote healthcare monitoring [1, 3] of the elder adults and patients suffering from chronic diseases. One of the basic building block of such monitoring systems is Wireless Body Area Network (WBAN) [5]. It is a network of wearable/implantable sensors that sense and transmit the sensed body vitals to a sink, Local Monitoring Unit (LMU). The sink in turn transmits the data from the WBAN to a central server over the Internet. IEEE 802.15.4 standard is generally used for data communication within a WBAN configuration. Here, the

S. Roy
Asian International School, Howrah, West Bengal, India

C. Chowdhury (✉)
Jadavpur University, Kolkata, India
e-mail: chandreyee.chowdhury@gmail.com

J. Tavares et al. (eds.), *Emerging Technologies in Data Mining and Information Security*, Lecture Notes in Networks and Systems 164,
https://doi.org/10.1007/978-981-15-9774-9_47

nodes first contend for slot(s) and then the LMU allocates Guaranteed Time Slots (GTSs).

Few works could be found on developing a slot allocation scheme for WBAN. Albi et al. [1] proposed an algorithm to improve the superframe structure to minimize the delay. If there is a request for GTS allocation, then the Contention free period starts immediately after the beacon. Otherwise, contention access period starts after the beacon. Kajima et al. [4] proposed an algorithm for periodic data transmission in a superframe to conserve energy of the sensors. This algorithm optimizes the active period of a sensor depending on the data amount it has to send. However, in a single superframe, only one node can get the chance to access the GTS. Zhang et al. proposed a priority guaranteed MAC protocol [7] which actually divides the active portion of the superframe into five sub sections to support collision free data communication. This protocol also considers two separate channels for medical and non-medical data transmission. Allocating GTSs to the sensor devices prevents data loss during transmission. However, it is more effective with multi-criteria fuzzy system for sensor prioritization that the existing work hardly considered. In [6], a multi-criteria prioritization scheme is presented which is based on Analytical Hierarchy Process (AHP). However, a WBAN configuration may contain up to 15–20 sensors. So, allocating GTSs to all sensors in a single superframe is not feasible. So, the selection of important sensors should be performed before slot prioritization while taking care of starvation.

This has motivated us to propose a sensor node prioritization scheme for GTS allocation based on a decision making process named as Fuzzy Analytical Hierarchy Process (FAHP). The severity of a data packet depends on a few factors like emergency conditions and buffer space as sensors have a limited amount of memory to store data before transmission. Battery life is another important factor as recharging is not feasible always. Interestingly, a sensor may ask for more than one slot to the LMU while assigning in that way may lead to starvation of some nodes. This is taken care of by the scheme. The proposed scheme facilitates the GTS management based on the mutual preference among the criteria while increasing the number of sensors.

The remaining section of the paper is organized as follows. State of the art works are summarized in Sect. 2, a brief description of the system model. The proposed scheme is detailed in Sect. 3 followed by discussion of results in Sect. 4. Finally, Sect. 5 concludes.

2 System Model

The system model represents a WBAN configuration consisting of N physiological body sensor devices, and one LMU. The LMU is so placed that all the physiological body sensor nodes are connected to it. The nodes sense the physiological parameters and sends the raw sensed data to the LMU. The LMU receives all the data from the N sensor nodes and aggregates it as described in Fig. 1. IEEE 802.15.4 protocol standard is in place. In the beacon enabled mode of IEEE 802.15.4 standard [2],

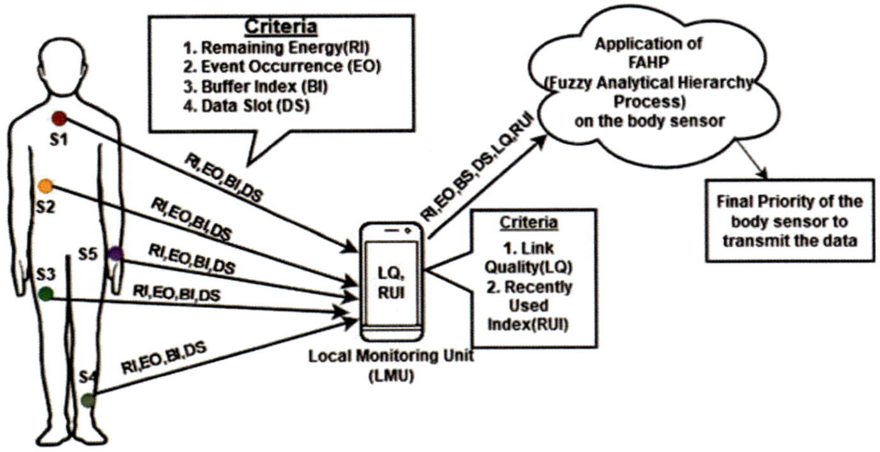

Fig. 1 Working model of the proposed scheme

LMU sends the beacon in certain time intervals to communicate with the connected nodes. The interval between two beacons is called the superframe.

Generally, the superframe is divided into two main parts active and inactive. The inactive mode is the low power consumption state for each of the connected devices whereas, the active section can also be divided into two sub sections- Contention Access Period (CAP) where the connected devices contend with each other to get access to the channel and Contention Free Period (CFP) where a non-preemptive guaranteed time slot is allocated to each of the nodes to transmit its data to the LMU. The proposed algorithm is focused on the CFP part of the superframe. In general, the sensor devices initiate the process by requesting for GTS slots to the LMU. The LMU then dynamically allocates a time slot to the sensor node. After that, the body sensor nodes transmit its buffered data to the LMU in that prescribed time slot.

3 Proposed Prioritization Scheme

The proposed scheme first selects a number of sensors that are important over the others to transmit data in the present superframe. Then, FAHP is applied to prioritize the sensors on the basis of multiple criteria. The GTS slots are allocated according to their priority. The generalized framework of FAHP consists of three parts- *goal*,*criteria*, and *alternatives*. In this problem, *goal* is to prioritize the connected body sensors. The *criteria* are network-specific or sensor-specific attributes based on which the prioritization would take place. Lastly, *alternatives* are the sensor nodes whose priority value would get computed by this model. Here, six attributes and its significance are described in Table 1.

Table 1 Significance of each criteria and corresponding values

Term	Significance	Values
Remaining energy (RE)	The remaining energy of a sensor	Full (00), almost full (01), less (10), very less (11)
Event occurrence (EO)	The type of sensing event occurring at a time instant	No. event (00), Regular event (01), High priority event (10)
Buffer index (BI)	The buffer status at a time instant	Empty (00), almost empty (01), almost full (10), and full (11)
Data slot (DS)	The no. of data slots solicited by a sensor at a time instant	One (01), two (10), three (11)
Link quality (LQ)	The quality of the link between a sensor with its LMU	Strong (00), average (01), weak (10), very weak (11)
Recently used index (RUI)	The number of superframe intervals a sensor is waiting to get a slot	Zero (00), one (01), two (10), three (11)

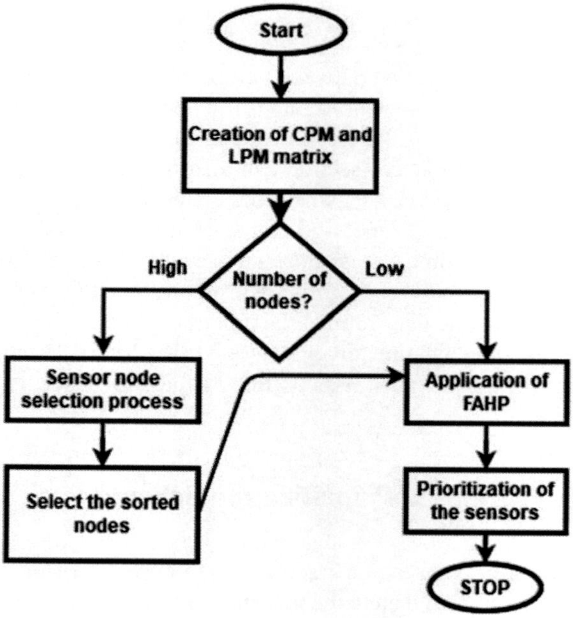

Fig. 2 Flowchart of the proposed algorithm

After receiving RI, EO, BI and DS values from all sensors, and LQ and RUI from itself, the LMU uses FAHP model to provide Guaranteed Time Slots (GTS) to all the sensors. The flow chart of the algorithm is shown in Fig. 2.

One criterion can be more important over the others and this preferences can change dynamically in nature. Although, the preference of one criteria over the other

Table 2 Significance of criteria preference value and the corresponding scale

Value	Explanation according to the preference
(1, 1, 1)	Preference are same
(2, 3, 4)	Moderately high preference
(4, 5, 6)	Strongly preferred
(6, 7, 8)	Very strongly preferred
(9, 9, 9)	Extremely high preference
(1, 2, 3), (3, 4, 5), (5, 6, 7), (7, 8, 9)	Intermediate values between two consecutive judgement

criteria cannot be crisp value always. So, the mutual preference of the R criteria are quantified as a $R \times R$ Criteria Preference Matrix, CPM = $[\alpha_{ij}]$. The preference of ith criteria over jth is represented as α_{ij}. Where α_{ij} is the fuzzy set of having lower, middle and upper limit, represented as (L_{ij}, M_{ij}, U_{ij}). The relationship between the two is maintained as $\alpha_{ji}= \frac{1}{\alpha_{ij}} = (\frac{1}{U_{ij}}, \frac{1}{M_{ij}}, \frac{1}{L_{ij}})$.

The possible values of α_{ij} and its assessments are summarized as Table 2.

α_{ij} is not static, rather it is subject to change in every turn, which makes the proposed algorithm even more dynamic in nature.

To check for consistency of the given preferences, LMU executes few more steps given by the generic algorithm of FAHP. At the very first step, the LMU calculates fuzzy geometric mean value,which is Rth root of row wise multiplication of each cell element for each row. It can be represented as follows.

$$\beta_i = (\prod_{j=1}^{r}(\alpha_{i,j}))^{\frac{1}{R}} \tag{1}$$

where, $i, j \in 1, 2, 3, \ldots, R$ represent the row and column of the CPM matrix respectively. Before doing column-wise normalization, calculation of fuzzy wight(W_i) is needed. It can represented as:

$$W_i = \beta_i \otimes (\beta_1 \oplus \beta_2 \oplus \cdots \oplus \beta_R)^{-1} \tag{2}$$

By using geometric mean defuzzification of W_i can be done. Which can be represented as $W_i = \dfrac{(L_i + M_i + U_i)}{3}$.

Now, the ith element of the normalized column vector can be represented as:

$$\gamma_i = \frac{W_i}{\Sigma_{i=1}^{R} W_i} \tag{3}$$

$$\lambda_{\max} = \frac{(\Sigma_{i=1}^{R} \gamma_i)}{R} \tag{4}$$

Here, to check the consistency of $\lambda_{\max}$ LMU needs to compute the Consistency Index (CI) as $\mathrm{CI} = \frac{\lambda_{\max}-R}{R-1}$.

Here, R represents the total number of criteria of the system. Finally, the LMU calculates the Consistency Ratio (CR) which can be represented as $\mathrm{CR} = \frac{\mathrm{CI}}{\mathrm{RCI}}$. Here, RCI represents the Random Consistency Index (RCI). The RCI value corresponding to a particular R is fixed.

If $\mathrm{CR} \leq 0.1$, then the system is consistent, otherwise, the value needs to be changed. The CR value shows the robustness of the system which can tolerate upto 10% of inconsistency. The mutual preference of one criteria over the other as well as the values of individual criterion may vary from one node to the other for a certain period of time. So, it can be represented as a Local Preference Matrix (LPM) as described below.

$$\mathrm{LPM} = \begin{array}{c} \\ S_1 \\ S_2 \\ \vdots \\ S_N \end{array} \begin{array}{c} \begin{array}{cccc} C_1 & C_2 & \cdots & C_R \end{array} \\ \begin{pmatrix} \mathrm{lpm}_{11} & \mathrm{lpm}_{12} & \cdots & \mathrm{lpm}_{1R} \\ \mathrm{lpm}_{21} & \mathrm{lpm}_{22} & \cdots & \mathrm{lpm}_{2R} \\ \vdots & \vdots & \ddots & \vdots \\ \mathrm{lpm}_{N1} & \mathrm{lpm}_{N2} & \cdots & \mathrm{lpm}_{NR} \end{pmatrix} \end{array} \quad (5)$$

Here R and N represent the number of criteria and the number of body sensors of the system respectively. FP_{ij} represents the local preference of the ith node about the jth criteria. From the existing LPM matrix, the column wise average value would fix the level of each criteria. If a sensor's 50% criteria cross the level, then that sensor participates in FAHP. Likewise, LMU generates the revised LMP and then it is normalized. However, if the number of sensors is less (all can be accommodated in the present superframe), then node selection is not needed. The final priority of ith node, FP_i can be calculated based on the normalized LPM matrix as $\mathrm{FP}_i = \Sigma_{k=1}^{R}(\mathrm{LPM}_{ik} * W_i)$.

FP matrix stores the final priority value of each of the sensor nodes. LMU allocates the GTSs to the sensor nodes following descending order of the FP matrix values.

4 Analytical Results

A WBAN configuration typically consists of 5–20 nodes [5]. The experimental setup considers a WBAN configuration with 15 nodes. This setup is executed in PYTHON 3.7. The criteria preference of the nodes are denoted as a tuple [RI EO BI DS LQ RUI]. Thus, S_1 : [3 1 2 1 2 1], S_2 : [1 2 3 2 1 1], S_3 : [3 2 0 1 0 1], S_4 : [3 1 3 3 1 1], ... upto S_{15}. The corresponding CPM is as follows.

$$\begin{array}{c} \\ RI \\ EO \\ FL \\ DS \\ LQ \\ RUI \end{array}\begin{array}{c} \begin{array}{cccccc} \text{RI} & \text{EO} & \text{BI} & \text{DS} & \text{LQ} & \text{RUI} \end{array} \\ \begin{pmatrix} (1,1,1) & (\frac{1}{4},\frac{1}{3},\frac{1}{2}) & (2,3,4) & (\frac{1}{4},\frac{1}{3},\frac{1}{2}) & (2,3,4) & (\frac{1}{3},\frac{1}{2},\frac{1}{1}) \\ (2,3,4) & (1,1,1) & (4,5,6) & (1,1,1) & (2,3,4) & (\frac{1}{3},\frac{1}{2},\frac{1}{1}) \\ (\frac{1}{4},\frac{1}{3},\frac{1}{2}) & (\frac{1}{6},\frac{1}{5},\frac{1}{4}) & (1,1,1) & (\frac{1}{4},\frac{1}{3},\frac{1}{2}) & (2,3,4) & (\frac{1}{3},\frac{1}{2},\frac{1}{1}) \\ (2,3,4) & (1,1,1) & (2,3,4) & (1,1,1) & (2,3,4) & (\frac{1}{3},\frac{1}{2},\frac{1}{1}) \\ (\frac{1}{4},\frac{1}{3},\frac{1}{2}) & (\frac{1}{4},\frac{1}{3},\frac{1}{2}) & (\frac{1}{4},\frac{1}{3},\frac{1}{2}) & (\frac{1}{4},\frac{1}{3},\frac{1}{2}) & (1,1,1) & (\frac{1}{3},\frac{1}{2},\frac{1}{1}) \\ (1,2,3) & (1,2,3) & (1,2,3) & (1,2,3) & (1,2,3) & (1,1,1) \end{pmatrix} \end{array} \quad (6)$$

Figure 3 represents the priority values of the body sensors according to FAHP and AHP solution after eliminating less important sensors. As, discussed before that

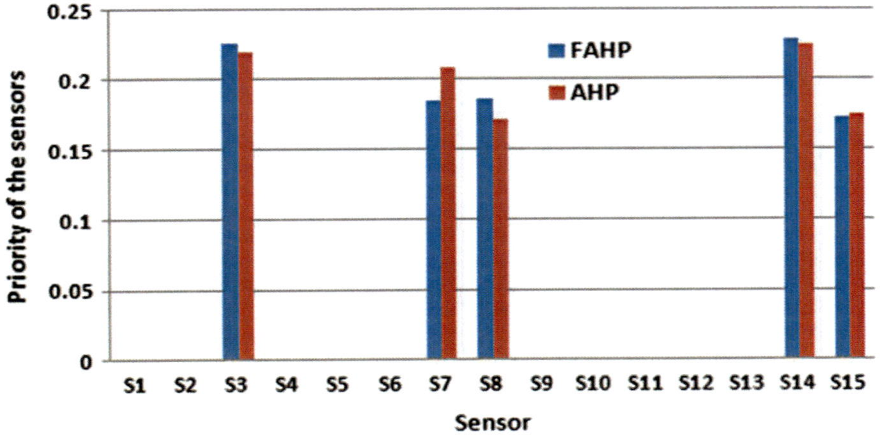

Fig. 3 Priority of each sensor nodes after applying the CPM and LPM

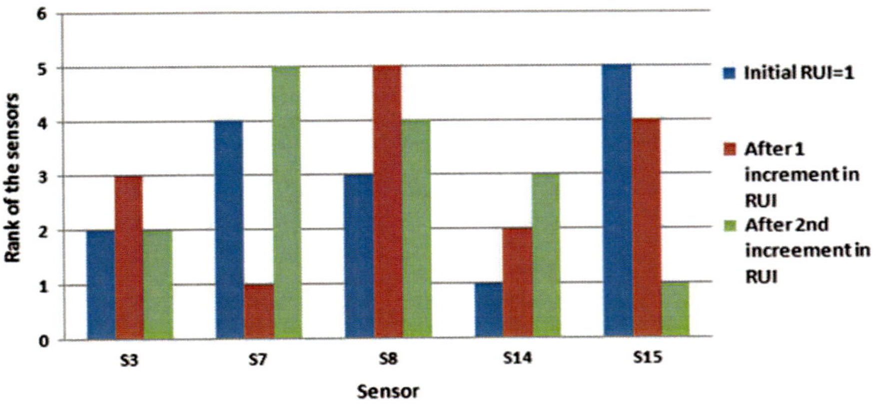

Fig. 4 Priority of the sensor after increment on the RUI value in every superframe structure

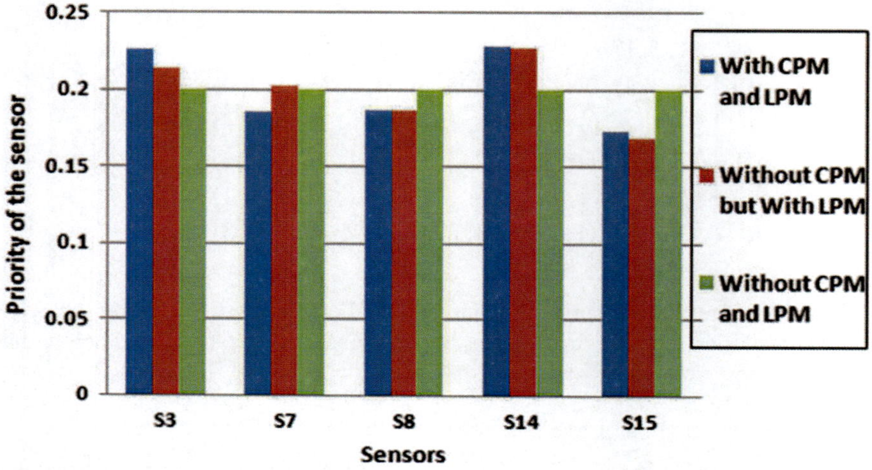

Fig. 5 The effect of criteria preference

criteria preference can not be crisp value all the time, so fuzzyness is applied. As both nodes S_3 and S_{14} contain emergency data, so they are assigned higher priority by both AHP and FAHP. As, S_8 has very less RI value, after comparing all six criteria, FAHP has given it the third slot to minimize data loss which AHP does not consider.

The next experiment explores the effect of RUI to minimize starvation. As can be observed from Fig. 4, S_{15} got the fifth position initially, in terms of priority of the nodes. But after increment in the RUI value in the next superframe, it has the fourth position and in the next slot first position. Thus, eventually it gets a chance to transmit its data packet.

Figure 5 shows the importance of the preferences among the criteria. S_3 has emergency data packet as indicated by the CPM. However, in absence of mutual criteria preference, it would not get a chance to transmit before the others. Thus, relative priorities are changed depending on the mutual preference of different criteria. To ascertain the effectiveness of the proposed solution, it is compared with another strategy, Random FCFS. In case of Random FCFS, the priority of all the sensors are same, so the average waiting time are same for all the sensors. However, almost 40% of the body sensors' waiting time (Fig. 5) is found to get reduced through the proposed scheme thus reducing the starvation possibilities.

5 Conclusion

In this paper, a multi-criteria prioritization scheme is proposed for body sensor nodes of a WBAN for transmission slot allocation through GTSs according to IEEE 802.15.4 standard. FAHP is applied to solve the multi-criteria decision problem along with mutual preference among the criteria. The LMU of a WBAN configuration is supposed to execute the scheme to assign node priorities. GTS slots are allocated in ascending order of priority, that is, more priority leads to allocation of more GTS slots as solicited by the sensors. Node starvation is also taken care of.

We plan to extend this work to inter-BAN communication where several WBANs in vicinity may compete and/or collaborate to transmit data to a medical server. This work will be challenging due to inter-BAN interference and relative mobility of the users wearing WBAN sensors.

References

1. A.N. Alvi, S.H. Ahmed, M.A. Yaqub, N. Javaid, S.H. Bouk, D. Kim, An improved ieee 802.15.4 superframe structure with minimum delay and maximum cfp link utilization. Adhoc Sens. Wirel. Netw. **35**, 151–171 (2017)
2. L., De Nardis, M.G. Di Benedetto, Overview of the ieee 802.15. 4/4a standards for low data rate wireless personal data networks, in *2007 4th Workshop on Positioning, Navigation and Communication*, pp. 285–289 (IEEE, 2007)
3. J.Y. Khan, M.R. Yuce, G. Bulger, B. Harding, Wireless body area network (wban) design techniques and performance evaluation. J. Med. Syst. **36**(3), 1441–1457 (2012)
4. F. Kojima, H. Harada, Long-lived smart utility network management using modified ieee 802.15. 4 mac, in*2010 IEEE International Conference on Communications Workshops*, pp. 1–5 (IEEE, 2010)
5. M. Roy, C. Chowdhury, N. Aslam, Designing transmission strategies for enhancing communications in medical iot using markov decision process. Sensors **18**(12), 4450 (2018)
6. S. Roy, I. Mallik, A. Poddar, S. Moulik, S-mac: Prioritized allocation of gtss in ieee 802.15. 4 mac protocol-a dynamic approach based on analytic hierarchy process, in *2017 14th IEEE India Council International Conference (INDICON)*, pp. 1–6 (IEEE, 2017)
7. Y. Zhang, G. Dolmans, A new priority-guaranteed mac protocol for emerging body area networks, in *2009 Fifth International Conference on Wireless and Mobile Communications*, pp. 140–145 (2009)

A Secure Lightweight Authentication Protocol for Mobile Payment

Gungun Sharma and Mohona Ghosh

Abstract In recent times, the numbers of mobile commerce (m- commerce) users have increased due to multiple benefits and convenience m-commerce applications offer. As m-commerce purchases and payments are done in online mode through mobile devices, security aspects of these applications are of utmost concern. In the last decade, many protocols like secure electronic transaction (SET) protocol, iKP etc. have been proposed to address these concerns. LPMP (Lightweight Protocol for Mobile Payment) is one of them which fulfills the basic requisites of security which are confidentiality, integrity, authentication, and non-repudiation and have been shown to be better in terms of memory, limited processing power, key generation and computation time in comparison to other existing protocols. In this work, we analyses LPMP protocol and show multiple flaws and drawbacks in this protocol. The first flaw involves usage of multiple secret keys between client and issuer due to which both (client and issuer) have to maintain and keep the order of the set of secret keys which requires memory space on both sides. The next flaw which we have observed involves forwarding of digital certificate by unauthorized users. LPMP protocol also lacks the anonymity and unlinkability properties of transaction and replay attack (which is done by the merchant) is not handled by this protocol. In this work, we discuss various mitigations that can be done and propose an improved version of LPMP which is more secure and memory efficient.

Keywords Mobile payments · M-Commerce · Lightweight protocol for mobile payment

G. Sharma · M. Ghosh (✉)
Department of Information Technology, Indira Gandhi Delhi Technical University for Women, New Delhi, Delhi, India
e-mail: mohonaghosh@igdtuw.ac.in

G. Sharma
e-mail: gungunsharma61195@gmail.com

J. Tavares et al. (eds.), *Emerging Technologies in Data Mining and Information Security*, Lecture Notes in Networks and Systems 164,
https://doi.org/10.1007/978-981-15-9774-9_48

1 Introduction

Mobile commerce, also known as m-commerce, is a way of buying and selling goods and products with the help of handheld devices such as smart phones and tablets etc. Examples of m-commerce include mobile banking; in-app purchasing and virtual marketplace apps such as Amazon mobile app or digital wallet such as Apple pay, and Samsung pay [1, 2]. Through this, users do not need to go to the store or do not need to carry a physical wallet or credit card, instead they can buy and sell the product anytime and anywhere. It provides more security and privacy than the physical credit card. As m-commerce applications have become one of the driving revenue generators and all purchases and payments are done in online mode, security of these applications is of paramount importance.

Mobile commerce has so many advantages in terms of ordering goods or services and time, but it has some limitations, i.e., restricted bandwidth and constrained resources when applied to mobile devices. There are many protocols proposed which are based on the public-key infrastructure or private-key infrastructure like secure electronic transaction(SET) protocol [3], iKP [4], modified SET [5], mobile pay center protocol [MPCP] [6], LPMP with two payment gateway [7] and many more [8–11].

Since SET protocol requires high computational operation and limited capacity of mobile devices, it is not suitable for mobile devices. Like SET, iKP cannot be used for mobile devices because of its high computational operation. Modified SET is an improved version of SET which uses the symmetric key encryption/decryption instead of public key encryption/decryption but it does not provide non-repudiation property.Limited power processing and memory space of low-end mobile devices are some of the challenges while designing payment protocol for constrained devices. While high end mobile devices have been proposed that have good power processors and memory space, they are expensive and hence could not be very popular.

Keeping these aspects in mind, authors in [12] proposed an efficient lightweight protocol for mobile payment system (LPMP) which covers all these flaws. LPMP is a secure lightweight protocol which is based on the secret key between client and issuer. Many security requirements that are satisfied by the mobile payment are privacy, integrity, authentication, and non-repudiation. It does not require any public key operation and set up or registration step for merchants. Moreover, this protocol uses less memory space at the client-side which makes it suitable for low end mobile devices. However, anonymity (i.e., deals with the secrecy of the user identity) and unlikability (i.e., the property that considers whether different communications can be attributed to the same agent) is not satisfied by the protocol. It is unable to handle the replay attacks which are done by merchants. It uses the set of secret keys which requires memory space at client side and issuer side. Here, in this work, we propose a secure lightweight authentication protocol for mobile payment which overcomes all these flaws.

2 Related Work

In this section, we review some of the important existing protocol that have been proposed in the literature as follows:

2.1 Secure Electronic Transaction(SET) Protocol [3]

SET is a secure payment protocol which is based on the public-key cryptography. Client(C), Merchant (M), Payment Gateway (PG), Issuer (I) and Acquirer (A) are the participants in SET protocol. In this, we can make a payment with a credit card through the internet. It is not suitable for mobile because of its high computational operation and its limited capacity of mobile devices.

2.2 Internet Key Protocol(IKP) [4]

i-Key Protocol is also based on public key principal. In this protocol, three entities have participated- Client, Merchant, and Payment Gateway. Here, i stand for how many participants have public-key. It can be 1KP, 2KP and 3KP. The higher value of i provides more security.

2.3 Modified SET Protocol for Mobile Payment [5]

As SET is not suitable for mobile payment; an improved version of it was proposed which is known as a Modified SET protocol. This protocol is suitable for mobile payment.

2.4 New Mobile Payment Protocol: Mobile Pay Center Protocol [6]

Mobile pay center protocol is an improved version of LPMP which uses the digital certificate. This protocol has two sub protocol which are registration and payment protocol. This protocol provides the non-repudiation property.

2.5 Lightweight Protocol for Mobile Payment with Two Payment Gateways [7]

In this protocol, a user may be able to make the payment via using the two credit cards through two payment gateway. Four entities have participated- Client, Merchant and two the Payment Gateway. It fulfils many security features like accountability, untractability, unlinkability, and prevents from the double-spending attack and Denial of Service attack.

2.6 Light-Weight and Privacy-Preserving Authentication Protocol for Mobile Payment in the Context of IOT [8]

This is a first mobile payment protocol with the certificate-less unidirectional signature scheme and provides the user anonymity and unforgeability.

2.7 Strong Anonymous Mobile Payment Against Curious Third Party Provider [9]

SATP provides the strong anonymous mobile payment by using the ticket which is blindly signed by the third-party provider (TPP). SATP does not require any public key certificate.

Later in Sect. 5, we compare the various security properties that are satisfied and not satisfied by these above mentioned protocols (shown in Table 4).

3 Background

3.1 Lightweight Protocol for Mobile Payment

In this section, we describe the LPMP protocol [12] in brief. General notations used in subsequent sections are listed in Fig. 1.

The LPMP protocol consists of 14 steps as follows (Fig. 2):

Fig. 1 Notation used in LPMP [12]

- {C, M, PG, I, A}: This tuple stands for client, merchant, payment gateway, issuer and acquirer respectively.
- OD: order description.
- Amount: price of good or service.
- DC_p: digital certificate of principal P.
- ID_p: identity certificate of principal P.
- TID: transaction identifier.
- TIDReq: TID request.
- MIDReq: request for ID_m.
- H(msg): hash of msg.
- PU_p: public key of principal P.
- PR_p: private key of principal P.
- E[msg, k]: encryption with key k.
- D[msg, K]: decryption with key k.
- KeyReq: request for session key.
- Y: session key between client and merchant.
- X_i: A key from the list of shared secret keys between the issuer and the client.
- P1→P2: message passing from principal P1 to P2.

3.2 Security Analysis and Discussion of LPMP

On analyzing the LPMP protocol, we found some errors and various security issues that are not addressed by this protocol. In this section we present and discuss them in detail as follows:

3.3 Key Management Protocol

In this protocol, the session key is shared between client and issuer via the set of secret keys.

$Xi = H(i$ bit cyclic shift of-(X, MPIN)), where $i = 1, 2\ldots, n$, $X =$ Shared Secret and MPIN = Mobile Personal Identification Number.

As both client and issuer have to maintain and keep the set of secret keys, this requires memory space on both sides. Moreover, the protocol does not mention the threshold limit of n, i.e., how many keys will be shared between client and issuer. As such, chances of repetitions of the secret key after a certain time exists which may be exploited by the adversary and poses security risk.

Counter measure Client can make the use of a single key instead of a set of keys by using the temporary identity and nonce of client [13]. Nonce is a random number that can be used only once. It is used to ensure that old messages cannot be sent again in

1) **C ->M :** IDc, i, DCi, TIDReq, MIDReq, H(Xi)
2) **M->PG:** E[((IDc,i,DCi,H(Xi),KeyReq),H(IDc,i,DCi,H(Xi),KeyReq),PRm],DCm),PUpg]
3) **PG->I:** E[((IDc,i,DCm,H(Xi),KeyReq), E[H(IDc,i,DCm,H(Xi),KeyReq, PUpg],DCpg), PUi]
4) **I->PG:** E[(E[Y, PUm],E[Y, Xi+l],E[H(E[Y, PUm],E[Y,Xi+l]), PRi],DCi), PUpg]
5) **PG->M:** E[(E[Y, PUm], E[Y,Xi+l], E[H(E[Y, PUm], E[Y,Xi+l]), PRpg], DCpg), PUm]
6) **M->C:** E[(TID, IDm),Y],E[Y, Xi+l],H(E[(TID, IDm),Y], E[Y, Xi+l])
7) **C->M:** E[(TID,Timestamp,OD,DCi,E[(TID,IDm,Timestamp,amount,H(OD,amount), Xi), Xi+1],H(TID,Timestamp,OD,DCi,E[(TID,IDm,Timestamp,amount,H(OD,amount),Xi),Xi+1])), Y], IDc
8) **M->PG:** E[(amount, E[TID,H(OD, amount),PUi],E[(TID, Timestamp, amount, H(OD, amount),Xi),Xi+1],E[H(amount,E[TID,H(OD,amount),PUi],E[(TID,Timestamp,amount,H(OD, amount), Xi), Xi+l]), PRm], DCm), PUpg]
9) **PG->I:** E[(E[TID,H(OD, amount),PUi],E[(TID, Timestamp, amount, H(OD, amount),Xi), Xi+l],E[H(E[TID,H(OD, amount),PUi],E[(TID, Timestamp, amount, H(OD, amount),Xi), Xi+ 1]),PRpg],DC pg),PUi]
10) **I->PG:** E[(TID,amount/no,E[((amount/no),IDm),Xi+l],E[H(TID,amount/no,E[((amount/no), IDm), Xi+l]) , PRi], DCi). PUpg]
11) **PG ->A:** E[(TID, amount, DCm, E[H(TID, amount, DCm), PRpg], DCpg), PUa]
12) **A -> PG:** E[(amount, DCm, E[H(amount, DCm), PRa],DCa), PUpg]
13) **PG->M:** E[(amount/no, E[((amount/no),IDm),Xi+l],E[H(amount/no, E[((amount/no),IDm),Xi+l]), PRpg], DCpg) ,PUm]
14) **M->C:** E[amount/no, E[((amount/no), IDm), Xi+ I],H(Yes/no, E[((amount/no), IDm) ,Xi+l]),y]

Fig. 2 Steps in LPMP [12, 16]

replay attacks. Temporary identity is used for several purposes. It provides privacy as the temporary identity is only known to client and issuer. So that merchant and payment gateway cannot map the temporary identity to the real identity. It also saves memory on both sides as only one key is now used between client and issuer.

3.4 Error in LPMP's Payment Protocol Step

One of the major flaws in this protocol is in its payment protocol. Digital certificates that are required for authenticating the owner's public key and also used when one wants to send and receive encrypted data always reside on the owner's own machine. Thus, it can only be shared by the owner of the digital certificate and not others. However, in this protocol, digital certificate has been forwarded by unauthorized users. As shown above in Sect. 3, in step 1 client forwards issuer's digital certificate (DCi) to merchant (M) and in step 2, merchant forwards issuer's digital certificate (DCi) to payment gateway PG. Similarly, in step 3, payment gateway forwards the merchant's digital certificate (DCm) to issuer(I).

(1) **C-> M**: IDc, i, DCi, TIDReq, MIDReq, H(Xi)
(2) **M-> PG**: E[((IDc, i, DCi, H(Xi), KeyReq), H(Xi), KeyReq), PRm], DCm), PUpg]
(3) **PG-> I**: E[((IDc, i, DCm, H(Xi), KeyReq), E[H(IDc, i, DCm, H(Xi), KeyReq), Rpg], DCpg), PUi].

Counter measure Client must send the identity of issuer (IDi) instead of issuer's digital certificate (DCi) to merchant (in the step 1) and also merchant must forward the issuer's identity (IDi) to payment gateway (in the step 2). Payment gateway verified the key using the DCm received from the merchant and must forward the merchant's public key (PUm) instead of digital certificate (DCm) to the issuer.

Assumption: Payment gateway is trusted by the issuer that it will not provide the merchant's false public key.

3.5 Anonymity

Anonymity means not revealing the real identity of the user. It deals with the secrecy of the user. It ensures that users may use the services or resources without being identified [1]. However, LPMP protocol does not use the anonymous identity in order to protect the client/merchant's identity, because client shares his/her real identity (IDc) with merchant and acquirer and merchant shares his/her real identity (IDm) with client and issuer.

Counter measure Client must send the temporary identity TIDc instead of his real identity IDc [12]. So, the client is hidden from the merchant. Similarly, merchant must use the temporary identity TIDm instead of their real identity IDm. Both Client and Issuer will periodically update the TIDc by concatenating the previous TIDc, IDc and increment of session key ($Y + 1$) and then take the message authentication code (MAC) of the entire message. MAC is a one-way hash function with the addition of a secret key [14]. Similarly, Merchant and Acquirer will periodically update the TIDm by concatenating the previous TIDm, and IDm and then take the MAC of the entire message.

3.6 Replay Attack by Merchant

In this protocol, merchant can resend the message to the payment gateway along with the DCm.

M-> PG: E[(Amount, E[TID, H(OD, amount), PUi], E[(TID, Timestamp, amount, H(OD, amount),Xi), Xi + 1], E[H(Amount, E[TID, H(OD, amount), PUi], E[(TID, Timestamp, amount, H(OD, amount), Xi), Xi + 1]), PRm], DCm), PUpg]

Payment gateway decrypts the message and removes the amount from the message. And digitally signed the message and encrypted it with the issuer's public key and sent it to the issuer.

PG-> I: E[(E[TID, H(OD, amount), PUi], E[(TID, Timestamp, amount, H(OD, amount), Xi), Xi + 1], E[H(E[TID, H (OD, amount), PUi], E[(TID, Timestamp, amount, H(OD, amount), Xi), Xi + 1]), PRpg], DC pg), PUi]

Issuer decrypts with their private key and performs payment gateway's signature verification. Then the issuer decrypts the remaining message with the secret key which is shared between client and itself and detects the timestamp. Further, issuer checks to guarantee that the timestamp is still valid and thinks that the amount is not transferred in the merchant's account. He thus sends the amount to the merchant's bank again. This happens because in this protocol clients do not send any acknowledgment to the issuer that the merchant has received the payment. Now the merchant hasfooled the issuer successfully. In fact, till the timestamp expires, merchant can fool everyone on the network [14].

Counter measure One way to counter replay attack is that clients can also add the lifetime with timestamp so that timestamp is valid until the lifetime has expired. The timestamp and lifetime ensure the freshness of the message and prevents the client as well as issuerfrom replay attack. Because the issuer can verify the message by checking the lifetime of the message. But we have to make sure that the client's clock is synchronized with the issuer's clock.

3.7 Linkability

Linkability means an attacker may be able to distinguish whether these items of interest (i.e., entities, messages, action, etc.) are related to each other or not [15]. In this protocol, the client uses the real identity which gives an attacker an opportunity to link the different transactions to a user.

Counter measure To eradicate this flaw we use the concept of unlinkability [15]. Unlinkability means attackers cannot be able to distinguish whether these items of interest (i.e., entities, messages, action, etc.) are related to each other or not. Client must send the unique temporary identity TIDc for every new transaction, so that merchants are unable to link the transaction with the same identity.

4 The Proposed Protocol

In this section, we present our protocol which saves memory, addresses the security concerns of LPMP and incorporates all the countermeasures discussed in Sect. 4. Our proposed protocol mitigates replay attack, is memory efficient as only one key is used, and provides anonymity and unlinkability of client to merchant as well as

Table 1 Additional notation used in our protocol

Symbol	Description
TIDp	Temporary identity of principal
Np	A random number or nonce generated by principal
X	Secret key share between client and issuer
Hx (ID, Nonce)	Keyed-hash function
Lifetime	Validity of timestamp

merchant to client. This protocol provides these properties by using the TIDc instead of IDc and TIDm instead of IDm. The additional notations used in our protocol apart from those used in Sect. 2 are given in Table 1.

Our protocol consists of 14 steps shown in below:

Step 1. C-> M: TIDc, Nc, IDi, TIDReq, MIDReq, Hx(TIDc, Nc)

In this step, the client makes a request to the merchant with his temporary identity (TIDc) insteadof his real identity (IDc) and generates a nonce (Nc). The client requests the session key by passing the keyed hash of TIDc and Nc between the issuer and himself Hx(TIDc, Nc). As TID is only known to client and issuer, merchant and payment gateway will not be able to know the user's real identity. As a result, it fulfills the anonymity and unlinkability property.

Step 2. M-> PG: E[((TIDc, Nc, IDi, Hx(TIDc, Nc), KeyReq), H(TIDc, Nc, IDi, Hx(TIDc, Nc), KeyReq), PRm], DCm), PUpg]

Merchant signs and encrypts the client's request and sends it tothe payment gateway along with his/her digital certificate.

Step 3. PG-> I: E[((TIDc, Nc, Hx(TIDc, Nc), KeyReq, PUm), E[H(TIDc, Nc, Hx(TIDc, Nc), KeyReq, PUm), PUpg], DCpg), PUi]

Payment gateway decrypts the message using his private key to recover the issuer identity. Then, it sends the message along with the public key of the merchant which was received from the merchant's digital certificate to the specified issuer.

Step 4. I-> PG: E[(E[Y, PUm], E[(Y, Ni, Nc + 1), x], E[H(E[Y, PUm], E[(Y, Ni, Nc + 1), x]), PRi], DCi), PUpg]

Issuer calculates the keyed-hash of TIDc and Nc and matches the calculated value with the value sent by the payment gateway. If the value has matched, issuer generates the session key and make two messages one for client which contains the session key, a new nonce (Ni) generated by issuer and Nc + 1 and encrypts it with the secret key and the other for merchant which contains session key and encrypts it with the merchant's public-key and passes it to payment gateway.

Step 5. PG-> M: E[(E[Y, PUm], E[(Y, Ni, Nc + 1), x], E[H(E[Y, PUm], E[(Y, Ni, Nc + 1), x]), PRpg], DCpg), PUm]

Message is forwarded as it is to the merchant by the payment gateway.

Step 6. M- > C: E[(TID, TIDm),Y],Ex[Y, Ni, Nc + 1],H(E[(TID, TIDm),Y], E[(Y, Ni, Nc + 1),x])

The merchant after deriving the session key from the message sends the transaction TID and his temporary identity(TIDm) instead of real identity encrypted with the session key, to the client.

Step 7. C-> M: E[(TID, Timestamp, Lifetime, OD, IDi, E[(TID, TIDm, Timestamp, Lifetime, amount, H(OD, amo unt), Ni + 1), x], H(TID, Timestamp, OD, IDi, E[(TID, TIDm, Timestam, Lifetime, amount, H(OD, amount), Ni + 1])), Y], TIDc

Client decrypts the message to extract the session key. Then, the client generates two messages by using the dual signature property [14], one for issuer that consists of a TID, TIDm, timestamp, lifetime, amount, order description(OD) and increment of nonce(Ni + 1) encrypted using the secret key and other for merchant which is encrypted using the session key. Thus,bank and other private details shared between issuer and client is not revealed to the merchant and issuer or payment gateway does not get to know the services or resources purchased by the client.

The remaining steps are same as described by the LPMP [12].

When the 14 steps are done, client and issuer update the TIDc in their database as follows:

$$\text{TIDc} + 1 = \text{MAC}(\text{TIDc}, \text{IDc}, \text{Y} + 1),$$

where MAC stands for Message Authentication Code [14] and Y is the session key.

Similarly, merchant and acquirer also update the TIDm in their database as follows:

$$\text{TIDm} + 1 = \text{MAC}(\text{TIDm}, \text{IDm}).$$

In the next transaction, client and merchant send TIDc + 1 and TIDm + 1 instead of their real TIDc and TIDm.

5 Analysis and Discussion of Proposed Protocol

In this section, we perform a comparative analysis of our protocol with the LPMP protocol.

5.1 Comparison Between LPMP and Proposed Protocol

See Table 2.

Table 2 Comparison between LPMP and proposed protocol

Step	LPMP [12]	Proposed Protocol
1	C-> M: IDc, i, DCi, TIDReq, MIDReq, H(Xi)	C-> M: TIDc, Nc IDi, TIDReq, MIDReq, Hx(TIDc, Nc) Our protocol uses the TIDc and nonce of the client. So, it removes the need for a set of keys and anonymity and unlinkability of the client also is satisfied. Error in LPMP is solved by sending the identity of issuer (IDi) instead of issuer's digital certificate (DCi) to merchant
2	M-> PG: E[((IDc, i, DCi, H((Xi), KeyReq), H(Xi), KeyReq), PRm], DCm), PUpg]	M-> PG: E[((TIDc, Nc, IDi, Hx(TIDc, Nc), KeyReq), H(TIDc, Nc, IDi, Hx(TIDc, Nc), KeyReq), PRm], DCm), PUpg] Merchant forwards the issuer's identity to the payment gateway
3	PG-> I: E[((IDc, i, DCm, H (Xi), KeyReq), E[H(IDc, i, DCm, H(Xi), KeyReq), PRpg], DCpg), PUi]	PG-> I: E[((TIDc, Nc, Hx(TIDc, Nc), KeyReq, PUm), E[H(TIDc, Nc, Hx(TIDc, Nc), KeyReq, PUm), PUpg], DCpg), PUi] Similarly, payment gateway forwards the merchant's public key (PUm) instead of digital certificate (DCm) to the issuer
4	I-> PG: E[(E[Y, PUm], E[Y, Xi +l], E[H(E[Y, Pm], E[Y, Xi + l]), PRi], DCi), PUpg]	I-> PG: E[(E[Y, PUm], E[(Y, Ni, Nc + 1), x], E[H(E[Y, PUm], E[(Y, Ni, Nc + 1), x]), PRi], DCi), PUpg] Issuer generates the Nc + 1 and encrypts with a secret key, so that client ensures that this session key is generated by the issuer.
5	PG-> M: E[(E[Y, PUm], E[Y, Xi + l], E[H(E[Y, PUm], E[Y, Xi + l]), PRpg], DCpg), PUm]	PG-> M: E[(E[Y,PUm], E[(Y, Ni, Nc + 1), x], E[H(E[Y, PUm], E[(Y, Ni, Nc + 1),x]), PRpg], DCpg), PUm]
6	M-> C: E[(TID, IDm), Y], E[Y, Xi + l], H(E[(TID, IDm), Y], E[Y, Xi + l])	M-> C: E[(TID, TIDm), Y], Ex[Y, Ni, Nc + 1], H(E[(TID, TIDm), Y], E[(Y, Ni, Nc + 1), x]) Anonymity and unlinkability of a merchant is also satisfied as the merchant uses his temporary identity (TIDm) instead of his real identity (IDm)
7	C-> M: E[(TID, Timestamp, OD, DCi, E[(TID, IDm, Timestamp, amount, H(OD, amount), Xi), Xi + l], H(TID, Timestamp, OD, DCi, E[(TID, IDm, Timestamp, amount, H(OD, amount), Xi), Xi + 1])), Y], IDc	C-> M: E[(TID, Timestamp, Lifetime, OD, IDi, E[(TID, TIDm, Timestamp, Lifetime, amount, H(OD, amount), Ni + 1), x], H(TID, Timestamp, OD, IDi, E[(TID, TIDm, Timestamp, Lifetime, amount, H(OD, amount), Ni + 1])), Y], TIDc Replay attack by merchant is handled by using the lifetime with timestamp. Since timestamp is valid within the limited period, replay attack will be mitigated

Table 3 Security Analysis

Security/protocol requirements	[12]	Our protocol
Size of secret key for making the payment with N merchant	No	1
Client's anonymity	No	Yes
Merchant's anonymity	No	Yes
Client's unlinkability	No	Yes
Merchant's unlinkability	No	Yes
Replay attack prevention	No	Yes
Keyed-hash function at client side	No	Yes

Table 4 Security feature comparison

Security properties	[3]	[4]	[5]	[6]	[7]	[8]	[9]	[12]	Our protocol
Confidentiality	✓	✓	✓	✓	✓	✓	✓	✓	✓
Authentication	✓	✓	✓	✓	✓	✓	✓	✓	✓
Integrity	✓	✓	✓	✓	✓	✓	✓	✓	✓
Non-repudiation	✓	✓	×	✓	✓	✓	✓	✓	✓
Client's anonymity	×	×	×	✓	✓	✓	✓	×	✓
Merchant's anonymity	×	×	×	×	×	×	×	×	✓
Client's unlinkability	×	×	×	✓	✓	✓	✓	×	✓
Merchant's unlinkability	×	×	×	×	×	×	×	×	✓

5.2 Security Analysis of LPMP and Our Protocol

Here, we show the security comparison between LPMP and our protocol (Table 3).

5.3 Security Feature Comparison

In Table 4, we provide a comparative analysis of our protocol with other existing protocols and show that it satisfies all the security properties unlike others.

6 Conclusion

In this work, we present an improved version of existing lightweight protocol for mobile payment (LPMP) which satisfies the anonymity and unlinkability property. This proposed protocol also prevents clients from merchant's replay attack, enables the client and issuer to share the session key via using the single secret key and also stops an entity from forwarding other's digital certificate.

References

1. F. Zamanian, H. Mala, A new anonymous unlinkable mobile payment protocol, in *6th International Conference on Computer and Knowledge Engineering (ICCKE)* (IEEE, 2016), pp. 117–122
2. https://searchmobilecomputing.techtarget.com/definition/m-commerce
3. C. Techapanupreeda, R. Chokngamwong, A secure payment protocol with accountability properties, in *2016 International Computer Science and Engineering Conference (ICSEC)* (IEEE, 2016), pp. 1–6
4. K. Ogata, Model checking the iKP electronic payment protocols. J. Inf. Secur. Appl. **36**, 101–111 (2017)
5. S.M. Shedid, M. Kouta, Modified SET protocol for mobile payment: an empirical analysis, in *2010 2nd International Conference on Software Technology and Engineering*, vol. 1, (IEEE, 2010), pp. V1–350
6. M.V.A. Dizaj, R.A. Moghaddam, S. Momenebellah, New mobile payment protocol: mobile pay center protocol (MPCP), in *Proceeding of 3rd International Conference of Electrons Computer Technology*, 2011, pp. 74–78
7. H. Lee, J. Kim, J. Moon, D. Kang, D. Won, A security enhanced lightweight mobile payment scheme based on two gateways. Biometrics **3**(9) (2017)
8. Y. Chen, W. Xu, L. Peng, H. Zhang, Light-weight and privacy-preserving authentication protocol for mobile payments in the context of IoT. IEEE Access **7**, 15210–15221 (2019)
9. C. Cao, X. Zhu, Strong anonymous mobile payment against curious third-party provider. Electron. Commer. Res. **19**(3), 501–520 (2019)
10. V.C. Sekhar, M. Sarvabhatla, A secure account-based mobile payment protocol with public key cryptography. Mrudula Sarvabhatla, ACEEE Int. J. Netw. Secur **3**(01), (2012)
11. R.A. Abouhogail, A new secure lightweight authentication protocol for NFC mobile payment. Int. J. Commun. Netw. Inf. Secur. (IJCNIS) **11**(2), 283–289 (2019)
12. D.M. Tripathi, A. Ojha, LPMP: an efficient lightweight protocol for mobile payment, in *3rd National Conference on Emerging Trends and Applications in Computer Science* (IEEE, 2012), pp. 41–45
13. J.T. Isaac, S. Zeadally, An anonymous secure payment protocol in a payment gateway centric model. Procedia Comput. Sci. **10**, 758–765 (2012)
14. B. Schneier, Applied Cryptography, 2nd edn, Wiley (1996)
15. A. Pfitzmann, M. Hansen, Anonymity, unlinkability, undetectability, unobservability, pseudonymity, and identity management-a consolidated proposal for terminology. Version **31**(15) (2008)
16. B. Narwal, Security analysis and verification of authenticated mobile payment protocols, in *4th International Conference on Information Systems and Computer Networks (ISCON)*, (IEEE, 2019), pp. 202–207

Create EC2 Instance, Install Splunk and Configure Using Ansible

Rishabh Shah and Vatsal Shah

Abstract In today's world, there is burgeoning of the projects in IT firms which unfortunately leads to tedious work, and human errors rise to a great extent. To take the above situation under control, computer automation is the best way. The behoof of this study is to lower the human task by the help of automation scripts. Considering this, there is an idea to automate the industrial task. Whenever there are large number virtual systems and if Splunk is required to install on each of them, it would be tedious. For this Ansible, the tool provided by the Red Hat for automation, will be used. The script will be written in the YAML file which will be executed and a number of infrastructures will be created and Splunk will be installed on it. After the installation and configuring the whole infrastructure by the automation script, it will be further used for data monitoring and analysis. Basically, the whole automation script is divided into several tasks which will execute step by step to become a great infrastructure.

Keywords Ansible · Automation · AWS EC2

1 Introduction

1.1 What Is Ansible

Ansible is an open-source IT engine developed by Red Hat that automates application deployment and many IT tools. We just need to write the playbook to install any application or any other things to automate. Just need to mention IP addresses of the systems in which automation is to be performed in the inventory. We have to run a playbook in control machine, and automation will be done in all the specified

R. Shah (✉) · V. Shah
IT Department, Birla Vishwakarma Mahavidyalaya, Anand, Gujarat, India
e-mail: rishabhshah1602@gmail.com

V. Shah
e-mail: vatsal.shah@bvmengineering.ac.in

J. Tavares et al. (eds.), *Emerging Technologies in Data Mining and Information Security*, Lecture Notes in Networks and Systems 164,
https://doi.org/10.1007/978-981-15-9774-9_49

machines. Ansible runs the playbook in the other systems via SSH by default and exits after the task is completed. After assigning the playbook and inventory Ansible, node management does SSH to the other machines and performs tasks. It is built on Python, so many Python functionalities are provided in Ansible. Playbook is a YAML file which runs on the system and SSH on remote machine to execute the mentioned tasks in playbook. Ansible is agentless and temporarily connects to remote machine via SSH. Many modules of Python are a part of Ansible. It can be run easily on UNIX-like operating systems.

1.2 Overview of the System

The proposed system would be able to automate the creation of EC2 instances, installing Splunk and configuring it. The automation tool which is going to be used is Ansible which is provided by Red Hat. Ansible code will be run on the instance which will create the instances how many we want. This will be used to do the tedious task of creating EC2 instances. The system will be having different parts like creating instances first, then installing Splunk and configure at last. In this system, we just need to enter the number of instances to be created, and all other things will be automated. This task is not tedious when there are a smaller number of systems or instances, but what if there are more than 50 systems in the infrastructure, then it will be really difficult to configure everything in each of the systems. Thus, a script will be created which will make the whole process easy saving time and also less chances of human error.

2 Literature Survey

2.1 "Unleashing Full Potential of Ansible Framework" by Brno University of Technology, Brno, Czech Republic. (2018) [1]

In this research paper, they have talked about managing the large-scale university laboratory infrastructure by a simple platform. This platform itself has been written in Python and allows users to script commands in YAML Ain't Markup language, which seems like a markup language, but it is not a markup language. They have implemented a system so they can manage the university laboratory from both local and public network infrastructures. They have created a Web application which enables configuration changes with the help of smart devices like smart mobile phones and tablets.

2.2 "Automated Provisioning of Application in IAAS Cloud Using Ansible Configuration Management" by Nishant Kumar Singh in [3]

This research is about management of cloud services with the help of Ansible automation. As nowadays there is a huge rise in the cloud services as it has a large amount of storage, and also, the main advantage is that we can access it from anywhere remotely. This paper has emphasized on the automation of the customer applications which are hosted on cloud right from environment provisioning to the deployment stage. The whole structure of the automated application deployment is assembled using Amazon as IAAS provider and Ansible as orchestration engine.

2.3 "Cloud Automation with Configuration Management Using CHEF Tool" by Ramandeep Singh in [2]

This is again the best example of automation in the IT infrastructure. Like Ansible, CHEF is the tool which can be used for making automation scripts which reduces the tedious human tasks. The cloud operation for the customer starting from provisioning the space for work to deployment can be scripted right away and reduce the human work with the help of CHEF tools. He has also focused on the DevOps part which is also the combination of Development and Operations. A system can be made which can monitor the DevOps operations regularly which at the end provides the efficient result.

2.4 "Survey of Automated Software Deployment for Computational and Engineering Research" by Benson in [4]

His team had stated about how whole cloud services can be moved or automated with the help of external tools such as CHEF, Salt Stack and Ansible based on some defined model. Their study has addressed the challenge of building an effective multi-cloud application deployment controller as a customer add-on outside of cloud utility service itself. Such external controllers must function within the constraints of cloud providers' API.

3 Implementation

3.1 Workflow of the System

No. of Instances we want to be active [Task 1]: Firstly, the script asks us about the number of instances we want to create for our infrastructure. It will generate the number of instances mentioned (Fig. 1).

Creation of EC2 Keypair [Task 2]: For creation of instance, the main thing is creation of EC2 keypair which is also called pem file. To SSH into the remote machine, one need pem file in the local system (Fig. 2).

Adding Key to SSH Agent [Task 3]: Adding the key to SSH agent, this is used to enable the single sign-on with SSH (Fig. 3).

Create a VPC [Task 4]: Virtual private cloud enables us to launch AWS resources in the virtual we have created. In this task, the script creates the virtual private cloud for the instances. This whole process will be tedious for the many number of virtual machines, but the script will automatically create a separate VPC (Fig. 4).

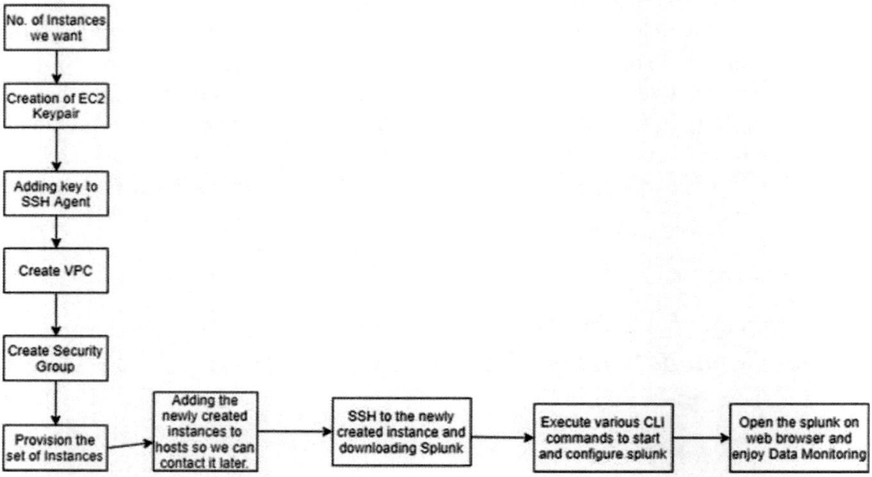

Fig. 1 Workflow of the system

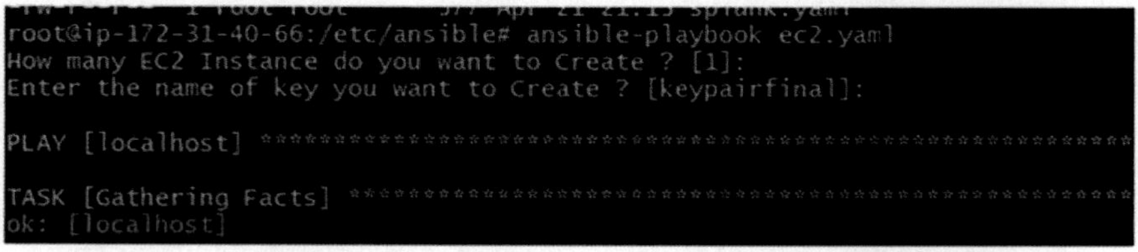

Fig. 2 No. of instances to be created

Fig. 3 Creation of EC2 keypair

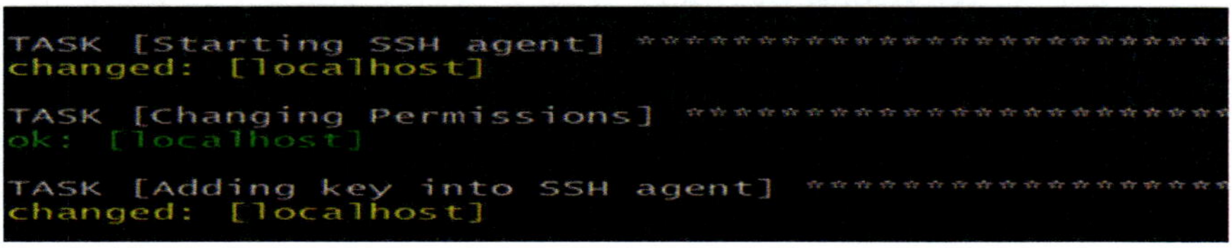

Fig. 4 Adding key to SSH agent

Creating a security group [Task 5]: This is used to control the inbound and outbound traffics that can control the AWS instances. Only the mentioned ports in the security group are available to use (Fig. 5).

Provision the set of Instances [Task 6]: At this stage, new instances are created, and at output, we can also see the image of instance ids. The number of instances is mentioned at the beginning of script that many numbers of instances will be active (Fig. 6).

Adding the newly created instances to the hosts so that we can contact them later [Task 7]: The public IPs of the newly created instances will be added to the host file in the ansible folder. In the host file, the IPs of the remote instances are stored jo that they can be contacted or SSH later on (Fig. 7).

SSH to the newly created Instance and Download Splunk [Task 8]: For downloading the Splunk application, we need to SSH into the remote instances which will be done by the script itself and procedure for downloading the Splunk will be initiated (Fig. 8).

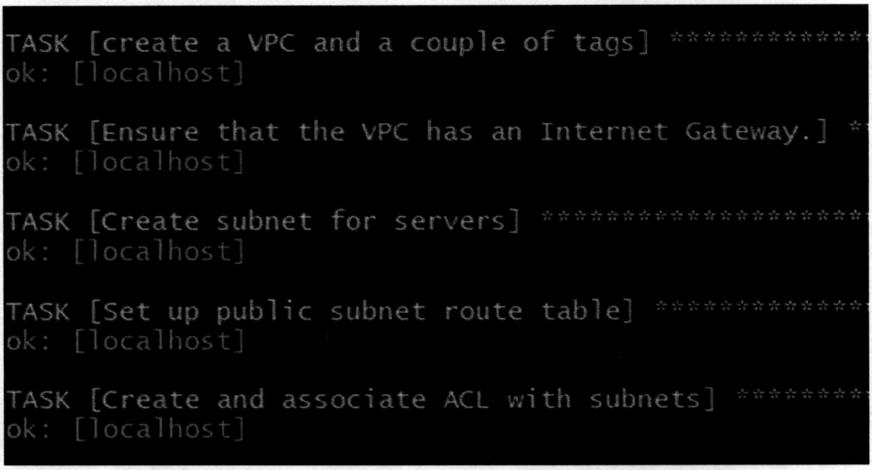

Fig. 5 Creation of VPC and subnet

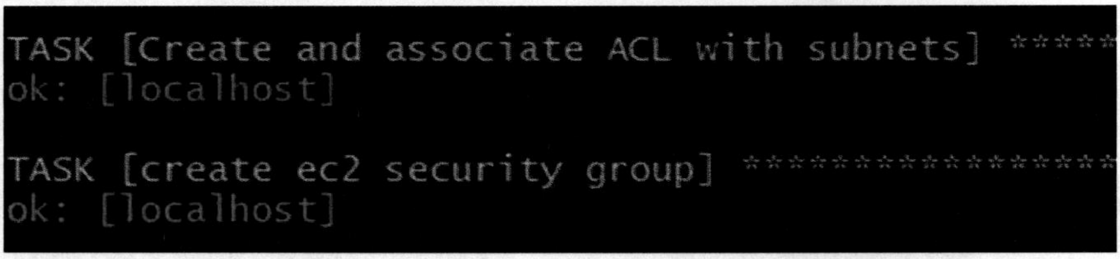

Fig. 6 Creating security group

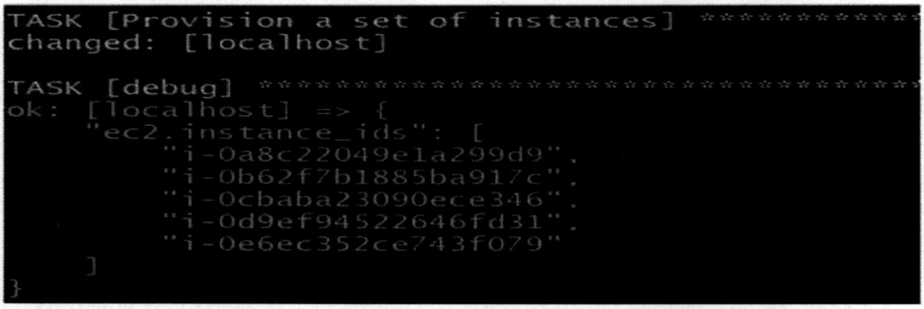

Fig. 7 EC2 instance provision

After the execution of the above automation task, we need to SSH into the instance we need and start the Splunk service. After starting the Splunk, we can access the UI of the Splunk by the public IP of the particular instance with 8000 port number

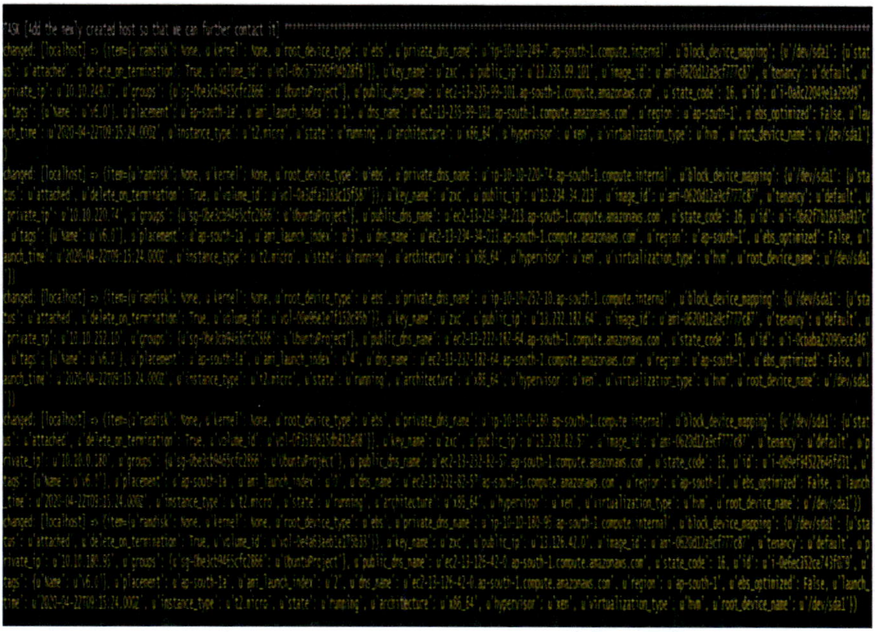

Fig. 8 Adding instances to host file of Ansible

which is the port number for management of Splunk. The UI of the Splunk will look like the below image (Figs. 9 and 10).

4 Conclusion

In the above paper, various research papers already published were, studied, and to summarize at the last, this paper was created considering the above studies. Some of the above researches mentioned about moving the cloud services to the automated platform using the external tool such as CHEF, Salt Stack, Puppet, Ansible and Terraform. Thus, this system was implemented with the help of Ansible automation tool. Splunk can be used majorly for the data monitoring and analysis tool in many of the organizations and also in government sections. Thus, the above system can be used to lower down the hard work required to set up an infrastructure having a lot of virtual systems.

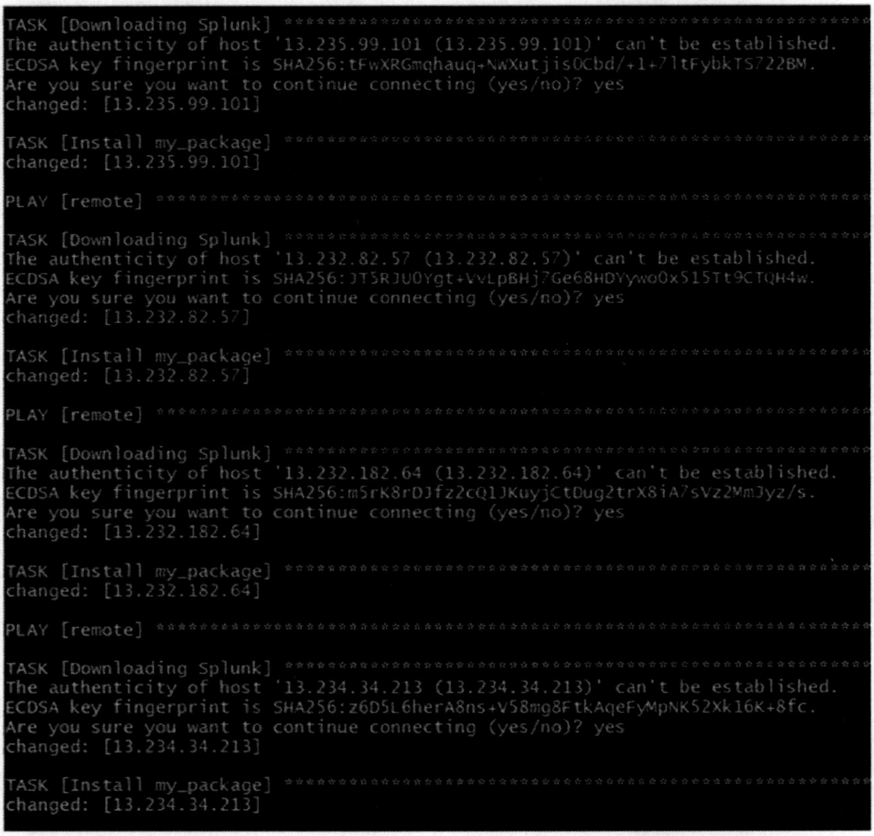

```
TASK [Downloading Splunk] ************************************************************
The authenticity of host '13.235.99.101 (13.235.99.101)' can't be established.
ECDSA key fingerprint is SHA256:tFwXRGmqhauq+NWXutjisOCbd/+1+7ltFybkTS722BM.
Are you sure you want to continue connecting (yes/no)? yes
changed: [13.235.99.101]

TASK [Install my_package] ************************************************************
changed: [13.235.99.101]

PLAY [remote] ************************************************************************

TASK [Downloading Splunk] ************************************************************
The authenticity of host '13.232.82.57 (13.232.82.57)' can't be established.
ECDSA key fingerprint is SHA256:JT5RJUOYgt+VvLpBHj7Ge68HDYywoOx515Tt9CTQH4w.
Are you sure you want to continue connecting (yes/no)? yes
changed: [13.232.82.57]

TASK [Install my_package] ************************************************************
changed: [13.232.82.57]

PLAY [remote] ************************************************************************

TASK [Downloading Splunk] ************************************************************
The authenticity of host '13.232.182.64 (13.232.182.64)' can't be established.
ECDSA key fingerprint is SHA256:m5rK8rDJfz2cQ1JKuyjCtDug2trX8iA7sVz2MmJyz/s.
Are you sure you want to continue connecting (yes/no)? yes
changed: [13.232.182.64]

TASK [Install my_package] ************************************************************
changed: [13.232.182.64]

PLAY [remote] ************************************************************************

TASK [Downloading Splunk] ************************************************************
The authenticity of host '13.234.34.213 (13.234.34.213)' can't be established.
ECDSA key fingerprint is SHA256:z6D5L6herA8ns+V58mg8FtkAqeFyMpNK52Xk16K+8fc.
Are you sure you want to continue connecting (yes/no)? yes
changed: [13.234.34.213]

TASK [Install my_package] ************************************************************
changed: [13.234.34.213]
```

Fig. 9 Automatically SSH into newly created instances

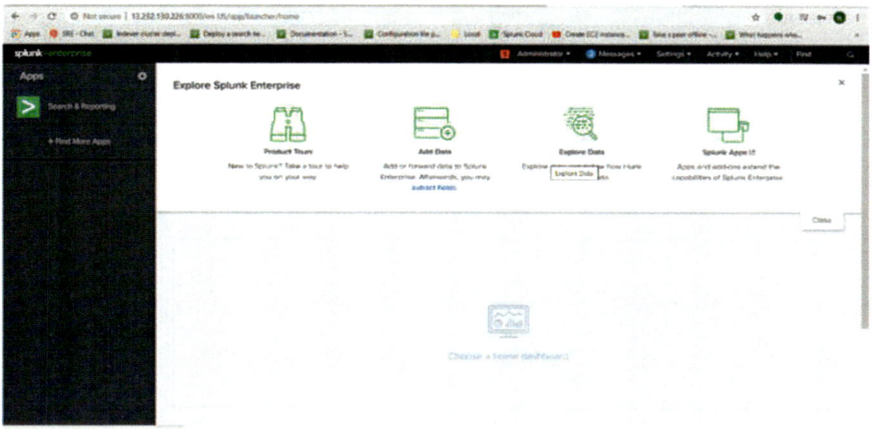

Fig. 10 Splunk UI

References

1. P. Masek, M. Stusek, J. Krejci, K. Zeman, J. Pokorny, M. Kudlacek, Unleashing full potential of ansible framework: university lab administration. ResearchGate (2018)
2. R. Singh, R.K. Purwar, Cloud automation with configuration management using CHEF Tool. Int. J. Eng. Res. Technol. (IJERT) (2019)
3. N.K. Singh, S. Thakur, H. Chaurasiya, H. Nagdev, Automated provisioning of application in IAAS cloud using ansible configuration management, ResearchGate (2015)
4. J.O. Benson, J.J. Prevost, P. Rad, Survey of automated software deployment for computational and engineering research. IEEE, June 2016

Quantum Computation-Based Secured Key Distribution Protocols

Priyanka Jojan, Kapil Kumar Soni, and Akhtar Rasool

Abstract The cryptographic system ensures authentication between the communicating parties through keys, and it cannot be sent over insecure communication channel due to eavesdropping adversary. Distribution approaches are required to support either private or public key cryptosystem that allows key sharing secretly. We represent the existing classical key distribution protocols and relevant quantum methods that overcome the challenges in classical protocols. The keys must be shared at the end of distribution process, and continuum to this, we have quantum key distribution that uses quantum principles which guarantee that key cannot be intercepted. The objective of this paper is to explore quantum key distribution protocols like BB84 and Eckert's protocol in detail and then to analyze them with classical approaches in comparative parameters and lastly concludes the quantum protocol to be highly secure, advantageous and superior to classical existing protocols.

Keywords Cryptography · Quantum computation · Key distribution · Superposition · Entanglement

1 Introduction

Cryptography is the process in which information is exchanged between two parties through the use of codes, so that only those for whom the information is intended can read and process it. In classical cryptography, key is used for encryption and decryption of data. Decryption can be performed by the third party only if he knows

---

P. Jojan (✉) · K. K. Soni · A. Rasool
Department of Computer Science and Engineering, Maulana Azad National Institute of Technology, Bhopal 462003, India
e-mail: priyankajojan@gmail.com

K. K. Soni
e-mail: prof.kapilsoni@gmail.com

A. Rasool
e-mail: akki262@gmail.com

J. Tavares et al. (eds.), *Emerging Technologies in Data Mining and Information Security*, Lecture Notes in Networks and Systems 164,
https://doi.org/10.1007/978-981-15-9774-9_50

the analogous key. The strength of any cryptographic system is decided by its key distribution mechanisms used. Thus, establishment of powerful key between two authorized parties is the main problem in cryptography and is known as the key distribution problem. Key distribution by symmetric cryptography requires two parties and a single key which can be obtained from the key distribution center.

Symmetric cryptographic algorithms require "$n\ (n - 1)/2$" unique keys for pair of hosts that want to communicate, and management of such a large key is not feasible, while asymmetric algorithm needs "$2n$" keys pair like public key and private key used for encryption and decryption, respectively.

Classical key distribution [1] can be categorized into two parts (1) distribution of public key (2) distribution of private key. In cryptography, it is important that the key distribution protocols should provide confidentiality and authentication. Confidentiality can be provided by encrypting message by receiver's public key, so that only intended receiver can decrypt it with its private key. Authentication mechanisms are applicable either over message or between communicating entities, whereas message authentications are provided through the existing protocol-based algorithm such as SHA1 or MD5, and entity authentication is done by the verifiable protocol Kerberos. There is no need of authentication in case of private key distribution but needed in case of public key distribution.

Quantum computing is a field which is constantly flourishing and is becoming an immense challenge for technology and science both. Enormous efforts and progress are been done to know maximum capacity of a quantum system since its discovery which was initially done by Richard Feynman in 1982 [2]. From the beginning of research done until now, quantum computers can provide dramatic speedups for some applications like for breaching cryptographic codes. Classical computers are not best fit for this job. To break the cryptographic code, even the most powerful supercomputers would take decades or even centuries. But with quantum computers, it could break them in a second.

As the field quantum computing is developing, researchers have devised a highly secure way of distributing key among two parties which provides security against the eavesdropper [3]. Thus, quantum key distribution has been developed, part of quantum cryptography which allows two parties to exchange a random secret key securely [4]. Quantum key distribution uses uncertainty principle devised by Stephen Wiesnar in 1969 [5] and photon polarization which ensures that only legitimate users get transmitted information. This provides quantum key distribution explicit security. Peev et al. state that "quantum key distribution provides unconditionally secure communication that cannot be compromised by any eavesdropping technique" [6].

Section 2 reviews some of the quantum fundamental concepts. Section 3 discusses the protocols that are used in classical key distribution. Section 4 presents quantum key distribution protocols. Section 5 gives a comparative analysis between classical key distribution and quantum key distribution. Section 6 talks about variations in quantum key distribution, and finally, Sect. 7 provides a conclusion (Fig. 1).

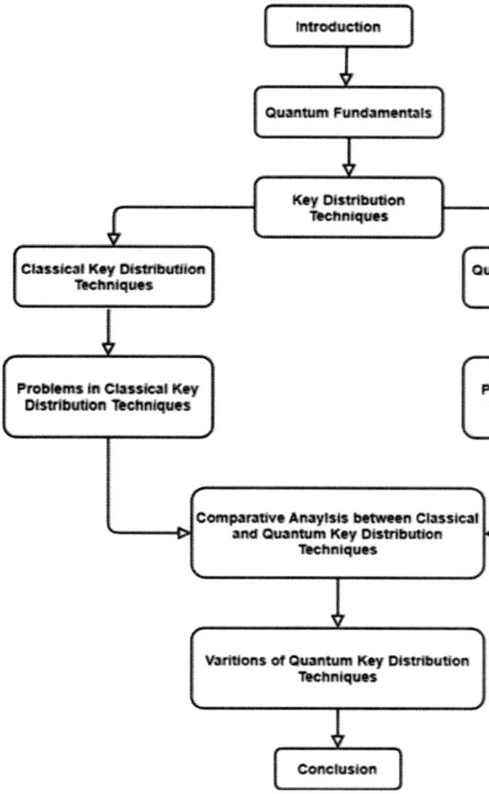

Fig. 1 Organization of paper

2 Quantum Fundamentals

Quantum key distribution mainly relies on quantum mechanics principle that provides unconditional security for the parties communicating with each other. Our normal computers can understand classical bit that is it can be 0 or 1, but in quantum computing, we have quantum bits or qubits; they can be 0 or 1 at the same time which is known as superposition [7]. This factor gives quantum computers superior computing power. Many tangible objects can be used as qubits like a single photon, nucleus and electrons.

Superposition is a phenomenon in which an object can be a present in more than one state at once. The other important principle is quantum entanglement. Suppose there are entangled particles, then irrespective of distance between them if we measure the particular property in one particle, then, at the same time exactly, opposite results will be shown by the other particle. So, it is impractical to communicate via entangled particles without first discussing the observations over a classical channel. "The process of communicating using entangled states by using classical communication channel is known as quantum teleportation, and this forms the basics of Ekert's protocol" [6, 8].

The two attributes of quantum mechanics with which Heisenberg's uncertainty principle deals are position and momentum. The principle says that identifying the position of the particle in the small region of space will make the momentum of the particle uncertain and while measuring the momentum of a particle makes the position uncertain. So this means that you cannot measure any attribute of quantum particle without changing the quantum attributes of the particle. If detection of attributes of any attributes is done, then disturbance will be introduced which will randomly alter that state of the photon in the undeterminable way.

One more important principle that is applied is the no-cloning theorem [9] according to which a quantum state cannot be copied. There will be destruction of original quantum state if any attempts to copy that state are made.

3 Classical Key Distribution

See Fig. 2.

3.1 Distribution of Private Key

Key Distribution Center (KDC)

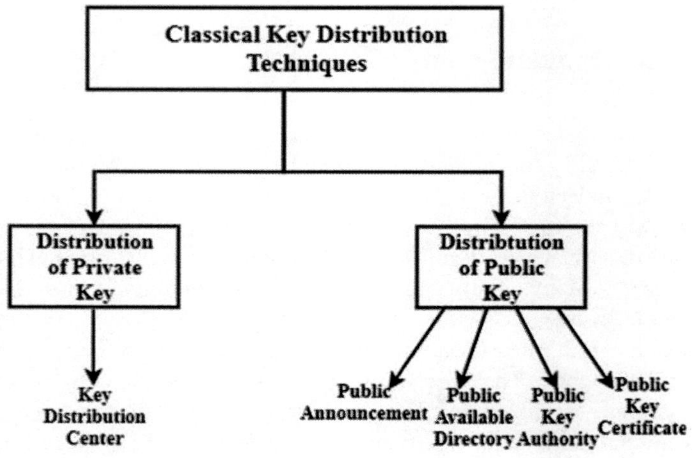

Fig. 2 Key distribution methods

1. A requests for session key to KDC.
2. KDC responds with message, encrypted with K_A for A, and message for B is encrypted with K_B
   $M_A \leftarrow E\ (M, K_A)$
   M: Message which contains two items, i.e., K_S, that is one-time session key, and it also contains the original message that is sent.
   K_A: It is the master key that A shares with KDC.
3. A sends M_B to B which was the intended receiver B that sends nonce (N_B) which is encrypted with K_S (Fig. 3).
4. A responds with some transformation on N_B which is again encrypted with K_S [10].

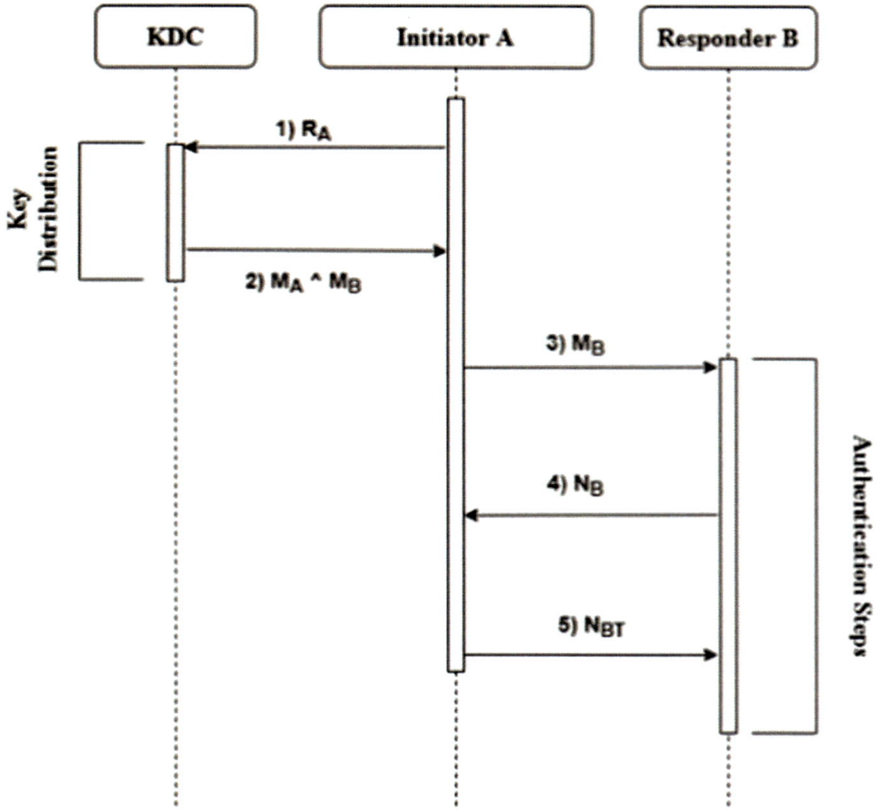

Fig. 3 Symmetric key distribution scenario

3.2 Distribution of Public Key

Public Announcement

1. P_A: *A* sends its public key or broadcasts it (Fig. 4)
2. P_B: *B* sends its public key or broadcasts it

Public Available Directory

1. Reg_A: *A* registers to public key directory with its public key.
2. Reg_B: *B* registers to public key directory with its public key.
3. Ask_A: *B* asks for public key of *A* to directory authority. (Fig. 5)
4. Ask_B: *A* asks for public key of *B* to directory authority.

Public Key Authority

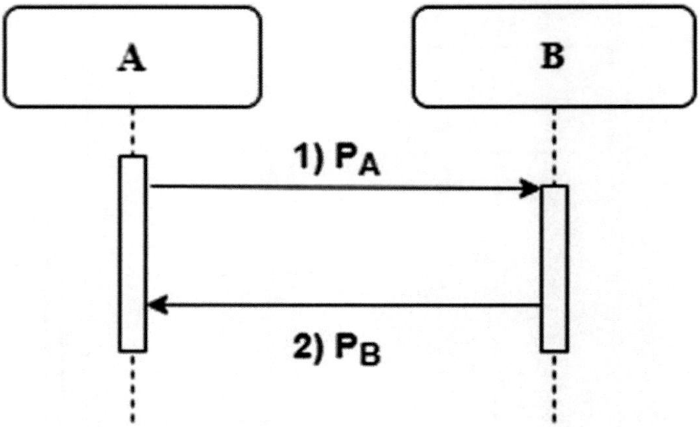

Fig. 4 Uncontrolled public key distribution

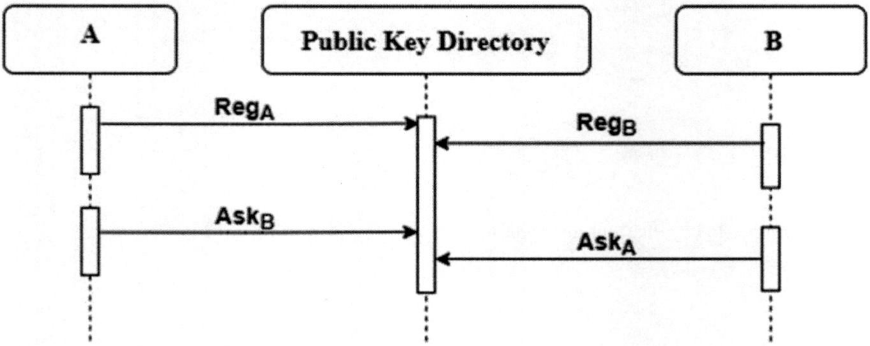

Fig. 5 Public key publication

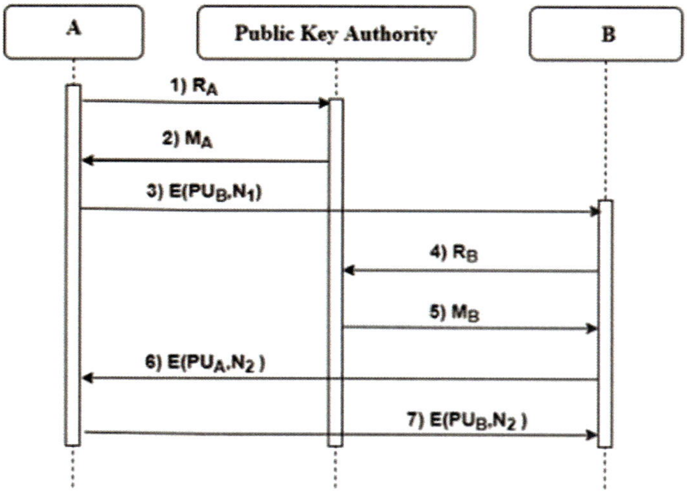

Fig. 6 Public key distribution

1. R_A: A requests for public key of B to public key authority.
2. Authority responds with message (M_A) which contains public key of B, encrypted with authority private key.

$$M_A \leftarrow E(\text{PR}_{\text{auth}}, \text{PU}_B \wedge R_A)$$

   PR_{auth}: Private Key of authority
   EN_1: A sends nonce and its ID which is encrypted with B's public key.
3. R_B: B requests for public key A to public key authority (Fig. 6).
4. Authority responds with message (M_B) which contains public key of A, encrypted with authority private key.
5. EN_2: B sends nonce which is encrypted with A's public key.
6. A sends same nonce again which is encrypted with public key of B to assure B that it corresponds to A.

Public Key Certificate

1. R_A: A requests for certificate by supplying its public key (PU_A).
2. C_A: Authority provides certificate to A. (Fig. 7)
3. B applies to certificate by supplying public key and requests for certificate.
4. Authority provides certificate to B [11].

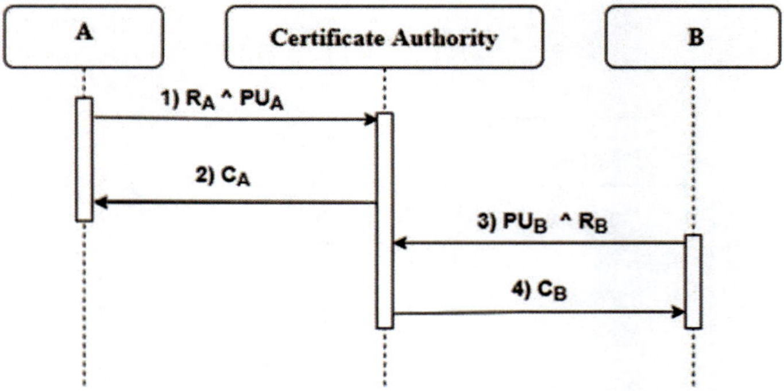

Fig. 7 Exchange of public key certificates

3.3 Problems in Classical Key Distribution Techniques

The main problem in classical key distribution is that it cannot detect the presence of the adversary. The public key authority could become a hindrance to the system. The public key authority maintains directory which contains names and public keys, and it can be tampered due to which privacy is violated. Besides, it is not possible to verify the medium through which parties are communicating which was intercepted, and if the medium is snooped, the contents can be copied and modified.

4 Quantum Key Distribution Protocols

The main problem in classical key distribution is that it cannot identify the presence of adversary, but in quantum key distribution, we can identify the presence of eavesdropper. This main challenge is overcome with help of the no-cloning algorithm. The basic model of quantum key distribution contains two parties Alice and Bob who wish to exchange keys, and they both have access to classical channel and quantum communication channel (Fig. 8). An eavesdropper Eve is also assumed to have access to both classical channel and quantum communication channel.

4.1 BB84 Protocol

Bits are encoded in the polarization state of photon in BB84 [12]. In rectilinear bases, Binary 0 is represented by 0° and Binary 1 is represented by 90°. And in diagonal bases, Binary 0 is represented by 45° and Binary 1 is represented by 135° (Fig. 9). Bit can be represented by polarizing photon in either one of two bases [13].

1. Alice sends randomly chosen bits, i.e., polarized photons on quantum channel.

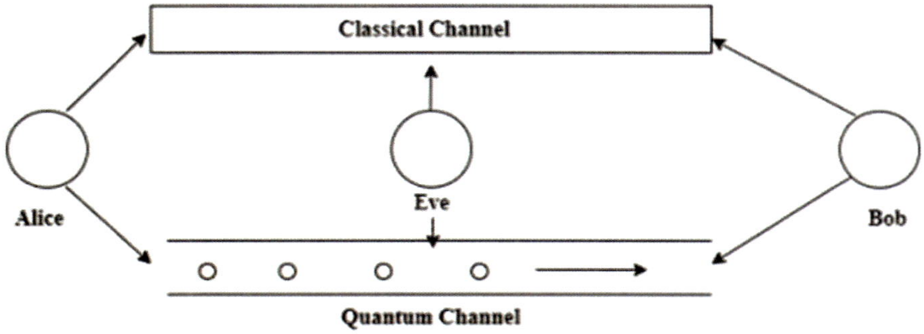

Fig. 8 Quantum key distribution model

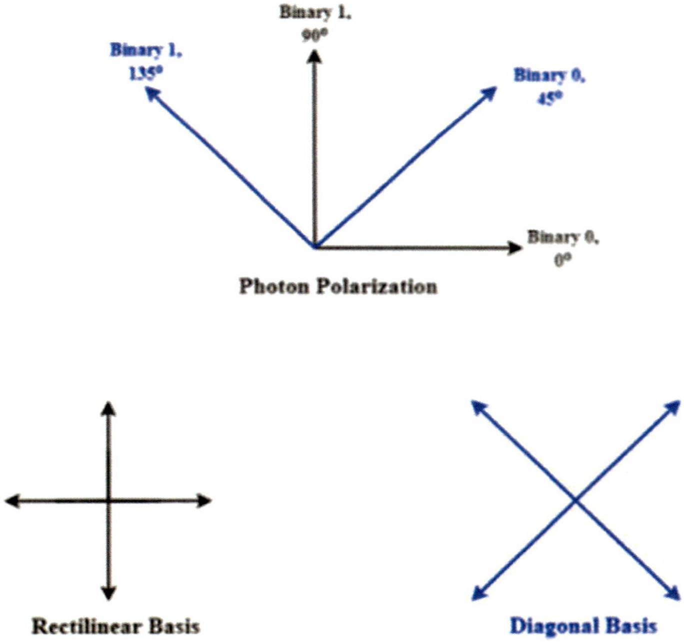

Fig. 9 Bit encoding in BB84

2. Eve can also read the bits that have been sent by Alice, but no-cloning algorithm cannot intercept the bits.
3. Bob reads the bits from the quantum channel. (Fig. 10)
4. Bob sends the basis he randomly chose to Alice via public channel.
5. Alice reads those bases.
6. Alice then compares its bases with those randomly chosen by Bob.

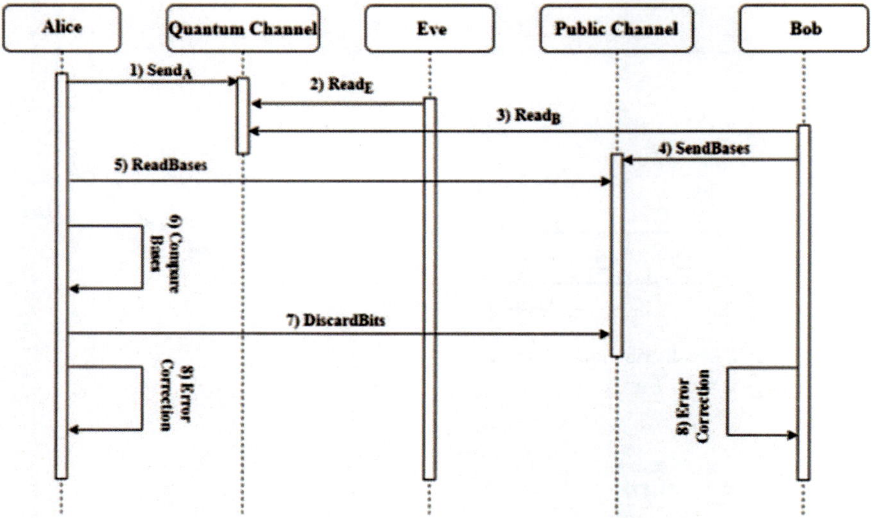

Fig. 10 BB84 key sharing

7. Alice reports back whether he choose correct basis. If both the bases are not same, then that corresponding bit is discarded.
8. Alice and Bob both calculate error (e) to detect the presence of eavesdropper. They compare their error with maximum bit error (e_{max}), and if it is greater ($e > e_{max}$), then communication is suspended, and it starts all over again.
9. Finally, they both share the secret key also known as sifted key.

4.2 *Eckert's Protocol*

In Eckert's protocol [8],

1. P: Single source that emits pairs of entangled polarized photon particles.
2. S_A: Alice shares its bases with Bob.
3. C_B ^ R: Bob compares its bases with Alice bases (C_B) and finds correlation between bases.
4. S_B: Bob shares its bases with Alice.
5. C_A ^ R: Alice compares its bases with Bob (C_A) and finds correlation between them.
6. If same bases have been used by Alice and Bob, then the result generated will be opposite due to the principle of quantum entanglement. This means that if Alice and Bob both interpret their measurements, then they would have a string which is binary complement of the other. So, either party could invert their key.
7. Then, both the parties share a secret key. (Fig. 11)

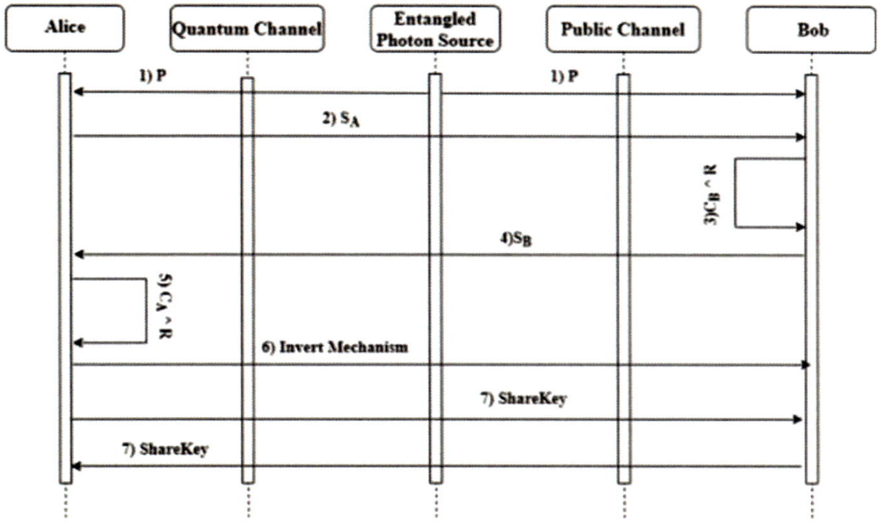

Fig. 11 Eckert's protocol

4.3 Challenges of Quantum Key Distribution

1. Photon Number Splitting (PNS) Attack

PNS is launched by eavesdropper, and BB84 is prone to PNS attack [14]. In BB84, it is assumed that Alice has perfect single-photon source which produces polarized photon for transmission. But practically, it is not possible to have perfect single-photon source. This means that most of the pulses will contain more than one photon.

As the pulses contain several photon, so Eve can spilt the photon and then collect these extra photons and then collect these extra photons and transmit the single photon to Bob. And Eve stores these extra photons in quantum memory until the stage where Bob and Alice compare their bases [15]. Eve may obtain the particulars on the key without producing an error by measuring her photons using correct basis.

2. Quantum Bit Error Rate

The ratio of an error rate to key rate is known as quantum bit error. It also gives information on whether the eavesdropper is present or not. Even though the eavesdropper is present in quantum channel, he does not know the bases which were used to decode the photon. BB84 has relatively higher QBER [16] as it may leak lot of information during communication while sharing secret key.

3. Raw Key Efficiency

Raw key efficiency is ratio of length of the session key which is shared by two communicating parties and length of the original random bits generated by the sender.

Secret key which is later used for encryption/decryption process should be to immune to exhaustive key search attacks and brute force attacks [17]. By taking secret key length as long as possible, the chances of getting any details related to the key will be reduced. The key efficiency rate for BB84 is very low. Because receiver may choose same basis as the sender, and this might happen at least 50% of bits that are generated and that are shared between the two communicating entities.

And the remaining at least 50% of the bits will be removed from the original bits due to unmatched basis [16]. So the resulted key generated will be small in size which affects the overall security of the information that is transmitted.

5 Comparative Analysis of Key Distribution Protocols

Table 1 gives a comparative observation between classical and quantum key distribution Protocols. The strength of a key distribution mechanism can be evaluated by considering the parameters such as identification of the adversary presence during key distribution process and the system vulnerability to passive attack, along with the inclusion of authentication mechanism which is required to be additionally implemented in quantum protocols. Once the parties are authenticated, the next step is the generation of a session key that can be done either through bit generation in classical protocols using pseudo-random number generator or through qubit by utilizing random polarized photon behavior. In classical protocol, there exists simultaneous, frequent accesses of third party which is been involved in key distribution and hence leads to network congestion, but the quantum protocols assure key exchange mechanism to be done by the parties themselves. The quantum protocol uses the principles which are evolved by quantum properties, and its behavior is probabilistic and hence needs measurement with high probability. These all aspects are ensuring quantum key distribution protocols highly secured over classical protocols.

6 Variations of Quantum Key Distribution Protocol

1. B92 Protocol

This protocol was put forth by Bennet in 1992, and this protocol uses two non-orthogonal as compared to four-state BB84 protocol. Out of two non-orthogonal states, one state will be chosen by Alice and another state will be sent to Bob. As the states are not orthogonal to each other, Bob cannot always get decisive result. So instead, he can use measurement called positive operator valued measure or POVM [23, 24]. So Bob can perform test which will sometimes fail to give answer, and other times, it will give answer. Instead of using binary system, Bob uses ternary system which has possible results like 0, 1 or inconclusive result. For example, if Alice sends

Table 1 Comparison between quantum and classical key distribution [18–22]

		Comparative parameters							
		Security aspects	Authentication requirement	Key generation randomness	Adversary identification	Attack identification	Third party requirement	Protocol behaviour	Principle machanism
Distribution protocol	Classical protocols								
	Symmetric key distribution	Less secure	No	Pesudo random	Yes	No	Yes	Deterministic key distribution	Key distribution centre
	Asymmetric key distribution	Less secure	Yes (default)	Pesudo random	Yes	No	Yes	Deterministic key Distribution	Public key certificate
	Quantum protocols								
	BB84 Key distribution	Highly secure	Yes (additional)	Polarized random	No	Yes	No	Probablilistic key distribution	Heisenberg's uncertainty principle
	Eckert's key distribution	Highly secure	Yes (additional)	Polarized random	No	Yes	No	Probablilistic key distribution	Quantum entanglement

1, Bob will not get 0, but he may either get either a 1 or some undeceiving result. And this protocol is also based on Heisenberg's uncertainty principle [25].

2. 4 + 2 Protocol

It was proposed by Huntier et al. and is combination of the merits of both two state B92 protocols four and state BB84 protocol [26]. The principle concept behind this protocol is that rather than using two orthogonal bases, two non-orthogonal bases should be used in BB84 protocol. This makes eavesdropping tougher. In B92 protocol, phase encoding is used as compared to polarization in BB84 protocol. The presence of eavesdropper can be detected even when the weak pulses containing multiple photons are used due to the use of phase encoding.

3. Six-State Protocol

This protocol was suggested by Brub, and it is an abstraction of BB84 protocol with six states. Its working is exactly the same as BB84 protocol, but now Alice is sending one of the six polarization states [27].

7 Conclusion

The article describes classical and quantum key distribution mechanism and realizes that interception of key is not possible in quantum protocols. In fact, the BB84 and Eckert's protocols are significant due to principle of no-cloning and quantum entanglement, and they both enhance the security aspects. Quantum protocols are enough capable to identify the presence of adversary as after the end of distribution process, the communicating parties ensure finding the computed error which should be less than certain threshold; otherwise, they discard the key and iteratively does the same. The parameters considered for comparing the protocols are already justified, and they prove that quantum key distribution is highly secured as by utilizing quantum computational properties and can be effectively used with less vulnerable possibilities.

References

1. W. Diffie, M.E. Hellman, New directions in cryptography. IEEE Trans. Inf. Theory, **IT-22**(6), 644–654 (1976). (IEEE)
2. R.P. Feynman, Simulating physics with computers. Int. J. Theor. Phys. 467–488
3. L. Lace, O. Scegulnaja-Dubrovska, R. Usovs, A. Zalcmane, Quantum cryptographic key distribution protocols, in *The European Social Fund (ESF)* (2008)
4. C. Elliot, Quantum cryptography. IEEE Secur. Priv. J. 57–61 (2004)

5. M. Peev, C. Pacher, T. Lorünser, M. Nölle, A. Poppe, O. Maurhart, M. Suda, A. Fedrizzi, R. Ursin, A. Zeilinger, Response to vulnerability of a novel protocol authentication algorithm ruling out a man-in-the-middle attack in quantum cryptography. Int. J. Quantum Inf. **07**(07), 1401–1407 (2009)
6. S. Wiesner, Conjugate coding. Sigact News **15**(1), 78–88 (1983)
7. P.K. Amiri, Quantum computers. IEEE Potentials **21**(5), 6–9 (2003)
8. A.K. Ekert, Quantum cryptography based on Bell's theorem. Phys. Rev. Lett. **67**, 661–663 (1999)
9. W.K. Wootters, W.H. Zurek, A single quantum cannot be cloned. Nature **299**, 802–803 (1982)
10. S. William, Cryptography and network security principles and practice. 2nd Edn, Prentice Hall (1998)
11. C. Pfleeger, Security in computing. Prentice Hall PTR (2001)
12. C.H., Bennett, Quantum cryptography: public key distribution and coin tossing, in *International Conference on Computer System and Signal Processing* (IEEE, 1984), pp. 175–179
13. V. Padamvathi, B.V. Vardhan, A.V.N. Krishna, Quantum cryptography and quantum key distribution protocols: a survey, in *International Conference on Advanced Computing (IACC)*, Bhimavaram (2016), pp. 556–562
14. G. Brassard, N. Lütkenhaus, T. Mor, B. Sanders, Limitations on practical quantum cryptography. Phys. Rev. Lett. **85**, 1330–1333 (2000)
15. N. Lütkenhaus, M. Jahma, Quantum key distribution with realistic states: photon-number statistics in the photon-number splitting attack. New J. Phys. **4**(1), 44 (2002)
16. D. Jin, P.K. Verma, S.K. Kartalopoulos, Fast convergent key distribution algorithms using a dual quantum channel. Secur. Commun. Netw. (2008)
17. K.H. Hong, O.M. Foong, T.J. Low, Challenges in quantum key distribution: a review, in *ICINS '16*, Dec 28–31, Kuala Lumpur, Malaysia (2016)
18. D. Bruss, G. Erdelyi, T. Meyer, T. Riege, J. Rothe, Quantum cryptography: a survey. ACM Comput. Surv. **39**(2), Article 6 (2007)
19. F. Yang, Y. Hao, The formal study of quantum cryptography protocols, in *10th International Computer Conference on Wavelet Active Media Technology and Information Processing (ICCWAMTIP)*, Chengdu (2013), pp. 29–33
20. R.C. Merkle, Protocols for public key cryptosystems, in *IEEE Symposium on Security and Privacy*, Oakland, CA, USA, pp. 122–122 (IEEE, 1980)
21. D.N. Kartheek, G. Amarnath, P. Venkateswarlu, Security in quantum computing using quantum key distribution protocols, in *International Multi-Conference on Automation, Computing, Communication, Control and Compressed Sensing(iMas4s)*, Kottayam (2013), pp. 19–25
22. P.W. Shor, J. Preskill, Simple proof of security of the BB84 quantum key distribution protocol. Phys. Rev. Lett. **85**(2), 441–444 (2000)
23. C.H. Bennett, Quantum cryptography using any two nonorthogonal states. Phys. Rev. Lett. **68**(21), 3121–3124 (1992)
24. M. Javed, K. Aziz, A survey of quantum key distribution protocols, in *FIT'09*, CIIT, Abbottabad, Pakistan (2009)
25. M. Elboukhari, M. Azizi, A. Azizi, Quantum key distribution protocols: a survey. Int. J. Univ. Comput. Sci. **1**, 59–67 (2010)
26. B. Huttner, N. Imoto, N. Gisin, T. Mor, Quantum cryptography with coherent states. Phys. Rev. Lett. **51**, 1863–1869 (1995)
27. D. Bruß, Optimal eavesdropping in quantum cryptography with six states. Phys. Rev. Lett. **81**, 3018 (1998)

A Detailed Survey on Epilepsy Detection Using the Classifications of EEG Signals

M. K. Mariam Bee and K. Vidhya

Abstract Epilepsy is a neurological disorder of the human brain, and it can be caused due to cells damage in brain. This disorder can be detected by identifying the signals from these affected regions in brain. These signals are captured by electroencephologram (EEG) sensor which places number of electrodes on brain. The signals which are acquired by affected regions of brain are focal, and the signals which are acquired by normal regions of brain are non-focal. Hence, there is a need for automatic classifications of EEG signals. This paper discusses various conventional methods for EEG signals detection and classifications and states its limitations.

Keywords Focal · Non-focal · EEG signals · Brain · Epilepsy

1 Introduction

Epilepsy is a typical interminable neurological issue, which is set apart by seizures. It is normally caused by unnecessary release of cortical cells from the brain. It is imperceptible and may prompt paroxysms. Epileptic seizures cannot be relieved, however it can be controlled through appropriate restorative medications. Since seizures happen once in a while and cannot be anticipated, programmed identification of seizures by EEG signals is much suggested.

EEG is a regularly utilized demonstrative instrument for epilepsy conclusion. Visual portrayal of EEG by brain mapping or quadrature EEG (QEEG) will be useful in distinguishing the seizure inception which can be utilized for clinical finding. An epileptic seizure is an electrical release inside the brain that spreads rapidly crazy all through the focal sensory system. There are two kinds of seizures to be specific generalized and focal seizures. Summed up seizures are aftereffects of anomalous

M. K. Mariam Bee (✉) · K. Vidhya
Department of Electronics and Communication Engineering, Saveetha School of Engineering, SIMATS, Chennai, India
e-mail: mariamresearch09@gmail.com

K. Vidhya
e-mail: kvidhyasri@gmail.com

J. Tavares et al. (eds.), *Emerging Technologies in Data Mining and Information Security*, Lecture Notes in Networks and Systems 164,
https://doi.org/10.1007/978-981-15-9774-9_51

Fig. 1 EEG sensor

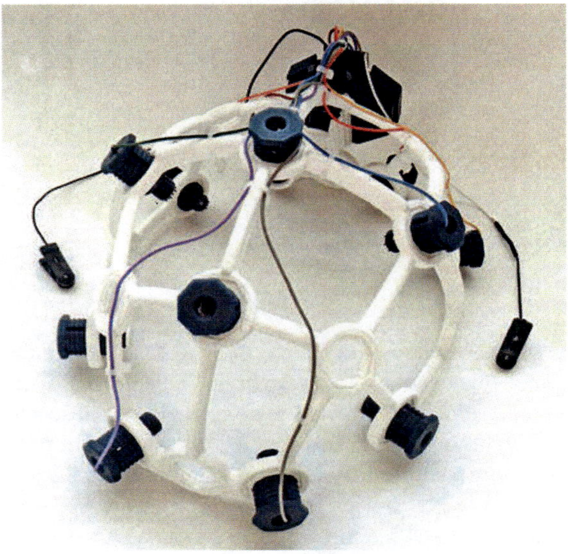

neuronal movement on the two sides of the brain. It influences both cerebral halves of the globe of brain. These seizures may cause loss of cognizance, falls or monstrous muscle fits. Once in a while, it is hard to group a few cases as either focal or summed up. Typically, focal or incomplete seizures influence a little territory of the brain. This kind of seizures happens when this electrical action stays in a constrained region of the brain. The signs of a focal seizure rely upon the piece of the brain required with the strange brain cell action. Seizure action starts as focal seizures and afterward spreads to the whole brain which winds up and summed up in nature. Figure 1 shows the EEG sensor setup which produces EEG signals.

Figure 2 shows the types of focal and non-focal signals for epilepsy detection and diagnosis. The non-focal signal is the normal signal, and focal signal is the abnormal signal. The focal signals exhibit high level of nonlinearities which are the main reasons for epilepsy disease as depicted in Fig. 2.

Confined focal seizures influence a little region of the brain which is a class of focal seizures. The starting point of irregular movement is typically limited particularly to a flap or an area in a specific side of the brain. Strange electrical action is typically shown as abating, spikes or sharp wave design saw on a specific projection or at a point where the seizures begin in a side of the equator of the brain. A man may encounter one or a wide range of sorts of seizures.

An information choice calculation relying upon stage congruency to decide interictal spikes from the foundation EEG was proposed by Logesparan and Rodriguez-Villegas [1]. Yusof et al. [2] proposed another transformation task for fast element determination by GA relying upon elitism of the allele. The fittest chromosomes were

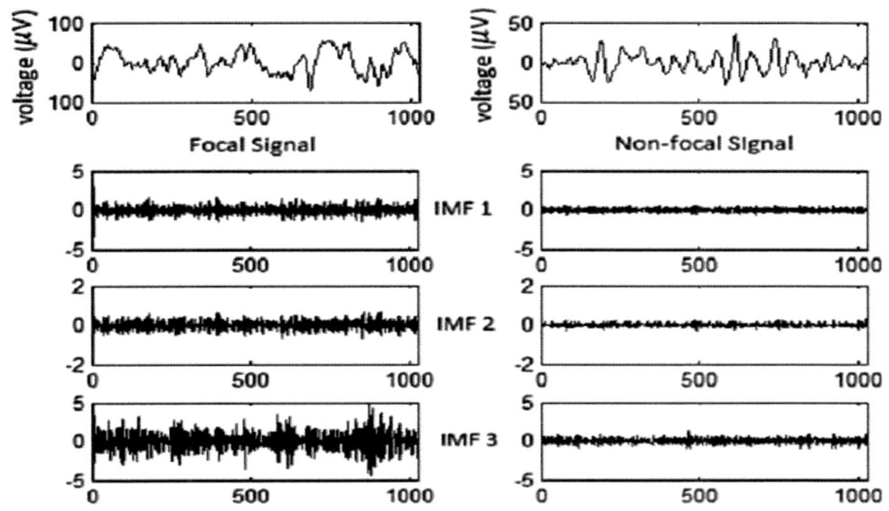

Fig. 2 Focal and non-focal EEG signals

saved by ordinary elitism in GA, which were then assessed by using the wellness work.

2 Survey Based on Preprocessing and Decomposition

Noise and artifacts are occurred in EEG signals during its signal acquisition stage. These disturbances should be detected and removed from EEG signals in order to improve the classification accuracy for disease detection and diagnosis. Preprocessing is used to remove the noises and distortions in an EEG signal for improving the classification accuracy of EEG signals classification system.

Tzyy-Ping et al. [3] have utilized independent component analysis (ICA) for transforming the spatial domain EEG signals into frequency domain signals. The authors separated the spatial coefficients from the transformed matrix coefficients. The decomposition rate was not optimum for further processing.

Singular value decomposition (SVD) is utilized by Trejo et al. [4] for pre-handling of EEG signals. This transform decomposed the EEG signals into three different submatrix coefficients. Lemm et al. [5] used common spatio-spatial patterns (CSSP) methodology as preprocessing stage for the detection of EEG signals for epilepsy disease detection and classifications.

Nathan et al. [6] used dry contact electrodes for removing the noise patterns from the obtained EEG signals. FurioBartoli et al. [7] used optimal linear filter for detecting and removing the noises from EEG signals from various channels in EEG sensors. The authors tested their proposed noise reduction filtering methodology on

Table 1 Survey of preprocessing for epilepsy disease detection

Authors	Methodology	Limitations
Tzyy-Ping et al. (2001)	Independent component analysis (ICA)	Signal recovery rate is low
Trejo et al. (2003)	Singular value decomposition (SVD)	High error
Lemm et al. (2005)	Common spatio-spatial patterns (CSSP)	Not suitable for real-time EEG signals for epilepsy detection
Nathan et al. (2014)	Dry contact electrodes	High distortion effect
FurioBartoli et al. (1983)	Optimal linear filter	Signal recovery rate is low
Garcés Correa et al. (2007)	Adaptive filters	Nonlinearities are high

different patients on real-time environment, and the results were compared with a conventional method Kalman filter. Garcés Correa et al. [8] applied adaptive filters on EEG signals for detecting and removing artifacts for improving the classification accuracy of epilepsy disease diagnosis.

Table 1 shows the survey of different conventional preprocessing methods for epilepsy disease detection and diagnosis.

3 Survey Based on Feature Extraction

Bajaj et al. [9] used EMD decomposition methodology to decompose the EEG signals into various modes. Further, the authors extracted Pearson product-moment correlation coefficient features and Spearman rank correlation features from focal and non-focal EEG signals for differentiating the normal EEG signals from abnormal EEG signals.

Pushpendra Singh et al. [10] detected and extracted mean-frequency (MF) and root mean-square (RMS) features from EEG signals. Khushnandan Rai et al. [11] developed feature extraction-based EEG signals classification methodology. The authors extracted Renyi entropy and the average entropy from EEG signals with respect to both normal and abnormal EEG cases. Mangala Gowri et al. [12] applied wavelet transform on EEG signals for epilepsy disease detection and diagnosis. The authors used Daubechies wavelet transform on EEG signals in order to produce wavelet coefficients matrix. Cecchin et al. [13] evaluated a semi-programmed strategy for lateralization of fleeting flap seizures utilizing crude scalp EEG signals.

Guerrero-Mosquera et al. [14] have proposed a technique for distinguishing seizures in EEG signals. For this, highlight extraction in time frequency distributions (TFD) was utilized. Highlights were removed from the smoothed Pseudo Wigner–Ville conveyance utilizing tracks from McAulayquatieri Sinusoidal model. Direito et al. [15] have looked at the three element determination techniques for minimum redundancy maximum relevance, recursive component elimination and elitist non-commanded sorting genetic algorithm. Stamoulis et al. [16] have proposed a subspace flag handling approach for enhancing the scalp EEG-based limitation of wide concentration epileptic seizures and for assessing the bearing of the source landing. Guo et al. [17] proposed a novel approach for programmed epileptic seizure discovery that used line length highlights relying upon wavelet change multiresolution disintegration. Fake neural network was utilized to arrange the EEG signals with respect to the nearness or non-attendance of seizure. The database was preprocessed by evacuating the antiques.

4 Survey Based on Classification Approaches

Krishnaprasanna et al. [18] developed EEG signals classification methodology using radial basis function (RBF) classifier. The authors extracted various intrinsic mode functions which were obtained from EEG signals for the classification of EEG signals into focal and non-focal signals for epilepsy disease diagnosis. Sriraam et al. [19] proposed a classification methodology for classifying focal signals from non-focal signals using SVM classification approach. The authors achieved 94.56% of sensitivity, 89.74% of specificity and 92.15% of overall classification accuracy on open access dataset. Taqi et al. [20] applied deep neural network classification approach for classifying the EEG signals into focal and non-focal signals. The authors achieved 100% classification accuracy for LeNet model in order to classify the EEG signals for epilepsy disease detection and diagnosis.

Anindya Bijoy et al. [21] classified EEG signals using empirical mode decomposition (EMD) and DWT transform. The authors extracted entropy features from EEG signals for both normal and abnormal cases. The authors tested their algorithm on 7500 EEG signals. The method incorporated KNN classification approach obtained 89.4% of classification accuracy and 90.7% of sensitivity.

Mammone et al. [22] have proposed automatic wavelet independent component analysis (ICA) model for consequently evacuating ancient rarities in EEG signals. This technique proposed depended on the joint utilization of wavelet transform and ICA. Bhattacharyya et al. [23] projected a component-based characterization way to deal with recognized redundant and transient antiquities. Boubchir and Boashash (2013) [24] have proposed another nonparametric Bayesian estimator for flag and picture denoising in wavelet area. The creators have outlined another hyper-parameter estimator in light of expectation maximization (EM) calculation. This was utilized to evaluate the parameters of the Bessel K form (BKF) thickness and contrasted and a cumulants-based estimator.

Acharya et al. [25] proposed a procedure to separatehigher-order spectra highlights from typical, interictal and seizure EEG sections and to utilize includes in the classifiers for identifying these classes. D. Najumnissa et al. [26] proposed a procedure with fuzzy logic and fuzzy set theory which are employed to describe human thinking and reasoning in a mathematical framework. Fuzzy-rule based modeling

Table 2 Comparisons of various conventional classification approaches for EEG signals classification with respect to its limitations

Authors	Methodology	Limitations
Krishnaprasanna et al. (2018)	Radial basis function (RBF) classifier	Low classification accuracy
Sriraam et al. (2017)	SVM classification approach	Low sensitivity and specificity
Taqi et al. (2017)	LeNet model	Limited to number of EEG signals
Anindya Bijoy et al. (2016)	KNN classification approach	Low classification accuracy
Bhattacharyya et al. (2013)	Component-based characterization	Low detection rate
Boubchir and Boashash (2013)	Nonparametric Bayesian estimator	Low sensitivity and specificity
Acharya et al. (2013)	SVM classification approach	Low classification accuracy
Hekim (2012)	ANFIS	High elapsed time

is a qualitative modeling scheme where the system behavior is described using a natural language. Hekim [27] classified EEG signals by utilizing two artificial neural networks (ANN) models and normal forces of power spectral densities (PSD) of EEG subgroups. Discrete wavelet transform (DWT) was utilized in this paper in order to deteriorate EEG signals into subgroups. Orhan et al. [28] have proposed out the EEG flag order by two ML techniques, for example, k-implies grouping and MLPNN. Highlight extraction has been performed utilizing DWT. Kousarrizi et al. [29] have built up a maximum likelihood (ML)-based procedure for for antiques expulsion from the EEG flag which have been utilized and after that order was performed on the information removed from wavelet change utilizing neural systems and support vector machine (SVM). Skinner et al. [30] have proposed hereditary-based ML classifier for EEG flag order. The authors constructed auto regressive (AR) models and fast Fourier transform (FFT) models where they proposed a hereditary-based learning classifier framework. The authors achieved overall precision of 99.3% for abnormal cases and overall precision of 88.9% for normal cases.

Liang et al. [31] have proposed a methodology for grouping of mental undertakings utilizing EEG motion with the proficiency of extreme learning machine (ELM) approach. Table 2 shows the comparisons of various conventional classification approaches for EEG signals classification with respect to its limitations.

5 Conclusions

This paper introduces the EEG signal and its types in detail with illustrations. This work mainly focuses on the survey of different research works performed in different aspects of EEG signal analysis. In this paper, various existing approaches and algorithms for epileptic seizure detection in EEG signals are discussed. In most of the research works, the reported methods did not select the attributes using a suitable technique and did not observe statistical significance of the attributes, although the parameters significantly affect the classification performance. From the literature, some techniques took more time for computation work, some of them were computationally complex and some of them had a limited success rate.

References

1. L. Logesparan, E. Rodriguez-Villegas, A novel phase congruency based algorithm for online data reduction in ambulatory EEG systems. IEEE Trans. Biomed. Eng. **58**(10), 2825–2834 (2011)
2. R. Yusof, U. Khairuddin, M. Khalid, A new mutation operation for faster convergence in genetic algorithm feature selection. Int. J. Innov. Comput. Inf. Control **8**(10(B)), 7363–7379 (2012)
3. J. Tzyy-Ping, S. Makeig, M.J. Mckeown, A.J. Bell, L. Te-Won, T.J. Sejnowski, Imaging brain dynamics using independent component analysis. Proc. IEEE **89**(7), 1107–1122 (2001)
4. L.J. Trejo, K.R. Wheeler, C.C. Jorgensen, R. Rosipal, S.T. Clanton, B. Matthews, A.D. Hibbs, R. Matthews, M. Krupka, Multimodal neuroelectric interface development. IEEE Trans. Neural Syst. Rehabil. Eng. **11**(2), 199–204 (2003)
5. S. Lemm, B. Blankertz, G. Curio, K.R. Muller, Spatio-spectral filters for improving the classification of single trial EEG. IEEE Trans. Biomed. Eng. **52**(9), 1541–1548 (2005)
6. V. Nathan, R. Jafari, Reducing the noise level of EEG signal acquisition through reconfiguration of dry contact electrodes, in *2014 IEEE Biomedical Circuits and Systems Conference (BioCAS) Proceedings,* Lausanne, 2014, pp. 572–575
7. S. Cerutti, F. Bartoli, An optimal linear filter for the reduction of noise superimposed to the EEG signal. J. Biomed. Eng. **5**(4), 274–280 (1983)
8. A. Garcés Correa, E. Laciarl, H.D. Patiño, M.E. Valentinuzzi, Artifact removal from EEG signals using adaptive filters in cascade. J. Phys. Conf. Ser. **90** (2007)
9. V. Bajaj, K. Rai, A. Kumar, D. Sharma, G.K. Singh, Rhythm-based features for classification of focal and non-focal EEG signals. IET Sig. Process. **11**(6), 743–748 (2017)
10. P. Singh, R.B. Pachori, Classification of focal and nonfocal EEG signals using features derived from Fourier-based rhythms. J. Mech. Med. Biol. **17**(4) (2017)
11. K. Rai, V. Bajaj, A. Kumar, Features extraction for classification of focal and non-focal EEG signals, in *2015, Information Science and Applications*, Lecture Notes in Electrical Engineering, pp. 599–605
12. S.G. Mangala Gowri, C.P. Raj, EEG feature extraction using daubechies wavelet and classification using neural network. Int. J. Pure Appl. Math. **118**(18), 3209–3223 (2018)
13. T. Cecchin, R. Ranta, L. Koessler, O. Caspary, H. Vespignani, L. Maillard, Seizure lateralization in scalp EEG using hjorth parameters. Clin. Neurophysiol. **121**(3), 290–300 (2010)
14. C. Guerrero-Mosquera, A.M. Trigueros, J.I. Franco, A. Navia-Vázquez, New feature extraction approach for epileptic EEG signal detection using time-frequency distributions. Med. Biol. Eng. Comput. **48**(4), 321–330 (2010)
15. B. Direito, F. Ventura, C. Teixeira, A. Dourado, Optimized feature subsets for epileptic seizure prediction studies, engineering in medicine and biology society, EMBC, in *2011 Annual International Conference of the IEEE* (2011), pp. 1636–1639
16. C. Stamoulis, I. Snchez Fernndez, B.S. Chang, T. Loddenkemper, Signal subspace integration for improved seizure localization, engineering in medicine and biology society (EMBC), in *Annual International Conference of the IEEE*, 2012, pp. 1016–1019

17. L. Guo, D. Rivero, J. Dorado, J.R. Rabunal, A. Pazos, Automatic epileptic seizure detection in EEGs based on line length feature and artificial neural networks. J. Neurosci. Methods **191**(1), 101–109 (2010)
18. A.R. Krishnaprasanna, V. Vijaya Baskar, Focal and Non-Focal EEG signal classification by computing area of 2D-PSR obtained for IMF. J. ICT **5**(2), 171–186 (2018)
19. N. Sriraam, S. Raghu, Classification of focal and non focal epileptic seizures using multi-features and SVM classifier. J. Med Syst. **41**(10), 160 (2017)
20. M. Taqi, F. Al-Azzo, M. Mariofanna, J.M. Al-Saadi, Classification and discrimination of focal and non-focal EEG signals based on deep neural network, in *2017 International Conference on Current Research in Computer Science and Information Technology (ICCIT)*, Slemani, 2017, pp. 86–92
21. A. Bijoy, D. Mohammed, I.H. Bhuiyan, Discrimination and classification of focal and non-focal EEG signals using entropy-based features in the EMD-DWT domain. Biomed. Signal Process. Control **29**, 11–21 (2016)
22. N. Mammone, F. La Foresta, F.C. Morabito, Automatic artifact rejection from multichannel scalp EEG by wavelet ICA. Sens. J. IEEE **12**(3), 533–542 (2012)
23. S. Bhattacharyya, A. Biswas, J. Mukherjee, A.K. Majumdar, B. Majumdar, S. Mukherjee, Detection of bursts from high energy bursts in neonatal EEG. Comput. Biol. Med. **43**(11), 1804–1814 (2013)
24. L. Boubchir, B. Boashash, Wavelet denoising based on the MAP estimation using the BKF Prior with application to images and EEG signals. IEEE Trans. Sig. Proc. **61**(8), 1880–1894 (2013)
25. U.R. Acharya, S.V. Sree, G. Swapna, R. Joy Martis, J.S. Suri, Automated EEG analysis of epilepsy: a review. Knowl. Based Syst. **45**, 147–165 (2013)
26. D. Najumnissa, T. Rangaswamy, Detection and classification of epileptic seizures using wavelet feature extraction and adaptive neuro-fuzzy inference system. Int. J. Comput. Eng. Res. **2**(3), 755–761 (2012)
27. M. Hekim, ANN-based classification of EEG signals using the average power based on rectangle approximation window. Electr. Rev. **88**(8), 210–215 (2012)
28. U. Orhan, M. Hekim, M. Ozer, EEG signals classification using the K-means clustering and a multilayer perceptron neural network model. Expert Syst. Appl. **38**(10), 13475–13481 (2011)
29. M.R.N. Kousarrizi, A.A. Ghanbari, Feature extraction and classification of EEG signals using wavelet transform, SVM and artificial neural networks for brain computer interfaces, in *The Proceedings of the International Joint Conference on Bioinformatics, Systems Biology and Intelligent Computing* (2009), pp. 352–355
30. B.T. Skinner, H.T. Nguyen, D.K. Liu, Classification of EEG signals using a genetic-based machine learning classifier, in *The Proceedings of the IEEE Engineering in Medicine and Biology Scociety* (2007), pp. 3120–3123
31. N. Liang, P. Saratchandran, G. Huang, N. Sundararajan, Classification of mental tasks from EEG signals using extreme learning machine. Int. J. Neural Syst. **16**(1), 29–38 (2006)

A Comparative Study of Machine Learning and Deep Learning for Prediction of COVID-19

Shaswata Saha and Sarmistha Neogy

Abstract In December 2019, a novel coronavirus was found in Wuhan, China. The disease spread due to it was named COVID-19 (Coronavirus Disease 2019). Since then, more than 28 lakh people have been affected worldwide by the virus at the time of writing this paper. The virus has claimed the lives of close to 2 lakh people. It has also affected the lives and livelihoods of millions of people around the globe. Thus it is important to predict and analyse how the disease will affect us in the future. This work uses various machine learning and deep learning models to predict the number of COVID-19 cases in the near future and compare the accuracies of the different results thus obtained.

1 Introduction

Wuhan, a province in China, experienced a surge in pneumonia patients in December 2019. Upon further investigation, it was announced that a novel coronavirus has been found. At the time of writing this paper, the virus has spread to 210 different nations. The world has seen more than 2.8 million COVID-19 patients. Around 200 thousand people have lost their lives. Countries with the best medical facilities have failed to curb the onslaught. The world has come to a standstill due to a microscopic being.

The magnanimity of the effects of the virus has made people and governments curious to know what lies ahead. There are various infectious diseases. This will help them know how the steps that have been taken are working and what steps must be taken in the near future.

There are various traditional models that predict the outcome of infectious diseases. These can be broadly classified into two types: differential models and time series models [1]. The differential models aim at representing the dynamics of the

S. Saha (✉) · S. Neogy
Jadavpur University, 188, Raja S.C. Mallick Road, Kolkata, West Bengal 700032, India
e-mail: shaswata97@gmail.com

S. Neogy
e-mail: sarmisthaneogy@gmail.com

J. Tavares et al. (eds.), *Emerging Technologies in Data Mining and Information Security*, Lecture Notes in Networks and Systems 164,
https://doi.org/10.1007/978-981-15-9774-9_52

infectious diseases by a differential equation. Examples of these are SI model, SIR model, SIS model, SIER model, etc,., where S, E, I and R stand for Susceptible, Exposed, Infected and Recovered respectively.

Time series prediction models predict the outcome of infectious diseases by analyzing time series progression of the infectious disease incidence. This is mainly done with the help of AutoRegressive Integrated Moving Average (ARIMA) model. It uses multiple differences to make the variable time series sequence into a constant one and then traces it back to a point in the past. Other time series prediction models include Exponential Smoothing model, Markov Chain model, etc.

This paper aims at checking the possibilities of reducing this problem to a machine learning or a deep learning problem. We have used various learning algorithms to train the forecasting model using the available data and predict values for the future. Section 2 of this paper familiarises the reader with the learning algorithms used here. Section 3 analyses the problem statement, the dataset used and the experiments conducted. Section 4 shows the accuracy of each predictive model and compares them with each other and analyses what the results mean for India. Section 5 concludes the paper by saying what is achieved in the paper and what is left to achieve.

2 Related Works

In this section, we review some of the existing techniques that are used for prediction purposes.

2.1 Machine Learning

Machine Learning refers to the process of a computer algorithm teaching itself through experiences given as input. There are various types of machine learning models like Regression Analysis, Support Vector Machines (SVMs) and various others. In this work, we have used models based on the above two Machine Learning techniques.

2.1.1 Regression Analysis

Regression analysis [2] means using statistical methods like linear functions, polynomial functions, etc., to understand the relationship between input and output variables. These methods use linear or polynomial curves to best fit the given data based on a loss function.

2.1.2 Support Vector Machines

SVMs [3] are supervised machine learning models used for classification and regression. Since this paper deals with a regression or prediction problem, this technique has also been considered. It is mostly used as a linear classification model.

2.1.3 Bayesian Networks

Bayesian networks [4] are a type of probabilistic directed acyclic graphical models which use Bayesian Inference for probability computations. Bayesian Regressors reduce regressions problem to a probabilistic model.

2.2 *Deep Learning*

Deep learning [5] is a class of machine learning. Deep learning algorithms take raw data as input and use multiple layers to extract features. Thus, unlike in machine learning, feature extraction module can be done away with in deep learning. Although deep learning is a class of machine learning, it is often considered to be a separate class of study because it is widely used in multiple fields and researchers are coming up with state-of-the-art algorithms. There are various types of deep learning models like Convolutional Neural Networks (CNN), Recurrent Neural Networks (RNN), among others. CNN is mostly used for image data. However, the data used here is text-only. Thus, the present work only uses RNN models.

2.2.1 Recurrent Neural Networks

An RNN [6] is a deep learning model that can take input sequences of arbitrary length. It processes this input by recursively applying a transition function to an internal hidden state vector ht of the input sequence. The output of the activation function of ht at time step t is a function of the previous hidden state ht-1 and the current input at time step t.
Long Short-term Memory (LSTM) LSTM [6] is a type of RNN that contains special memory blocks in the hidden layer of the RNN. These memory blocks are made up of memory cells that are connected to each other. These connections store the temporal state of the network. This nature of LSTM makes it useful for working with time series data.

3 Proposed Method

In this section we state the problem statement, the data available, and the approaches adopted to achieve a prediction. This paper aims at developing machine learning and deep learning models for predicting COVID-19 the number of patients and fatalities at a given date in the future. The problem statement and the datasets used are explained in details below.

3.1 Problem Statement and Dataset

The paper attempts to develop machine learning and deep learning models which take previous count of COVID-19 patients and fatalities in various countries as input and predict the same for future dates. The models may also use various other input parameters like quarantine dates, age ratio, hospital facilities, etc. in each country, so that more information can be obtained for better prediction.

Three datasets have been used for this task. The nature of the datasets and the information they provide have been given below. All the datasets used here are public datasets available in Kaggle.

1. *COVID19 Global Forecasting (Week 4)* [7]: This dataset contains daily data of COVID-19 patients and deaths from 313 different states or countries. The dataset contains train data from 22nd January, 2020 to 11th April, 2020. It also asks participants to predict data from 12th April, 2020 to 14th May, 2020. However, Kaggle has not yet released the ground truths for the test dates. The train data has the following fields: Country, Province/State, Date, Confirmed cases, Fatalities. The test data only has the first 3 fields.
2. *COVID19/lockdown* [8]: This dataset contains the lockdown dates of every state or country present in the above dataset, the type of the lockdown: full, partial or none, and the reference for related information.
3. *COVID 19 Enriched Dataset* [9]: This dataset contains various demographic details of most of the states or countries present in the train set of COVID 19 Global Forecasting dataset. The data contains the percentage share of age groups from 0–4 to 95–99 in every state or country. It also contains the population density, number of hospital beds, percentage of people having lung diseases, the percentage of smokers, etc for each area.

3.2 Machine Learning Approaches

This section describes the various machine learning methods that have been used to predict the future COVID-19 data using the above datasets.

1. *Regression models*: In these models, two input variables are used, namely, number of days since the first case d_f and number of days since lockdown d_l. For each province or country the date when the first case is detected is found. If any input has $d_l < 0$, then the output will be zero, otherwise the input will be fed into the regression model. The following regression models have been used.
   a. In the *Linear Regression (LR)* model, a linear function is fit to the input variables and the output variables.
   b. A cubic polynomial function is fit to the input variables and the output variables in *Polynomial Regression (PR)* model. Polynomial functions of different degrees from 2 to 10 were tried and tested on the model. The cubic polynomial model gave the best results.
   c. In the *Exponential Regression (ER)* model, a logarithmic function is applied on the output variables and a linear function is fit to the input variables and the log of the output variables, thus making it an exponential regression.
2. *Support Vector Machines (SVM)*: In this model, the same input variables have been used as above. An SVM is trained to fit to the given dataset and is used later for prediction.
3. *Bayesian Regression models (BR)*: A Bayesian Ridge is fit to the same input and output variables as said before. The advantage of such a model is that it uses a regularization parameter that can be used in the estimation procedure.

3.3 Deep Learning Approaches

This section sheds light on the deep learning methods that have been used to predict the future COVID-19 data using the datasets mentioned. LSTM based models and Hybrid models using both LSTM and ANN have been used in this work.

1. *LSTM based models (LSTM)*: In this approach, LSTM has been used for COVID-19 forecasting. This model takes only two input variables, i.e., the new number of Confirmed Cases and Fatalities that have come up in each of the past 7 days. Experiments were also conducted taking sequences of past 5, 10 and 14 days too. However, the model taking sequences of the past 7 days gave the best realistic result.
2. *Hybrid model (HM)*: In this approach, at first an LSTM based model is trained on the following input variables: Confirmed Cases, Fatalities, Quarantine and Schools. Quarantine is a binary input field that stores 1 if the place is under quarantine on that day and 0 otherwise. Similarly, School is another binary input variable that stores 1 if schools are open in that area on the specific date and 0 otherwise. The outputs of this LSTM model is appended to a 1-D vector that stores demographic information, like, age group, lung diseases, etc., at that place and time. This appended 1-D vector is finally fed to an ANN model. This ANN model gives the Confirmed Cases and fatalities as output.

4 Results

The results from the above methods are then compared using root-mean-squared-log-error (RMSLE). Equation 1 shows the formula of RMSLE where y is the predicted output and ŷ is the ground truth.

$$\text{RMSLE} = \sqrt{\frac{1}{n}\sum_{i=1}^{n}(\log(y_i + 1) - \log(\hat{y}_i + 1))^2} \quad (1)$$

The RMSLE between the outputs and the real-life figures is obtained for each of the two output variables. Then the average of these RMSLEs is obtained. Let this be known as the Column-Wise Average RMSLE (CWA-RMSLE). Table 1 shows the (CWA-RMSLE). Detailed results from all models were submitted to Kaggle. However, the ground truths have not yet been released by Kaggle. Thus all the figures in Table 1 are the Kaggle scores (CWA-RMSLE) obtained from running the above models.

This shows that the LSTM model beats other conventional machine learning models by a large margin. Also, although the hybrid model variant conceptualized here is not a conventional model, it performs better than state-of-the-art machine learning models. It only falls short to the LSTM model. To put the results into perspective, the best score on Kaggle's private leaderboard is 0.19695 at the time of writing.

This success of LSTM can be attributed to the fact that the data given here is a time series data. Although other datasets are available and used in the other models, the LSTM model obtains the result by using the daily count of the COVID-19 patients and deaths and thereby contradicts the general notion that 'more the information, better the results'.

Table 1 Comparison of various models

Model	CWA-RMSLE
LR	2.95006
PR	10.32774
ER	2.95006
SVM	34.17189
BR	4.16549
LSTM	**0.10203**
HM	1.26248

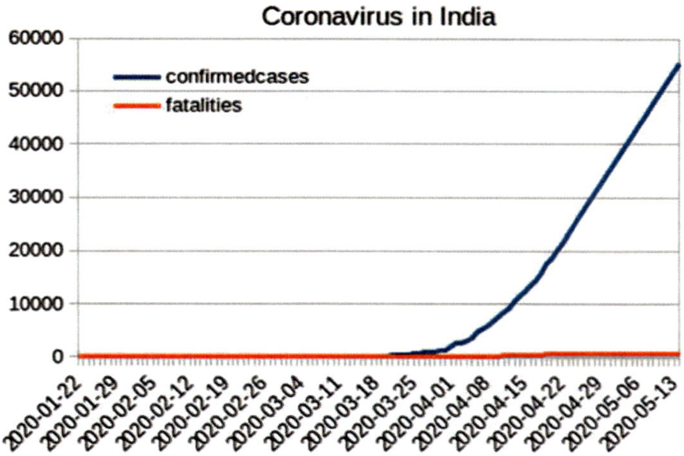

Fig. 1 Coronavirus situation in India

4.1 Analysis of the Results Obtained for India

At the time of writing this paper, 26,917 have been affected by the novel coronavirus. The virus has also claimed the lives of 827 of them [10]. The LSTM model predicted that by this time, 27,534 Indians would have been affected by the virus and 681 would have lost their lives. Thus, the model slightly overestimates the number of patients while it underestimates the deaths caused by it. Figure 1 shows the impact of coronavirus in India from 22nd January, 2020 to 14th May, 2020. X-axis denotes the dates and Y-axis denotes the no. of cases. The figure shows the predicted values for dates from 12th April to 14th May.

This figure shows that the total number of coronavirus cases in India is expected to be around 55,000 in mid May. The model is optimistic that the virus would not be able to take the lives of many in this period. Although we know that the model underestimates the count of fatalities, we sincerely hope death count stops ticking in the near future.

5 Conclusion

Forecasts or predictions have their uncertainties. However, in such times of unprecedented crisis, it is important to get an estimate of the size of the problem at hand and the size to which it can grow. This is because, this global pandemic not only affects the people who are affected by the virus but also raises severe health concerns, like the possibility of the entire health system getting overwhelmed and imposing risks on the economy. Thus this work attempts to develop a model (hybrid variant) that can predict future patients affected by the novel coronavirus. It uses various pre-existing learning models for this forecasting purpose and then compares their results. These results show that instead of interpreting the data as a classic regression dataset, it can be considered to be a time series dataset. The work also conceptualizes a hybrid variant model that beats the machine learning models but falls short of the recurrent network.

The models used in this paper do not take any information about the virus as input. Thus, the models are independent of the type of the disease or the microbe causing the disease. So, governments and researchers can use these models to predict future data if similar outbreaks of infectious diseases take place in the future and take steps accordingly. The paper also contains enough future scope like predicting for a larger range of future dates using a smaller dataset. However, the authors hope that the condition only gets better as soon as possible.

References

1. L. Jia, K. Li, Y. Jiang, X. Guo, T. Zhao, *Prediction and Analysis of Coronavirus Disease 2019* (2020). https://arxiv.org/abs/2003.05447
2. A. Gupta, A. Sharma, A. Doel, Review of regression analysis models. Int. J. Eng. Res. Technol. **6**(08), 58–61 (2017)
3. H. Drucker, C.J. Burges, L. Kaufman, A. Smola, V. Vapnik, Support vector regression machines. Adv. Neural Inf. Process. Syst. **9**, 155–161 (1996)
4. D. Margirits, *Learning Bayesian Network Model Structure from Data*. Doctor of Philosophy Thesis (Carnegie Mellon University, Pittsburgh, USA, 2003)
5. B.-B. Benuwa, Y. Zhan, B. Ghansah, D.K. Wornyo, F.B. Kataka, A review of deep machine learning. Int. J. Eng. Res. Africa **24**, 124–136 (2016)
6. H. Sak, A. Senior, F. Beaufays, Long short-term memory recurrent neural network architectures for large scale acoustic modeling, in *15th Annual Conference of the International Speech Communication Association* (Singapore, 2014)
7. COVID 19 Global Forecasting Week 4 dataset. https://www.kaggle.com/c/covid19-global-forecasting-week-4/data
8. COVID 19 dataset. https://www.kaggle.com/shaswata97/covid19
9. COVID 19 Enriched Dataset. https://www.kaggle.com/amoghjrules/covid-19-enriched-dataset-week-2
10. Ministry of Health and Family Welfare, Government of India. https://www.mohfw.gov.in/

Epileptic Seizure Severity Analysis Using Artificial Neural Network and CANFIS Classification Approach

M. K. Mariam Bee and K. Vidhya

Abstract Epilepsy is a disorder of the neurology characterized by epileptic seizures. Epileptic seizure can be analyzed via the normal and abnormal brain activity. This anomalous activity can only be observed by using an efficient algorithm. An effective algorithm method also uses signal processing, in which an epileptic signal can be considered as an input signal. This paper introduces an epileptic signal detection technique and compares the characteristics of the brain signals at different stages. Our algorithm is based on the signal processing techniques used in the EEG signal to detect epilepsy. We present a computer-aided automatic detection and classification method for the focal and non-focal EEG signal in this article. Dual Tree Complex Wavelet Transform (DT-CWT) decomposes the EEG signal, and the characteristics are determined from the decomposed coefficients. These functions are trained and classified using classifier ANN and CANFIS. The proposed system achieves 99% accuracy for EEG signal classification. The experimental results are presented to demonstrate the effectiveness of the proposed classification method for classifying the EEG focal and non-focal signals.

Keywords Epilepsy · CANFIS · Decomposition · EEG

1 Introduction

A seizure is a brief case, marked by a temporary occurrence of symptoms attributable to irregular neuronal synchronous activity in the brain, and epilepsy is a brain disorder involving repeated unprovoked seizures [1]. Epilepsy disease is a neurological disorder affecting around 1–2% of people around the world [2]. Epileptogenesis may be the main reason for seizures occurring in epileptic patients [3] Neurologists

M. K. Mariam Bee (✉) · K. Vidhya
Department of Electronics and Communication Engineering Saveetha School of Engineering (SIMATS), Chennai, Tamilnadu, India
e-mail: mariamresearch09@gmail.com

K. Vidhya
e-mail: kvidhyasri@gmail.com

J. Tavares et al. (eds.), *Emerging Technologies in Data Mining and Information Security*, Lecture Notes in Networks and Systems 164,
https://doi.org/10.1007/978-981-15-9774-9_53

generally use electroencephalogram (EEG) signals to understand human brain's electrical activity to diagnose disorders such as epilepsy, autism, sleep-related disorders, alcoholism, and so on. Epileptic patients who suffer from generalized and partial epilepsy during their treatment procedure may be resistant to drugs [4] In order to treat these patients, it is important to locate the affected region of the brain to conduct the surgery. Signal processing-based techniques can play an important role in identifying EEG recordings corresponding to the brain area affected by focal epilepsy [5].

Previous studies have shown that the use of a k-nearest neighbor (KNN) classifier based on the self-assessment manikin model for EEG attentiveness recognition will yield an average accuracy of 57.03% [6], and the use of the power spectral density model (PSD) support vector machine (SVM) has led to an average accuracy of 76.82% [7]. Furthermore, when taking into account approximate entropy using fuzzy entropy [8], the accuracy can be increased to 81%. KNN combining correlation based feature selection (CFS) to identify attention during the learning process yields a classification rate of 80.84% [9].

On a single subject level, the accuracy of common spatial patterns is up to 89.4% And the non-linear system of shared information [10]. Taken together, frequency specific and nonlinear features derived from the EEG are important for the identification of attentiveness.We present a computer-aided automatic detection and classification method for the focal and non-focal EEG signal in this article. Dual Tree Complex Wavelet Transform (DT-CWT) decomposes the EEG signal, and the characteristics are determined from the decomposed coefficients. Such features are qualified and categorized using CANFIS (Classifier for Adaptive Neuro Fuzzy Inference System). The proposed system achieves accuracy of 99% for the classification of EEG signal. The experimental results are presented to demonstrate the efficacy of the proposed classification method for classifying the focal and non-focal EEG signal.

2 Dataset Description

The proposed model for the detection and prevention of EEG signal ocular artifacts uses BCI Competition IV Dataset 2a, which was recorded at the University of Technology at the BCI Laboratory in Graz. The data set generally consists of 3-channel monopolar electrooculography data (EOG) and 22-channel EEG data during four class motor imaging tasks. This data-set information was recorded from a total of nine subjects. The ground electrodes were positioned on the right mastoids while the signals are being recorded, and the reference electrodes were positioned on the left mastoids. The sampling of the signals was done at 250 Hz, and the filtering was done using the 4–40 Hz band pass filter.

3 Methods

This article proposes automatic detection and analysis of the severity of epilepsy disease using a hybrid classification algorithm. The proposed method consists of stage of decomposition, stage of extraction of features, and stage of classification. Using wavelet transform, the EEG signals are decomposed and features are then extracted from those coefficients. These characteristics are then classified using an approach to NN classification to classify the EEG signals into either focal or non-focal EEG signals. The severity of the focal EEG signal is further analyzed using the classification approach of CANFIS. The proposed detection of epilepsy disease using a hybrid classification approach is shown in Fig. 1.

3.1 Complex Wavelet Transform

Transforms are used to convert the signal, suitable for analysis, from one format to another. In this issue,

Transform multiresolution as CWT is used to decompose the EEG signals in multi-resolution mode. For further research, discrete wavelet transform (DWT) is often used in traditional methods to decompose the EEG signals. CWT's shift invariant

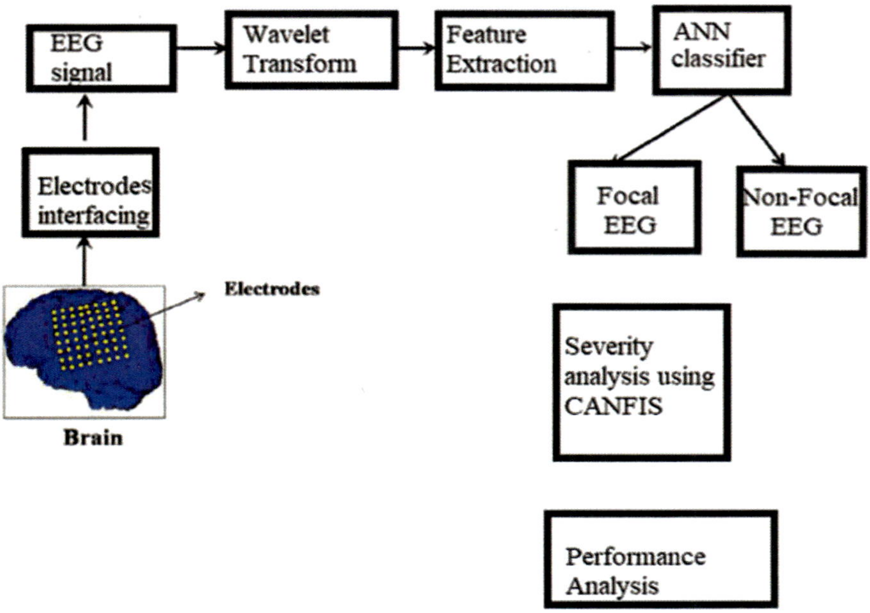

Fig. 1 Proposed epilepsy detection using ANN classification

and directional selectivity property is weak, and for severity analysis, it can not be used to decompose the EEG signals. Overcomes this limitation by proposing DT-CWT to decompose the EEG signals.

The EEG signal is decomposed using its scaling factor and wavelet factor, as shown in the following equation:

However, the scaling factor is noted as $s(n)$ and it is given in the following equation:

$$x(t) = \sum_{n=-\infty}^{\infty} s(n) * \varphi(t-n) + \sum_{i=0}^{\infty} \sum_{n=-\infty}^{\infty} \sum_{n=-\infty}^{\infty} w(i,n) * 2^{i/2} * \vartheta(2^{i}t-n). \quad (1)$$

$$s(n) = \int_{-\infty}^{\infty} x(t) * \varphi(t-n).dt. \quad (2)$$

The wavelet factor is noted as $w(i, n)$ and it is given in the following equation:

$$w(i,n) = 2^{i/2} * \int_{-\infty}^{\infty} x(t) * \vartheta(2^{i}t-n).dt. \quad (3)$$

3.2 Feature Extraction

The features discriminate the focal EEG signal contents from the contents of non-focal EEG signals. In this article, the following features are extracted from the test EEG signal (DT-CWT coefficients) and described in the section below.

3.2.1 Mean

The average number of samples in each signal is represented by its average characteristic and is defined as follows.

$$\text{Mean}(\mu) = \sum_{i=0}^{N-1} ip(i) \quad (4)$$

where $p(i)$ is the EEG signal test and I is the EEG signal sample at separate, finite intervals. The total number of samples present in the EEG test signal is shown as N.

3.2.2 Variance

The average number of samples in each signal is represented by its average characteristic and defined as follows.

$$\text{Variance}(\sigma^2) = \sum_{i=0}^{N-1} (i - \mu)^2 p(i). \tag{5}$$

3.2.3 Skewness

The skewness attribute describes the third-order statistical attribute in the EEG test signal and it is measured using mean signal function:

$$\text{Skewness}(\mu_3) = \sigma^{-3} \sum_{i=0}^{N-1} (i - \mu)^3 p(i). \tag{6}$$

where the signal SD is interpreted as σ.

3.2.4 Standard Deviation

The EEG test signal SD is defined according to the following equation:

$$\sigma = \sqrt{\frac{\sum (x - \mu)^2}{N}} \tag{7}$$

The multivariate properties of the EEG test signal are defined by its moment characteristic and the following equation is given:

$$\text{Moment}(m_k) = E(x - \mu)^k. \tag{8}$$

3.2.5 Correlation Coefficients

The pixel relationship between different components (x and y components in the EEG signal) is defined by their correlation coefficients, and is given by the equation that follows.

$$CC = \frac{C(x_1, x_2)}{\sqrt{C(x_1, x_1) * C(x_2, x_2)}}. \tag{9}$$

Table 1 Extracted feature set for focal and nonfocal EEG signals

Statistics	Focal EEG signal	Non-Focal EEG signal
Mean	−0.710	0.222
Variance	3438.435	11,188.444
SD	149.074	244.317
Skewness	−1.544	−1.547
Kurtosis	130.944	158.948
Correlation coefficients	16.000	16.000
Moment	3433.291	11,171.418
Covariance	3438.435	11,188.444

Table 1 displays the features derived for both the focal and non-focal EEG signals.

3.2.6 Kurtosis

The fourth-order statistical feature in the EEG test signal is defined by skewedness and is calculated using the mean signal function

$$\text{Kurtosis}(\mu_4)\sigma^{-4}\sum_{i=0}^{N-1}(i-\mu)^4[p(i)-3]. \tag{10}$$

3.2.7 Covariance

Covariance defines the pixel ratio between the x-component and the y-component of the EEG signal. And it is equated with the following

$$\text{Covariance} = E[x_1 - \mu_1] * (x_2 - \mu_2)], \tag{11}$$

where x_1 is the x-component of the test EEG signal; x_2 is the y-component of the test EEG signal; μ_1 is the mean of signal with respect to the x-component; and μ_2 is the mean of signal with respect to the y-component.

3.3 Classifications

To discriminate the focal EEG signals from nonfocal EEG signals, the extracted EEG signal features are fed into the classifier architecture. This article uses hybrid classification methodology to classify EEG signals for the process of detecting and diagnosing epilepsy disease. The classification algorithm integrates classification of neural networks with the method of classification by CANFIS. The NN classification algorithm classifies the EEG signals into focal and nonfocal based on the EEG signals' extracted discrimination characteristic. In addition, the EEG focal signal severity is detected through the CANFIS classification algorithm proposed in this article.

3.3.1 Neural Network Classifier

This classification algorithm functions on the basis of the neurons present in the human brain. It is known as a radial NN and it feeds NNs of propagation backwards. The radial NN has no feedback to train the EEG signals samples, and can't be used in this article to distinguish EEG signals for epilepsy detection due to the lack of feedback when the error rate is intolerable. This limitation of this radial NN is overcome by feed forward propagation NNs, which have feedback and when there is an intolerable error the generated output is feedback to the input. This feed back propagation NN has six internal layers with single layer input and output. With this unsupervised network every internal layer has a number of neurons to be set automatically.

This feed forward backpropagation NN is trained through extracted feature set derived from both focal and non-focal EEG signals, as shown in Fig. 2. Then, the same features are extracted from the EEG test signal, and these features are discriminated against with respect to qualified feature patterns by the unsupervised NN classification mode. The reaction from this feed back propagation NN is either negative or positive. The positive value is that the EEG test signal is nonfocal, and the negative value is the focal point of the EEG test signal.

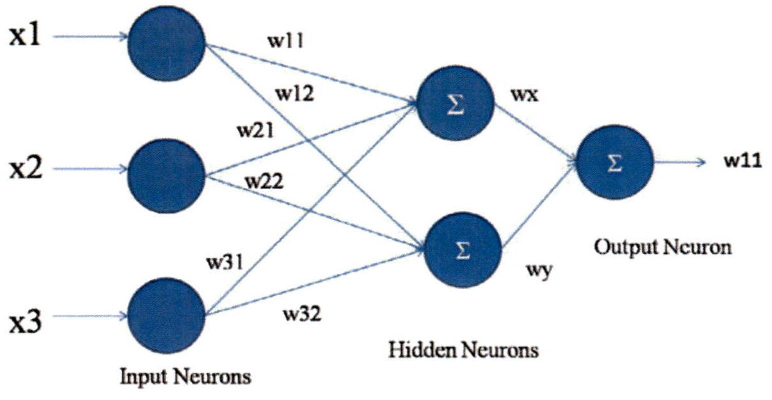

Fig. 2 Feed forward back propagation neural networks

3.3.2 CANFIS Classifier

In a two-step procedure, CANFIS combines Classification and Regression Trees (CART) with the Neuro-Fuzzy Inference System (NFIS). CART is a tree-based algorithm used to optimize the selection process from a large pool of potential predictors to select relevant predictors. Using the predictors selected, NFIS develops a model for the predict and the continuous output. In this sense, CANFIS migrates different degrees of Neuro-fuzzy spectrum between the two extremes: a fully comprehensible FIS and a black-box NN, which is at the other end of interpretability. Neuro-fuzzy models can be defined by a Neuro-fuzzy continuum, taking into account linguistic consistency and precision mapping of input-output.

For its severity detection in the focal EEG signal, the classified focal signal is diagnosed into different categories as class 1 and class 2. The CANFIS classifier is a fuzzy logic hybridization incorporated by NN algorithm. The classified pattern from feedforward backpropagation NN is trained with extracted feature set from the EEG signals by the classifier CANFIS. The CANFIS classificator response is either class 1 or class 2. If the focal signal severity is high, the CANFIS classifier will generate class 1. If the focal signal severity is low then CANFIS classifier generates Class 2. Figure 3 illustrates the proposed CANFIS classificator architecture.

4 Conclusions

EEG signals are used for detecting epileptogenic region in the epileptic patients' brain. The focal and non-focal EEG signals are automatically ranked for epileptic detection in this article. In this function, to decompose the signals into transformed coefficients, dual tree complex wavelet transform is implemented on the EEG signals. Additionally, the DT-CWT coefficients derive mean standard deviation and energy characteristics. Based on the extracted feature collection, these features are trained and categorized using ANFIS classifier to classify the obtained EEG signals into either focal or non-focal signals. The average accuracy of the EEG signal classification achieved is around 99%. Therefore, the method proposed in this article can be used in diagnostic epilepsy centers to detect and diagnose the patients' epileptic characteristics through a computer-aided approach. The future research in this field is focused on predicting the occurrence of seizure points in patients' EEG signals.

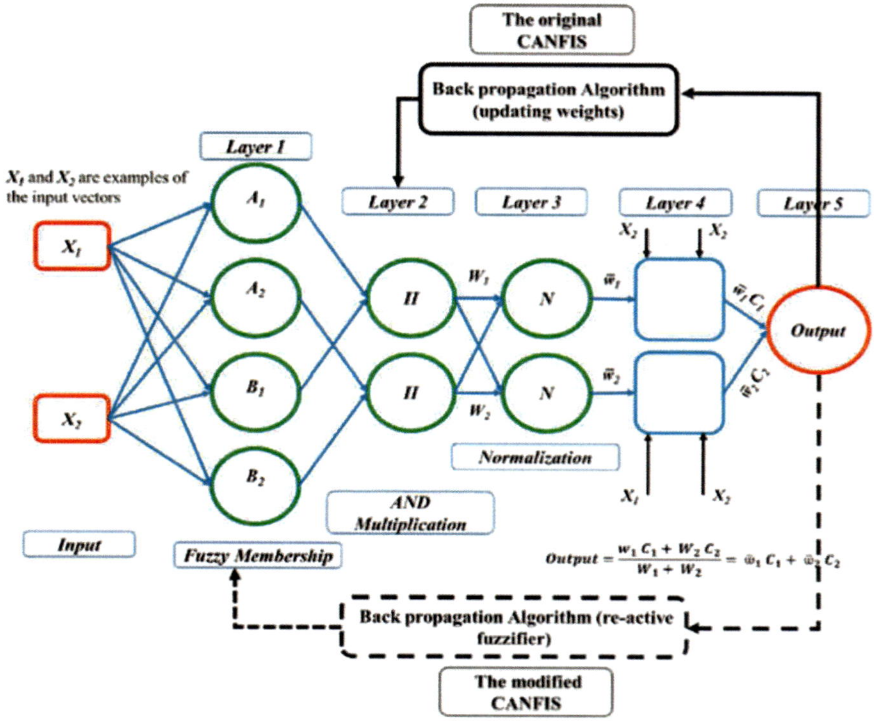

Fig. 3 Architecture of the proposed CANFIS classifier

References

1. R.S. Fisher, C. Acevedo, A. Arzimanoglou, A. Bogacz, J.H. Cross, C.E. Elger, L. Forsgren, J.A. French, M. Glynn, D.C. Hesdorffer, B.I. Lee, G.W. Mathern, S.L. Moshe, E. Perucca, I.E. Scheffer, T. Tomson, M. Watanabe, S. Wiebe, A practical clinical definition of epilepsy. Epilepsia **55**(4):475–482 (2014). https://doi.org/10.1111/epi.12550
2. L.D. Iasemidis, Epileptic seizure prediction and control. IEEE Trans. Biomed. Eng. **50**(5), 549–558 (2003)
3. D.J. Cross, J.E. Cavazos, *The Role of Sprouting and Plasticity in Epileptogenesis and Behavior; Behavioral Aspects of Epilepsy* (Demos Medical Publishing, New York, NY, USA, 2007)
4. S. Pati, A.V. Alexopoulos, Pharmacoresistant epilepsy: from pathogenesis to current and emerging therapies. Clevel Clin. J. Med. **77**, 457–467 (2010)
5. R. Sharma, R.B. Pachori, U.R. Acharya, An integrated index for the identification of focal electroencephalogram signals using discrete wavelet transform and entropy measures. Entropy **17**(8), 5218–5240 (2015)
6. Y. Li, X. Li, M. Ratcliffe, L. Liu, Y. Qi, Q. Liu, A real-time EEG-based BCI system for attention recognition in ubiquitous environment, in *Proceedings of 2011 internationalworkshop on ubiquitous affective awareness and intelligent interaction*, 2011 (pp. 33–40); N.-H. Liu, C.-Y. Chiang, H.-C. Chu, Recognizing the degree of human attention using EEG signals from mobile sensors. Sensors **13**(8), 10273–10286 (2013)

7. L. Xu, J. Liu, G. Xiao, W. Jin, Characterization and classification of EEG attention based on fuzzy entropy, in *Third international conference on digital manufacturing and automation(ICDMA)*, pp. 277–280. IEEE
8. B. Hu, X. Li, S. Sun, M. Ratcliffe, Attention recognition in EEG-based affective learning research using CFS + KNN algorithm. IEEE/ACM Trans. Comput. Biol. Bioinf. **15**(1), 38–45 (2016)
9. B. Hamadicharef, H. Zhang, C. Guan, C. Wang, K.S. Phua, K.P. Tee, et al., Learning EEG-based spectral-spatial patterns for attention level measurement, in *IEEE international symposium on circuits and systems*, pp. 1465–1468. IEEE
10. C.-F. Lin, J.-D. Zhu, Hilbert-Huang transformationbased time-frequency analysis methods in biomedical signal applications, in *Proceedings of the Institution of Mechanical Engineers Part H—Journal of Engineering in Medicine,* vol. 226, no. (H3), pp. 208–216 (2012)

Inferring Political Preferences from Twitter

Mohd Zeeshan Ansari, Areesha Fatima Siddiqui, and Mohammad Anas

Abstract Sentiment analysis is the task of automatic analysis of opinions and emotions of users towards an entity or some aspect of that entity. Political Sentiment Analysis of social media helps the political strategists to scrutinize the performance of a party or candidate and improvise their weaknesses far before the actual elections. During the time of elections, the social networks get flooded with blogs, chats, debates and discussions about the prospects of political parties and politicians. The amount of data generated is much large to study, analyze and draw inferences using the latest techniques. Twitter is one of the most popular social media platforms enables us to perform domain-specific data preparation. In this work, we chose to identify the inclination of political opinions present in Tweets by modelling it as a text classification problem using classical machine learning. The tweets related to the Delhi Elections in 2020 are extracted and employed for the task. Among the several algorithms, we observe that Support Vector Machines portrays the best performance.

Keywords Sentiment analysis · Text classification · Support vector machines

1 Introduction

Analyzing people's opinion on political views via surveys and polls is a time consuming and expensive task. It is impossible for humans to read and summarize all relevant documents and in terms of the expressed sentiments. Therefore, an automatic political sentiment analysis is required to deal with a large amount of text data [1]. People engaging in political issues actively are relying more on online

M. Z. Ansari · A. F. Siddiqui (✉) · M. Anas
Department of Computer Engineering, Jamia Millia Islamia, New Delhi 110025, India
e-mail: areeshaf7@gmail.com

M. Z. Ansari
e-mail: mzansari@jmi.ac.in

M. Anas
e-mail: anas.1633.m@gmail.com

J. Tavares et al. (eds.), *Emerging Technologies in Data Mining and Information Security*, Lecture Notes in Networks and Systems 164,
https://doi.org/10.1007/978-981-15-9774-9_54

sources to get informed about the latest news and events [2]. The use of social media such as Twitter has changed the way people express their views, feelings, opinions and this user-generated content in form of blogs, posts, tweets etc. are easily available publicly in an unstructured format. Twitter provides a platform to share and express political thoughts which have a huge impact on the political sphere in India. Twitter users mainly comprise of the educated class, political figures, prominent personalities which are influencers, are fewer in number when compared to the population of India. Their sentiments and opinion are considerably able to affect their follower's belief and opinions [3].

In this work, we queried the Twitter for scraping tweets that are related to Delhi Elections 2020 by using specific hashtags. We analyzed the tweets and labelled the polarity of sentiments of each tweet with respect to the major political parties taking part in elections. A total of 6060 tweets were collected and preprocessed into the relevant dataset including removal of all the noise that is present in the social media text. We carried out annotations on the extracted tweets by categorizing them into distinct sentiments classes as in *favour* or *against* a party or an alliance. Subsequently, we modelled the problem as a classical text classification problem and investigated the performances of several algorithms on the prepared dataset. The structure of the paper is as follows, Sect. 2 consists of the study of work that been has done in the past related to our topic of research. Section 3 elaborates the corpus acquisition method and theoretical dimensions to specify the adopted approach for our research. Sections 4 and 5 consists of the procedure of experiments and the results obtained. Finally, in Sect. 5 we conclude the work done.

2 Related Work

There is a large increase in the availability of data in the form of text and documents that expresses the opinion of people which can be used for sentiment analysis and work in this field has grown rapidly [4]. The interest in analyzing a large amount of data produced around the world is generated due to the research on network and social media analysis and these data collected can be examined to discover interactions and behavioural patterns and have a better understanding of issues which are unrelated [5, 6]. The sentiment in Twitter has been done in the domain of stock markets, politics and social movements for prediction and measurement [7, 8]. With the increase in the involvement of citizens in the electoral process empowers the democratic process on different levels and brings forward a new environment [9]. Prati and Hung (2017) [10] evaluated the exchanging of texts having a "defined ideological load", and the citizen contribution on Twitter during the Spanish electoral process in which they analyzed classes of segregation noticed in political orientations in messages that were posted digitally and they also considered event timelines [11, 12]. The volume of tweets on Twitter about the political affiliations was found out to be a good estimator in the 2009 German Elections while to detect the candidates in Singapore's 2011 national election by Twitter sentiment analysis was a failure [8, 13]. Colleoni et al.

(2014) used classical learning and social media analysis approach to predict party preferences in an American Democratic and Republican voters database [14]. Several statistical approaches for democratic ideology estimation based on ideological stance and alignment of the traditional approaches to Twitter as a new source of data over the political campaign and keeping a track of people's behaviour and perception as the campaign developed over the time.

Sentiment classification is in a two-step task wherein the primary step, the data (tweets) relevant to our work is being collected and subsequently, in the second step, the sentiment from the data collected is being extracted. Relevant tweets contain words from a list of target keywords compiled either manually or semi-automatically from expanding a seed set [13]. Once the set of such messages has complied, several approaches are applied which are used to obtain the sentiment of the text. Unsupervised methods depend on a list of 'positive' and 'negative' keywords, which estimate a sentiment on the basis of the ratio of occurrences of keywords with respect to one another or just by counting the occurrence of each word concerning each other [8, 15]. Advanced approaches employ supervised learning techniques and train prediction models on either tweet classified manually or on the tweets which have an emotional context [16–18]. Jain and Dandannavar (2017) research study focuses on the combination of a lexicon-based method and a machine learning-based algorithm to define a mixed approach for performing sentiment analysis [19]. Sahayak et al. (2015) research studies intended to make it effortless for the companies to collect the feedback about the products they sell and for the customers who want to get other's opinions about a product prior to purchasing it [20]. Other research studies were focused on the conventional authorities of media, campaigns for elections, voter's social engagement and movement [21, 22].

3 Proposed Method

3.1 Political Corpus Preparation

The scraping of the politically inclined tweets from Twitter is carried using the automatic tool. The extraction of tweets is based on the hashtags and twitter account of popular politicians of respective political parties focused on Delhi Elections 2020. The hashtags that were used are: #DelhiElections2020 #DelhiPolls #delhielections etc. Total 6060 tweets related to Delhi Elections 2020 were obtained in the process for this task. The collection of tweets is significant to three major political parties which were mainly participating in Delhi Elections. We labelled the polarity of sentiment in tweets for every party and incorporated the inclination of a tweet towards or against the political party. All the tweets under consideration are updated and tweeted from Sept 2019 to January 2020 (Fig. 1).

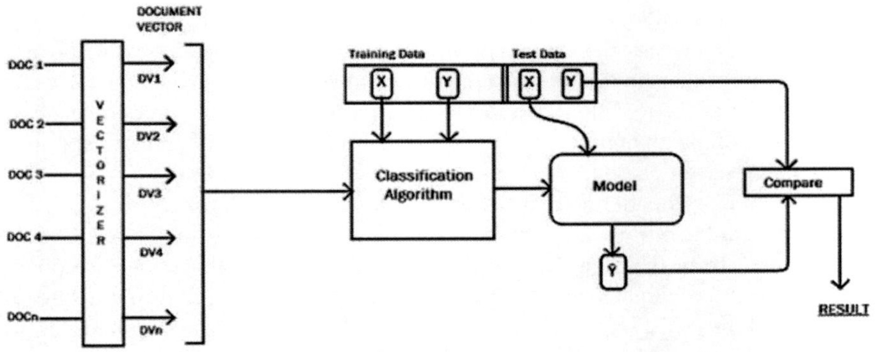

Fig. 1 The text classification framework

3.2 *Preprocessing*

Prior to analyzing the political inclination of sentiments in extracted tweets, we cleaned them into proper form for retrieving the relevant features required for successful classification. Several steps were taken to clean the tweets and remove the unwanted data such as (1) removing the username of the user who tweeted, (2) all phone numbers, (3) RT stands for retweets, (4) all punctuation, (5) all trailing spaces, (6) replacing every email id with "emailaddr", (7) all break and white spaces with single space, (8) every URL with "urladdr", (9) all currency symbol with "moneysymb", (10) all numbers with "numbr" and (11) converting all text into lowercase.

3.3 *Annotation*

After the preprocessing of tweets, annotation of tweets was performed in which each tweet was given a label according to its polarity of sentiment. Being considerably mindful of the electoral politics, the sentiment either positive or negative towards the respective political party or leader was identified for each tweet. We considered three political parties i.e. P1, P2 and P3 as primary parties because the majority of tweets were focused on these three parties for Delhi Elections 2020. The classes to which the cleaned tweets are labelled are present in Table 1.

Taking into consideration the approach similar to Wilson et al. (2005) we prepare the sentiment analysis model (Fig. 1) [23]. The tweets were labelled to eight different classes which are defined as follows: the tweet that supported and opposed party P1 was given class 1 and class 2 respectively. Similarly, classes 3 and 4 were given to tweets that supported and opposed party P2 and classes 5 and 6 were given to tweets that supported and opposed party P3. Instead of discarding the non-relevant annotations, they were assigned to a proper class (class 7) to analyze the dissemination

Table 1 Sentiment classes

Class	Sentiment
1	Support P1
2	Oppose P1
3	Support P2
4	Oppose P2
5	Support P3
6	Oppose P3
7	Non relevant
8	None

of political and non-political tweets. The tweets that did not support or oppose any political party but were related to Delhi Elections were given class 8. There are some tweets which can belong to more than one class so for our annotation we labelled the tweet into the supporting sentiment taking the standard that support has a higher preference than opposing. The manually annotated corpus achieved the inter-annotator agreement over 97.3%. Each cleaned tweet is considered as a document for text classification. After using NGrams, the document vectors (DV1, DV2,... DVn) are formed. The vectorized data is then sent for training different classification algorithms such as Random Forest, Support Vector Machines, Logistic Regression and Naïve Bayes. After running classification algorithms on training data, models are built and their performance is compared.

The number of tweets extracted is 6060 and after categorizing the tweets into the respective classes according to their polarity of sentiment, the total number of tweets that belonged to each class is presented in Table 2. The analysis of annotated corpus shows that 17.06% of total tweets belong to a class that supports party P1, while 14.46% of tweets belong to a class that oppose party P1. The percentage of tweets that oppose party P2 is 13.04% and that support party P2 is drastically less percentage of 4.85%. The percentages of tweets that support and oppose party P3 are very less to a combined percentage of 1.48%.

Table 2 Corpus statistics

Sentiment classes	#Tweets per class	% of tweets in each class
1 (Support P1)	1034	17.06
2 (Oppose P1)	876	14.46
3 (Support P2)	294	4.85
4 (Oppose P2)	790	13.04
5 (Support P3)	37	0.61
6 (Oppose P3)	53	0.87
7 (Non relevant)	623	10.29
8 (None)	2353	38.82
Total	6060	100

Table 3 Count of N-grams

N-grams	Total number of N-grams
Unigrams	6081
Bigrams	22,468
Trigrams	42,879
Tetragrams	50,236

4 Simulation and Experimental Results

4.1 Feature Extraction

N-gram language modelling is a significant probabilistic technique to extract essential features from a sample of text. We used unigrams, bigrams, trigrams and tetragrams of the tweets and created N-grams as presented in Table 3. The Term Frequency-Inverse Document Frequency being a significant method to convert a collection of raw documents to relevant features. It is denoted as tf-idf, tf-idf1, tf-ifd2, tf-idf3 respectively.

4.2 Classification Algorithms

The classification algorithms employed for this task encompass the different approaches to classification such as probabilistic models, ensemble methods, kernel methods etc.

Support Vector Machine. Support Vector Machine was first put forward by Cortis and Vapnik in 1992. It is a supervised learning model that can be used for classification, regression and outlier detection. Support vector machine is highly preferred as it has a simple structure and produces results with significant accuracy and less computation power. SVM also does not require a large number of features.

Random Forest. A Random Forest is a classification model that consists of a collection of tree-structured classifiers. It is an ensemble learning method that operates by constructing multiple decision trees at training time and giving the output as the class that is the most suitable for input x.

Logistic Regression. Logistic Regression is widely used to examine and define a relationship between a response variable, which is binary in nature, and a set of predictor variables. Logistic Regression fits data to a logit function and predicts the probability of occurrence of an event; therefore it is a type of predictive analysis. It is used for classification problems.

Table 4 Performance of classifiers over various features

	N = 1			N = 2			N = 3			N = 4		
	P	*R*	*F*	*P*	*R*	*F*	*P*	*R*	*F*	*P*	*R*	*F*
Random forest	0.84	0.83	0.83	0.81	0.79	0.79	0.80	0.77	0.77	0.81	0.76	0.76
Naïve bayes	0.79	0.78	0.77	0.80	0.78	0.77	0.81	0.78	0.77	0.81	0.78	0.77
SVM	**0.85**	**0.85**	**0.85**	0.83	0.82	0.82	0.82	0.80	0.80	0.82	0.80	0.80
Logistic regression	0.83	0.82	0.81	0.81	0.78	0.77	0.81	0.76	0.76	0.80	0.75	0.74

Bold indicates highest values

Naïve Bayes. Naïve Bayes is a collection of probabilistic classifiers that are based on Bayes theorem. In Naïve Bayes, every pair of feature classified is independent of each other and gives an equal contribution to the outcome. It is used for the classification problem. It requires fewer amounts of training data to estimate the necessary parameters.

Table 4 shows a comparison of performance for different classifiers Random Forest, Naïve Bayes, Support Vector Machine and Logistic Regression with respect to different features like bigrams, trigrams and tetragrams. The performance of Support Vector Machine is best when compared to other classifiers in unigram. Support Vector Machine with tf-idf-unigram exhibit a precision of 0.85 while with tf-idf-bigram, the precision is 0.83. The precision of SVM with tf-idf-trigram and tf-idf-tetragram is 0.82. The recall and F1-score of Support Vector Machine are highest compared to others which is equal to 0.85 in tf-idf-unigrams. The performance of Naïve Bayes classifier is worst when compared to all classifiers. The precision for Naïve Bayes is least for unigram which is 0.79. Precision for Naïve Bayes is better than the precision of Logistic Regression for tetragram model whereas the same for trigrams.

5 Conclusion

In this paper, several classification methods for sentiment analysis are examined on political tweets obtained from active users. We exploited the sentiments keywords significant to major political parties in the tweets with respect to Delhi Elections 2020 and successfully annotated the corpus. The prepared annotated corpus from Twitter is used for political analysis. The data analysis significantly shows both, the supporting and opposing opinions of a substantial amount. The Support Vector Machine reports the highest precision, recall and F1 score among all the algorithms. The tweets collected at large undoubtedly represent a fraction of the actual population participating in the election process. Therefore, enormous attention is required for the sampling of data and generalization of models developed using the social media text. The performance of the models and the overall result can be enhanced by expanding the corpus with the incorporation of a large number of tweets and learning domain-specific keywords.

References

1. M. Ringsquandl, D. Petković, Analyzing political sentiment on twitter, in *2013 AAAI Spring Symposium* vol. SS-13-01 (2013)
2. D. Najar, S. Mesfar, Opinion mining and sentiment analysis for Arabic on-line texts: application on the political domain. Int. J. Speech Technol. **20**(1), 1–11 (2017). https://doi.org/10.1007/s10772-017-9422-4
3. M.Z. Ansari, M.B. Aziz, M.O. Siddiqui, Analysis of political sentiment orientations on twitter. Procedia Comput. Sci. **167**(2020), 1821–1828 (2019)
4. B. Pang, L. Lee, Opinion mining and sentiment analysis. Found. Trends Inf. Retrieval. **2**(1–2), 1–135 (2008)
5. G. Xu, L. Li, Social media mining and social network analysis: emerging research and information, in *Science reference. IGI Global, Hershey* (2013). https://doi.org/10.4018/978-1-4666-2806-9
6. M.Z. Ansari, S. Khan, T. Amani, A. Hamid, S. Rizvi, Analysis of part of speech tags in language identification of code-mixed text, in *Advances in Computing and Intelligent Systems. Algorithms for Intelligent Systems.* (Springer, Singapore, 2020)
7. J. Bollen, H. Mao, X. Zeng, Twitter mood predicts the stock market. J. Comput. Sci. (2011). https://doi.org/10.1016/j.jocs.2010.12.007
8. M. Choy, L.F.M. Cheong, N.L. Ma, P.S. Koo, A sentiment analysis of Singapore presidential election 2011 using twitter data with census correction, in *Research Collection School of Information Systems* (2011)
9. K.M. Wagner, J. Gainous, Digital uprising: the internet revolution in the middle east. In. J. Inf Technol. Politics. (2013). https://doi.org/10.1080/19331681.2013.778802
10. R. Prati, E. Said-Hung, Predicting the ideological orientation during the Spanish 24 M elections in Twitter using machine learning, in *AI & Society 2017* (2017). https://doi.org/10.1007/s00146-017-0761-0
11. A. Gruzd, J. Roy, Investigating political polarization on twitter: a canadian perspective. Policy Internet **6**(1), 28–45 (2014)
12. G. Elmer, Live research: Twittering an election debate. New Media Soc. (2012). https://doi.org/10.1177/1461444812457328
13. A. Tumasjan, T.O. Sprenger, P.G. Sandner, I.M. Welpe, Predicting elections with twitter: what 140 characters reveal about political sentiment, in *Proceedings of the Fourth International AAAI Conference on Weblogs and Social Media* (2010)
14. E. Colleoni, A. Rozza, A. Arvidsson, Echo chamber or public sphere? Predicting political orientation and measuring political homophily in Twitter using big data. J. Commun. https://doi.org/10.1111/jcom.12084 (2014)
15. B. O' Connor, R. Balasubramanyan, B.R. Routledge, N.A. Smith, From tweets to poll: linking the text sentiment to public opinion time series, in *Proceedings of the ICWSM Conference* (2010)
16. H. Wang, D. Can, A. Kazemzadeh, F. Bar, S. Narayanan, A system for real-time twitter sentiment analysis of 2012 U.S. presidential election cycle, in *Proceedings of the ACL 2012 System Demonstrations*, pp 115–120 (2012)

17. M. Conover, B. Gonc alves, J. Ratkiewicz, A. Flammini, F. Menczer, Predicting the political alignment of Twitter users, in *Proceedings of SocialCom/PASSAT conference*, pp. 192–199 (2011)
18. B. Chambers, Learning for micro blogs with distant supervision: political forecasting with Twitter, in *Proceedings of the 13th Conference of the European Chapter of the Association for Computational Linguistics* (2012)
19. P. Dandannavar, A. Jain, Application of machine learning techniques to sentiment analysis, in *IEEE 2nd International Conference on Applied and Theoretical Computing and Commuication Technology ICATCCT* (2017)
20. V. Sahayak, A. Pathan, V. Shete, Sentiment analysis on twitter data. Int. J. Innovative Res. Adv. Eng. **2**, 178–183 (2015)
21. K. Koc-Michalska, R. Gibson, T. Vedel, Online campaigning in France, 2007–2012: political actors and citizens in the aftermath of the Web.2.0 evolution. J. Inf. Technol. Politics (2014). https://doi.org/10.1080/19331681.2014.903217
22. C.F. Fominaya, *Social movements and globalization: how protests, occupations ad are changing the World* (Palgrave Macmillan, New York, 2014)
23. T. Wilson, P. Hoffmann, J. Wiebe, Recognizing contextual polarity in phrase-level sentiment analysis, in *Proceedings of the 2005 Conference on Empirical Methods in Natural Language Processing (EMNLP)* (2005)

Implementation of an Intelligent Online Job Portal Using Machine Learning Algorithms

Zarrin Tasnim, F. M. Javed Mehedi Shamrat, Shaikh Muhammad Allayear, Khobayeb Ahmed, and Naimul Islam Nobel

Abstract Business intelligence and analytics are data management solutions implemented in companies and enterprises to collect historical and present data, while using statistics and software to analyze raw information, and deliver insights for making better future decisions. In the circumstances of today's world, to survive and established own business need an analytical and find an easiest way or intelligence business model. The main objective is to examine the performance of various Machine Learning algorithms in order to perform with the system of an online job portal. This proposed module integrated with three phase such as, the Clusters similar kind of job search phase (CSK) is a way of knowing the demand is to create a visual graph showing clusters of similar kinds of job searched by the job seekers in the website of the job portal, the email notifications send phase (ENS) is responsible to send email notifications to the job seekers when a job circular is posted in the website, extract the job circular phase (EJC) is the way to extract the job circular post from the career section of each of the company's website. The result shows the successful clustering of similar job search, email notification send to specific people and extracts the information from the web.

Z. Tasnim (✉) · F. M. Javed Mehedi Shamrat · N. I. Nobel
Department of Software Engineering, Daffodil International University, Dhaka, Bangladesh
e-mail: zarrint25@gmail.com

F. M. Javed Mehedi Shamrat
e-mail: javedmehedicom@gmail.com

N. I. Nobel
e-mail: nobel775@diu.edu.bd

S. M. Allayear
Department of Multimedia and Creative Technology, Daffodil International University, Dhaka, Bangladesh
e-mail: drallayear.swe@diu.edu.bd

K. Ahmed
Department of Computer Science and Engineering, Daffodil International University, Dhaka, Bangladesh
e-mail: khobayeb15-5200@diu.edu.bd

J. Tavares et al. (eds.), *Emerging Technologies in Data Mining and Information Security*, Lecture Notes in Networks and Systems 164,
https://doi.org/10.1007/978-981-15-9774-9_55

Keywords Machine learning algorithms · Intelligent online job portal · K-means algorithm

1 Introduction

The intelligent enterprise utilizes connectivity (between things, people, and enterprises), data, cloud applications, algorithms, and advanced analytics (instead of hard-coded rules and rigorous procedures). This enables them to come to the right decisions with minimal human involvement even in turbulent, fast-changing environments. For a machine to do any work as well or even better than human, it need a combination of technologies to collect and process the data (e.g., IoT and cloud applications), artificial intelligence and Machine Learning [1]. We are working on topics that is a business intellectual model that will optimize the process of job seek seeking, skill development for job seekers and for human resource department (HRD) to hire candidates. Within three phase we accomplished the whole process.

It is an essential method for collecting data on, and keeping in touch with the rapidly increasing Internet. This Paper briefly reviews the concepts of web crawler, its architecture and its various types [2].

The paper discusses the Bayesian decision tree algorithm, its structure and computation. In the paper, the authors gave a detailed calculation of the efficiency of the algorithm with the partition space. It describes the structure and formation of the greedy-modal tree (GMT) is given with numerical example [3].

The main focus of the paper is to make clusters of the verses of the Holy Qur'an. In the paper, the authors talked about mining the text from the Holy Qur'an and applying the K-means algorithm to determine the number of steamed and unsteamed words in each cluster. The final visualization shows the different densities in each cluster [4].

The objective of the paper is to give a fast and efficient seeding method for text document clustering using k-means algorithm. The authors suggested vectorizing the text document. After that the initial seed point are select as far away from one another as possible to get the best result. Furthermore they compared the system with Points, K-means++, and KMC2 seeding methods [5].

From the research gap a system is proposed that contains three modules. The web crawler is used to collect data from the web automatically and efficiently. The decision tree algorithm is used to select candidates for sending email notification and finally the K-means clustering algorithm is used to study he job market demand.

2 Methodology

2.1 Proposed System

For the proposed system a number of machine learning algorithms [6] are implemented. This algorithms calculate the arrangement of the data to exhibit efficient performance. The proposed system incorporates three critical phase that is, automatically extracting and gathering data from the web is optimize and enrich the database automatically for the effectiveness of the machine learning algorithms [7] to be applied. The second phase is sending email notifications to the job seekers based on the data extracted from the first phase comparing it to the data of the job seekers available in the database. Finally the phase where clustering similar kinds of jobs is done using a machine learning algorithm in order to study and illustrate the proportions of demand of job in the job market at present. This proposed system enhance the roundabout time of an organization's site. The outcome demonstrates the fruitful bunching of comparative pursuit of employment, email notice sent to explicit individuals and concentrates the data from the web (Fig. 1).

2.2 System Overview

Our proposed system contains three (3) phases. Those are given below:

1. Extract job circular phase (EJC)
2. Cluster similar kind of job search phase (CSK)
3. Email notifications sending phase (ENS).

3 Implementation

3.1 Extraction of Job Circular Phase (EJC)

In this phase, to extract job circular information from the web, a web crawler is implemented [8]. The URLs of websites are set in a queue that contain job circulars from where the crawlers extract the data and saves in the parameters in the database. The crawler is implemented in order to automatically updating the database and to make to system efficient.

We assume that T is the set of relevant web pages in test dataset; U is the set of relevant web pages assigned by classifier. Therefore, we define Precision and Recall as follows [9]:

$$\text{Precision} = \frac{|U \cap T|}{|U|} \times 100\% \tag{1}$$

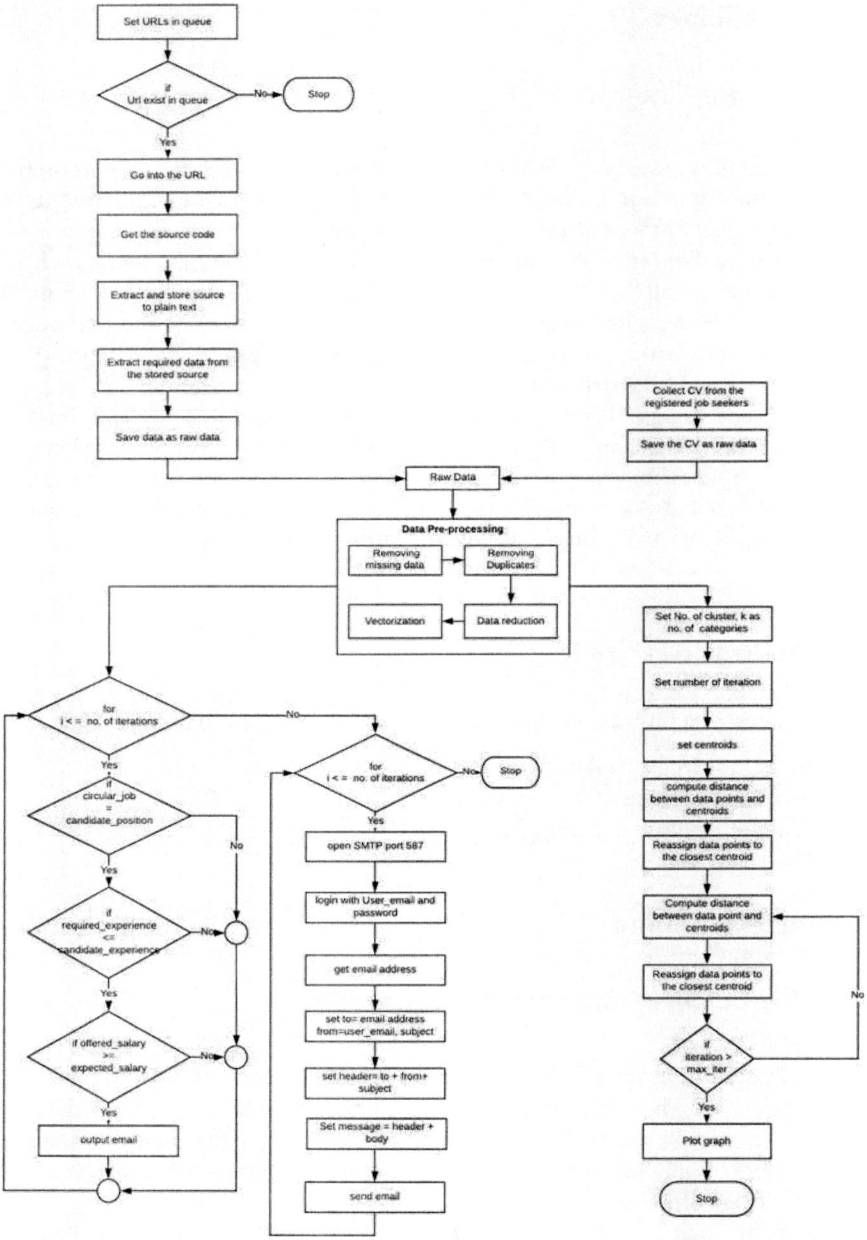

Fig. 1 System diagram

$$\text{Recall} = \frac{|U \cap T|}{|T|} \times 100\% \quad (2)$$

To implement the proposed phase, Algorithm 1 is followed. Setting *i* number of pages in the web crawler, the process is run until all the expected pages are crawled into and all the required data is gathered.

Algorithm 1: Web crawler

Input: Number of pages = i, URL link
Output: Source, URL link, Job Circular Data
Step 1: Set i;
Step 2: **for** iterate till i
Step 3: Set the URL link to be crawled;
Step 4: Extract the source of the set URL;
Step 5: Download and store the source into text format;
Step 6: Extract data of URL links from the text source;
Step 7: **end for**
Step 8: **for** all the extracted URLs are crawled;
Step 9: Crawl in to the extracted links;
Step 10: Download and store the source of the URL into text format;
Step 11: Extract required data of job, required experience and offering salary;
Step 12: **end for**
Step 13: Save the retrieved data into a CSV file;

3.2 Data Preprocessing

Once the raw data is extracted by the web crawler and stored, the data needs to be preprocessed for further use. Both the extracted data of job circular and he job seekers' data has to be preprocessed before applying them in the decision tree and

K-means clustering machine learning algorithm [10]. The preprocessing of the raw data is done following the Algorithm 2.

Algorithm 2: Data preprocessing

```
Input: Job titles = JT, salary = Sal, experience = Ex and email = Em of job
       seekers' data;
Output: Job Vector = JV, X vector of job title (JT) = XJV and Y vector of job
       title (JT) = YJV
Step 1:  Drop null values;
Step 2:  Remove duplicate values;
Step 3:  Converts string to lower case;
Step 4:     for all rows in data frame
Step 5:        if JT exists in dataframe then
Step 6:           set current JV = previous occurrence;
Step 7:           set XJV = previous occurrence of XJV + ( 0.01 X Random number );
Step 8:           set YJV = previous occurrence of YJV + ( 0.01 X Random number );
Step 9:        end if
Step 10:       else
Step 11:          set current JV = random generated number;
Step 12:          set XJV = Random number + ( 0.01 X Random number );
Step 13:          set YJV = Random number + ( 0.01 X Random number );
Step 14:       end else
Step 15:    end for
```

3.3 *Email Notification Sending Phase (ENS)*

In this part, a decision tree algorithm is implemented that algorithm gives a list of suitable candidates for a certain job circular based on three attributes i.e. job position, experience and expected salary from the job circular and job seekers' data. The data must be preprocessed such that the categorical variables are based on same degrees. The job position data is considered as the root node of the tree. The tree will provide the emails of the job seekers' once the set conditions are fulfilled. This email addresses will receive to notification in order to be relevant.

Entropy is the measure of uncertainty of a random variable, it characterizes the impurity of an arbitrary collection of examples. The higher the entropy more the information content. [10]. Entropy is defined as below [11]:

$$E(T) = -\sum_{i=1}^{j} p_i \log_2 p_i \tag{3}$$

In Algorithm 3, the steps of implementation of the decision tree is described where the job circular data is compared with the data of all the job seekers' in order to find a match. The comparison is done based on the conditions set for each case.

Algorithm 3: Decision tree

Input: Job position =JP, Job position vector= JPV, expected experience = EEx, Offering salary = OSal, Job title = JT, Job Vector =JV, experience = Ex, salary= Sal
Output: Email

```
Step 1:   for all JT in dataset
Step 2:   |    if JP = JT then
Step 3:   |    |    set JPV as JV;
Step 4:   |    end if
Step 5:   |    else
Step 6:   |    |    set JPV as random generated number
Step 7:   |    end else
Step 8:   end for
Step 9:   for all row in dataset
Step 10:  |    if JPV = JV then
Step 11:  |    |    if EEx < = Ex then
Step 12:  |    |    |    if OSal > = Sal then
Step 13:  |    |    |    |    Increment flag;
Step 14:  |    |    |    |    output Email ;
Step 15:  |    |    |    end if
Step 16:  |    |    end if
Step 17:  |    end if
Step 18:  end for
Step 19:  if flag = 0 then
Step 20:  |    output "No candidate";
Step 21:  end if
```

In Algorithm 4, the steps taken to send the email notifications to the candidates are given. The email is sent to the candidates those email is received as the output of the decision tree implementation.

Algorithm 4: Implementing the decision tree algorithm

Input: Job position =JP, Job position vector= JPV, expected experience = EEx, Offering salary = OSal, Job title = JT, Job Vector =JV, experience = Ex, salary= Sal
Output: Email

```
Step 1:   for all JT in dataset
Step 2:   |    if JP = JT then
Step 3:   |    |    set JPV as JV;
Step 4:   |    end if
Step 5:   |    else
Step 6:   |    |    set JPV as random generated number
Step 7:   |    end else
Step 8:   end for
```

3.4 Clustering Similar Kind of Job Search Phase (CSK)

In the CSK phase, an unsupervised K-means algorithm is implemented that gives the cluster of similar kinds of jobs in the dataset. The algorithm retrieves the data of the job titles from the dataset and categorize them in similar types. This data is then vectorized based on their categories. The vectorized values are then plotted into a scattered graph. The K-means identifies clusters from the scattered graph. Each cluster represent a certain type of job. The density of each cluster can be analyzed to identify the demand of job in the job market.

To calculate the new cluster center the following equation is used [12]:

$$v_i = \frac{1}{c_i} \sum_{J=1}^{c_i} x_i \tag{4}$$

To calculate the distance between the data points, p and q, Euclidean distance equation is used. The equation is as follows [13]:

$$d(p, q) = \sqrt{\sum\nolimits_{i=1}^{n} (q_i - p_i)^2} \tag{5}$$

The implementation of Algorithm 5 gives the K-means clustering algorithm. The algorithm gives the cluster of data points as output in a scattered graph which uses the preprocessed data of current job position of job seekers'.

Algorithm 5: K-means Clustering

Input: X Job Vector = XJV, Y Job Vector = YJV, number of clusters = K, number of centroids = C

Step 1: Set K;

Step 2: Set C = K;

Step 3: Plot data points, XJV and YJV;

Step 4: for iterating until no change occurs to the centroids

Step 5: Compute distance between XJV, YJV and centroids with Euclidean distance [13],

$$\boldsymbol{dist}((\boldsymbol{x}, \boldsymbol{y}), (\boldsymbol{a}, \boldsymbol{b})) = \sqrt{(\boldsymbol{x} - \boldsymbol{a})^2 + (\boldsymbol{y} - \boldsymbol{b})^2}$$

Step 6: Assign each data point to the closest cluster;

Step 7: Compute the centroids for the clusters by taking the average of the all data points that belong to each cluster;

Step 8: end for

Fig. 2 Extracted data by web crawler

4 Result

4.1 Extracted Data Using Web Crawler

A HTTP link of a website is set in the web crawler using which the web crawler goes in the website and extracts the required data from the website's source, copy it and stores it for further use. Such data is extracted using the implemented web crawler in order to gather data for the system.

In Fig. 2, the data that is extracted from the website is shown. The web crawler copied the information from the web page [15] and stored it in the database as raw data to enrich the database. This data can be processed for further use in the following phases.

4.2 Email Notifications

The phase of sending email notification consists of two parts. First the decision tree algorithm is implemented. The algorithm gives as a decision of the suitability job seekers to a job circular. As output, the algorithm provides the email addresses of the job seekers'. Next an email notification is send out to the address obtained from the result of the decision tree algorithm using an smtp server.

In Fig. 3, if suitable candidates are found for any job circular, a list of email addresses of those candidates will be provided. This email addresses will only get the notification of that particular circular.

In Fig. 4, a test email is send to a recipient. This process can be used to send email notification of a job circular to a job seeker who is suitable for the job.

Fig. 3 Output (email addresses) of decision tree

```
Email addresses:
nqqj820u@gmail.com
l0wfsg7l@gmail.com
7cxvtrs1@gmail.com
6k1m65e1@gmail.com
sqhsrd9s@ymail.com
tv2edn3m@ymail.com
47pljawl@gmail.com
28czccsc@yahoo.com
nymh2fk6@hotmail.com
k03oxxxf@gmail.com
gu55hbvp@gmail.com
pe278lug@gmail.com
hyhmp4jn@gmail.com
996r5rec@ymail.com
ciiv3f3t@yahoo.com
8rqr1wy4@ymail.com
jqknuk2n@yahoo.com
sqibagfv@ymail.com
di86m5ms@gmail.com
i9nwafhd@yahoo.com
```

Fig. 4 Test email notification

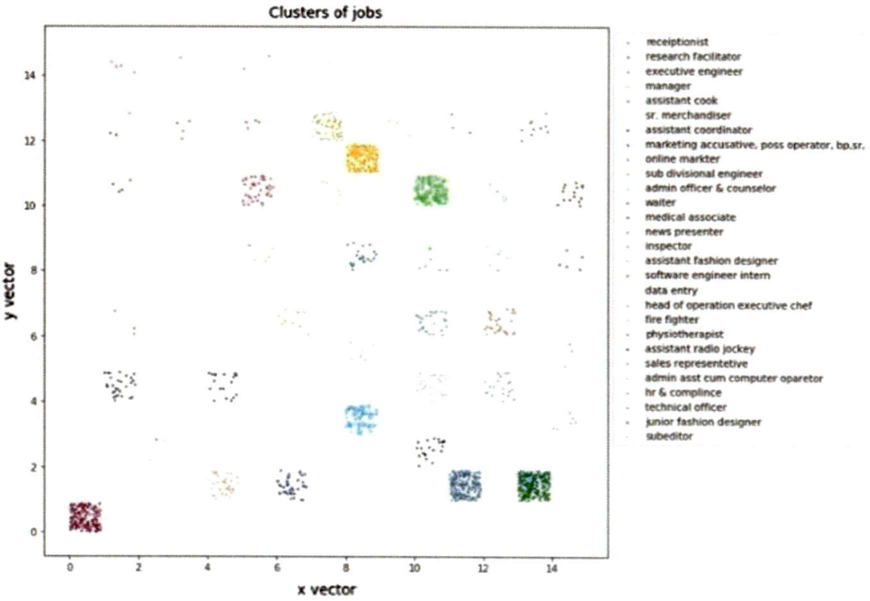

Fig. 5 K-means cluster of types of jobs

4.3 *Clusters of Similar Kinds of Jobs*

Analyzing the clusters, the market demand can be realized as the high density of cluster has higher demand compared to the lower density clusters [14]. The final scatter graph (Fig. 5) is shown that indicates 28 different clusters in along with a color code legend. Analyzing this clusters and comparing the color legend, we can identify the demand of the jobs.

4.4 *Efficiency Measurement*

To test the implemented system, a number of test cases were used. In Table 1, ten test cases were tested and the result are demonstrated. The test cases contain URLs for web crawler to extract data and job seekers data for decision tree and clustering algorithms. For Test 1, it is seen that the crawler fails to gather any data. From Test 2 to onwards, the crawler successfully extract data and completed all the processes with different accuracy rate with up to 100% accuracy. A graphical interface is illustrated in Fig. 6.

Table 1 Accuracy measurement for web crawler

Input (test case)	Output	Response time (s)	System accuracy (%)
Test 1	Fail	3.7	0
Test 2	Pass	2	87
Test 3	Pass	2.1	91
Test 4	Pass	1.7	91
Test 5	Pass	1.8	94
Test 6	Pass	0.8	98
Test 7	Pass	0.7	98
Test 8	Pass	0.7	100
Test 9	Pass	0.7	100
Test 10	Pass	0.7	100

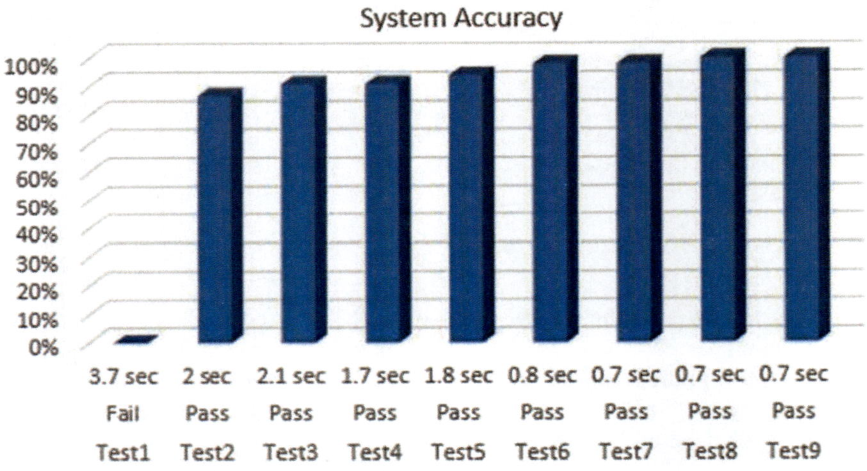

Fig. 6 Accuracy graph for web crawler

5 Conclusion

The system is proposed for online job portals. The system contains three different phases. Each phase is responsible to improve the efficiency of the system. First of all, EJC phase is responsible for extracting the job circulars from different company's website automatically using a web crawler. The web crawler extracts the job circular from the company website's career section and store it in the database of skill.jobs. Next is the CSK phase that is tasked to give a graph containing the cluster of same kind of jobs from the data using the K-means algorithm. Analyzing the density of the cluster, skill enhancement training programs can be organized to help jobseekers get better jobs. Finally in ENS phase a decision tree is implemented to make decision about to whom email notification of a certain job circular should be send. This phase uses the data stored in the database, provided by the jobseekers in order to match the job seekers' job position, salary and experience to find a suitable job from the job. When job circular is matched with a suitable job seeker, the job seeker receives an email allowing him to know about the job circular. This phases together makes an efficient system that automatically gathers job circulars and allow job seekers to know if any job matches their requirement which helps job seekers to apply for jobs more efficiently. At the same time, analyzing the demand job seekers, training programs can be arranged that helps enhance skills to get better jobs.

References

1. F.M.J.M. Shamrat, M.A. Raihan, A.K.M.S. Rahman, I. Mahmud, R. Akter, An analysis on breast disease prediction using machine learning approaches. Int. J. Sci. Technol. Res. **9**(02), 2450–2455 (2020). ISSN: 2277-8616
2. T.V. Udapure, R.D. Kale, R.C. Dharmik, Study of web crawler and its different types. IOSR J. Comp. Eng. (IOSR-JCE) **16**(1), Ver. VI (2014). e-ISSN: 2278-0661, p-ISSN: 2278-8727
3. G. Nuti, L.A. Rugama, A. Cross, *A Bayesian Decision Tree Algorithm* ArXiv, abs/1901.03214 (2019)
4. C. Slamet, A. Rahman, M.A. Ramdhani, W. Darmalaksana, Clustering the verses of the Holy Qur'an using K-means algorithm. Asian J. Inf. Technol. **15**(24), 5159–5162 (2016)
5. E. Sherkat, J. Velcin, E.E. Milios, in *Fast and Simple Deterministic Seeding of K-Means for Text Document Clustering*. 9th International conference of the CLEF Association, CLEF, vol. 11018 (2018)
6. F.M.J.M. Shamrat, M. Asaduzzaman, A.K.M.S. Rahman, R.T.H. Tusher, Z. Tasnim, A comparative analysis of Parkinson disease prediction using machine learning approaches. Int. J. Sci. Technol. Res. **8**(11), 2576–2580 (2019). ISSN: 2277-8616
7. A.K.M.S. Rahman, F.M.J.M. Shamrat, Z. Tasnim, J. Roy, S.A. Hossain, A comparative study on liver disease prediction using supervised machine learning algorithms. Int. J. Sci. Technol. Res. **8**(11), 419–422 (2019). ISSN: 2277-8616
8. F.M.J.M. Shamrat, Z. Tasnim, A.K.M.S. Rahman, N.I. Nobel, S.A. Hossain, An effective implementation of web crawling technology to retrieve data from the world wide web (www). Int. J. Sci. Technol. Res. **9**(01), 1252–1256 (2020). ISSN: 2277-8616
9. H. Lu, D. Zhan, L. Zhou, D. He, In *An Improved Focused Crawler: Using Web Page Classification and Link Priority Evaluation*. Hindawi Publishing Corporation, Mathematical Problems in Engineering, vol 2016. Article ID 6406901, 10 p (2016). http://dx.doi.org/10.1155/2016/6406901
10. Decision Tree Algorithm, https://www.geeksforgeeks.org/decision-tree-introduction-example/. Last Accessed 12 April 2020
11. Decision Tree Algorithm, https://en.wikipedia.org/wiki/Decision_tree_learninglast. Accessed 13 April 2020
12. K =Means Algorithm, https://sites.google.com/site/dataclusteringalgorithms/k-means-clustering-algorithm. Last Accessed 15 April 2020
13. Euclidean Distance, https://en.wikipedia.org/wiki/Euclidean_distance. Last Accessed 15 April 2020
14. F.M.J.M. Shamrat, Z. Tasnim, I. Mahmud, M.N. Jahan, N.I. Nobel, Application of K-means clustering algorithm to determine the density of demand of different kinds of jobs. Int. J. Sci. Technol. Res. **9**(02), 2550–2557 (2020). ISSN: 2277-8616
15. https://jobs.waltonbd.com/last. Accessed 03 April 2020

A Review Paper on Reconfigurable Antenna Technique and Methodology

Mritunjay Srivastava and Abhijeet Kumar

Abstract The numerous features of the reconfigurable antenna are discussed in this paper from the basic idea which will give us the broader perception into the fundamental design approach to the advance methodology. These are the modern technological antennas which are capable of analyzing the surrounding and perform necessary adjustment in order to attain and maintain maximum performance and these adjustments are performed by applying different technologies such as varactors, microelectromechanical systems (based on radio frequency), PIN diodes, and photoconductive plates. This kind of antenna is also capable of avoiding noise sources and capable of improvising to gain of the system and security due to their special properties, they consume less power and saves the energy by directing signals only toward the intended direction. Due to this phenomenal property of self-adjustment of with respect to the environments condition, reconfigurable antenna gets a wide-open field of operation in cognitive radios, satellite communications, multi-input multi-output systems (MIMO), and other various fields.

Keywords MIMO · UWB · NBM · WLAN · RF-MEMS · Wi-MAX and composite right-/left-handed transmission line (CRLH-TL)

1 Introduction

The most phenomenal change that occurs in next generation of antenna technology is the reconfigurable antenna. The ability to adapt dynamically to the changes by observing the surrounding and environment condition to achieve highest efficiency and maximize the gain of antenna and get a better performance is the kind of property we need. These are the kind of property that can bring an enormous amount

M. Srivastava · A. Kumar (✉)
Department of Electronics and Communication Engineering, HMR ITM, Delhi, India
e-mail: abhijeetkmr07@gmail.com

M. Srivastava
e-mail: srivastavamritunjay01@gmail.com

J. Tavares et al. (eds.), *Emerging Technologies in Data Mining and Information Security*, Lecture Notes in Networks and Systems 164,
https://doi.org/10.1007/978-981-15-9774-9_56

of change in the device working methodology and performance efficiency [1]. It is the most promising next-generation idea in the radio networks. The reconfigurable antenna can be of many types according to the situation demands and in what scenario what kind of change do we want to make accordingly to ensure better efficiency and performance. As in the frequency reconfigurable reception, apparatus can obtain dynamic change to different recurrence bands also the best spectrum allocation [2]. Bandwidth reconfigurable antenna: these have ability to shift from narrow band to wide band; Pattern reconfigurable antenna: these are capable of producing radiation patterns with various directivities at a similar sort of activity at different frequency; Polarization reconfigurable antenna: these types of antenna have a very unique property of offering exceptional multipath fading rejection [3]. Over the last few decades, there is a huge amount of changes occur in the wireless communication and radar system.

On the contrary, in [5], suggested hierarchical quadrature amplitude modulation (HQAM) for better protection of high-priority data during image transmission. Nevertheless, no channel model such as AWGN is considered in this work.

We have developed a wide range of antenna like loop antenna, helical antenna, log-periodic antenna, horn antenna, reflector antenna, microstrip antenna, and dipoles/monopoles and all these types of antenna have a very unique type of property which makes them suitable for various fields of communication and radar system. At the point when confronted with another framework configuration, engineers change and adjust these essential reception apparatuses, utilizing hypothetical information and general plan rules as beginning stages to grow new structures that regularly produce satisfactory outcomes [4]. But apart from their various qualities, these kinds of antenna have a very strong drawback which is vulnerability toward one's environmental condition which has very enormous effect on its performance and the efficiency [5]. To have a clear solution and to avoid these performance and efficiency issues, we have a concept of reconfigurability property. Radio waves are reconfigured by modifying the transmitting fields of the usable distance in the receiving waves. It depends on a purposely revamped antenna flows or a reconfiguration of the radiating edges of the receiving system. This redistribution of properties results in a modification of the functionality of the reception antenna [6]. This differential technology allows users in multiple remote communications stages to offer reconfigurable antenna. The designers of radio wire are faced with tough issues about the attempt to reconfigure antennas. The way toward making a reconfigurable reception antenna is intense undertaking; we have to take care of many factors such as gaining of a good gain, stabilizing the extents of radiations and also the good impedance which will match through all the antenna's operation states [7]. To change the functionalities of operation of reception antenna apparatus, switches like PIN diodes, varactor diodes and RF MEMS [8], even as optically driven switches of optic fiber connections, are being exploited. A receiving antenna apparatus is connected through a varactor diode, the capacitance of which is controllable by a fluctuating predisposition voltage, may provide a broad tuning range. The reconfigurable configuration of antenna receiving the feed through biasing device with varactor diodes [9].

2 Designing Mechanism of Reconfigurable Antenna

The four reconfiguration properties can be achieved by a reconfigurable receiving antenna system. A reception antenna may display a configurable operation recurrence, a reconfigurable configuration of the radiation, or a reconfigurable conduct of polarisation or a combination of these characteristics. With respect to the form of reconfiguration, it relies on the radio wave property that can be reconfigured. A model chooses a strategy that satisfies the imposed restrictions and productively performs the antenna configuration process at the same time [1]. With the advent of the reconfigurable antenna, a few reconfiguration techniques have been suggested. In four major categories, the proposed reconfiguration methods are isolated: electrical, optic, mechanical, and material change. Electrical reconfiguration strategies depend on using antenna parts receiving changes or disengaging them, and redistribute them through the adjustment of the powerful opening emanated fields of the radio wire. Micromechanical frameworks (RF-MEMS), varactor diodes, FET, or PIN diodes for the purpose of diverting the source flow are coordinated into the recycling wire. RF-MEMS switches put together reception apparatuses depend with respect to the mechanical development of the changes to accomplish reconfiguration. Numerous radio wire plans have depended on RF-MEMS to reconfigure their exhibition (Figs. 1 and 2).

2.1 *Reconfiguration Properties*

Depending on the situation, the reconfigurable antenna can be of many kinds and in what scenario do we want to make changes, what will help? During the last decades, the wireless networking and radar infrastructure have experienced a great deal of change. The variety of antenna helical antennas, log periodical antennas, horn antennas, microstrip antennas, and dipoles/monopolies has now been created and all

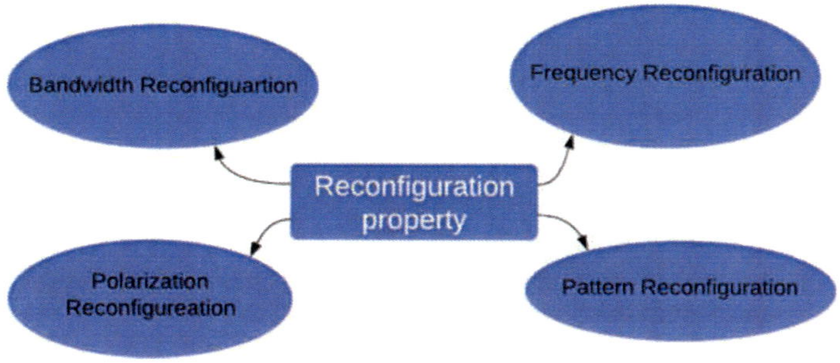

Fig. 1 Different properties exhibited by the reconfigurable antenna

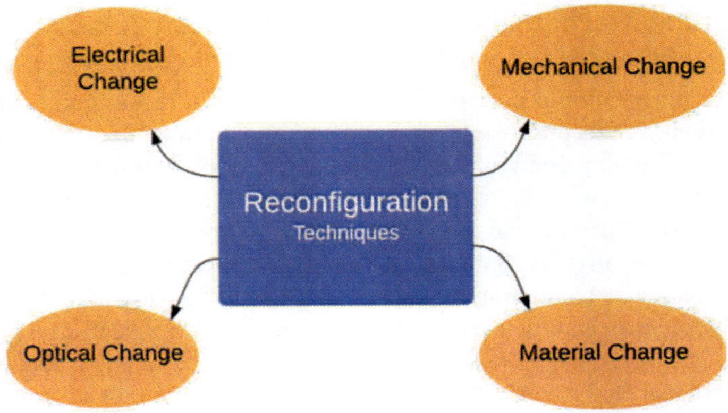

Fig. 2 Different type of reconfiguration techniques

these antenna types have a very distinctly different set of property, which makes such antennas ideal for various fields of communications and for the radar network.

2.1.1 Frequency Reconfigurable Antennas

Two components: electric and mechanical can create the antenna. The electric system uses continuous and discrete tuning techniques. By using RF switches, discrete tuning can be achieved and varactor diodes can perform continuous tuning. For example, liquid crystals and metasurfaces are used to conduct frequency reconfiguration by the mechanical device for impedance stacking tunable materials. [6] For most cases, they are used in multi-band/multi-service systems which involve efficient dynamic range re-allocation: various types of switches are used. The best-known technique is to adjust the antenna radiator's electrical length with PIN diodes. Varactor diode switches will set the band in UWB mode and select a great narrow band mode (NBM). There are many plane approaches for accomplishing antenna reconfigurability, which are as per the following:

- Integration of electronic switches (PIN diodes/varactor diodes)
- Switching by means of RF MEMS switches
- Substrate tunable materials (permittivity/substrate height changes).

The theory of reconfigurable frequency antenna structure can be describing the radio wave configuration is based on the CRLH-TL, which requires a shunt inductance/capacitance and series capacitance/inductance. By piling the varactor in the meander line slot, the radio wave operating band can be balanced.

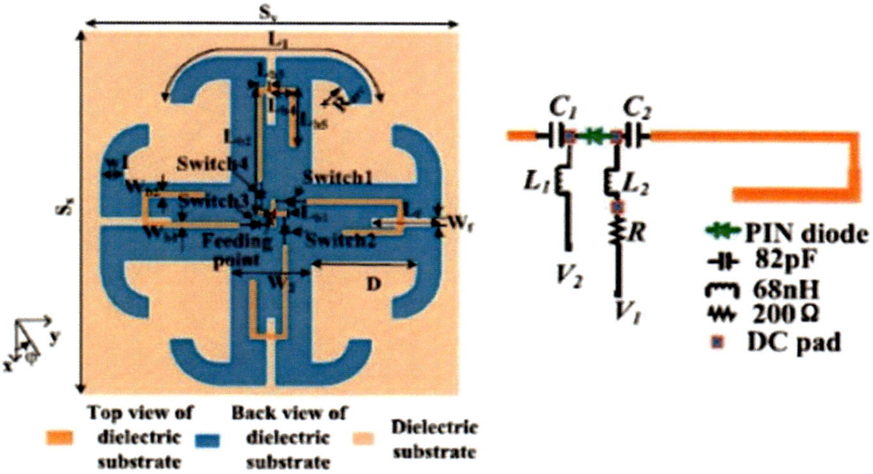

Fig. 3 Antenna configuration for different diode state

2.1.2 Pattern Reconfigurable Antennas

These antennas use mobile/rotatable devices, such as the metasurface or the remembering switch to adapt the circular propagation of radiation patterns to reactivate capacitive parts. The reaccessions patterns of the antenna are appealing for use in reconnaissance and monitoring as they deliver the radiation pattern for the various guidelines in this area. Regularly either structure is planned with the possibility to create design assorted variety or the takings care of arrangements are reconfigurable by using of array theory. In Fig. 3, on shifting sides of a Rogers 4350B layer, there tend to be four identical circular section dipoles next to the reconfigurable device treatment system. By modifying the conditions of diodes, a 90-degree identifying antenna radiation example in the azimuthal plane can be managed to cover all headings. The normal reflection (S11) for the antenna in each diodes state is steady and covers the frequency of the 4G and 5G operating bands from 2.3 to 3.2 GHz. This antenna is also used in MIMO wireless communication to improve system performance by reducing noise.

2.1.3 Polarization Reconfigurable Antennas

These transmission antenna systems are interchanged between various polarizations, such as direct polarization, LHCP and RHCP, using several modes configuration or metasurface. In portable devices, a swap between horizontal, vertical, and circular polarization is required to minimize uncoordinated polarization [9]. With this type, an Antenna can be modified into several modes and the antenna can be used in many different ways. Most inquiries in reconfigurable polarization antennas involve

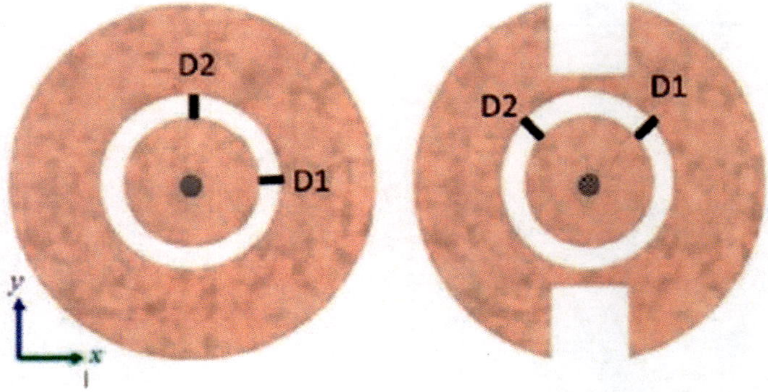

Fig. 4 A circular radiation patch with a C-shaped slot

exchanges between RHCP and LHCP at an ideal bandwidth, while linear polarization (vertical or horizontal modes) is also considered. The most common methods for polarization reconfiguration include the use of adjusted structures such as hole, slits, parasite structures, and shortened corners of the main radiator with mounted dynamic sections. A polarization structure is describing for 4G applications, Fig. 4 which is reconfigurable with a C-shaped slot. The patch consists of a spherical radiation patch with a C-shaped opening and two diodes on the radiation patch over a concentric circular slot. The antenna will shift between vertical and even horizontal (VP/HP) modes as in the circular modes LHCP and RHCP.

2.1.4 Bandwidth Reconfigurable Antennas

This type of antenna uses the concomitant tuning of different antenna parameters, such as radiation design and frequency, for independent reconfiguration of working frequency, pattern of radiation and polarizing, by means of a layer of parasitic pixels. The operational bandwidth can also be changed, usually via a matched network or reconfigurable from a narrow to a large band. [3] The antenna needs to be very well done throughout the whole spectrum to protect both the narrow band and the broad band. A modified ground plane, particularly for a patch antenna, is another technique. In a reconfigurable antenna for the wireless local area network (WLAN) and global interoperativeness of the world wide interoperability for microwave access (WiMAX) application was presented. The antenna scheme proposed in Fig. 5 is delineated. It features a rectangular radiation patch with pivoted F-formed slot, which is combined with a feed line for a microstrip and a shorter ground plane behind a FR4 substratum. The current flow is regulated by a few PIN diodes in the space of the radiation patch. The bandwidth feature of the antenna can be reconfigured from narrow band to broadband by exchanging the operating state of diodes. The frequency of the impedance varies from 22 to 78%, making the antenna ideal in different wireless communications systems. Moreover, the antenna can also be repository to switch to

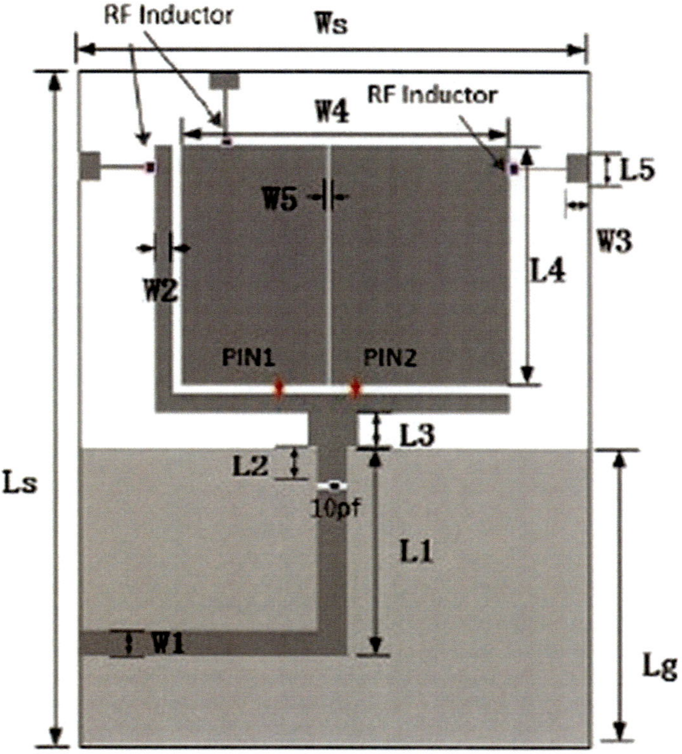

Fig. 5 Bandwidth reconfigurable antenna

a wideband/UWB property with a reconfigurable trademark. This should be achieved to avoid intrusion from UWB networks and other wireless services such as 4G, Wi-Fi, WiMAX, and satellite communications. It improves the flexibility of antennas combining diverse wireless networking systems into a single network.

2.2 Reconfiguration Technique

There are four primary types of reconfiguration methods, electrical, optic, mechanical, and material change. In compliance with the antenna reconfiguration requirements, the part of electrical techniques is RFMEMS, Varactor diodes, FETs or PIN diodes. Just like electrical with the help of photoconductive materials, the reconfigurable property gained is known as optical, structural alteration comes under the mechanical property and the metasurface liquid crystal, ferrites, etc., is the part of material change (Fig. 6).

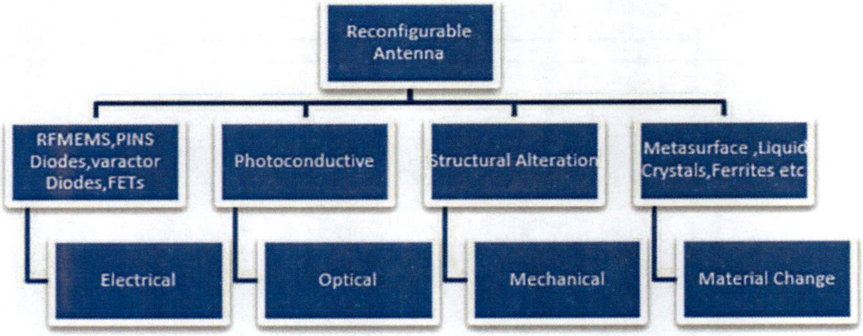

Fig. 6 Reconfigurable antenna techniques

2.2.1 Electrical Technique

An antenna for transmitting structures or the emitting edges may be modified for electronic parts such as RF-MEMS, varactor diodes, PIN diodes, or FETs in electrically reconfigurable surface flow circuits. The incorporation of switches into the antenna structure simplifies the development of the perfectly customizable usability. The electrical system uses discrete tuning and continuous tuning techniques [6]. The diodes varactor, PIN, and field effect transistors (FETs) can be used. Direct current (DC) source and biasing circuits are essential for the operation of these electronic parts in the antenna circuit. In this way, an electrically reconfigurable antenna is conditioned by a DC electric source and segments of the electronic exchange which affect adverse activity and the antenna execution. A circular slot antenna is used to reconfigure frequency and pattern. Reconfigurability of the antenna is retained by the PIN diodes that monitor the input matches, while diode switches are placed at slot positions to regulate the bearing of a pattern null normal to specific antenna operation.

2.2.2 Optical Technique

Optically reconfigurable antenna is known as radiation components with the ability to adjust radiation properties by using switches that could be optical triggering of reactive element silicon switches. Metal wires that interfere with the radiation properties of the antenna can be separated in the case of optically powered devices. Additional metal microstrip or wired partition lines are used to allow complicated antenna interferences and find it difficult to track the necessary radiation pattern by using DC-controlled microstrips. In this case, two silicon-photo conducting switches have been fitted with a balanced dipole built out of high resistivity silicon as seen in Fig. 7 Light was used to power the switches from infrared laser diodes guided by fiber-optic cables. The antenna performed at a smaller frequency of 2.16 GHz

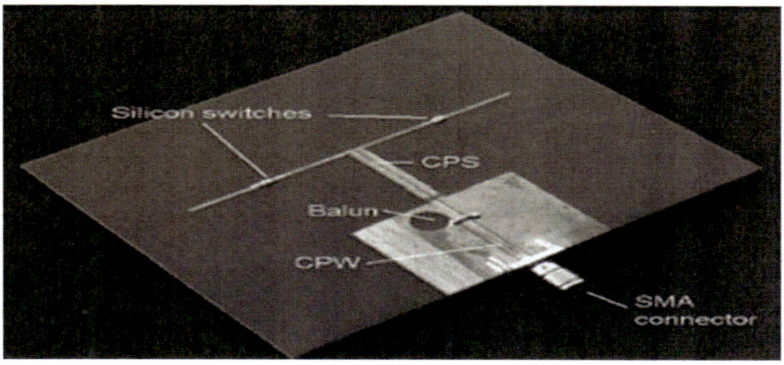

Fig. 7 An optically switched dipole antenna that provides frequency reconfigurability

and the antenna worked at 3.15 GHz with both switches closed. The variability in antenna gain as a function of the optical power used to activate the switches. A degree of pattern reconfigurability is also achieved with this design by activating only one switch at a time, resulting in a shift of the pattern null in the direction of the longer dipole arm by approximately 50°.

2.2.3 Mechanical Technique

It is also possible to reconfigure antennas by manually adjusting the antenna radiation function. The internal adjustment of the radiant antenna elements means that the antenna is adjusted. What is critical is that it does not relay any switching processes, biasing lines, or the incorporation of optical fiber/laser diodes. However, this approach depends on the device's mechanical weakness. A piezoelectric actuator device was demonstrated in 1998 as a mechanical tuning model antenna to modify the distance between the antenna and a parasite radiator in order to adjust the antenna operating frequency. Figure 8 reveals a snapshot of the antenna [8]. While the parasite dimension usually has a very tiny bandwidth (1%), it has a controlled

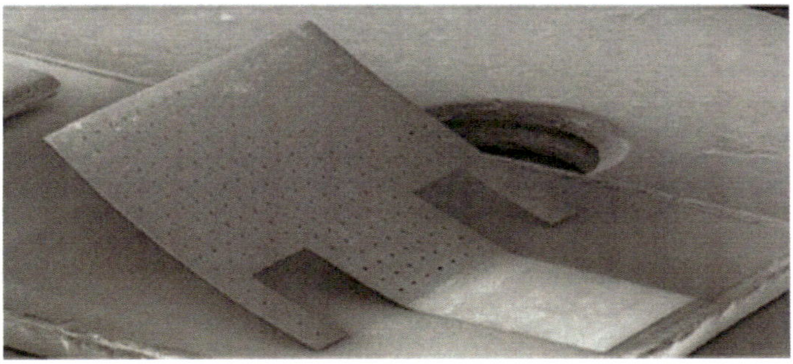

Fig. 8 Magnetically actuated reconfigurable microstrip antenna

movement of about 9%. This example shows that it is difficult to reconfigure one type without modifying the other characteristics of the antenna; the bandwidth and configuration gains are often changed as a function of the parasite aspect size, but cannot be isolated. A magnetically operated microstrip antenna is another example of constant frequency shift that can be made possible by mechanical changes. A thin film of magnetic material coated and removed from the substratum a microstrip antenna designed for operation of around 26 GHz. The addition of an additional magnetic DC field allows the plastic antenna, when attached to the feed line of the microstrip, to be deformed at the boundary point to obtain an angle over the ground. With the micro-machine method called the plastic deformation assemblage.

2.2.4 Material Change

In addition, antennas are reconfigured by modifying their substrates with materials such as dielectric fluids, liquid crystals, ferrites, or metasurfaces. The change in substance is achieved through changes in relative electrical admissibility or magnetic permeability. Indeed, liquid crystal is a nonlinear substance that, by a change in the orientation of liquid crystal molecules, is able to alter the dielectric constant at various voltage levels. With regard to a ferrite material, the relative permittivity/permeability may be modified by statically applied electric/magnetic field. The metasurface is positioned directly on top of the patch antenna; the antenna beam can be steered continuously. This influences the relative primitiveness of the system with which the resonant frequency of the antenna is tuned [5]. The frequency tuning can also take place by electrical controllable properties of liquid crystals, yttrium iron garnet (YIG), barium-strontium titanate (BST), dielectric fluids, and artificial fluids. A frequency-tuned ferrite-based antenna in which the variable static magnetic field on the substratum plane and the resonant aspect of the patch are equipped with a 40% continuous tuning range. However, the radiation efficiency of the system was satisfactory, with rates of cross polarization considerably higher than expected in a conventional rectangular microstrip antenna. Several groups have recently produced thin film ferroelectric material in order to reduce the losses added into the circuit and have a degree of tunability. The majority of applications also use tunable materials in the feed structure or in parasitic elements rather than in the antenna itself because of planar constraints and the realistic standardization of the films.

3 Conclusion

This paper introduced a comprehensive investigation into different types of antennas, such as frequency reconfigurable antennas, which can dynamically adjust their operating frequency, polarization reconfigurable antennas which allow switching between different modes of polarization. Method of reconfigurable antenna: this is based on the purposeful change in the spherical radiation frequency distribution and bandwidth reconfigurable antenna: these are capable of changing their bandwidth from narrow-band to wide-band along with the thorough study of design and reconfigurable methods, such as electrical: electrically reconfigurable RF-MEMS, PIN diodes, varactor diodes or FET antennas, Optical: Photoconductive, Mechanical: Structural alteration and material based: Meta surface, liquid crystals, ferrites, etc. A detailed and descriptive study of these designing techniques is presented here.

References

1. C.G. Christodoulou, Y. Tawk, S.A. Lane, S.R. Erwin, 'Reconfigurable antennas for wireless and space applications'. Proc. IEEE **100**(7), 2250–2261 (2012)
2. E.R. Brown, RF-MEMS switches for reconfigurable integrated circuits. IEEE Trans. Microw. Theory Tech. **46**(11), pt. 2, 1868–1880 (1998)
3. E. Erdil, K. Topali, M. D'Unlu, O.A. Civi, T. Akin, Frequency tunable patch antenna using RF-MEMS technology. IEEE Trans. Antennas Propag. **55**(4), 1193–1196 (2007)
4. N.O. Parchin, Y. Al-Yasir, A.M. Abdulkhaleq, I. Elfergani, A. Rayit, J.M. Noras, J. Rodriguez, R.A. Abd-Alhameed, in *Frequency Reconfigurable Antenna Array for mm-Wave 5G Mobile Handsets*. Proceedings of the 9th International Conference on Broadband Communications, Networks, and Systems,Faro, Portugal, 19–20 Sept 2018
5. R. Hussain, M.U. Khan, M.S. Sharawi, *An integrated dual MIMO antenna system with dual-function GND-plane frequency-agile antenna* (IEEE Antennas Wirel. Propag., Lett, 2018)
6. S. Sam, H. Kang, S. Lim, Frequency reconfigurable and miniaturized substrate integrated waveguide interdigital capacitor (SIW-IDC) antenna. IEEE Trans. Antennas Propag. **62**, 1039–1045 (2014)
7. S.L. Chen, P.Y. Qin, W. Lin, Y.J. Guo, Pattern-reconfigurable antenna with five switchable beams in elevation plane. IEEE Antennas Wirel. Propag. Lett. **17**, 454–457 (2018)
8. Y. Tawk, J. Costantine, C.G. Christodoulou, A varactor based reconfigurable filtenna. IEEE Antennas Wirel. Propag. Lett. **11**, 716–719 (2012)
9. S.Y. Chen, Q.X. Chu, N. Shinohara, in *A Bandwidth Reconfigurable Planar Antenna for WLAN/WiMAX Applications*. Proceedings of the Asia-Pacific Microwave Conference (APMC), New Delhi, India, 5–9 Dec 2016

An Exploratory Study for Process Optimization in IT Industry

H. Aditya Pai, Piyush Kumar Pareek, M. S. Narasimha Murthy, Sunanda Dixit, and Sneha Karamadi

Abstract Software defect prediction (SDP) is the method essential for testing during life cycle of the software development (LCSD). It finds out those stages of the software which are more prone towards defect and requires extensive testing. In this way, the testing can be without usage of extra resources. Though SDP is an essential step in the testing, it is not always easy to do the prediction of which stage of the software is having defects. There are various reasons which barriers for the smooth performances of defect prediction. In this paper, we carried out the survey on different IT companies and analyzed their software process model by performing SWOT chart. The chart gave us the idea to perform value stream mapping (VSM) that identifies the non-value added process activities in the IT companies. Also failure effective analysis model (FEAM) to know the outcome or threats over the defects identified. We also came up with the hypothesis for the delay in software development process. ANOVA analysis was carried out to understand the turbulence in the business environment.

H. Aditya Pai (✉) · S. Karamadi
Department of Computer Science and Engineering, K. S. Institute of Technology, Bangalore, India
e-mail: adityapaih2007y@gmail.com

S. Karamadi
e-mail: snehakaramadi@ksit.edu.in

P. K. Pareek · M. S. Narasimha Murthy
Department of Computer Science and Engineering, East West Institute of Technology, Bangalore, India
e-mail: piyushpareek88@gmail.com

M. S. Narasimha Murthy
e-mail: china.mur@gmail.com

S. Dixit
Department of Computer Science and Engineering, BMS Institute of Technology and Management, Bangalore, India
e-mail: sunandadixit_cse@bmsit.in

J. Tavares et al. (eds.), *Emerging Technologies in Data Mining and Information Security*, Lecture Notes in Networks and Systems 164,
https://doi.org/10.1007/978-981-15-9774-9_57

1 Introduction

Life cycle of the software development (LCSD) is the processes that performs the basic activities in the development stages. The software developers can then study the result of the LCSD in order to develop an application in the considerable amount of budget and time. The software developers can use various strategies for their single project. Thus, there is blend of various techniques to be use that suits the particular project based on the scale of the project [1].

On the contrary, in [5], suggested hierarchical quadrature amplitude modulation (HQAM) for better protection of high-priority data during image transmission. Nevertheless, no channel model such as AWGN is considered in this work.

Here, only salt and pepper noise is taken into account. A raised cosine filter is used in the AWGN channel is introduced in [6]. Here, authors evaluate the performance of the communication channel through the bit error rate (BER). In [7], M array QAM is used in AWGN channel for evaluating the polar coding technique. In this method, the characteristic of the band limited and channels inter symbol interference not considered.

In this paper, a raised cosine filter is proposed before transmission of compressed image.

1.1 Background

In the era of 1960s, the SDLC was created to build the application software in the humongous scale. Every software programming requires assembling the software, module wise that are used to create models which solves complex defects. This helped the IT companies easier to develop the software. For example, it is the role of the business analyst to ensure the client's requirements are properly being gathered. This is followed by the designers carrying out the plan to carry out the requirements in the form of software architectures and so on. The process is followed by continuous testing to keep up the quality. This helps in optimization of the processes. There has been great advancement being made in the era of 1970s in understanding software application's requirements. For this, many software engineers were being trained to be developers.

2 Literature Survey

The literature survey is divided into various segments: (1) Agile Software Development; (2) Six sigma to reduce lead time; (3) Value stream mapping; (4) Failure mode effective analysis

2.1 Agile Software Development

As indicated by the examination did by Kupiainen et al. [2]. The Agile Software Development strategy has been generally utilized by the product business. The paper focuses on which are the measurements being utilized by the software engineers in Lean and Agile Software Development approach [19]. To discover this, the creator utilizes the philosophy of doing systematic literature review of around 774 papers which was brought down to 30 for essential examinations. The outcomes from the investigation show that the measurements like in five classes—Sprint and Project Planning, Sprint and Project Progress Tracking, Understanding and improving Quality, Fixing Software Process Problems, and Motivating People are the territories being engaged.

Subsequently, the creator finishes up with the utilization of Agile it has cut down the issues as there were different dashes being utilized to find and fix these issues.

As indicated by the investigation did by Brian Fitzgerald et al. [3]. The ubiquity of the nimble advancement has been expanding according to the study being completed where 80% of the associations are currently following dexterous methodology. The nimble strategies are most appropriate for those frameworks which are little and non-basic in nature. Anyway, there is a few requirements of use of dexterous methodology in the well-being basic frameworks. The paper examines about the qualities of the light-footed methodology where there is requirements utilizing in the security basic condition.

2.2 Six Sigma to Reduce Lead Time

As per Athul Jayaram et al. [4], LSS wipes out superfluous procedures and imperfections which can profit a venture from diminished expenses and less wastage of assets. Industry 4.0 alludes to robotization of enterprises by the trading of information between the store network and coordinations.

Modern Web of things IIoT is mechanical machines associated with the undertaking distributed storage zone for information stockpiling just as information recovery.

The discoveries of the paper is that LSS when applied to the Global Supply Chain wipes out deformities in the delivered great along these lines expanding generally speaking effectiveness. Industry 4.0 and IIoT assist with improving the proficiency of creation. It helps in quicker, effective, and efficient administration of SCM exercises all inclusive.

The creator Man Mohan Siddh et al. [5] in the paper clarifies that lean assembling [31] is one of the activities taken by significant organizations all around the globe that have been attempting to embrace so as to stay serious in the undeniably worldwide market and six sigma [21, 22] was an American "creation". The focal thought behind six sigma [24] is to gauge the "abandons" that are there all the while. The paper tells how incorporating store network with the lean practices [33] and six sigma [23] procedures can create prevalent items with ideal outcomes where the procedures can be changed from delayed to quick by lean assembling [38].

2.3 Value Stream Mapping

As per Lin et al. [6] aims in improving money to trade process duration out the machines by utilizing the VSM that distinguishes the loss all the while and by utilizing Discrete Event Simulation. The creator shows the exhibition through the contextual investigation on how the decrease is done and the outcomes show that through the mix of the VSM and DES improves the C2C process duration altogether.

As per Liu et al. [7], the fundamental point is to see lean creation instruments that are being applied by numerous organizations for assembling, has a preferred position of keeping an edge from its rivals. The paper utilizes the VSM instrument to distinguish the ways to deal with take out the losses during the creation procedure [10–12].

2.4 Failure Mode Effective Analysis

As indicated by Huang et al. [8], the paper points in arranging the purposes behind programming disappointment modes. The paper appears there are two arrangement of disappointment modes one is general disappointment and another is extraordinary disappointment. The creator has clarified the grouping with the assistance of the contextual investigations [13–15]. As indicated by Nguyen et al. [9], the paper presents able method for taking care of the issue that is FEAM [17]. This methodology gives all the data [20, 32, 36] that is causing the disappointment and their impact on the product framework during each procedure. FMEA [35] additionally checks what the likely disappointments are at each stage and what the remedial moves to be made.

3 Research Methodology

The methodology followed in the research is as shown in Fig. 1. The methodology explains that the research is carried out by the survey carried out through questionnaire, and from the sample, we can analyze what are the non-value added activities via VSM [18]. To understand the causes and effects of the longer lead time, we use FEAM [39] technique. Next, we do ANOVA [25, 26] analysis to prove the hypothesis [27–30].

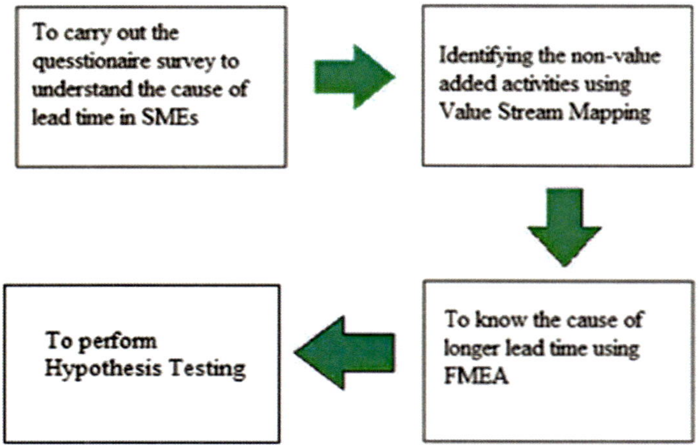

Fig. 1 Research methodology used

4 Identify the Non-value Added Activities and FEAM

The team of software developers first understand the software process and then analyzes the potential failure which are the non-value added activities by using VSM. Through VSM, the following are the non-value added activities as shown in Table 1 [16].

Table 1 Non-value added activities

NV code	Non-value added activities
NV01	Delay in requirement collection
NV02	Number of requirement is different than the actual requirement
NV03	Bugs fixed versus bugs pending
NV04	Testing done during unit tests less than executing stage
NV05	Pending of unit testing
NV06	Number of testing done
NV07	Errors found after testing
NV08	Developers lack of understanding the requirement
NV09	Total number of bugs fixed by the developer
NV10	Gaps between different stage of software process
NV11	Lack of skill from the client
NV12	Requirements number increase
NV13	Lack of information to newer techniques
NV14	Requirement not up to the market standards
NV15	Defects unrecognized at one stage is carried over to next stage
NV16	Lack of participation from the stake holders

Table 2 Failure effective analysis model solution

NV code	Non-value added activities
NV04	The activities that are slowing the process such as the errors are required to be fixed at the same stage rather than carrying over the next stage. If the testing is done at the initial period, the discrepancy in the data being collected will be understood. For this, there is a need of ensuring right requirement being collected by analyzing the practical applicability of the data
NV09	It is important the engineers who collect the right requirement must also do the testing for the errors being found and must report the same to the developers for bug fix
NV12	The software process used in the IT companies are mainly Agile, thus incremental approach is being followed. To tackle too many new requirements. The whole process must be split into the Sprints
NV15	There is a need of experienced tester within the team to ensure that each software module is tested and accuracy is maintained, and for this, re-testing is required to make sure the software is error free

The team of software developers first understand the software process and then analyzes the potential failure which are the non-value added activities by using VSM [18]. Through VSM [34, 37] the following are the non-value added activities as shown in Table 1.

After listing the activities in Table 1, we perform FEAM [40] to find out the causes and effects of the failures listed. Some of them are listed in Table 2 where the solutions to the said highly risked failures are described [16].

5 Hypothesis Testing

Statement: To find out the reason for turmoil in the Business prospects in IT company.

Null Hypothesis H_0: There is absolutely no co-relationship between the turmoil in the business prospects and what type of IT company it is.

Alternate Hypothesis H_1: There exists strong co-relationship between the turmoil in the business prospects and what type of IT company it is.

5.1 Independent Variable

- Unknown market trends
- Tough competition within Indian market
- Tough completion across global market
- Introduction to newer technology

- Reliance on automobile market
- Affect due to changes made by government policies
- Need of working capital
- Capital's high cost
- Lot of import of junk machinery done by Indian IT companies
- Threat from China, Korea, and Taiwan due to low cost of machines (Tables 3 and 4).

Degrees of freedom 1: 4
Degrees of freedom 2: 238
Probability level: 0.05
Critical F-value: 2.40957301

As observed Avg. $\mathbf{F}_{\textbf{(statistic Value)}} = 7.23 > \mathbf{F}_{\textbf{(Critical Value)}}$ 2.4095 at probability level 0.05, 95% confidence level, and avg. level of significance 0.01005 < 0.05, Hence, we reject null hypothesis and accept alternate hypothesis.

There exists strong relationship between turbulence of business environment on types of company.

6 Conclusion

The sample size of 243 was taken for the survey of IT firms in Bengaluru, where we were able to find out the reason for turmoil in the business prospects in IT company. Mainly due to threat from Global Market, Introduction to newer technology, Frequent Changes in Government Policies etc.

After this, potential failure modes were found using FMEA techniques for the selected IT firms in Bengaluru where we could find out what are the causes of the failure modes and effects for the same. The hypothesis testing was carried out bases on the survey which says.

Hypothesis Statement: To find out the reason for turmoil in the business prospects in IT company—There exists strong co-relationship between the turmoil in the business prospects and what type of IT company it is.

Table 3 ANOVA analysis—I

		N	Mean	Std. deviation	Std. error	95% confidence interval for mean		Minimum	Maximum
						Lower bound	Upper bound		
The reason for turmoil in the business prospects in IT company is unknown market trends	Sole proprietary	20	2.45	0.605	0.135	2.17	2.73	2	4
	Private Ltd.	72	2.67	1.343	0.158	2.35	2.98	1	5
	Public Ltd.	24	2.29	1.233	0.252	1.77	2.81	1	4
	PSU's	25	2.60	1.118	0.224	2.14	3.06	1	5
	SME's	102	2.70	1.348	0.133	2.43	2.96	1	5
	Total	243	2.62	1.265	0.081	2.46	2.78	1	5
The reason for turmoil in the business prospects in IT company is tough competition within Indian market	Sole proprietary	20	2.60	1.046	0.234	2.11	3.09	1	4
	Private Ltd.	72	2.89	1.251	0.147	2.59	3.18	1	5
	Public Ltd.	24	2.33	1.404	0.287	1.74	2.93	1	4
	PSU's	25	3.24	1.508	0.302	2.62	3.86	1	5
	SME's	102	2.70	1.406	0.139	2.42	2.97	1	5
	Total	243	2.77	1.354	0.087	2.59	2.94	1	5
The reason for turmoil in the business prospects in IT company is tough completion across global market	Sole proprietary	20	1.80	1.196	0.268	1.24	2.36	1	4
	Private Ltd.	72	3.11	1.338	0.158	2.80	3.43	1	5
	Public Ltd.	24	3.00	1.445	0.295	2.39	3.61	1	4
	PSU's	25	3.20	1.225	0.245	2.69	3.71	1	4
	SME's	102	3.15	1.285	0.127	2.89	3.40	1	5

(continued)

Table 3 (continued)

		N	Mean	Std. deviation	Std. error	95% confidence interval for mean		Minimum	Maximum
						Lower bound	Upper bound		
	Total	243	3.02	1.345	0.086	2.85	3.19	1	5
The reason for turmoil in the business prospects in IT company is introduction to newer technology	Sole proprietary	20	2.40	1.392	0.311	1.75	3.05	1	4
	Private Ltd.	72	3.22	1.236	0.146	2.93	3.51	1	5
	Public Ltd.	24	1.67	0.761	0.155	1.35	1.99	1	3
	PSU's	25	2.88	1.092	0.218	2.43	3.33	1	4
	SME's	102	3.01	1.324	0.131	2.75	3.27	1	5
	Total	243	2.88	1.308	0.084	2.71	3.04	1	5
The reason for turmoil in the business prospects in IT company is reliance on automobile market	Sole proprietary	20	2.80	1.196	0.268	2.24	3.36	1	4
	Private Ltd.	72	3.56	1.174	0.138	3.28	3.83	2	5
	Public Ltd.	24	1.50	0.511	0.104	1.28	1.72	1	2
	PSU's	25	1.64	0.952	0.190	1.25	2.03	1	3
	SME's	102	2.73	1.145	0.113	2.50	2.95	1	5
	Total	243	2.74	1.286	0.082	2.58	2.91	1	5
The reason for turmoil in the business prospects in IT company is affect due to changes made by government policies	Sole proprietary	20	2.40	0.503	0.112	2.16	2.64	2	3
	Private Ltd.	72	3.00	1.300	0.153	2.69	3.31	1	5
	Public Ltd.	24	2.67	1.404	0.287	2.07	3.26	1	4
	PSU's	25	3.88	1.236	0.247	3.37	4.39	2	5
	SME's	102	2.74	1.455	0.144	2.45	3.02	1	5
	Total	243	2.90	1.370	0.088	2.72	3.07	1	5

(continued)

Table 3 (continued)

		N	Mean	Std. deviation	Std. error	95% confidence interval for mean		Minimum	Maximum
						Lower bound	Upper bound		
The reason for turmoil in the business prospects in IT company is need of working capital	Sole proprietary	20	2.60	0.503	0.112	2.36	2.84	2	3
	Private Ltd.	72	2.83	1.311	0.154	2.53	3.14	1	5
	Public Ltd.	24	2.83	1.090	0.223	2.37	3.29	1	4
	PSU's	25	3.56	1.417	0.283	2.98	4.14	1	5
	SME's	102	2.71	1.432	0.142	2.42	2.99	1	5
	Total	243	2.84	1.326	0.085	2.67	3.00	1	5
The reason for turmoil in the business prospects in IT company is capital's high cost	Sole proprietary	20	4.00	0.918	0.205	3.57	4.43	3	5
	Private Ltd.	72	3.44	1.174	0.138	3.17	3.72	1	5
	Public Ltd.	24	2.50	1.532	0.313	1.85	3.15	1	4
	PSU's	25	3.12	1.236	0.247	2.61	3.63	2	5
	SME's	102	3.48	1.384	0.137	3.21	3.75	1	5

(continued)

Table 3 (continued)

		N	Mean	Std. deviation	Std. error	95% confidence interval for mean		Minimum	Maximum
						Lower bound	Upper bound		
	Total	243	3.38	1.329	0.085	3.21	3.55	1	5
The reason for turmoil in the business prospects in IT company is lot of import of junk machinery done by Indian IT companies	Sole proprietary	20	2.60	0.503	0.112	2.36	2.84	2	3
	Private Ltd.	72	3.61	1.015	0.120	3.37	3.85	2	5
	Public Ltd.	24	2.50	1.532	0.313	1.85	3.15	1	4
	PSU's	25	4.04	1.020	0.204	3.62	4.46	3	5
	SME's	102	3.74	1.185	0.117	3.50	3.97	1	5
	Total	243	3.51	1.207	0.077	3.36	3.67	1	5
The reason for turmoil in the business prospects in IT company is threat from China, Korea, and Taiwan due to low cost of machines	Sole proprietary	20	2.60	0.503	0.112	2.36	2.84	2	3
	Private Ltd.	72	3.61	1.015	0.120	3.37	3.85	2	5
	Public Ltd.	24	2.50	1.532	0.313	1.85	3.15	1	4
	PSU's	25	4.04	1.020	0.204	3.62	4.46	3	5
	SME's	102	3.74	1.185	0.117	3.50	3.97	1	5
	Total	243	3.51	1.207	0.077	3.36	3.67	1	5

Table 4 ANOVA analysis—II

		Sum of squares	df	Mean square	F	Sig.
The reason for turmoil in the business prospects in IT company is unknown market trends	Between groups	3.921	4	0.980	0.608	0.657
	Within groups	383.487	238	1.611		
	Total	387.407	242			
The reason for turmoil in the business prospects in IT company is tough competition within Indian market	Between groups	12.247	4	3.062	1.689	0.153
	Within groups	431.383	238	1.813		
	Total	443.630	242			
The reason for turmoil in the business prospects in IT company is tough completion across global market	Between groups	32.829	4	8.207	4.822	0.001
	Within groups	405.105	238	1.702		
	Total	437.934	242			
The reason for turmoil in the business prospects in IT company is introduction to newer technology	Between groups	50.088	4	12.522	8.183	0.000
	Within groups	364.208	238	1.530		
	Total	414.296	242			
The reason for turmoil in the business prospects in IT company is reliance on automobile market	Between groups	115.130	4	28.782	24.031	0.000
	Within groups	285.052	238	1.198		
	Total	400.181	242			
The reason for turmoil in the Business prospects in IT company is Affect due to changes made by government policies	Between groups	33.802	4	8.450	4.781	0.001
	Within groups	420.626	238	1.767		
	Total	454.428	242			
The reason for turmoil in the business prospects in IT company is need of working capital	Between groups	15.946	4	3.986	2.317	0.058
	Within groups	409.470	238	1.720		
	Total	425.416	242			

(continued)

Table 4 (continued)

		Sum of squares	df	Mean square	F	Sig.
The reason for turmoil in the business prospects in IT company is capital's high cost	Between groups	29.290	4	7.323	4.380	0.002
	Within groups	397.879	238	1.672		
	Total	427.169	242			
The reason for turmoil in the business prospects in IT company is lot of import of junk machinery done by Indian IT companies	Between groups	53.976	4	13.494	10.751	0.000
	Within groups	298.724	238	1.255		
	Total	352.700	242			
The reason for turmoil in the business prospects in IT company is threat from China, Korea, and Taiwan due to low cost of machines	Between groups	53.976	4	13.494	10.751	0.000
	Within groups	298.724	238	1.255		
	Total	352.700	242			

References

1. S. Kader, M.M.K. Akter, Analysis of the factors affecting the lead time for export of readymade apparels from Bangladesh; proposals for strategic reduction of lead time. Eur. Sci. J. **10**(33) (2014). ISSN: 1857–7881 (Print) e-ISSN: 1857–7431
2. E. Kupiainen, M.V. Mäntylä, J. Itkonen, Using metrics in agile and lean software development—a systematic literature review of industrial studies. Inf. Softw. Technol. **62** (2015)
3. B. Fitzgerald, K.J. Stol, R. O'Sullivan, D. O'Brien, in *Scaling Agile Methods to Regulated Environments: An Industry Case Study*. ICSE '13 Proceedings of the 2013 International Conference on Software Engineering, pp. 863–872
4. A. Jayaram, in *Lean Six Sigma Approach for Global Supply Chain Management using Industry 4.0 and IIoT*. 2nd International Conference on Contemporary Computing and Informatics (IC3I), pp. 89–94 (2016)
5. M.M. Siddh, G. Soni, G. Gadekar, R. Jain, in *Integrating Lean Six Sigma and Supply Chain Approach for Quality and Business Performance*. 2nd International Conference on Business and Information Management (ICBIM), pp. 53–57 (2014)
6. W.D. Lin, E.S. Chan, L.F. Kwan, in *Integrated Value Stream Mapping and Simulation for Cash-To-Cash Cycle Time Improvement of a Machining Facility*. IEEE International Conference on Industrial Engineering and Engineering Management (IEEM) (2017)
7. Q. Liu, H. Yang, in *Lean Implementation Through Value Stream Mapping: A Case Study of a Footwear Manufacturer*. 29th Chinese Control and Decision Conference (CCDC) (2017)
8. B. Huang, H. Zhang, M. Lu, in *Software FMEA Approach Based on Failure Modes Database*. 8th International Conference on Reliability, Maintainability and Safety (2009)
9. D. Nguyen, in *Failure Modes and Effects Analysis for Software Reliability*. Annual Reliability and Maintainability Symposium. 2001 Proceedings. International Symposium on Product Quality and Integrity (Cat. No. 01CH37179), Jan 2001

10. S.M. Zahraee, A. Hashemi, A.A. Abdi, A. Shahpanah, J.M. Rohani, Lean manufacturing implementation through value stream mapping: a case study. Jurnal Teknologi **68**(3), 119–124 (2014)
11. M. Poppendieck, *Principles of lean thinking.* Poppendieck.LLC (2012)
12. M. Rother, J. Shook, *Learning to See–Value-Stream Mapping to Create Value and Eliminate Muda* (Lean Enterprise Institute, Cambridge (US), 2009), pp. 1–4
13. P.L. Goddard, in *Software FMEA Techniques.* Annual Reliability and Maintainability Symposium (2000)
14. J.B. Bowles, C. Wan, in *Software Failure Modes and Effects Analysis for a Small Embedded Control System.* Proceedings Annual Reliability and Maintainability Symposium (2001)
15. N. Ozarin, in *Failure Modes and Effects Analysis During Design of Computer Software.* RAMS (2004)
16. H. Aditya Pai, H.S. Sameena, S. Soman, P.K. Pareek, Analysis of causes and effects of longer lead time in software process using FMEA. Int. J. Res. Adv. Technol. **7**(6S), 71–77 (2019)
17. P.K. Pareek, T.V. Nandikolmath, P. Gowda, FMEA implementation in a foundry in Bangalore to improve quality and reliability. Int. J. Mech. Eng. Robot. Res. **1**(2), 81–87 (2012)
18. T. Nandikolmath, P.K. Pareek, S.A. Vasantha Kumara, Implementation of a lean model for carrying out value stream mapping in manufacturing industry. Int. J. Mech. Eng. Robot. Res. **1**(2), 88–95, (2012)
19. P.H. Aditya, H.S. Sameena, S. Sandhya, D. Pareek, P. Kumar, *ROC Structure Analysis of Lean Software Development in SME's Using Mathematical CHAID Model* (May 17, 2019). Available at SSRN: https://ssrn.com/abstract=3508577 or http://dx.doi.org/10.2139/ssrn.3508577
20. N. Geetha, K.D. Menon, A. Raj Jain, V. Janardhan, P.K. Pareek, Quantitative Analysis of Student Data Mining (May 17, 2019). Available at SSRN: https://ssrn.com/abstract=3510975 or http://dx.doi.org/10.2139/ssrn.3510975
21. A. Pai, V.S. Veesam, B.S. Babu, P.K. Pareek, Six sigma approaches used in implementing in supply chain management: a review. Res. Appl. Web Develop. Des. **1**(2), 12–16 (2018)
22. P.K. Pareek, Questionnaire survey using CHI-square test in six sigma in SME's in Bengaluru. Adv. Image Process. Pattern Recognit. **1**(1), 20–24 (2018)
23. P.K. Pareek, Analysing tools in six sigma in SME's in Bengaluru. J. Adva. Softw. Eng. Test. **1**(1), 19–29 (2018)
24. P.K. Pareek, Six sigma approaches used in implementing in supply chain management: a review. J. Adv. Softw. Eng. Test. **1**(1), 14–18 (2018)
25. A. Pai, V.S. Veesam, P.K. Pareek, B.S. Babu, Challenges in SME's anova analysis part-2 in Bengaluru. Res. Rev. Adv. Robot. **1**(1), 9–15 (2018)
26. A. Pai, V.S. Veesam, B.S. Babu, P.K. Pareek, Anova analysis part one of challenges in SME'S in Bengaluru. Res. Rev. Adv. Robot. **1**(1), 1–8 (2018)
27. A. Pai, V.S. Veesam, B.S. Babu, P. K. Pareek, Anova analysis part one of challenges in SME'S in Bengaluru. Research and reviews, understanding the adaptability of SME'S in Bengaluru. Res. Rev. Adv. Robot. **1**(1), 16–22 (2018)
28. A. Pai, V.S. Veesam, B.S. Babu, P.K. Pareek, Anova analysis part one of challenges in SME'S in Bengaluru, Research and Reviews:,,Comparative Analysis of Conven-tional Approach and Automated Approach Used in Lean Six Sigma While Implementing in Supply Chain Management: A Review, Research and Applications of Web Development and Design 1 (1), 7–13 (2018)
29. D Chandramma Piyush Kumar Pareek, ANOVA Analysis Part One of Challenges in SME'S in Bengaluru, research and reviews:, fast and accurate parts of speech tagging for Kannada-Telugu pair. Int. J. Appl. Eng. Res. **13**(10), 7857–7867 (2018)
30. P.K. Pareek, in *Anova Analysis Part One of Challenges in SME'S in Bengaluru, Research and Reviews*, ed. by K. Swathi, P. Shetteppanavar. An Efficient Machine Translation Model for Dravidian Language, 2017 2nd IEEE International Conference on Recent Trends in Electronics
31. P.K. Pareek, in *An Adoptive Model for Lean Software Development in Small and Medium Level Firms in Bengaluru* (2016). https://shodhganga.inflibnet.ac.in/handle/10603/120481

32. K.V. Rao, R. Balakrishna, H.A. Pai, P.K. Pareek, in *Data Mining for Healthy Tomorrow with the Implementation of Software Project Management Technique, Artificial Intelligence and Evolutionary Computations in Engineering Systems*. https://link.springer.com/chapter/10.1007/978-81-322-2656-7_31
33. P.K. Pareek, A.N. Nandakumar, Lean software development Survey on Benefits and challenges in agile and lean usage in small and medium level firms in Bangalore. Int. J. Adv. Res. Comput. Sci. Softw. http://ijarcsse.com/Before_August_2017/docs/papers/Volume_4/12_December2014/V4I12-0474.pdf
34. T. Nandikolmath, P.K. Pareek, Implementation of a lean model for carrying out value stream mapping in a manufacturing industry. Int. J. Mech. Eng. Robot. Res. http://www.ijmerr.com/show-123-120-1.html
35. P.K. Pareek, T.V. Nandikolmath, P. Gowda, FMEA implementation in a foundry in bangalore to improve quality and reliability. Int. J. Mech. Eng. Robot. Res.
36. P.K. Pareek, Education data mining–perspectives of engineering students. Int. J. Innov. Res. Comput. Sci.
37. P.K. Pareek, S.A. Vasantha Kumara, Implementation of a lean model for carrying out value stream mapping in a manufacturing industry. Int. J. Mech. Eng. Robot. Res.
38. P.K. Pareek, Reduction of cycle time by implementation of a lean model carried out in a manufacturing industry. Int. J. Eng. Soc. Sci. ISSN: 2249–9482
39. A. Pai, B.N. Ramesh, P.K. Pareek, P. Prasoon, in *Failure Mode Effective Analysis for Software Processes*. Int. J. Res. Eng., IT, Soc. Sci. **8**(10), October 2018, Impact Factor (SJIF) 6.452
40. Failure Mode Effective Analysis of Requirements Phase in small software Firms. Paper ID: ICSTM. http://ijates.com/images/short_pdf/1421736687_292.pdf

Self-driving Car in Virtual Simulation: A Non-AI Approach

Debkumar Chowdhury, Sujoy Paul, Sreejit Bagchi, Trisha Das, and Bipasha Mukhopadhyay

Abstract Over the last few decades, the self-driving car industry has come a long way and is going toward a future where human drivers might no longer exist. Researchers and engineers in motor companies are trying to innovate new ways to make and train self-driving cars to reduce the number of collisions. So, a lot of progress has been made but the cost of making them grew up with them. In this paper, we propose a simpler version of making and training a self-driving car in a simulated environment using a non-AI approach.

Keywords Virtual simulation · PID controller · Self-driving car · Reinforcement learning · Imitation learning

1 Introduction

While making a self-driving car inside a simulated environment there are three ways we can do that:

1. PID controller
2. Imitation learning
3. Reinforcement learning.

D. Chowdhury · S. Paul (✉) · S. Bagchi · T. Das · B. Mukhopadhyay
University of Engineering and Management, Kolkata, India
e-mail: psujoy09@gmail.com

D. Chowdhury
e-mail: debkumar.cse@gmail.com

S. Bagchi
e-mail: sreejitbagchi@gmail.com

T. Das
e-mail: trisha.uemkcse@gmail.com

B. Mukhopadhyay
e-mail: bipasha.mukherjee@uem.edu.in

J. Tavares et al. (eds.), *Emerging Technologies in Data Mining and Information Security*, Lecture Notes in Networks and Systems 164,
https://doi.org/10.1007/978-981-15-9774-9_58

In the normal PID controller, we detect the error and reduce that error using a predefined function that will use mathematical models to reduce the error.

In the imitation learning method, the problem is solved by training the policy to make decisions using demonstrations. So, it is by far a subset of supervised learning with the only key difference is that the predictions are made sequentially.

On the other hand, reinforcement learning is to maximize the reward and minimize the penalty. In this project, we have a unity 3D game engine to simulate the environment and the unity of ML agents to make self-driving cars.

2 Literature Survey

In this section, we are going to shuffle a brief comparison between heuristic erudition (PID controller), imitation learning and reinforcing stimulus learning (Table 1).

In the below section, we are going to discuss the previous researches and related work proposed for the said topic [4, 5] (Table 2).

Table 1 Classification of approaches for implementation of a self-driving car inside a simulated environment

S. No.	Method name	Year proposed	Implementation
1.	PID controller	1911	Calculates the error and reduces using predefined mathematical functions [1]
2.	Imitation learning	1952	Clones the user behavior which is a subset of supervised learning [2]
3.	Reinforcement learning	1954	Reinforcement learning learns by trial and error [3]

Table 2 Chronological chart related to research that has been conducted by various companies using simulation

S. No.	Year	Organization	Results
1.	2015	Amazon	It developed an self-governing conveyance robot known as "Amazon Scout." It is functioning as a multi-operational self-governing moveable object in Washington State, America
2.	2016	Apple	A self-governing system was developed by the Apple in 2017 in association with researchers and experts of NASA
3.	2017	Baidu	Built "Android of autonomous driving" to support the development of self-driving technology
4.	2018	Audi	3 M autonomous kilometers driven in urban environments Opened AI research laboratory in Silicon Valley
5.	2020	Aptiva	It is the first company to establish free hand driving Flagship self-runnable vehicle which can be moved on the streets of Europe

3 Proposed Methodology

3.1 Introduction to the Proposed Methodology

We have implemented a PID controlling mechanism to make a self-driving car drive it. The proportional–integral–derivative controller is a control loop feedback mechanism widely used in industrial control systems and a miscellanea of other applications requiring continuously modulated control. This PID controller is widely used in other industrial control systems as a control loop feedback mechanism [6].

3.2 Algorithm

Step 1: The car detects objects and localizes to get desired position $r(t)$
Step 2: The position is located using sensors represented by $y(t)$.
Step 3: Calculate an error value $e(t)$ will be difference between $r(t)$ & $y(t)$.
Step 4: Add $u(t)$ as steering angle the PID controller will try to minimize $e(t)$.
Step 5: Set $u(t)$ proportional to the error value.
Step 6: Set integral value to reduce small error overtime where the error is too small to detect by the proportional mechanism.
Step 7: Predicts the error value and looks into the future state of the system.

3.3 Flowchart

See Fig. 1.

3.4 Mathematical Equation Establishment

A new mathematical equation can be established w.r.t our proposed methodology. It is as follows:

$$u(t) = C_{\mathrm{p}} \cdot e(t) + C_{\mathrm{i}} \int_{0}^{t} e(\tau) \cdot d(\tau) + C_{\mathrm{d}} \frac{\mathrm{d}r(t)}{\mathrm{d}(t)} \quad (1)$$

where
C_{p}: Proportional coefficient
C_{i}: Integral coefficient
C_{d}: Derivative coefficient

Fig. 1 Flowchart of the proposed methodology

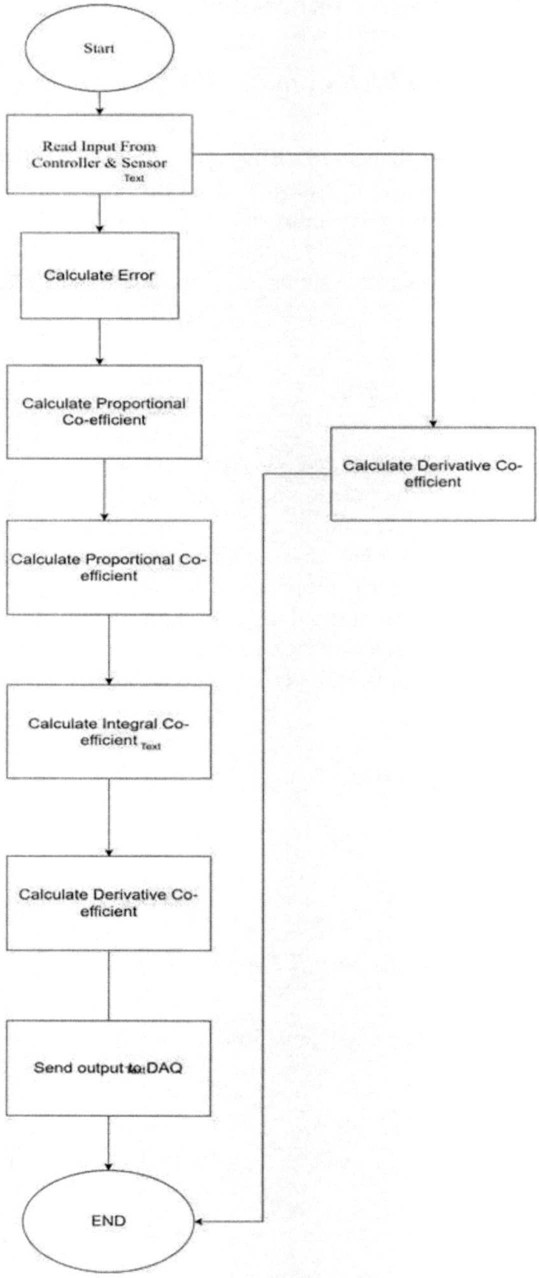

Output = Proportional + Integral + Derivative.

3.4.1 Mathematical Explanation

Let us say we are 50 m behind the line, so in this case, the error = 50. Now, we assume for sake of simplicity that the gains are 1. Now, we will find the P and D where P is proportional to the error and D will be preemptive control.

Here, $P = 50$ and $D = 0$ as we are not moving.

$$1.50 + (-1) = 50 \tag{2}$$

Now, we can find that the gain for the derivative is negative. This is because it wants to resist the proportional gain. The near one gets to the point the slower one goes. Thus, we assumed that the speed to negatively turn the acceleration value down.

So, acceleration is now,

$$1.40 + (-1).5 = 35 \tag{3}$$

Then, we go for integral gain. An integral is known as a summation over some time. If an equation maps the velocity of an object the integral of that equation will be the position of the object as it records the sum of the velocity at each moment.

4 Experimental Result and Analysis

4.1 Comparison Between the Proposed Approach and Machine Learning Approach

4.1.1 Proposed Approach

PID controllers are useful for finding optimal policies in continuous dynamical systems, and often these domains are also used as benchmarks for RL, precisely because there is an easily derived optimal policy. However, in practice, you would prefer a PID controller for any domain in which you can easily design one: The controller's behaviors are well understood, while RL solutions are often difficult to interpret.

If we wanted to make an agent maneuver a plane from in front of an enemy plane with known movement patterns to behind it, using the least amount of fuel, we would like to prefer to use a PID controller.

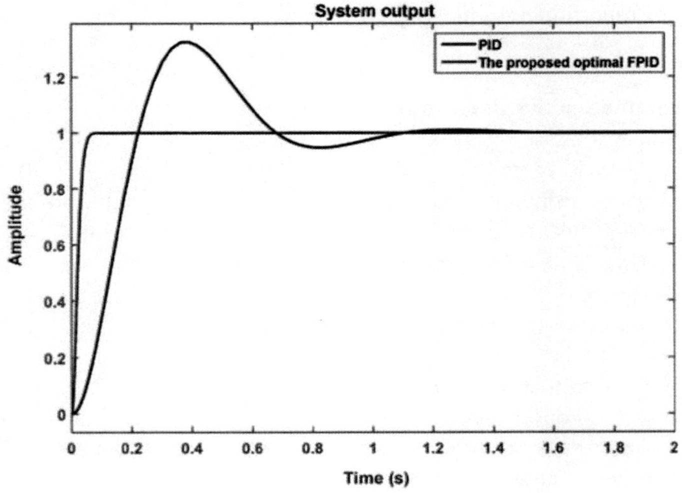

Fig. 2 Comparison graph for the establishment of better performance of our newly proposed technology

4.1.2 Machine Learning Approach

Where RL shines is in tasks where we know what good behavior looks like (i.e., we know the reward function), and we know what sensor inputs look like (i.e., we can completely and accurately describe a given state numerically), but we have little or no idea what we want the agent to do to achieve those rewards [7].

If we wanted to make an agent control a plane and shoot down an enemy plane with enough fuel left to land, but without a formal description of how the enemy plane might attack (perhaps a human expert will pilot it in simulations against our agent), we would much prefer RL (Fig. 2).

4.2 Result Analysis of Proposed Methodology and Experimentation

We get a good result from our proposed method. These are as follows:

1. Feasible and easy to implement.
2. Use lower resources.
3. Easy to tune mismatches.
4. Easier to implement by trial and error.
5. Better response to unmeasured disturbances.
6. It works better than model-based controllers who have only integral systems.

4.3 Experimental Results

See Figs. 3 and 4.

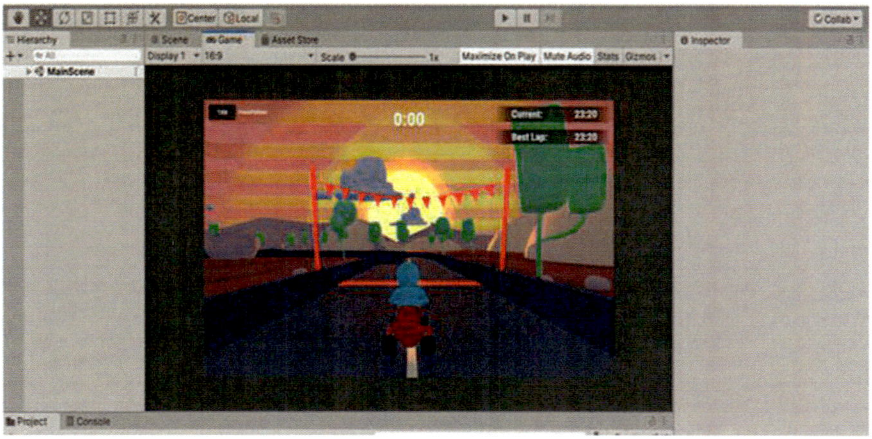

Fig. 3 Initializationof the proposed system

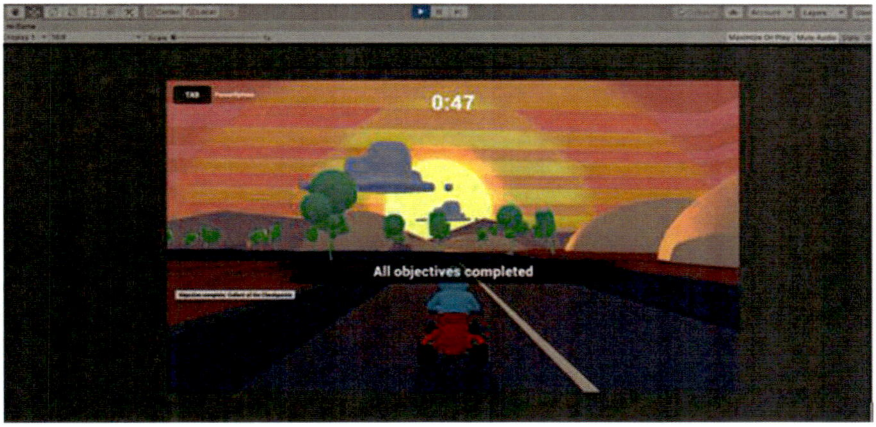

Fig. 4 Completion of the proposed system

5 Conclusion

In the world of robots, there are a huge number of controllers. Depending on the type of robot you want to control, it can be more or less simple. The environments can also drama a function.

In an autonomous railway car, pinch braking may occur involving a more sophisticated controller than an indoor robot. The control stage is mandatory for robots and monotone that we want to make autonomous. This step relies, however, on all the previous faculty (perception, location, preparation) which must be developed perfectly. Controllers are difficult to export from one car to another since they are unique to a vehicle.

References

1. K.H. Ang et al., PID control system analysis, design, and technology. IEEE Trans. Control Syst. Technol. **13**(4), 559–576 (2005)
2. B. Fang, S. Jia, D. Guo et al., Survey of imitation learning for robotic manipulation. Int. J. Intell. Robot. Appl. **3**, 362–369 (2019)
3. A.S. Polydoros, L. Nalpantidis, Survey of model-based reinforcement learning: applications on robotics. J. Intell. Robot. Syst. 153–173 (2017). https://doi.org/10.1007/s10846-017-0468-y
4. E. Yurtsever et al., A survey of autonomous driving: common practices and emerging technologies. IEEEAcess. arXiv:1906.05113v3 [cs.RO], 2 Apr 2020
5. W.K. Wolterink, G. Heijenk, G. Karagiannis, in *Constrained Geocast to Support Cooperative Adaptive Cruise Control (CACC) Merging*. 2010 IEEE Vehicular Networking Conference, pp. 41–48 (2010). https://doi.org/10.1109/vnc.2010.5698268
6. H.O. Bansal, R. Sharma, P.R. Shreeraman, PID controller tuning techniques: a review. JCET **2**(4), 168–176 (2012)
7. A.R. Fayjie et al., in *Driverless Car: Autonomous Driving Using Deep Reinforcement Learning in Urban Environment*. IEEE 15th International Conference on Ubiquitous Robots (2018)

Exploration Study to Study the Relationships Between Variables of Secure Development Lifecycle (SDL)

Sandhya Soman, Piyush Kumar Pareek, Sunanda Dixit, R. M. Chethana, and Veeranna Kotagi

Abstract Secure development lifecycle consists of a detailed plan that defines the process organizations use to build an application from inception until decommission. In this research paper, our focus was to conduct a survey with the help of a well-structured questionnaire to understand the relationship between various variables considered for the study with the help of a correlation test. Summary of the relationships is portrayed in the findings with respect to various levels. Data was analysed with the help of SPSS software. Nearest neighbour analysis also confirms the relationships among the variables considered.

Keywords Secure development lifecycle · Nearest neighbour analysis etc.

S. Soman
Department of Computer Science, Shri JJT University, Jhunjhunu, Rajasthan, India

(✉)
Department of Computer Science, Kristu Jayanti College, Bangalore, India
e-mail: sandhyasomancg@gmail.com

P. K. Pareek · R. M. Chethana · V. Kotagi
Department of Computer Science and Engineering, East West College of Engineering, Bangalore, India
e-mail: piyushpareek88@gmail.com

R. M. Chethana
e-mail: chethana.12@gmail.com

V. Kotagi
e-mail: veerkotagi21@gmail.com

S. Dixit
Department of Computer Science and Engineering, BMS Institute of Technology and Management, Bangalore, India
e-mail: sunanda.bms@gmail.com

J. Tavares et al. (eds.), *Emerging Technologies in Data Mining and Information Security*, Lecture Notes in Networks and Systems 164,
https://doi.org/10.1007/978-981-15-9774-9_59

1 Introduction

Varieties embrace code review, style review, formal qualification review, needs review, and take a look at readiness review' (IEEE 1990). IEEE commonplace says that the aim of a technical review is to judge a wares by a team of qualified personnel to work out its quality for its meant use and determine discrepancies from specifications and standards. Following square measure a number of the inputs to the technical review:

- A statement of objectives for the technical review (mandatory).
- The wares being examined (mandatory).
- Software project management set up (mandatory).
- Current anomalies or problems list for the wares (mandatory).
- Documented review procedures (mandatory).
- Relevant review reports (should).
- Any laws, standards, guidelines, plans, and procedures against that the wares are to be examined (should).
- Anomaly classes (See IEEE Std 1044-1993 [B7]) (should).

Inspection: A static analysis technique that depends on visual examination of development standards, and alternative issues. Varieties embrace code inspection; style inspection' (IEEE 1990). Scrutiny has several names, some referred to as software system scrutiny that might cowl style and documentation, and a few others can decision it as code scrutiny that relates additional ASCII text file written by the developer. Fagan scrutiny is another name that came because the name of the one that made-up QA and testing technique. Code scrutiny could be a time intense task however statistics telling that it should conceal to ninetieth of the contained errors if we tend to apply that in a very systematic means. Figure 1 shows time, employee relationship in a software development process.

IEEE Standard for Software Reviews (IEEE 1028-1997 standard) is talking about manual static testing methods like inspections, reviews, and walkthroughs.

2 Review of Literature

As part of the pilot study and other case studies, we are done the evaluation of many automation frameworks, tools, automation documents, etc. All tools and frameworks have its own advantages and disadvantages [1, 2]. In this research, we completed the pilot study and case studies using rational build forge and test-all tools (Proprietary tools of IBM). IBM Rational build forge V 7.13 information centre document will give an overall details of the framework. Build forge is a process execution framework tool that automates the process of creating a deployable application from source artifacts [3–5]. It has the features like build and release process management, source control and deployment; automate manual and repetitive task allowing focus on more

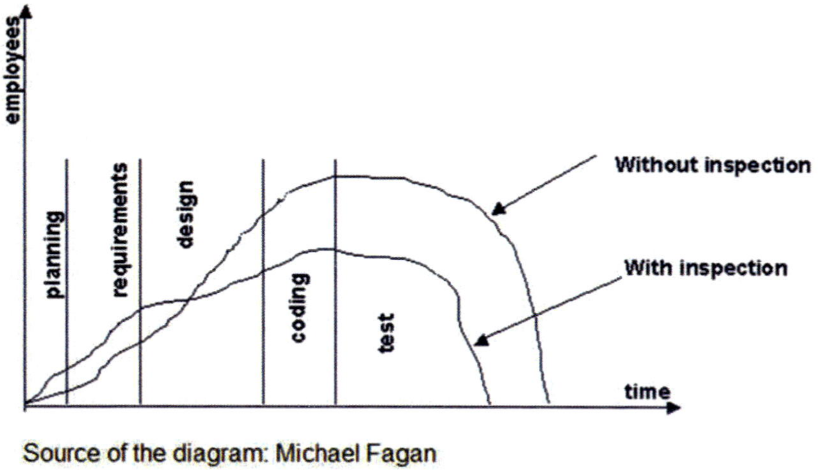

Fig. 1 Employee relationship in a software development process

strategic functions etc. Build forge tool will help to streamline software delivery process, automate, accelerate, coordinate, track and analyse the system [6–8]. The tool accelerates the software delivery cycles by enabling rapid iterations and rapid feedback, improves the quality by managing the product assembly tasks for consistent results each time, practices solutions that leverage existing scripts and tools for rapid implementation and results. It also helps to reduces global development cost by automating processes, handoffs, and communication and provides traceability control for compliance management and better software governance [9–11].

References [12–14] worked on an industrial case study for improving the testing efficiency of a data indexing tool and came out with some recommendations that benefits the testing team for reducing the testing time and resources. Author suggesting the importance of knowing the current efficiency for knowing what to improve. Also the paper talks about the importance of bringing risk based testing, lightweight test automation, usage of proper testing tools, install verification automation, etc. Paper also showing the improvements that the team achieved after introducing the new framework for regression testing [15].

3 Results and Discussions

3.1 Reliability Test

See Table 1.

Table 1 Reliability test summary statistics

Case processing summary			
		No.	%
Cases	Valid	235	92.9
	Excluded[a]	18	7.1
	Total	253	100

[a]List wise deletion based on all variables in the procedure

Reliability statistics	
Cronbach's alpha	Number of items
0.864	52

From the case processing summary and reliability statistics, it can be observed that, for a total of 235 respondents size, 52 number of items, the reliability achieved was 0.864, which indicates questionnaire is valid and reliable.

3.2 Nearest Neighbour Analysis

See Fig. 2.

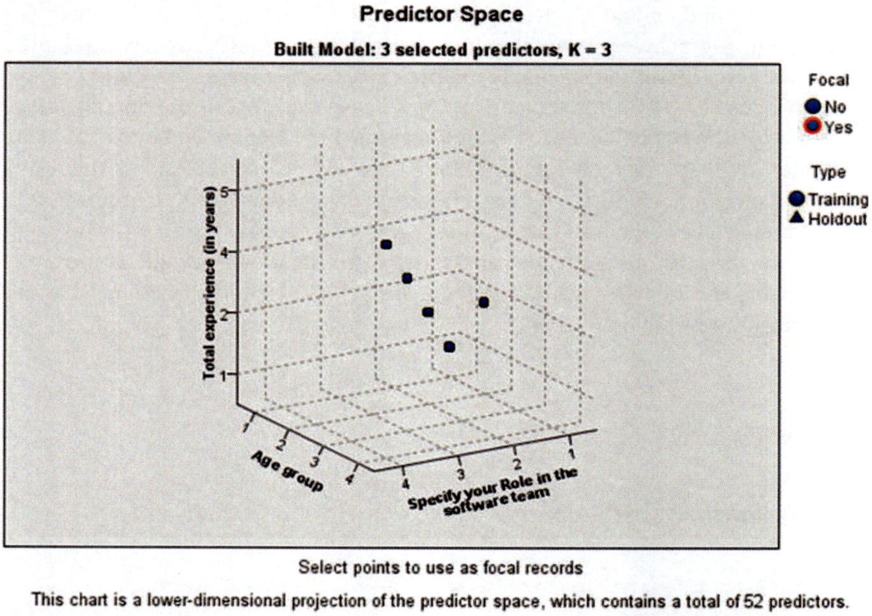

Fig. 2 Nearest neighbour analysis plot

Case processing summary			
		No.	Percent (%)
Sample	Training	169	71.9
	Holdout	66	28.1
Valid		235	100.0
Excluded		18	
Total		253	

Table 2 significant relationships between variables are indicated as follows
**Correlation is significant at the 0.01 level (2-tailed).
*Correlation is significant at the 0.05 level (2-tailed).

4 Conclusion

Members detailed that finding a defencelessness or encountering a security break direct is another factor that supported security practices and mindfulness in their groups. Regardless of broad exposure around security vulnerabilities, attention to and duty to security stays low. A few members said their group chooses their security rehearses dependent on the accessible spending plan and additionally representatives who can perform security undertakings. Another persuasive factor showed by members is the groups' discernment of security and whether it is a piece of the organization culture.

Table 2 Correlation test analysis

Correlations

		Depth of inheritance (DIT)	Weighted methods per class (WMC)	Code complexity (Complexity)	Coupling between objects (CBO)	Change in code	File or class size (LoC)	Number of previous bugs	Number of modified lines	More number of revisions (releases)
	Sig. (2-tailed)	0.000	0.000	0.000	0.276	0.178	0.000	0.115	0.006	0.000
	N	253	253	253	253	253	253	253	253	253
Depth of inheritance (DIT)	Pearson correlation	1	0.799**	0.794**	0.266**	0.140*	−0.472**	−0.100	−0.288**	−0.304**
	Sig. (2-tailed)		0.000	0.000	0.000	0.026	0.000	0.113	0.000	0.000
	N	253	253	253	253	253	253	253	253	253
Weighted methods per class (WMC)	Pearson correlation	0.799**	1	0.609**	−0.071	0.024	−0.418**	0.123	−0.268**	−0.275**
	Sig. (2-tailed)	0.000		0.000	0.257	0.704	0.000	0.050	0.000	0.000
	N	253	253	253	253	253	253	253	253	253
Code complexity (Complexity)	Pearson correlation	0.794**	0.609**	1	0.372**	0.395**	−0.113	0.168**	−0.410**	−0.435**

(continued)

Table 2 (continued)

Correlations

		Depth of inheritance (DIT)	Weighted methods per class (WMC)	Code complexity (Complexity)	Coupling between objects (CBO)	Change in code	File or class size (LoC)	Number of previous bugs	Number of modified lines	More number of revisions (releases)
	Sig. (2-tailed)	0.000	0.000		0.000	0.000	0.073	0.007	0.000	0.000
	N	253	253	253	253	253	253	253	253	253
Coupling between objects (CBO)	Pearson correlation	0.266**	−0.071	0.372**	1	0.523**	−0.092	−0.369**	−0.047	−0.364**
	Sig. (2-tailed)	0.000	0.257	0.000		0.000	0.146	0.000	0.454	0.000
	N	253	253	253	253	253	253	253	253	253
Change in code	Pearson correlation	0.140*	0.024	0.395**	0.523**	1	0.051	0.296**	−0.337**	−0.085
	Sig. (2-tailed)	0.026	0.704	0.000	0.000		0.416	0.000	0.000	0.176
	N	253	253	253	253	253	253	253	253	253
File or class size (LoC)	Pearson correlation	−0.472**	−0.418**	−0.113	−0.092	0.051	1	0.101	−0.261**	0.207**

(continued)

Table 2 (continued)

Correlations

		Depth of inheritance (DIT)	Weighted methods per class (WMC)	Code complexity (Complexity)	Coupling between objects (CBO)	Change in code	File or class size (LoC)	Number of previous bugs	Number of modified lines	More number of revisions (releases)
	Sig. (2-tailed)	0.000	0.000	0.073	0.146	0.416		0.110	0.000	0.001
	N	253	253	253	253	253	253	253	253	253
Number of previous bugs	Pearson correlation	−0.100	0.123	0.168**	−0.369**	0.296**	0.101	1	−0.148*	0.273**
	Sig. (2-tailed)	0.113	0.050	0.007	0.000	0.000	0.110		0.019	0.000
	N	253	253	253	253	253	253	253	253	253
Number of modified lines	Pearson correlation	−0.288**	−0.268**	−0.410**	−0.047	−0.337**	−0.261**	−0.148*	1	−0.166**
	Sig. (2-tailed)	0.000	0.000	0.000	0.454	0.000	0.000	0.019		0.008
	N	253	253	253	253	253	253	253	253	253
More number of revisions (releases)	Pearson correlation	−0.304**	−0.275**	−0.435**	−0.364**	−0.085	0.207**	0.273**	−0.166**	1
	Sig. (2-tailed)	0.000	0.000	0.000	0.000	0.176	0.001	0.000	0.008	
	N	253	253	253	253	253	253	253	253	253

**Correlation is significant at the 0.01 level (2-tailed)
*Correlation is significant at the 0.05 level (2-tailed)

References

1. P.K. Pareek, T.V. Nandikolmath, P. Gowda, FMEA implementation in a foundry in Bangalore to improve quality and reliability. Int. J. Mech. Eng. Robot. Res. **1**(2), 81–87 (2012)
2. T. Nandikolmath, P.K. Pareek, S.A. Vasantha Kumara, Implementation of a lean model for carrying out value stream mapping in manufacturing industry. Int. J. Mech. Eng. Robot. Res. **1**(2), 88–95 (2012)
3. P.K. Pareek, P. Gowda et al., Ergonomics in a foundry in Bangalore to improve productivity. Int. J. Eng. Soc. Sci. **2**(5), 1–6 (2012). ISSN: 2249- 9482
4. S. Soman, P.K. Pareek, J. Phys. Conf. Ser. 1427 012010 6 2020; P.K. Pareek, Optimization of the critical parameters affecting the fuel lid performance (2019) 2019-28-2413
5. P.K. Pareek, in *ROC Structure Analysis of Lean Software Development in SME's Using Mathematical CHAID Model* (17 May 2019)
6. K.D. Menon, A. Raj Jain, P.K. Pareek, in *Quantitative Analysis of Student Data Mining* (2019)
7. H.S. Sameena, S. Soman, P.K. Pareek, in *Fast and Efficient Parallel Alignment Model for Aligning both Long and Short Sentences* (2019)
8. M.R. Suresh, P.C. Behera, C. Dash, P.K. Pareek, Face recognition with convolutional neural network. J. Adv. Res. Mob. Comput. **1**(1)
9. A. Pai, V.S. Veesam, B.S. Babu, P.K. Pareek, Six sigma approaches used in implementing in supply chain management: a review. Res. Appl. Web Develop. Des. **1**(2), 12–16
10. S. Dixit, S. Ranjitha, H.N. Suresh, Segmentation of handwritten kannada text document through computation of standard error and weighted bucket algorithm. Int. J. Adv. Comput. Technol. **3**(2), 55-62 (2014). ISSN: 2319-7900
11. S. Dixit, H.N. Suresh, Mahesh, in *Kannada Text Line Extraction Based on Energy Minimization and Skew Correction*. Advance Computing Conference, IEEE International, pp. 62–67 (2014). ISBN: 978-1-4799-2571-1
12. S. Dixit, H.N. Suresh, in *Kannada Text Line Extraction and Skew Correction using State-of-the art Algorithms*. Elsevier, First Revision as per their Comments
13. S. Dixit, H.N. Suresh, Segmentation of kannada handwritten text line through computation of variance. Comput. Sci. Inf. Secur. **12**(2), 56–60 (2014). ISSN: 1947-5500
14. S. Dixit, S. Nilotpal, H.N. Suresh, Text line segmentation of handwritten documents in hindi and english. Int. J. Rec. Innovation Trends Comput. Commun. **2**(4), 733–739 (2014). ISSN: 2321-8169
15. S. Dixit, K. Agnani, V.G. Pai, S.R. Vaishna Priyan, V. Claud Rodrigues, in *3D Reconstruction of 2D X-Ray Images*. 4th IEEE International Conference on Computational Systems and Information Technology for Sustainable Solutions [CSITSS-2019]. RVCE, 22–23 Dec 2019

Recommendation-Based Interactivity Through Cross Platform Using Big Data

G. K. Suhas, S. N. Devananda, R. Jagadeesh, Piyush Kumar Pareek, and Sunanda Dixit

Abstract A recommendation is a suggestion or proposal as to the best course of action, especially one put forward by an authoritative body. Shopping is a necessity of every human being, and when we do shop, it is definitely either the product we like or our friends like. Aim of the research paper is to give recommendation to the user based on the user's interest in a single cross platform. Design a cross-platform layers. Currently provide the public centres. Provide a platform for mining user interest content across different social networks. In the proposed system, we develop a Web application where in the user has to subscribe by giving the users credentials along with his interest in different fields. Based on user's priority and interest, recommendations are displayed. A cross-platform Web application which is purely based on the user interest and interactivity obtain big data based on the user interactivity and use it for future recommendations. Content is recommend based on user interest.

Keywords Recommendation system · Big data · Cross platform

1 Introduction

We are influenced to buy merchandise recommended by agents because we trust the sales person. In this digital age, any online Web site, we visit employs a kind of recommendation tools like click-through rate conversions, and other important

G. K. Suhas (✉) · P. K. Pareek
East West Institute of Technology, Bengaluru, India
e-mail: suhask300@gmail.com

S. N. Devananda
PES Institute of Technology and Management, Shivamogga, India

R. Jagadeesh
CMR University, Bengaluru, India

S. Dixit
BMS Institute of Technology, Bengaluru, India

J. Tavares et al. (eds.), *Emerging Technologies in Data Mining and Information Security*, Lecture Notes in Networks and Systems 164,
https://doi.org/10.1007/978-981-15-9774-9_60

metrics, if it is initiated and configured accordingly, it can seriously hike profits. Furthermore, they can have a very good impact on the customer gratification as well, which in turn improves customer satisfaction and retention. This can be made easily possible with the help of recommendation engine. Recommendation engines essentially comprises of data filtering tools that make use of algorithms and data to recommend the most relevant items to the appropriate customer. In elementary terms, recommendation engines are nothing but an automated form of a "outlet counter person". We can inquire him for a product. Along with showing our desired product, he also recommends the related products based on our interest which we could consider buying. They are very much best trained in cross-selling and up-selling. With the expanding volume of data on the internet and with a compelling growth in the number of users, it is much essential for companies to search, map, and provide customers with the relevant chunk of information according to their choice and preferences.

2 Background

Zaharieva et al. [1], in their work, have proposed that detection and retrieval of media related to social events have gained increasing attention in the research community. The researchers have differentiated between known and unknown social events. linking of metadata and heterogeneous media from various platform itself is a major challenge in a given social event. Supervised learning approach based on machine learning algorithms to detect cross-platform social event was employed in the existing approach. To develop the retrieval model to recognize unknown event, in this approach, it requires a training data set to learn the properties from the known social events.

Yan et al. [2] the researches have focused on the problems of single platform and have catered a extensive possibilities to overcome challenges and problems of one single platform. To handle the problem of cold-start recommendation, they have investigated into behaviour information and cross-platform social relation. Their research mainly focuses on thorough investigation on data analysis to explore what information can be exceptionally transferred from one platform to another platform; the results they have achieved indicate a firm interrelation for the common contact behaviour and bidirectional relation between test platforms. Their approach is better in gathering social information to boost the performance of friend recommendation.

Niveditha et al. [3] in their research focused on two aspects: First, to handle the difficulties of cross platform user experience theoretically. Second, by using a mobile application based on android and a cloud architecture is mapped for an account of theoretical specification of big data user centric approach and interactivity. To address the challenges of cross platform architecture, they have gone through the study of cross platform using big data tools and techniques, for intensive operations such as data mining and data processing at large scale.

Zelenkauskaite et al. [4] have proposed an adaptive user-centric model to enhance user experience. Their model focuses on facility of offline and online territory and self-governing situation that are keen to the changing data fluctuations. They developed this model as a prototype for an ad hoc media group, that company was focusing on enhancing their user experience through social media through many years. In their proposed prototype, they have used big data tools for interlinking varying social media streams and external Web data with the media company's data.

Shyam Sundar et al. [5], G.K. Suhas et al. [6] have emphasized the importance of customization. According to them, customization is so appealing because the content is tailored or because the user feels greater agency. They have conducted number of tests to check the preferences of different types of users. In study 1, they came to a conclusion that power users rated content quality higher when it had a customizable interface, whereas non-power users preferred personalized content.

3 Proposed Methodology

3.1 Information Extraction

The interests are extracted from different Web sites like YouTube, Facebook, twitter by using the Web API's such as YouTube data API protocol, Public content solution API, Twitter JSON API, respectively, and stored in the information extraction repository. The repository contains id's of URL of different Web service provider like YouTube, Facebook, google, etc., then the different categories like sports, comedy, movies, news, etc., and unstructured data such as videos, audios, images, etc., and contains URL's from multiple sites like YouTube, Facebook, google, etc.

3.2 Interest Mining Based

On user browsing behaviour on contents, this module learns the user interest and constructs user profiles grouping user of similar interest. First the training set is built for the respective user by using rating on a particular item. Based on the ratings provided by user, the predictive model is generated. Rank filtering method is applied to the prescient model, then top-positioned items are incorporated in the list of proposal. Users interests change rapidly so the profile learner is updated automatically.

3.3 Content Recommendation

This module will match the contents to user interest based on meta data matching and also collaborative recommendation and provides content recommendation to the user. Feature vector is built and stored as data model by extracting content form metadata. Consistency of the data model is maintained by removing outliers. The contents meeting the threshold value is recommended to the user. If a particular user has not viewed an item, then the neighbour user will provide recommendation for that particular item.

3.4 Providing Recommendation and Calculating Similarity Measure

Content recommended to the user for which the user has not viewed yet. This is achieved by calculating the cosine similarity measure between the neighbour URL ratings. All the information regarding User, URL, Rating, Location will be stored in a file. Data model is built based on the ratings provided by the users in a particular location.

Similarity between the products can be calculated by the using the formula

$$\mathrm{Sim}(i, j) = \cos(s_{\rightarrow}, y_{\rightarrow}) = s_{\rightarrow} \cdot y_{\rightarrow} / \,||s||2 * ||y_{\rightarrow}||2 \tag{1}$$

System Architecture

System architecture is the conceptual design that defines the structure and behaviour of a system. Content/subscription management of user is done in the registration phase. The interests are extracted from different websites like YouTube, Facebook, and Instagram by using the Web API's such as YouTube data API protocol, Public content solution API, Instagram Graph API, respectively, and stored in the information extraction repository. User browsing behaviour is extracted and stored in the interest mining repository. Content matching can be done between the information extraction and the interest mining repository and finally content recommendation is given to the user.

The system architecture is shown below (Fig 1):

In this work, we address the question of how to predict fine particulate matter given a combination of weather conditions. Therefore, this study aims to elaborate a statistical model to predict the pollution levels from the meteorological conditions. This statistical approach is based on data mining, which is searching for some patterns in raw big data in order to extract regularities that can be used to build a predictive model. In the proposed system, by implementing the collaborative filtering algorithm, mining all the previous data about the pollution details and predicting the next day's fine particulate matter of the environment.

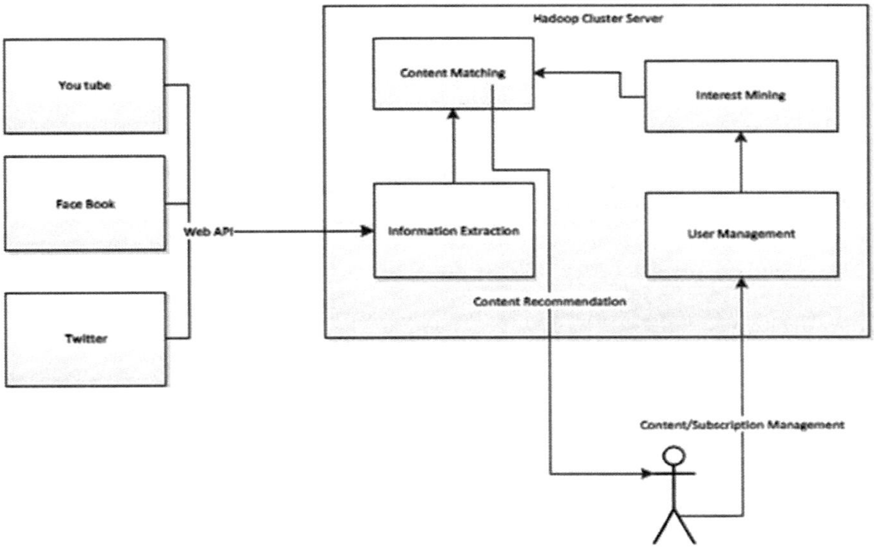

Fig. 1 Diagrammatic representation of proposed methodology

Proposed methods is based on following algorithms.

Algorithm : Interest Mining

Step1: The training set TRa for Ua must be characterized.
Step2: TRa is a set of pairs <Ik, Rk>
where Rk is the rating gave by Ua on the item representation Ik.
Step 3: Given an arrangement of item representation named with ratings, the profile learner generate a prescient model—the client profile.
Step 4: Filtering part executes Rank Filtering to rank fascinating things as indicated by significance concerning client profile. Highest ranked items are included in the list of recommendation La that is given to Ua.
Step5: User premiums generally changes as time slip by; subsequently progressive data ought to be kept up and given to the profile learner to redesign the client profile consequently.
Step 6: Feedback is tackled suggestions produced by letting clients express their like or dissimilar to with things La.
Step 7: After the getting the criticism, the learning procedure is completed again on the new preparing information and the subsequent profile is adjusted to the upgraded client premiums. Keeping in mind the end goal to account the dynamic way of client inclination

Algorithm: Content Recommendation

Step 1: Each content extracted from social networking site is expressed in form of feature vector

$$D = (W1, W2 \ldots WN)$$

Where $W1$, $W2$ are wight vectors for the Feature items. Feature items are taken from the meta data of the content.

$$Wi = tf * idf = (\text{freq}/\text{max freq}) * \log N/ni \quad (2)$$

Step 2: Store the features vectors whenever a user browses a particular content and store it as interest feature vector.
Step 3: Remove outliers and cluster the items in the interest feature vector
Step 4: Any new content extracted from the social site, compare the similarity of feature vector of content to cluster centres of the interest feature vector created, if similarity value is less than threshold, the content is recommended to the user.
Step 5: Group the groups of similar cluster centres of interest feature vector and whenever the user is group watches a content recommend the same content to another user in the group if they have not watched it.

4 Simulation and Experimental Results

Test Case 1:

Table 1 shows the test cases for interest profiler and content download. The parameters are test functions, input, expected output, actual output, and remarks. The testing is done on the interest profiler, content download by taking the inputs as browsing history and information retrieved from interest vector, respectively. The expected output is interest vector has to build and retrieve the information and list of url's based on the user interest, respectively, and testing gives the same result hence it states successful.

Table 1 Testing table for interest profiler and content download

S. No	Test function	Input	Expected output	Actual output	Remarks
1	Interest profiler	Browsing history	Interest vector has to build and retrieve the information	Interest vector built and Information retrieved	Success
2	Content download	Information retrieved from interest vector	List of URL's based on the user interest	List of URL's based on the user interest	Success

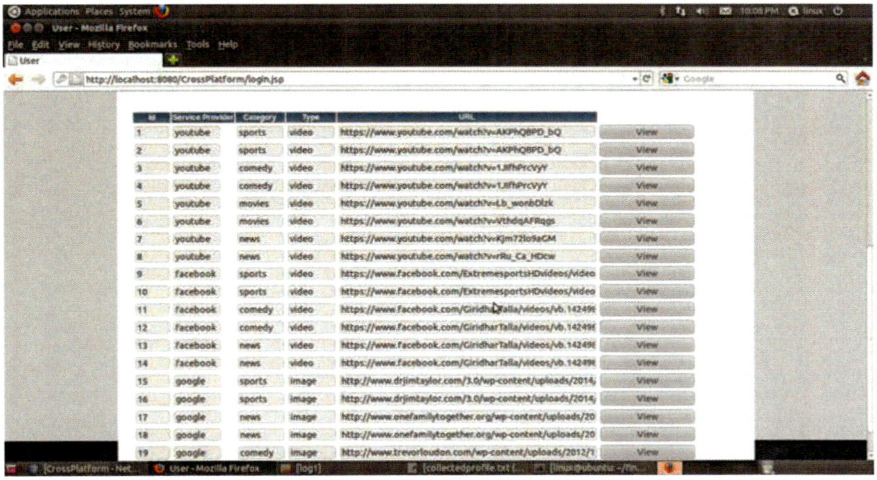

Fig. 2 Snapshot of cross platform Web portal

Result for test case 1:
Figure 2 shows that cross platform Web portal contains id's of url, different Web service provider like youtube, facebook, google, etc., different categories like sports, comedy, movies, news etc., type such as videos, audios, images, etc., and contains url's from multiple sites like youtube, facebook, google, etc.

Test case 2:

Table 2 shows the test cases for provide rating. The parameters are test functions, input, expected output, actual output, and remarks. The testing is done on the provide rating by taking input as rate item from radio button. The expected output is rating given to a url should store in choose profiler folder and testing gives the same result as expected hence it states successful.

Result for test case 2:
See Fig. 3

After viewing different url by the user, the user has to provide rating to those url in the button provided at the bottom. All the ratings provided by the respective user is stored in the choose profiler folder.

Table 2 Testing table for provide rating

S. No	Test function	Input	Expected outputt	Actual output	Remarks
1	Provide rating	Rate item from radio button	Rating given to a URL should store in choose profiler	Rating provided by user is stored in choose profiler	Success

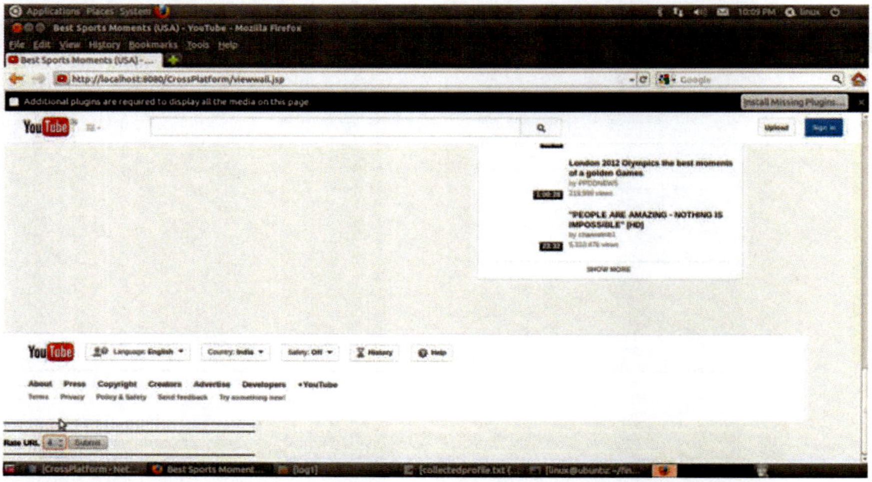

Fig. 3 Snapshot of ratings given by user after watching respective URL

Table 3 Testing table for recommendation

S. No	Test function	Input	Expected output	Actual output	Remarks
1	Recommendation	User has to recommend through feature vector	Content recommended	Recommendation done	Success

Test Cases 3:

Table 3 shows the test cases for providing recommendation. The parameters are test functions, input, expected output, actual output, and remarks. The testing is done on the recommendation by taking inputs as user has to recommend through feature vector. The expected output is content recommended and testing gives the same result. Hence it states successful.

5 Conclusion

To provide a cross-platform based on user experience. Implement an application which contains information retrieval, information filtering, and rank filtering processes. Design virtual cloud to account for theoretical parameters of big data user centric approach and interactivity. Provide user friendly data based on the user history and profile.

References

1. M. Zaharieva, M. Del Fabro, M. Zeppelzauer, Cross-platform social event detection. IEEE MultiMedia, 1–1 (2015). https://doi.org/10.1109/mmul.2015.57
2. M. Yan, J. Sang, T. Mei, C. Xu, in *Friend Transfer: Cold-Start Friend Recommendation with Cross-Platform Transfer Learning of Social Knowledge.* 2013 IEEE International Conference on Multimedia and Expo (ICME) (2013). https://doi.org/10.1109/icme.2013.6607510
3. N.M. Niveditha, Design an interest based interactivity platform for social media through big data. Int. J. Adv. Eng. Res. Technol. (IJAERT) **3**(2) (2015). ISSN: 2348–8190
4. A. Zelenkauskaite, B. Simões, in *The Big Data User-Centric Model*
5. S.S. Sundar, S.S. Marathe, Personalization versus customization: the importance of agency, privacy, and power usage. Hum. Commun. Res. **36**(3), 298–322 (2010). https://doi.org/10.1111/j.1468-2958.2010.01377.x
6. G.K. Suhas et al., An exploration on recommendation based interactivity through multiple platforms in big data. IOSR J. Comput. Eng. (IOSR-JCE) **22**(1), 31–36 (2020)
7. M. Latif, Y. Lakhrissi, E.H. Nfaoui, N. Es-Sbai, in *Cross Platform Approach for Mobile Application Development: A Survey.* 2016 International Conference on Information Technology for Organizations Development (IT4OD) (2016). https://doi.org/10.1109/it4od.2016.7479278
8. A. Labriji, S. Charkaoui, I. Abdelbaki, A. Namir, in *User Interest Center Based on a Semantic User Profile.* 2016 5th International Conference on Multimedia Computing and Systems (ICMCS) (2016). https://doi.org/10.1109/icmcs.2016.7905556
9. A. Pai, V.S. Veesam, B.S. Babu, P.K. Pareek, Six sigma approaches used in implementing in supply chain management: a review. Res. Appl. Web Develop. Des. **1**(2), 12–16

A Comprehensive Survey: Toward Providing Personal Safety Services in Emergency Situations

Ramesh M. Kagalkar and Lokesh B. Bhajantri

Abstract Human security has become one of the most challenging issues for the researchers, owning it to its grave significance and the expanded challenge in the real world environment. Hundreds or thousands of human security gadgets are being created due to the quick progression in the field of Internet of things (IoT) that include detecting advancements, implanted frameworks, remote correspondence innovations, assortment of sensors, and so forth. A basic capacity of these gadgets is human activity recognition. The present human security gadgets are ceaselessly track human exercises with the assistance of sensors. In this work, we have presented the comprehensive survey on smart ring device as compared to existing systems in real world environment. Finally, we have proposed the safety device system during emergency situations. This system is an integration of three apparatus mainly wearable finger ring device, smart mobile phone device, and four service systems. The services systems are police, hospital, ambulance, and health services systems. Regularly, each services system will communicate to each other for exchanging information with one another. This will be monitored and accessed by an agency service system.

Keywords Internet of things · Wearable finger ring device · Agency service system · Hospital service system

---

R. M. Kagalkar
KLE College of Engineering and Technology, Affiliated to VTU, Belagavi, Chikodi, Karnataka, India
e-mail: rameshvtu10@gmail.com

L. B. Bhajantri (✉)
Basaveshwar Engineering College (Autonomous), Affiliated to VTU, Belagavi, Bagalkot, Karnataka, India
e-mail: lokeshcse@yahoo.co.in

J. Tavares et al. (eds.), *Emerging Technologies in Data Mining and Information Security*, Lecture Notes in Networks and Systems 164,
https://doi.org/10.1007/978-981-15-9774-9_61

1 Introduction

The Internet of things (IoT) world view gives connectivity to devices or nodes for data sharing between the nodes or devices which are connected by the high speed of network. These nodes sense the data from the physical world environment and process the sensed data to its respective base station. Consequently, the IoT is utilized in a number of applications such as e-health monitoring, transportation, environment monitoring, logistics management, and so on. Typically, IoT is expected to offer advanced connectivity of devices, systems, and services that goes beyond machine-to-machine communications [1]. The rapid growth of IoT technology and infrastructure has made our lives easier. The advent of technology has also increased the traffic hazards and road accidents take place frequently which causes huge loss of life and property because of the poor emergency facilities like difficult to detect the exact location of the victim. There is no technology in current times to address in real-time human emergency. The IoT has the following characteristics [2]: dynamic and self-adapting, self-configuring, interoperable communication protocols, unique identity, integrated into the information network, context awareness, and intelligent decision-making capabilities.

In this work, we have planned to propose a new model for safety services in emergency situations. The proposed system is fully automated, as it finds the accident spot, identify the person details, sending message to his/her relatives, sending message to police station to lodge the compliant, providing medical treatment to the patient, paying of medical bill for the hospital automatically, and helping by saving the life of patients to reach the hospital in time.

Accordingly, the proposed framework helps in lessening the loss of lives of human which occur by the mishaps. It provides safe and secure to the victim in a very reasonable cost. The principle objective of the proposed system is to limit the time hole between the event of mishap and time required for emergency vehicle to reach at the area of mishap for offering treatment to the person. The proposed system provides security service to human beings in various situations like road accidents, kidnap cases, health issues, attention toward the aged individuals, person in trouble, emergency alert, and explains new or unknown places in the real world environment.

The rest of the paper organized is as follows: Sect. 2 describes the detailed literature survey toward the proposed work. Section 3 discusses the system overview of the proposed work. Finally, it concludes the proposed work in Sect. 4.

2 Literature Survey

In this work, some of the related works are carried out towards the problem statement of the proposed work, which is shown in the below Table 1.

Based on the above literature survey, we have identified the boundary lines issues and limitations of the existing systems in terms of text messages or call, health care

Table 1 Approaches and issues of related works

Citation	Approach	Issues
[3]	***Aim***: To verify, effective, and easy to use validation framework for older individuals or patients in basic conditions during restorative administrations ***Methodologies***: The proposed methodology has created five unique elements framework, User (U), Registration Center (RC), Access Points (APs), Medical Sensors or End Nodes (ENs) offering well-being-related administrations and the Medical Server (MS) for Privacy-protecting biometrics based confirmation conspire created Performed formal security confirmation utilizing CDVT/AD (Cryptographic-convention Development and Verification Tools with Attack Detection ***Results***: Better execution regarding secure, calculation and correspondence costs contrasted, and the current biometrics plans	• User visit every time to hospital • More complexity
[4]	***Aim***: To improve the performance of the IoT-based healthcare systems (accuracy and power consumption) **Methodologies**: The proposed scheme is implemented with optimization algorithms such as cloud handling, big data handling, security algorithm, and power consumption ***Results***: Consumes a low power to achieve considerable efficiency and also achieved high performance	• High power consumption • Security issues due to the utilization of many devices
[5]	***Aim***: To feature the difficulties of crisis administrations and its past arrangements. To spare lives through improving interchanges among patient and crisis administrations ***Methodologies***: Implemented an algorithms for smart hospital emergency system and also implanted a service request algorithm for utilizing health data held within a personal cell phone, and interior followed information (GPU, Accelerometer, Gyroscope, and so on.) ***Results***: It gives a good usability, reliability, and communication performance	• Real-time video communication protocols are to be improved

(continued)

Table 1 (continued)

Citation	Approach	Issues
[6]	***Aim***: To build a virtual neighborly gadget and shield themselves while in a tough situation ***Methodologies***: It supports to hardware components such as GPS, GSM modem, Microcontroller, RF handset, Temperature Sensor, and Voice Recognizer. GPS Receiver to gets the data from satellites. This work has additional features like alarm sound and voice recognition. The voice recognizer circuit is used to communicate with voice by using the codes of SAVE ME and HELP ME ***Results***: Designed as a smart a portable kit that can be easily performed in the real world environment	• For only women safety and security system
[7]	***Aim***: To build a wearable shrewd band that communicated over the telephone networks and continuously update the emergency information into the Web page ***Methodologies***: The proposed module is used to control and communicate and use the hardware components such as trigger, microcontroller, GSM module, GPS module, IoT module, Neuro Stimulator, Buzzer, and Vibrating Sensor ***Results***: It provides safety and security to the women in danger situation	• For only women safety • Hardware complexity
[8]	***Aim***: A shrewd IoT gadget for child safety to help the parents for locating and monitoring their children's in the real world environment ***Methodologies***: The proposed system using LinkIt ONE board module with temperature, heartbeat, touch sensors and also GPS, GSM & digital camera modules. It automatically alerts the parent/care taker by sending SMS during emergency situations ***Results***: The parameters such as touch, temperature, and heartbeat of the child are used for parametric analysis. It ensures the safety and tracking of children	• For only child safety

(continued)

Table 1 (continued)

Citation	Approach	Issues
[9]	***Aim***: To build a productive half breed cloud structure system combined with Li-Fi communication for a human-driven IoT network ***Methodologies***: It calculates the neighborhood cloud to diminish the dormancy deferral and transmission capacity cost and improves the proficiency, security, unwavering quality, and accessibility ***Results***: The proposed scheme reduces the performance of system in terms of bandwidth costs and latency delay by using Li-Fi communication scheme. Also, supports the communication medium at the edge of the network	• Energy harvesting technique to build green cloud computing
[10]	***Aim***: To develop a smart wearable device like a smart jewelry bracelet which is used to automatically sense, detect, and identify physical assault ***Methodologies***: It utilizes the following modules like sensor module, analyze, and classifier unit module, communication module, native presentation module, and a phone-based App agent module. The Ada Fruit controller has implanted 3-pivot accelerometer, spinner, flex sensor, and temperature sensor. It utilizes sensors like large number and furthermore utilizes AI strategy to identify an ambush as it happens and afterward continues to contact crisis administrations ***Results***: It gives great convenience and adequacy in furnishing a minimal effort with aiding of old clients	• For only women safety
[11]	***Aim***: To develop human safety devices to follow the human exercises with the assistance of sensors and furthermore track down any surprising movements using sensor and machine learning algorithms ***Methodologies***: Implemented with a machine learning algorithms for sensor data analysis. It uses the various sensors for human wellbeing gadgets to identify strange exercises ***Results***: It provides good performance and also it gives a solitary answer for some issues from various areas	• Requires human intervention to serve better

(continued)

Table 1 (continued)

Citation	Approach	Issues
[12]	**Aim**: To present a wearable IoT sensor node to monitor the hurtful ecological conditions for security applications by means of LoRa remote innovation ***Methodology***: A LoRa door used to interface sensors for the Internet access. The framework center is used to check the carbon monoxide, carbon dioxide, bright, and some broad ecological parameters. Poor condition quality could make serious medical issues people ***Results***: It demonstrates the well-being observing system can work dependably with low force utilization	• Smart wearable device for little children

applications, and gesture-based controller with less efficient for communications in emergency situations. The proposed device over comes the current device limitations and issues that provides the new way of insights toward the emergency services to humans.

3 Proposed Work

The proposed system is used to provide emergency services to the users in the real world environment. This system is integrated with three apparatus mainly wearable finger ring (WFR) device, smart mobile phone device, and emergency service systems as shown in Fig. 1. The emergency service systems like the centralized agency, police,

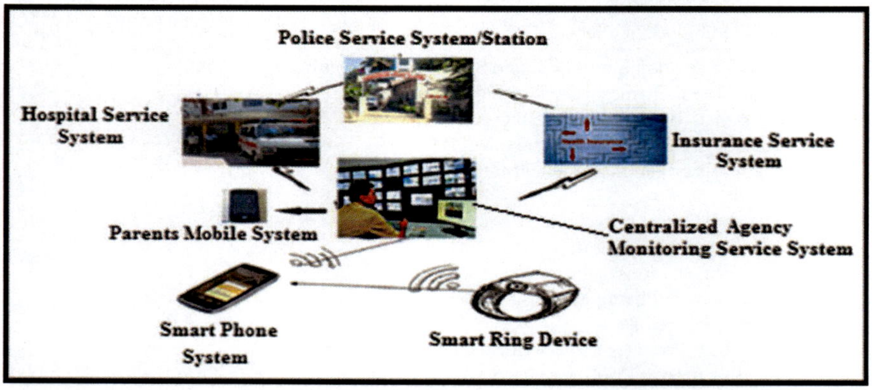

Fig. 1 Overview of the proposed system architecture

hospital, and health insurance service system. In this architecture, each service system will communicate with each other for exchanging emergency information with one another regularly. Also it has knowledge base for storing emergency information for further processing of data in a real-time environment. The proposed agency service system is the main service system that controls all the service systems during emergency situations. In this work, the ring device is an interface between all service systems, as and when required information is to be sent and receive to and from in real-time environment.

In this work, the smart ring device has a panic button that has been used in emergency situations such as kidnap cases, health issues, accidents, women harassments, etc., by pressing the panic button on the device to start sending the emergency information to the agency service systems during an emergency situation. The proposed agency system will get an alert message (emergency) by the device. Based on the context data, the proposed agency system will analyze, if the agency team has found the context data or situation is an emergency, then it will send the notification along with the location of the person to the respective sub-system as well as an alert message to the parents or registered dependent users phone numbers to know the situation. Also, at the same time, the device sends the emergency context data to nearby police station and ambulance service system in the area. The agency system continuously monitors the progress of the emergency services by the other service systems. Finally, the agency service system will get acknowledgment from all the service systems once they have taken the action on the particular emergency situation of a user.

Also, the proposed agency system sends the message to the nearest police service system to automatically register the complaint on behalf of an emergency user. If any emergency situation happens in the real world environment, for example major road accident, meanwhile the agency service system will send location details to the hospital for emergency services. This will help doctors to be in ready position to handle the situation quickly. All services will take place through online within a minute and which helps to prepare the required thing in advance and the ambulance service will pick up the patient from location and rush into the hospital, where a special doctor team will handle and operate to save the life of patients or victims. Parallelly, the agency service system verifies whether the victim or patient has made a health insurance policy if so, it claims and automatically settles the medical expenses without knowledge of victim or patient. Therefore, all these service will be provided within very less time from the time of the incident to hospital for saving the lifeline of individuals. Also, the proposed system is used to handle health issues, attention toward the aged individuals, etc., in the real world environment.

The proposed ring device sends a signal for an emergency alert that advises family, companions, and crisis service systems when the user is in the trouble. The overview of the system shows, how it interacts with all the service sub-systems? and also communicates with each other. Regardless of whether you are in a tough situation, encountering a medical problem or being assaulted, it is there to call for help when you need it most.

The objectives of the proposed system is as follows: (1) To develop a ring device for humans in various emergency situations; (2) To develop an agency system for real-time monitoring; (3) To provide urgent notification to parents, police, and ambulance in a real-time environment and (4) To identify the location and records the video and audio. In this work, the agency service system will be the main service system or central service system which sends the emergency alert to the other service systems as well as parents, relatives or loved ones. The proposed system consists of mainly five modules such as: Wearable Finger Ring Device System (WFR device), Agency Service System, Police Service System, Hospital Service System (Ambulance service), and Health Insurance Service System. The descriptions of each of system are described below:

- *Wearable Finger Ring Device System*—This ring device system contains a button which is used to send the context aware data to the agency service system along with the location of the user. Also, this device consists of two main sub-modules, camera module and compact module system. The camera module consists of a camera that provides live streaming of the video. Compact module system consists of various hardware components such as Arduino, Wi-Fi module, GPS module, LCD display, power supply, etc. This system also sends the emergency text message to its emergency service systems.
- *Agency Service System*—This system receives the real-time context data which is sent by the ring device system. It analyses the context data and correspondingly sends the alert message and location of the emergency user to other service systems as well as parents. This is the main service system among all the service systems and whenever the user is in an emergency, this system first receives the alert notification from the device. This system receives the location of the user along with the information of the user who is in an emergency, information such as name, address, mobile number, E-mail ID, picture of the user, and the location in terms of latitude and longitude.
- *Police Service System*—This system receives the emergency alert only when the user is in an emergency. It can view the details and location of the user so that they can take action and send the acknowledgement to the agency system. Also, it receives the alert notification from the agency service system whenever the users in emergency situations like harassments, kidnap cases, and accidents.
- *Hospital Service System*—This system receives the alert message from the agency service system only when the user has undergone some severe health problems. It can view the details and location of the user so that they can take action to save the user. The hospital system can send the acknowledgment to the agency system. This is one of the service systems for the proposed system. It receives the alert notification from the agency service system whenever the user is having health problems as well as accident situations. It receives the information and current location of the user who is in an emergency situation.
- *Insurance Service System*—The health service system receives the alert message from the agency service system. This system checks the policy status of the health insurance of the emergency user. If found such information of the user, the system

will be paid the hospital charges who are in under emergency situations immediately through the online system. Also send the acknowledgment to the agency system.

4 Conclusion

The proposed system is used to provide real-time emergency services to different users like children, women, and aged individuals. The system acts like a "Life-Saving System". Hence the proposed system is efficient, reliable, and secure for all services. As per the literature survey, we found that no such similar type of working mechanism devices available in the market. The proposed system is used to provide multiple services to users during emergency situations. As a future work, using artificial intelligence and machine algorithms are used to provide emergency services to the users effectively.

Acknowledgements This research work was financially supported by the Visvesvaraya Technological University (VTU), Belagavi under the TEQIP Competitive Research Grant Scheme (RGS). The authors would like to thank Vice-Chancellor, Registrar (Academics), Registrar, Registrar (Evaluation), TEQIP Coordinator, Management, Principal of KLE College of Engineering and Technology, Chikodi, and Basaveshwar Engineering College, Bagalkot for their help, support, and guidance during the research work.

References

1. P.P. Ray, A survey on internet of things architectures. J. King Saud Univ. Comput. Inf. Sci. **30**, 291–319 (2018)
2. L. Atzori, A. Iera, G. Morabito, The internet of things: a survey. J. Comput. Netw. **54**(15), 2787–2805 (2010)
3. T. Kumar, A. Braeken, A.D. Jurcut, M. Liyanage, M. Ylianttila, AGE: authentication in gadget—free healthcare environments. Int. Inf. Technol. Manage. **12**(6), 1–22 (2019)
4. S. Selvaraj, S. Sundaravaradhan, Challenges and opportunities in IoT healthcare systems: a systematic review. Int. J. SN Appl. Sci. **2**, 1–8 (2019)
5. M. Al-khafajiy, H. Kolivand, T.B D. Tully, A. Waraich, Smart hospital emergency system via mobile-based requesting services. Int. J. Multimedia Tools and Appl. **22**(21), 20087–20111 (2019)
6. B. Sumathy, P.D. Shiva, P. Mugundhan, R. Rakesh, S. Sai Prasath, in *Virtual Friendly Device for Women Security*. International Conference on Physics and Photonics Processes in Nano Sciences, pp. 1–13 (2019)
7. B. Sathyasri, U.J. Vidhya, G.V. K. JothiSree, T. Pratheeba, K. Ragapriya, Design and implementation of women safety system based on IoT technology. Int. J. Rec. Technol. Eng. **7**(63), 177–181 (2019)
8. M.N. Priyanka, S. Murugan, K.N.H. Srinivas, T. D.S. Sarveswararao, E. KusumaKumari, Smart IoT device for child safety and tracking. Int. J. Innov. Technol. Expl. Eng. **8**(8), 1791–1795 (2019)
9. P. Sharma, J. H. Ryu, K.Y. Park, J. Ho, J. Park, H. Park, Li–Fi based on security cloud framework for future IT environment. Int. J. Hum. Centric Comput. Inf. **18**(15), 1–13 (2018)

10. J. Patel, R. Hasan, in *Smart Bracelets: Towards Automating Personal Safety using Wearable Smart Jewellery*. 15th IEEE Annual Consumer Communications & Networking Conference (CCNC), pp. 1–2 (2018)
11. K. Sharma, D.D. Londhe, in human safety devices using IoT and machine learning: a review. 3rd International Conference for Convergence in Technology (I2CT), pp. 1–7 (2018)
12. F. Wu, C. Rüdiger, J. Redouté, M.R. Yuce, in *WE-Safe: A Wearable IoT Sensor Node for Safety Applications via LoRa*. IEEE 4th World Forum on Internet of Things (WF-IoT), pp. 144–148 (2018)

Classification of Bharatnatyam and Kathak Dance Form Through Audio Signal

Sanghamitra Das, Suchibrota Dutta, Debanjan Banerjee, and Arijit Ghosal

Abstract This work proposes a simple method to classify two forms of Indian classical dance—Bharatnatyam and Kathak. These two dance forms are one of the popular dance forms of India. Most of the prior endeavors for classification of different dance forms were based on pose or facial expression of artists involving processing of lots of images. Some works were based on video of dance. These are time-taking process. To minimize classification time, this effort relies on audio signal. Processing of audio signal does not require much amount of time. Spectral flux is a very good aural facet which points toward the fact that how fast power spectrum of an audio signal changes from one frame to its adjacent frame. Power spectrum is dependent on the energy stored. Theory of co-occurrence matrix generation has been used on spectral flux to clear-cut study its nature. Statistical facet extorted from the co-occurrence matrix also considered in this work. Bharatnatyam and Kathak dance forms maintain certain breed of unlikeness between them. Skewness is an aural facet which principally points out aural dissimilarity present within different kinds of aural data set. Skewness is also applied in this exertion to incarcerate the certain breed of unlikeness between Bharatnatyam and Kathak. Support vector machine or SVM,

S. Das · A. Ghosal (✉)
Department of Information Technology, St. Thomas' College of Engineering and Technology, Kolkata, West Bengal 700 023, India
e-mail: ghosal.arijit@yahoo.com

S. Das
e-mail: dassanghamitra1708@gmail.com

S. Dutta
Department of Information Technology and Mathematics, Royal Thimphu College, Thimphu, Bhutan
e-mail: suchibrota@gmail.com

D. Banerjee
Department of Management Information Systems, Sarva Siksha Mission Kolkata, Kolkata, West Bengal 700 042, India
e-mail: debanjanbanerjee2009@gmail.com

J. Tavares et al. (eds.), *Emerging Technologies in Data Mining and Information Security*, Lecture Notes in Networks and Systems 164,
https://doi.org/10.1007/978-981-15-9774-9_62

RepTree (a kind of Decision Tree), and Naïve Bayes supported classifier have been made use of to differentiate Bharatnatyam and Kathak dance forms. This work has been judged against some preceding exertions to show up the pros of the proposed idea.

Keywords Classical dance · Bharatnatyam · Kathak · Classification · Spectral flux · Co-occurrence matrix · Skewness

1 Introduction

India possesses varied types of culture from long ago better to say from very prehistoric age. India is formed with many numbers of states each having diverse types of culture, lingo. Even performing arts is also diverse among these states. India is also known to be a land of the most ancient and traditional performing arts. In the domain of classification of different dance forms of India, not very much research efforts were done so far. Bharatnatyam and Kathak are two well-admired classical dance forms of India. Bharatnatyam is originated from South India and Kathak is originated from North India. So, if India is divided into two geographical divisions—North and South, Bharatnatyam and Kathak are the best dance forms to represent the performing arts of these two geographical regions. Instruments supporting Kathak dance falls under Hindustani category of Indian classical music [1]. Kathak dance has links with Moghuls [2]. These dance forms have distinct *Tala* (beats), *Raga* (musical framework) and diverse musical instruments.

Bharatnatyam is almost 2000 years old [3]. This dance expresses various South Indian religious and spiritual themes. Bharatnatyam dance is accompanied by music and a singer. This dance is supported by several musical instruments—'*Mridangam*', '*Ghatam*', '*Flute*', '*Violin*' and '*Veena*' [4]. Dancers use '*Ghungroos*' [4]. Rhythmic foot movement is the major emphasis in Kathak dance. Kathak mainly describes stories basically dependent on incidents from the myths, epics, and legends [1]. Kathak dance is accompanied by quite a few numbers of musical instruments—'*Tabla*' (a pair of hand drums), '*Paquwaj*', '*Sarangi*' and '*Tanpura*' [2]. Occasionally, '*Flute*' or a '*Clarinet*' is also used in Kathak dance [2].

Bharatnatyam and Kathak dance, representing performing arts of two large geographical regions of India, further can be sub-divided into different '*Gharanas*' (school) once they are classified. Previous efforts dealing with classification of classical dance forms involved processing of images and videos. These are time-eating process. Execution of those methods for huge number of images or video is wearisome. For faster classification of classical dance form, audio-based approach is much more reliable as it does not take much time to extract aural facet from each of the data. As this work is planned for categorization of Bharatnatyam and Kathak dance, not all aural facets can be employed here. Well-judged aural facets need to be used here. Facets which can distinguish Bharatnatyam and Kathak dance only those need to be applied here. Dance form discrimination is enormously inspirational and excellent research domain to survey as it finds its usage in search engines, digital documentation of dance forms comprising smart human/computer interactive structures. Progress in the domain of data mining and signal processing is an added motivation of research bustles in the area of aural data practicing.

This article comprises of some sections—after introduction section preceding research works are conferred in next section. Section 3 illustrates methodology of the scheme. Investigational consequences in conjunction with relative examination with preceding endeavors are conferred in Sect. 4 and subsequently the conclusion segment.

2 Literature Analysis

Previously, many researchers have paid their endeavors to put forward a good facet set to catalogue different forms of Indian dances. Many of these exertions were based on either by pose of the performers or by facial expressions generated by them during dance. Very few efforts involved classification of dance forms through audio signal.

Indian classical dance actions have been categorized by Kishore et al. [5] by making use of convolutional neural network (CNN)-based approach. Pattnaik and Samantaray [4] have performed a comparative analysis between Odissi and Bharatnatyam dance. Indian classical dance is separated with Adaboost multiclass classifier by Kumar et al. [6]. Bisht et al. [7] have recognized Indian dance form videos. Bhatt and Patalia [8] have discriminated Indian folk dance song employing mel-frequency cepstral coefficients (MFCC) and linear predictive coding (LPC). Their work were limited to only 4 categories of Indian folk dance—'Garba', 'Lavani', 'Ghoomar' and 'Bhangra'. Tiwary [9] has classified Indian classical dance forms based on pose. Bharatnatyam, Odissi and Kathak were considered under Indian classical dance form. Mohanty et al. [10] have also worked with Indian classical dance. Devi et al. [11] have carried out a review to be aware of dance gestures. Samanta and Chanda [12] have used local spatiotemporal feature model on manifold system for discrimination of diverse styles of Indian classical dance.

Different hand gestures of Bharatanatyam dance have been identified by Saha et al. [13]. Kapsouras et al. [14] have worked with Greek folk dances through video. For classification purpose, support vector machine or SVM along with clustering was applied in their approach. Saha et al. [15] have spotted different hand gestures of Bharatnatyam dance depending on fuzzy L membership function. Lalli [2] has shared knowledge on Kathak, especially Lucknow Kathak. Statistical facet extraction process from co-occurrence matrix was suggested by Haralick [16]. Center for Cultural Resources and Training (functioning under the auspices of Ministry of Culture, Government of India) has discussed on Kathak [1] and Bharatnatyam [3].

3 Suggested Design

Preceding study bustles indicate that prior research ventures were largely derived from video and skeletal information like body pose and facial appearances. Facet mining from skeletal information necessitates image processing. This makes the whole process slower as well as complicated. Facet mining from video data has several shortcomings. Firstly, it entails hefty storage space which is costly. Next, if facet mining has done during runtime from websites, it is too much intricate and possibilities of unforeseen faults are huge. Also facet mining from video records is time captivating. There are very few moves which have done facet mining from aural signal. Facet mining from aural data has several benefits. Firstly, it requires small storage compared to video records. Also it is simple and speedy compared to image processing and video data processing as it requires less number of moves. This is one impetus for preferring aural facet set in this work.

Bharatnatyam dance employs '*Mridangam*', '*Ghatam*', '*Flute*', '*Violin*' and '*Veena*' with '*Ghungroos*' [4]. But Kathak dance involves '*Tabla*' (a pair of hand drums), '*Paquwaj*', '*Sarangi*' and '*Tanpura*' along with sporadic use of '*Flute*' or a '*Clarinet*' [2]. Power spectrum variation of the instruments linked with Bharatnatyam and Kathak dance form are different. This variation is well monitored through spectral flux.

Presentation of Bharatnatyam and Kathak dance as well as instruments related with these two dance forms makes these two dance forms dissimilar when it is heard. This dissimilarity is properly reflected through Skewness. Aural facet mining and categorization of aural signal into two diverse dance forms are explicated in the ensuing sub-segments. The steps of the advised approach are elucidated in Fig. 1.

3.1 Computation of Features

Any classifier can exhibit high classification performance if it is fed with fine facet set. A facet set is said to be fine when it holds the following characteristics:

1. any fine facet set should be invariance to irrelevancies like noise

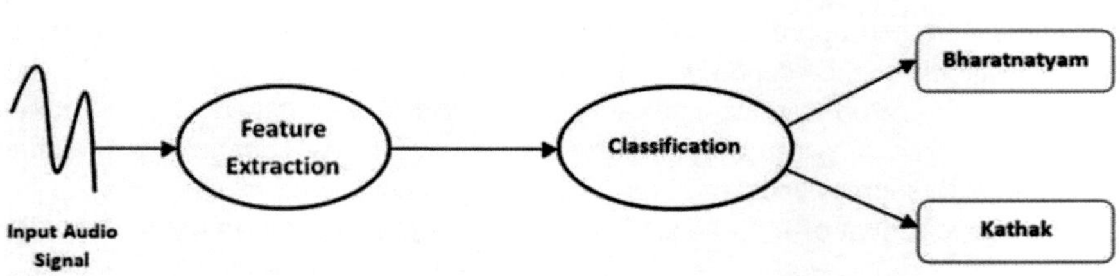

Fig. 1 Steps of the advised approach

2. high discriminative power means it should preserve alike values inside identical category but dissimilar values across different categories. It should not contain unnecessary facets which have low discriminative power or no discriminative capacity
3. length of the facet set should not be too large to minimize computational expense.

3.1.1 Facets Based on Spectral Flux

Power spectrum of a signal alters from one spectrum to its next spectrum. How rapid this alteration arises that is assessed by spectral flux, sometimes termed as spectral variation. Actually difference of the assessment of spectrum amid two successive frames is assessed by spectral flux. It is calculated by applying Eq. (1):

$$\mathrm{SpF}(h) = \sum_{j=0}^{n-1} s(h, j) - s(h-1, j) \tag{1}$$

SpF(h) symbolizes spectral flux of hth spectrum. $s(h, j)$ symbolizes the assessment of the hth bin in the hth spectrum. $s(h - 1, j)$ symbolizes the identical thing corresponding to a spectrum which arises before h. Assessments of every bin of the earlier spectrum are deducted against the value of the analogous bin in present spectrum and then addition of those variations takes place to quantify spectral flux. In this fashion, spectral flux of the entire spectrum is assessed. Mean and standard deviation are judged to capture the overall idea of spectral flux.

To accurately assess the characteristics of spectral flux, a co-occurrence matrix of spectral flux is needed as this matrix is capable to return repetitive character of its successive assessments. Statistical facets recommended by Haralick [16] are assessed from the matrix. A co-occurrence matrix, CO_SF_m, having dimension $T \times T$ (where, $T = \max\{\mathrm{SpF}_i\} + 1$) is summarized. Any element in the matrix $CO\_SF_m(r, v)$ echoes repetition figure of spectral flux r and v in successive cases. Entropy, energy, correlation, contrast, along with homogeneity—these statistical facets [16] are assessed from the co-occurrence matrix CO_SF_m.

Spectral flux co-occurrence nature for Bharatnatyam and Kathak dance are portrayed in Figs. 2 and 3 respectively. Figures 2 and 3 discloses that happening nature of spectral flux is diverse for Bharatnatyam and Kathak dances.

3.1.2 Facet Based on Skewness

Skewness is a statistical facet which assesses the dissimilarity present for a standard distribution in contrast to its mean location. Bharatnatyam dance maintains this dissimilarity when judged against Kathak dance. This dissimilarity is produced owing to different supporting instruments. Skewness skw is enumerated using Eq. (2).

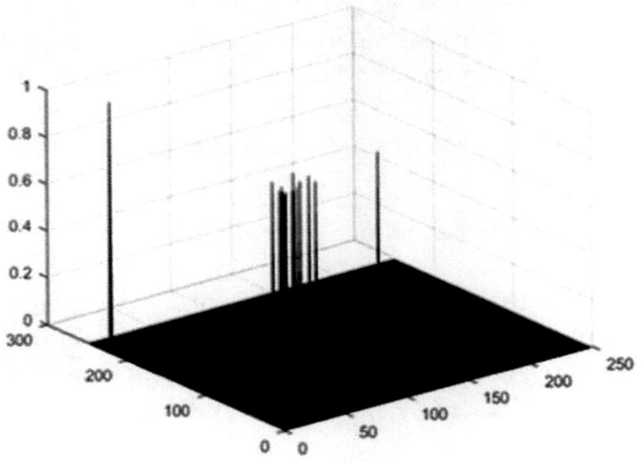

Fig. 2 Co-occurrence nature of spectral flux plot for Bharatnatyam

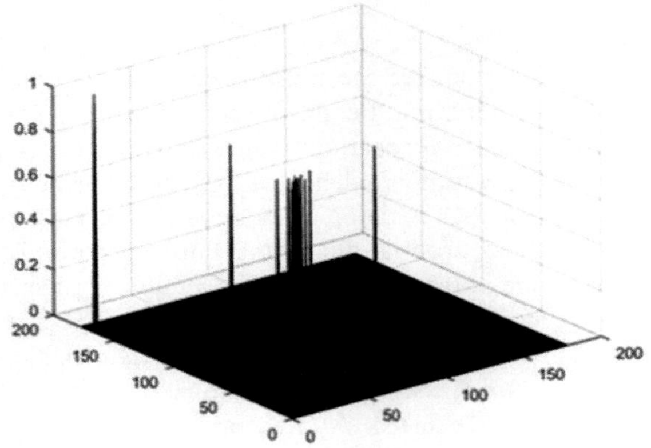

Fig. 3 Co-occurrence nature of spectral flux plot for Kathak

$$\mathrm{skw} = \frac{E(da - \mu)^3}{\sigma^3} \tag{2}$$

Here, mean of sampled data *da* is symbolized by μ, standard deviation of *da* is symbolized by σ, and expected assessment of quantity z is symbolized by *E(z)*.

3.2 Classification

Main endeavor of this work is to stratify Bhatarnatyam dance forms from Kathak dance forms by means of easy and tiny dimensional yet powerful facet set. Hence, plain classification design comprising of support vector machine or SVM, RepTree (a kind of decision tree), and Naïve Bayes has been used here. Stratification power of the advised eight-dimensional facet set (mean and standard deviation of spectral flux

+ 5 statistical facets extracted from spectral flux co-occurrence matrix + 1 skewness depended facet) is examined by feeding it to the classifiers. A data set has been constructed having 600 audio files—300 files for Bharatnatyam dance form and 300 files for Kathak dance form. Half of the dataset is used to coach the classifiers and rest is used for testing resulting coaching dataset consisting of 300 files where 150 files each for Bharatnatyam and Kathak dance forms. While implementing SVM, 'Quadratic' kernel is used. tenfold cross-validation is considered for both Naïve Bayes and RepTree which is a kind of decision tree.

4 Experimental Results

To accomplish the job of classification of Bharatnatyam and Kathak dance form using aural signal, a data set has been prepared. This dataset cuddles 600 aural files—300 files for Bharatnatyam dance forms and 300 files for Kathak dance forms. All files in the dataset are mono kind and all of them have 90 s length. These aural files are mounted up from assorted sources like recordings in compact disc, aural recordings of various live performances in addition to YouTube. To judge the capacity of the advised facet set, a few data of this data set is noisy to symbolize realistic circumstances.

All sound files in the dataset are segregated into 50% overlie frames just to curtail loss of any periphery characteristic of a certain frame. For coaching support vector machine or SVM, RepTree (a kind of decision tree), and Naïve Bayes supported classifier half of data set is acquired. Rest of the data set is acquired for the purpose of testing those classifiers. Once categorization task is completed, coaching and investigating dataset are overturned and categorization task is accomplished for a second time. Average of these two rounds of categorization task is judged as ultimate categorization outcome. This ultimate categorization outcome is put into Table 1.

Table 1 Bharatnatyam dance form and Kathak dance form stratification accurateness

Categorization method	Stratification accurateness (in %) for advised facet set
RepTree (a kind of decision tree)	80.50
SVM	79.00
Naïve Bayes	78.50

Table 2 Comparative examination of advised work with former exertion in % correctness

Precedent method	Categorization precision (in %)
Bisht et al. [7]	75.83
Advised method (RepTree based)	80.50

4.1 Comparative Examination

This exertion has been contrasted with previous effort. Proposed method has been evaluated against the endeavor done by Bisht et al. [7]. They have recognized Bharatnatyam, Odissi, Kathakali, Kathak, Kuchipudi, Sattriya, Manipuri, and Mohiniyattam dance form using 626 videos where 82 videos were for Bharatnatyam and 99 videos were for Kathak. They have used dataset collected from YouTube similar to this work. They have used multi-class SVM but in this work linear SVM has been used as there are only categories—Bharatnatyam and Kathak. The proportional performance is tabularized in Table 2.

Experimentally, it is observed that average facet computation time by the method proposed by Bisht et al. [7] is 3.92 s for a single video file whereas for proposed method, it is 1.16 s for a single audio file.

Hence, it is noticeable that advised facet set surpasses the scheme suggested by Bisht et al. [7].

5 Conclusion

For discriminating Bharatnatyam and Kathak dance forms through audio signal, a simple idea has been proposed for good feature set based on spectral flux and skewness. The concept of co-occurrence matrix has been wielded in the proposed work for analyzing the facet set more precisely. Investigational consequences prove the capability of this facet set compared to other preceding exertions. To highlight the strength of the intended facet set, an easy classification design based on support vector machine or SVM and Naïve Bayes has been taken into deliberation. In the future, further sub-classification of both Bharatnatyam and Kathak based on '*Gharana*' may be explored.

References

1. http://ccrtindia.gov.in/kathak.php
2. G. Lalli, A North Indian classical dance form: Lucknow Kathak. Visual Anthropol. **17**(1), 19–43 (2004)
3. http://ccrtindia.gov.in/bharatnatyam.php

4. S. Pattnaik, S.P. Samantaray, Comparative study on classical dances: odissi & bharatanatyam. World Wide J. Multi. Res. Develop. **3**(12), 147–150 (2017)
5. P.V.V. Kishore, K.V.V. Kumar, E. Kiran Kumar, A.S.C.S. Sastry, M. Teja Kiran, D. Anil Kumar, M.V.D. Prasad, Indian classical dance action identification and classification with convolutional neural networks. Adv. Multimedia **2018**(5141402), 10 (2018). https://doi.org/10.1155/2018/5141402
6. K.V.V. Kumar, P.V.V. Kishore, D. Anil Kumar, Indian classical dance classification with adaboost multiclass classifier on multifeature fusion. Math. Prob. Eng. **2017**(6204742), 18p (2017). https://doi.org/10.1155/2017/6204742
7. A. Bisht, R. Bora, G. Saini, P. Shukla, in *Indian Dance Form Recognition from Videos*. 2017 13th International Conference on Signal-Image Technology & Internet-Based Systems (SITIS). IEEE, pp. 123–128, Dec 2017
8. M. Bhatt, T. Patalia, Neural network based indian folk dance song classification using MFCC and LPC. Int. J. Intell. Eng. Syst. **10**(3), 173–183 (2017)
9. U.S. Tiwary, in *Classification of Indian Classical Dance Forms*. International Conference on Intelligent Human Computer Interaction. Springer, Cham, pp. 67–80, Dec 2016
10. A. Mohanty, P. Vaishnavi, P. Jana, A. Majumdar, A. Ahmed, T. Goswami, R.R. Sahay, Nrityabodha: towards understanding indian classical dance using a deep learning approach. Sig. Process. Image Commun. **47**, 529–548 (2016)
11. M. Devi, S. Saharia, D.K. Bhattacharyya, Dance gesture recognition: a survey. Int. J. Comput. Appl. **122**(5) (2015)
12. S. Samanta, B. Chanda, in *Indian Classical Dance Classification on Manifold Using Jensen-Bregman Logdet Divergence*. 2014 22nd International Conference on Pattern Recognition. IEEE, pp. 4507–4512, Aug 2014
13. S. Saha, A. Konar, D. Gupta, A. Ray, A. Sarkar, P. Chatterjee, R. Janarthanan, in *Bharatanatyam Hand Gesture Recognition Using Polygon Representation*. Proceedings of The 2014 International Conference on Control, Instrumentation, Energy and Communication (CIEC). IEEE, pp. 563–567 (2014)
14. I. Kapsouras, S. Karanikolos, N. Nikolaidis, A. Tefas, in *Folk Dance Recognition Using a Bag of Words Approach and ISA/STIP Features*. Proceedings of the 6th Balkan Conference in Informatics, pp. 71–74, Sept 2013
15. S. Saha, L. Ghosh, A. Konar, R. Janarthanan, in *Fuzzy L Membership Function Based Hand Gesture Recognition for Bharatanatyam Dance*. 2013 5th International Conference and Computational Intelligence and Communication Networks. IEEE, pp. 331–335, Sept 2013
16. R.M. Haralick, L.G. Shapiro, in *Computer and Robot Vision*, vol. I (1992)

SWOT Analysis of BYOD (Bring Your Own Device)

P. Soubhagyalakshmi and K. Satyanarayan Reddy

Abstract Bring your own device (BYOD) grants delegates to use their own gadgets, application, and framework for business or work reason, with an idea of extending productivity, reduces the IT costs, and adjusting undertaking system. A SWOT assessment of BYOD is furthermore a result to provide a succinct audit of favorable circumstances with downsides of BYOD technology. Furthermore, an examination of a modification concept is given which explains a comprehensive diagram of various needs of IT industry and their unmistakable security level over their data and structure for executing BYOD. At the present time, we are focusing on conduction of SWOT examination of BYOD.

Keywords Bring your own device (BYOD) · SWOT · Technology · Strategy · Security

1 Introduction

Bring your own device (BYOD) is another example that is being picked by a creating number of IT firms for reducing costs and extending benefit. Extremely various IT associations raised an admonition for executing BYOD concept, essentially in light of in any event one security concerns [1]. Regardless, starting late, IT firm required to have guidelines with the need of BYOD technology and is setting ready in realizing BYOD approaches or head-ways. Close to the end, it is about end customers having the choice to use enrolling and furthermore specific devices of their tendency to fabricate effectiveness and adaptability [2, 3].

P. Soubhagyalakshmi (✉)
Visvesvaraya Technological University, Belagavi, India
e-mail: slakshmi.p@gmail.com

K. Satyanarayan Reddy
ISE Department, Cambridge Institute of Technology, Bengaluru, India
e-mail: ksatyanreddy@yahoo.com

J. Tavares et al. (eds.), *Emerging Technologies in Data Mining and Information Security*, Lecture Notes in Networks and Systems 164,
https://doi.org/10.1007/978-981-15-9774-9_63

1.1 What is BYOD Technology?

Bring your own device (BYOD) is an idea where end users can customize their own gadget(s) or program(s) or communication channel(s) for business/work reason. The IT circumstances with resources are contributed to the beginning plan of IT structures fuse PCs, frameworks, Wi-Fi, etc. Alternately, undertaking agent uses the own gadgets to work, the system cost of the endeavor is decreased and laborer feel satisfied while using their own devices just as PC applications. Suppose, an endeavor using a default program for picture changing is Microsoft paint. A specialist uses another picture changing program as he is convenient in using it. Directly, when the delegate fathoms that the standard program used in the undertaking is not more powerful, the specialist differences the his own picture modifying program, and may push toward the endeavor to use his own picture adjusting program for work; thus, the laborer uses his own picture changing program for work purpose through executing authentic security measures and implied as bring your own application(BYOA).

1.2 Advantages of Bring Your Own Device (BYOD)

Extended productivity: More productivity achieved in less time as BYOD grants agents to use their own devices or download application(s) for business reason.
Improved specialist solace and affirmation: As the laborer can uses their gadgets, the time needed for alteration is basically diminished.
Diminished IT establishment cost: As the agent of an endeavor uses their gadgets which seriously decrease fast IT system cost.
Versatile Working Culture: An endeavor supports a laborer to use their gadget(s)/program(s), then delegates posses a versatile circumstance to work in business society.

2 Literature Survey

BYOD is a mechanical example that proposes understudy centered learning. Phones in a BYOD circumstance enable tutors to assess understudies by creating for all intents and purposes fast information [4]. BYOD licenses understudies the tutors to use their imaginative devices that support to understand study objectives and demonstrating logically capable and beneficial [5]. In customary preparing, the principle wellspring of data inside an examination lobby was the educator, without his quality the investigation corridor is only an unfilled live with no resource for learn. With BYOD, each understudy approaches information from this world, for instance, events, places, affiliations, people, and various understudies authentically [6]. Studying about the BYOD circumstance will by and large be straightforward and capable since customers think about their own devices, rather than making sense of how to function another device. This segment can add to progressively conspicuous study in less time [7–9]. With the help of BYOD, educators and understudies are learning together, thus making correspondence, collaboration, essential thinking, and imagination [5]. Regardless of the way that the BYOD design got vitality in affiliations and their delegates, on account of various central focuses, for instance, extended customer productivity, diminished costs in obtaining of hardware and programming, extended transport ability, versatility, specialist work satisfaction [10]. Starting at now, various exercises that incorporate the BYOD example to give new and better strategies for training and study objectives learning [11, 12]. A couple of assessments ensure that BYOD may benefit understudy responsibility and participation [13]. Understudies invite the affiliation that BYOD offers among tutors for development to their learning [14]. Similarly, BYOD provides propelled open entryways to self-ruling studying, all through class and supports advancement in training and studying [5]. A noteworthy activity in guidance, teachers notwithstanding everything miss the mark on the fundamental capacities to execute a constructive approach maintained by technology and this leads understudies to get reinforce compelled to make essential aptitudes [15, 16]. Disregarding the way that the preferences are obvious, there are a couple of associations of cutting edge training power a hard and fast limitation on the usage of PDAs by understudies inside the homerooms. These gadgets interference an issue for the organization [17]. Moreover [17] shows that couple of instructors concede as no information on the focal points in learning of educating capabilities. Regardless of the impressive number of central focuses showed in the assessment presented, there are in like manner imperatives and hindrances that avoid the apportionment of BYOD in education and learning circumstances reliant on m-learning. Subsequently, SWOT examination incorporating personal gadgets in m-learning is required.

3 SWOT Analysis

SWOT Analysis of BYOD is discussed as follows:

3.1 Strengths

1. More productivity
2. More work satisfaction and employee ethics
3. Convenient for work field
4. Decreased IT cost of association
5. Simple IT infrastructure
6. Agent settlement
7. Beneficial business gadgets/programs.

3.2 *Weakness*

1. Portability issue.
2. Security threats.
3. Affects affiliation as significant laborer leaves the firm with their gadgets.
4. Support for various sorts of programs/orchestrate accessibility/working gadgets with huge business.

3.3 *Opportunity*

1. Seeking the business information for new considerations
2. Opportunity in middleware and gadget the board technology
3. Growth in programming progression for BYOD gadgets compact/PDA/PC
4. Huge enthusiasm for memory the board, against contamination and unfriendly to spy-ware programming improvement technology
5. Huge enthusiasm for united synergistic instrument for managing various sort of application(s)/arrange accessibility/working structure with enormous business system(s) similarly as agent device(s).

3.4 *Threat*

1. Data/Gadgets stolen
2. Habitual attitude impact production benefits
3. Work in unsafe zone can be impacted (Fig. 1).

4 Data Analysis and Results

4.1 *Model Fit Summary*

Baseline Comparisons

Model	NFI	RFI	IFI	TLI	CFI
	Delta1	Rhol	Delta2	Rho2	
Default model	0.031	−0.615	0.031	−0.619	0.029
Saturated model	1.000		1.000		1.000
Independence model	0.000	0.000	0.000	0.000	0.000

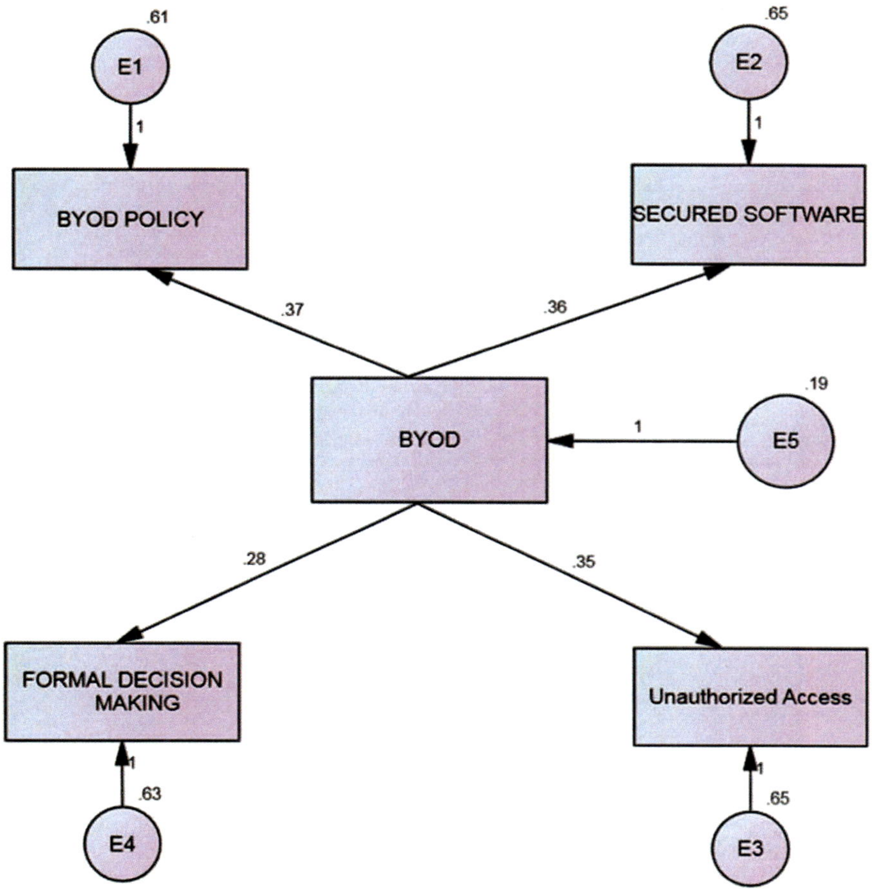

Fig. 1 SEM model

Parsimony-Adjusted Measures

Model	PRATIO	FNFI	PCFI
Default model	0.600	0.019	0.017
Saturated model	0.000	0.000	0.000
Independence model	1.000	0.000	0.000

Table 1 Summarry of SWOT analysis

Results of SWOT analysis of BYOD			
Strengths	Opportunities	Weaknesses	Threats
1. Mobility	1. High mobile penetration	1. Lack of training	1. Security of personal and institutional information
2. Ubiquity	2. Reduced costs of access to mobile internet	2. Excessive number of applications	2. Wide variety of devices and platforms
3. Accessibility			
4. Collaboration	3. Great amount of educational app based on games	3. technological obsolescence of institutional mobile infrastructure	3. Android OS security threat
5. Utility			
6. Privacy	4. Learning independence, improved focus in the classroom and the link with the teacher	4. Low educational support, testing, and applications	4. Costly, complex and incompatible applications on some mobile devices
7. Adaptability			
8. Portability			
9. Multiplatforni		5. Low acceptance of the interested parties	5. Culture change of traditional teaching
10. Flexibility	5. Interaction and instant feedback		
	6. Improvement of commitment and academic participation	6. Economic, technological, and institutional policy factors	
	7. Satisfaction of the mobile device user	7. Applications not compatible with multiple mobile devices	
		8. Usability (screen, battery)	

RMSEA (Root Mean Square Error of Approximation)

Model	RMSEA	LO 90	HI 90	PCLOSE
Default model	0.844	0.811	0.878	0.000
Independence model	0.664	0.638	0.690	0.000

See Table 1.

5 Conclusion and Future Work

A survey of the writing is a moment photo of the field examined at a given time. In spite of the fact that the writing search was a thorough procedure, numerous pertinent articles may have been disregarded. In spite of this, the exploration created pointers that show the significance of estimating the effect of BYOD on instructors and tending to the effect of academic methodologies vital, because of the assortment of devices, usefulness, working frameworks, and access to advanced apparatuses. The utilization of cell phones and the new advancements of data and correspondence, both by understudies and educators, creates critical changes, and better approaches for producing information. The utilization of BYOD is invaluable for some reasons, most importantly, improving access to Internet assets and computerized devices on the side of educating and learning. Regardless of the quantity of arrangements and versatile learning applications at present accessible, there are still a few constraints as far as ease of use, related costs, innovative foundation, intricacy, and so forth. For instance, there are numerous instructive applications accessible that solitary work on explicit convenient devices. This can expand the boundaries in the selection of BYOD in instructive establishments. What has been said above opens new developing standards that ought to be considered for future research. Understudies ought to be associated with instructive encounters that utilize cell phones, for instance a mix of advances (BYOD) and mobile crowd-sensing (MCS).

References

1. F. Giannakas et al., in *CyberAware: A Mobile Game-Based App for Cybersecurity Education and Awareness*. Proceedings of International Conference on Interactive Mobile Communication Technologies and Learning, Thessaloniki, Greece, pp. 54–58 (2015)
2. A.I. Han et al., in *Crowd Sourced Testing Through End Users for Mobile Learning Application in the Context of Bring Your Own Device*. Proceedings of 7th IEEE Annual Information Technology, Electronics and Mobile Communication Conference, Vancouver, BC, Canada, pp. 1–6 (2016)
3. T. Mahalingam, A.V. Rajan, in *Cloud and Mobile Computing: Affordances of the 21st Century Teaching and Learning*. Proceedings of International Conference on Current Trends in Information Technology, Dubai, United Arab Emirates, pp. 125–128 (2013)
4. S. Soman, P.K. Pareek, J. Phys. Conf. Ser. **1427**, 012010 (2020)
5. M. Stork et al., Global perspectives on BYOD: a reflection on the EdMedia 2015 conference. TechTrends **59**(6), 13–14 (2014)
6. G.K.W. Wong, Engaging students using their own mobile devices for learning mathematics in classroom discourse: a case study in Hong Kong. Int. J. Mob. Learn. Organ. **8**(2), 143 (2014)
7. I.T. Sanusi, S.S. Oyelere, in *Exploring Students and Teachers Activities, Experiences and Impact of Opón Ìmò Mobile Learning Device on Teaching and Learning*. IEEE Africon 2017 Proceedings, pp. 809–814 (2017)
8. M. Zappatore et al., in *Mobile Crowd Sensing-Based Noise Monitoring as a Way to Improve Learning Quality on Acoustics*. Proceedings of International Conference on Interactive Mobile Communication Technologies and Learning, pp. 96–100 (2015)
9. P.K. Pareek, T.V. Nandikolmath, P. Gowda, FMEA implementation in a foundry in Bangalore to improve quality and reliability. Int. J. Mech. Eng. Robot. Res. **1**(2), 81–87 (2012)
10. N. Zahadat et al., BYOD security engineering: a framework and its analysis. Comput. Secure. **55**, 81–99 (2015)
11. T. Nandikolmath, P.K. Pareek, S.A. Vasantha Kumara, Implementation of a lean model for carrying out value stream mapping in manufacturing industry. Int. J. Mech. Eng. Robot. Res. **1**(2), 88–95 (2012)
12. P.K. Pareek, P. Gowda et al., Ergonomics in a foundry in Bangalore to improve productivity. Int. J. Eng. Soc. Sci. **2**(5), 1–6 (2012). ISSN: 2249-9482

13. Y. Song, Y. Wen, Integrating various apps on BYOD (Bring Your Own Device) into seamless inquiry-based learning to enhance primary students' science learning. J. Sci. Educ. Technol. **39**(17), 1–12 (2017)
14. C. Dobbins, P. Denton, MyWallMate: an investigation into the use of mobile technology in enhancing student engagement. TechTrends **61**(6), 541–549 (2017)
15. P.K. Pareek, in *Optimization of the Critical Parameters Affecting the Fuel Lid Performance* (2019). 2019-28-2413
16. P.K. Pareek, in *ROC Structure Analysis of Lean Software Development in SME's Using Mathematical CHAID Model*, 17 May 2019
17. R. Olasoji et al., in *Requirement Engineering for Effective Mobile Learning: Modelling Mobile Device Technologies Integration for Alignment with Strategic Policies in Learning Establishments.* Federated Conference on Computer Science and Information Systems, vol. 2, pp. 851–860 (2014)

A Novel Approach for Improving Security in Software Development in Small Software Firms: A Literature Review

Prakash Chandra Behera, Chinmaya Dash, and Piyush Kumar Pareek

Abstract Developing secure software is really a challenge now for all types of software industry like small or big industry. Security is an essential part of any type of software development. There is a number of drawbacks to our existing system of software development starting from conceptualization to the last phase of an SDLC Software security does not depend on maintaining security in each step of software development but also it depends on the technology we are using for maintaining the security of the software. In this paper, we did a literature review to identify the basic security challenges during development of a software and also pointing out research gap. Here, around sixty research articles are reviewed related to software issues or software security.

Keywords Software security · SDLC · Agile method · Information security

1 Introduction

A software development is a list of interrelated steps for the developing a software or system. It is also a flow of interrelated steps where each step depends on its previous steps output completely. Each step follows a list of predefined activity as their stage role, like beginning from requirement analysis to the last step of software

P. C. Behera (✉)
St. Claret College, Bangalore, India
e-mail: prakasbehera@gmail.com

C. Dash
St. Claret College, Bangalore, India
e-mail: da.chinmaya.sh@gmail.com

P. K. Pareek
East West Group of Institution, Bangalore, India
e-mail: Piyushkumarpareek88@gmail.com

J. Tavares et al. (eds.), *Emerging Technologies in Data Mining and Information Security*, Lecture Notes in Networks and Systems 164,
https://doi.org/10.1007/978-981-15-9774-9_64

maintenance. There are many models are available for the development of software like the waterfall model, iterative model, spiral model, agile and lean, etc.

SDLC Phases:

1. Conceptualization
2. Requirements (elicitation, analysis and specification)
3. Design of systems (architectural and infrastructure design, detailed design, etc.)
4. Module development or coding
5. Testing
6. Deployment
7. Maintenance.

1.1 System Engineering

The software to be developed must always work as part of a larger system. Software engineering is the step where many interactions of customers and systems took place. As per the requirement of software, hardware should work according to the instruction from the software and it should behave or responds in time and order. In this stage, the system should ensure that the software will work properly with people, hardware and other system linked to the hardware and software.

1.2 Software Requirements Analysis

In this stage, all the requirements related to developing of software are taken care, and it gathers all the data or information which is required to the particular function of the software. Here, the software engineer plays a big role, who will analyse the requirement and understand what type of information is required. Software engineer also has to analyse the functional and performance of the software and how it will interlink each other.

1.3 Design

After the requirement stage, the most important stage is the design stage. Here, an actual design will be made, and the whole project will divide into the different modules as per their functionality. And for each module, many characteristics will be added like software architecture, database management, data flow diagram and interface of the different modules will design.

1.4 Coding (Build)

After the design stage, the next important stage of the development process is the coding stage which is called the backbone stage of software development. In this stage, each module will again be divided into sub-modules based on the functional requirement started coding using different sure development language. Here, the functionality is translated into machine understanding form.

1.5 Testing

After the development of each module or once the coding is done, the next important stage is testing each module. In this, each module will be tested that is it working as per the requirement or its assign functionality. There are two testing namely unit testing and integration testing, wherein unit testing each module will be tested individually and in integration testing the different module is combined for a specific function is tested and making sure that it is fulfilling the assigned requirement.

1.6 Maintenance

After the software is installed and running, there are many causes make demand to change to this system. There is much reason to maintain the software in the system such as system or software up-gradation, changes of functionality as per new requirements of the customer or technological advancement, etc., so maintained means doing some alternation in the existing system as per the customer requirement.

Software is an essential part of our day-to-day life, and each and every activity of world are anyhow depending on software functionality. We interact with software not only on home or office computers but in our day-today everyday activities, the recent increase in cybersecurity attacks and another misuse of software is a bad situation to handle. The best way to maintain software is sure means, maintain security in each and every stages of software development. The SDLC is a set of steps necessary to develop any type of software which includes the following stages planning and design, development, testing and implementation phases. The best way to maintain software is sure means, maintain security in each and every stages of software development. Software security is an essential part of the software development life cycle, for developing secure software, security should be maintained from every stage of the software starting from requirement gathering to implement stage.

2 Meaning of Software Security

Software security is an idea to protect the software or the system from malicious attack or from hackers, so that software can work properly without any problem till the software existence in the system. Secure software means it maintains security goal of a software like integrity, confidentiality and availability.

- Confidentiality: Protect the entrance control and exposure limitations on data and ensure that nobody will disrupt the guidelines of individual security and restrictive data
- Integrity: Maintaining a strategic distance from unapproved data adjustment or obliteration.
- Availability: It implies all the data must be accessible at whatever point required.

A protected programming consistently gives a quality item which must have accessibility, uprightness, interoperability, rightness, viability, execution, dependability, reusability, adaptability, security, testability and ease of use.

3 Software Security in SDLC

The best way to maintain software is sure means, maintain security in each and every stages of software development. The only secure method to ensure that software is developed in secure is by maintaining a security in the entire software development life cycle (SDLC). Before software developer are not serious in development phase, so problems are identifyed easily in the next stage of the development. Software security begins from requirement stage and carries forward till the maintenance phase.

4 Background and Related Work from 2011 to 2020

In this research paper, we reviewed around 20 research articles, text book and some white paper, but here, we consider only 16 research articles for our reference where all the authors highlight different issues related to maintaining security during software development. Many authors highlight some models to maintain more security during SDLC.

In the year 2018, the authors Assal and Chaiasson have written a paper on, "Security in the Software Development Lifecycle", [1]. In this article, author explores different software security such as security knowledge, operational strategies. Division of labour, company culture, external pressure and experiencing a security incident are factors need to address for maintaining more secure in development of software. In the year 2016, the author Sanjay Kumar has written a paper on," Security Enhancement in Software Development Life Cycle" [2]. In this paper, creator said that the outcome acquired by secure programming is of very significance in light of the fact that numerous significant capacities are altogether subject to the product. With least expense and time to form the product into the market, numerous engineers begin thinking security after that. In the year 2017, the authors Alam et al. publish a paper on "Security Estimation Framework for Development of Secure Software and Cyber Attacks", [3]. In this paper, as indicated by authors to configuration, assemble and convey secure programming implies it must apply security in SDLC and follow the ongoing new advancement innovation identified with programming security during improvement procedure, for example, spry and others.

In the year 2012, the authors Ruchi Dabey, Vishnu Kumar, Suresh publish a paper on, "Improve Security in a Software Development Life Cycle", [4] in this paper, author said effective software security management means it must apply new technique which is simple to apply and maintain and easy to understand. Software security means all the new methodology used for secure development must understand all security issue of development process or SDLC. In the year 2012, the author Dejan Baca publishes a paper on, "Developing Secure Software—in an Agile Process", [5] in this paper, author said software security can be maintained more if the company follows the agile process because to create secure programming, numerous organizations use security designing procedures that are plan substantial and inflexible and all the things are effectively feasible in agile procedure. In the year 2019, the authors Tamimi et al. publish a paper on, "A systematic snapshot review of custom-made software enterprises from the development perspectives", [6] in this paper, author made a review around 102 research papers from different source and made a study furthermore, this study exhibited that there was a higher research enthusiasm for their own made programming. In the year 2017, the authors Saleh et al. publish a paper on, "Comparative Study on the Software Methodologies for Effective Software Development", [7] in this paper, author made a comparative study on waterfall model, XP model, the objective of the study is to find out different functionality of each model and going through thoroughly. In the year 2019, the authors, Alenezi and Almuairfi, have written a paper on "Security Risks in the Software Development Lifecycle", [8] in this paper, authors make a study on nature and phases of secure software development and the security risk assessment models and practices involved in it. They present a qualitative study looking at real-life practice employed towards software security risk and also on how well current risk practices follow best practice, identify pitfalls and explore why these occur and mitigate. In the year 2013, the authors Sundararajan et al. have written a paper on, "An Empirical Study of Industry Practice in Software Development Risk Management", [9] in this paper, authors propose a model for changed administration pertinent to non-in-house ventures and furthermore clarify the unmistakable highlights of our model in examination with the current writing on software hazard, in that model the key centre zones of software hazard the board in redistributed undertakings so as to accomplish the best result. In the year 2017, the authors, Al-Zewairi et al., have written a paper on, "Agile Software Development Methodologies: Survey of Surveys", [10] in this paper, author made a comparative study on various tools used by different author on agile process, they made study by comparer and review method (CR), and the proposed agile methods that have not been tended to till now in some other writing survey were looked into and thought about as far as where the progressions that they proposed lay on the SDLC. In the year 2014, the authors, Hijazi et al., have written a paper on, "Risk Factors in Software Development Phases", [11] in this paper, authors made a comparative study of many major risk factor associated with software development in SDLC. Around 100 risk factors were addressed in the paper which are mostly common risk factors in all software development and in all software industry. In the year 2014, the authors, Sahu et al., have written a paper on, "Risk Management Perspective in SDLC", [12] in this paper, authors have tried to combine between software security and software risk in a solitary string. It is a finished procedure that will assist a designer with choosing most suitable risk the board plan for giving software greater security and build up a real existence cycle is proposed which will assist with recognizing and moderate dangers at the beginning phase of development.

In the year 2018, the author, Siddiqui, has written a paper on, "Significance of Security Metrics in Secure Software Development", [13] in this paper, author makes a study on security metrics of software development phases and some standardized criteria for validation. In the study, every stage of metrics is compared to each other, and the final product is noted. The software derived from the proposed security metrics of the software will be secured and more qualitiable. In the year 2014, the authors, Praveen et al., have published a paper on," Software Security Issues: Requirement Perspectives", [14] in this paper, the author examines 21 issues identified with security with separate to prerequisite which can evacuate all the security threats in start of improvement life cycle. By remembering these issues, working individuals in the zone of software security can be ready to assemble increasingly secure software, and furthermore, the objectives of security can without much of a stretch be distinguished. In the year 2015, the authors, Manoj and Sangwan, have written a paper on, "Information Security requirement in Project Development Life Cycle", [15] in this paper, the author discovers how data security could be incremented and built up another system and new philosophy that can help coordinate data security in venture improvement life cycle procedures and information security contribution in software project development life cycle. In the year 2017, the author, Kapur, has written a research paper on, "A stochastic model to enhance information security in software development through risk management", [16] in this paper, author develops a model to identify different threats in overall information security management. In the year 2015, the authors, Rafi et al., have written a research paper on, "US-Scrum: A Methodology for Developing Software with Enhanced Correctness", [17] in this paper, authors develop a model on FDD and scrum principle to improve quality and security of a software, and also, this model provides more quality production using the concept verification and validation process. In the year 2016, the authors, Sharma and Mehta, have written a research paper on, "Implementation of Fuzzy Logic for Introducing Security at Process Level in Agile Development", [18] in this paper, the authors highlight different concern of security requirement of the project using agile process using fuzzy logic technique which uses Java language and GUI.

5 Observation from Literature Review

After reviewing above article published by different author, here are list of some secure software development approach.

1. Security knowledge and training to developer [19].
2. Security goal should be clear to all development team [20, 21].
3. Requirement analysis from security as part of it [22].
4. Formal risk assessment at each stage of development.
5. Thirty party verification of software related to security level after development [23, 24].
6. Security level must be verifying pre and post of development [25].
7. Security test and evaluation must maintain [26].
8. Each software development should maintain security goal, confidentiality, integrity and authenticity [27, 28].
9. Cryptography.
10. Identifying and handling sensitive data [29].
11. Authentication mechanism and authorization [30, 31].
12. Data validation.
13. Identifying and handling sensitive data [32].
14. Do not rely on single testing process.
15. Implement centralize security in both client and server side [33].
16. Object reuse or code reuse concept should utilize [34, 35].
17. Software item was recognized, chosen and executed in a true situation [36, 37].
18. Adopting new methodologies for more secure in development like agile or lean [38, 39].

6 Problem Statement

After studying, an existing propose model and research on security in software development process, in all big software industry are following different methodology for developing secure software, but in case small software industry it quite difficult to adopt new technology or tools for developing secure software.

Here are few problem statements for secure software development in small software firms are as follows:

1. Will the agile procedure assist with improving security in software development in small software firms?
2. Will the hybrid model of waterfall and agile assist with improving security in software development?
3. Would Agile-Lean model be appreciated and adopted by the small software firm to develop secure?
4. Would Microsoft SDL, lightweight application security (LASP) security approach improve software security in the software development in small firms.

7 Conclusion

Development of secure software is a challenge to all types of industries like big or small software firms. In small software firms developing a secure software a big challenge, so small software firms may face different type of attack during the development process or after the development of the software which will lead to many expenses and for making that secure it will be so costly and that's really a big challenge for small software firms. So all small software firms should maintain high security in their development process.

8 Suggestion for Future Research and Research Gap

All the research on software security enhancement in the software development process is based on research methodology, research development process and risk factor identification, but till now there is no such type of research is there for enhancing security level during software development life cycle in small software farm [40–42].

1. All the research is based on methods or process of secure software development, but no one addresses cost factor, if a software is not secure.
2. All the research is on the overall development process, but there is no such research on "How to enhance software security in software development in small scale software company".
3. There is no such research to identify different issues in secure software development for small industry.

9 Future Work on My Research

Finalizing my research problem, framing a new methodology, questioner formation for my research topics. Developing one model for small software industry.

Acknowledgements The author is thankful to all the authors whose references are included.

References

1. H. Assal, S. Chiasson, in *Security in the Software Development Lifecycle*. Proceedings of the Fourteenth Symposium on Usable Privacy and Security, Baltimore, 12–14 Aug 2018). ISBN 978-1-939133-10-6
2. S. Kumar, Security enhancement in software development life cycle. IJSTE—Int. J. Sci. Technol. Eng. **3**(06) (2016). ISSN (online): 2349-784X

3. S.M.S. Alam, S.K. Singh, S.A. Khan, Security estimation framework for development of secure software and cyber attacks. Int. J. Eng. Manage. Res. **7**(3), 146–154 (2017). ISSN (ONLINE): 2250-0758, ISSN (PRINT): 2394-6962
4. R. Dabey, V. Kumar, Improve security in a software development life cycle. Int. J. Eng. Sci. Res. Technol., 399–405. ISSN: 2277-9655, [Dabey, **1**(7), Sept 2012], pp. 399–405
5. D. Baca, in Developing Secure Software - in an Agile Process, School of Computing. Blekinge Institute of Technology, SE-371 79 Karlskrona, Sweden Printed by Printfabriken, Karlskrona, Sweden (2012). ISBN: 978-91-7295-229-4, ISSN 1653
6. M. Tamimi, F. Alghamdi, A. Yaseen, A systematic snapshot review of custom-made software enterprises from the development perspectives. Int. J. Inf. Syst. Manage. Res. Develop. (IJISMRD) **9**(1), 1–24 (2019). ISSN (P): 2250-236X; ISSN (E): 2319-4480
7. S.M. Saleh, M. Ashikur Rahman, K. Ali Asgor, Comparative study on the software methodologies for effective software development. Int. J. Sci. Eng. Res. **8**(4) (2017). ISSN: 2229-5518
8. M. Alenezi, S. Almuairfi, Security risks in the software development lifecycle. Int. J. Rec. Technol. Eng. (IJRTE). **8**(3) (2019). ISSN: 2277-3878
9. S. Sundararajan, M. Bhasi, K.V. Pramod, An empirical study of industry practice in software development risk management. Int. J. Sci. Res. Publ. **3**(6) (2013). ISSN: 2250-3153
10. M. Al-Zewairi, M. Biltawi, W. Etaiwi, A. Shaout, Agile software development methodologies: survey of surveys. J. Comput. Commun. **5**, 74–97 (2017). ISSN Online: 2327-5227; ISSN Print: 2327-5219
11. H. Hijazi, S. Alqrainy, H. Muaidi, T. Khdour, Risk factors in software development phases. Eur. Sci. J. **10**(3) (2014). ISSN: 1857-7881 (Print); e-ISSN: 1857-7431
12. K. Sahu, Rajshree, R. Kumar, Risk management perspective in SDLC. Int. J. Adv. Res. Comput. Sci. Softw. Eng. **4**(3) (2014). ISSN: 2277-128X
13. S.T. Siddiqui, Significance of security metrics in secure software development. Int. J. Appl. Inf. Syst. (IJAIS). **12**(6) (2017). ISSN: 2249-0868 Foundation of Computer Science FCS, New York, USA
14. N. Parveen, M.R. Beg, M.H. Khan,"Software security issues: requirement perspectives. Int. J. Sci. Eng. Res. **5**(7) (2014). ISSN: 2229-5518
15. Manoj, S. Sangwan, Information security requirement in project development life cycle. Int. J. Adv. Res. Comput. Sci. Manage. Stud. **3**(3) (2015). ISSN: 232 7782 (Online)
16. R. Kapur, A stochastic model to enhance information security in software development through risk management. Int. J. Eng. Sci. Invention Res. Develop. **IV**(IV) (2017), www.ijesird.com, E-ISSN: 2349-6185
17. U. Rafi, T. Mustafa, N. Iqbal, W.U.I. Zafar, US-Scrum: a methodology for developing software with enhanced correctness, usability and security. Int. J. Sci. Eng. Res. **6**(9) (2015). ISSN: 2229-5518
18. R. Sharma, P. Mehta, Implementation of fuzzy logic for introducing security at process level in agile development. Int. J. Latest Trends Eng. Technol. (IJLTET) **7**(2) (2016). ISSN: 2278-621X
19. N.S.A. Karim, A. Albuolayan, T. Saba, A. Rehman, The practice of secure software development in SDLC: an investigation through existing model and a case study. Security Comm. Netw. **9**, 5333–5345 (2016). Published online 16 November 2016 in Wiley Online Library (wileyonlinelibrary.com). https://doi.org/10.1002/sec.1700
20. N. Davis, in *Secure Software Development Life Cycle Processes*. Software Engineering Institute Carnegie Mellon University, PA, pp. 15213-2612 (2013)
21. E. James, in *Building Security into the System Development Life Cycle (SDLC) A Case Study*. SANS software Institute, Last Updated 29 Jan 2020
22. "E-Devlet Uygulamalarının Çevik Geliştirme Süreçlerinde Güvenli Yazılım," Secure software development in agile development processes of e-government applications. J. Int. Sci. Res. **3**(1) (2018)
23. M.I. Daud, in *Secure Software Development Model: A Guide for Secure Software Life Cycle* (2010). Research Gate

24. S.R. Ahmed, Secure Software Development - Identification of Security Activities and Their Integration in Software Development Lifecycle. Master Thesis Software Engineering Thesis no: MSE-2007–12 March 2007, School of Engineering Blekinge Institute of Technology Box 520 SE—372 25 Ronneby, Sweden
25. M.M. Morana, in *Building Security into The Software Life Cycle a Business Case*
26. Fortify an HP Company, a white paper, in *Seven Practical Steps to Delivering More Secure Software* (2011)
27. N. Davis, in *Secure Software Development Life Cycle Processes: A Technology Scouting Report*. Technical Note CMU/SEI-2005-TN-024
28. B.M. y Parareda, J. Streit, in *Software Quality Modelling Put into Practice*. TUM-I0811 April 08, TUM-INFO-04-I0811-0/1.-FI
29. K.M. Gentzel, in *Software Security Assurance*. State-of-the-Art Report (SOAR) July 31, 2007. Information Assurance Technology Analysis Center (IATAC) Data and Analysis Centre for Software (DACS) Joint endeavour by IATAC with DACS
30. S. Preda, N. Cuppens-Boulahia, F. Cuppens, J. Garcia-Alfaro, L. Toutain, in *Model-Driven Security Policy Deployment: Property Oriented Approach*. IT TELECOM Bretagne CS 17607, 35576 Cesson-S´evign´e, France, hal-00540842, version 1, 29 Nov 2010
31. S.S. Dadhwal, in *Secure Software Policy*. CSCI – E41 Secure Software Development, ©ACME Retail – All Rights Reserved
32. A. Hudaib, M. AlShraideh, O. Surakhi, M. Khanafseh, A survey on design methods for secure software development. Int. J. Comput. Technol. **16**(7). ISSN: 2277-3061
33. M. Dawson, D.N. Burrell, E. Rahim, S. Brewster, Integrating software assurance into the software development life cycle (SDLC). J. Inf. Syst. Technol. Plann. JISTP **3**(6), 49–53 (2010). ©2010 Published by Intellect Base International Consortium, USA
34. C. Banerjee, S.K. Pandey, in *Research on Software Security Awareness: Problems and Prospects*. ACM SIGSOFT Software Engineering Notes, September 2010, vol. 35, Number 5
35. M. Rizwan Jameel Qureshi, in *Agile Software Development Methodology for Medium and Large Projects*. Published in IET Software Received on 22nd June 2011 Revised on 14th August 2011. https://doi.org/10.1049/iet-sen.2011.0110. ISSN: 1751-8806
36. S. Mahmudova, in *Some Methods for Software Security*. Society for Science and Education United Kingdom. ISSN: 2054-7420
37. D. Chinmaya, D. Ramesh, C.B. Prakash, Software quality management: a process to improve the quality of software project using SQA. Int. J. Comput. Sci. Inf. Technol. (2016). ISSN: 0975-9646
38. M. Upendra Kumar, Literature survey on agile security architecture model. SSRG Int. J. Mob. Comput. Appl. (SSRG – IJMCA) **1**(1) (2014)
39. N. Newton, C. Anslow, A. Drechsler, in *Information Security in Agile Software Development Projects: A Critical Success Factor Perspective*, ed. by Newton et al. Information Security in Agile Development
40. M.W. Mwadulo, Suitability of agile methods for safety-critical systems development: a survey of literature. Int. J. Comput. Appl. Technol. Res. **5**(7), 465–471 (2016). ISSN: 2319–8656
41. A. Sharma, R. Sharma, R.K. Bawa, An integrated framework for security enhancement in agile development using fuzzy logic. IJCST **7**(2) (2016). ISSN: 0976-8491 (Online); ISSN: 2229-4333 (Print)
42. J. Ahola, C. Frühwirth, M. Helenius, L. Kutvonen, J. Myllylahti, T. Nyberg, A. Pietikäinen, P. Pietikäinen, J. Röning, S. Ruohomaa, C. Särs, T. Siiskonen, A. Vähä-Sipilä, V. Ylimannela, in *Handbook of the Secure Agile Software Development Life Cycle*. University of Oulu, 2014. ISBN: 978-952-62-0341-6

Literature Survey on Exploratory Analysis of Waste in IT Industries During Project Development Using Agile–Lean Practices

Chinmaya Dash, Prakash Chandra Behera, and Piyush Kumar Pareek

Abstract In the last two decades, Implementing and maintaining the standard of IT projects is a critical task. IT projects have not successfully facilitated standard due to cost involved in project completion, required resources along with inefficient risk and stakeholder. Agile approach is a customer-oriented approach which enhances the efficiency of software development process. This paper focuses on need of agile–lean model which focuses on two major concepts that are eliminating waste and improving flow of productivity during software development and maintenance and to be acknowledged positively by the IT organizations during project development.

Keywords Agile · Lean · Software waste · Agile–lean model

1 Introduction

Agile methodology contains three repetitive processes such as experiment, inspect, and adapt to generate more productivity which is adopted in various organizational sectors such as software development, finance, sales, and manufacturing. Lean practice continuously reduces the waste using rapid experimentation process. The general root causes for project failure are as follows: [1]

- Insufficient planning finally ends with a project failure.
- Ineffective stakeholder management, inexperienced user, and their involvement.
- Unrealistic estimated time.

C. Dash (✉) · P. C. Behera
Shri JJT University, Rajasthan, India
e-mail: da.chinmaya.sh@gmail.com

P. C. Behera
e-mail: prakasbehera@gmail.com

St. Claret College, Bangalore, India

P. K. Pareek
East West Group of Institution, Bangalore, India
e-mail: Piyushkumarpareek88@gmail.com

J. Tavares et al. (eds.), *Emerging Technologies in Data Mining and Information Security*,
Lecture Notes in Networks and Systems 164,
https://doi.org/10.1007/978-981-15-9774-9_65

- Budget, project goal along with change in technology.
- Elimination of quality assurance mechanism.
- Unqualified resource and environment and Underutilization of employee.
- Poor and ineffective communication between stakeholders.
- Planning and managing risk.
- Poor project management and managing changes in control.
- Undefined project development process.
- Waste during software development and time pressure to increase productivity.

Software project maintains its standard by dealing with risk, waste, and quality during the development stage with the help of agile methodologies and lean practices.

2 Agile Practice

Agile methodology is a lot of standards and conventions for software improvement where in prerequisites and arrangements develop through cooperation between self-sorting out and cross-useful groups. It is adaptable, lightweight, and group arranged software improvement approach. The agile techniques center around individuals, working software, client coordinated effort and change rather than forms, instruments, agreements, and plans. The usually utilized structures are Extreme Software, Scrum, Dynamic System Development Method, Feature Driven Development, Adaptive Software Development, Kanban, Bimodal, Hybrid, Crystal, and Lean [2, 3].

2.1 Agile Approach

It can be defined as [4]

- It is acceptable to change the necessities, even late being developed.
- Deliver working software modules in time.
- Stakeholders and engineers must cooperate all through the venture.
- Build ventures with roused people.
- Convey data inside an improvement group ought to be eye-to-eye discussion.
- Sustainable way of improvement must be kept up in a consistent pace.
- Technical greatness and great structure must be ceaselessly joined in.
- It is basic to know the measure of work not done.
- The prerequisites and plans can be risen up out of self-sorting out groups.
- The group ought to be increasingly viable during venture improvement.

2.2 *Agile Features*

A set of features which shortens the development life cycles of projects: [5]

- Short iterative modules ranging from 1 to 6 weeks that enable rapid corrections and verifications.
- Modularity is required for development process.
- Adaptive with current potential risks.
- An incremental process method allows creating and functioning applications and eliminates the entire unneeded activities which attempts to reduce the risk.
- The agile processes help users to adequate any technologies and processes.
- Communicative and collaborative working style at the workplace throughout the development.

3 Lean Practice

Lean software development approach helps in optimizing the software development process by eliminating the waste which means removing not essential things from the final product and helps deliver a product in refined quality. The lean improvement group centers around learning and conveying the item as quick as conceivable to expand productivity. Lean spotlights on client and their fulfillment. It targets interfacing with client as often as possible, understanding their worth and spotlights on its key procedures to improve it which actually focuses on two major concepts that are eliminating waste and improving flow of production. The principles of lean are: [6].

3.1 *Eliminating Waste*

It means eliminating the things which are not useful for the customers, termed as waste. These wastes must be eliminated to reduce the unnecessary complications, developing the wrongly featured product to reduce stress or workload.

3.2 *Amplify Learning*

The ways through which work can be improved and not on adding more and more features. It promotes continuous learning in each iteration which involves discovering newer ways/ideas to write a code or build a product in an optimal manner. In most of the cases, user feedback is also taken to understand their perspective and requirements.

3.3 Decide as Late as Could Reasonably Be Expected

This rule defers the decision until we get especially portrayed fact which is the greater delay in decisions and duties. This alters more forward-thinking changes which can be costly.

3.4 Deliver as Fast as Possible

This also helps in delaying the decision so that customer gets enough time to discover their requirements properly. This helps in getting the product as fast as possible which further helps in improving the product in the next iteration. This also leads a complete product which satisfies customer's requirement.

3.5 Empower the Team

In this, manager helps the team member in taking any decision, assigning tasks to them, providing them with the decision, suggestion, and also help in removing the obstacle.

3.6 Build Integrity

It can be further classified into two terms: perceived integrity which is related to the overall experience of the customer with the production and how the product is useful; completed integrity focuses on how separate component of the system are integrated to make it more efficient, maintainable, and flexible product can be designed.

3.7 See the Whole

Most of the defect in the software system can be eliminated by developing the system into subsystem and developing them separately using well-defined standards. The larger the system or the larger the organization involved in the development which results in the development of complex, efficient, and powerful system. But using lean, the defects can be identified in each stage of software designing.

4 Category of Waste

There are eight categories of wastes [7].

4.1 *Overproduction*

Production in huge scope in limited capacity to focus time brings about poor progression of data or merchandise and utilization of overabundance stock.

4.2 *Defects*

Frequent blunders in material/item prompts adjust to create the quality just as poor conveyance execution.

4.3 *Unnecessary Stock*

Delay of data or items lead overabundance expenses and poor client support.

4.4 *Inappropriate Preparing*

Using an inappropriate arrangement of instruments, methodology, or frameworks, the efficiency can be diminished.

4.5 *Excessive Transportation*

Excessive development of individuals, data, or merchandise results wastage of time and cost.

4.6 *Waiting*

Long times of latency for individuals, data, or merchandise result poor stream and long lead times.

4.7 Unnecessary Movement

Poor work environment association leads ineffectual development of the association and its partners.

4.8 Underutilization of Worker

Employee's innovativeness and aptitudes ought to be improved for the procedures and practices which allude to squandering of accessible information, experience or expertise of the staff/workforce.

5 Process of Eliminating Waste

The main objective of developing software using agile and lean approaches is eliminating waste. The process of eliminating waste is: [8].

5.1 Minimizing the Partially Work Done

Partially software development work can carry financial risks. Minimizing this, the risk factor will reduce and become an effective waste reduction strategy.

5.2 Reduction of Additional Processes

Lot of documentation occurs during software improvement which may expand the loss in software advancement process. Plan documentation assists with limiting the additional documentation during improvement which can be comprehended by all partners of the venture.

5.3 Avoid to Add Extra Code and Features

Developers have the practice of keeping some code in comment for future references but extra code and features are not necessary, putting those into a development practice is a waste.

5.4 Avoid Visit Task Exchanging

The quickest method to create and finish two undertaking that utilizes same assets is to do them each in turn and second in the wake of completing initial one.

5.5 Delay Decision as Late as Possible

Delay the decision until the last possible moment. It is the best way to deal with the uncertainty.

5.6 Team Should Work in a Solitary Workroom

Team work in a solitary workroom offers consent to access to partners which gives the stream to the turn of events.

5.7 Detect the Defect as Early as They Occur

The way to reduce these wastes is to find and detect defect followed by test immediately, integrate and release in the form of product immediately.

6 Survey from 2011 to 2020

Jeffry, Edward and Sutherland in 2020 [9] derived the success rate of software projects using agile and lean methodologies along with related approaches. They also derived that social and economic growth may be produced using agile methods. Kuutila et al. [10] investigated how time pressure can affect software quality and productivity through different experiments and case studies. They concluded with the time pressure; the productivity may increase but the quality of the product will be reduced. Yadav et al. [11] investigated how leagile is implemented in different software development projects. They discussed and derived various application of lean and also analyzed the performance of lean on same. Sedano et al. [12] conducted in depth study on software development waste and wasteful practices. They also described different types of waste and different process for eliminating those. Atawneh [13] demonstrated how agile methods can be implemented in developing software projects which produce significant benefits over other conventional methods. He also represented the agile software development methods into community of software engineering and derived quick development process along with the modification which is the need of dynamic environment. Qureshi et al. [14] proposed lean–agile integration LScrum can be useful for developing large size project. They reasoned that LScrum can assist with acquiring properties incline toward scrum and to achieve enormous size undertakings as big business asset arranging frameworks. Poth et al. [15] introduced the outline and standpoint of agile and lean methodologies and how the constant software process improvement can be acquired from these methodologies. They also highlighted the lean–agile software process improvement approached depending on target environment. Kiss and Rossi [16] performed systematic literature review on agile to lean changes in wording advantages, difficulties, and measurements. Thus, they discovered advantages as non-stop improvement and imperfection fix rate; challenges as distinguishing waste and scaling adaptability; and metric as lead time. Alahyari et al. [17] conducted case study using semi-structured open interviews to investigate how waste is drawn closer in agile software advancement associations. They ordered the loss in 10 unique sorts and inferred that a joint and all-encompassing perspective idea or advancement the entire association and entire item is required to recognize and take out waste. Matkovic et al. [18] determined how particular architectural practices are incorporated into agile development processes. They referred traditionalization of agile processes which is associated with incorporation of elements of traditional development into agile process. Kettunen and Lanti [19] recognized the design principles of future of software organizations to synthesize a new development impact mapping grid for upcoming digitization process. They experimented resource-based and computer-based views through real-time examples. Shaga and Shinde [8] estimated how agile methodology can be benefitted for small software companies. They proposed that agile methodology is customer-oriented approach which enhances the software development process along with software developers of small software companies. Yaman et al. [20] conducted multiple case study to derive the process of continuous experimentation into organization which reduce risk of wasting the invested effort. They concluded that the benefits of using this process can be reduced development effort and better development decisions; challenges could be complex stakeholder structure and exploring success criteria. They also enlightened that this process requires a change in various operations of organization. Todd et al. [21] conducted participants observation study to distinguish and portray various kinds of waste in software improvement. They iterated between theoretical and analysis sampling and introduced empirical waste taxonomy which insights different types of waste and its cause along with how it is related with lean software development. Deokate and Shaga [22] focused on agile–lean software development methodology which underpins the waste end procedure to decrease the non-esteem include exercises. They also highlighted how to ignore wastes while developing a software product.

Razzak [23] conducted empirical study to join lean and agile practices which scale and oversee appropriated software advancement and worldwide software improvement. He also derived that the complex management structure and very sized team can be attended by lean and agile combination. Behroozi and Kamandi [24] reviewed different agile methodologies for Web development. They also introduced different patterns in Web engineering through which waste can be affected and eliminated by using verities of agile methodologies. Nataraja and Ramesh [25] focused on lean software development, and different practices are used lean with agile, usage of lean for reducing wastes during software development and maintenance. They also derived continuous improvement model which insights stable development model for different organizations. Shahzeydi and Gandomani [26] conducted case study for adding lean thinking principles with agile software development scrum practices to gain more business values compared to competitors. They enlightened the scrum team cannot fulfill few lean practices. Almasri [27] derived the agile methodologies that are most suitable mechanism for mobile development projects because of few metrics such as flexibility and reduced wastes. He also concluded that a process model is required for mobile application which can overcome different challenges such as business requirements and technical constraints for mobile system. Kundu [28] derived how to distinguish and sort non-esteem include exercises in various industry settings. He also highlighted eliminating wastes can be processed through lean thinking and applying lean principles. Power and Conboy [29] reframed lean concept as impediments of complex human system. They also enlightened how to categorize the impediments faced by organizations as well as software development and how to reduce those in their system. Al-Baik and Miller [30] proposed another model to characterize waste happened in IT associations. They built up another model of waste arrangement for medium-scale IT associations. Likewise, they talked about various waste disposal techniques by expelling the underlying driver of waste and accentuated the need of lean think. Auerbach and McCarthy [31] used phenomenological approach which concentrated how systems are chosen for agile and lean. They reasoned that how lean and agile procedures are important and furthermore can be viable being developed and execution of data framework ventures. Pernstal et al. [32] inferred distinctive lean holes during lean item advancement and agile software improvement in enormous scope software framework advancement. Swaminathan and Jain [33] conducted case study to apply ideas of constant improvement and stream to agile software advancement. They derived from their study that using those concepts, there is a significant benefit and can be implemented in different industry product development process. Ebert et al. [34] addressed how lean principles delivers value and introduced to manage the change. Wang [35] analyzed set of experience reports to understand the use of lean and agile approaches. He derived lean can be applied in guiding principles for agile development furthermore, stream-based lean procedures lead to substitute time boxed agile procedure.

7 Research Findings

- Social and economic growth are produced using agile methods.
- With the time pressure, the productivity may increase but the quality of the product will be reduced.
- Various applications of lean and its performance.
- Different types of waste and different process for eliminating those.
- Agile methods can be implemented in developing software projects which produce significant benefits over other conventional methods which are a need of dynamic environment.
- Lean–agile integration LScrum can be useful for developing large size project.
- The continuous software process improvement can be inherited from agile and lean approaches which depends on target environment.
- Agile to lean transformations in terms benefits as continuous improvement and defect fix rate; challenges as identifying waste and scaling flexibility; and metric as lead time.
- A joint and holistic view concept or optimization of the whole organization and whole product is required to identify and eliminate waste.
- Traditionalization of agile processes which is associated with incorporation of elements of traditional development into agile process.
- The design principles of future of software organizations to synthesize a new development impact mapping grid for upcoming digitization process.
- Agile methodology is customer-oriented approach which enhances the software development process along with software developers of small software companies.
- Empirical waste taxonomy which insights different types of waste and its cause along with how it is related with lean software development.
- Combining lean and agile practices scale and oversee disseminated software advancement and worldwide software improvement.
- Different patterns can be created in Web engineering through which waste can be affected and eliminated by using verities of agile methodologies.
- Lean thinking principles with agile software development scrum practices can be used to gain more business values compared to competitors.
- Agile methodologies are most suitable mechanism for mobile development projects because of few metrics such as flexibility and reduced wastes.
- Model of waste categorization for medium-scale IT organizations which manages waste end procedures by evacuating the main driver of waste and underscored the need of lean think.
- Different lean holes during can be found during lean item improvement and agile software advancement in huge scope software framework advancement.
- Concepts of non-stop improvement and stream to agile software advancement can be executed in various industry item advancement process.
- Lean can be applied in core values for coordinated agile events and stream-based lean procedures lead to substitute time boxed agile procedure.

8 Problem Statement

The current investigations lead toward the combination of lean and agile methodologies. This paper attempts to propose the agile–lean model by combining agile and lean exercises to produce various advantages. [36] Agile–lean model leads two major concepts that are eliminating waste and improving the production flow during software development and maintenance and to aware about the wastes occurring while developing a product and how it can be eliminated.

RQ1: Which combination of agile method and lean principle could be explored as agile–lean model which could be fruitful for IT industries and be valued by the product advancement associations for the improvement on flow of productivity along with social and economic growth during the software development and maintenance?

9 Conclusion

Agile software improvement strategies are viewed as lightweight techniques that could utilize a gradual and iterative life cycle went with short necessities and emphasis, which could be altered inside the advancement with expansive support by the customer [13]. Agile methodology is a client arranged methodology which upgrades the proficiency of software improvement process. Agile and lean software advancement forms are utilized to expel the loss in the product improvement. In this paper, we have proposed the need of agile–lean model which centers around two significant ideas that are disposing of waste and improving stream during software improvement and upkeep and to mindful about the squanders happening while at the same time building up an item and how it very well may be dispensed with. We also need to do more studies for eliminating research gaps: [37]

RG1: How can be the agile–lean model implemented in software development organizations?
RG2: What are the implications of agile–lean model on requirements engineering and product release and on the coordination mechanisms within agile teams?

Acknowledgements As a part of my academic affiliation with the Shri JJT University for the doctoral degree program in Computer Science, through this review, we have attended to enlighten the positive features of lean through which wastes have been eliminated with improved productivity in software development organizations. We are satisfied and appreciative to all creators whose references are incorporated. I am also pleased and grateful to my co-author and research supervisor for their constant guidance.

References

1. D. Chinmaya, D. Ramesh, C.B. Prakash, Software quality management: a process to improve the quality of software project using SQA. Int. J. Comput. Sci. Inf. Technol. (2016). ISSN 0975–9646
2. M.A. Awad, A comparison between agile and traditional software development methodologies (2005)
3. K. Anupama, A literature review on agile software development. Int. J. Adv. Res. Comput. Commun. Eng. (2016). https://doi.org/10.17148/ijarcce.2016.5971
4. K. Beck, Principles behind the agile manifesto. Agile Alliance (2001)
5. G.G. Miller, The characteristics of agile software processes, in *Tools* (IEEE, Piscataway, 2001)
6. S. Bhabya, K.S. Ashish, J. Gitanjali, Comparative study and analysis of scrum and lean methodology. Int. J. Res. Appl. Sci. Eng. Technol. 2321–9653 (2018)
7. J.P. Womack, D.T. Jones, *Lean thinking: banish waste and create wealth in your corporation*, 2nd edn. (Free Press, New York, 2003)
8. V. Shaga, L. Shinde, Agile project estimation: an imperative approach for small software industries. Int. J. Comput. Sci. Trends Technol. **5**, 54–57 (2017)
9. J. Saltz, E. Anderson, A. Sutherland, Agile and lean: organizations, products and developments, in *Proceedings of the 53rd Hawaii International Conference on System Sciences* (2020). 978-0-9981331-3-3
10. M. Kuutila, M. Mäntylä, U. Farooq, M. Claes, Time pressure in software engineering: a systematic literature review (2019)
11. RamKaran Yadav, M.L. Mittal, Rakesh Jain, Lean practices in software development projects: a literature review. AIP Conf. Proc. **2148**, 030044 (2019). https://doi.org/10.1063/1.5123966
12. T. Sedano, P. Ralph, C. Peraire, Removing software development waste to improve productivity, in *Rethinking Productivity in Software Engineering*, eds. by C. Sadowski, T. Zimmermann (Apress, New York, 2019). https://doi.org/10.1007/978-1-4842-4221-6_19
13. S. Atawneh, The analysis of current state of agile software development. J. Theor. Appl. Inf. Technol. 1992–8645 (2019)
14. M.R. Qureshi, M. Alassafi, H. Shahzad, Lean agile integration for the development of large size projects. Int. J. Mod. Educ. Comput. Sci. (2019). https://doi.org/10.5815/ijmecs.2019.05.03
15. A. Poth, S. Sasabe, A. Mas, A.-L. Mesquida, Lean and agile software process improvement in traditional and agile environments. J. Softw. Evol. Process (2019). https://doi.org/10.1002/smr.1986
16. F. Kiss, B. Rossi, Agile to lean software development transformation: a systematic literature review, in *Proceedings of the Federated Conference on Computer Science and Information Systems* (2018). https://doi.org/10.15439/2018f53
17. H. Alahyari, T. Gorschek, R.B. Svensson, An exploratory study of waste in software development organizations using agile or lean approaches: a multiple case study at 14 organizations. Inf. Softw. Technol. (2018). https://doi.org/10.1016/j.infsof.2018.08.006
18. P. Matkovic, M. Maric, P. Tumbas, M. Sakal, Traditionalization of agile processes: architectural aspects. Comput. Sci. Inf. Syst. **15**, 79–109 (2018). https://doi.org/10.2298/CSIS160820038M
19. P. Kettunen, M. Laanti, Future software organizations—agile goals and roles. Eur. J. Futures Res. (2017). https://doi.org/10.1007/s40309-017-0123-7
20. S. Yaman, M. Munezero, J. Munch, F. Fagerholm, O. Syd, M. Aaltola, C. Palmu, T. Mannisto, Introducing continuous experimentation in large software-intensive product and service organizations. J. Syst. Softw. (2017). https://doi.org/10.1016/j.jss.2017.07.009
21. S. Todd, R. Paul, P. Cecile, Software development waste, in *ICSE '17: Proceedings of the 39th International Conference on Software Engineering* (2017). https://doi.org/10.1109/icse.2017.20
22. P. Deokate, V. Shaga, Agile lean development for waste management: a review (2017). https://doi.org/10.13140/rg.2.2.18694.68167

23. M.A. Razzak, An empirical study on lean and agile methods in global software development, in *IEEE 11th International Conference on Global Software Engineering Workshops (ICGSEW)* (Orange County, CA, USA, 2016). https://doi.org/10.1109/icgsew.2016.22
24. N. Behroozi, A. Kamandi, Waste elimination of agile methodologies in web engineering, in *Second International Conference on Web Research (ICWR). Transaction on Engineering and Management* (2016). https://doi.org/10.1109/tem.20 https://doi.org/10.2081675
25. P. Nataraja, B. Ramesh, A review on lean thinking for software process improvement. Int. J. Adv. Res. Comput. Commun. Eng. (2016). https://doi.org/10.17148/ijarcce.2016.51128
26. M. Shahzeydi, T.J. Gandomani, Adding lean principles to agile software development: a case study report. Int. J. Softw. Eng. Technol. **2**, 12–15 (2016)
27. A. Almasri, A proposed hybrid agile framework model for mobile applications development. Int. J. Softw. Eng. Appl. (2016). https://doi.org/10.5121/ijsea.2016.7201
28. G.K. Kundu, Lean wastes: classifications from different industry perspectives. ICTACT J. Manag. Stud. (2015). https://doi.org/10.21917/ijms.2015.0007
29. K. Power, K. Conboy, Impediments to flow: rethinking the lean concept of 'waste' in modern software development, in *Agile Processes in Software Engineering and Extreme Software. XP 2014*. Lecture Notes in Business Information Processing, vol 179 (Springer, Cham, 2014). https://doi.org/10.1007/978-3-319-06862-6_14
30. O. Al-Baik, J. Miller, Waste identification and elimination in information technology organizations. Empirical Software Eng. (2014). https://doi.org/10.1007/s10664-014-9302-3
31. B. Auerbach, R. McCarthy, Does agile + lean = effective: an investigative study. J. Comput. Sci. Inf. Technol. 2334–2374 (2014)
32. J. Pernstal, R. Feldt, T. Gorschek, The lean gap: a review of lean approaches to large-scale software systems development. J. Syst. Softw. **86**, 2797–2821 (2013). https://doi.org/10.1016/j.jss.2013.06.035
33. B. Swaminathan, K. Jain, Implementing the lean concepts of continuous improvement and flow on an agile software development project: an industrial case study, in *Proceedings—Agile* (India, 2012). https://doi.org/10.1109/agileindia.2012.12
34. C. Ebert, P. Abrahamsson, N. Oza, Lean software development. Softw. IEEE. **29**, 22–25 (2012). https://doi.org/10.1109/MS.2012.116
35. X. Wang, The combination of agile and lean in software development: an experience report analysis. in *Proceedings—2011 Agile Conference, Agile* (2011). https://doi.org/10.1109/agile.2011.36
36. H. Soltan, S. Mostafa, Lean and agile performance framework for manufacturing enterprises. Procedia Manuf. **2**, 476–484 (2015). https://doi.org/10.1016/j.promfg.2015.07.082
37. R.K. Surbhi, S. Ritesh, Literature survey on agile information systems development. Int. Res. J. Eng. Technol. **2**(8) (2015). ISSN 2395-0056

Classifying Medical Histology Images Using Computationally Efficient CNNs Through Distilling Knowledge

Aakash Garg, Karan Aggarwal, Manan Saxena, and Aruna Bhat

Abstract Breast cancer is one of the most prevalent cancer types among women, and early detection of carcinogenic tissue from histology images can go a long way in effective treatment as well. Recent approaches to solving this problem of classification of tissues utilize heavy deep learning-based architectures, which take higher inference time and storage space for model parameters. In our work, we utilize a CNN architecture with a lesser number of parameters that can be effectively deployed on a resource-constrained device. For this, we use two different approaches, namely the knowledge distillation and a hybrid technique combining attention transfer and knowledge distillation techniques. We evaluate our approach on high-resolution Breast Cancer Histology slides of the BACH 2018 dataset.

Keywords Deep learning · Residual networks · Patchwise extraction · Pattern recognition · Stain normalization · Knowledge distillation · Attention transfer

1 Introduction

In 2019, according to the American Cancer Society [12], approximately 268,600 new cases of breast cancer were reported, out of which 41,760 cases resulted in the death of the patient. Diagnosis of breast cancer involves histopathology; it is the foremost

A. Garg (✉) · K. Aggarwal · M. Saxena · A. Bhat
Delhi Technological University, Shahbad Daulatpur, Main Bawana Road, Delhi 110042, India
e-mail: aakashgarg_bt2k16@dtu.ac.in

K. Aggarwal
e-mail: karanaggarwal_bt2k16@dtu.ac.in

M. Saxena
e-mail: manan_bt2k16@dtu.ac.in

A. Bhat
e-mail: aruna.bhat@dtu.ac.in

J. Tavares et al. (eds.), *Emerging Technologies in Data Mining and Information Security*, Lecture Notes in Networks and Systems 164,
https://doi.org/10.1007/978-981-15-9774-9_66

technique for distinguishing malignant and benign breast cancer tissue and thereby also determining the patients possessing in situ or invasive carcinogenic cells in the body tissues of the patients.

The laboratory procedure for determining the nature of the biopsy tissue consists of analyzing the tissue under a microscope wherein the pathologist can look for specific features such as the spatial arrangement of the cells, morphometric characteristics of the nuclei, etc. [4]. However, this manual examination of cancerous tissue is prone to errors, and two highly trained pathologists may have different opinions on the same biopsy tissue. As a solution, computer-aided diagnosis (CAD) has proven to be quite useful for giving accurate information, which is quite close to what a trained pathologist would provide.

Breast tissue slides stained with certain chemical compounds (H&E) are used to aid the computer-aided diagnosis of biopsy slides. These slides typically determine a crucial rule in determining the nature of the breast tissue, namely if it is non-cancer (i.e., normal or benign) or a malignant (in situ or invasive carcinoma) tissue. When the breast tissue is stained with hematoxylin, the nuclei of the tissue in question are stained purple, whereas eosin turns the cytoplasm surrounding the nuclei to a pinkish color.

To develop a solution to this problem, we have chosen to move forward with The Breast Cancer Histology Challenge (BACH) 2018 dataset, which consists of high-resolution chemically stained breast cancer histology microscopy images [1]. The images contained in this dataset are RBG images of the size 2048×1536 pixels. Each pixel covers $0.42\,\mu\text{m} \times 0.42\,\mu\text{m}$ of tissue area. The images in this dataset are labeled by two professional experts in the field of medicine, and in cases of discord, the picture is not added to the final dataset. The images are classified into four categories: (i) normal tissue, (ii) benign lesion, (iii) in situ carcinoma, and (iv) invasive carcinoma as per the judgement of these two professionals.

Many researchers have worked on the issue of early diagnosis of breast cancer using the tissue biopsy slides and have achieved significant success in this area [3, 10, 13]. The images present in the BACH 2018 dataset, however, are very high resolution in comparison to other datasets, like MNIST [9]. The model architectures proposed in these papers generally have very high computational resource requirements in terms of inference and training time and require high storage space for model parameters. In this paper, we propose a lower parameter model, trained separately via the knowledge distillation [6] and a hybrid of knowledge distillation [6] and attention transfer [14] techniques that perform reasonably well when compared to a high parameter model.

The motivation behind using a deep learning architecture with lower computational requirements instead of one with a higher number of parameters stems from the fact that deployment of such a model would be troublesome, especially if the model is designed to cater to a large number of people. This smaller model would also be suitable for deployment in resource-constrained devices, which cannot support larger models with high resource requirements, thereby making our methods commercially viable.

To summarize, our main contributions are:

1. We utilized both knowledge distillation [6] and a hybrid of knowledge distillation [6] and attention transfer [14], for training a low resource requirement model for Medical Histology Image Classification.
2. We demonstrate the effectiveness of both techniques for the task consisting of very high-resolution images and a low number of classes, compared to standard datasets such as CiFAR-10 and MNIST.

2 Background and Related Work

Deep learning has become the principal tool to achieve high-level performance for many vision and non-vision problems. But such high accuracy is usually achieved by utilizing a very deep neural network or their ensemble. Also, using these deep neural networks requires significant memory space and computational resources. Hence, over the years, many techniques for model compression have been devised, such as knowledge distillation [6] and attention transfer [14].

Hinton et al. introduced knowledge distillation with neural networks. [6] drawing focus toward "distilling" the knowledge gained from training a powerful teacher network to train a weaker student network. Distillation can be done by using "soft targets" as class probabilities for the student model implying usage of the last layers of the teacher model as "feature representation" to train the student model [7].

Artificial attention mechanisms have been developed for problems that could benefit from paying more "attention" to a particular scene. Over the years, attention transfer [14] became a mechanism for transferring knowledge between two networks, with the use of spatial attention maps that utilize either the activation approach or the gradient approach. Research [14] suggests that in both cases, attention transfer works better as compared to a full activation transfer.

Classification and detection of breast cancer in microscopic tissue images at early stages can significantly help in devising the course of action used to treat the patient; hence, it has been a crucial area of research for a significant amount of time. Recently, several works aimed at usage of deep learning models for breast cancer histology image classification have been published [3, 10, 13]. Differences in these works lie in the model and the patch extraction technique used. Patches are extracted either by convolutional-based split [10, 13] or nuclei-based split [3]. The deep learning models used vary from being a standard CNN architecture [10] to using huge models like InceptionV3 for transfer learning [3, 13] (Fig. 1).

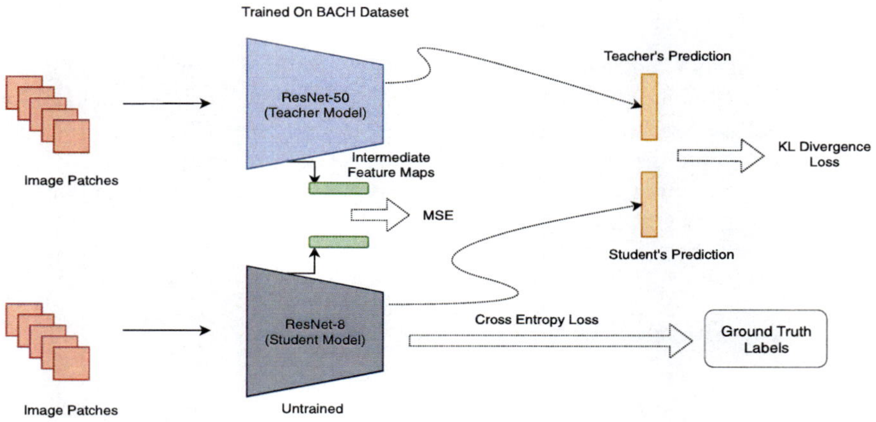

Fig. 1 Model architecture

3 Methodology

In this paper, we solved this problem by using a ResNet-50 network as a teacher and a ResNet-8 network as a student. For this paper, we are making the use of high-resolution breast cancer histology images; we preprocess the images using stain normalization and a patchwise image split strategy, as mentioned in [10]. After the images are preprocessed, they are used in the training phase of the ResNet-50 network wherein the classes assigned to the slides are used as the ground truth label. Post-training, the accuracy of the network is measured with respect to the assumed ground truth. Once ResNet-50 has been trained, we use knowledge distillation and attention transfer to train the ResNet-8 model and compare the performance of the ResNet-8 model in both cases.

3.1 Stain Normalization

A well-known problem in histological image analysis is of variations present in the stained images, which are mainly caused by unquantified staining procedures for preparing tissues before microscopic imaging for cancer diagnosis. Therefore, stain normalization is used to make the images usable for computer-based analysis. In this paper, we have utilized the approach proposed by Reinhard et al. [11], which maps the color histogram of the image to that of the target image, following the transformation of RGB colorspace to the decorrelated LAB colorspace via a series of linear transformations. It is done by using linear transforms to match the standard deviation and mean of each color channel of the source and target images in LAB colorspace. This is necessary because the variations present in the histology slide do not cope well with the in-depth learning networks used in this project, thereby giving inaccurate results.

3.2 Patchwise Image Split

Training deep learning models on high-resolution images of the dataset used requires ample memory space or downsampling of the images. However, neither a large memory is readily available, nor downsampling can be done as it leads to loss of discriminative features in the image. Also, if we train on the whole high-resolution images, then the models might learn the most distinctive features and disregard the intricate details of the image. Hence, feeding images in the form of patches to the network is the simplest solution. In this paper, we have used the patchwise image generation mentioned by [10]. For an image, we extract fixed-sized patches by sliding a window (patch) of size $m \times m$ over the image with stride n. With this, we would get a total number of $[1 + \frac{I_x - m}{n}] \times [1 + \frac{I_y - m}{n}]$ patches where I_x and I_y are image height and width, respectively. In our experiments, we take a patch size of 512×512 with a stride of 256, which gives us 35 patches of each image. The patches generated here are overlapping, which is beneficial as it helps in learning of features shared between patches.

3.3 Training Residual Networks

In this paper, we have used the ResNet [5] as our base deep learning model. ResNet is based on residual learning framework, which enables the network to retain the learning of previous residual blocks by doing an identity mapping weight function where the output is equal to the input, preserving by adding on to what the network already learned. Such a network is easier to optimize and, therefore, enables the learning of deeper networks. For our teacher model, we have used ResNet-50, which has 23.5 million parameters and 151 layers, whereas, for student model, we have used ResNet-8, which had only 94.7k parameters and merely 37 layers. The low number of parameters makes the model less computationally expensive. To train the model, we use the cross-entropy loss function, which is described as:

$$H_{y'}(y) = -\sum_i y'_i \log(y_i) \tag{1}$$

3.4 Knowledge Distillation

Knowledge distillation [7] is a technique that aims to "distill" or transfer the knowledge contained in a cumbersome model to a smaller model that is more suitable to be deployed in resource-constrained devices. As compared to traditional knowledge transfer techniques, knowledge distillation aims to make the smaller models generalize better to the training data and reflect the real objective of the user rather than to optimize the performance on the training data. Training a model using this approach would, however, require information about the generalizations, patterns, and the relations inherent in the data to make the generalizations possible. In deep learning applications, the class probabilities for a particular case depict a lot of hidden information and not just the label to be chosen. Analyzing the class probabilities sometimes reveals hidden patterns among the data that may prove to be very useful. However, these relationships have very little influence on the cross-entropy cost function used during the learning phase due to the small probabilities associated with them. Knowledge distillation addresses this problem by raising the temperature of the softmax until the cumbersome model produces a sufficiently soft set of targets. This temperature is later used in the training phase of the small model. The loss function used is the KL divergence function given by the following equation:

$$\begin{aligned} D_{\mathrm{KL}}(M\|N) &= \int_{a_1}^{a_2} M(a)\log\left(\frac{M(a)}{N(a)}\right)\mathrm{d}a \\ &= \int_{b_1}^{b_2} M(b)\log\left(\frac{M(b)\frac{\mathrm{d}b}{\mathrm{d}a}}{N(b)\frac{\mathrm{d}b}{\mathrm{d}a}}\right)\mathrm{d}b \\ &= \int_{b_1}^{b_2} M(b)\log\left(\frac{M(b)}{N(b)}\right)\mathrm{d}b \end{aligned} \tag{2}$$

3.5 Attention Transfer

Attention transfer [14] is a technique that aims to help processes with different knowledge and different goals (thereby different attentional strategies) to focus their "attention" on that part of the "scene" or input, which would be beneficial in terms of attaining their specific objective functions. In the case of artificial neural networks employed in image classification problems, attention transfer defines attention with the help of spatial maps that try to encode on which specific areas in the input the neural network should focus on in terms of deciding the output. These maps, in turn, are defined to capture attentional information at various levels of detail with respect to the input image. In our work, we use the activation-based spatial maps to transfer the attention information from a teacher to the student network. The transfer losses in activation-based attention transfer [14] are placed between student and teacher attention maps of the same spatial resolution; however, interpolation of the attention maps helps to match their shapes. The total loss is defined as:

$$\mathcal{L}_{AT} = \mathcal{L}(\mathbf{W}_S, x) + \frac{\beta}{2} \sum_{j \in \mathcal{I}} \| \frac{Q_S^j}{\|Q_S^j\|_2} - \frac{Q_T^j}{\|Q_T^j\|_2} \|_p \ , \tag{3}$$

Moreover, we are using a hybrid technique that aims to combine attention transfer with knowledge distillation [7]. To do so, we include an additional term corresponding to the cross-entropy loss between the softened distributions over the labels of the teacher and the student model to the knowledge distillation loss.

4 Experiments and Results

The BACH 2018 dataset is used in this project,[1] which contains 400 high-resolution breast cancer histology tissue slides, stained with hematoxylin and eosin. The images in the dataset have one of the four labels: normal, benign, invasive, and in situ, with 100 images of each category. We took 70 samples, 15 samples, and 15 samples from each class for training, validation, and testing, as to avoid class imbalance. Since the images are high resolution, it is impractical to train a CNN using whole photographs. Therefore, we employ the technique of patchwise generation in this case. From each image, 35 patches are generated, thereby covering the entire surface of the input image. Each patch was assigned a class label of the image from which it is extracted.

We train the networks on a single NVIDIA Tesla P-100 GPU with 12 GB memory using the Adam optimizer [8], with a mini-batch size of 16 and an initial learning rate of 0.001, with a decay of 0.1 every 20 epochs. The results reported in Tables 1 and 2 are reported on patchwise classification of data. From Table 1, we can see that the ResNet-50 (teacher model) has the best precision on majority of classes. This is because a higher parameter model has a better generalization performance [2]. The precision on ResNet-8 (without distillation training) showed poor results, but after using both distillation and attention transfer, the model showed good generalization over the data. Similarly in Table 2, we report the F1 score of all the classes and accuracy of the prediction to demonstrate the effectiveness of using knowledge distillation technique and the hybrid of knowledge distillation and attention transfer for Medical Histology Image Classification, for the task of improving upon the base performance of our student model.

[1] https://iciar2018-challenge.grand-challenge.org/dataset/.

Table 1 Precision and recall

Model Architecture	Precision				Recall			
	Benign	In-situ	Invasive	Normal	Benign	In-situ	Invasive	Normal
ResNet-50	0.85	0.83	0.92	0.72	0.62	0.84	0.91	0.90
Resnet-8	0.68	0.79	0.82	0.74	0.74	0.77	0.76	0.75
ResNet-8 with KD	0.81	0.81	0.80	0.79	0.76	0.78	0.82	0.85
ResNet-8 with AT + KD	0.75	0.77	0.81	0.83	0.72	0.85	0.80	0.79

Table 2 F1 score and accuracy

Model Architecture	F1 score				Accuracy (%)
	Benign	In-situ	Invasive	Normal	
ResNet-50	0.72	0.84	0.92	0.80	82.00
ResNet-8	0.71	0.78	0.79	0.75	75.71
ResNet-8 with KD	0.79	0.79	0.81	0.82	80.24
ResNet-8 with AT + KD	0.74	0.80	0.81	0.81	78.95

5 Conclusion and Future Work

In this paper, we have demonstrated the fact that architectures with lower parameters can perform well with the help of knowledge distillation and hybrid of knowledge distillation and attention transfer. We also intend to further research on the vast array of applications of our paper in the industrial sector. Moreover, for further study, we can use several novel methods proposed in medical image classification by replacing the base models with low parameter models. Also, we can utilize semi-supervised techniques on these to improve the generalizability of our models.

References

1. G. Aresta, T. Araújo, S. Kwok, S.S. Chennamsetty, M. Safwan, V. Alex, B. Marami, M. Prastawa, M. Chan, M. Donovan, G. Fernandez, J. Zeineh, M. Kohl, C. Walz, F. Ludwig, S. Braunewell, M. Baust, Q.D. Vu, M.N.N. To, E. Kim, J.T. Kwak, S. Galal, V. Sanchez-Freire, N. Brancati, M. Frucci, D. Riccio, Y. Wang, L. Sun, K. Ma, J. Fang, I. Kone, L. Boulmane, A. Campilho, C. Eloy, A. Polónia, P. Aguiar, Bach: grand challenge on breast cancer histology images. Med. Image Anal. **56**, 122–139 (2019). https://doi.org/10.1016/j.media.2019.05.010
2. A. Brutzkus, A. Globerson, Why Do Larger Models Generalize Better? A Theoretical Perspective via the XOR Problem. arXiv e-prints arXiv:1810.03037 (2018)
3. A. Golatkar, D. Anand, A. Sethi, Classification of Breast Cancer Histology using Deep Learning. arXiv e-prints arXiv:1802.08080 (2018)

4. Group DIA, Breast Histology (2019). http://diagnijmegen.nl/index.php/Breast_Histology. Accessed 15 Mar 2020
5. K. He, X. Zhang, S. Ren, J. Sun, Deep Residual Learning for Image Recognition. arXiv e-prints arXiv:1512.03385 (2015)
6. G. Hinton, O. Vinyals, J. Dean, Distilling the Knowledge in a Neural Network. arXiv e-prints arXiv:1503.02531 (2015)
7. J. Karlekar, J. Feng, Z. Sian Wong, S. Pranata, Deep Face Recognition Model Compression via Knowledge Transfer and Distillation. arXiv e-prints arXiv:1906.00619 (2019)
8. D.P. Kingma, J. Ba, Adam: A Method for Stochastic Optimization (2014)
9. Y. LeCun, C. Cortes, MNIST Handwritten Digit Database (2010). http://yann.lecun.com/exdb/mnist/
10. K. Nazeri, A. Aminpour, M. Ebrahimi, Two-Stage Convolutional Neural Network for Breast Cancer Histology Image Classification. arXiv e-prints arXiv:1803.04054 (2018)
11. E. Reinhard, M. Adhikhmin, B. Gooch, P. Shirley, Color transfer between images. IEEE Comput. Graphics Appl. **21**(5), 34–41 (2001). https://doi.org/10.1109/38.946629
12. R.L. Siegel, K.D. Miller, A. Jemal, Cancer statistics, 2019. CA: Cancer J. Clin. **69**(1), 7–34 (2019). https://doi.org/10.3322/caac.21551
13. S. Vesal, N. Ravikumar, A. Davari, S. Ellmann, A. Maier, Classification of Breast Cancer Histology Images Using Transfer Learning. arXiv e-prints arXiv:1802.09424 (2018)
14. S. Zagoruyko, N. Komodakis, Paying more attention to attention: improving the performance of convolutional neural networks via attention transfer. CoRR abs/1612.03928. http://arxiv.org/abs/1612.03928 (2016)

Analyzing the Spread of COVID-19 in India Through PageRank and Diffusion Techniques

Sawon Bhattacharya, S. Siva Sathya, and S. Sharmiladevi

Abstract From time immemorial, the world has faced and overcome several epidemics claiming millions of lives among which the corona virus disease-2019 (COVID-19) has become a global threat with nations trying to figure out the techniques to analyze, mitigate, and control the spread of the pandemic. Various mathematical and statistical models have been utilized to understand the spread of epidemics in the past. This paper tries to use the concepts and models used in social network analytics like page rank, diffusion, and cascading effects to understand the spread of COVID-19 in the Indian subcontinent. The identification of hot spot is done through the page rank algorithm, and the spread of the disease is explained through the process of diffusion and cascading effects. The concept has been implemented on the real-life dataset collected day-to-day from Kaggle using python to visualise the results and draw a conclusion.

Keywords COVID-19 · Social network · PageRank · Pandemic · Epidemic · Diffusion · Cascading · Tablighi Jamaat

1 Introduction

The prime concern of any nation today is the fight against COVID-19, which has created a colossal havoc all over the world. Here, for this work, only cases in India have been considered. The Government is battling through the disease by introducing the concept of social distancing and complete lockdown to curtail the severity of the

S. Bhattacharya · S. Siva Sathya · S. Sharmiladevi (✉)
Department of Computer Science, Pondicherry University, Puducherry 605014, India
e-mail: sharmiladevi94@gmail.com

S. Bhattacharya
e-mail: sawon17081997@gmail.com

S. Siva Sathya
e-mail: ssivasathya@gmail.com

J. Tavares et al. (eds.), *Emerging Technologies in Data Mining and Information Security*, Lecture Notes in Networks and Systems 164,
https://doi.org/10.1007/978-981-15-9774-9_67

spread as it has happened in rest of the world. Compared to other countries, it is reported that the spread in India is much less and has been attributed to the efforts of the government through social distancing. Even then, the minimal increase that India sees today could be explained through the concepts of page rank, diffusion, and cascading effects predominantly studied with social network analytics. Though India witnessed mild increase for the first few weeks, the event called "Tablighi Jamaat," a huge congregation that took place in Delhi (the capital of India) has been claimed by several sources to have spurred the increase in the spread.

To substantiate the claims and analyze the nature of the spread, this paper has created a network of all states and union territories based upon the number of cases found in the dataset. Every node in the network is a state or a Union Territory like Delhi, West Bengal Tamil Nadu, etc. Each edge between a pair of state represents a direct connection between them. By analyzing this network, the pagerank of each state could be calculated.

After pagerank, we will discuss how the Delhi congregation has become a major factor in diffusing the COVID-19 virus all over the India. Also, the infectiousness of the corona virus and how fast it has spread through the network has been analyzed by implementing the cascading concepts in the network. Finally, the number of confirmed cases (State wise) based upon the date of "Tablighi Jamaat" has been drawn up.

The rest of the paper is organized as follows: Sect. 2 gives a brief report on the literature related to page rank, diffusion and cascading. Sect. 3 discusses about the dataset used for the study; Sect. 4 details the use of pagerank for determining the hotspot of COVID-19. Section 5 on the spread of the disease using the concept of diffusion and its results, Sect. 5.1 on the cascading effects of the disease, and Sect. 7 concludes the paper.

2 Related Work

The COVID-19 outbreak has also seen a surge in the number of research articles trying to study the disease in various perspectives. For instance, one article has used ARIMA model to study the COVID-19 dataset [1]. Some other papers like [2–4] have employed machine learning and time series prediction techniques for the analysis. But this work focuses on the concepts of social network analytics techniques like page rank, diffusion, and cascading behavior to analyze the COVID-19 dataset [5–12]. Hence, a brief review of the above-mentioned techniques is presented in this section.

2.1 Page Rank

The page rank algorithm was invented in 1998 by the founder of Google search engine, Larry Page, and Sergey Brin to prioritize or quantify pages with respect to the others in the World Wide Web assisting search engines to list them out based on relevance to a search query. The basic concept of page rank is that the importance of a page is directly proportional to the number of web pages linking to that page. If a page is linked by a large number of other pages that are themselves very important or carry a large pagerank, then the page is given a higher page rank meaning that it is more relevant to a given search query.

Hence, to rank the web pages, the page rank algorithm works by counting the number and quality of links to a page to determine a rough estimate of how important the website is.

The basic definition of Page rank as given in [13] assumes page rank as a kind of "fluid" that circulates through the network, passing from node to node across edges, and pooling at the nodes that are the most important. Specifically, page rank is computed as follows:

$$[X] = 1 - d + [\mathrm{PR}(Y1)/C(Y1) + \ldots + \mathrm{PR}(Yn)/C(Yn)]$$

2.2 Diffusion and Cascading

Diffusion and cascading concepts are used to study the spread of something (message, new product, technology, disease, etc.) across a network [5, 6, 9, 11]. Most often, an understanding of the mode and speed of spreading is required to improve or combat the spread based on whether it is a wanted or unwanted phenomenon. For instance, in viral marketing, the spread of a new product is an essential phenomenon whereas in the spread of an epidemic, it is an unwanted phenomenon which needs to be brought to a halt before infecting other nodes. Hence whatever may be the case, understanding the concept of diffusion and cascading becomes mandatory to act accordingly. Diffusion and cascading are determined by a number of parameters of which the important ones pertaining to the COVID-19 dataset are listed below:

Primary Parameters for Diffusion

Infectiousness metric. Infectiousness basically defines the power of the news/ideas or disease. It basically implies how fast these ideas or news or disease can spread through the network. In this paper, the infectiousness of the virus COVID-19 is used to calculate the payoff of the nodes in the network. In fact, this virus has a very high infectiousness metric.

Resistance. The word "Resistance" itself defines the meaning and that is the ability not to be affected by anyone or anything. In this concept, basically resistance of a node means the power to oppose against anything whether it is an idea or disease. In this paper, resistance is not discussed broadly but to understand the concept of how to stop the diffusion of the virus it can be helpful.

Topology. Topology is an important parameter in the concept of diffusion and cascading behavior. It basically depicts the idea of connectivity among nodes across the network. Better connectivity will help the ideas/news or disease to spread faster across the network. Just like infectiousness topology also plays an important role in calculation of the payoff from which the idea of the cascading behavior of the network will be better to understand.

Dissemination Strategy. Last but not the least parameter is dissemination strategy. Basically, by observing the action or fact of spreading some news/idea or disease one has to draw the conclusion whether the strategy is random or through some defined logic. In this paper, the main topic is COVID-19 virus and the virus has spread randomly.

Modeling Diffusion through a Network. The spread of information or technology or disease through a network called as diffusion could be modeled mathematically in order to understand the dynamism of the spread and to make intelligent decisions appropriate to the scenario. The spread may follow a particular pattern and may depend on the topology and the adoption strategy of the nodes surrounding the node which starts the spread. The adoption strategy may depend on certain payoffs, for instance, if its a new technology, the gains reaped out of adopting it may be considered as the payoff. But in the case of a new disease, it's not the adoption of the nodes, but the resistance or immunity of the nodes surrounding an infected node. If the resistance of the surrounding nodes is less or the infectiousness metric of the affected node is too high, the disease spreads rapidly to the other regions and this may result in a cascading effect throughout the entire network.

In [13], a game theoretic approach to model diffusion has been proposed in which each node in a network is considered to be an individual in the population. There is an edge (i, j) if i and j can interact with each other. The payoffs that they get are defined as follows: if i and j both adopt behavior A, they each get a payoff of $a > 0$; and if they both adopt B, they each get a payoff of $b > 0$ and if they adopt opposite behaviors, they each get a payoff of 0.

Now, if a node j has multiple neighbors say "d" and a p fraction of them adopts behavior A and $(1 - p)$ fraction of them adopts B, then node j wishes to maximize its payoff. The strategy is defined through an inequality which states that if at least a $q = b/(a + b)$ fraction of its neighbors follow behavior A, then j should also follow the same.

3 Dataset

The datasets for this study and analysis come from an acclaimed website "Kaggle" [14] which is an online community of data scientists and machine learning practitioners. The chronology of the events that created the accumulation of this dataset is as follows:

(a) January 30: The first case of the COVID-19 in India was reported, originating from China.
(b) March 14: Central government declares COVID-19 as a "notified disaster"
(c) March 15: The number of positive cases crosses 100
(d) March 23: People participated in "Tablighi Jamaat" started flying from Delhi to several states and cities of India [14]
(e) March 30 and April 10th indicates the weekly updates of the number of positive cases in the various states.

The data has been pre-processed to create a dataset suitable for the analysis. A csv file contains the links or paths between states or union territories. If there is a path, then it will be represented by 1 else it will be 0. From this file, the network to analyze the page rank, diffusion, and cascading effects have been created. Another csv file contains the number of total confirmed cases of each states or union territories.

4 Hotspot Detection in COVID-19 Dataset Using Page Rank

Though page rank algorithm is used to find the rank or importance of the web pages as discussed in Sect. 2.1, the concept has been successfully employed here in the network constructed to derive the most important node based on the connectivity of the node through the network. By this, we can say that if a virus spread through the network, then the node with highest pagerank can play a crucial role. The pagerank has been computed for each node and given in Table 1. From this network shown in Fig. 1, the page rank of each node where each node represents a state or union territory has been calculated. From Table 1, it is seen that Delhi has the highest pagerank of 0.04058. Hence, Delhi could be considered as most dangerous state in terms of becoming a vector to carry the virus.

5 Diffusion and Cascading

Once the hotspot has been identified, the next task is to study the diffusion of the virus within the network using the definitions given in Sect. 2.2. Basically here, Delhi has been identified as the hotspot after the conduction of the congregation and the result ascertained by the pagerank algorithm. The node marked in red in Fig. 2 is

Table 1 Page rank of states and union territories in India

State/UT	ANI	AP	ASSM	Bihar	Chandi	CHG	DLH	Goa	Guj	Har
Page rank	0.032542	0.040475	0.039227	0.033943	0.030063	0.039221	0.040585	0.02991	0.032594	0.028893
State/UT	HP	JK	Jhar	KNTK	Ker	LDK	MP	Mah	Man	Miz
Page rank	0.02765	0.030101	0.03895	0.036401	0.032622	0.031353	0.03895	0.035517	0.014977	0.027772
State/UT	OD	PDC	Pun	Raj	TN	Tel	Trp	UP	UK	WEB
Page rank	0.028824	0.02521	0.03259	0.0364	0.036342	0.036681	0.040537	0.032651	0.036392	0.032633

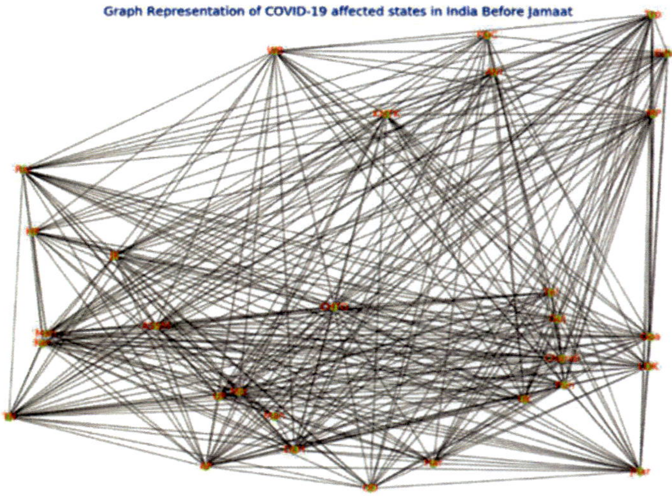

Fig. 1 Network of states and union territories in India

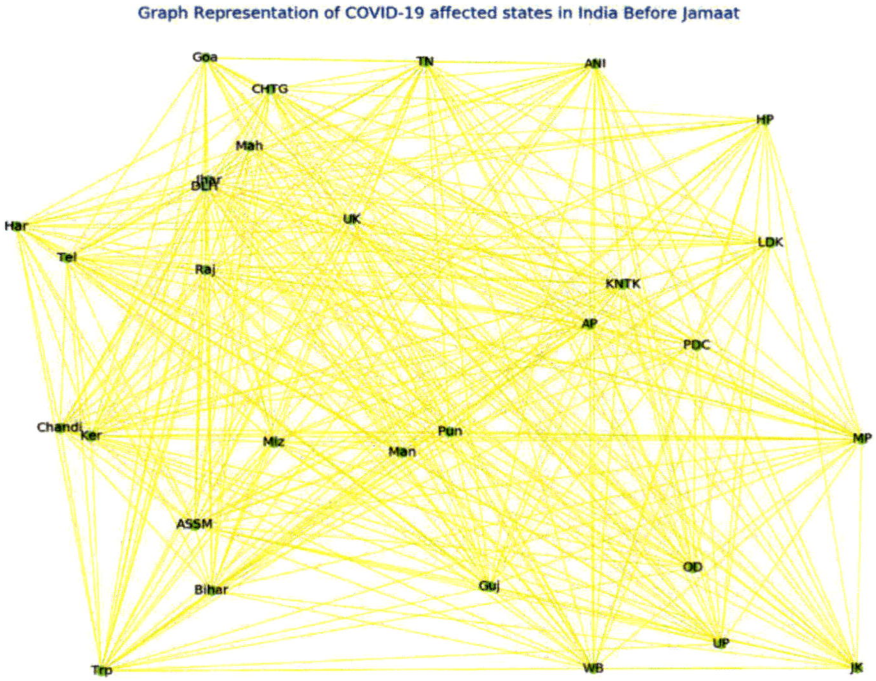

Fig. 2 Cascading started from Delhi (red node)

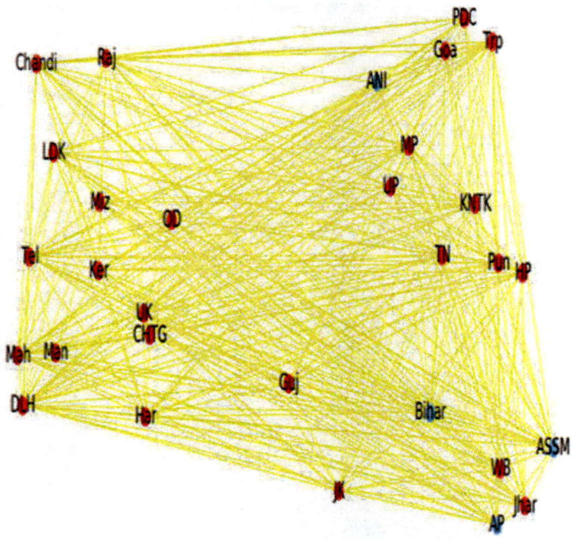

Fig. 3 Cascading step 2 (red nodes have increased)

Delhi which represents that it has adopted a particular behavior. Hence, applying the concepts of diffusion elaborated in Sect. 2.2 to visualize the next round of cities to which the virus spreads, Fig. 3 shows that from the initial adopter namely Delhi, which hosts the virus, the next set of states to which the disease has diffused. Except for 4 states shown in blue, all other states have adopted the new behavior which means the infectiousness metric is high and resistance is low in these cities.

5.1 *Cascading Behavior*

Most of the time, cascading behavior is exhibited in a network wherein, step by step, either all the nodes adopt a certain behavior or the diffusion might stop at some instance. This results in some nodes following behavior A while others following behavior B such that both A and B coexist in the network, i.e., the cascading effect percolates for a while and stops due to various reasons-maybe topology, infectiousness metric, resistance, etc., of the nodes in the network. To study the cascading effect of the COVID-19 to all the states in India, the parameters were adjusted and the iterations continued.

In this network, all nodes initially were not that much affected and symbolized by "sky-blue" color, except the node "DLH" (Delhi). That node with "red" color represents the event "Jamaat". After that event, the novel virus had increased rapidly from Delhi across the network. The payoff parameter, which is infectiousness metric here, was empirically analyzed to model the real scenario (i.e., the actual data obtained about the covid-19 positive cases) a payoff of each "red" node = 100 and blue node

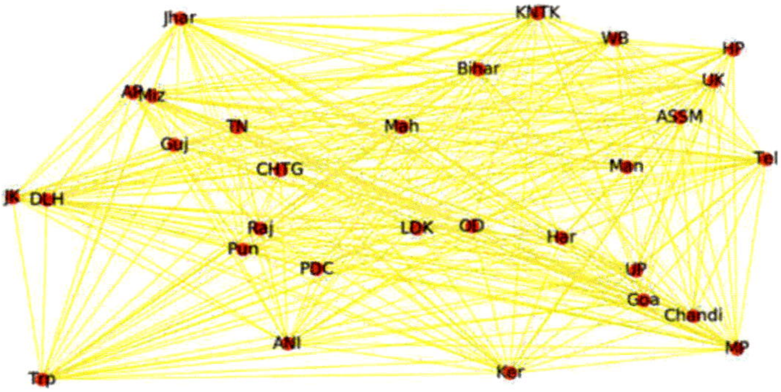

Fig. 4 Cascading step 3 (red nodes have increased)

= 5 set as the infectiousness metric was found to project the real scenario as per the data collected.

From the images, we can see that cascading behavior had started from Fig. 2 and in Fig. 4 full cascade model is obtained. This depicts that just in two iterations the virus had diffused through the network. After the first iteration, it is visible that only 4 states are not affected. Those are: "Assam," "Arunachal Pradesh," "Bihar," and "Andaman and Nicobar Islands." From this, it can be observed not only the high payoff (infectiousness parameter), but also the topology of the network (connectivity) is also an important factor to diffuse the virus through the network. Because when the cascading is being calculated for each node, the number of neighbors for each node plays a huge role. Geographically places like Assam, Arunachal Pradesh, Andaman and Nicobar Islands are little isolated than the other states or union territories. In first level, those are not affected.

But in the second iteration, from Fig. 4, it can be noticed that all of the states are being colored "red" which indicates the full cascading. It takes only two iterations to complete the full cascading through the network. This indicates the deadliness and the connectivity between the edges because earlier it is mentioned that the payoff and connectivity play important roles in analyzing the cascading behavior. This is where the social distancing and lockdown of hotspots play a key role in containing the virus, which India has followed precariously to combat the COVID-19.

So from here, we can say that not only because the payoff of red node was high, but also the structure and connectivity of the network were also the other factors.

6 Result Analysis

Based on the hotspot identification and the depiction of the spread of the disease to all the states through the diffusion model, the actual data collected from Kaggle were analyzed. For simplicity, the weekly data has been taken commencing from 23rd

Table 2 Weekly updates of COVID-19 cases in top 10 states

Name of state/UT	23-03-2020	30-03-2020	10-04-2020
Mah	74	193	1364
TN	9	50	834
Delhi	29	53	720
Raj	28	57	463
Tel	32	69	442
UP	31	75	410
Ker	67	194	357
Andhra	7	19	348
MP	6	33	259

March which represents the day that the congregation was held in the hot spot Delhi. The data was analyzed thereafter as before the event and after the event. From the graph, we can see the epic difference in the total number of confirmed cases. Table 2 holds only top ten affected states though by this time span, more than 90% states were affected by the virus. Though the number of confirmed cases were relatively lesser than the other countries like USA, Italy, Japan, etc. But as time has increased, the numbers have increased exponentially, i.e., after the "Tablighi Jamaat" took place in Delhi and from 23rd March, people from Delhi have gone to their places. So, after week 3, there is a huge difference in the number of cases confirmed for each state or union territory. Had the congregation been held in any of the other place whose page rank is less, instead of the hotspot Delhi, then the spread would not have been to this extent. This has led to the contact tracing and isolation of members who attended the congregation which has substantially resulted in a decrease after the 21 days lockdown.

7 Conclusion

As the entire world is facing a tough crisis in the name of COVID-19, it is essential that appropriate techniques be used for understanding the disease spread to mitigate further disaster. This paper proposed the adoption of techniques used for social network analytics namely page rank, diffusion, and cascading effects to analyze the pandemic. On the one hand, it is proved that the use of these techniques could yield useful insights about the spread of the pandemic. On the other side, the results showed that the spread of the disease in India is attributed to the congregation held in Delhi, which became the hotspot initially and from which the virus slowly spread to other states which wouldn't have reached far and wide if the social distancing was adhered to strictly.

Acknowledgements This work was supported by Council of Scientific & Industrial Research, India under the scheme Direct- SRF with grant File No:09/559(0141)/19-EMR-I.

References

1. D. Benvenuto, M. Giovanetti, L. Vassallo, S. Angeletti, M. Ciccozzi, Application of the ARIMA model on the COVID-2019 epidemic dataset. Data Brief **29**, 105340 (2020)
2. Y. Ma et al., Effects of temperature variation and humidity on the death of COVID-19 in Wuhan, China. Sci. Total Environ. 138226 (2020)
3. D. Fanelli, F. Piazza, Analysis and forecast of COVID-19 spreading in China, Italy and France. Chaos, Solitons Fractals **134**, 109761 (2020)
4. D. Kang, H. Choi, J.-H. Kim, J. Choi, Spatial epidemic dynamics of the COVID-19 outbreak in China. Int. J. Infect. Dis. (2020)
5. Y. Zhang, J. Zhao, Social network analysis on data diffusion in delay tolerant networks, in *Proceedings of the International Symposium on Mobile Ad Hoc Networking and Computing (MobiHoc)* (2009), pp. 345–346
6. M. Kimura, K. Saito, Tractable models for information diffusion in social networks, in *Lecture Notes in Computer Science* (including subseries Lecture Notes in Artificial Intelligence and Lecture Notes in Bioinformatics), vol. 4213 (LNAI, 2006), pp. 259–271
7. T. Haveliwala, Topic-sensitive pagerank: a context-sensitive ranking algorithm for web search. IEEE Trans. Knowl. Data Eng. **15**(4), 784–796 (2003)
8. P. Cui, S. Jin, L. Yu, F. Wang, W. Zhu, S. Yang, Cascading outbreak prediction in networks: a data-driven approach, in *Proceedings of the ACM SIGKDD International Conference on Knowledge Discovery and Data Mining*, vol. Part F1288 (2013), pp. 901–909
9. M. Sattari, K. Zamanifar, A cascade information diffusion based label propagation algorithm for community detection in dynamic social networks. J. Comput. Sci. **25**, 122–133 (2018)
10. C. Tong, W. He, J. Niu, Z. Xie, A novel information cascade model in online social networks. Physica A **444**, 297–310 (2016)
11. Y. Tonta, H.R. Darvish, Diffusion of latent semantic analysis as a research tool: a social network analysis approach. J. Informetrics **4**(2), 166–174 (2010)
12. B. Yu, H. Fei, Modeling social cascade in the flickr social network, in *6th International Conference on Fuzzy Systems and Knowledge Discovery, FSKD 2009*, vol. 7 (2009), pp. 566–570
13. M. Aalabaf-Sabaghi, Networks, crowds and markets: reasoning about a highly connected world. J. Roy. Stat. Soc. A (Stat. Soc.) **175**(4), 1073 (2012)
14. COVID-19 Corona Virus India Dataset. (2020), Available https://www.kaggle.com/imdevskp/covid19-corona-virus-india-dataset. Accessed 19 Apr 2020

Graph-Assisted Attention for Path Finding in Question Answering Task

M. Guruprasad, Jai Agarwal, T. N. Lokesh Kumar, and Bhaskarjyoti Das

Abstract Attention-based memory networks, a class of deep learning algorithms in Natural Language Processing (NLP), capture long-range dependencies present in text data and is a popular recipe in currently available question answering (QA) systems. However, multi-hop QA systems pose additional challenges that these memory networks cannot comfortably handle with their attention spans. Path-finding tasks are a flavor of such multi-hop QA, and it does not have the additional complexity of implicit reasoning. A directed knowledge graph of entities, built from any text corpus, can capture both direction and implicit reasoning. This work attempts to address the bottleneck faced by sophisticated memory network implementations on a dataset that is appropriate for solving such path-finding tasks over multiple hops. It uses the knowledge graph approach for the directional information to address the search challenge in path finding. The experiments conducted using this approach have produced impressive results and can be extended in the future for higher-order challenges that consist of implicit reasoning.

Keywords Question answering · Attention · Knowledge graph · Graph linearization · End to end memory network · Dynamic memory network · bAbI dataset · Path finding task

M. Guruprasad (✉) · J. Agarwal · T. N. Lokesh Kumar · B. Das
PES University, Bengaluru, India
e-mail: mguru1998@gmail.com

J. Agarwal
e-mail: jai.bhageria@gmail.com

T. N. Lokesh Kumar
e-mail: lokeshkumartn30@gmail.com

B. Das
e-mail: bhaskarjyoti01@gmail.com

J. Tavares et al. (eds.), *Emerging Technologies in Data Mining and Information Security*,
Lecture Notes in Networks and Systems 164,
https://doi.org/10.1007/978-981-15-9774-9_68

1 Introduction

The main focus of question answering (QA) discipline within Natural Language Processing (NLP) is to develop systems that are to answer questions posed to them. This is mainly achieved by training machine learning systems through which they are taught how to answer the questions. Most NLP tasks can be converted to a question answering task, like sequence-to-sequence translation, co-reference resolution, etc. These systems are used majorly in chatbots and for information retrieval as a sub-task of another task like text summarization. QA systems can be classified into various types. Based on the range of topics on which a system can answer a question, they can be classified as open-domain QA systems, which can answer questions on a wide variety of topics and closed-domain QA Systems, which can answer questions on a particular domain captured in a given text corpus. Another type of classification is into single-hop QA systems, which can answer questions by referring to just one paragraph or sentence, and multi-hop QA systems, which would have to refer multiple paragraphs or sentences to answer a question. Multi-hop QA is more challenging since the required information to answer a given question is usually spread across multiple paragraphs. Co-reference resolution is key in multi-hop QA.

A multi-hop QA system can also come in various flavors. In the first scenario, the task involved is tracing the relevant answer by traversing across multiple paragraphs of a document and answer being explicitly present in the document. This scenario can be thought of as some kind of search problem over a directed graph of entities present in the set of paragraphs or documents. This is essentially a path-finding task. Path-finding QA refers to computing answers for path-finding tasks. A path-finding task contains a set of locations and their locations in the form of inter-related sentences and the objective is to determine the path between any two given locations. A flavor of this problem can exist as a knowledge-base search problem where the knowledge base itself is composed of the set of documents over which the QA system is constructed. In the second scenario, additionally, there can be reasoning involved and this is essentially a complicated multi-hop QA scenario that requires a fair bit of multi-hop reasoning.

In this work, the particular flavor of multi-hop QA systems that consists of "path-finding tasks" is targeted. Almost all the present-day QA systems rely on deep learning systems in NLP that utilize sequence to sequence architectures coupled with an attention scheme. However, as indicated in the existing work section, the efficacy of various attention mechanisms has reached a bottleneck. The work described in this paper proposes an approach that works around this bottleneck for the path-finding tasks of multi-hop QA systems.

2 Existing Work

2.1 Question Answering (QA) Systems and Existing Datasets

Text-based question answering system has been a popular topic in NLP as QA provides an easy way to judge any NLP system's capability in understanding as well as reasoning natural language texts. As mentioned earlier, open-domain versus closed-domain is a broad categorization of QA systems. Similarly, the other broad categorization is essentially about the breadth of the context that the answers in a QA system have. Single-hop QA is mostly about the case when the context of the answer (reasoning or fact) is in a single paragraph or document. In case the context is multiple paragraphs or documents, the QA system is multi-hop. As the capability of QA systems has evolved, multiple QA datasets have come into existence. Each dataset in turn has supported multiple tasks and few tasks of single-hop learning datasets have required multi-hop reasoning as well. Hence, first-generation datasets such as SQuAD [13] and bAbI [18], at least for their fact-finding tasks, are single-hop reasoning types. Task 19 (path finding task) of bAbI is not strictly a single-hop learning task. The focus of this work is multi-hop reasoning in QA. Recently, researchers have formed few multi-hop QA datasets as well. For example, SearchQA [6] and TriviaQA [9] have a multi-document focus, whereas COMPLEXWEBQUESTIONS [16] has the scope of existing knowledge bases. HotpotQA [20] has a focus on reasoning in a multi-paragraph context.

2.2 Memory Network in Sequence-to-Sequence Learning

Modern QA systems in NLP are most commonly engineered by deep learning techniques that employ sequence-to-sequence learning using LSTM and GRU with attention mechanisms. A common issue with such pure attention [1]-based network is that the learning network does not parse the input multiple times, which in turn hinders the scope of attention. Memory networks tend to address this issue by incorporating memory in the design and have remained the most popular choice in NLP QA systems. There are two main approaches that employ sophisticated techniques:

1. **End-to-end memory network (MemN2N)** [15]: upgrades the existing attention mechanism in recurrent neural network (RNN) model with large external memory for performing multiple reads. End-to-end memory network is available as a single-layer or multi-layer variant. The single-layer variant can be used to implement single-hop tasks, which essentially means that a single context from the memory representations is used to generate the answer. This can be stacked multiple times, resulting in a multi-layer variant that can be used to achieve multiple hops over the memory. Two different types of sentence vectorization can be applied to the given set of input sentences before feeding them to a MemN2N

model. The first is a bag-of-words model, and the second is a positional encoding model. The latter is preferred over the former since the bag-of-words model cannot capture the order of words in a given sentence. On top of this, a temporal embedding is also generated for sentences that have an inherent notion of time since the model would have to understand the temporal context.
2. **Dynamic memory network (DMN)** [10]: This is a deep learning network that is mainly used to train models for tasks like question answering. The highlight of a dynamic memory network is that it can form episodic memories of the given set of input sentences and use them to compute the required answer. This is carried out by refining the mechanism of attention, wherein, when a question is given to the model, an iterative attention process is triggered, hence resulting in the model tuning its attention on the current input and the result of previous iterations. The architecture of DMN consists of four main components. The first is an Input Module which transforms any given sentence or a set of sentences into a normalized vector representation to account for the varying lengths of sentences. The second is a Question Module which generates an embedding for a given question. The Episodic Module follows this, and it updates the hidden state based on the embeddings generated by the Input Module. Finally, the Answer Module generates the answer to a given question.

There have been multiple pieces of research [5, 14] that have augmented the idea of such memory networks. However, the memory network is not a panacea in every kind of QA task. Chen et al. [2] analyze the same in three multi-hop reasoning scenarios. In two out of three scenarios, memory networks failed to perform the reasoning and did not perform well unless supplemented with additional supervision. Hence, in this work, the focus has been on additional supervision in terms of directional knowledge.

2.3 *Graph-Based Approaches in Question Answering Systems*

There have been two kinds of graph-based approaches toward QA systems in NLP. One is a graph-based parallel to neural techniques, and the other is around the knowledge that can be represented as a graph. De Cao et al. [4] addresses reading comprehension in a multi-document or multi-paragraph context by building a graph of entities in the document where nodes are entities and edges are "across" or "within" relations. Graph convolution networks are then used on these graphs as a supervised learning task for multi-hop reasoning. Recently, Xiao et al. [19] extended this graph approach by building this entity graph dynamically based on the query. Zhang et al. [22] also use a deep learning approach on the graph but addresses the query matching using a graph-based attention network.

In a recent work, Fan et al. [7] use a graph-structured knowledge base in open-domain tasks such as multi-document summarization and shows that it performs better than traditional TF-IDF-based approaches. Collarana et al. [3] break the QA system into two parts, i.e., information extraction where they utilize the graph and then the subsequent part of usual attention-based QA on the extracted paragraphs. Essentially the approaches based on knowledge indicate that pure sequence and memory-based approaches can indeed take some advantage from the implicit graph inside the document. Kumar et al. [11] proved the reverse hypothesis as well by generating multi-hop questions from a knowledge graph.

2.4 Open-Domain QA Systems

Yu et al. [21] in their work have presented a novel approach of sentence selection for answering a given question. Semantic similarity-based encoding has been used to match a given question with the appropriate answer. The focus of the authors has been on open-domain question answering and various information retrieval techniques that can fetch relevant snippets from various sources to construct the final answer have been experimented with. A two-step approach has been used wherein, initially, the relevant set of documents is extracted, followed by the extraction of the relevant paragraphs and sentences. Experimentation was performed on the TREC QA dataset [17], and it was found that a CNN-based bi-gram model was successful in capturing long-range dependencies compared to a normalized bag-of-words approach. Their model, however, failed to pick the correct answers when reference to background information was required to answer a question.

The challenges of large knowledge bases and context-less open-domain questions have been addressed by Feldman and El-Yaniv [8] in their work. A joint vector representation of the paragraphs and the question has been used to iteratively retrieve the relevant paragraphs with the help of multi-hop paragraph retrieval and maximum inner product search algorithms. Using a CNN for word embeddings and a bidirectional GRU to capture more context, sentence embeddings are generated. Sentence-level embedding was preferred over paragraph-level embedding for better reasoning. It was established that the proposed model underperformed in the presence of noise and could perform at optimal levels with a maximum of two iterations only.

The above two pieces of research provide interesting examples of open-domain QA using information extraction as a step and taking advantage of knowledge. The work described in this paper is closed-domain QA but similarly benefits from a knowledge graph approach.

3 Dataset

3.1 The bAbI Dataset

The primary focus of this work is knowledge-assisted directional search in attention-driven question answering system in a multi-hop setup. Decoding the implicit reasoning in the dataset in a multi-hop context is not the focus in this work, and this will only introduce an additional level of complexity over the path-finding task. Hence, after carefully considering the HotpotQA [20] and task 19 of bAbI [18], it was decided to use the bAbI task 19 data. The bAbI dataset is a first-generation dataset where almost all the tasks are designed around single-hop tasks. The existing baselines are state-of-the-art performances on these tasks. However, task 19 of bAbI is a multi-hop QA task. Predictably, existing baselines on task 19 have plenty of scope to improve. Hence, this fitted the requirement in this work perfectly as bAbI did not have reasoning components in its dataset (Fig. 1).

The bAbI dataset is developed by the Facebook AI research team. The dataset comprises a collection of twenty well-defined tasks, which are used for textual understanding and reasoning of a model. Each task is designed in such a manner that it requires a unique aspect of text and reasoning, as a result of which the different capabilities of the learned models are tested. For each task, there are two versions: The 1k dataset comprises 1000 training questions and 1000 testing questions along with the respective stories, and the 10k dataset, which contains 10,000 training questions and

```
1 Mary moved to the bathroom.
2 John went to the hallway.
3 Where is Mary?          bathroom          1
4 Daniel went back to the hallway.
5 Sandra moved to the garden.
6 Where is Daniel?        hallway 4
7 John moved to the office.
8 Sandra journeyed to the bathroom.
9 Where is Daniel?        hallway 4
10 Mary moved to the hallway.
11 Daniel travelled to the office.
12 Where is Daniel?       office  11
13 John went back to the garden.
14 John moved to the bedroom.
15 Where is Sandra?       bathroom          8
```

Fig. 1 An example of a bAbI single supporting fact task. The number after the answer to a question denotes the sentence number needed to answer the given question

1000 testing questions along with the respective stories. A story is defined as a set of sentences corresponding to a particular scenario. The idea is to create models that can answer a given question and provide an explanation by identifying the referenced sentences. As the name bAbI suggests, the tasks are intended to be simple and not entirely realistic. The idea behind this is that, if researchers use a common set of tasks as defined in bAbI, the comparison of the results can be done easily.

3.2 Task 19 of bAbI Dataset

Task 19 of the bAbI dataset corresponds to a path-finding task, in which, given a set of sentences in the form of stories containing locations of places, the path between two locations based on the question by the user needs to be found. These questions effectively emulate a search problem. Referring to Fig. 3, it can be seen that there is a story with five sentences where each sentence portrays a directional relationship between two entities. The goal is to answer the question mentioned in the sixth line. In this task, all the questions can be answered with two-directional hops. This can be achieved by making the model learn the layout of all the locations and their relative directions. The image also shows the answer in the form of two comma-separated directions followed by two comma-separated numeric values, which denote the supporting sentences using which the answer can be computed. Shorthand notations for the directions have been used in the dataset, that is, n for north, e for east, and so on.

4 Methodology

The approach used in this work consists of five modules as shown in Fig. 2.

1. **Input module**: The role of the first module, which is the Input Module, is to read the training data from the dataset file and feed the data to the Graph Creation module.

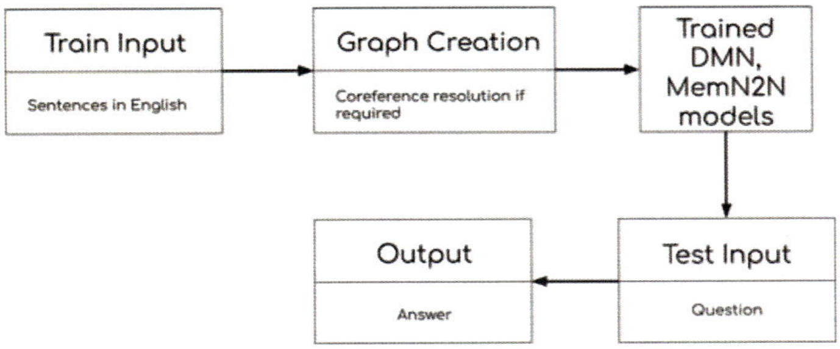

Fig. 2 Methodology represented in a block diagram

2. **Graph Creation module**: Here a graph is constructed by doing entity extraction, etc. and then the graph data is linearized as described in Sect. 4.2.
3. **Training module**: This is the "learning" phase in the machine learning approach used. Here, appropriate training models such as an end-to-end memory network and a dynamic memory network are trained, and the trained models are then passed onto the testing module.
4. **Testing module**: This is the "generalization" phase in the machine learning approach used. This module takes in a question from the user or generates a random question from a set of questions. The answer is then computed here and sent to the Output Module.
5. **Output module**: This displays the answer on an interface.

4.1 Linearized Graph as an Additional Supervision in Seq2seq Attention

A "directional sentence" is one that describes the location of a given entity with respect to another entity. Graph linearization can be defined as the process of converting a given "directional sentence" into a concise form of itself while adding necessary supporting facts.

The idea of graph linearization originated from the results obtained from the experiments conducted on single-hop questions. The model could perform satisfactorily because of the attention mechanism, which however did not help the model perform well on multi-hop path finding questions. In the case of a single-hop question, the model was able to pick the right words to which more attention had to be provided, thereby ignoring the irrelevant information. This was partly because the model had enough information from the input sentences to predict the correct answer.

The drawback discovered with the model on path-finding tasks was that the model had to implicitly learn the relations between the directions. For example, if Place A is to the south of Place B, the model had to internally learn that Place B is to the north of Place A. This information was absent in the input stories. The attention scores were computed based on the relation between the question, present input and the internal memory state of the network, hence performing graph linearization on a given set of directional sentences is expected to greatly improve the performance of a model on path-finding tasks since necessary information has been incorporated into the input and irrelevant information has been removed.

4.2 Implementation of Graph Linearization

Knowledge is usually described in knowledge bases in the form of "subject, predicate, object" triple as per Resource Description Framework (RDF). For textual data, the subject and object can be entities, whereas predicate can be sort of relation between the two of them. Given a piece of textual data, such RDF triples can be extracted and a knowledge graph can be built. For a knowledge-based question answering, this knowledge graph can be queried.

In a directional path-finding task, the predicates between any two entities are the directions between them. One can adopt a convention that the "entity1, direction, entity2" means a direction of entity2 starting from entity1. The main purpose of graph linearization in path-finding tasks is defined as the extraction of such triples from the sentences in the training dataset (Fig. 3).

The sentences were first pre-processed by removing the stop words and out-of-context words. Post this, directional representations for each sentence were created by extracting the two entities and relation (which in this case is the direction) from each sentence and joining them as "entity1, direction, entity2". The direction in every directional representation represents the direction in which one needs to move to get to entity2 from entity1. Additional representations were created as "entity2, opposite direction, entity1". This additional representation was essential since it provides enough information to the model to ensure that the attention mechanism works accurately.

The questions which are part of the stories were also treated in the same manner and represented as "entity1, entity2?". The answers were left unmodified, and the numbers indicating the supporting sentences were modified accordingly to accommodate the additional representations.

```
1 The office is east of the hallway.
2 The kitchen is north of the office.
3 The garden is west of the bedroom.
4 The office is west of the garden.
5 The bathroom is north of the garden.
6 How do you go from the kitchen to the garden? s,e      2 4
```

```
1 hallway office east.
2 office hallway west.
3 office kitchen north.
4 kitchen office south.
5 bedroom garden west.
6 garden bedroom east.
7 garden office west.
8 office garden east.
9 garden bathroom north.
10 bathroom garden south.
11 kitchen garden?      s,e     4 8
```

Fig. 3 A sample story from bAbI Task 19 dataset before and after graph linearization

Finally, the above was written to a file, which was then passed as an input to the end-to-end memory network and dynamic memory network models. It is expected that, when the linearized graph is fed as input to a memory network and the parameters tuned accordingly, higher accuracy can be achieved since more attention can be given to the relevant words.

4.3 Learning Using End-to-End Memory Network (MemN2N)

End-to-end memory network consists of multiple modules, and each module plays an important role in ensuring that the input is transformed into an appropriate form and that learning happens effectively. The three main modules of this memory network are Input Memory Representation Module, Output Memory Representation Module and the Final Prediction Module. In the Input Memory Representation Module, the sentences from the input stories are converted to story vectors of dimension 'd' with the help of an embedding matrix 'A'. Similarly, the given question is transformed into a question vector using an embedding matrix 'B'. Once done, a match between the story vectors and the question vector is made by considering the inner products followed by a Softmax, which generates probability vectors that are fed into the next module.

The role of the Output Memory Representation Module is to compute response vectors and pass them into the final module. This computation is nothing but the combination of the output vectors of input sentences represented by the embedding matrix 'C' and the probability vectors from the previous module.

In the Final Prediction Module, the sum of the response vectors and story vectors is passed through a final weight matrix 'W' followed by a Softmax to produce as output the predicted label.

The weights for the matrices A, B, C, and W are learned using the backpropagation algorithm to minimize the binary cross-entropy loss.

For the experiments, the activation functions used were Softmax and ReLU, respectively, and the embedding method used was positional encoding. The training was performed over 60 epochs with a batch size of 32. The initial learning rate used was 0.01, which decayed by half after every 20 epochs.

4.4 Learning Using Dynamic Memory Network (DMN)

Dynamic memory network comprises four modules, which are Input, Question, Episodic and Answer modules.

The Input Module embeds the given input sentences using an embedding algorithm as well as a gated recurrent unit (GRU) to capture the context of each story.

A similar process takes place for the given question inside the Question Module. Using the context from the Input Module and the question embedding, the Episodic Module updates the internal memory state of the network over multiple iterations with the help of attention. Finally, the Answer Module generates the required answer based on the internal memory state as well as the question embedding.

As part of this work, we trained a dynamic memory network using GloVe [12] vectors and an embedding dimension of 300. For the hidden layers of attention, the tanh activation function was used, while for the final layer of attention and Answer Module, Sigmoid and Softmax, respectively, were the activation functions used.

5 Results and Conclusions

This section presents the results obtained along with the various inferences and conclusions drawn from them.

It can be seen that, when the 1k dataset is used (Table 1), there is a large difference between the training accuracy and the validation accuracy. This is a clear indication of overfitting. This usually can be resolved by training the model using more data so that it can effectively learn the patterns. Another observation that can be made is that there is a clear increase in the training and validation accuracies when linearized graph-based preprocessing is applied to the 1k dataset, this is likely because the model was able to capture the context better with the help of attention modification by the graphs. When the embedding size was increased from 20 to 100 and activation function changed to ReLU, a small increase in the training and validation accuracies can be seen, although overfitting persists. Hence, it can be said that an increase in the embedding dimension enables the model to capture more information and predict the answer more accurately.

Table 1 Results of MemN2N for 1k dataset

Method	Embedding dimension	Activation function	Train accuracy (%)	Validation accuracy (%)
Trained model with original task19 stories	20	Softmax	36.28	8.33
Trained model with original task19 stories	100	ReLU	61.61	9.48
Trained model with linearized graph for task19 stories	20	Softmax	60.38	15.63
Trained model with linearized graph for task19 stories	100	ReLU	73.98	19.81

Table 2 Results of MemN2N for 10k dataset

Method	Embedding dimension	Activation function	Train accuracy (%)	Validation accuracy (%)
Trained model with original task19 stories	20	Softmax	35.48	31.1
Trained model with original task19 stories	100	ReLU	73.21	68.55
Trained model with linearized graph for task19 stories	20	Softmax	67.22	61.98
Trained model with linearized graph for task19 stories	100	ReLU	**100**	**99.98**

From Table 2, it can be observed that there is a sharp increase in the validation accuracy and a low difference between the training and validation accuracies when the 10k dataset is used for training. This shows that overfitting has reduced considerably when the 10k dataset is used since a large amount of training data provides more context to the model to predict the answers. When a combination of linearized graph-based preprocessing, a higher embedding dimension of 100 and ReLU activation unit, the model provides near-perfect accuracy on the validation set, thereby reiterating on the fact that linearized graph-based preprocessing can help in improving the attention and larger embedding dimension coupled with larger training set can help in gathering more context and achieving high accuracies.

The results obtained by training a DMN model on the 1k dataset are shown in Table 3. The first conclusion that can be drawn from the table is that providing a linearized graph of the stories and questions to the model rather than the original stories largely helps in increasing the training and validation accuracies and reducing the binary cross-entropy error. Similar to MemN2N, overfitting is observed when training with the 1k data which can be avoided by using more training data.

Table 3 Results for DMN for 1k dataset

Method	Train accuracy (%)	Validation accuracy (%)	Binary cross-entropy error
Trained model with original task19 stories	84.5	34.5	13.49
Trained model with linearized graph for task19 stories	93.2	68.17	8.40

Table 4 Results for DMN for 10k dataset

Method	Train accuracy (%)	Validation accuracy (%)	Binary cross-entropy error
Trained model with original task19 stories	96.9	89.26	0.89
Trained model with linearized graph for task19 stories	**98.1**	**93.17**	**0.62**

From Table 4, it can be said that providing supporting facts to the model while generating the linearized graph coupled with the usage of 10k dataset results in an increase in both training and validation accuracies and a reduction of overfitting and binary cross-entropy error, thereby implying that providing more training data with appropriate pre-processing is key in achieving the desired results.

The baseline results that are used here are the work on MemN2N by Sukhbaatar et al. [15] and DMN work by Kumar et al. [10]. The MemN2N baseline was 17.2% accuracy on 1K dataset (ours is 19.81%) and 33.5% on 10K dataset (ours is 99.98%). The DMN baseline was 34.5% on 1K dataset (ours is 68.17%) and 89.26%(ours is 93.17%) on 10k dataset. The results conclusively establish the advantage provided by the extraction of a knowledge graph containing directional information and linearization of the same for path-finding question answering tasks.

6 Future Work and Further Steps

The work described in this paper conclusively establishes the efficacy of using graph-based directional knowledge in path-finding question answering task. The general principle followed is the knowledge graph-assisted attention mechanism in seq2seq learning. The graph-based directional knowledge can be generalized further for any other type of knowledge hierarchy in the dataset. One possibility can be to investigate the effect of this technique in other more complicated multi-hop question answering datasets by extending the knowledge triples to cover types other than direction. A natural next step will be the HotpotQA dataset that has a reasoning component.

The other possible direction of the future work is to further refine the current approach without changing the scope itself. Letting the model dynamically decide the type of weight tying mechanism to use (between adjacent and linear weight tying) for the encoding matrices can also be explored. Making the model implicitly understand that north is the complementary direction of south, that of the east is west, and so on is key in reducing the amount of training data after graph linearization by 50% and hence can be a potential extension to this work. Having multiple memory states in the training network coupled with hierarchical attention to the memory states for longer stories is also something that can be considered in the future.

References

1. D. Bahdanau, K. Cho, Y. Bengio, Neural machine translation by jointly learning to align and translate. arXiv preprint arXiv:1409.0473 (2014)
2. J. Chen, G. Durrett, How to learn (and how not to learn) multi-hop reasoning with memory networks (2018)
3. D. Collarana, T. Heuss, J. Lehmann, I. Lytra, G. Maheshwari, R. Nedelchev, T. Schmidt, P. Trivedi, A question answering system on regulatory documents, in *JURIX* (2018), pp. 41–50
4. N. De Cao, W. Aziz, I. Titov, Question answering by reasoning across documents with graph convolutional networks. arXiv preprint arXiv:1808.09920 (2018)
5. B. Dhingra, H. Liu, Z. Yang, W.W. Cohen, R. Salakhutdinov, Gated-attention readers for text comprehension. arXiv preprint arXiv:1606.01549 (2016)
6. M. Dunn, L. Sagun, M. Higgins, V.U. Guney, V. Cirik, K. Cho, Searchqa: a new q&a dataset augmented with context from a search engine. arXiv preprint arXiv:1704.05179 (2017)
7. A. Fan, C. Gardent, C. Braud, A. Bordes, Using local knowledge graph construction to scale seq2seq models to multi-document inputs. arXiv preprint arXiv:1910.08435 (2019)
8. Y. Feldman, R. El-Yaniv, Multi-hop paragraph retrieval for open-domain question answering. arXiv preprint arXiv:1906.06606 (2019)
9. M. Joshi, E. Choi, D.S. Weld, L. Zettlemoyer, Triviaqa: a large scale distantly supervised challenge dataset for reading comprehension. arXiv preprint arXiv:1705.03551 (2017)
10. A. Kumar, O. Irsoy, P. Ondruska, M. Iyyer, J. Bradbury, I. Gulrajani, V. Zhong, R. Paulus, R. Socher, Ask me anything: dynamic memory networks for natural language processing, in *International Conference on Machine Learning* (2016), pp. 1378–1387
11. V. Kumar, Y. Hua, G. Ramakrishnan, G. Qi, L. Gao, Y.F. Li, Difficulty-controllable multi-hop question generation from knowledge graphs, in *International Semantic Web Conference* (Springer, 2019), pp. 382–398
12. J. Pennington, R. Socher, C.D. Manning, Glove: global vectors for word representation, in *Proceedings of the 2014 Conference on Empirical Methods in Natural Language Processing (EMNLP)* (2014), pp. 1532–1543
13. P. Rajpurkar, J. Zhang, K. Lopyrev, P. Liang, Squad: 100,000+ questions for machine comprehension of text. arXiv preprint arXiv:1606.05250 (2016)
14. Y. Shen, P.S. Huang, J. Gao, W. Chen, Reasonet: learning to stop reading in machine comprehension, in *Proceedings of the 23rd ACM SIGKDD International Conference on Knowledge Discovery and Data Mining* (2017), pp. 1047–1055
15. S. Sukhbaatar, J. Weston, R. Fergus, et al., End-to-end memory networks, in *Advances in Neural Information Processing Systems* (2015), pp. 2440–2448
16. A. Talmor, J. Berant, Repartitioning of the complexwebquestions dataset. arXiv preprint arXiv:1807.09623 (2018)
17. E.M. Voorhees, D.M. Tice, Building a question answering test collection, in *Proceedings of the 23rd Annual International ACM SIGIR Conference on Research and Development in Information Retrieval (2000)*, pp. 200–207
18. J. Weston, A. Bordes, S. Chopra, A.M. Rush, B. van Merriënboer, A. Joulin, T. Mikolov, Towards AI-complete question answering: a set of prerequisite toy tasks. arXiv preprint arXiv:1502.05698 (2015)
19. Y. Xiao, Y. Qu, L. Qiu, H. Zhou, L. Li, W. Zhang, Y. Yu, Dynamically fused graph network for multi-hop reasoning. arXiv preprint arXiv:1905.06933 (2019)
20. Z. Yang, P. Qi, S. Zhang, Y. Bengio, W.W. Cohen, R. Salakhutdinov, C.D. Manning, Hotpotqa: a dataset for diverse, explainable multi-hop question answering. arXiv preprint arXiv:1809.09600 (2018)
21. L. Yu, K.M. Hermann, P. Blunsom, S. Pulman, Deep learning for answer sentence selection. arXiv preprint arXiv:1412.1632 (2014)
22. T. Zhang, B. Liu, D. Niu, K. Lai, Y. Xu, Multiresolution graph attention networks for relevance matching, in *Proceedings of the 27th ACM International Conference on Information and Knowledge Management* (2018), pp. 933–942

Towards Open Ended and Free Form Visual Question Answering: Modeling VQA as a Factoid Question Answering Problem

Abhishek Narayanan, Abijna Rao, Abhishek Prasad, and S. Natarajan

Abstract Visual Question Answering (VQA) is a multi disciplinary challenging problem involving various fields such as Natural Language Processing, Computer Vision, Deep Learning and Knowledge Representation, which has garnered much interest among researchers, especially with the recent advancements in machine perception. The problem at hand not only involves reasoning over visual elements present in the image or natural language understanding of the input query, but also may involve outside world knowledge in order to infer the answer. In this paper, we convert the VQA problem to a factoid question answering task over a set of natural language facts extracted from images in the Visual Genome Dataset [1]. Recent literatures in textual question answering have established the effectiveness of End to End Memory Networks (MemN2N) over the standard LSTMs. Inspired by the approaches incorporated by researchers in this direction, as a first step to create an explainable VQA model, this paper proposes the incorporation of MemN2N with soft attention for inferring the answer from a set of regional facts extracted from the image. We also experiment with the addition of a Bayesian Neural layer for posterior reasoning of the answer from a fixed vocabulary, as proposed in [2] which shows a significant improvement in the accuracy score compared to the other models tested.

Keywords Visual question answering · Factoid question answering · End to end memory networks · Bayesian inference

A. Narayanan (✉) · A. Rao · A. Prasad · S. Natarajan
Department of Computer Science and Engineering, PES University, Bangalore , India
e-mail: abhishek.1010n@gmail.com

A. Rao
e-mail: abijnarao@gmail.com

A. Prasad
e-mail: abhishek.pes2016@gmail.com

S. Natarajan
e-mail: natarajan@pes.edu

J. Tavares et al. (eds.), *Emerging Technologies in Data Mining and Information Security*,
Lecture Notes in Networks and Systems 164,
https://doi.org/10.1007/978-981-15-9774-9_69

1 Introduction

In the recent years, there have been significant advancements in research in various domains under AI, such as computer vision, deep learning and natural language understanding. This largely motivated researches to address an important problem in machine perception: visual question answering which largely involves all these domains. Intuitively, the problem at hand can be defined as follows: the input which is fed to the model essentially comprises of an image along with a natural language open ended and free form question about that image, involving but not limited to its visual contents. The model is supposed to subsequently infer and output a correct answer to the input query with respect the image fed.

Though the VQA problem was introduced recently, various researchers have experimented on improving the effectiveness of feature extraction of the image channel and the question channel and have applied a wide variety of deep learning architectures. However due to the vastness and complexity involved in the problem, the existing state-of-art is still constrained with certain limitations and there still exists, a large scope for further improvement in this domain .Some of the existing literatures are often also limited to only a certain variety of questions such as counting, object detection, etc. Most approaches are also not "Explainable", i.e., it is not always evident how the model arrived at a particular prediction and are hence "black box" approaches. Motivated by this problem, in this paper, we propose a novel approach as a preliminary step to make VQA explainable and also achieve reasonable good performance for answering open ended and free form visual question answering using the baseline visual genome dataset. We convert the VQA problem to factoid question answering on textual region-level captions extracted from the image and train deep learning architectures to learn to infer a relevant answer from a vocabulary of answers using a single supporting fact. Extensive experimentation in this direction indicated an accuracy score, comparable with the existing state-of-art approaches.

The remaining part of the paper has been organized as follows. Section 2 provides an insight into the existing literatures and state-of-art and the limitations involved in some of these approaches. The proposed methodology and experimentation performed have been elaborated in detail in Sect. 3. Section 4 portrays the results obtained in our experiments and presents a comparison of all models experimented with. Section 5 concludes the paper with a brief insight into the future directions of research in order to extend the current work.

2 Related Work

A large number of research experiments targeted at VQA have been proposed in the past few years. A large fraction of these existing literatures have portrayed VQA as a classification problem, where each answer is treated as a distinct category, rather than generating natural language answers of variable length and essentially comprise of (1) extraction of features from image channel, (2) extraction of features for the question channel, and (3) an algorithm to combine the aforementioned features. The combined feature set is then fed to a classifier, traditionally a multi layer perceptron for classification of the answer. Forthe purpose of image feature extraction, most approaches use pre-trained variants of CNNs In the question encoding channel, a large variety of text feature extraction techniques have been explored, starting with simple approaches such as bag-of-words (BOW) which generate sparse high dimensional feature vectors and gradually evolving into more complex approaches for extracting dense feature vectors using long short term memory (LSTM) encoders [3], word2vec [4] and glove vectors [5]. However the pre-dominant CNN/LSTM-based approaches to VQA, are often limited, as portrayed in recent literatures, because of the monolithic vector representations which largely ignore structure in the scene and semantics of the question. CNN feature vectors are incapable of effectively capturing the presence of multiple instances of objects, while LSTMs process questions as sequence of words, which may ot often reflect the actual complexity which might be involved in the language structure. Many of the aforementioned pre-trained text feature extraction techniques like word2vec and glove have the disadvantage of grouping together words which are meant to be distinguished, such as "left" and "right". Apart from this, out of vocabulary words appearing in test question samples cannot be represented by word2vec or glove pre-trained models which hinders learning of the semantic representation of the entire question.

The aforementioned approaches differ significantly in how they combine the question and image feature vectors. Some of the approaches include the following:

- Combination of the image and question feature vectors using basic techniques such as concatenation, element wise addition or element wise multiplication [6–9]. However these approaches are incapable of capturing complex inter-relationships between the image and question channels.
- Usage of bilinear pooling or related schemes to combine features [10, 11]. However, it was argued by the authors of [12] that this operation is computationally too expensive for real time performance.
- Computation of spatial attention maps for visual features using question features and adaptively scaling local features based on their relative importance. Various literatures have explored spatial visual attention for VQA [10, 13–16]. Interestingly, though attention mechanisms seemed to have improved performance compared to approaches using global features, a closer inspection based on type of questions fed have shown relatively poor performance on binary (yes/no) questions. Related literatures hypothesize that this is because open-ended and free form questions often require identifying and naming only one concept from the image whereas binary questions typically require reasoning over longer chains of words.

Most of the aforementioned approaches are black box algorithms, as it is close to impossible to determine the inference steps taken by the model to output a conclusive answer. As a first step to make VQA more explainable or interpretable, we propose modeling the task at hand as a factoid question answering problem, wherein we convert the images to a series of facts using region-level captions and infer the answer using a single supporting fact using state-of-the-art deep learning approaches for factoid textual question answering. However, the scope of the current work does not include re-tracing the hops taken by the model to retrieve the single supporting fact or chain of facts from which the answer to a question is inferred and hence, this step has been proposed as future work to extend this research.

3 Proposed Methodology

Inspired by the textual Question-Answering systems, we adopt the following approach. We first extract the regional descriptions corresponding to images present in the visual genome dataset, which serve as the supporting facts for reasoning out the answer. The question answer pairs corresponding to these images available in the dataset are appended to the sequence of supporting facts and a training corpus is formed. Since such ready made regional descriptions are not available for real life test images, for the purpose of testing, we employ a dense captioning algorithm, DenseCap [17] to extract region level scene descriptions or captions to be used as supporting facts. The advantage of using DenseCap is the fact that it has been trained on the visual genome dataset as well, which enables it to generalize well on the data. The training corpus thus created is subsequently fed to memory based end to end deep learning architectures which have been shown to specialize in textual question answering task. Since most of the existing literatures are focussed at predicting one word answers or answering multiple choice questions, this research is focussed at one word answers in order to establish a fair comparison based on accuracy metric and hence natural language generation for outputting multi word phrases or long answers is beyond the scope of the current work.

3.1 Training Corpus Procurement

For the purpose of experimentation and training a VQA model, we use the visual genome dataset due to the large size of the dataset and convenient availability of regional descriptions for each image along with multiple question answer pairs and the mapping from these question answer pairs to the image region from where the answer has been inferred. Using the aforementioned information, we generate a training corpus wherein each image data and the question answer pair corresponding to that image is represented in the following format. Each line contains a unique id starting with 1 initially, a space, followed by a regional description of the image which acts as the supporting fact from which the answer is inferred. This set of facts is followed by a line which indicates the end of the particular image-question-answer set. This line also comprises of an id, which is sequentially next number relative to that corresponding to the last supporting fact, a space, followed by a question, answer and the id of the single supporting fact separated by tabs. This format is repeated consistently for 50,000 randomly sampled image, question, answer sets with the id re-initialized to 1 for each new data instance. The number 50,000 was chosen emperically based on the system memory constraints for training. The questions chosen are primarily those which contain "what" in them because this category spans a wide variety of questions which can be asked which could be pertaining to finding colour of objects, actions, type of objects, object properties, etc. We filter out questions based on object counting as the regional captions may fail to capture count of objects correctly. Also, since the answer is predicted from a vocabulary formed from the training corpus, counting based questions are not appropriate as the model may tend to predict a numeral value from the vocabulary, which might result in anomalous answers.

3.2 LSTM Based Baseline Model

We set up a simple baseline model as a preliminary first step to approach the problem. We model the factoid question answering task as follows. Each of the data instances in the training corpus generated using visual genome possesses a story component comprising a set of supporting facts and a query component. An LSTM encoder network is run over both of these components, with the motive of transforming long sequences of words into a fixed vector representation. Since neural models are incapable of processing raw text directly, the words themselves are first converted to semantic preserving word embeddings using the embedding layer provided by keras before the LSTM encoder captures sequential information. Various related literatures in text processing have shown that the fixed feature vector representations learnt by the recurrent neural network are capable of effectively encapsulating and capturing relevant information from the raw input fed should hopefully encapsulate all of the relevant input. Finally, the two fixed vector representations are concatenated and fed into a traditional dense neural network, which outputs the most probable classification among a set of answers as the answer to the query. However, as seen in section IV, the baseline LSTM model does not perform well and it was observed through our experiments that the model does not substantially improve even upon scaling up of the network tremendously. This could be due to the fact that the model is incapable of taking advantage of knowing where the supporting facts are, and therefore is unable to learn where to focus attention in the input, or performing multiple "lookups" to track relevant information. In order to incorporate these features in the model, we employ a state-of-the art memory based deep learning architecture termed as MemN2N or end to end memory networks [18], which achieves performance comparable to other existing approaches for vqa using the visual genome dataset.

3.3 End to End Memory Network (MemN2N) Model

The proposed model takes as input, a discrete input set $x_1, \ldots, x_n$ which are stored in memory, an input question q, and produces as output, an answer a. Each of the x_i, q, and a are composed of words from a fixed vocabulary comprising V words. The model stores in memory, the inputs x up to a fixed buffer size, and subsequently generates a continuous representation for the x and q. These vector representations are then processed into an output a via multiple hops. This facilitates backpropagation of the error computed, through multiple accesses to memory back to the input while training the model. The model has been described considering a single layer, focused at implementing a single memory hop. In practice, as proposed in [18], we then extend it to a stacked architecture to facilitate multiple memory hops. We consider a set of inputs $x_1, \ldots, x_i$ which are meant to be stored in memory. The set of $\{x_i\}$ are transformed into memory vectors $\{m_i\}$ of d dimensions, which are computed by transforming each x_i it to embeddings in a continuous space using an embedding matrix A. The question q is also transformed into an embedding using an embedding matrix B having the same dimensions as A to generate an internal state u. Subsequently, the match between this internal state u and each of the memory vectors m_i is computed by computing their inner products followed by application of the softmax function as shown below. The scores obtained by computing the inner product intuitively represents the similarity between the question and a regional caption and a higher score value would be obtained for the regional captions or supporting facts which are most relevant to the question. Applying the softmax activation scales these scores to values between 0 and 1 corresponding to probability scores which act as the attention weights, wherein the relevant facts are assigned more weightage.

p is a vector of probabilities over the given inputs. Corresponding to each x_i, another embedding matrix C is employed in order to generate an output vector c_i. The sum over the transformed inputs c_i, weighted by the probability vector from the input is computed which gives the response vector o. Since the function from input to output is smooth, the gradients can easily be computed and backpropagated. When a single layer is used, the sum of the output vector o and the input embedding u is multiplied with a final weight matrix W and then the softmax activation is applied to produce the predicted label:

$$\begin{aligned} p_i &= \text{Softmax}(u^T m_i) \\ o &= \sum_i p_i c_i \\ \hat{a} &= \text{Softmax}(W(o+u)) \end{aligned} \tag{1}$$

The overall architecture of the model has been portrayed in Fig. 1. Al lof the three three embedding matrices A,B and C, as well as W are jointly learnt during training while minimizing a standard cross-entropy loss between the actual answer a and $\hat{a}$. Training is performed using stochastic gradient descent.

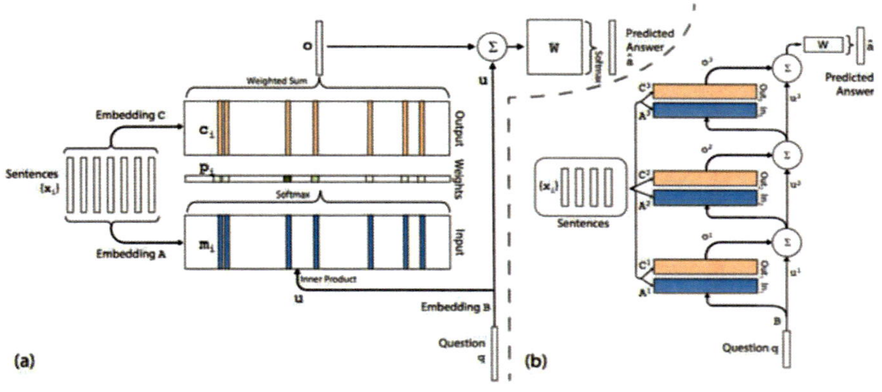

Fig. 1 **a** Model with single layer, **b** a three layered architecture of the model [18]

3.4 *Bayesian Inference*

We replace the artificial neural network layer with a bayesian neural network layer which combines the properties and advantages of artificial neural networks with stochastic or probabilistic modeling and has been found to prevent overfitting and improve performance in classification tasks in previous literatures [2], especially when data is scarce. Bayesian neural networks differ from traditional neural networks in that the weights are assigned a probability distribution rather than a point estimate or a single value. The uncertainty in weights described by these probability distributions may be used to estimate uncertainty in predictions. Instead of learning the weights, variational inference is employed to learn the parameters of these distributions.

A neural network may be portrayed as probabilistic model $p(y|x, w)$. A likelihood function $p(D|w) = \prod_i p(y(i)|x(i), w)$ which is a function of parameters w can be constructed when given a training dataset $D = x(i), y(i)$. Maximizing this likelihood function results in the maximum likelihood estimate (MLE) of w. However MLE can lead to severe overfitting.

However both MLE as well as MAP provide point estimates of parameters. Instead if a full posterior distribution over parameters is computed, weight uncertainty can be taken into account, which can be covered by the posterior predictive distribution $p(y|x, D) = \int p(y|x, w)p(w|D)dw$. This would theoretically be equivalent to averaging of predictions from an ensemble of neural networks which have been weighted by the posterior probabilities of their parameters w.

However, an analytical solution for computing the posterior probability $p(w|D)$ in neural networks is practically intractable. Therefore, a variational distribution $q(w|\theta)$ of a known functional form whose parameters we intend to estimate is incorporated in order to approximately estimate the posterior. This is done by minimizing the Kullback-Leibler or K L divergence between $q(w|\theta)$ and the true posterior $p(w|D)$

with respect to θ. The optimization objective correspondingly is as follows:

$$F(D,\theta) = \mathrm{KL}(q(w|\theta)||p(w)) - E_{q(w|\theta)} logp(D|w) \tag{2}$$

The first term refers to the Kullback-Leibler or K L divergence between the variational distribution $q(w|\theta)$ and the prior $p(w)$ and the second term is the expected value of the likelihood with respect to the variational distribution. Since all of terms in the equation above may be expressed as expectations with respect to the variational distribution $q(w|\theta)$, the optimization objective may be approximated by drawing samples $w(i)$ from $q(w|\theta)$.

$$F(D,\theta) \approx \frac{1}{N}\sum_{i=1}^{N}[\mathrm{log}q(w^{(i)}|\theta) - \mathrm{log}p(w^{(i)}) - \mathrm{log}p(D|w^{(i)})] \tag{3}$$

The neural network is parameterized with μ and σ rather than being parameterized by the weights w directly, and therefore the number of parameters are double as compared to a traditional neural network.

A training iteration comprises of a forward pass and a backward pass. During the forward pass, from the variational posterior distribution, a single sample is drawn, which is then used to evaluate an approximation of the cost function. The gradients of μ and σ are calculated during the backward pass using the backpropagation algorithm

Re-parameterization is performed for backpropagation to work because the forward pass involves stochastic sampling step. This essentially involves sampling from a parameter free distribution and subsequently transforming the sampled ε with a deterministic function $t(\mu, \sigma, \varepsilon)$ for which a gradient can be defined. In our task, ε is drawn from a standard normal distribution i.e. $\varepsilon\ N(0, 1)$ and the function $t(\mu, \sigma, \varepsilon) = \mu + \sigma \circ \varepsilon$ essentially scales the sample μ with σ and shifts the sample by the mean μ, where $\circ$ refers to point wise multiplication.

The network is parameterized with ρ instead of σ directly and ρ is transformed using the softplus function in order to obtain $\sigma = \log(1 + \exp(\rho))$, so as to maintain numerical stability and ensure that σ is always positive.

Since the task at hand is classification, categorical cross entropy is used as the loss function. The optimizer chosen was Adam, which was found to perform well relative to other optimizers during hyperparameter tuning. The models were trained for 100 epochs. Early stopping and dropout with a dropout rate of 0.2 have been employed to prevent over-fitting of the model. An 80–20% split up was used for training and testing respectively. All of the experimentations were performed on Google Collaboratory research platform, with a GPU enabled backend and 25 GB RAM environment.

4 Results and Discussion

A detailed analysis of the proposed algorithms experimented with, in this research has been presented in terms of accuracy score (Table 1). However there is a scope for improvement in results as accuracy score or string matching of the predicted and actual answers may penalize too much in some cases where one is a synonym or different form of the other. For instance, the model would be flagged to be wrong if predicted output is "mobilephone" while actual output is "phone". Though the usage of word similarity metrics as evaluation metrics rather than exact character by character matching could prove effective in such cases, this has been left as a future experimentation task. Due to memory limitations while training, the scope of this research has currently been limited to questions having 'What' in them because such questions tend to span a variety of question types in the training set as discussed earlier. The remaining question types have been proposed to be experimented with as future work as an extension of this work. However the models perform poorly when faced with questions based on counting of objects present in an image because of the fact that the regional descriptions used as supporting facts may not be capable of counting the number of instances of an object globally in an image as they focus on only specific regions. Also since the output of the models is a probability distribution over a large vocabulary formed using the training corpus, the model may assign a large probability score to numerical values in the vocabulary, which may not correspond to the count of objects and may be an irrelevant answer. Hence such questions were filtered out during creation of the training corpus. Figure 2 shows few sample test data instances with some open ended questions which our model was able to answer reasonable well. It is also observed that the proposed methodology has reasonably fast inference times for real time testing which is critical for any application, which was about 81 s on an average, though there is more scope for improvement with respect to accuracy.

5 Conclusion and Future Work

It is evident that Visual Question Answering is a significantly complex problem, especially when the questions involved are open ended and free form. In this work, we have not only proposed a detailed procedure for performing VQA by modeling it as a textual question answering problem, but also obtained a reasonable accuracy in predicting answers for free form questions. Since the dataset used for training is a subset of the large visual genome dataset, the comparison of our models with the state-of-approaches has not been portrayed as there is no base for a fair comparison. The future direction of work in order to improve Visual Question Answering may be categorized into three prospective areas to explore: (1) Extending the model to generalize to other suitable question types and experiment with other state-of-the art models in textual factoid question answering in an attempt to improve upon

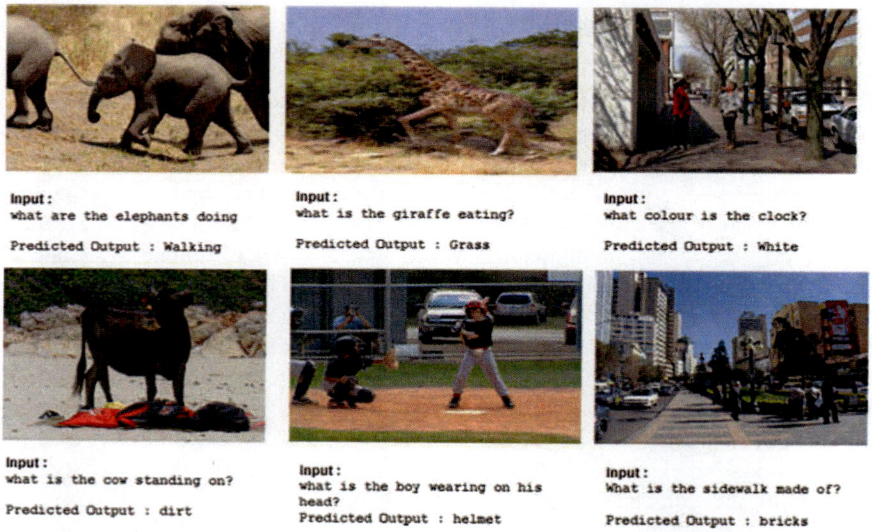

Fig. 2 Some examples of the output obtained using the proposed model

Table 1 Comparison of proposed algorithms in terms of accuracy score

Model	Accuracy (%)
LSTM baseline	41.07
MemN2N	43.56
MemN2N with Bayesian	48.93

the existing model (2) Re-trace the inference steps and retrieve the relevant single supporting fact or chain of facts used to reach the answer predicted in order to make the approach explainable (3) constructing a scene graph and reasoning answers to questions using novel recent developments in research on graph neural networks and graph embeddings.

References

1. R. Krishna, Y. Zhu, O. Groth, J. Johnson, K. Hata, J. Kravitz, S. Chen, Y. Kalantidis, L.J. Li, D.A. Shamma et al., Visual genome: connecting language and vision using crowdsourced dense image annotations. Int. J. Comput. Vis. **123**(1), 32–73 (2017)
2. C. Blundell, J. Cornebis, K., Kavukcuoglu, D. Wierstra, *Weight Uncertainty in Neural Networks* (2015). arXiv preprint arXiv:150505424
3. S. Hochreiter, J. Schmidhuber, Long short-term memory. Neural Comput. **9**(8), 1735–1780 (1997)
4. T. Mikolov, I. Sutskever, K. Chen, G.S. Corrado, J. Dean, Distributed representations of words and phrases and their compositionality, in *Advances in Neural Information Processing Systems*, pp. 3111–3119 (2013)

5. J. Pennington, R. Socher, C. Manning, Glove: Global vectors for word representation, in *Proceedings of the 2014 Conference on Empirical Methods in Natural Language Processing (EMNLP)*, pp. 1532–1543 (2014)
6. K. Kafle, C. Kanan, Answer-type prediction for visual question answering, in *Proceedings of the IEEE Conference on Computer Vision and Pattern Recognition*, pp. 4976–4984 (2016)
7. B. Zhou, Y. Tian, S. Sukhbaatar, A. Szlam, R. Fergus, *Simple Baseline for Visual Question Answering* (2015). arXiv preprint arXiv:151202167
8. S. Antol, A. Agrawal, J. Lu, M. Mitchell, D. Batra, C. Lawrence Zitnick, D. Parikh, Vqa: Visual question answering, in *Proceedings of the IEEE International Conference on Computer Vision*, pp. 2425–2433 (2015)
9. H. Gao, J. Mao, J. Zhou, Z. Huang, L. Wang, W. Xu, Are you talking to a machine? dataset and methods for multilingual image question, in *Advances in Neural Information Processing Systems*, pp 2296–2304 (2015)
10. A. Fukui, D.H. Park, D. Yang, A. Rohrbach, T. Darrell, M. Rohrbach, *Multimodal Compact Bilinear Pooling for Visual Question Answering and Visual Grounding* (2016) . arXiv preprint arXiv:160601847
11. J.H. Kim, S.W. Lee, D. Kwak, M.O. Heo, J. Kim, J.W. Ha, B.T. Zhang, Multimodal residual learning for visual qa, in *Advances in Neural Information Processing Systems*, pp. 361–369 (2016a)
12. J.H. Kim, K.W. On, W. Lim, J. Kim, J.W. Ha, B.T. Zhang, *Hadamard Product for Low-Rank Bilinear Pooling* (2016b). arXiv preprint arXiv:161004325
13. K.J. Shih, S. Singh, D. Hoiem, Where to look: focus regions for visual question answering, in *Proceedings of the IEEE Conference on Computer Vision and Pattern Recognition*, pp. 4613–4621 (2016)
14. Z. Yang, X. He, J. Gao, L. Deng, A. Smola, Stacked attention networks for image question answering, in *Proceedings of the IEEE Conference on Computer Vision and Pattern Recognition*, pp. 21–29 (2016)
15. H. Xu, K. Saenko, Ask, attend and answer: Exploring question-guided spatial attention for visual question answering, in *European Conference on Computer Vision*, Springer, pp. 451–466 (2016)
16. I. Ilievski, S. Yan, J. Feng, *A Focused Dynamic Attention Model for Visual Question Answering* (2016). arXiv preprint arXiv:160401485
17. J. Johnson, A. Karpathy, L. Fei-Fei, Densecap: fully convolutional localization networks for dense captioning, in *Proceedings of the IEEE Conference on Computer Vision and Pattern Recognition*, pp. 4565–4574 (2016)
18. S. Sukhbaatar, J. Weston, R. Fergus et al., End-to-end memory networks, in *Advances in Neural Information Processing Systems*, pp. 2440–2448 (2015)

Integrating Secured Crypto System with Cloud for Enhancing Cloud-Based Encrypted Data Sharing Services

Ravi Verma, Chetan Agrawal, and Neelesh Jain

Abstract Cloud services are most growing technology for future business system and information technology applications, as recent advancement has proofed themselves the power and success of cloud computing network, which offers more advance services to its users on demand, such advancement for storage and computational activities motivate many users to take part into cloud computing services. The nature of cloud services model make real-time services to be processed more effectively for best streamed-based communication services. However, along with this benefits, in addition security issues are consider to be a long term concern for different types of cloud computing service models, the chances of data hijacking, loss of information putting such technology on high risk over cloud service model although many researchers and scientist has invented many algorithm and techniques to get secure and provide trustworthy cloud computing services based on demand to its users. In this paper, we proposed a data encryption technique called crypto system to overcome the weakness of existing attribute-based encryption (ABE) technique; ABE is not cooperative with cloud computing service models and other services of cloud; in this paper, we focus on the weakness of attribute-based encryption (ABE) techniques and propose a integrated crypto system for resolving data sharing and security concerns which brings efficiency, reliability, and security for cloud services model.

Keywords Attribute based encryption · Cloud services · Crypto system

---

R. Verma (✉)
Department of Computer Science and Engineering, Radharaman Institute of Technology and Sciences, Bhopal, India
e-mail: ravi.verma0099@gmail.com

C. Agrawal
Department of Computer Science and Engineering, University Institute of Technology, Rajiv Gandhi Proudyogiki Vishwavidyalaya, Bhopal, India
e-mail: chetan.agrawal12@gmail.com

N. Jain
Department of Computer Science and Engineering, Sagar Institute of Research Technology and Science, Bhopal, India
e-mail: neeleshcmc@gmail.com

J. Tavares et al. (eds.), *Emerging Technologies in Data Mining and Information Security*, Lecture Notes in Networks and Systems 164,
https://doi.org/10.1007/978-981-15-9774-9_70

1 Introduction

This section address about proposed technology (crypto system) which reduces the chances of losing data confidentiality, security, and reliability over different level of data sharing among different cloud services models. It is always a big challenge for maintaining open data with security and reliability; the openly available data can be hacked and lost anytime, all such concerns losses users trust over cloud services, this section introduces the concept of crypto system technique for the betterment of future of cloud computing and data sharing services. However, the proposed model is emerging its services with API session in order to get implement secure crypto system in between the process of sharing data from node to node, crypto system entitle the concept of making data lightweight using encryption technique and integrating crypto layer among all the services layer of existing cloud network, using the concept of secure key of 256 bit size, cloud user who are authorized for services are only applicable for doing demanded communication through determined.

API at crypto layer system in this case remaining attackers or non authorized users cannot get access the data services at run time, even if one can get access with API keys, lightweight encrypted data can't be understand by others, the secure API-based cloud network improves data sharing services on demand and protecting its original users. Figure 1 illustrates the concept of crypto system.

In [1], describes the various aspect of attribute-based encryption (ABE) system, as discussed ABEs having lots of vulnerabilities during the process of data sharing with software as services (SaaS) cloud computing interface, ABEs is specially designed for ensure application level services with one to many relationship among the cloud objects but ABE lacks the security issues such as it lack down with threshold value

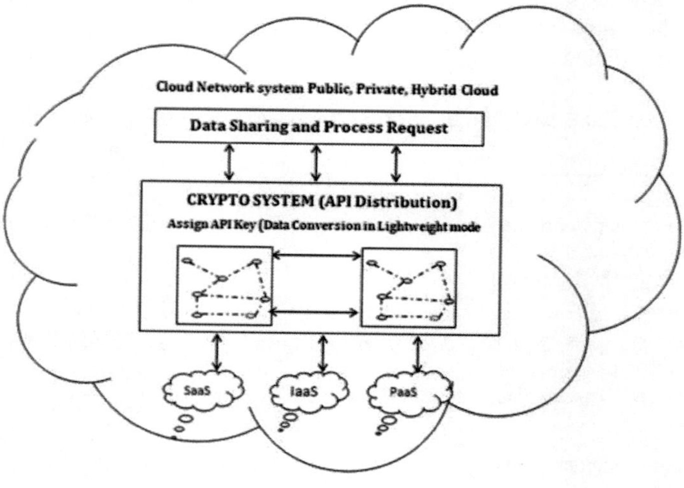

Fig. 1 Integration of crypto system with cloud architecture

Table 1 Details of vulnerability in various version of attribute-based encryption (ABE)

Policy	Vulnerability description
Attribute-based encryption with key	Vulnerable in encryption services for sophisticated communication, it is not flexible for many applications, decrypt authorities not know by sender
Attribute-based encryption with expressive key	Vulnerable with the applications which flexible and efficient encryption services
Attribute-based encryption with cipher text	Lacking the security services with prevention of data from outside eavesdropper and attacker is the real challenge
Hierarchical attribute-based encryption with key	Lacking the security services when in hierarchical communication a lower level entity is not working or if absent, then all system will shut down
Weighted attribute-based encryption with key	Vulnerable to manage cost of computation and operation

with computation overheads during and in between the process along with this Table 1 describes major limitations of ABE.

2 Related Work

When we go through detailed study of cloud network, there are numerous security concerns which affect the performance of overall system and encompass all such parameters to other computing resources like operating system, applications, storage interfaces, and devices; therefore, security will be the major consideration for all types of cloud computing services for example as virtual cloud network interconnect various communication services also may lead for security concern, for example, as in virtualization paradigm, all the virtual machine are interconnected with other physical nodes they should also lead to security considerations such security concern will make network more immune from attackers and hackers but for the purpose of data privacy and security concerns, one should carried out data security with enhanced crypto system which allow system to interconnect from one and to any authorized node within network with enforced data sharing policies. In [2], describes that security major are taken care for resource allocation and memory management with secure methodology and algorithms. In [2] defines that cloud uses *s* various mining tools and techniques to get detect virus like malware from the database, such technique is specially used for the intrusion detection in secure cloud net work [3]. Cloud network deals with large area of application, where cloud keep variety of data which leads variation at run time, therefore, it is not easy to manage open environment data with security concern, Fig. 2, define some common working area of cloud on demand services environment.

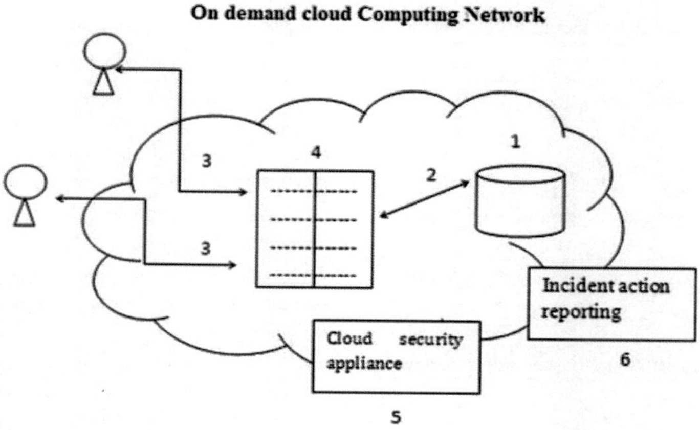

Fig. 2 Define some common working area of cloud on demand services environment that consider six area are: (1) Security of open environment data, (2) security of data during communication, (3) authentication and authorization of users, process and cloud resources, (4) mapping, and classification of different data set to different users, (5) cloud security appliance, (6) incident action reporting

As we knew that security issues are always considered as an important content whenever one talk about the security parameters of network likes (confidentiality, integrity, non-repudiation, availability, accountability, privacy and authenticity) as in [4]. In [5], author have been presented an investigation report on various methods of secret communication in cloud such as entry point to detect intrusion in cloud network, attacks likes denial of service (DoS), side channel attack, man in the middle attack. According [6], present discussion about security attribute as key to get safety and privacy and to overcome the limitation of cloud computing using the concept of data sharing and communication through key exchange scheme. In [7], describes about the challenges on latest cloud computing model used for human health and monitoring called electronic health reword system, author presented different challenges for keeping data with privacy and safety over cloud network. In [8], authors presented a summary on uses of encryption technology in cloud using different types of searchable key-based encryption, public key-based encryption technique uses has been explained with their limitations, in Table 2, we present a comparative literature study of different articles presented in last few years.

3 Designing Secure API Keys for Crypto System

This paper presenting the concept of using application programming interface keys in order to create a protection layer that should be integrated with cloud services (SaaS, IaaS, PaaS) providers, according to [14], author implementation of application key for different operation will improve result with cloud SaaS infrastructure, in addition

Table 2 Detail of various security concerns in difference research findings

Reference	Publication year	Problem area	Research findings	Cloud category
[7]	2013	Cloud computing applications	Defines the communication policy with normalized encryption	Public
[8]	2015	Cloud computing applications	Suffering with encryption-based computing at application level	Public
[9]	2017	Cloud computing applications	Application level securities are limited with side channel security issues	Public
[10]	2018	Fog computing	Fog cloud having limited services due to lack of poor encryption scheme	Hybrid
[11]	2018	Healthcare cloud	Focused on searchable encryption communication techniques	Small group of users
[12]	2019	Internet of things	Cloud with trustful services	Small group of users
[13]	2019	Block chain	Focused on various accessing and controlling technique	Public

with this, it also provide a way to check log details as well as system are now able to backtrack the past events. However, usage of key is limited with SaaS services and later on it should be integrated with IaaS also as in [7]. As experienced in some case that API keys are also used by some system admin for the maintenance of data services and permission as defines in [15], but designed key announced by mailing services to one and other, it may lead to stolen the key through massage, the popular services provider like Yahoo, Google also experience the same but they need to address the solution by implementing API keys using some intermediate process through some algorithm which can encapsulate the deigned key to specific purpose with transmission request; here, it is also essential to design API key to ensure administrative operation for inserting, deleting, and updating data values in lightweight encrypted manner, such resulted encrypted data is called lightweight data which can easily managed with efficiency, whenever any transmission request initiated, it should be a part of key-based crypto system that brings security concerns which we has discussed in Sect. 2.

Algorithm1: Discovering API Keys for Data Sharing

1. Initialized all the available nodes, user input (a_i & b_i) by initial values.
2. Initialize N_i, E_i, D_i (int), $P_i = a_i$, $Q_i = b_i$;
3. Star event as Open_event initialize();
4. Initialize timmer T_i=timmer();
5. Do
   a. $N_i = P_i \times Q_i$;
   b. $\varnothing(n) = (P_i - 1)(Q_i - 1)$;
   c. $E_i = \frac{1[\mathrm{MOD}\varnothing(n)]}{Di}$;
   d. $D_i = \frac{1+K\varnothing(n)}{Ei}$;
   e. Distribute API as $[N_i, E_i]$ and $[N_i, D_i]$.
   f. Calculate $C_i = M^{Ei}\sqrt{n}$ and $M_i = C^{Di}\sqrt{n}$;
6. Until (T_i> = timmer());
7. End

The above algorithm works as intermediate layer for maintaining sensitive data in lightweight mode with the help of API key, which is going to be change time to time whenever a new request has been input into the system; using such pseudo code, one can get a level of security with moving data with high overheads, as we have been studying yet that in [1] for security concern past research defines that still cloud are not reliable for maintaining security services as per customer perspective, cloud need to ensure its reliability and confidentiality according to the need of its user so that one can go with cloud-based communication, however, this paper present a new concept to encompass the drawback of existing cloud network regarding data loss and data theft issues, the entire proposed concept is working logically creating and sharing data based on intermediate data conversion and crypto transmission layer. As in [16], storing sensitive data values like credit card details over cloud network encryption is always considered as a powerful technique and tool but the potential drawback of existing cloud network is safety of keys which are sharing via physical channel because as we knew a hacker can easily access and control over the sharing system which my result loss over many times as in URL [17].

4 Simulation and Result Analysis

As discussed in Sect. 3, the primary objective of author is to design and securely distribute the API keys among the cloud object, the methodology adopted over the vulnerabilities of attribute-based encryption technique as discussed in Table 1 (Sect. 2), we need to use the services of proposed crypto system with the concept of API distribution as discussed in algorithm 1; to realize the benefits of proposed encryption technique, we implement the concept using the simulation environment of ns2 as a wireless networking tool to represent and compare the result for flexibility, performance, and scalability parameters with existing encryption policies.

4.1 Setup Preparations

Figure 3 describes the communication story between authorized and non-authorized nodes (attackers) as multiple active nodes are showing in ns2 simulation environment for establishing cloud computing network, when session starts, the initial server is serving the service to each active node as authorized node; meanwhile, the nodes like attackers such as DoS attack can easily get entered into the system as an authorized node but in progressive way, our proposed crypto system will separate the non-authorized attacker nodes because every valid node should have their own real key which is not even determined in attribute-based encryption technique, crypto policy separate such node from communication zone and similarly the attackers like man in the middle also want to participate during communication that time our policy again protect the communication system, it make an active session of all the authorized active nodes to accomplish the task with trustful and reliable cloud computing environment.

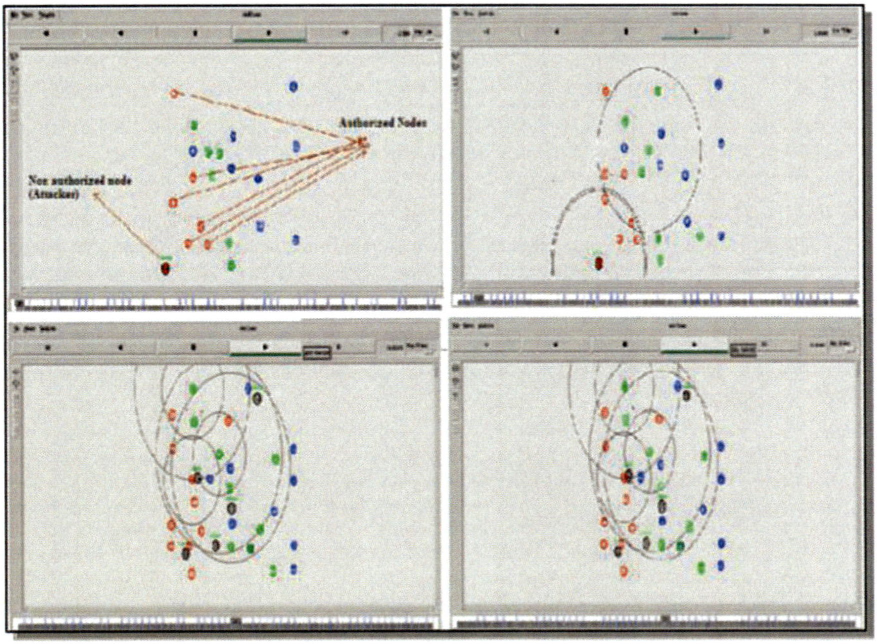

Fig. 3 Simulation of crypto system

$$D_{ms} = \frac{\sum Nb}{Rms} \times Tn$$

Data Packets in Byte	No. Of Bits	Proposed	SAE	ABE
60	61440	0.3	1.2288	0.6
90	92160	0.5	1.8432	0.9
120	92160	0.5	1.8432	0.9
150	122880	0.6	2.4576	1.2
180	153600	0.8	3.072	1.5
200.0	184320	0.9	3.6864	1.8
230	204800	1.0	4.096	2.0

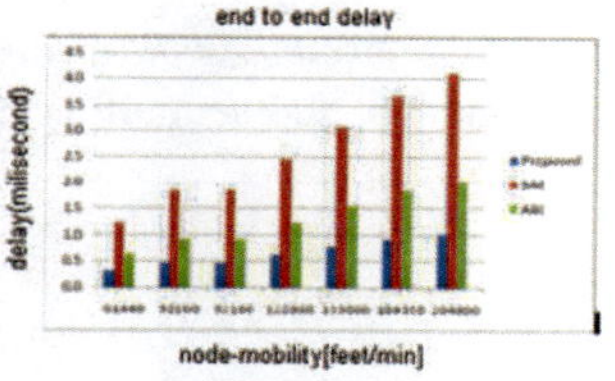

$$T = \frac{\sqrt{Ws}}{RTTxi \ldots\ldots xn} \times N$$

MAX WS	RTT	nodes	Proposed	KTT	ABE	KTT	SAE
1224	25	2	97.9	28.0	87.4	26.0	94.2
1224	51	4	96.0	61	80.3	54.0	90.7
1224	77	6	95.4	85.6	83.5	82.0	89.6
1224	104	8	94.7	120.0	81.6	115.0	85.1
1224	131	10	93.4	138.0	78.5	130.0	81.6
1224	159	12	92.4	192.0	76.5	183.0	80.3
1224	180	14	95.2	198.0	86.5	190.0	90.2

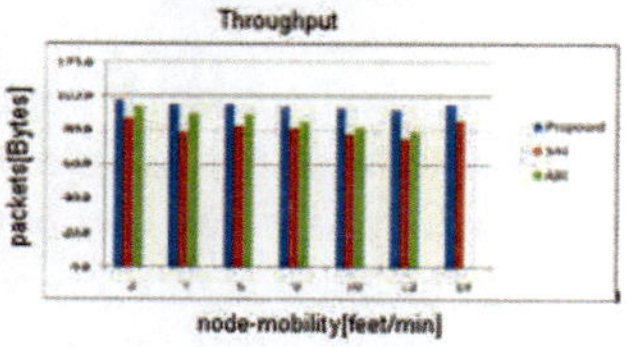

$$PDR = \frac{\sqrt{(ws - Tn) + Dt}}{\sum Sxi \ldots\ldots xn} \times 100$$

MAX WS	nodes	Data Packets in Byte	ta Transmit	Proposed	ABE	SAE
1224	2	60	97.9	138[illegible]	1356.6	60.0
1224	4	90	90	132[illegible]	14.0	94.0
1224	6	120	95.4	1324	1265.1	120.0
1224	8	150	94.7	1294	1215.2	147.6
1224	10	180	93.4	1276	1180.3	178.2
1224	12	200.0	92.4	1262	1164.1	198.8
1224.0	14	230.0	95.2	1255	1145.4	225.5

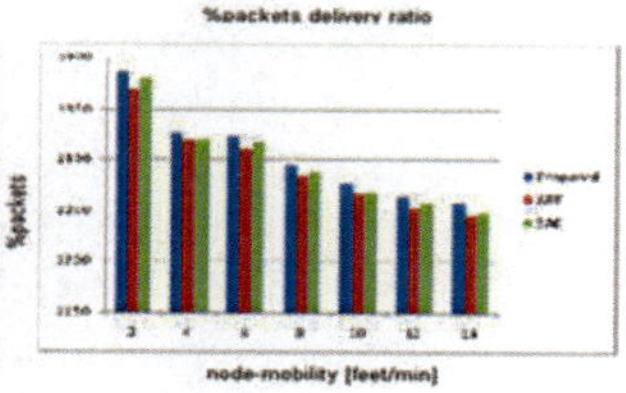

Fig. 4 Comparison between proposed, ABE, SAE policies

4.2 Result Analysis

Finally, this section justify the objectives of this paper, to overcome the vulnerabilities defined in Table 1 as attribute-based encryption (ABE), searchable asymmetric encryption, and other like searchable symmetric encryption technique was not applicable for other applications of cloud computing, many times system need to face high computing cost and overhead issues. In result, we try to compare the performance of our proposed policy with existing, based on mathematical metrics used for quality of services and security consideration in cloud computing environment. Figure 4 illustrates the computing metric for proposed, ABE, SAE policy along with mathematical construct for delay, flow control, and packet delivery ratio calculation in Fig. 4, we find proposed policies working well under security consideration compare to existing policies.

5 Conclusion

This paper has reviewed and analyzed different security concerns with existing encryption policy such as attribute-based encryption and other approaches along with its functionality and limitations. The weakness of ABE and SAE policies as discussed (Sect. 2) are motivating us to bring a policy for cloud computing called crypto system to overcome these limitations and bring the cloud data sharing services to its users with more reliability, efficiency, and security; after simulating the proposed approach (Sect. 4), we found our proposed approach is performing well compare to other policies as discussed in (Sect. 4), Fig. 3 illustrating the comparisons between proposed and existing encryption technology, using mathematical construct, proposed cloud policy incorporating with the application services of cloud for sharing data and managing it in lightweight mode for efficient storage and operation of data on demand.

References

1. M.Y. Shabir, A. Iqbal, Analysis of classical encryption techniques in cloud computing. Trans. Cloud Comput. **21**(1), 11 (2016)
2. G. Batista, C.H.G. Ferreira, D.C.M. Segura, D.M.L. Filho, M.L.M. Peixoto, A QoS-driven approach for cloud computing address-ing attributes of performance and security. Future Gener. Comput. Syst. **68**, 260–274 (2017)
3. X. Zhao, Q. Lin, J. Chen, X. Wang, J. Yu, Z. Ming, Optimizing security and quality of service in a real-time database system using multi-objective genetic algorithm. Expert Syst. Appl. **64**, 11–23 (2016)
4. Z. Xiao, Y. Xiao, Security and privacy in cloud computing. IEEE Commun. Surv. Tutorials **15**, 843–859 (2013)
5. D. Gonzales, J.M. Kaplan, E. Saltzman, Z. Winkelman, D. Woods, Cloud-trust—a security assessment model for infrastructure as a service (IaaS) clouds. IEEE Trans. Cloud Comput. **5**(3), 523–536 (2017)
6. A. Nieto, J. Lopez, A context-based parametric relationship model (CPRM) to measure the security and QoS tradeoff in configurable environments, in *Proceedings IEEE International Conference Communication (ICC)*, vol 21 (2014), pp. 755–760
7. R.K. Kalluri, C.V.G. Rao, Addressing the security, privacy and trust challenges of cloud computing. Int. J. Comput. Sci. Inf. Technol. **5**(5), 6094–6097 (2014)
8. D. Stevenson, J. Pasek, Privacy concern, trust, and desire for content personalization, in *Proceedings Research Conference Communication, Information Internet Policy*, vol. 08, no. 2 (2015) pp. 1–30
9. G. Shynu, K.J. Singh, A comprehensive survey and analysis on access control schemes in cloud environment. Inf. Technol. **16**(1), 19–38 (2016)
10. J. Ni, K. Zhang, Securing fog computing for internet of things applications: challenges and solutions. IEEE Commun. Surv. Tutorials **20**(1), 601–628 (2018)
11. R. Zhang, R. Xue, L. Liu, Searchable encryption for health-care clouds: a survey. IEEE Trans. Serv. Comput. **11**(6), 978–996 (2018)
12. S. Chenthara, K. Ahmed, H. Wang, F. Whittaker, Security and privacy-preserving challenges of e-Health solutions in cloud computing. IEEE Access **7**, 74361–74382 (2019)
13. T. Salman, M. Zolanvari, Security services using blockchains: a state of the art survey. IEEE Commun. Surv. Tutorials **21**(1), 858–880 (2019)
14. The Statistics Portal. Number of smartphone users world-wide from 2014 to 2020. Online. Available, vol. 12, no. 3 (2014)
15. J. Chen, Q. Zhu, Security as a service for cloud-enabled inter-net of controlled things under advanced persistent threats: a contract design approach. IEEE Trans. Inf. Forensics Secur. **12**(11), 2736–2750 (2017)
16. Cloud Security Alliance. Home page URL https://cloudsecurityalliance.org
17. E. Bertion, F. Paci, R. Ferrini, Privacy-preserving digital identity management for cloud computing. IEEE Comput. Soc. Data Eng. Bull. **12**(1), 1–4 (2009)

Performance Analysis of UAV Relay in Vanets against Smart Jamming with Q-Learning Techniques

L. Rajesh, K. Bhoopathy Bagan, P. Thayal Sankar, and V. Suchitra

Abstract Due to the high mobility of on-board units (OBUs) and the large-scale network topology frequency hopping-based anti-jamming techniques are not applicable in vehicular ad hoc networks (VANETs). In this paper, to improve the communication performance of VANETs against smart jamming, unmanned aerial vehicles (UAVs) are used. These unmanned aerial vehicles observe the ongoing OBU and UAV communication status and induce the UAV to use a specific relay strategy. The UAV relays the OBU message to another roadside unit (RSU) with a better radio transmission condition if the serving RSU is heavily jammed or interfered. The interactions between a UAV and a smart jammer are formulated using an anti-jamming UAV relay game. The UAV decides whether or not to relay the OBU message to another RSU based on the relay game, the jammer observes the UAV and the VANET strategy it chooses the jamming power accordingly. The proposed relay strategy can reduce the bit-error rate of the OBU message efficiently and increasing the benefit of the VANET Using Q-learning analysis.

Keywords VANET · On-board unit · Roadside unit · Unmanned aerial vehicle · Q-learning

L. Rajesh (✉) · K. Bhoopathy Bagan · P. Thayal Sankar · V. Suchitra
Department of Electronics Engineering, Madras Institute of Technology, Anna University, Chennai 600044, India
e-mail: virudhairaj@gmail.com

K. Bhoopathy Bagan
e-mail: bhoopathy02@yahoo.com

P. Thayal Sankar
e-mail: p.t.s.sharma@gmail.com

V. Suchitra
e-mail: suchitra98406@gmail.com

J. Tavares et al. (eds.), *Emerging Technologies in Data Mining and Information Security*, Lecture Notes in Networks and Systems 164,
https://doi.org/10.1007/978-981-15-9774-9_71

1 Introduction

Vehicular ad hoc networks (VANETs) are scalable and cost-effective solutions for applications such as road safety, traffic efficiency and entertainments through multi-hop vehicle-to-vehicle wireless communications [1]. Vehicular ad hoc networks (VANETs) help to build an unmanned-driving, and support booming applications of on-board units (OBUs). The high mobility of OBUs and the large-scale dynamic network with fixed roadside units (RSUs) reduces the utility of networks and they are vulnerable to jamming. VANET can be used by a group of vehicles to set up and maintain a communication network among them without using any central base station or any controller. Infrastructureless VANET has additional responsibilities on vehicles because every vehicle becomes part of the network. These VANETs also manages and controls the communication on this network along with its own communication requirements [1, 2]. Unmanned aerial vehicle (UAV) technology is being used increasingly for military and civilian purposes. The primary reason for increase is that UAVs eliminate the risk to human life in difficult and dangerous missions, are cost effective, and easily are deployed [3]. Jamming occurs while there is a higher mobility of OBUs and the large-scale dynamic network with fixed roadside units (RSUs). A jammer sends duplicated signals which aim to block the ongoing transmissions between OBUs and the serving RSUs. A smart jammer has flexible control over the jamming frequencies and signal strengths and this induces the VANET to use a specific communication strategy and then attacks it accordingly. By using unmanned aerial vehicles (UAVs) to relay the OBU message the anti-jamming communication of VANETs can be significantly improved. UAVs generally have more desirable channel states due to the line-of-sight (LOS) links and smaller path-loss exponents. When UAV communicate with OBUs and RSUs at a fixed location on the ground it might be severely blocked by a smart jammer. To improve the signal-to-interference-plus-noise-ratio (SINR) of the OBU signals, and to reduce the bit-error-rate (BER) of the OBU message relaying the OBU message is done with the help of UAVs. However, the jamming resistance in the UAV-aided VANETs is still a problem. We propose an anti-jamming UAV relay in this paper which the UAV decides whether or not to relay an OBU message to another RSU, and the smart jammer chooses its jamming power under a UAV channel model with the path loss, log-normal shadowing, and Rayleigh fading.

2 Related Works

Vehicular ad hoc network (VANET) is emerging technology past few years. Sparse VANETs have more realistic significance compared with vehicular networks which are densely distributed. There are some challenges faced by a sparse vehicular network. The first most common challenge of the sparse VANET system is that the network suffers from frequent disconnections. To adapt the transmission route to the dynamic mobility pattern of the vehicles is the second most challenge faced by network. Some infrastructural requirements are hard to meet when deploying a VANET widely [1] the author has proposed an infrastructureless unmanned aerial vehicle (UAV) assisted VANET system called vehicle-drone hybrid vehicular ad hoc network to face the above difficulties. This utilizes quad rotor drones, to boost vehicle-to-vehicle data message transmission under circumstances conducted by our distributed vehicle location prediction algorithm. VDNET considers the geographic information. Vehicles observe the location information of other vehicles to predict the location of a destination vehicle and to construct a transmission route. To deliver data message directly to destination, some vehicles in VDNET equips an on-board UAV. This can relay messages in a multi-hop route, and collect location information while flying above the traffic. VDNET achieves high efficiency and low end-to-end delay with controlled communication overhead.

Vehicular ad hoc networks (VANETs) are the most promising and fastest growing subsets of mobile ad hoc networks (MANETs). VANETs are the combination of smart vehicles and roadside units (RSU) which communicate through unreliable wireless medium. They are very susceptible to attacks which may result in life-endangering situations. It is essential to develop security mechanisms in order to detect such attacks against VANETs. To survey from possible attacks, the corresponding detection mechanisms [4] have proposed in the paper. The attacks can be classified and explained along with their effects.

Vehicular ad hoc networks (VANETs) can be used to improve transportation security, reliability, and management. This proposed system investigates security aspects of VANETs within a game-theoretic framework. Here, defensive measures are optimized with respect to threats posed by malicious attackers. The formulations are chosen in order to maximize the utility of the models and solutions to future systems. The security games are taken as input centrality measures and they are computed by mapping the centrality values of the car networks to the underlying road topology. The above strategies help to locate the most valuable or vulnerable points (e.g., against jamming) in vehicular network [5]. The author have proposed an optimal deployment of traffic control and security infrastructure is investigated both in the static (e.g., fixed roadside units) and dynamic cases (e.g., Mobile law enforcement units). Multiple assumptions for the players, leads to fuzzy game and fictitious play formulations in addition to classical zero-sum games. The effectiveness of the security game is evaluated numerically using realistic simulation data that is obtained from traffic engineering system.

In LTE, networks are used which show vulnerable to denial-of-service (DOS) and loss of service attacks from smart jammers. This may employ simple narrowband jamming techniques to attack without any need to hack the network or its users. The utilities of jamming and anti-jamming actions are modelled and played by the jammer in the network under the framework of single-shot and repeated Bayesian games. In a single-shot game formulation, the Nash equilibria (NE) is the applicable strategy equilibria for which network usage is severely compromised. A repeated-game learning and strategy algorithm for the network is proposed [6] that outperforms single-shot games by a significant margin.

Unmanned aerial vehicle (UAV) systems can be exposed to smart attackers, who are subjective end-user. They use smart radio devices to change their attack types and policies based on the ongoing UAV transmission and network states. The author [7] has proposed the excepted theory to formulate an individualized smart attack game for the UAV transmission, in which a smart attacker makes individualized decisions in order to choose the type of attack such as jamming, spoofing, and eavesdropping without knowing the attack detection accuracy of the UAV system. The UAV transmit power on multiple radio channels is selected to resist smart attacks. To achieve the optimal power allocation against smart attacks without knowing the attack model and the channel model in the dynamic game reinforcement learning-based UAV power allocation strategies are proposed. A deep Q-learning-based UAV power allocation strategy combines Q-learning and deep learning. This helps to accelerate the learning speed for the case with a large number of channel states and attack modes. Our proposed system of UAV power allocation strategy can suppress the attack motivation of subjective smart attackers and increase the secrecy capacity and the utility of the UAV system can proved with stimulation results [7].

Due to the vision to adopt technology information and communication in this context and to facilitate access to urban services. Security requirement being a permanent challenge which affects most smart cities applications. Vehicular ad hoc networks (VANETs) are one which can be used for smart mobility axis. VANETs are certainly affected by security risks faced to the users. The global positioning system (GPS) which is widely used in several applications by human is vulnerable to different attacks like jamming, blocking, and spoofing. The last attack tries to provide to the receiver fake information, and because of this, it computes a wrong time or location. In paper [8], author studied the impact of spoofing attack on VANETs communications. The paper deals with the vulnerabilities produced by the signals to the fake GPS that can distract drivers. This can affect people security and congestion in roads of the cities. By the application of smart radio devices, a jammer can dynamically change its jamming policy based on opposing security mechanisms. This induces the mobile device to enter a specific communication mode and then launch the jamming policy accordingly. This can exploit spread spectrum and user mobility to address both jamming and interference in mobile devices. A two-dimensional anti-jamming mobile communication scheme is proposed in which a mobile device leaves a heavily jammed/interfered-with frequency or area. By applying reinforcement learning techniques, a mobile device can achieve a required communication policy without the need to know the jamming and interference model and the radio channel model in a dynamic game framework. Author have the proposed hot booting deep Q-network-based two-dimensional mobile communication scheme exploits simulation in similar scenarios to reduce the exploration time at the beginning of the game. This is applied to the deep convolutional neural network and macro-action techniques which accelerates the learning ability in dynamic situations. The simulation results show that there is an improvement in both the signal-to-interference-plus-noise ratio of the signals and the usage of the mobile devices against cooperative jamming compared with existing schemes [9].

In [10], the author has proposed the intelligent transportation systems (ITS), vehicular ad hoc networks (VANETs). It mainly deals with the prevention, detection, and mitigation of denial-of-service attacks in a vehicle networks. The evaluation of the physical workings of a vehicle networks and the wireless communication between the vehicles and the possibility of malicious interference. Defense methods against jamming attacks are implemented and tested. These methods include interference reduction, data redundancy, and warning systems based on on-board vehicle sensors. The results are positive and the defenses are successfully increasing vehicle networks resilience to attacks.

Smart jammers can analyze the ongoing radio transmission with flexibility and powerful control on jamming signals through serious threats on cognitive radio networks. Game theory [11] provides a powerful approach to study the interactions between smart jammers and secondary users (SUs). In [12] have proposed the power control strategy of an SU against a smart jammer under power constraints is designed using Stackelberg game. The jammer as follower of the game chooses the jamming power based on the observed ongoing transmission, while the SU as the leader determines it transmit power based on the estimated jamming power. The impact of the observation accuracy of the jammer regarding the transmit power of the SU is obtained and studied. The Stackelberg equilibrium of the anti-jamming game is derived and compared with the Nash equilibrium of the game. The transmission of an SU benefits from the observation error of the jammer with a higher signal-to-interference-plus-noise ratio and utility is obtained.

The communication between a secondary user (SU) with frequency hopping and a jammer with spectrum sensing in heterogenous cognitive radio networks is investigated. The power control interactions are formulated as a multi-stage anti-jamming game, in which the SU and jammer repeatedly choose their power allocation strategies over multiple channels simultaneously without interfering with primary users. To achieve the optimal transmission power and channel with unaware parameters such as the channel gain of the opponent. In [13], the author has proposed a power allocation strategy for the SU based on reinforcement learning algorithms including Q-learning and WoLF-Q. The proposed power allocation strategy can improve the secondary user's performance against both sweeping jammers and smart jammers with learning in heterogeneous CRN. A two-dimensional anti-jamming communication scheme for cognitive radio networks is developed, in which a secondary user (SU) exploits both spread spectrum and user mobility to address jamming attacks, while not interfering with primary users [14]. By application of the deep Q-network algorithm, this scheme determines whether to recommend the SUs to leave the area of heavy jamming. This helps to choose a frequency hopping pattern to defeat smart jammers. The SU derives an optimal anti-jamming communication policy using Q-learning in a proposed dynamic game, without knowing the jamming model and the radio channel model. This can be applied to a deep convolution neural network to accelerate the learning speed with a large number of frequency channels. The proposed scheme can increase the signal-to-interference-plus-noise ratio and improve the utility of the SU against cooperative jamming, compared with a Q-learning-only based benchmark system. Unmanned aerial vehicles (UAVs) are used as relays between ground-based terminals and a network base station. An algorithm is developed for optimizing the performance of the ground-to-relay links to control of the UAV heading angle. In [10] defined the author has ergodic normalized transmission rate (ENTR) for the links between the ground nodes and the relay to quantify link performance. The derivation of a closed-form expression in terms of the eigen values of the channel correlation matrix is obtained to increase the performance. The current UAV relay assignments cannot meet the minimum link requirements. So, there is a need to investigate the deployment and heading control problem for new UAV relays as they are added to the network. We propose a smart handoff algorithm that updates node and relay assignments as the topology of the network evolves.

In [15], the author has proposed unmanned aerial vehicles (UAVs) deal with smart jamming that updates the jamming policy based on the ongoing communication status of the VANET. The UAV relays the data of OBUs to another roadside unit (RSU) with a better transmission condition if the serving RSU is located in a heavily jammed area. The interactions between the UAV and the jammer are formulated by the anti-jamming UAV relay game. UAV decides whether or not to relay the data of the OBU to another RSU that is far away from the jammer with the above experiences, and thus it chooses the jamming power. The Nash equilibrium (NE) of the game is derived to reveal how the best UAV relay strategy depends on the transmission cost and the radio channel model. To address jamming in the dynamic UAV-aided VANET game without the knowledge of network model and jamming model, a hot booting policy hill climbing (PHC)-based UAV relay strategy is proposed.

3 Proposed Work

An OBU that moves along the road at a speed denoted by $e^k \in [0, V]$ at time slot k, where V is the maximum speed and time is partitioned into slots of a constant duration. The OBU aims to send a message to a server via several RSUs and a UAV in a time slot, as illustrated in Fig. 1. We assume that both the UAV and RSU1 receive the message from the OBU and then the UAV decides whether to connect to the server via RSU2. For simplicity, the constant channel power gains are assumed to be constant in each time slot. Let $\mathbf{d}(k) = [d^{(k)}, d^{(2k)}, d^{(3k)}, d^{(4k)}, d^{(5k)}]$ denote the topology vector of the network at time slot k, where $d^{(k)}$ denotes the distance between the OBU and RSU1, $d^{(2k)}$ is the distance between the OBU and the UAV, $d^{(3k)}$ corresponds to the distance of the jammer-RSU1 link, $d^{(4k)}$ is the distance between the jammer and the UAV, and $d^{(5k)}$ is the distance between the UAV and RSU2. The distances $d^{(2k)}$ and $d^{(k)}$ depend on the speed of the OBU at time slot k. The OBU sends a message to a server via RSU1 and the UAV that is connected with RSU2 at time slot k with a fixed transmit power $P^{(k)}$. The SINR of the signals received by r ($r = 1$ for RSU1 or $r = 2$ for the UAV) sent from the OBU and the SINR of the signals received by RSU2 sent from the UAV at time slot k are denoted by $\rho r^{(k)}$ and $\rho 3^{(k)}$, respectively. The BER of a signal denoted by $pe^{(\rho)}$ depends on the SINR per bit, ρ, of the received signal for a given modulation mode. The BER of

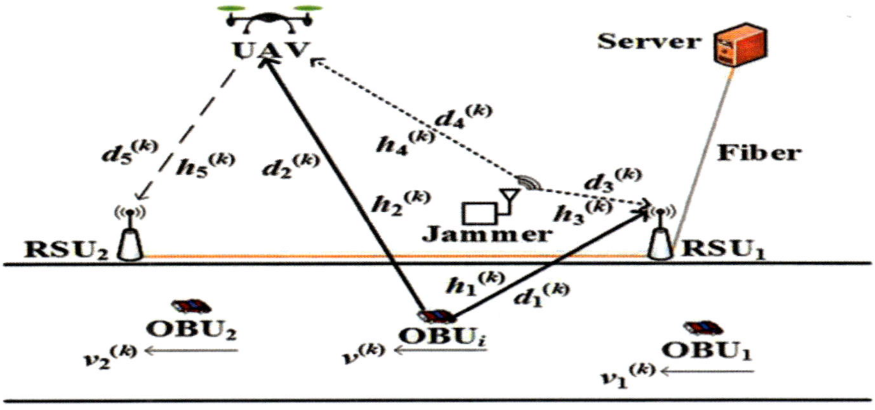

Fig. 1 UAV relay in VANETs

the OBU message at time slot k denoted by $pe^{(k)}$ depends on the OBU-RSU1 link, the OBU-UAV link and the UAV-RSU2 link, and is given by

$$P_e^{(k)} = \min\left(p_e\left(\rho_1^{(k)}\right), p_e\left(\min\left(\rho_2^{(k)}, \rho_3^{(k)}\right)\right)\right). \tag{1}$$

The BER of the message depends on the minimum of the BER of the OBU-RSU1 signal as shown in the first term, and the BER of the weaker signal of the OBU UAV link and the UAV-RSU2 link at that time as represented in the second term. According to the channel quality and the BER of the OBU message, the UAV decides whether or not to relay the OBU message to RSU2, which is denoted by $x \in \mathbf{A} = \{0, 1\}$, where $\mathbf{A}$ is the feasible action set of the UAV. The UAV relays the OBU message at time slot k with a fixed transmit power $PU^{(k)}$ and a transmit energy cost U if $x = 1$, and keeps silent otherwise. The system server can gather the sensing message sent from the $C^{(k)}$ OBU via RSU1 or the UAV. The jammer at a fixed location sends faked or replayed signals at time slot k to block RSU1 with a jamming cost $CJ^{(k)}$ estimates the transmit power of the OBU and the UAV. In addition, the smart jammer applies smart radio devices to eaves drop the control channel of the VANET and estimate the VANET transmission policy. According to the estimated VANET communication, the smart jammer changes its jamming power $y \in$ [0, P J M], where P M J is the maximum jamming power. For simplicity, we assume a constant noise power in the received signal denoted by σ and the jammer is too far away from the UAV and RSU2 to block them.

The channel power gain vector of the system denoted by $h^{(k)} = h1^{(k)}, h2^{(k)}, h3^{(k)}, h4^{(k)}, h5^{(k)}$ [at time slot k consists of the channel power gain of the OBU-RSU1 link $h1^{(k)}$, the OBU-UAV link $h2^{(k)}$, the jammer-RSU1 link $h3^{(k)}$, the jammer-UAV link $h4^{(k)}$, and the UAV-RSU2 link $h5^{(k)}$ Similar to [16], the channel gain is modeled as

$$h_i^{(k)} = \theta_0 \Delta_i^{(k)} \left(\frac{d_i^{(k)}}{d_0} \right)^{-\alpha_i} \tag{2}$$

where θ_0 is the channel power gain at the reference distance d_0, the channel time variation $\Delta^{(k)}$ i depends on the Doppler shift due to the node mobility. The path-loss exponent α_i, for example, $\alpha_i = 2$ for the OBU-UAV, jammer-UAV, and UAV-RSU2 links and $\alpha_i = 4$ otherwise. The path loss of the jammer-UAV radio link is assumed to be much higher than that of the jammer-RSU1 link due to a longer distance. Similarly, the radio link between the OBU and the UAV has a higher path loss compared with that of the OBU-RSU1 link. Due to the uncorrelated locations of RSU1 and the UAV, $h1^{(k)}$ is independent with $h2^{(k)}$. Time slot k denoted by p^k i, m, n depends on the OBU speed given by the interactions between the UAV and the jammer are formulated as an anti-jamming relay game.

BER of the UAV Relay in VANET

$$P_e = \frac{1}{2} \text{erfc}\left(\sqrt{\frac{Ph_2}{2\sigma}} \right). \tag{3}$$

4 Result and Discussion

The simulations have been performed to evaluate the performance of the proposed UAV relay strategy in the dynamic game against smart jamming. In the simulations, both the UAV and the jammer were stationary and observed the ongoing VANET communication. The jammer applied greedy strategy to choose the jamming power regarding the expected reward uJ in (10) during the VANET communication in each time slot. Vehicles can communicate with themselves through RSU. VANET scenario show in Fig. 2.

Numerical results with $P = 8$, PU $= 1$, $\sigma = 0.2$, $h1 = 0.3$, $h3 = 0.6$, $h4 = 0.3$ and $h5 = 0.4$ in Fig. 3a, show that the BER of the OBU message at the NE of the game decreases with the jamming cost CJ. For instance, the BER decreases from

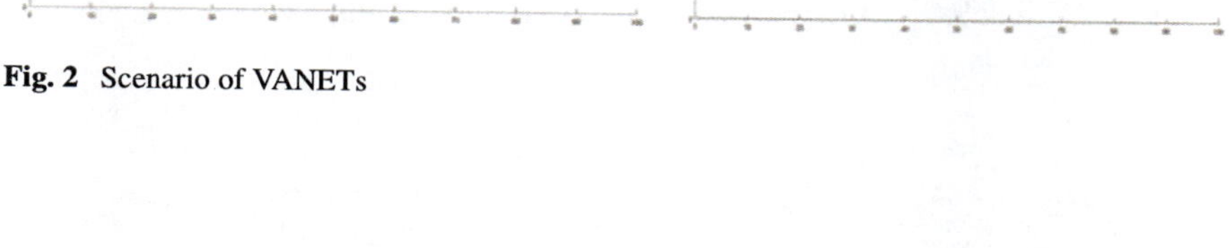

Fig. 2 Scenario of VANETs

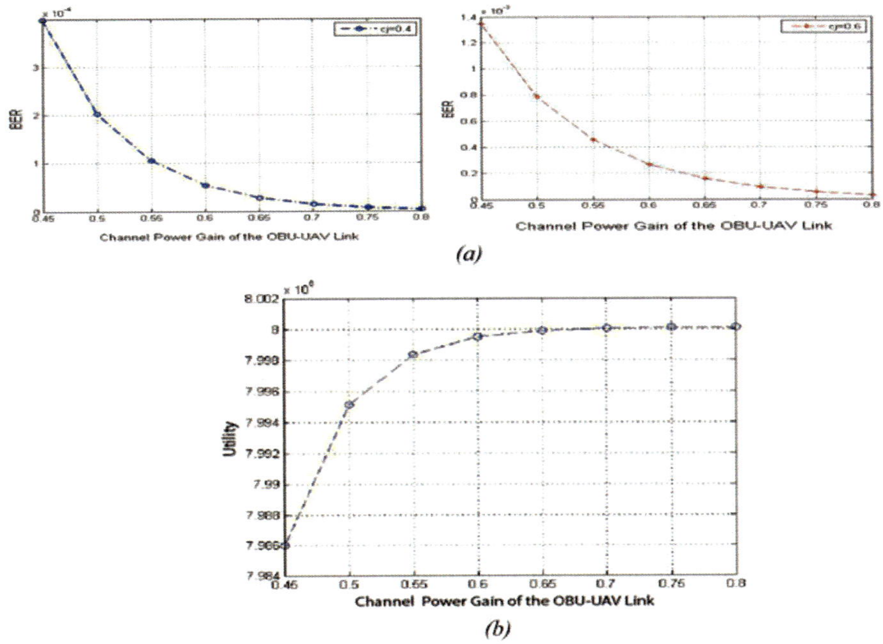

Fig. 3 **a** BER of the OBU message in UAV relay-VANET. **b** Utility

3.7% to 1.6% and the utility of the UAV increases by 43%, as CJ changes from 0.4 to 0.6, because the smart jammer is less motivated to attack the VANET under a higher jamming cost. The utility of the UAV increases with the channel gain $h2$, and the utility of the jammer decreases with it. For example, the utility of the UAV increases by 56.5% as the channel gain $h2$ changes from 0.35 to 0.8 is shown in Fig. 3b.

Numerical results with $P = 8$, $\mathrm{PU} = 1$, $h = [h1, h2, 0.3, 0.5, 0.7]$, $\mathrm{CU} = 2$ and $\mathrm{CJ} = 0.6$, Fig. 4 show that the BER of the OBU message at the NE of the stochastic game increases with the transition probability of $h1$ and that of $h2$. For instance, the BER increases by 82% by 14.6% as p2,1,2 changes from 0.4 to 0.6, because the UAV has more trouble to gain the optimal strategies with the unstable channel conditions.

The repeated interactions between the UAV and the smart jammer in the VANET can be formulated as a dynamic game, in which the jammer determines its jamming power based on the previous VANET transmission, and the UAV chooses its relay strategy based on the system state, which consists of the radio channel state and the BER of the OBU message observed in last time slot. The next system state observed by the UAV is independent of the previous states and actions, for a given system state and UAV relay strategy in the current time slot. The communication performance of the UAV-aided VANET in the dynamic game for utility of the UAV. For the sake of tinkering, let us first change EPISODES 1–4000, just to keep things quicker to

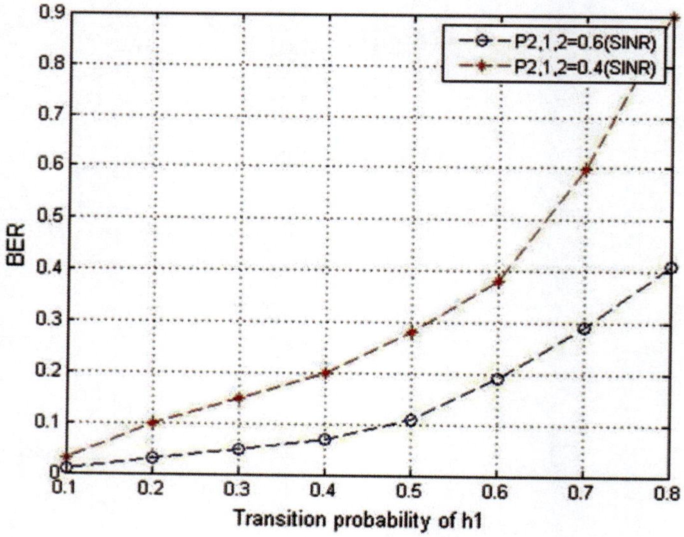

Fig. 4 BER of the OBU message in UAV relay-VANET using SINR

iterate for now. Then we will add a *new parameter STATS_EVERY, and set that to 100*. Result shown on Fig. 5.

To change our Epsilon decay policy. Let us set that to decay to the very end: END_EPSILON_DECAYING = EPISODES. To raise the episodes to 10,000. Result shown on Fig. 6.

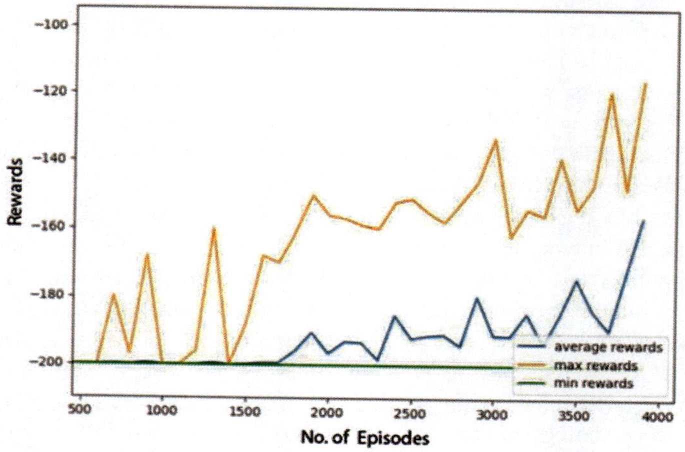

Fig. 5 Performance of the UAV relay-VANET in Q-learning episodes 1–4000 for utility of the UAV

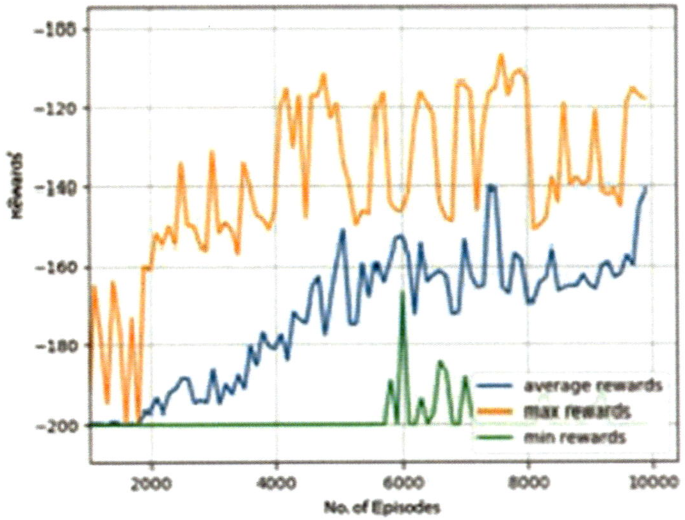

Fig. 6 Performance of the UAV relay-VANET in Q-learning episodes to 10,000 for utility of the UAV

To raise the episodes to 20 k. since it had high overall rewards and to curtail the size 100 mb. Result shown on Fig. 7

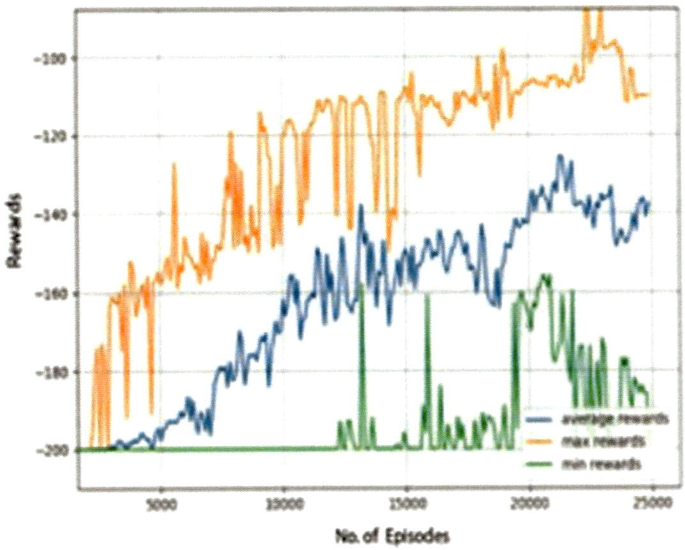

Fig. 7 Performance of the UAV relay-VANET in Q-learning episodes to 20 k for utility of the UAV

5 Conclusion

To reveal the impacts of the UAV, transmit cost, and the radio channel condition on communication performance of the VANET against smart jamming a UAV relay in VANET. Simulation results have shown that the proposed relay strategy can improve the anti-jamming transmission performance of the VANET efficiently. The proposed scheme reduces the BER of the OBU message in the VANET by 76.8% and increases the usage of the UAV by 62% compared with the various Q-learning-based relay strategy. In future work that will carry out a more in-depth game-theoretic study which incorporates more jamming models, including a smart jammer that can choose both the jamming strength and the mobility. To protect VANETs at the beginning stage of the learning process, a backup anti-jamming transmission mechanism can be used. To improve the anti-jamming communication performance of UAV relay in VANETs.

References

1. X. Wang, L. Fu, Y. Zhang, X. Gan, X. Wang, VDNet: an infrastructure-less UAV-assisted sparse VANET system with vehicle location prediction. Wireless Commun. Mob. Comput. **16**, 2991–3003 (2016). https://doi.org/10.1002/wcm2727
2. L. Xiao, X. Lu, D. Xu, Y. Tang, L. Wang, W. Zhuang, UAV relay in VANETs against smart jamming with reinforcement learning. IEEE Trans. Veh. Technol. **99**, 1 (2018). https://doi.org/10.1109/tvt.2018.2789466
3. E.E. Ulku, B. Dogan, O. Demir, I. Bekmezci, Sharing location information in multi-UAV systems by common channel multi-token circulation method in FANETs. Elektronika Ir Elektrotechnika **25**(1), (2019) ISSN 1392-1215. https://doi.org/10.5755/j01.eie.25.1.22738
4. F. Sakiz, S. Sevil, A survey of attacks and detection mechanisms on intelligent transportation systems: VANETs and IoV. Ad Hoc Netw. **61**, (2017). https://doi.org/10.1016/j.adhoc.2017.03.006
5. G. Noubir, R. Rajaraman, B. Sheng, B. Thapa, On the robustness of IEEE 802.11 rate adaptation algorithms against smart jamming, in *Proceedings of the Fourth ACM Conference on Wireless Network Security, WISEC 2011*. https://doi.org/10.1145/1998412.1998430
6. F.M. Aziz, J.S. Shamma, G.L. Stuber, Resilience of LTE networks against smart jamming attacks: wideband model, in *IEEE 26th International Symposium on Personal, Indoor and Mobile Radio Communications* (PIMRC 2015). https://doi.org/10.1109/pimrc.2015.7343507
7. L. Xiao, C. Xie, M. Min, W. Zhuang, User-centric view of unmanned aerial vehicle transmission against smart attacks. IEEE Trans. Veh. Technol. **99**, 1 (2017). https://doi.org/10.1109/tvt.2017.2785414
8. A. Boudhir, M. Benahmed, A. Ghadi, M. Bouhorma, Vehicular navigation spoofing detection based on V2I calibration, Colloquium calibration, in *Colloquium in Information Science and Technology, 4th IEEE CIST* (2017). https://doi.org/10.1109/cist.2016.7805006
9. L. Xiao, D. Jiang, D. Xu, H. Zhu, Y. Zhang, P.H. Vincent, Two-dimensional antijamming mobile communication based on reinforcement learning. IEEE Trans. Veh. Technol. **99**, 1 (2018). https://doi.org/10.1109/tvt.2018.2856854
10. G. Patounas, Y. Zhang, S. Gjessing, Evaluating defence schemes against jamming in vehicle platoon networks, in *IEEE Conference on Intelligent Transportation Systems, Proceedings, ITSC Electronic* ISSN: 2153-0017 (2015). https://doi.org/10.1109/itsc.2015.348

11. N. Namvar, W. Saad, B. Niloofar, B. Kelley, Jamming in the internet of things: a game-theoretic perspective, in *2016 IEEE Global Communications Conference, GLOBECOM 2016—Proceedings* (2016). https://doi.org/10.1109/glocom.2016.7841922
12. L. Xiao, T. Chen, J. Liu, H. Dai, Anti-jamming transmission stackelberg game with observation errors. IEEE Commun. Lett. **19**(6), 1 (2015). https://doi.org/10.1109/lcomm.2015.2418776
13. T. Chen, J. Liu, L. Xiao, L. Huang. Anti-jamming transmissions with learning in heterogenous cognitive radio networks, in *2015 IEEE Wireless Communications and Networking Conference Workshops (WCNCW)* (2015). https://doi.org/10.1109/wcncw.2015.7122570
14. G. Han, L. Xiao, H.V. Poor, Two-dimensional anti-jamming communication based on deep reinforcement learning, in *IEEE International Conference on Acoustics, Speech and Signal Processing (ICASSP)—Proceedings* (March 2017). https://doi.org/10.1109/icassp.2017.7952524
15. X. Lu, D. Xu, L. Xiao, L. Wang, W. Zhuang, Anti-jamming communication game for UAV-aided VANETs, in *IEEE Global Communications Conference, Electronic* ISBN: 978-1-5090-5019-2 (2017). https://doi.org/10.1109/glocom.2017.8253987
16. Y. Gwon, S. Dastangoo, C. Fossa, H.T. Kung, Competing mobile network game: embracing anti jamming and jamming strategies with reinforcement learning, in *2013 IEEE Conference on Communications and Network Security (CNS) Electronic* (2013) ISBN:978-1-4799-0895-0. https://doi.org/10.1109/cns.2013.6682689
17. G. Zheng, E.A. Jorswieck, B. Ottersten, Cooperative communications against jamming with half-duplex and full-duplex relaying, in *IEEE Vehicular Technology 77th Conference* (2013). https://doi.org/10.1109/vtcspring.2013.6692490
18. X. Zhou, D. Niyato, A. Hjørungnes, Optimizing training-based transmission against smart jamming. IEEE Trans. Veh. Technol. **60**(6), 2644–2655 (2011). https://doi.org/10.1109/tvt.2011.2151890
19. F.M. Aziz, J.S. Shamma, G.L. Stüber, Resilience of LTE networks against smart jamming attacks, in *2014 IEEE Global Communications Conference, GLOBECOM* (2014). https://doi.org/10.1109/glocom.2014.7036895
20. M.N. Kurt, Y. Yilmaz, X. Wang, Real-time detection of hybrid and stealthy cyber-attacks in smart grid. IEEE Trans. Inf. Forensics Secur. **99** (2018). https://doi.org/10.1109/tifs.2018.2854745
21. Y.A. Abdalla, A.K. Siti-Nabiha, A.S. Md Shahbudin, Cellular network traffic scheduling with deep reinforcement learning. Int. J. Econ. Manage. (2017)
22. Y. Fan, X. Xiao, W. Feng, An anti-jamming game in VANET platoon with reinforcement learning, in *2018 IEEE International Conference on Consumer Electronics-Taiwan, ICCE-TW Electronic* ISSN: 2575-8284 (2018). https://doi.org/10.1109/icce-china.2018.8448435
23. F.M. Aziz, Farhan Muhammad, Resilience of LTE networks against smart jamming attacks: A game- theoretic approach. Georgia Institute of Technology (2017)
24. C. Xie, L. Xiao, Liang, User-centric view of smart attacks in wireless networks, in *2016 IEEE International Conference on Ubiquitous Wireless Broadband* (ICUWB 2016). https://doi.org/10.1109/icuwb.2016.7790439
25. T. Yu, H.Z. Wang, B. Zhou, K.W. Chan, J. Tang, Multi-agent correlated equilibrium Q(λ) learning for coordinated smart generation control of interconnected power grids, in *IEEE Transactions on Power Systems Electronic* ISSN: 1558-0679 (2015). https://doi.org/10.1109/tpwrs.2014.2357079

Recent Trends in IoT–Based Smart Healthcare Applying ML and DL

Sohail Saif, Mayurakshi Jana, and Suparna Biswas

Abstract Internet of Things (IoT) is an interconnected network of different sensors, communication channels, cloud data storage, and application software. Applying machine learning algorithms in IoT-based smart healthcare systems can achieve a significant growth in prediction and detection of diseases. This review aims to provide recent research trend in various directions of smart healthcare application. A systematic methodology based on extensive search has been followed to identify most relevant works for further detailed review and investigation. A total number of (n = 836) papers are screened published during 2014–2019 from most popular databases such as IEEE Xplore, Science Direct, Springer Link and Semantic Scholar. After two stage filtering, only 24 most relevant papers are identified and categorized in 4 different categories. In first category is disease identification and diagnosis where 54.16% (n = 13/24) paper belongs, 29.16% (n = 7/24) papers belongs to second category, patient monitoring using IoT. Rests 12.50% (n = 3/24) and 4.18% (n = 1/24) papers belongs to category 3 and 4, drug discovery, personalized treatment and smart health record, respectively.

Keywords Smart healthcare · Machine learning · Deep learning · Systematic review

S. Saif · S. Biswas (✉)
Department of Computer Science and Engineering, Maulana Abul Kalam Azad University of Technology, Kolkata, West Bengal, India
e-mail: mailtosuparna@gmail.com

S. Saif
e-mail: sohailsaif7@gmail.com

M. Jana
Department of Computer Science, Bijay Krishna Girls' College, University of Calcutta, Howrah, WB, India
e-mail: mjcompsc@gmail.com

J. Tavares et al. (eds.), *Emerging Technologies in Data Mining and Information Security*, Lecture Notes in Networks and Systems 164,
https://doi.org/10.1007/978-981-15-9774-9_72

1 Introduction

The Internet of Things (IoT) is an emerging technology which consists of a set of intelligent sensor devices, real-time transmission medium, and cloud computing facility. IoT spans many areas, i.e., smart healthcare, smart transportation, smart parking, smart traffic management, smart environment, smart home, etc. With the advancement of IoT and cloud computing, there is a significant increase in smart healthcare domain. The interconnected network system can offer specialized doctors, healthcare centers, hospitals, etc. But in recent times rapid increase in population, it is not only limited to find healthcare centers, but to monitor patients both in hospitals and at home. In IoT-cloud scenario, medical signals are collected from various sensor devices which is obviously very much complex to handle. Generally, the sensor devices are used in this paradigm are—temperature sensor, pressure sensor, blood-oxygen level measurement sensor, blood-glucose level measurement sensor, accelerometer, pulse oximeter, electrocardiogram (ECG), electroencephalogram (EEG), altimeter, GPS location tracker, etc. The sensors can be wearable or can be implanted in the patients' body for collection of data. Then, the data can be transmitted in real time using cloud server so the doctors can take necessary actions [1]. 3G, 4G, Wi-Fi, bluetooth technologies are used to maintain smooth delivery of this complex heterogeneous data. The cloud platform is used to store and analyze these data. Since cloud servers have more computation ability, machine learning or deep learning are deployed there for sensor data analysis, knowledge extraction, and classification. This can help doctors to identify or diagnosis a disease easily.

Finally, doctors can see the prediction or the raw health data in a handheld device like a smart phone. Figure 1 shows the architecture for smart healthcare using machine learning.

Smart healthcare includes various applications such as diseases identification and diagnosis, drug discovery and production, medical imaging diagnosis, personalized medicine, machine learning-based behavioral modification. Smart health records, clinical trial and research, crowd sourced data collection, better radiotherapy, outbreak prediction, etc.

This paper discusses recent research works on the aforementioned smart health care applications. Section "Introduction" presents a brief introduction of smart healthcare. Section 2 "Preliminaries" describes the basics of machine learning, electronic health record (EHR), and deep learning techniques. Section 3 "Methodologies" represents the process behind the searching, filtering, and collecting relevant papers. Section 4 "Results" describes the outcome of the systematic review; finally, this review has been concluded in Sect. 5.

Fig. 1 Architecture of smart healthcare

2 Preliminaries

In this section, we have briefly discussed about the basics of machine learning techniques, deep learning approaches and role of electronic health record in healthcare.

2.1 Machine Learning

With the advancement of artificial intelligence, machine learning technique is used to handle "Big health data." This technique mainly focuses on complex computing tasks and is used to reduce computational complexity, resources with an impressive accuracy. Generally, machine learning techniques are used for diagnosis and disease prediction. Also, machine learning algorithms successfully identify relapse of any disease or transition from one disease to another [2]. Any machine learning algorithm has two parts—a training data set which is used to train the system with existing data and their outcomes and a test data set based on which the machine learning algorithm is used to predict or detect diseases. Machine learning algorithms are broadly classified into two groups: supervised learning and unsupervised learning. The main difference between these two types of machine learning is in supervised learning; there is a training dataset which is used to instruct the machine and test dataset which is used test the machine with accuracy whereas in unsupervised learning, the data points are clustered which have same features.

2.2 *Deep Learning*

Deep learning has gained popularity as sometimes machine learning algorithms cannot solve all complex classifications. It uses hidden layers more than classical neural network to extract features from data. It works on complex data with diverse structures. Some popular deep learning algorithms are:

- Stacked Autoencoder (SAE)—Stacked autoencoder consists multiple hidden layers of autoencoders where output of each layer is the input of the next layer.
- Convolution Neural Network (CNN)—CNN consists of convolutional or pooling layers. Each layer contains a set of neurons which is used to detect patterns. These patterns are further formed into feature map due to their similarity.
- Recurrent Neural Network (RNN)—RNN consists of input and output sequential data. The input sequences are mapped to output sequences based on the structure of the network.

2.3 *Electronic Health Record (EHR)*

Electronic health record contains patient's medical history that is disease, clinical reports, diagnosis results, medications, treatment methods, etc. The record should be in the form of natural text or images. Therefore, natural language processing (NLP) and named entity recognition (NER) techniques are used to map between English language to intelligent systems.

3 Methodology

The critical keywords related to smart healthcare covered in this study are "disease identification," "drug discovery," "medical images analyses," "smart health records." A general search was made to identify the papers related to smart healthcare in the popular database such as, (1) IEEE Xplore [3] (2) Science Direct [4] (3) Springer Link [5] (4) Semantic Scholar [6]. These are all well-known database where technical articles in technology, medical domain are available. The strategy behind choosing appropriate papers involves searching the databases in two rounds. In first round, papers published during 2014–2019 have been considered, then duplicate and not related to healthcare papers are excluded by reading the abstract and title of the papers. Next filtering is conducted to exclude the papers which are not related to key-words by reading the full text of the paper. In second round, again filtering is conducted by accurate full paper reading of the examined papers of first round based on detail application of the keywords. Finally, the papers are categorized based on the application of smart healthcare as mentioned earlier. This search was conducted at the end of April, 2020 by using search engines and different keywords. The combination of the various

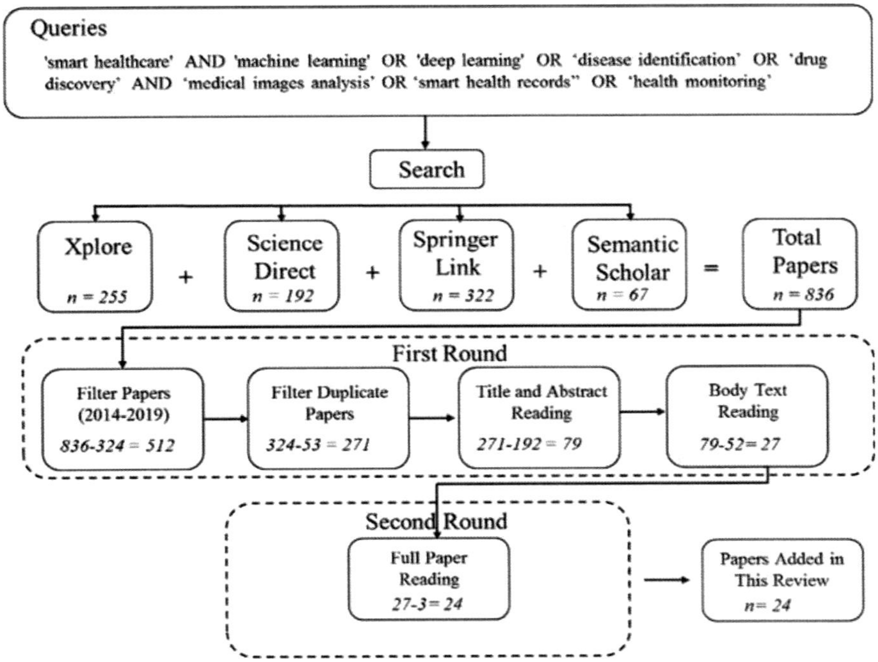

Fig. 2 Workflow of the paper search and filtering process

keywords including "disease identification using deep learning," "drug discovery in healthcare," "healthcare," "IoT," "ML in e-health," "medical imaging analysis," etc., in several syntax of logical keywords were searched performing "AND" and "OR" operation, as shown in Fig. 2. This search includes book chapters and various reports rather than searching only journals and conference proceedings. The paper chosen based on the criteria given in Fig. 2 has been included in this review.

4 Results

The search using the above-mentioned queries resulted 836 papers, where 255 papers found in IEEE Xplore, 92 papers found in science direct, 322 in springer link and 67 in semantic scholar. In the first round, 512 papers out of the 836 published papers during 2014–2019 have been identified, then 53 papers found as duplicate and excluded. After reading titles and abstract, 192 papers have been excluded, since those are not relevant. Next, body texts were read and another 52 papers excluded. Now, total number of papers after the screening in first round is 27. In second round, full papers were read and 24 papers found which are highly related with the keywords. Thus,

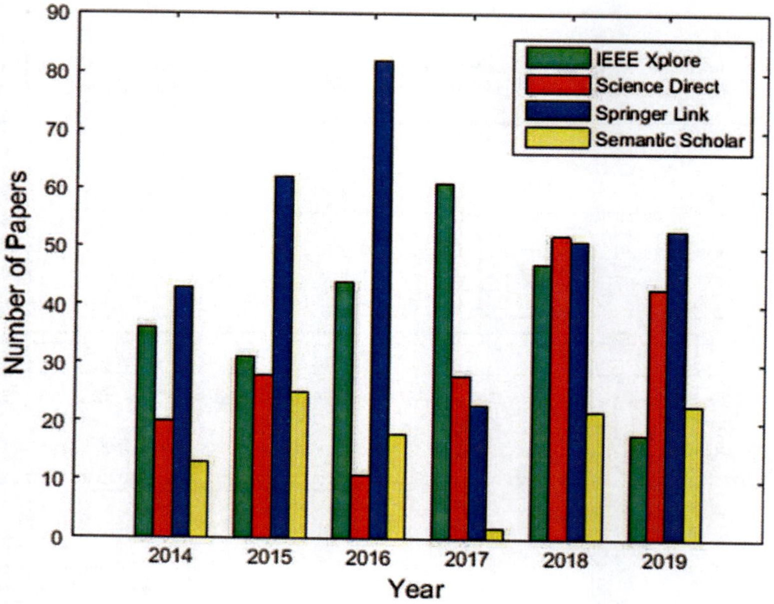

Fig. 3 Year-wise distribution of the papers found in various database before filtration

after completion of two rounds of rigorous processing, 24 papers have been selected for the detailed study in this review.

Figures 3 and 4 show the distribution of papers published during 2014–2019 before filtration and after filtration, respectively. Finally, identified 24 papers [7–30] as most relevant have been categorized in 4 categories, based on the application in smart health which is reflected in Fig. 5. Summary of the research works studied and reviewed here have been briefly discussed below.

4.1 Disease Identification and Diagnosis

One of the successful applications of machine learning in healthcare is the identification and diagnosis of some chronic diseases which are sometime considered difficult to diagnose. Various cancers which are difficult to be detected in early stages, genetic diseases, tumor diagnosis using machine intelligence are found trending now days.

Two very common heart diseases—Myocardial ischemia and cardiac arrhythmias detection have been shown in [7]. The electrocardiogram (ECG)-based MIT-BIH data-bases are preprocessed using LMS algorithm where delayed LMS algorithm is used to calculate—updated weights. The discrete wavelet transform (DWT) technique is used for feature extraction and *R*-peak feature is used to determine heart rate variability (HRV). Out of 200 extracted features, 180 (90%) is training dataset and 20

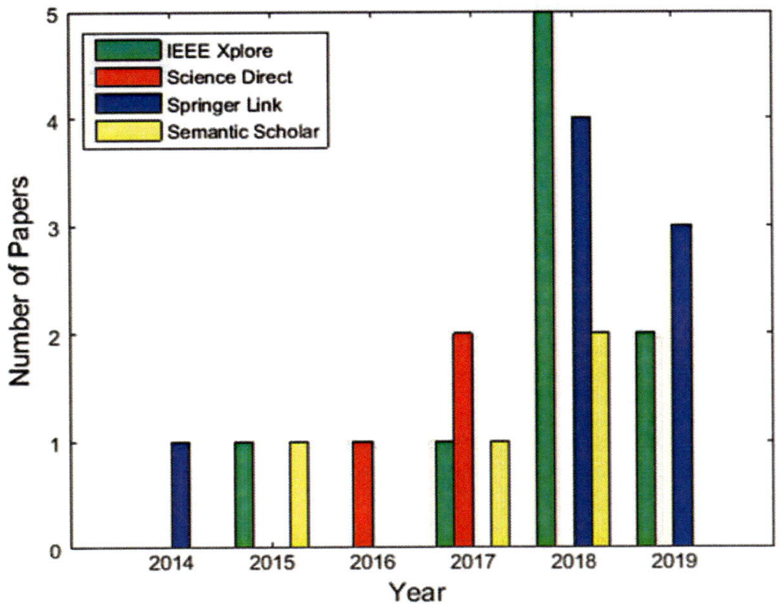

Fig. 4 Year-wise distribution of the papers found in various database after filtration

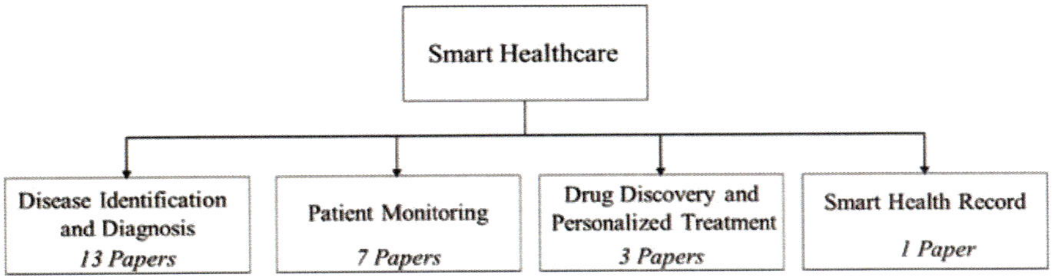

Fig. 5 Taxonomy of research literatures on smart healthcare

(10%) is test dataset. Experimental results shows that the HRV-based SVM classifier and based Neuro-fuzzy technique obtain the accuracy of 96% and 94.2%, respectively, Fuzzy-decision and kNN classifier obtain the accuracy of 91.8% and 90.4%. Also, the ECG-based artificial neural network (ANN) and principal component analysis (PCA) have been done which obtains accuracy of 88.5% and fuzzy-kNN obtains 92.5% of accuracy.

Breast cancer detection is one of the useful applications in smart healthcare [8]. Well known Breast cancer Wisconsin diagnostic dataset which is available openly is mostly used by researchers. It is a fine-needle aspiration (FNA) of cell nuclei taken from Breast masses. Authors have used this dataset and applied linear programming, pattern reorganization, and decision tree techniques to extract the features. SVM classifier has been used to create a hyper plane to divide into two classes—malignant and benign. Features are selected using LASSO, and generalized linear model (GLM) has been used to reduce some features as number of features are greater than number of instances. Artificial neural network (ANN) has been used for error minimization and deep neural network (DNN) is used for image processing.

Researchers in [9] have used medical images to identify possible Melanoma. The International Skin Imaging Collaboration (ISIC): 2018 [31] and the International Skin Imaging Collaboration (ISIC): 2019 [32] are used as input in convolutional neural network (CNN). Both this dataset contains images in JPEG format. Various data augmentation techniques have been applied, and images are normalized before the learning operation. Various image recognition models have been used like—Inception V3, ResNet50, VGG 16, Inception Resnet, MobileNet. Based on measurement parameters, the Inception Resnet CNN model achieves accuracy of 91%.

Epileptic seizure detection and monitoring with the help of CHB-MIT dataset is done by researchers in [10]. Training dataset of CHB-MIT contains 686 EEG samples of 23 epilepsy patients out of which 198 has one or more seizures. They have used deep CNN algorithm for feature extraction. Filters or kernels are used to take 2-D input and represents output as 3-D feature maps. To train the input as well as output, three layer neural network or auto-encoder has been employed. Experimental results that average value of sensitivity is 93%, average accuracy of recognition is 99.2%, overall accuracy of deep CNN-stacked auto-encoder is 99.5%.

One of the successful applications of medical imaging diagnosis and prognosis is Alzheimer's disease diagnosis. Many researchers are working on this to introduce a high accurate algorithm which can classify Alzheimer's disease and normal cases efficiently. Here, in this work [11], researchers have used deep learning-based Siamese convolutional neural network with three branches of ResNet-34. Objective of this research is to classify Alzheimer's disease and normal cases from structural magnetic resonance imaging. Experimental result shows that the proposed method achieved the accuracy of 98.72% which is best compared to the similar state of art methods.

4.2 *Patient Monitoring*

IoT has many heterogeneous domains. Among them, most important domain is smart healthcare [12]. The intelligent sensor devices—Electrocardiography (ECG), electromyography (EMG), temperature sensor, respiratory sensor, blood-glucose level are responsible data collection, and machine learning algorithms are useful to deal with missing data value, streaming of real-time data and changing of dimensionality of data. Visualization between heterogeneous numerical data and continuously monitored graphical data should be performed accurately using machine learning algorithms so the patients can be monitored remotely and efficiently.

For a developing country scenario, the whole population cannot be covered under existing medical facilities. Therefore, a low-cost, highly reliable health monitoring system [13] is needed to provide medical facility to a large section of people. Researchers proposed a wrist band-based method to collect physiological data from patients' body. Two types of sensor devices; i.e., pulse oximeter and temperature sensor are utilized in this architecture. For data transmission, IEEE 802.15.4 protocol which consumes less power than Bluetooth or any other proprietary protocols have been used to transmit data to Raspberry Pi 3. After receiving data, Raspberry Pi 3 can directly send to the doctors by emails using its inbuilt IEEE 802.11 (Wi-Fi) protocol. The overall accuracy is achieved 98% and 95% for temperature and pulse rate readings, respectively. This system costs only $8.4; Hence, it is a low-cost system can serve a large number of populations.

In IoT-based smart healthcare paradigm, the continuous monitoring of medical data is necessary but the protection of this data is another important task. In [14], temperature sensor (LM-35), heartbeat sensor (AD8232), ECG sensor (AD8232), accelerometer sensor (ADXL335), Raspberry Pi, ADC 0808 for interfacing all microprocessors and for converting analog signal to digital signal has been used. Support vector machine (SVM) and radial basis function (RBF) have been employed along with patient's health record to train the machine for possible heart disease detection. AES encryption technique ensures confidentiality and privacy of those medical data. The encrypted data is kept on data server and keys associated with are kept on a key server using Shamir's secret key sharing algorithm.

With a growing number of populations, more people are affected with chronic diseases like diabetes, cardiovascular diseases, obesity, etc. In this paper [15], a real-time health monitoring system "WISE" (Wearable IoT-based health monitoring system) has been introduced which collects data from body area sensor network (BASN). BASN is a network of interconnected sensor devices which are placed on human body or covering immediate surroundings. Support vector machine (SVM) algorithm has been used to retrieve knowledge from heartbeat, body temperature, and blood pressure data to make important decisions based upon these parameters. Only authorized users can access these information from cloud as they have to login in the WISE-Cloud network.

In [16], an IoT-based health monitoring system is designed such that sudden cardiac arrest of patients can be reduced. Therefore, a wireless sensor network has been used to collect body vitals. This network consists of temperature sensor (LM-35) and ECG sensor. The collected data is sent to the low-cost Arduino uno microcontroller using GSM or wireless LAN. The Wi-Fi module ESP8267 can send data to doctors. Therefore, patients do not need to visit doctor regularly.

Contactless real-time patient monitoring can stop the spread of disease like Covid'19. Patients in intensive care unit (ICU) are monitored with extra attention. But when huge number of patients are admitted to ICU, it is really difficult to pay attention on each patient personally. Therefore, when some medical parameters like heart rate, respiratory rate, pulse rate, saturated percentage of oxygen in blood exceeds or goes below, then an alarm can be used to identify that particular patient for immediate care. But incorrect value may cause false alarms. Therefore it is very much difficult to identify true and false alarms. In [17] this research work support vector machine and its two incremental versions LASVM and ISVM (SVM) are being applied on multi-parameter intelligent monitoring for intensive care (MIMICII) dataset for detection of true and false alarm. It is observed that the incremental and modified version of SVM (LASVM) is faster and uses less memory than SVM. Experimental results show that a considerable change in monitoring patients using SVM with incremental versions LASVM has reduced the rate of false alarm up to 99.98%.

In [18], a cloud-assisted IoT-based student health monitoring framework has been proposed. This system can be used for regular health monitoring which can help in early detection of on-going and potential disease without visiting the hospitals. Researchers have considered detection of waterborne disease in work. Body vitals such as heart rate, blood pressure, ECG, EEG have been collected through wearable sensors and sent to cloud using internet. Using Bayesian belief network the data are divided into sensitive event and non-sensitive event. Data is captured using data mining algorithms in a particular time series and based on decisions, alerts are generated and sent to the doctors or healthcare professionals. Ten-fold stratified cross validation method is used to classify the performance of machine learning algorithms based on measurement parameters accuracy. kNN and decision tree classifiers both give best results based on response time and measurement parameters, respectively.

4.3 Drug Discovery and Personalized Treatment

In a recent study, authors [19] have introduced machine learning-based framework to identify the medication pattern in electronic medical record (EMR). Techniques like Jaccard similarity evaluation, spectral clustering, and cross-matching among multiple features have combined together to find the medication patterns. Their proposed framework outperforms other similar approaches.

In [20], researchers used machine learning-based biological structure activity relation (QSAR) modeling for drug discovery. Using this model, essential biological active modules can be identified from a lot of candidate compounds. Machine learning-based drug discovery resulted few commercials drugs such CCT244747, PTC725, RG7800, and GDC-0941.

Another QSAR–based drug discovery has been introduced by few researchers in the year 2018 [21]. In this work, first chemical features have been extracted from compound database by characterizing them in chemical substructure fragments. Depending on the presence of absence of substructure fragments, Chemical fingerprints are constructed and using QSAR approach training has been performed, so the compound property can be identified during testing. Melanoma is a type of skin cancer occurring in both men and women has significantly risen over the years.

4.4 Smart Health Record

Maintaining up-to-date medical records is a complex process, data entry process is much easier due to advancement in technology, but the truth is that even now, most of these processes need a long time to complete. Machine learning technique has big contribution which makes the medical record documentation process easy. Document classification with the help of vector machine and OCR recognition has played key role in this purpose. With the advancement of artificial intelligence, smart health record consists of all the health related data of a patient. Therefore, a disease should be easily predictable. Various machine learning algorithms are used in this purpose but for accurate result, recurrent neural network (RNN) has been used by authors [22]. Patient medical history has been given as sequential input to the RNN, which predicts the risk of heart failure. MIMICIII dataset has been treated as patient medical history here. Table 1 shows a comparative analysis of various smart health applications.

Table 1 Application of ML and DL in smart healthcare

Authors year	Application	Technique	Dataset	Result
Amin et al. [23]	Brain tumor detection	Support vector machine	Harvard, rider and local	Accuracy—97.1% auc—0.98 Sensitivity—91.9% Specificity—98.0%
Chen et al. [24]	Chronic disease of cerebral infarction detection	Convolutional neural network	Real-life hospital data	Accuracy—98.5%
Kumar et al. [25]	Malaria disease detection	Convolutional neural network	National library of medicine (NLM)	Accuracy—96.62%
Ramsundar et al. [26]	Drug discovery	Pyramidal multitask neural network	PCBA, MUV and Tox2	auc(PCBA)—0.873 auc(MUV)—0.841 auc(MUV)—0.818
Bar et al. [27]	Chest pathology detection	Convolutional neural network	Chest X-ray images	auc—0.94
Alam et al. [28]	Lung cancer detection	Multiclass support vector machine	Lung CT scan images	Detection accuracy—97% Prediction accuracy—87%
Ahmed et al. [29]	Breast cancer classification	Deep belief networks	Wisconsin breast cancer dataset	Accuracy—99.68% Sensitivity—100% Specificity—99.47%
Gupta and Kaur [30]	Head and neck cancer diagnosis and classification	Multilayer feed forward neural network	CT scan images (cancer imaging archive)	Accuracy—98.8% Sensitivity—99.5% Specificity—98.3%

5 Conclusion

This systematic review explores potential applications of smart healthcare system. A rigorous method has been conducted to find the most relevant and recent papers during 2014–2019. Papers have been searched from major database like IEEE explore, science direct, springer link, and semantic scholar. Out of 836 papers, we have found 24 most relevant papers after two stage filtering. These papers have been divided in 4 categories where 13 papers belong to disease identification and diagnosis category, 7 papers belong to patient monitoring, 3 papers belong to drug discovery and personalized treatment, and 1 paper belong smart health record category. Interested researchers can find research directions in this domain through detailed review of most significant and relevant recent works. It has been observed from this systematic investigation that machine learning and deep learning are widely used in recent smart healthcare applications.

Acknowledgements This work has been carried out with grant received from WBDST sanctioned research project on secure remote healthcare with project sanction no. 230(Sanc)/ST/P/S&T/6G-14/2018.

References

1. S. Saif, R. Gupta, S. Biswas, Implementation of cloud-assisted secure data transmission in WBAN for healthcare monitoring, advanced computational and communication paradigms. Adv. Intell. Syst. Comput. **706**, 665–6740 (2018)
2. S. Saif, S. Biswas, S. Chattopadhyay, Intelligent, secure big health data management using deep learning and blockchain technology: an overview, deep learning techniques for biomedical and health informatics. Stud. Big Data **68**, 187–209 (2019)
3. IEEE Xplore, https://ieeexplore.ieee.org
4. Science Direct, https://www.sciencedirect.com
5. Springer Link, https://link.springer.com
6. Semantic Scholar, https://www.semanticscholar.org
7. C. Venkatesan, P. Karthigaikumar, A. Paul, S. Satheeskumaran, R. Kumar, ECG signal preprocessing and SVM classifier-based abnormality detection in remote healthcare applications. IEEE Access **6**, 9767–9773 (2018)
8. J. Sidey-Gibbons, C. Sidey-Gibbons, Machine learning in medicine: a practical introduction. BMC Med. Res. Methodol. **19**, 64 (2019)
9. H. Nahata, S.P. Singh, Deep learning solutions for skin cancer detection and diagnosis, machine learning with health care perspective. Learn. Anal. Intell. Syst. **13**, 159–182 (2019)
10. M. Alhussein, G. Muhammad, M.S. Hossain, S.U. Amin, Cognitive IoT-cloud integration for smart healthcare: case study for epileptic seizure detection and monitoring. Mob. Netw. Appl. **23**, 1624–1635 (2018)
11. M. Amin-Naji, H. Mahdavinataj, A. Aghagolzadeh, Alzheimer's disease diagnosis from structural MRI using Siamese convolutional neural network, in *2019 4th International Conference on Pattern Recognition and Image Analysis (IPRIA)* (Tehran, Iran, 2019), pp. 75–79
12. M. Sathya, S. Madhan, K. Jayanthi, Internet of things (IoT) based health monitoring system and challenges. Int. J. Eng. Technol. **7**(1.7), 175–178 (2018)

13. V.V. Garbhapu, S. Gopalan, IoT based low cost single sensor node remote health monitoring system. Procedia Comput. Sci. **113**, 408–415 (2017)
14. A. Kamble, S. Bhutad, IOT based patient health monitoring system with nested cloud security, in *2018 4th International Conference on Computing Communication and Automation (ICCCA)* (Greater Noida, India, 2018), pp. 1–5
15. J. Wan, M.A.A.H. Al-awlaqi, M. Li et al., Wearable IoT enabled real-time health monitoring system. J. Wirel. Commun. Networking **298** (2018)
16. D.S.R. Krishnan, S.C. Gupta, T. Choudhury, An IoT based patient health monitoring system, in *2018 International Conference on Advances in Computing and Communication Engineering (ICACCE)* (Paris, 2018), pp. 01–07
17. F. BenRejab, K. Nouira, A. Trabelsi, Health monitoring systems using machine learning techniques, intelligent systems for science and information. Stud. Comput. Intell. **542**, 423–440 (2014)
18. P. Verma, S.K. Sood, S. Kalra, Cloud-centric IoT based student healthcare monitoring framework. J. Ambient Intell. Human Comput. **9**, 1293–1309 (2018)
19. H. Huang et al., Discovering medication patterns for high-complexity drug-using diseases through electronic medical records. IEEE Access **7**, 125280–125299 (2019)
20. L. Zhang, J. Tan, D. Han, H. Zhu, From machine learning to deep learning: progress in machine intelligence for rational drug discovery. Drug Discovery Today **22**(11), 1680–1685 (2017)
21. Y.-C. Lo, S.E. Rensi, W. Torng, R.B. Altman, Machine learning in chemoinformatics and drug discovery. Drug Discovery Today **23**(8), 1538–1546 (2018)
22. S. Khedkar, P. Gandhi, G. Shinde, V. Subramanian, Deep learning and explainable AI in healthcare using EHR, deep learning techniques for biomedical and health informatics. Stud. Big Data **68**, 129–148 (2019)
23. J. Amin, M. Sharif, M. Yasmin, S.L. Fernandes, A distinctive approach in brain tumor detection and classification using MRI. Pattern Recogn. Lett. (2017)
24. M. Chen, Y. Hao, K. Hwang, L. Wang, L. Wang, Disease prediction by machine learning over big data from healthcare communities. IEEE Access **5**, 8869–8879 (2017)
25. A. Kumar, S. Sarkar, C. Pradhan, Malaria disease detection using CNN technique with SGD, RMSprop and ADAM optimizers, deep learning techniques for biomedical and health informatics. Stud. Big Data **68**, 211–230 (2019)
26. B. Ramsundar, S. Kearnes, P. Riley, D. Webster, D. Konerding, V. Pande, Massively multi-task networks for drug discovery, arXiv preprint arXiv:1502.02072 (2015)
27. Y. Bar, I. Diamant, L. Wolf, S. Lieberman, E. Konen, H. Greenspan, Chest pathology detection using deep learning with non-medical training, in *2015 IEEE 12th International Symposium on Biomedical Imaging (ISBI)* (New York, 2015), pp. 294–297
28. J. Alam, S. Alam, A. Hossan, Multi-stage lung cancer detection and prediction using multi-class SVM classifie, in *2018 International Conference on Computer, Communication, Chemical, Material and Electronic Engineering (IC4ME2)* (Rajshahi, 2018), pp. 1–4
29. A.M. Abdel-Zaher, A.M. Eldeib, Breast cancer classification using deep belief networks. Expert Syst. Appl. **46**, 139–144 (2016)
30. P. Gupta, A. KaurMalhi, Using deep learning to enhance head and neck cancer diagnosis and classification, in *2018 IEEE International Conference on System, Computation, Automation and Networking (ICSCAN)* (Pondicherry, 2018), pp. 1–6
31. International Skin Imaging Collaboration, ISIC 2018: skin lesion analysis towards melanoma detection (2018). Available https://challenge2018.isic-archive.com/
32. International Skin Imaging Collaboration, ISIC 2019 (2019). Available https://challenge2019.isic-archive.com/

House Price Prediction Using Machine Learning

Navia Kamal, Ekta Chaturvedi, Siddharth Gautam, and Shruti Bhalla

Abstract Machine learning plays an important role in image detection, voice recognition, speech to text conversion, weather forecasting, etc. Safe automobile systems and fine customer services are also provided by machine learning systems. It is a part of artificial intelligence which means making the computers think like humans and solve the problems in the way that we do. In this paper, the prediction of housing prices that are generated by machine learning algorithms are discussed. We compare various prediction methods for the good results. Linear regression is chosen as our model because of the linearity. It makes the prediction procedure simple. We use the model and train the data and then do the testing. Finally, a supervised data is produced which is very essential to estimate the housing price in the real estate business. Our approach of this issue should be successful, and it must predict correctly. In this way, the buyers will get an idea whether to buy a house or not according to their budget.

Keywords Artificial intelligence · Linear regression · Prediction

1 Introduction

In this era of globalization, most people have opted investment as their interesting business activity. Many people use different objects for making the investments

N. Kamal · E. Chaturvedi (✉) · S. Gautam · S. Bhalla
Department of Information Technology, HMR Institute of Technology and Management, Delhi, India
e-mail: ektachaturvedi0302@gmail.com

N. Kamal
e-mail: naviakamal@gmail.com

S. Gautam
e-mail: siddharthinfo92@gmail.com

S. Bhalla
e-mail: shrutibajaj826@gmail.com

J. Tavares et al. (eds.), *Emerging Technologies in Data Mining and Information Security*, Lecture Notes in Networks and Systems 164,
https://doi.org/10.1007/978-981-15-9774-9_73

such as gold, property, etc. Estimating the accurate prices in a real estate business is necessary for making the good profits. Human beings can predict the prices of houses on the basis of different standards but that prediction may become wrong. So, the best way to estimate the correct prices of houses is using machine learning models. Machine learning has different forms: supervised, unsupervised, and reinforcement learning (hit and trial). Supervised learning is a task of machine learning in which the input is mapped to the output pairs. Some examples of supervised learning include: KNN, linear regression, support vector machines, etc. [1]. In this learning, we predict from a labeled dataset. Unsupervised learning is a technique where there is no need to supervise the model. Instead, a model is allowed to work on its own to find the information. *k*-means for clustering problems is a very important example of unsupervised learning [2]. Reinforcement learning is the area of machine learning in which software agents take actions in the environment in order to enlarge the idea of cumulative reward [3]. It is a type of dynamic programming in which algorithms are trained using a system of reward and punishment [4]. There are many ways that we can use to determine the price of the house, one of them is the analysis by making predictions. There are two models of supervised learning: classification and regression.

Classification—It categorizes a given dataset into different classes. For example: The iris dataset is divided into three different classes—Sentosa, Versicolor, and Virginica. We can perform the classification on both structured and unstructured data. The classes are often referred to as targets or labels [5].

Regression—It is a statistical method which we use in finance, investing, etc., to decide the strength and character of the connection between one dependent variable and a group of independent variables. If the depending one changes, then the independent variable will also be affected [6]. There are linear and logistic regressions. But the reality is there are more than 10 types of regression algorithms which are especially designed for different types of analysis. Each form has its own significance [7].

A real estate deliberation firm has the data apartments costs in Boston. This data includes standards like crime rate, age, accessibility, population, etc. Depending on the extracted data, the company chooses to decide the price of the new apartments [8]. This issue can be figured through a linear regression model. Other regression techniques have also been used such as decision tree and random forest regressions [9]. These models will estimate the accurate house prices. We aim to build a model for predicting the housing prices using regression to acquire the absolute outcomes. The property investors will also be benefitted by the prediction analysis as they get to know the mode of housing prices in a particular location. Predicted house prices are supposed to guide people who are planning to purchase a home so that they know the price ranges in future, and they can also plan their finance well [10].

2 Background

Classification and regression are the two most complicated areas in obtaining accurate results through class labels. Usual conventional data mining such as neural network is an additional materialized approach which is basically executed in data mining [11]. The neural network has the capability to reveal all the potential communications within predictors. This was treated as the most useful method of prediction. While purchasing the houses, buyers are extremely interested in the dealings related to housing prices in their ups and downs [12]. Despite of the great observations of real estates, most of the buyers face huge losses to invest high prices. To figure out such types of problems, systematic prediction information has been specified [13]. The fundamental search of housing price primarily investigates the bearing of housing price at the price levels and rate of growth. Further, arising the perspective of statistical and analytical, the time series method, impacts the tendency of house pricing predictions [14]. This paper mainly functions in a few situations like to analyze the real estate information of Boston city that consequences the surroundings of the house, area, location, and facilities to value the housing price [15]. Based on the house price, information in Boston city a specific and practical prediction was given. A method using linear regression not only capture the feature attributes of the house, but also reflect the time series attributes exactly [16].

2.1 Steps of Machine Learning

1. Obtaining the Dataset

The first step in machine learning is to gather the data. This is a very crucial step as the standard and amount of data that will be collected decides the quality of your predictive model.

2. Data Preparation

Here, data is loaded into a satisfactory place and prepared for the use in our training. Sometimes, the data collected requires other forms of adjusting and manipulation [17].

3. Selecting a Model

The next step in this procedure—to choose a suitable model. Many models have been created by our data scientists in the previous years.

4. Training

The main step in our machine learning is training of the data. In some manner, this is same as someone learning how to drive in the first time. At first, they do not have any knowledge of how pedals, switches and knobs work or at which time they are utilized. But, by practicing a lot, they correct their mistakes and emerge as licensed drivers.

5. Evaluation

After training of data is completed, we should see if the model is suitable by using evaluation. In evaluation, our model is tested against the data which is never trained. This metric helps us to check the performance of the model against data which it has not seen [7].

6. Parameter Tuning

Once the evaluation has been done, it is possible that we can improve our training. This can be done by tuning the parameters [18]. One of the parameters is learning rate which estimates the step size at every iteration while heading towards a loss function which is minimum.

7. Prediction

Prediction is that step in which questions are answered by us. This is the main step of our work, here we realize the significance of machine learning. The model is now finally used to make the predictions [19] (Fig. 1).

3 Related Work

A model is trained and tested on data obtained from homes in Boston. Then, we will create and estimate the predictive ability and standard of performance of this model. The prediction has been carried out with real estate information. Once a good fit is obtained, this model will be used to guess financial amount of a house which is in Boston. This model is very important for an agent who is in real estate to guess the correct predictions for the house prices [20]. Implementation of deep learning predictions examine the feature attributes of the house pricing accurately. We have taken the dataset from the UCI machine learning repository [21]. This data was gathered in 1978, and it contains five hundred six entries (506) which show data (aggregated) about fourteen features for homes which are located in Boston.

- 16 data points have an 'MEDV' amount of 50.0. Missing values are also contained by these data points which have been erased.
- 1 data point has an `'RM'` value = 8.78. This data point has been removed because it might be observed as an outlier.
- The feature 'MEDV' is measured to account for thirty-five years of inflation in market.

Some features like 'MEDV,' 'LSTAT,' 'PTRATIO,' and 'RM' are important. Rest of the non-applicable features are eliminated.

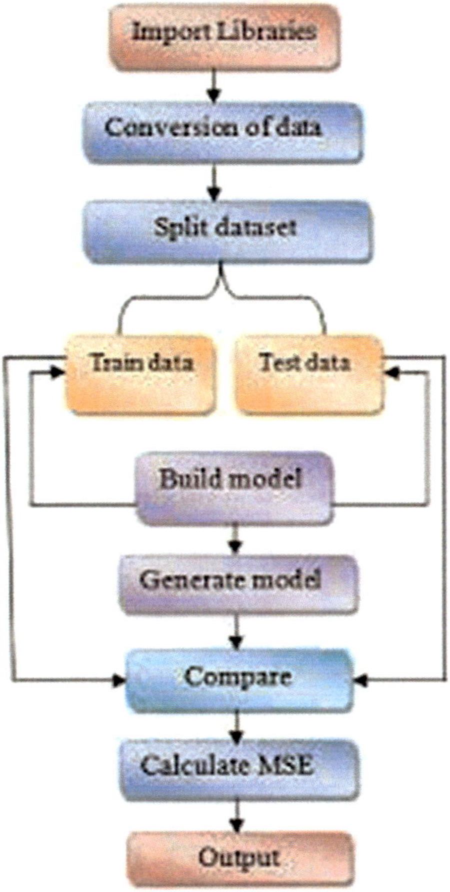

Fig. 1 Steps of machine learning

Our model must have the capacity to predict the amount of homes; the dataset is divided into target variables and features. All the non-essential features have been removed [22].

- 'MEDV' is the target variable which we will predict. It will be stored as prices.
- Quantitative information of each and every data point will be provided by the features like 'LSTAT,' 'RM,' and 'PTRATIO.'

4 Methodology

Step 1: Initially, import all the related libraries to function such as decision tree, random forest, and linear regressions.

Step 2: Convert tabular data to a data frame to obtain a format and check it for regression analysis.
Step 3: After conversion, load the data into python and isolate our dependent variables from independent variables. We will split the dataset into test and train sets.
Step 4: A regression model is built.
Step 5: Now, we will apply the respective model to train set.
Step 6: Execute predictions using the generated model and use the same model for testing.
Step 7: Compare both predicted values and check for the efficiency of the model.
Step 8: Finally, the mean square error is calculated and output is printed.

Features of Boston dataset are summarized below:

- ZN: It describes the land proportion which is residential and zoned for lots greater over 25,000 sq. ft.
- CRIM: It states the crime rate by each person in the town.
- CHAS: This is an autonomous variable of Charles River which is equal to 1 when the tract leaps river, or it will be 0.
- INDUS: It will be described as proportion of business acres (non-retail) in each city.
- NOX: It defines the concentration of Nitric Oxides which is in parts per ten million.
- RM: Number (average) of rooms in each residence.
- AGE: Quantity of units in which owners live and they are built prior to 1940.
- RAD: It is the accessibility token to highways arranged radially.
- DIS: Distances (weighted) to five employment centers in Boston.
- TAX: It tells us the full-amount property-tax rate.
- PTRATIO: Pupil-teacher proportion in Boston.
- B-1000 (Bk—0.63)^2—Bk is the amount of blacks in town.
- MEDV: Median proportion of homes preoccupied by owners (in $1000’s).
- LSTAT: It states the percentage of population which has lower status.

The respective dataset is loaded into a pandasdataframe using pd.DataFrame. Then, the first five rows of the data will be pinted by using head() (Fig. 2).

	CRIM	ZN	INDUS	CHAS	NOX	RM	AGE	DIS	RAD	TAX	PTRATIO	B	LSTAT
0	0.00632	18.0	2.31	0.0	0.538	6.575	65.2	4.0900	1.0	296.0	15.3	396.90	4.98
1	0.02731	0.0	7.07	0.0	0.469	6.421	78.9	4.9671	2.0	242.0	17.8	396.90	9.14
2	0.02729	0.0	7.07	0.0	0.469	7.185	61.1	4.9671	2.0	242.0	17.8	392.83	4.03
3	0.03237	0.0	2.18	0.0	0.458	6.998	45.8	6.0622	3.0	222.0	18.7	394.63	2.94
4	0.06905	0.0	2.18	0.0	0.458	7.147	54.2	6.0622	3.0	222.0	18.7	396.90	5.33

Fig. 2 Boston dataset

```
CRIM       0
ZN         0
INDUS      0
CHAS       0
NOX        0
RM         0
AGE        0
DIS        0
RAD        0
TAX        0
PTRATIO    0
B          0
LSTAT      0
PRICE      0
dtype: int64
```

Fig. 3 No missing values in the Boston dataset

Boston = pd.DataFrame(Boston_dataset.data, columns = Boston_dataset.feature_names)
Boston.head()

We will find 'MEDV' (target variable) missing in given data. A new column of target values is created and added in DataFrame [11].

Boston['MEDV'] = Boston_dataset.target
After loading dataset, we must see if the data contains any missing values. We count the number of missing values for each feature using isnull()
boston.isnull().sum()

There are no missing values in the Boston dataset (Fig. 3).

5 Design Approach

5.1 Linear Regression

This type of regression is a method of statistics that helps to study the relationship between two continuous quantitative variables. One variable (say x) is considered as the predictor or an independent variable. The other variable (say y) is considered to be the outcome or dependent variable [23] (Fig. 4).

Fig. 4 Equation of linear regression

$$y = \theta_1 + \theta_2 . x$$

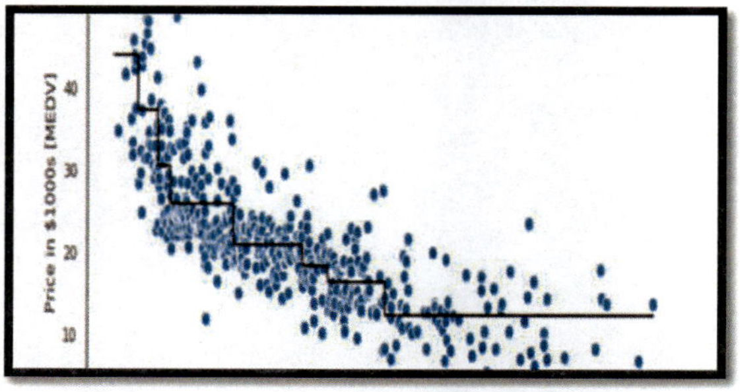

Fig. 5 Regression using decision trees

5.2 *Decision Tree Regression*

Classification and regression models are built in the form of a tree structure. A dataset is broken into smaller subsets, and an associated decision tree is built at the same time. We get a tree with leaf or decision nodes as our final result. Standard deviation is used for calculating the similarity of a numerical sample. A decision tree is developed top-down from a root, and data is partitioned into subsets that consist of instances with same values [24] (Fig. 5).

5.3 *Random Forest Regression*

It is one of the examples of supervised learning in which ensemble learning method is used for classification and regression. It is a bootstrap aggregation technique. There is no interaction between the trees and they run parallel. It aggregates multiple decision trees, with some helpful changes:

1. A random sample is drawn by each tree from the real dataset when it's splits are generated, thus, an element of randomness is also added that helps us to prevent the overfitting.
2. The number of total trees which can be divided at each node is limited to some percentage of whole. This is also called as the hyper parameter.

Ensemble learning is a mechanism in which multiple models are generated and combined to obtain better predictive performance than that which could be obtained from any of the separate constituent learning algorithms [25]. Bagging or bootstrap aggregation is a type of ensemble learning designed to improve the accuracy and

Fig. 6 Random forest simple explanation

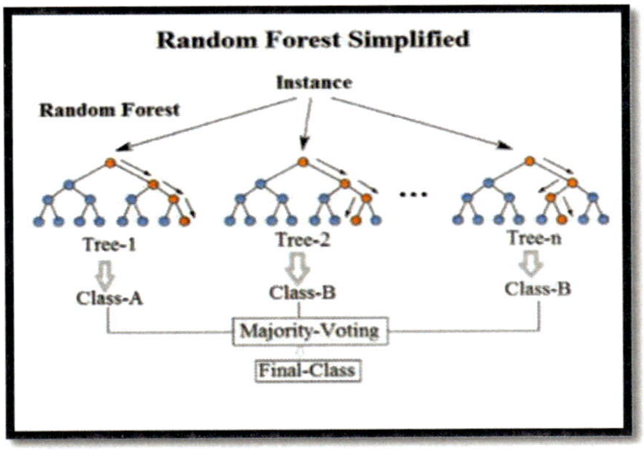

stability of machine learning algorithms. It also reduces overfitting and variance [26]. Here, multiple classifiers are modeled on distinct sub-samples of the same data set. A bagging classifier fits base classifiers one on subsets which are chosen randomly from the original dataset and now, each prediction is aggregated by voting for guessing a final prediction (Fig. 6).

6 Data Exploration

Data exploration is a necessary step before training the model. The distribution of target variable MEDV is plotted below.

Sns.set(rc = {'figure.figsize'(11.8,8.26)})
Sns.distplot(Boston['MEDV'], bins = 30)
plt.show() (Fig. 7)

Some of the plots are created using matplotlib python library. It is used to make different types of plots such as scatter, area, and pie plots. It provides an object-oriented API for making the plots using general purpose tools such as Tkinter, GTK+, etc. It is an amazing visualization library in python for creation of 2D plots of arrays. Similarly, different plots are created for the Boston dataset with the help of this library.

Many observations have been made after making the plots.

1. The prices tend to fall as the LSTAT increases.
2. There is an increase in prices as there is a linear increase in the value of RM (Fig. 8).

After plots are made, mean square value is calculated for the dataset. MSE is calculated by finding the regression line, then, the X-values are inserted into the linear regression equation to find the new Y-values, and the new y value is subtracted

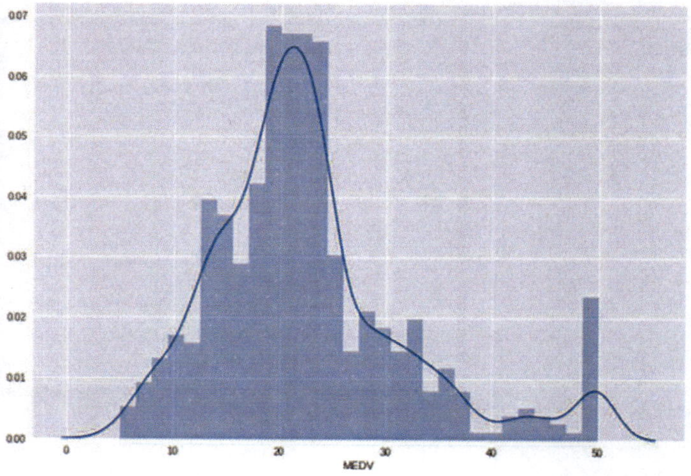

Fig. 7 MEDV distribution

from the original one to obtain the errors, square the errors, sum the errors, and mean is found (Fig. 9).

A greater MSE value means the data values have been dispersed widely around it's mean. So, the smaller MSE value is preferred because data values are dispersed closely to it's central moment. After MSE value calculation, a graph has been plotted between the measured and predicted values (Fig. 10).

7 Conclusion

A machine learning regression project is created and we observed various insights about regression models and learnt how they have been developed. We have performed several tests by using linear, decision tree, and random forest regression methods to predict the house prices. These models will predict the house prices correctly so that the real estate company does not suffer from any losses. The buyers will get the idea of the house price so that they can decide whether they can buy it or not. More advanced learning rate has speed up the merging for the trained format. Based on learning rate, the optimal prediction performance progress on the high learning rate. In this paper, implementation of linear regression for Boston dataset is executed for predicting the house prices.

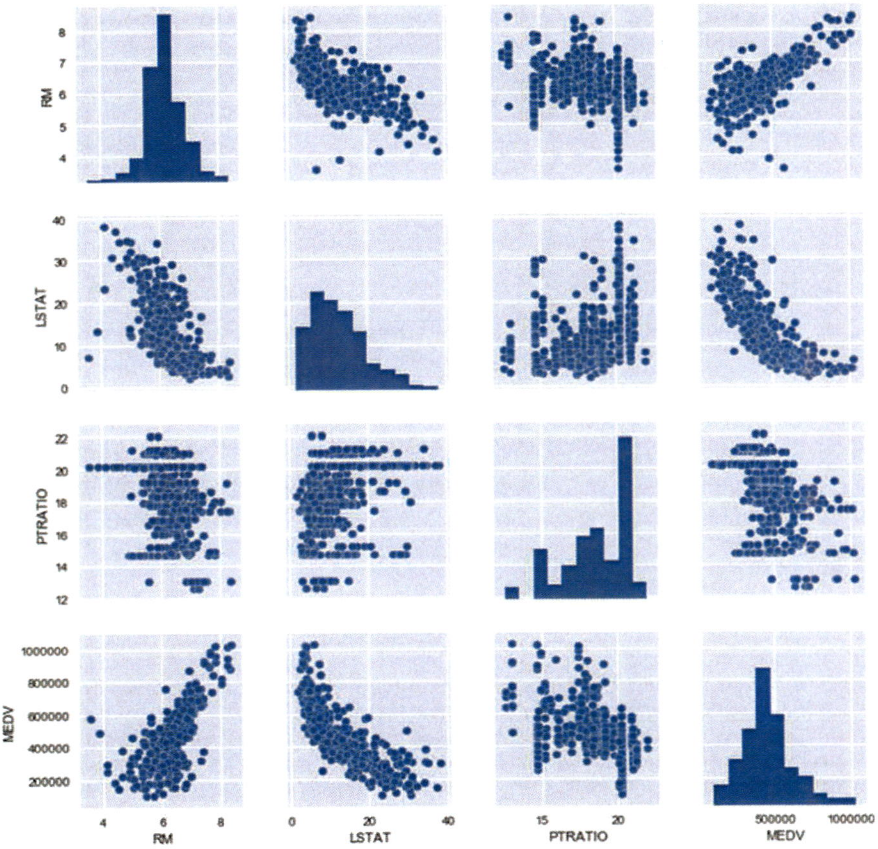

Fig. 8 Plots

Fig. 9 MSE value calculation

$$\text{MSE} = \frac{1}{n} \sum_{i=1}^{n} (y_i - \tilde{y}_i)^2$$

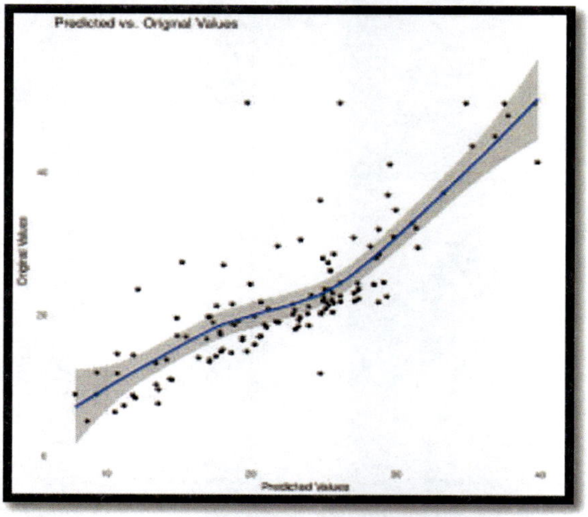

Fig. 10 Demonstrating the measured and predicted values

References

1. Rangan Gupta*, Periklis◆VasiliosPlakandaras + and TheophilosGogas. Forecasting the U.S. real house price index
2. L. Yu, C. Jiao, H. Xin, Y. Wang, K. Wang, Prediction on housing price based on deep learning. World Acad. Sci. Eng. Technol. Int. J. Comput. Inf. Eng. **12**(2) 2018
3. M. Grover, B. Verma, N. Sharma, I. Kaushik, Traffic control using V-2-V based method using reinforcement learning, in *2019 International Conference on Computing, Communication, and Intelligent Systems (ICCCIS)* (2019) https://doi.org/10.1109/icccis48478.2019.8974540
4. A.M. Shahiri, W. Husain, A review on predicting students performance using data mining techniques. Procedia Comput. Sci. **72**, 414–422 (2015). https://www.sciencedirect.com/science/article/pii/S1877050915036182
5. R. Gupta, Forecasting US real house price returns over 1831–2013: evidence from copula models
6. S. Jadon, A. Choudhary, H. Saini, U. Dua, N. Sharma, I. Kaushik, Comfy smart home using IoT. SSRN Electron. J. (2020). https://doi.org/10.2139/ssrn.3565908
7. N.M. Akbar, F.M. Ali, Z. Maryam, The dynamic effect of macroeconomic factors on functions of housing price in Iran (1990–2007) (J) (2010)
8. C. Manchanda, R. Rathi, N. Sharma, Traffic density investigation and road accident analysis in India using deep learning, in *2019 International Conference on Computing, Communication, and Intelligent Systems (ICCCIS)* (2019). https://doi.org/10.1109/icccis48478.2019.8974528
9. Z.B. Xian, L. Qiang, Arma-based traffic prediction and overload detection of network (J). J. Comput. Res. Dev. **39**(12), 1645–1652 (2002)
10. C. Patel, A. Patel, D. Patel, Optical character recognition by open source OCR tool tesseract: a case study. Int. J. Comput. Appl. **55**(10), 50–56 (2012). https://doi.org/10.5120/8794-2784
11. N. Sharma, I. Kaushik, R. Rathi, S. Kumar, Evaluation of accidental death records using hybrid genetic algorithm. SSRN Electron. J. (2020). https://doi.org/10.2139/ssrn.3563084
12. S. Li, L. Lin, C. Sun, Click-through rate prediction for search advertising based on convolution neural network(J). Intell. Comput. Appl. (2015)
13. A. Jain, A.R. Zamir, S. Savarese et al., Structural-RNN: deep learning on spatio-temporal graphs (J). 5308–5317 (2015)
14. D.E. Rapach, J.K. Strauss, Forecasting real housing price growth in the Eighth District states

15. A. Amberkar, P. Awasarmol, G. Deshmukh, P. Dave, Speech recognition using recurrent neural networks, in *2018 International Conference on Current Trends towards Converging Technologies (ICCTCT)* (Coimbatore, 2018), pp. 1–4
16. M. Harjani, M. Grover, N. Sharma, I. Kaushik, Analysis of various machine learning algorithm for cardiac pulse prediction. in *2019 International Conference on Computing, Communication, and Intelligent Systems (ICCCIS)* (2019). https://doi.org/10.1109/icccis48478.2019.8974519
17. S. Soni, B. Bhushan, Use of machine learning algorithms for designing efficient cyber security solutions, in *2019 2nd International Conference on Intelligent Computing, Instrumentation and Control Technologies (ICICICT)* (2019). https://doi.org/10.1109/icicict46008.2019.8993253
18. R. Tibshirani, Regression shrinkage and selection via the lasso. J. Roy. Soc. London. Ser. B (Methodol.) 267–288 (1996)
19. M. Chakarverti, N. Sharma, R.R. Divivedi, Prediction analysis techniques of data mining: a review. SSRN Electron. J. (2019). https://doi.org/10.2139/ssrn.3350303
20. N. Srivastava, G.E. Hinton, A. Krizhevsky, I. Sutskever, R.R. Salakhutdinov, Dropout: a simple way to prevent neural networks from overfitting. J. Mach. Learn. Res. **15**(1), 1929–1958 (2014)
21. D. Puri, B. Bhushan, Enhancement of security and energy efficiency in WSNs: machine learning to the rescue, in *2019 International Conference on Computing, Communication, and Intelligent Systems (ICCCIS)* (2019). https://doi.org/10.1109/icccis48478.2019.8974465
22. D. De Cock, Ames Iowa: alternative to the Boston housing data as an end of semester regression project. J. Stat. Educ. **19**(3) (2011)
23. P.A. Merolla et al., A million spiking-neuron integrated circuit with a scalable communication network and interface. Science **345**(6197), 668–673 (2014)
24. R. Sil, A. Roy, B. Bhushan, A. Mazumdar, Artificial intelligence and machine learning based legal application: the state-of-the-art and future research trends. in *2019 International Conference on Computing, Communication, and Intelligent Systems (ICCCIS)* (2019). https://doi.org/10.1109/icccis48478.2019.8974479
25. M. Chu et al., Neuromorphic hardware system for visual pattern recognition with memristor array and CMOS neuron. IEEE Trans. Ind. Electron. **62**(4), 2410–2419 (2015)
26. M. Jindal, J. Gupta, B. Bhushan, Machine learning methods for IoT and their future applications, in *2019 International Conference on Computing, Communication, and Intelligent Systems (ICCCIS)* (2019). https://doi.org/10.1109/icccis48478.2019.8974551

MatriVasha: A Multipurpose Comprehensive Database for Bangla Handwritten Compound Characters

Jannatul Ferdous, Suvrajit Karmaker, A. K. M. Shahariar Azad Rabby, and Syed Akhter Hossain

Abstract Bengali or Bangla language is native to Bengal. Bengali speaks their mother tongue, which they feel comfortable. In 1952, UNESCO announced that 21st February is the international mother language day. People remember this day with great respect. We achieved our mother languages for our martyrs. And, there are belongs to a different type of characters. Bangla's handwritten compound character is one of them. For helping Bengali people in different purposes, Bangla's handwritten compound character recognition performs a vital role. Such as Bangla character recognition, they are extracting data from handwritten paper and digitalization of any kind of documents, etc. In Bangla language, there are have ten numerical digits, 50 basic characters, greater than 200 compound characters, and ten modifiers. But there is no strong dataset for Bangla character recognition. Most of the researchers are working for only digit or basic characters or a few numbers of compound characters. But works with compound characters are complicated because of their different shapes and styles. For this kind of reason, MatriVasha has proposed a dataset designed to recognize 120 compound characters in Bangla that consist of 306,240 images written by 2552 distinct writers from Bangladesh. Our proposed dataset is the most comprehensive dataset for handwritten compound characters in Bangla; so far, MatriVasha datasets can be found on this link https://shahariarrabby.github.io/matrivasha.

J. Ferdous · S. Karmaker (✉) · A. K. M. S. A. Rabby · S. A. Hossain
Daffodil International University, Dhaka, Bangladesh
e-mail: suvrajit15-6944@diu.edu.bd

J. Ferdous
e-mail: jannatul15-7287@diu.edu.bd

A. K. M. S. A. Rabby
e-mail: shahariar.cse@diu.edu.bd

S. A. Hossain
e-mail: akhter.hossain@ulab.edu.bd

J. Tavares et al. (eds.), *Emerging Technologies in Data Mining and Information Security*, Lecture Notes in Networks and Systems 164,
https://doi.org/10.1007/978-981-15-9774-9_74

Keywords Computer vision · Bangla handwritten compound characters · Machine learning

1 Introduction

Researchers always want to research different types of applications, and many countries are exploring to introduce their handwritten recognition, text, and characters. Currently, automatic handwriting compound character recognition is one of the most critical research areas for various applications like OCR [4, 5] which helps to identify the character from the images, and that's process consumes the time and also research this application last few years, handwritten hardcopy information covert into a digital file format which is more reliable. And that's a result this system is comfortable to handle the document. But still now, for all Bangla characters, there is no dataset. For this reason, a significant number of older papers are written through typing. This handwritten document is not added to the digital file and this is the challenge that is trying to convert the documents by typing the manual and following the traditional method for copying them. Most of the time, we need to change the digital file from the handwritten file. But if we do it manually, then this works will waste our huge time. But we can do it using OCR [4, 5] using OCR, we can easily convert our handwritten file to digital file. It saves our huge time. But there is no strong dataset for Bangla character recognition. In Bangla language, there are ten numerical digits, 50 basic characters, greater than 200 compound characters, and ten modifiers. Most of the researchers are working for only digit [1, 6, 16] or basic characters [2, 10, 17] or a few numbers of compound characters [3, 11, 18]. But works with handwritten characters is challenging because of their unique shapes and patterns for the different types of people. And, the format of the Bengali letter is complex for its classification, and some of them are similar in addition to the complementary compound letters of the other basic characters.

The fundamental language of one language is from different languages like Latin scripts are distinct from Bengali because Bengali comes from the Sanskrit script. On the other hand, there is also every country's language that is different from other languages [7–9]. Many Bangla compound characters that are unknown to students and people. They do not know how to create a sentence using compound characters. So, the MatriVasha dataset help to know the compound character.

2 Literature Review

At present, For Bangla characters, there have four online datasets that play a vital role in recognizing Bangla's handwritten characters. These are the BanglaLekha-Isolated [12], the CMATERdb [13], and the ISI. [14], the Ekush [15]. But this kind of dataset has few numbers of compound characters.

Table 1 Comparison of the number of images with our proposed dataset

Dataset name	Unique compound characters	Total compound characters
Ekush [15]	50	150,840
CMATERdb [13]	160	42,248
BanglaLekha-Isolated [12]	24	47,407
ISI. [14]	None	None
MatriVasha	120	306,240

The MatriVasha dataset consists of 306,240 images that contain 120 unique compound characters and the largest dataset for Bangla compound characters where belong to different types of compound characters. A comparison is given in Table 1 between MatriVasha and the other four famous Bangla handwriting datasets are: BanglaLekha-Isolated dataset, CMATERdb, Ekush, and ISI handwritten dataset.

In Table 1, where we can see the comparison between the MatriVasha dataset and the other four famous datasets for Bangla handwritten characters (Ekush, CMATERdb, BanglaLekha-Isolated, and the ISI handwritten dataset). The MatriVasha dataset consists of 306,240 images that contain 120 unique compound characters and the largest dataset for Bangla compound characters where belong to different types of compound characters.

3 Dataset

MatriVasha is a Bangla compound character dataset. This dataset can be used in different ways. MatriVasha dataset consists of 120 unique compound characters. MatriVasha dataset collected from various kinds of students like university students, college students, school students in Bangladesh, and a total of 2600 people fill-up the forms where 50% from the male and the other 50% from the female. After the checking and processing the full dataset, we have 2552 people correct data where 1267 male and 1285 female. Finally, the MatriVasha has 306,240 images that contain 120 unique compound characters.

4 Dataset Processing

Initially, data are collected in the given form. After they collect all data, then we are checking those forms manually. Then we separate the correct form as male and female. Then, we scanned it in JPEG format with 300 dpi. After that, we rename the scanned form. Then, we have to crop individual character. For this, we have to follow three-step. Firstly, we cropped our data form as a square, which consists of 120 compound characters. Secondly, we crop 120 characters as ten rows which each row consists of 12 compound characters. And lastly, we crop each row. And finally, we get 120 individual characters. Then, after some process, we prepare data for character recognition.

Each step, we use a python script. We can easily and faster do our work using such a script. To scan our form, we used automated GUI in python. Also, for cropping the image using OpenCV, P.I.L., PyQT. After cropping, we transfer our image in different types for multiple purposes. Such as JPEG, grayscale, invert, and CSV.

4.1 Data Collection

There is more than 200 compound character. From those characters, we try to figure out which character is using most frequently. Ekush paper has the 20 most frequent compound characters. Without these 20 characters, we try to select the next 120 most frequent characters. To collect these 120 compound characters, we try to make a perfect form such that after collecting data, we can easily preprocess these data. We must maintain the same size for collecting individual characters. Otherwise, it's too tough to process these data. MatriVasha dataset collected from various kinds of students like university students, college students, school students in Bangladesh, and a total of 2600 people fill-up the forms where 50% from the male and the other 50% from the female (Fig. 1).

Fig. 1 Data collection form

4.2 Form Scanning

For scanning the form, we use the default app "IJ Scan Utility" for the Canon scanner. It takes 10 s to scan each form. But it has a big problem that this application cannot repeatedly scan the file automatically. The user needs to click the scan button for all images. For this reason, we need more time to scan a form. To solve this problem, we used an automated GUI in python. We define a specific position to click the button after a certain time for scanning the file by our python script. Using python script, we need 15 s to scan each page with 300 dpi in grayscale.

Algorithm: Automatically save files after the scan.

```
1.    Import necessary python library
2.    Find the perfect position for scan button and an exit button
3.    For all form do the certain number of times:
4.       Click scan button using python function
5.       Wait for a certain time
6.       Click exit button using python function
```

4.3 Correction and Crop

In Fig. 2, firstly, we can see the original image. And then, we can see the resized image. When we scan each form individually, then some form changes their angle for some issues. But for our next process, our all form has to be at the same angle. We try to solve our problem using canny edge detection. This algorithm tries to find the biggest contour. Our form has a big black boundary. This algorithm detects this boundary and changes its angle for all forms.

4.4 Separate Each Row

In Fig. 3, we can see the sample output after row-wise cropping. After the previous step, all our data will be of the same size and same angle. Now, we must separate all characters from the form data. Firstly, we have to crop as row-wise. In this step, we have to crop our form as row-wise. Each form has ten rows. We save these ten rows in 10 separate folders after each cropping. For each image, we can select the same distance for cropping as row-wise. As when we make our form, then we maintain the exact distance between each row. We can initialize the start and end value for each row. Then each time we can increase the beginning and end value concerning the distance value. And after cropping each row, we can save the same row in the same file.

Fig. 2 Before and after the cropping

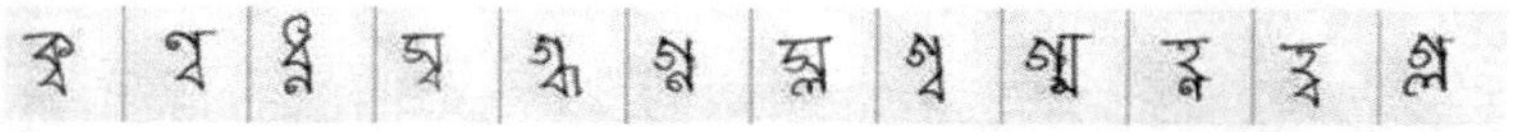

Fig. 3 Row cropping

Fig. 4 Example of after separates each character

4.5 *Separate Each Character*

In Fig. 4, we can see the example of the output image. After cropping as row-wise, we can see that each row has 12 characters as each small box has equal-sized; so ,we can maintain the same distance for cropping each box. Now, we have to crop each row image as column-wise. But now, we have to save 120 unique characters in a unique folder. We can easily handle this task in our python script.

4.6 *Noise Remove and Smooth Image*

The threshold applies to a single-channel array with fixed-level thresholding. In the grayscale image, this function used to get a binary level image. And for the filtering image, this function removes small or large noise values from the image. After

Fig. 5 Example image after noise remove and smoothing

Fig. 6 Example images after inverting the images

removing the noise, add a particular value OSTU. And Gaussian blur used to smooth the image. In Fig. 5, we can see the example of the output image.

4.7 Invert the Images

In this step, we just invert the current image bit. The white bit will be black, and the black bit will be white. That means the white part will be black, and the black part will be white. In Fig. 6, we can see the example of inverted images.

4.8 Data Labeling

We label our data as that we can use our dataset for multiple purposes. Every single character has a unique id. This id represents people's gender, hometown, age, education level, and serial number. We follow a format: gender_hometown_age_educationLevel_serialNumber for this ID. If the first number is 0, that means the writer was male, and the number is one that means the writer was female. The next one represents hometown names. Then, next come to age, here fill-up age which is given in the form. Then, they give their education or occupation level using the number to understand which level they belong. Here, 0 means the primary level (0–5), 1 means high school level (6–10), 2 means college level (11–12), 3 Means University, and the last one is the serial number of each form. Here, an example

1_BAR_18_2_00032

So, in this example, the first number represents it was written by the female who is from Barishal district, and her age is 18, and she is a college student, and the last number is denoted by serial number.

5 Conclusion and Future Work

This work initially resulted in the development of a comprehensive repository for computer vision and NLP analysis, and the dataset was known as MatriVasha. Initial tests using a sophisticated indicator proves the recognition problem. It leaves a great place to improve and helps the industry to look for new methods of accreditation. It provides a new and classical quality standard for the classification and learning systems becoming created at present.

In the future, we will build a robust model for recognizing Bangla's handwritten character. Also, we will develop an application such that the user can easily convert their handwritten paper to a digital version.

References

1. M.Z. Alom, P. Sidike, T.M. Taha, V.K. Asari, Handwritten bangla digit recognition using deep learning, (2017). arXiv preprint arXiv:1705.02680
2. M.M. Rahman, M.A.H. Akhand, S. Islam, P.C. Shill, M.H. Rahman, Bangla handwritten character recognition using convolutional neural network. Int. J. Image Graphics Signal Process (IJIGSP) **7**(8), 42–49 (2015)
3. S. Roy, N. Das, M. Kundu, M. Nasipuri, Handwritten isolated bangla compound character recognition: a new benchmark using a novel deep learning approach. Pattern Recogn. Lett. **90**, 15–21 (2017)
4. B.B. Chaudhuri, U. Pal, A complete printed Bangla OCR system. Pattern Recogn. **31**(5), 531–549 (1998)
5. U. Pal, On the developement of an optical character recognition (OCR) system for printed Bangla script (Doctoral dissertation, Indian Statistical Institute, Calcutta) (1997)
6. U. Pal, B.B. Chaudhuri, Automatic recognition of unconstrained off-line Bangla handwritten numerals, in *International Conference on Multimodal Interfaces* (Springer, Berlin, Heidelberg, 2000, October), pp. 371–378
7. B. Zhu, X.D. Zhou, C.L. Liu, M. Nakagawa, A robust model for on-line handwritten Japanese text recognition. Int. J. Doc. Anal. Recogn. **13**(2), 121–131 (2010)
8. F. Yin, Q.F. Wang, X.Y. Zhang, C.L. Liu, ICDAR 2013 Chinese handwriting recognition competition, in *2013 12th International Conference on Document Analysis and Recognition* (IEEE, 2013, August), pp. 1464–1470
9. H. Begum, M.M. Islam, Recognition of handwritten Bangla characters using gabor filter and artificial neural network. Int. J. Comput. Technol. Appl. **8**(5), 618–621
10. N. Das, S. Pramanik, Recognition of handwritten Bangla basic character and digit using convex hall basic feature, in *2009 International Conference on Artificial Intelligence and Pattern Recognition (AIPR-09)*
11. N. Das, B. Das, R. Sarkar, S. Basu, M. Kundu, M. Nasipuri, Handwritten Bangla basic and compound character recognition using MLP and SVM classifier, (2010). arXiv preprint arXiv:1002.4040
12. M. Biswas, R. Islam, G.K. Shom, M. Shopon, N. Mohammed, S. Momen, A. Abedin, Banglalekha-isolated: a multi-purpose comprehensive dataset of handwritten bangla isolated characters. Data brief **12**, 103–107 (2017)
13. R. Sarkar, N. Das, S. Basu, M. Kundu, M. Nasipuri, D.K. Basu, CMATERdb1: a database of unconstrained handwritten Bangla and Bangla-English mixed script document image. Int. J. Doc. Anal. Recogn. **15**(1), 71–83 (2012)

14. U. Bhattacharya, B.B. Chaudhuri, Handwritten numeral databases of Indian scripts and multi-stage recognition of mixed numerals. IEEE Trans. Pattern Anal. Mach. Intell. **31**(3), 444–457 (2008)
15. A.S.A. Rabby, S. Haque, M.S. Islam, S. Abujar, S.A. Hossain, Ekush: a multipurpose and multitype comprehensive database for online off-line Bangla handwritten characters, in *International Conference on Recent Trends in Image Processing and Pattern Recognition* (Springer, Singapore, 2018, December), pp. 149–158
16. S. Haque, A.S.A. Rabby, M.S. Islam, S.A. Hossain, ShonkhaNet: a dynamic routing for Bangla handwritten digit recognition using capsule network, in *International Conference on Recent Trends in Image Processing and Pattern Recognition* (Springer, Singapore, 2018, December), pp. 159–170
17. A.S.A. Rabby, S. Haque, S. Islam, S. Abujar, S.A. Hossain, Bornonet: Bangla handwritten characters recognition using convolutional neural network. Procedia Comput. Sci. **143**, 528–535 (2018)
18. A.S.A. Rabby, S. Haque, S. Abujar, S.A. Hossain, Ekushnet: using convolutional neural network for bangla handwritten recognition. Procedia Comput. Sci. **143**, 603–610 (2018)

Study on Efficient DRLSE-Oriented Edge-Based Medical Image Segmentation of Cardiac Images

Pramit Brata Chanda and Subir Kumar Sarkar

Abstract Today, medical image-based processing is very much used in recent research in medical domain very effectively. So, this technical approach helps the medical expertise person to take faster decision regarding earlier diagnosis and faster decision regarding operation and next stages treatments. Level set approach is a crucial method that is used domain as computer vision. So, the normal level set approach is more error prone and creates several irregularities for the evolution in the stages of stabilities of algorithms. Here, the potential function is built and create a distance regularized term for making solution of initiating level set function in periodical way. A energy function is created by combining internal and external energy function. The main purpose of this task is medical image-based detection. The DRLSE-based segmentation approach reduces the re-initialization problems and can minimize the errors. The problems of further initialization can hamper the accuracy or rate of performance. This approach can efficiently able to initialize the level set value which is flexible to changing the time steps in a difference scheme for elimination of amount of iterations. The approach provides more than 85% overall accuracy and sensitivity scores for proper segmentation of the cardiac images in medical imaging domain. This approach can effectively segment the area of contour region to the boundary with varying number of iterations with lesser error rate.

Keywords Computer vision · Medical image segmentation · Segmentation · MAE · RMSE · DRLSE · Level set

P. B. Chanda · S. K. Sarkar (✉)
Department of Electronics and Tele Communication Engineering, Jadavpur University, Kolkata, India
e-mail: su_sircir@yahoo.co.in

P. B. Chanda
e-mail: pramitcse@gmail.com

J. Tavares et al. (eds.), *Emerging Technologies in Data Mining and Information Security*, Lecture Notes in Networks and Systems 164,
https://doi.org/10.1007/978-981-15-9774-9_75

1 Introduction

Today, medical imaging become a crucial area of research which may be used for reducing the situation of surgery-related problem. It can help the planning of radiotherapy and tracking any disease earlier. It fully depends on the prior experience of the clinical expertise person [1, 2]. So the computerized algorithm or analytical methodology can be used to segment the several areas of image using some approaches of image segmentation. Now, medical image segmentation takes a vital role in the area of image processing [3]. It is based on the areas in engineering applications in real world data, satellite images, and videos surveillance-based system [4]. In pre-processing of images, it separates an image to separate several regions with consistency at certain level. Today, working in image processing domain, it becomes a challenge with dealing noise, lower contrast, non-homogeneous intensity level, and background level difficulties for segmentation [5]. So the image segmentation is of several categories as (i) traditional approach and (ii) soft computing-based approach [6]. So the previous approach is more easier for implementation and has been much used for getting appropriate solutions for several real world applications based on practical implementation. So, these are combined to several classes for clustering based, threshold level methods, segmentation based on region wise, edge wise segmentation among others [7]. Several clustering approach as k-means and FCM-based methods are used in image segmentation [3].

Wang et al. [8] show an method for detection of plant-based disease images using k-means clustering with higher accuracy level of classification and identification of disease area. Hu et al. [3, 9] have made a kernel-based FCM approach with adaptive thresholding where clustering spatial constraints model with control easily with the neighborhood pixels with the central pixels. The main problem of the method is it converges very slowly with number of iterations.

Singla with Patra [10] have illustrated a threshold-based selection algorithm for solving image segmentation-related problem with getting quicker solution. Ananthi et al. [11] show that L-interval-valued fuzzy sets with the lesser entropy as threshold for segmenting image. Sometimes, several approach used multi-threshold-based approach for segmentation which sometimes enhance computational time [12]. Gao et al. [13] show an artificial bee colony-based optimization tool for threshold image segmentation in multilevel. So there are several types of level set approach are there edge-oriented techniques [3, 14–16], region-wise algorithms [3, 17–20], mixed approaches [3]. However, level set approach is so much sensitive for the initial positions of level sets and Chan Vese (CV) model [18], is a type of crucial region-based level set approach, was studied. The cv approach depends on initial position and nondependent with gradients of edges with weaker edges. The another approach as hybrid level set-based approach with edge-oriented models with region models of image segmentation [3]. This method of global segmentation model has all advantages of region-based methods, but can not be able to segment heterogeneous images. Also several real world images mainly medical images use this methods of segmentation.

A lot of algorithms are there for performing the tasks of segmentation as level set-based method, edge-based method, Wavelet-driven detection, PSO-based segmentation, region-based approach, threshold-based methods, distance regularized level set(DRLSE)-based methodology.

2 Theoritical Background

Previously, the level set approach has been developed in 1987 for capturing the dynamic shapes and interfaces. Here, for level set approach, the level set function take as zero level set of more higher dimensional function and formulation of the function is based on the motion of the contours as the evaluation of these function. These approach can be used of tackling topological changes as merging and splitting in an efficient approaches, that is not been allowed in contour-based models unless extra algorithms are provided its implementations. So level set-based methods may be used to solve various critical scientific and engineering domain-related problems [21, 22]. Their applications are suffered with various irregularities of these function. In the traditional LSF approach, the LSFs are used to create irregularities during its evaluation which are responsible to create some errors and stability of LSF may be lost for it. So solving these difficulty, the re-initialization approach is deployed for restoring the regularity of the values of level set and maintain the appropriate stability of level set value evaluation. In case of re-initialization, there may be some theoretical and implementation level problems. DRLSE-based approach proposes several approach for solving these problems. DRLSE can be used to reduce the requirement of initialization and avoid the side effects which are not desirable [23–31]. These methods can reduce the number of steps used for execution of the problem and also reduce the time of computation. So for this, DRLSE-based method is useful to maintain the numerical accuracy.

In case of DRLSE approach, there are two stages are used. The primary stage is mainly required for the contour toward the boundary of the objects, enhancing the process of evolution when the contour is slightly situated outer area rather than boundary region, and reducing the process of the evolution when the area of the particular contour is more nearer toward boundary. While refinement of the results of segmentation is done on the later stages.

3 Proposed Method

The approach combined with several steps. The steps are:

Step 1: Start.
Step 2: Take input image.
Step 3: smoothing of image is done with Gaussian kernel.

Step 4: Take initial level set function.
Step 5: Edge indicator function of the actual image is found for partial difference.
Step 6: Distance regularization function is calculated.
Step 7: Dirac delta function, regularized term, and area length of differential function is calculated.
Step 8: Energy functional is deduced.
Step 9: If either due to consecutive iterations the crossing points variation closes or it checks whether it exceeds over ano of iterations of maximum value, then terminate iteration.
Step 10: Getting final segmented image.
Step 11: Image classification with calculating accuracy parameters are error matrix parameters.
Step 12: Stop.

4 Results and Analysis

Here, some statistical parameters are calculated for analysis of the algorithm. As like as mean, sd, etc. Here, the images of average intensity value are treated as mean value and standard deviation(SD) is the variation of pixels values toward a appropriate mean values (Table 1).

The mean and standard deviation values for first cardiac cases image are quite higher than the other cases. So the variation of the particular case 1 is quite higher than other cases.

Here, the results show that our method produces good accuracy values comparing to other method. So method produce near about 90% accuracy rates for level set-based segmentation. Here, for the particular cases of cardiac problems how below: (Fig. 1).

These case studies are shown that how the abnormalities of affected region are changed according to changing several no of iterations using the approach for starting level diagnosis of disease. The marked zone highlights the appropriate region of affected zone of cardiac MRI. Here, for computational time required for 1002 number of iterations is 488.59 s. Here, for computational time required for 1402 number of iterations is 773.253 s (Fig. 2).

Table 1 Statistical parameter measurement for analysis

Image No.	Mean	Standard deviation	Time for execution (s)	Accuracy score
Image 1 (cardiac case1)	24.4491	6.346	773.253	0.8764
Image 2 (cardiac case 2)	22.154	5.985	855.71	0.8563
Image 2 (cardiac case 3)	21.0871	5.1076	958.34	0.8431

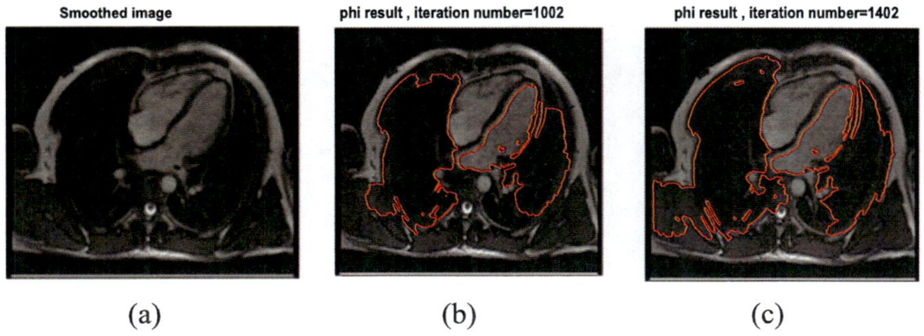

Fig. 1 **a** Input image. **b** Resultant image of 1002 iterations. **c** Resultant image of 1402 iterations

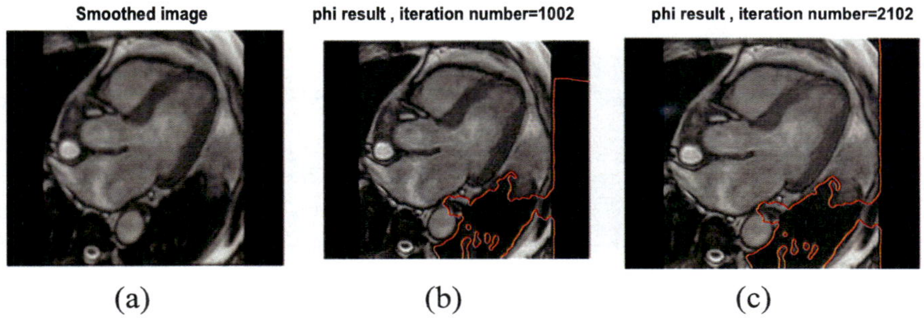

Fig. 2 **a** Input image. **b** Resultant image of 1002 iterations. **c** Resultant image of 2102 iterations

Here, for second cases, for cardiac MR images, the computational time required for 1002 number of iterations is 188.66 s. Here, for computational time required for 2102, number of iterations is 855.712 s. The resultant images are changed according to several number of iterations changes the segmentation performance and the classification accuracy rates or mean square error rates are varying with continuous changes of several number of steps of the particular images with level set and distance regularized terms.

Here, the comparison chart of several approaches is shown in given figure which shows the proposed DRLSE-based segmentation approach provides higher accuracy, sensitivity rates than the other threshold, adaptive approaches. Also the precision values are quite better than the other methodologies. In case of DRLSE-based approach, the false positive rates are comparatively lesser than the other segmentation methods (Fig. 3).

In case of DRLSE-based segmentation, it will be segmented with several iterations which are being very useful to segment the accurate contour region of image. So the segmentation performance is quite better than other approach. Also the precision rates, accuracy, sensitivity parameters are quite good as per comparing other segmentation and classification methods (Fig. 4).

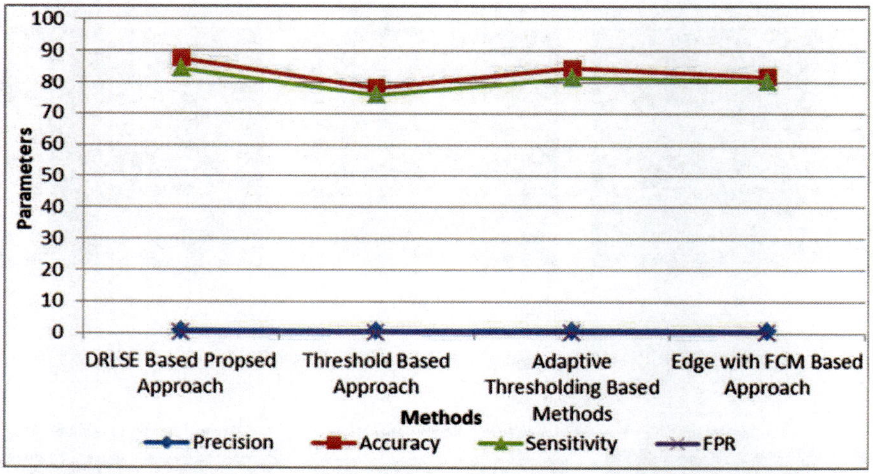

Fig. 3 Comparison of several approach in terms of performance parameter

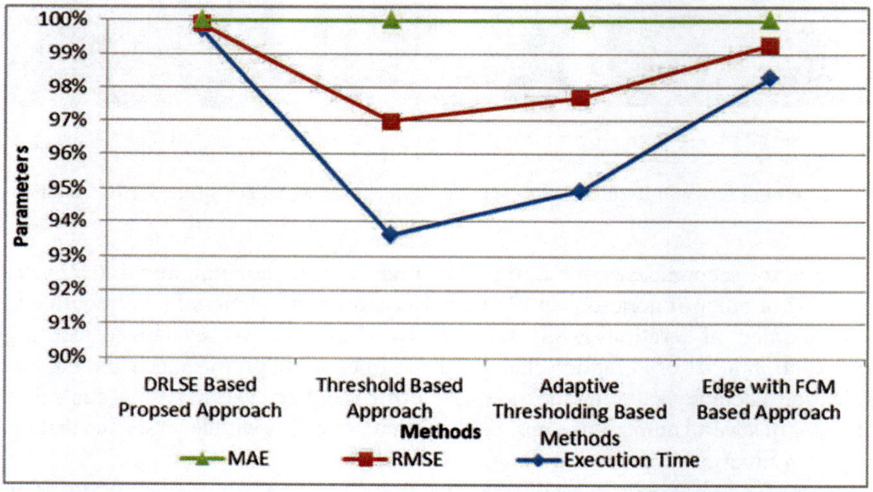

Fig. 4 Comparison of several types of error of performance parameter

Here, the results show that the DRLSE segmentation approach takes higher execution time period than the other methods as it requires several number of iterations for covering the total contour area of segmentation. As the results of the segmentation varies to the number of changing of iterations and value of distance regularized term with initial level set function values. So the mean absolute error or RMSE error rates are quite higher for the proposed methods as it requires several iterations for segmenting the actual disease affected region with better accuracy. So it requires higher mean square error (RMSE) rates, mean absolute error rate than the other segmentation approach for proper identification of the area than conventional algorithms. The rate of absolute error or mean square error rate is comparatively lower for adaptive thresholding and edge with fcm-based segmentation approach than the proposed DRLSE-based methodology.

5 Conclusion

DRLSE is a very simpler, efficient computational approach than other conventional methods. It is an edge-oriented active contour-based approach for segmentation. Here, after changing time steps, the number of iterations, time of computation can be reduced. Here, the used approach highly efficient as it reduces the requirements of re-initialization further and reduces the amount of errors due to use of distance regularization with level set function. In this approach, the segmentation quality changes according to varying number of iterations. Sometimes, the effective number of iterations may increase but the segmentation performance is better. So here after changing the number of iterations that will change the performance of segmentation. This approach provides better results of segmentation of partially effected image. After studying several papers, we have analyzed that this DRLSE-based segmentation is effective approach than other level set or other approach. It is based on active contour-based edge-oriented segmentation. The statistical outcomes are better for diagnosis of the patients effectively with more than 80% of accuracy score. Also the other statistical parameters as like as accuracy, sensitivity, precision rates, etc., are better for proposed level set methods than the approach as edge based, threshold-based approach. Also some error matrixes significantly showed that the performance rate is quite satisfactory with comparison to threshold, adaptive methods. As per the performance of DRLSE-based segmentation, we can expect that it can perform better in case of critical image segmentation-related applications in medical domain.

Acknowledgements Here we acknowledge DoveMed [32–34] which is based on healthcare based technology which provides several hospital based information as of taking several cardiac MRI images which are very crucial parts of our research work. Several patients images are taken from this organization which are basically one of the healthcare based company portal quite user-friendly. A lot of medical images are used from them which are really effective for working with cardiac or ventricular related studies. Also some image datasets are taken from the blog of Dr. Paul Babyn, Radiologist-in-Chief, and Dr. Shi-Joon Yoo, Cardiac Radiologist of hospital of Toronto.

References

1. P.B. Chanda, S.K. Sarkar, Detection and classification technique of breast cancer using multi Kernal SVM classifier approach ASPCON18, IEEE

2. P. Brata Chanda, S. Sarkar, Analysis on efficient brain MRI detection and classification strategy using medical image segmentation techniques (Elsevier SSRN, 28,2020)
3. L. Sun, X. Meng, J. Xu, S. Zhang, An image segmentation method based on improved regularized level set model. Appl. Sci. **8**, 23–93 (2018). https://doi.org/10.3390/app8122393
4. H. Min, W. Jia, X. Wang, Y. Zhao, R.X. Hu, Y.T. Luo, F. Xue, J.T. Lu, An intensity-texture model based level set method for image segmentation. Pattern Recogn. **48**, 1547–1562 (2015)
5. P. Telagarapu, V.J. Naveen, A.L. Prasanthi, G.V. Santhi, Image compression using DCT and wavelet transformations. Int. J. Sig. Process. Image Process. Pattern Recogn. **4**, 61–74 (2011); X.F. Wang, H. Min, L. Zou, Y.G. Zhang, A novel level set method for image segmentation by incorporating local statistical analysis and global similarity measurement. Pattern Recogn. **48**, 189–204 (2015)
6. S.S. Chouhan, A. Kaul, U.P. Singh, Soft computing approaches for image segmentation: a survey. Multimed. Tools Appl. **77**, 28483–28537 (2018)
7. M.A. Kader, F. Ghani, R. Badlishah, Development and performance evaluation of hierarchical quadrature amplitude modulation (HQAM) for image transmission over wireless channels, in *2011 3th International Conference on Computational Intelligence, Modelling and Simulation (CIMSiM)* (2011), pp. 227–232
8. Z. Wang, K. Wang, S. Pan, Y. Han, Segmentation of crop disease images with an improved K-means clustering algorithm. Appl. Eng. Agric. **34**, 277–289 (2018)
9. G. Hu, Z. Du, Adaptive kernel-based fuzzy c-means clustering with spatial constraints for image segmentation. Int. J. Pattern Recogn. Artif. Intell. **33**, 1954003 (2019)
10. A. Singla, S. Patra, A fast automatic optimal threshold selection technique for image segmentation. Sig. Image Video Process **11**, 243–250 (2017)
11. V.P. Ananthi, P. Balasubramaniam, P. Raveendran, A thresholding method based on interval-valued intuitionistic fuzzy sets: an application to image segmentation. Pattern Anal. Appl. **21**, 1039–1051 (2018)
12. J. Han, C.H. Yang, X.J. Zhou, W.H. Gui, A new multi-threshold image segmentation approach using state transition algorithm. Appl. Math. Model. 44, 588–601 (2017)
13. Y. Gao, X. Li, M. Dong, H.P. Li, An enhanced artificial bee colony optimizer and its application to multi-level threshold image segmentation. J. Cent. South Univ. **25**, 107–120 (2018)
14. S. Gao, T.D. Bui, Image segmentation and selective smoothing by using Mumford-Shah model. IEEE Trans. Image Process. **14**, 1537–1549 (2005)
15. A. Sarti, R. Malladi, J.A. Sethian, Subjective surfaces: a geometric model for boundary completion. Int. J. Comput. Vis. **46**, 201–221 (2002)
16. X. Xie, M. Mirmehdi, MAC: magnetostatic active contour model. IEEE Trans. Pattern Anal. Mach. Intell. **30**, 632–646 (2008)
17. D.B. Mumford, J. Shah, Optimal approximations by piecewise smooth functions and associated variational problems. Commun. Pure Appl. Math. **42**, 577–685 (1989)
18. F. Tony, T.F. Chan, L.A. Vese, Active contours without edges. IEEE Trans. Image Process. **10**, 266–277 (2001)
19. Z.X. Ji, Y. Xia, Q.S. Sun, G. Cao, Q. Chen, Active contours driven by local likelihood image fitting energy for image segmentation. Inf. Sci. **301**, 285–304 (2015)
20. Q. Ge, L. Xiao, H. Huang, Z.H. Wei, An active contour model driven by anisotropic region fitting energy for image segmentation. Digit. Signal Process. **23**, 238–243 (2013)
21. I. Rianto, Pranowo: distance regularized level set evaluation for medical image segmentation, CITACEE (2013). ISSN:2328-5154
22. G. Avinash, S. Bhalerao, M. Deepashree, Image segmentation using distance regularized level set method. IJARECE **5**(2) (2016). ISSN: 2278-909X
23. D.K. Chy, M. Khaliluzzaman, Comparative performance of BER in the simulation of digital communication systems using raised cosine filter, in *Third International Conference on Advances in Computing, Electronics and Electrical Technology-CEET* (2015), pp. 29–33
24. N. Mahata, S. Kahali, S.K. Adhikari, J.K. Sing, Local contextual information and Gaussian function induced fuzzy clustering algorithm for brain MR image segmentation and intensity inhomogeneity estimation. Appl. Soft Comput. **68**, 586–596 (2018)

25. C. Li, C. Xu, C. Gui, M.D. Fox, Distance regularized level set evolution and its application to image segmentation. IEEE Trans. Image Process. **19**(12) (2010)
26. R. Malladi, J.A. Sethian, B.C. Vemuri, Shape modeling with front propagation: a level set approach. IEEE Trans. Pattern. Anal. Mach. Intell. **17**(2), 158–175 (1995)
27. T. Chan, L. Vese, Active contours without edges. IEEE Trans. Image Process. **19**(12) (2010)
28. J. Sethian, *Level set methods and fast marching methods, Cambridge* (Cambridge University Press, U.K., 1999)
29. M. Sussman, P. Smereka, S. Osher, A level set approach for computing solutions to incompressible two-phase flow. J. Comput. Phys. **114**(1), 146–159 (1994)
30. B. Rajitha, A. Tiwari, S. Agarwal, Image segmentation and defect detection techniques using homogeneity, in *1st International Conference on Futuristic Trend in Computational Analysis and Knowledge Management (ABLAZE 2015)*
31. A.S. Baby, K. Balachandran, A parallel approach for region-growing segmentation, in *International Conference on Advances in Computer Engineering and Applications (ICACEA)* (IMS Engineering College, Ghaziabad, India, Year 2015)
32. https://www.dovemed.com/common-procedures/radiology-procedures/magnetic-resonance-imaging-mri-cardiac-heart/
33. http://www.cse.yorku.ca/~mridataset/
34. https://www.med-ed.virginia.edu/courses/rad/cardiacmr/Techniques/SSFP.html
35. B. Han, Y.Q. Wu, Active contours driven by median global image fitting energy for SAR river image segmentation. Digit. Signal Process. **71**, 46–60 (2017)

Big Data Analysis Using R and Hadoop

Vikas Rao Vadi, Varun Tiwari, and Shafiqul Abidin

Abstract Today very large amount of data is available in the world. Therefore, there is an immediate need for maintaining, managing and accessing the large amount of data. Major IT players Like Amazon, Google have been extensively working on cloud computing since year 2000. In year 2006 August, Amazon introduced its elastic compute cloud for Amazon Web services. Later in year 2008, April Google released Beta version of Google App Engine. The world is familiar with cloud and in some manner using cloud, so the companies now have abundance of data. There are many storage options available for storing million user's data such as hard disk, CD, DVD mobile phone, Internet cloud area, etc. Three types of data are available that is structured, semi-structured and unstructured. The challenge faced by the companies is of maintaining and analyzing the huge data. It is essential for the companies to retrieve the best processed information out of the data available. Now, the issue arises how to analyze, process large data in cloud. One of the finest solutions to resolve the above-mentioned problem is R programming. The authors in this research paper have tried to focus on R programming, as it is used for analyzing and accessing the data using R. R being at a developing stage utilizes familiar scripting platforms such as Python, Pig for reducing processing and generating faster results. This research paper will aim at identifying that how R programming used for data analysis. The paper will shed light on how R programming is beneficial for analyzing data.

Keywords Big data analysis · Hadoop · R programming

V. R. Vadi (✉)
Don Bosco Technical School, Bosco Technical Training Society, New Delhi, Delhi, India
e-mail: vikasvadi@gmail.com

V. Tiwari
Trinity Institute of Professional Studies, GGSIP University, New Delhi, Delhi, India
e-mail: varuntiw1984@gmail.com

S. Abidin
HMR Institute of Technology & Management, GGSIP University, New Delhi, Delhi, India
e-mail: shafiqulabidin@yahoo.co.in

J. Tavares et al. (eds.), *Emerging Technologies in Data Mining and Information Security*, Lecture Notes in Networks and Systems 164,
https://doi.org/10.1007/978-981-15-9774-9_76

1 Introduction

R programming is an open source, and it is widely used by data scientist for data analysis. R is a wonderful programming for displaying statistical data analysis represented easily in the form of graphics. R is preferred by maximum data scientists, but R has a major shortcoming as it loads all objects in main memory, and as it deals with big data, it is difficult to fit all data in RAM memory of a single machine [1]. Therefore, for better results on huge data, a need arises to integrate Hadoop with R programming. It is revolutionizing approach toward analysis, operation and maintenance of huge data. Information technology companies are adopting various scripting platforms with R to reduce the time for analyzing and structuring the gigantic data. R is the need of the hour for database companies which deal with enormous data at every step. Apache Pig is an important language of R which reduces the coding and development time for analyzing big data [2]. This research study furnishes information on big data and R using various challenges related to big data. This makes available information on R architecture and specifications of various components of different layers of R statistical processing. This study explicates the comparison of scripting platforms and provides metaphor between two platforms R and R Studio. Existing work of big data analysis using R integration with data has also been portrayed.

2 Working with Big Data in R

There are lots of packages and statistical computation providers of R language using big data. The paper depicts R programming using R Studio for analysis big data. Distributing, computing and data processing are accompanied by R. Data can be loaded from any source by providing the availability of vector, list, factor and array. Same as R Studio is a IDE which provide the facilities to show the chart, bar plot, etc.

Large amount of data can be handled efficiently with R and R studio. The combination of R and R Studio plays a wonderful result-oriented goal for data analysis [3].

3 R and Hadoop

As already mentioned above, R is one of the most chosen programming tools, and especially when combined with Hadoop, R pitfalls get converted into the best tool for statistical data analysis and to convert this data analysis to interactive graph, chart and plot. In R Programming, the objects are stored in the main memory on a single system. R is not scalable because only limited amount of data can be processed at a time [4]. Nowadays, Hadoop has become a choice of professional. In short, loading is being preferred by Hadoop over R. Keeping in view the functionalities and characteristics, SAS is being preferred jointly with R for storing large data.

Hadoop is a distributed processing framework. It is used to perform large operation and handle large datasets. It is a popular framework for big data using R, and this is highly scalable [5].

4 Big Data Analysis Using R and Hadoop

R packages and a variety of scripts are being used by data scientists for data processing during current professional scenario. Mainly, Java programming language is used to rewrite these packages and scripts. For data processing, Hadoop MapReduce algorithms are used. Hadoop and R support various techniques and methods for big data integration.

4.1 R-Hadoop

Hadoop and R provide the wonderful solution for big data. The user can access data directly from HBASE database systems and HDFS system. It includes five packages:

Rhbase: It uses database management functions for HBase within R; Rhdfs: provides connectivity to HDFS; Plyrmr: provides data manipulation operations for large datasets; Ravro: it is used to read and write Avro files from HDFS. It is a row-oriented and data serialization framework developed for Hadoop project and Rmr2: used for stored statistical analysis data on Hadoop [6].

4.2 Rhipe

Rhipe: It is R and Hadoop integrated environment. It is providing the facility to use R library within MapReduce. It also provides data distribution and integration within Hadoop.

4.3 Streaming with R and Hadoop

User directly runs MapReduce using an executable script. It reads data from standard input and writes data using mapper or reducer. R scripts is fully integrated with Hadoop Streaming.

4.4 RHive

RHive: This package is directly used to hive queries within R. It retrieves data using names, column names and table name. It provides libraries and algorithms in R to stored data in Hadoop.

4.5 Orch

This is Oracle connector for Hadoop. It used to run directly MapReduce program without help of new programming language.

5 R Techniques for Big Data Analysis

There are following techniques used for big data analysis with R:

5.1 Sampling

Sampling is used to decrease the size of data where data is too big.

5.2 Hardware

R stored all objects in a single memory. So that the problem comes if data is very large. So, this problem solves when you increase the memory size of machine.

5.3 Alternative Interpreters

Alternative Interpreters: There are two alternative interpreters such as pqR (pretty quick R) and Renjin which can easily run on the Java virtual machine.

5.4 *Integration of High Programming Language*

Integration of High Programming Language: R is high performing programming language. So, the small program is easily transferred from R to another language.

6 Processing by MapReduce

Processing of large dataset distribution on a large cluster is quite useful in MapReduce. Performing huge data processing of throughout thousands of servers is possible by Hadoop clusters. It also provides a secure platform for commercial applications where data is to be kept confidential. This Hadoop cluster concept is being overwhelmingly accepted and derived from Google MapReduce [7]. The processing of huge data in parallel mode and on large clusters of commodity hardware is reliable and fault tolerant. The conceptual technique has been divided into two forms. The first concept is mapping and another is reducing. Map deals with key, and reduce value pairs the data. These map and reduce techniques are processed in sequence on the cluster. The output of this sequential process becomes the input for reduce. These techniques and concepts have been explored as follows:

Map: In this phase, datasets are assigned to the task tracker, and data functional operation will be performed over the data, assign a map key and value pairs for the output.

Reduce: In this phase, master node will collect all the answer, subproblems and combines them for the output.

There are five steps used for map and reduce:

Preparing the Map () input: It will take input data row wise and emit key value pairs per rows. It changes explicitly according to requirements.

$$\text{Map input: list } (a1, b1) \tag{1}$$

$$\text{Execute user-provided Map () code: Map output: list}(a2,\ b2) \tag{2}$$

$$\text{Shuffle: it shuffles the similar keys and input them to the same reducer} \tag{3}$$

Execute the user given Reduce() code: In this phase, developed customized reducer code will be executed, a code which was written by developer to run on a shuffled data and emit key and value [8].

$$\text{Reduce input: } (a2, \text{list}(b2)) \text{ Reduce output:}(a3, b3) \tag{4}$$

Get final output: Finally

$$\text{the master node collects all reducer output and combines and writes them in a text file} \tag{5}$$

7 Role of Hive

It is Hadoop-based data framework. It is developed by Facebook. It performs task using Sql Query language such as HiveQL, which are highly abstracted to Hadoop MapReduce. It is easily integrated with business intelligence and visualization tools for real-time processing [9].

R functions used in Hadoop MapReduce scripts: There are some utility functions used in Hadoop Mapper and Reducer for data processing-:

File is used to perform or create a connection to a file for read and write operation. It uses stdin and stdout function for read and write. It is used for mapping and reducing phase.

$$\text{con} \leftarrow \text{file("stdin", "r")} \tag{6}$$

Write—It is used to write data in a file. It will be used after the key and value pair is set in the mapper.

$$\text{write(paste(city, pagepath, sep = "\textbackslash t"), stdout())} \tag{7}$$

print—Used to print or write data to a file.

$$\text{print(paste(city, pagepath, sep = "\textbackslash t"), stdout())} \tag{8}$$

close—Closing the connection of the file after reading and writing operation is completed.

$$\text{Close(con)} \tag{9}$$

stdin—It is standard input function. It provides text mode connection. It returns the connection object.

$$\text{conn} \leftarrow \text{file("stdin", open = "r")} \tag{10}$$

stdout—It is standard output function. It provides text mode connection. It returns the connection object.

$$\text{Print(list(city.key, page.value), stdout())} \tag{11}$$

sink—It drives the R output to the connection. If the connection is file or stream, it will return to file or stream. Sink can also find the errors in Mapper and Reducer.

$$\text{sink("abc.txt")}\backslash\backslash k \leftarrow 1:6\backslash\backslash \\ \text{for}(i \text{ in } 1{:}\,k)\{\backslash\backslash \\ \text{print(paste("value of } k\text{", } k))\backslash\backslash \\ \}\text{sink()}\backslash\backslash \text{ unlink(abc.txt)} \tag{12}$$

8 Case Study

In the education field, student examination result data is completely digitalized. Now, this data is stored in PDF, excel and csv file. This excel, csv file contains students result data [10].

The main purpose of this case studies is how to store data or read data and perform basic operation to csv file in R Studio. After uploading or reading this file, how to perform some calculation such as sum, average, max, min operation using R and how to represent this data in graph or plot form. The problem is that many people are not aware of calculation and representation of data in R [11].

Problem 1: Collecting Trinity Institute of Professional Studies (Affiliated to GGSIPU) BCA students result data and save as Trinity BCA Result.csv (see Fig. 1). The problem is that how to read csv file and perform operation in R [12].

Solution: There are some steps follow for completing this task:

Step 1 Firstly, this Trinity BCA Result.csv file store or read.

$$> \text{swastik} \leftarrow \text{read.csv ("Trinity BCA Result.csv") press enter}$$

This syntax used to read a csv file in R Studio.

$$\text{swastik} \leftarrow \text{transform(swastik, SUM} = \text{rowSums(swastik))}$$

This syntax is used to create SUM Field and transform sum of rows and store in SUM Fields.

Now, write these data into buffer.

$$> \text{write.csv(swastik, row.Names} = \text{FALSE)}$$

This is used to write data in csv file and then show the output of the file (see Fig. 2).

Step 2 Now, find the average of rows (see Fig. 3)

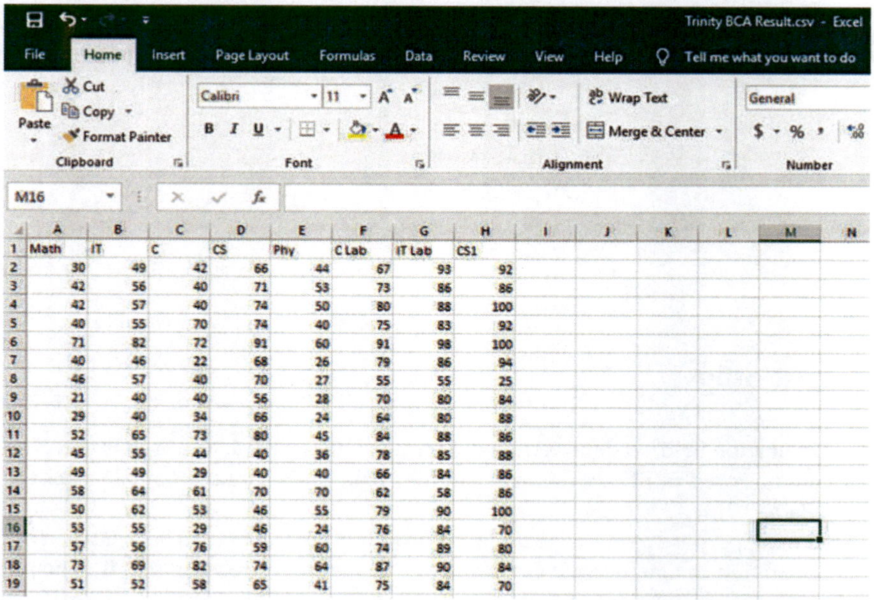

Math	IT	C	CS	Phy	C Lab	IT Lab	CS1
30	49	42	66	44	67	93	92
42	56	40	71	53	73	86	86
42	57	40	74	50	80	88	100
40	55	70	74	40	75	83	92
71	82	72	91	60	91	98	100
40	46	22	68	26	79	86	94
46	57	40	70	27	55	55	25
21	40	40	56	28	70	80	84
29	40	34	66	24	64	80	88
52	65	73	80	45	84	88	86
45	55	44	40	36	78	85	88
49	49	29	40	40	66	84	86
58	64	61	70	70	62	58	86
50	62	53	46	55	79	90	100
53	55	29	46	24	76	84	70
57	56	76	59	60	74	89	80
73	69	82	74	64	87	90	84
51	52	58	65	41	75	84	70

Fig. 1 Trinity BCA Result.csv

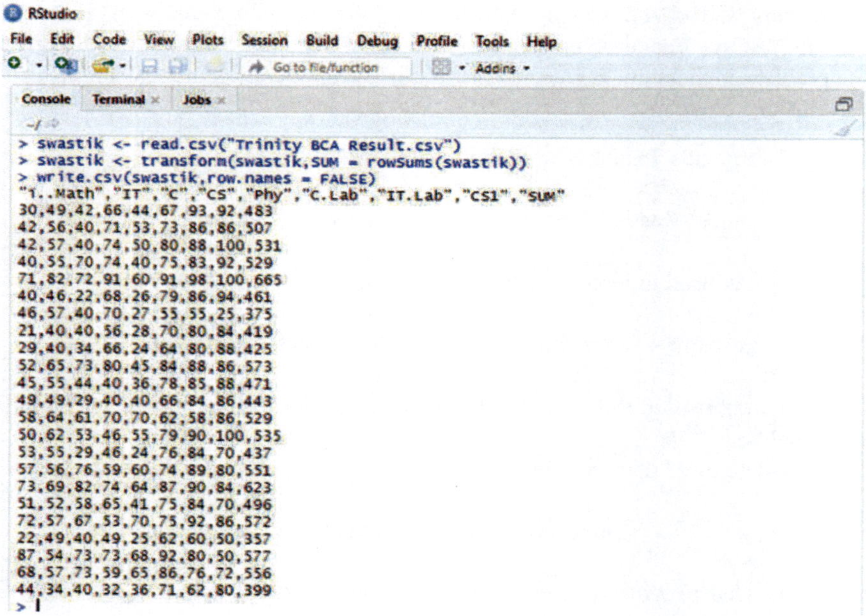

Fig. 2 Perform sum operations in csv

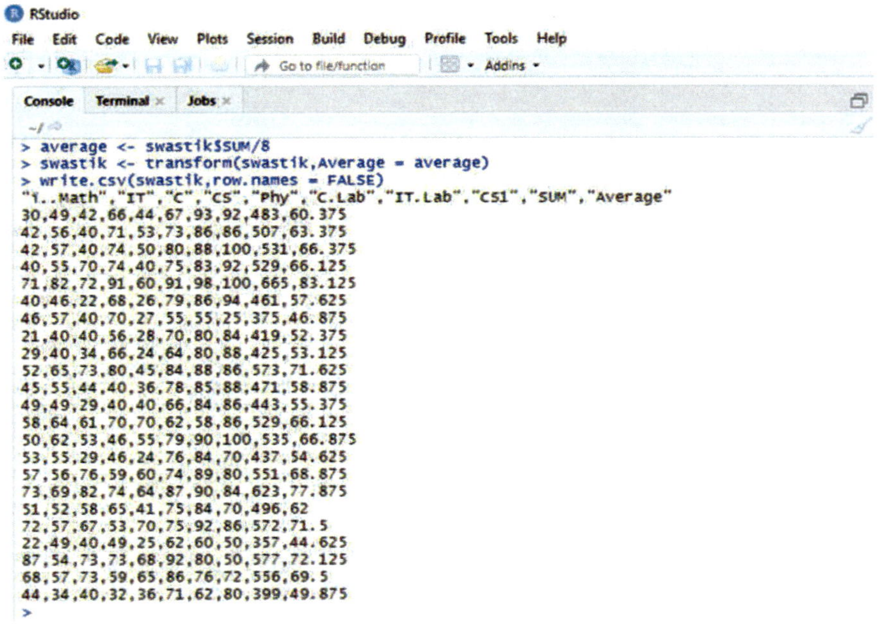

Fig. 3 Perform average operation in csv

> average ← swastik$SUM/8
> swastik ← transform(swastik, Average = average)
> write.csv(swastik, row.names = FALSE)

Step 3 Find the mean, median, min and max of marks.

> max(swastik$IT)
[1]82
> min(swastik$C.Lab)
[1]55
> median(swastik$IT.Lab)
[1]84
> mean(swastik$CS1)
[1]80.3913

Step 4 Now, find the graph according to average:

> plot(swastik$Average)

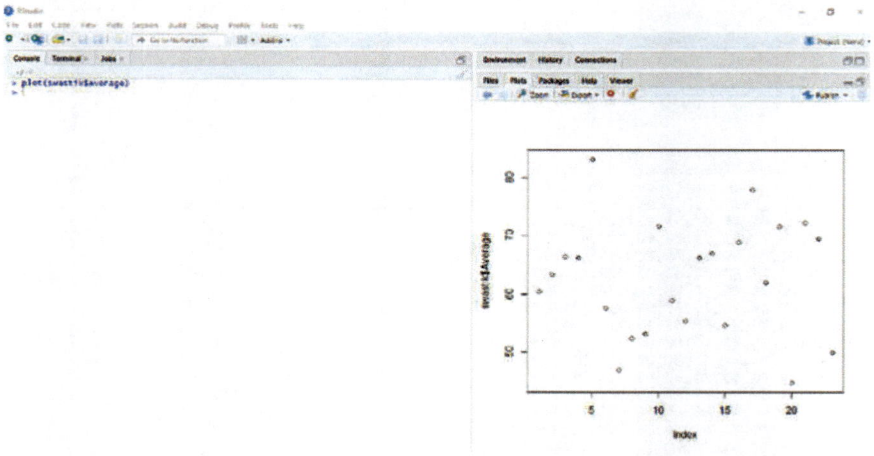

Fig. 4 Representation of graph for average

Problem 2: Collecting air quality data from Delhi save as aqi.csv. The problem is that how to read aqi.csv file, perform operation in R and how to represent in hist and box plot form [13] (Fig. 4).

On the contrary, in [5], hierarchical quadrature amplitude modulation (HQAM) is suggested for better protection of high-priority data during image transmission. Nevertheless, no channel model such as AWGN is considered in this work [14].

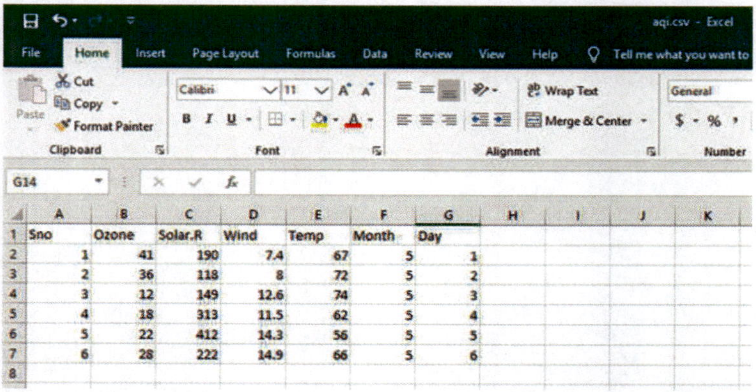

Sno	Ozone	Solar.R	Wind	Temp	Month	Day
1	41	190	7.4	67	5	1
2	36	118	8	72	5	2
3	12	149	12.6	74	5	3
4	18	313	11.5	62	5	4
5	22	412	14.3	56	5	5
6	28	222	14.9	66	5	6

Step 1 Now, show the result in histogram format.

$$> \text{hist}(\text{pqr\$wind})$$

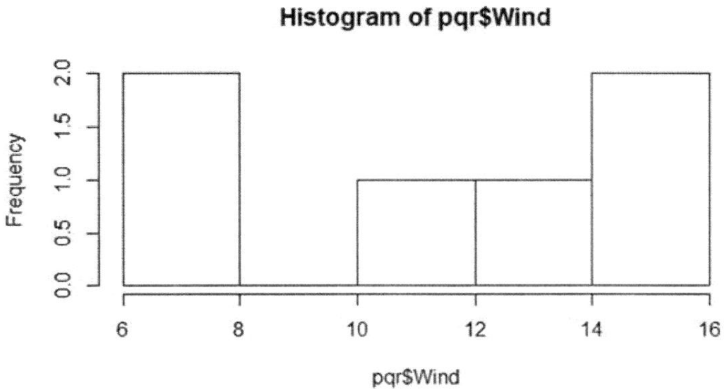

Step 2 Now, show the result in histogram format according to row and column.

$$> \text{par}(\text{mtow} = c(2, 2))$$
$$> \text{hist}\big(\text{pqr\$Temp, probality} = \text{TRUE}\big)$$
$$> \text{hist}\big(\text{pqr\$ozone, probality} = \text{TRUE}\big)$$

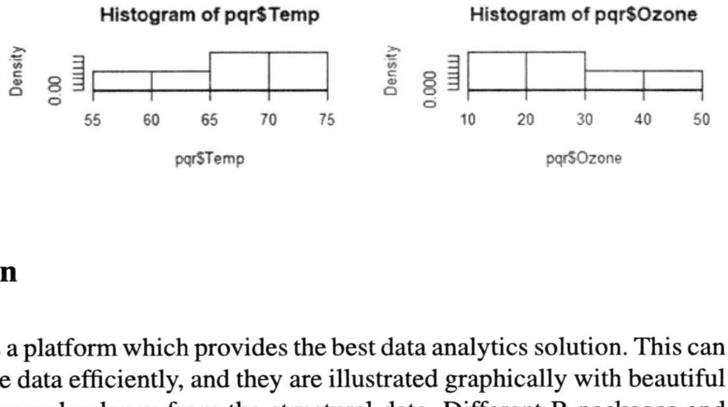

9 Conclusion

R with Hadoop is a platform which provides the best data analytics solution. This can process very large data efficiently, and they are illustrated graphically with beautiful visuals. It removes redundancy from the structural data. Different R packages and scripts are there and could be used in combination with Hadoop. R-Hadoop is the best approach of developing algorithm which is easily used in Hadoop. Linear or logistic regression can be shown in the form of graph and chart for better understanding of the task. Data analysis, manage data, store data all these tasks are easily performed by the Hadoop using R.

References

1. R. Buyya, Big data analytics = machinelearning cloud computing, in *Big Data* (Morgan Kaufmann Publisher, Massachusetts, USA, 2016), pp. 7–9
2. A.A. Chandio, N. Tziritas, C.-Z. Xu, Big-data processing techniques and their challenges in transport domain, in *ZTECommunications* (2015)
3. N. Kumar, V.R. Vadi, S.A. et al., Classifying time—bound hierarchical key agreement schemes, in 4th *Springer International Conference on Computer, Communication and Computational Sciences (IC4S 2019)*, Bangkok, Thailand, 11–12 October, 2019, Publication in Advances in Intelligent Systems and Computing (ISSN: 2194–5357).
4. T. Cheatham, A. Fahmy, D. Stefanescu, L. Valiant, Bulk synchronous parallel computing—a paradigm for transportable software, pp. 61–76. https://doi.org/10.1007/978-1-4615-4123-3_4.
5. Gartner, Gartner IT Glossary, Gartner Inc., S. Ghemawat, H. Gobioff, S.-T. Leung, The Google file system. ACM SIGOPS Oper. Syst. Rev. **37**(5), 29.https://doi.org/10.1145/1165389.945450
6. J. Ginsberg, M.H. Mohebbi, R.S. Patel, L. Brammer, M.S. Smolinski, L. Brilliant, Detecting influenza epidemics using search engine query data.Nature **457**(7232), 1012–4. https://doi.org/10.1038/nature07634
7. K. Hammond, A.S. Varde, *Cloud Based Predictive Analytics: Text Classification, Recommender Systems and Decision Support*, pp. 607–612. doi:https://doi.org/10.1109/icdmw.2013.95
8. S. Abidin, Enhancing security in WSN by artificial intelligence, in *Springer International Conference on Intelligent Data Communication and Internet of Things (ICICI—2018)*, Chapter 93 , Lecture Notes on Data Engineering and Communication Technologies Series ISSN : 2367-4512, ISSN: 2367-4520 (Electronic), ISBN: 978-3-030-03145-9, ISBN: 978-3-030-03146-6 (ebook), pp. 814–821, January 2019
9. I.A. Hashem, I. Yaqoob, N. BadrulAnuar, S. Mokhtar, A. Gani, S.U. Khan, The riseof -big data‖on cloud computing: review and open research issues. Inf. Syst. **47**, 98–115 (2015). https://doi.org/10.1016/j.is.2014.07.006
10. S. Harini, K. Jothika, K. Jayashree, A review of big data computing and cloud. Int. J. Pure Appl. Math. **118**(18), 1847–1855 (2018). ISSN: 1311-8080 (printed version); ISSN: 1314-3395
11. A. Gahlawat, Tata Consultancy Services Ltd., Big data analysis using R and Hadoop. IJCEM Int. J. Comput. Eng. Manage. **17**(5) (2014). ISSN (Online): 2230-7893
12. Bogdan OANCEA, University of Bucharest, Raluca Mariana The Bucharest University of Economic Studies, *Integrating R and Hadoop for Big DataAnalysis*
13. M. Wedel, R.T. Rust, T.S. Chung, Up close and personalized: a marketing view of recommendation systems. RecSys, 3–4 (2009)
14. S. Abidin, R. Kumar, V. Tiwari, A review report on cryptovirology and cryptography. Int. J. Sci. Eng. Res. **3**(11) (2012). ISSN: 2229-5518

Operating Systems Support and Network Optimization View of Internet of Things

Vikas Kumar Agarwal, Nitish Kumar, Rajesh Singh, Anuradha Pathak, and Bharat Bhushan

Abstract With the advent of Internet of things (IoT) in our daily life, the security issues and network optimization in such systems have gained significant attention both from industry and academia. IoTs connect various individual microcontroller-based embedded devices name "Things" that can receive and transmit information. These devices collect humungous amount of data thereby imposing huge overhead on the IoT network. Further, these devices are constrained in terms of resources, memory, and power. Hence, it is necessary to provide solutions to various IoT issues such as energy conservation, security, routing, heterogeneity, interoperability, and multicast. This brings forth the need for an adequate operating system (OS) suited for IoT applications. An appropriate OS makes the IoT devices flexible because of its kernel, networking, and real-time adaptability. This paper presents an extensive review of various OSs designed for IoT applications and compares these on the basis of their architecture, scheduling policy, memory management schemes, and the employed programming model. Finally, the paper outlines various issues and challenges associated with the IoT network optimization.

Keywords Internet of things · Operating system · Network optimization

---

V. K. Agarwal · N. Kumar · R. Singh · A. Pathak
Department of Computer Science and Engineering, RTC Institute of Technology, Ranchi, India
e-mail: vikashhagarwal@gmail.com

N. Kumar
e-mail: nitish1588@gmail.com

R. Singh
e-mail: rajesh_Singh@rtcit.ac.in

A. Pathak
e-mail: annu2712.pathak@gmail.com

B. Bhushan (✉)
Department of Computer Science and Engineering, Sharda University, Greater Noida, India
e-mail: bharat_bhushan1989@yahoo.com

J. Tavares et al. (eds.), *Emerging Technologies in Data Mining and Information Security*, Lecture Notes in Networks and Systems 164,
https://doi.org/10.1007/978-981-15-9774-9_77

1 Introduction

Recent advances in ubiquitous computing, Internet, and wireless communication led to the emergence of a new paradigm called Internet of things (IoT) dedicated to connect numerous physical devices named "Things" to the internet. These "Things" are connected to the Internet using varied technologies such as ZigBee (IEEE 802.15.4), Wi-Fi (IEEE 802.11), Bluetooth (IEEE 802.15.1), machine-to-machine technologies, and various cellular technologies [1–3]. In order to guarantee reliable communication, "Things" rely on various critical characteristics that encompasses efficient security services, protocols, architecture, and network optimization associated to discrete application types [4, 5]. Consequently, IoT enables smart functionality of the existing traditional devices by deriving technologies like embedded devices, artificial intelligence, pervasive computing, and different communication technologies. Since the IoT devices have energy constraints and limited memory, scalability and congestion in the network need to be accounted for efficient data delivery [6]. IoT applications include smoke detection, fire detection, healthcare monitoring, intelligent transportation system, and intrusion detection applications.

Veracity of the IoTs is ensured by enabling devices to communicate with the rest of the world via wireless technology. The fundamental role of the associated operating system (OS) is to provide management resources, opportune scheduling policies and hide the minor details of the "Things" [7]. OSs for IoT devices must achieve goals including interaction, security, and multitasking. Owing to the lack of interoperability among various IoT-specific solutions, there is need of a profound solution compatible for constrained IoT devices. OSs suited for IOT applications must support hardware agnostic operations, reliable communication, data storage, protocols support, real-time energy efficiency, end-device management, and small memory footprint [8]. OSs suited for IoT devices can be organized into various categories based on their architecture, scheduling policy, and programming model. As the devices are placed at remote locations, the reprogramming of these device is significant making the dynamic programming essential [9, 10].

Since IoT spans over various kinds of domains, the aforementioned factors challenge the efficient management of resources in IoT applications. Therefore, high priority must be assigned to optimize the network resource utilization and in turn enable efficient spectrum utilization. Hence, various OSs for IoT applications and efficient network optimization schemes for efficient data delivery in IoT systems. This paper addresses various IoT security constraints in terms of its OS, surveys various proposed OSs in terms of its applications and communication technologies, and outlines the issues and challenges related to efficient network optimization in IoT systems.

The remainder of the paper is organized as follows. Section 2 presents various IoT applications and the need for security therein. Section 3 outlines the architectural design requirement for IoT systems. Section 4 surveys various existing OSs for IoT applications and presents their comparison on the basis of their architecture memory management schemes and the employed scheduling mechanism. Section 5 outlines various issues and challenges associated with IoT network optimization followed by conclusion in Sect. 6.

2 IoT Application and Needs of Security

Security provisions are essential in an IoT network to guard all the confidential data and critical infrastructures. The variety of technologies and IoT devices create numerous vulnerabilities to cyber-physical security which are exploited by the attackers. Advanced cyber attacks, like daily attacks, have developed considerably which might lead to catastrophic consequences for the human ecosystem. Various types of IoT applications and the need for security therein are explored in the sub-sections below.

2.1 Smart Home

Smart homes are developing into very popular trend lately. Smart homes have an objective of integrating intelligence with daily used home gadgets or appliances (like decorative pieces, furniture's, locks of entrances, surveillance cameras setup and so on), and incorporating them with existing cyber-infrastructures. Adding intelligence to physical appliances/objects provides human life with a lot of advances that may include providing safety, techniques of utilizing natural resources, better way to ensure security and convenience to users. As an example, smart home may tune the temperature according to the heat and humidity outside and also according to the person living inside the house, turn the lights on and off when someone step on the entrance or leaves the arena saving the human effort, and inform emergency services whenever any emergency is identified. In smart homes, if the gadget is compromised its aftereffects might be harsh, for example, if the locks are manipulated by hackers, strangers may get in; manipulating a microwave might lead to fire in the building. Continuous surveillance and collection of data may rise privacy concerns and in the worst case, the data collected can be used against the owners [11].

2.2 Smart Grid

The main idea behind designing and implementing smart grid is to refine and upgrade reliability, optimize conventional grid systems, and to reduce the operational cost. With the integration of renewable and green resources (like solar power and wind power), the aim is to improve the management and reliability of conventional power grid in a more efficient way. Smart grids not only store the data related to energy usage but also keeps track of status of the whole system. This will also help to provide an outline for a scaled as well as fair pricing model which will consider most of the unbalanced energy consumption. Through making monitoring application for monitoring the status of smart grids, it also makes it easier to pin point the failures occurring in the grid system as fast as it occurs. This enables the design of a fault tolerant machinery for handling failures and miss happenings in a more efficient manner. A lot of methodologies and technologies which also include automated metering infrastructures (AMI) are being put forward to develop the network communication system of smart grid.

2.3 Smartly Connected Health

Smart health is one of the typical applications of IoT and was put forward to advance the functioning of systems in healthcare Arena and also to lower the treatment cost. Incorporating newly developed and advanced smart healthcare gadgets in ongoing health institutions, facilitate a proper and more convenient way for healthcare professionals to take care of patients. Ongoing researches are directed toward planting sensors into the patient's physic to facilitate a proper, detailed and continuous evaluation of their health conditions [12]. By doing a properly detailed analysis of the collected information from the gadgets, medical experts will also be able to record and identify behavioral changes of the specimens along with changing medication and level of disease during the course of treatment. With all these health systems, security and confidentiality are also major problem [6]. With these interconnected gadgets, gathering the information and checking the status of gadget are going to be very easy task but also full of risk as it eases the interruption and manipulation of the device.

3 Architectural Security Design for IoT

Novel security arrangements are required for achieving standard security in IoT. This can be achieved by designing proper, lightweight, and effective privacy algorithms capable of safeguarding physical assets. Various types of architectural security design for IoT are explored in sub-sections below.

3.1 End-to-End Security

End-to-end security of devices is an interesting and complex field where protocols like 6LoWPAN [13] and IPv6 [14] are developed only to backup End to End mechanism. A way to backup end-to-end security provision in IoT networks is by adding an extra security hardware (like physical Unclonable Functions 9PUF), that works on the concept of digital fingerprint and acts like a unique identification for the device [15]. They are mostly extended or updated versions of already deployed IP-based security arrangements. Even after lot of research attempts made to get a proper security mechanism, there still exist numerous open issues that needs to be considered. The light protocols that facilitate this mechanism of end-to-end security requires to be properly and precisely planned. Further, novel protocols are required for handling heterogeneity of IoT devices.

3.2 Edge Layer

Numerous end side terminals like RFID and smart bulbs tend not to acquire optimum facilities to backup end-to-end security. Rather than allowing these terminals to manage their security themselves, the security optimization work could be offloaded from less efficient thermal systems to highly strong systems. Authorization identification may be supervised and controlled by the edge terminals. Security methods that depend on edge layer are getting huge recognition these days. The ongoing researches attempts to find out how to prepare an efficient and secure edge. Various proposed methods rely on basic plan layout to guard edge devices, method of relating edge with end side terminals in a secure manner, and set up end side terminals to carry out complex collaborative security functions.

3.3 Distributed Security Model for IoT

Proper authentication is a way in which mutual trust can be developed between terminal devices and edge devices. Many of the current scalable verification protocols employ symmetric or public keys. However, the end side terminals do not have the required features or resources to cope up with these high computing operations. With good degree of faith in one other, cloud may give its sensitive information to edge devices and then they can further earn confidence of terminal devices by showing confirmable information from cloud. This plan may be applied in four steps. First of all, the edge devices ask for permissions to access a particular end term system from the cloud, by directing a request. Then, the cloud authenticates the edge terminal. Thirdly, edge terminal shows the information received through the cloud to the end side terminals for its authentication. If all these points are successfully adhered, the end side terminals start trusting the edge terminals.

4 Major Oss for IoT

In the recent past, huge range of IoT applications have been proposed that use varied OSs and software. OS architecture has enormous influence on kernel size. OSs in IoT can be classified into three types: *Monolithic architecture* combines OS applications and components and implements services separately resulting in underprivileged OS design. *Micro-kernel architecture* that provides mitigated kernel functionality and reduced kernel size. Virtual architecture that relies on virtual machines being exported to user programs. IoTs require an OS that possess small memory footprint, small kernel size and lends reliability and flexibility to the entire system. Some of the major proposed OSs used by IoT systems are explored in the subsections below.

4.1 Contiki

Contiki [16] is a reputed platform that was developed initially for WSNs but after some improvements, is also used as an IoT Platform. Contiki is licensed under Berkley software distribution (BSD) and possess numerous versions. Owing to its cleaver features, is has been declared as winner and is also the most referred OS for IoT. Contiki employs modular architecture that combines both event and thread model. The advantage of such modular architecture is that it provides less kernel functionalities resulting smaller kernel sizes. Kernel of Contiki comprise of lightweight event scheduler capable of dispatching events toward the running processes. Execution of process is initiated either by polling mechanism or event dispatched by the kernel. It also prevents the crashing of the entire system in case of failure of any single server. Contiki serves as an event-driven OS at the kernel level and supports multithreading. This architecture brings forth user-level server customization, ease of extension, and better reliability. The kernel of Contiki makes use of a shared single stack for all its executions.

Primarily being an event-driven OS, Contiki possesses hybrid programming model as it implements threads. It integrates pre-emptive multi-threading and supports context switching. Applications requiring multi-threading is linked with the library which is divided into two: Platform dependent (Implements pre-emption and stack switching) and platform independent (Event kernel Interface). Dynamic memory management is supported by Contiki which in turn supports dynamic program linking. It makes use of managed memory allocation that compacts the memory blocks in order to avoid issues of memory fragmentation. The library provided by Contiki includes powerful memory management functions for fixed lengths and blocks.

4.2 TinyOS

TinyOS [17] is an event-driven programmed model that supports monolithic architecture. TinyOs was first developed and conceived, at the California Berkeley University in the year 2000. TinyOS is an open-source OS that supports multiple applications because it is portable, flexible, and component-based. Earlier edition of TinyOS was implemented with FIFO as it was the only cooperative scheduler, while various scheduling algorithms were later added. Efficient memory management schemes are required for better performance. Owing to the limited space, TinyOS supports static memory allocation. Therefore, during the design time, resources are allocated. Dynamic memory management is not supported by the NesC language of TinyOS.

4.3 RIOT

The heterogenic IoT devices are resource constrained in terms of power consumption, memory, and other requirements needed for communication of real-time systems. RIOT [18] supports real-time scheduling and provides uniform programming platform that allows multi-threading aiming to be effective in terms of APIs, modulator, memory, and energy. RIOT uses a micro-kernel architecture that lends features such as synchronization, process communication, and scheduling. It ensures that the crashing of the file systems or failure of the device drivers do not cause complete harm to the system. This results in increased stability of the overall system. RIOT integrates a priority based, pre-emptive scheduler that aims to enhance the energy efficiency of the IoTs devices. These feature makes it highly reliable, developer friendly and robust against attacks or bugs.

4.4 Nano-RK

Carnegie Mellon University proposed a pre-emptive, reservation-based real-time operating systems (RTOS) known as Nano-RK [19] that supports multi-hop networking in WSNs. It employs real-time scheduling algorithms that use the assigned deadline priority. In addition, it also supports multitasking, priority-based scheduling, and time sensitive applications. It uses 18 KB of ROM and 2 KB of RAM that runs on micaZ platform and fire-fly sensor platform. Nano-RK introduces reservation techniques and implies static approach for efficient memory utilization. Applications specify resource requirements and timeliness. Nano-RK implements scheduling of tasks and definite privilege to system resources.

4.5 LiteOS

LiteOS is an open-source OS that implements both threads and events in its programming model making it suitable for wide range of IoT applications. It follows modular architecture that provides an effective method for minimizing learning complexities and enables user-friendly operations. LiteOS implements RR and priority-based scheduling for its scheduling purpose. LiteOS has a ready queue in which tasks are scheduled and executed based on their priority. Tasks must run to their completion in LiteOS. An interrupt will be generated which will go to its sleeping mode whenever a task requests for an unavailable resource. The task will resume itself once the resource is available. When the execution of a task is completed, it will come out of the kernel. The running system will go its sleeping mode, whenever there is no task. At the kernel level memory management is supported by the LiteOS. Kernels are compiled to the services which are separated from user application. The aim is to provide dynamic memory allocation to the users for which LiteOS has APIs that includes free and malloc for application manipulation [20].

4.6 MantisOS

MantisOS [21] implements thread as programming modelled system and has varied architectural layers. It has footprint which includes the scheduler, network stack, and the kernel of 500 bytes. Portability for varied platforms are supported by MantisOS and its portability could be examined on a personal computer. It attempts to provide support to resources with less consumption of intricate tasks as it is a type of multi-threading OS. It introduces the concept of atomic pre-emption by employing time sliced multi-threading of size 16 bytes and length of 10 ms by default. With this time-sliced multi-threading, buffer access could be avoided as the consumer can easily execute within the scheduled time period. Priority-based scheduling along with RR is implied which means execution of highest priority will be finished first and low priority tasks could starve. Middleware, virtual machine, OS, and algorithms must survive within a minimum amount of memory. Therefore, at run-time, dynamic memory allocation is a must for OS. However, memory is allocated statically in MantisOS.

4.7 RETOS

In 2007, an open-source OS named RETOS [22] was developed that implemented multi-threading as its programmed model and bears a modular architecture. Its network stack could either be dynamic or static. RETOS has been developed aiming to enhance resource utilization, vigorous activity, and reconfiguration. RETOS provides application code checking and dual-mode operation to ensure reliability and efficiency of the system. The dual model operation is achieved using stack switching. It uses user stack when the application is in the user area but switches to kernel stack upon receiving interrupts or system calls. The application code checking mechanism includes both static and dynamic checks. Static code checks are used during compile time to validate immediate addressing instructions. Dynamic checks are used to verify accurate usage of indirect addressing. RETOS demands stack reservation for every thread and it is a multi-threaded system where the relation of resources required is directly proportional to the thread reservation. Context saving, interrupt handler and system calls all require resources. Thus, for memory efficiency, single kernel stack method is adopted.

Table 1 compares the aforementioned types of operating systems in terms of their overview, architecture, memory management schemes, and the employed programming model.

5 Open Issues and Challenges for Network Optimization in IoT

Evolution of IoT has added new dimensions to various fields such as logistics, medical, smart city, home network, retail, and aeronautics. However, several issues and challenges related to smart and IoT networked objects are posed by this evolution. Such open issues and challenges are summarized in Table 2 and explored in the subsections below.

5.1 Network Routing

An effectual network depends on the network architecture and the associated topology. In order to transmit packets within mesh network topology, a competent routing mechanism must be used. Considering IEEE 802.15.4 technology, routing can take place either in link layer or in network layer inside a mesh network. For static routes, spanning tree construction is done for routing in link layer. Conversely, when the dynamic scenario of routing method comes into view, it poses a challenge. IPv6 is used in lossy and low power networks. It minimizes the overhead of sustaining the table at non-root nodes. However, overhead is caused because of slighter transmission maximum unit (TMU) when routing information is entered in the header of the packet.

Table 1 Comparison of various OSs

OSs	Overview	Architecture	Memory management	Programming model
Contiki [16]	Licensed under Berkley software distribution (BSD) and is the most referred OS for IoT	Employs modular architecture that combines both event and thread model	Dynamic memory management	Hybrid
TinyOS [17]	Open source OS that supports multiple applications because it is portable, flexible, and component-based	Event-driven programmed model that supports monolithic architecture	Supports static memory allocation	Event-based
RIOT [18]	Integrates a priority based, pre-emptive scheduler that aims to enhance the energy efficiency of the IoTs devices	Micro-kernel architecture that lends features such as synchronization, process communication, and scheduling	Supports real-time scheduling and allows multi-threading	Hybrid
Nano-RK [19]	Pre-emptive, reservation-based real-time operating systems (RTOS)	Supports priority-based scheduling and multi-hop networking that relies on pre-emption	Supports static memory allocation	Multi-threading
LiteOS [20]	Open-source OS that implements RR and priority-based scheduling for its scheduling purpose	Modular architecture	Dynamic memory allocation	Event driven
MantisOS [21]	Attempts to provide support to resources with less consumption of intricate tasks	Introduces the concept of atomic pre-emption by employing time-sliced multi-threading	Static memory allocation	Multi-threaded

(continued)

Table 1 (continued)

OSs	Overview	Architecture	Memory management	Programming model
RETOS [22]	Provides application code checking and dual-mode operation to ensure reliability and efficiency of the system	Implemented multi-threading as its programmed model and bears a modular architecture	Multi-threaded system	Event driven

Table 2 Summary of open issues and research challenges

Trends	Issues	Challenges
Network routing	• To offer effective routing • Selecting the most energy efficient algorithm	• Overhead caused due to TMU when routing information is entered into the packet header
Mobility	• To offer macro-mobility along with improved QoS in the network • Offering an augmented route	• High overhead rates of Mobile IP • To address the micro-mobility
Multicast	• In order to preserve energy, nodes tend to sleep thereby causing multicast packet loss • Recovering or tracing the overlooked packets at link layer	• Sending data packets at a consistent speed • Identify missing packets
Heterogeneity	• Energy conservation • Network's resource integration • Network's security	• Complex network utilization • Vulnerable to network failure • Limited flexibility
Interoperability	• Transmitting data to Internet from end devices • Creating interoperability between distinct kinds of devices	• Storage interoperability • Constrained network resources

5.2 *Mobility*

Network mobility refers to transition of mobile IP subnet from its attachment point at the IP backbone network. Common mobile structure contains multiple or singular mobile nodes along with mobile routers having defined topology. In a network, the fundamental necessity for mobility is handover which is defined as transition of point of attachment of mobile's node to the network. In network terminology, mobility is of two types: micro-mobility and macro-mobility. Micro-mobility is defined as the subscriber's movement inside any two points within a network whereas macro-mobility is defined as the movement between networks. In VoIP (Voice over IP), service degradation can occur because of recurrent handoff.

5.3 Multicast

IoT network employs multicast to request the concerned source for some resource or to notify their presence to other group nodes. The wireless protocol acknowledgment is mostly restricted for multicasting at the link layer. Hence, it is challenging for the sender to be able to identify packets that are missed out.

5.4 Heterogeneity

Heterogeneous network involves several assimilated networks wherein end user communicates with communication modalities. These modalities have distinct attributes like wireless, wired, and satellite modalities with distinct attributes and capacities. Whenever distinct devices are combined, heterogeneity becomes a serious concern. Various types of related issues and challenges are outlined in the subsections below.

5.5 Interoperability

Heterogeneity between communicating object stacks and protocols is referred to as interoperability. Distinct network technologies are used by distinct devices in applications of IoT. Network resource deployment and interoperability get impacted because of lesser standardization and documentation at the time of designing IoT devices.

6 Conclusion

Owing to its ubiquitous nature, IoT has been an emerging field that gained significant attraction from both industry and academia. IoT has been successfully implemented in huge range of applications related to smart city, smart home, intelligent transportation system, smart healthcare, and smart grid. The IoT devices are resource constrained and, therefore, an efficient and appropriate system is required for effectively managing these resources. Suitable solutions are needed to manage the system, provide secure communication, and prevent device failure. This paper presents various IoT security constraints and extensively reviews various OSs in terms of its applications and communication technologies. The study mainly focused on open source, real-time OSs involving latest networking technologies and compared these on the basis of their architecture, scheduling mechanism, and the employed programming model. RIOT and Contiki are the most widely accepted OSs for IoT applications as they meet majority of the application requirements. OSs for IoTs must employ 6LoWPAN protocol for communication and provide real-time functionalities. Finally, the paper outlines the issues and challenges related to efficient network optimization in an IoT system. The outcome of this paper explores various OSs suited for IoT applications and highlights the need for network optimization in IoT.

References

1. A. Nikoukar, S. Raza, A. Poole, M. Gunes, B. Dezfouli, Low-power wireless for the internet of things: standards and applications. IEEE Access **6**, 67893–67926 (2018). https://doi.org/10.1109/access.2018.2879189
2. E. Sisinni, A. Saifullah, S. Han, U. Jennehag, M. Gidlund, Industrial internet of things: challenges, opportunities, and directions. IEEE Trans. Industr. Inf. **14**(11), 4724–4734 (2018). https://doi.org/10.1109/tii.2018.2852491
3. P. Sinha, A.K. Rai, B. Bhushan, Information security threats and attacks with conceivable counteraction. in *2019 2nd International Conference on Intelligent Computing, Instrumentation and Control Technologies (ICICICT)* (2019). https://doi.org/10.1109/icicict46008.2019.8993384
4. H. Tran-Dang, D. Kim, An information framework for internet of things services in physical internet. IEEE Access **6**, 43967–43977 (2018). https://doi.org/10.1109/access.2018.2864310
5. A. Al-Fuqaha, M. Guizani, M. Mohammadi, M. Aledhari, M. Ayyash, Internet of things: a survey on enabling technologies, protocols, and applications. IEEE Commun. Surv. Tutorials **17**(4), 2347–2376 (2015). https://doi.org/10.1109/comst.2015.2444095
6. A.K. Goel, A. Rose, J. Gaur, B. Bhushan, Attacks, countermeasures and security paradigms in IoT, in *2019 2nd International Conference on Intelligent Computing, Instrumentation and Control Technologies (ICICICT)* (2019). https://doi.org/10.1109/icicict46008.2019.8993338
7. M.O. Farooq, T. Kunz, Operating systems for wireless sensor networks: a survey. Sensors **11**(6), 5900–5930 (2011). https://doi.org/10.3390/s110605900
8. Q. Jing, A.V. Vasilakos, J. Wan, J. Lu, D. Qiu, Security of the internet of things: perspectives and challenges. Wireless Netw. **20**(8), 2481–2501 (2014). https://doi.org/10.1007/s11276-014-0761-7
9. J. Fu, Y. Liu, H. Chao, B.K. Bhargava, Z. Zhang, Secure data storage and searching for industrial IoT by integrating fog computing and cloud computing. IEEE Trans. Ind. Inf. **14**(10), 4519–4528 (2018). https://doi.org/10.1109/tii.2018.2793350
10. R. Oliver, C. Science, U. Kingdom, Reprogramming embedded systems at run-time. 2–4 (2014)
11. A. Arora, A. Kaur, B. Bhushan, H. Saini, Security concerns and future trends of internet of things, in *2019 2nd International Conference on Intelligent Computing, Instrumentation and Control Technologies (ICICICT)* (2019). https://doi.org/10.1109/icicict46008.2019.8993222
12. M. Hartman, Tools for the vision impaired [opinion]. IEEE Technol. Soc. Mag. **34**(2), 16–17 (2015). https://doi.org/10.1109/mts.2015.2425812
13. Z. Shelby, C. Bormann, 6LoWPAN: the wireless embedded internet, vol. 43 (John Wiley & Sons, 2011)
14. C. Huitema, *IPv6: the new internet protocol* (Prentice Hall PTR Upper Saddle River, NJ, USA, 1996)
15. C. Herder, M. Yu, F. Koushanfar, S. Devadas, Physical unclonable functions and applications: a tutorial. Proc. IEEE **102**(8), 1126–1141 (2014). https://doi.org/10.1109/jproc.2014.2320516
16. A. Dunkels, Contiki—a lightweight and flexible operating system for tiny networked sensors
17. N. Tsiftes, J. Eriksson, A. Dunkels, Low-power wireless IPv6 routing with ContikiRPL, in *Proceedings of the 9th ACM/IEEE International Conference on Information Processing in*

Sensor Networks, IPSN '10 (Stockholm, Sweden, 12–16 April 2010, ACM Press: New York, 2010), p. 406

18. E. Baccelli, O. Hahm, M. Gunes, M. Wahlisch, T. Schmidt, RIOT OS: towards an OS for the internet of things, in *2013 IEEE Conference on Computer Communications Workshops (INFOCOM WKSHPS)* (2013). https://doi.org/10.1109/infcomw.2013.6970748
19. A. Eswaran, A. Rowe, R. Rajkumar, Nano-RK: an energy-aware resource-centric RTOS for sensor networks, in *26th IEEE International Real-Time Systems Symposium (RTSS'05)* (n.d.). https://doi.org/10.1109/rtss.2005.30
20. J. Polastre, R. Szewczyk, D. Culler, Telos: enabling ultra-low power wireless research, in *IPSN 2005. Fourth International Symposium on Information Processing in Sensor Networks, 2005* (n.d.). https://doi.org/10.1109/ipsn.2005.1440950
21. S. Bhatti, J. Carlson, H. Dai, J. Deng, J. Rose, A. Sheth, R. Han, MANTIS OS: an embedded multithreaded operating system for wireless micro sensor platforms. Mob. Netw. Appl. **10**(4), 563–579 (2005). https://doi.org/10.1007/s11036-005-1567-8
22. H. Cha, S. Choi, I. Jung, H. Kim, H. Shin, J. Yoo, C. Yoon, RETOS: resilient, expandable, and threaded operating system for wireless sensor networks, in *Proceedings of the 6th International Conference on Information Processing in Sensor Networks—IPSN '07* (2007). https://doi.org/10.1145/1236360.1236381

Information and Data Security Model: Background, Risks, and Challenges

Bhuvan Taneja and Bharat Bhushan

Abstract Information security and data security are the processes that are mainly based on preventing information from unauthorized access or in other words to secure information and data corruption throughout its lifecycle be it the information that is stored as physical format such as files which are subject to thefts or vandalism, etc. Since the very beginning of communication, everyone understood it was important to have a mechanism to preserve the private information. In the nineteenth century, more complex systems were made to let the government manage the information effectively. During the First World War, multi-tier systems were introduced to communicate info from one place to another. This study shows the details of different models of information and data security and also the management of risk and different challenges is possessed by security and different threats and also the importance of security.

Keywords Information security · Data security · Software life cycle · Security approach

1 Introduction

Organizations wherever in the world are heavily making a move toward the information technology (IT) sector to assure their critical assets [1]. Julius Caesar is believed to invent Caesar cipher which was to prevent his messages from being read by anyone else. In the nineteenth century, more complex systems were made to let the government manage the information effectively. During the First World War, multi-tier systems were introduced to communicate info from one place to another [2]. Whenever a company needs to defend a brand or their capital or the information

B. Taneja (✉)
Department of Computer Science and Engineering, HMRITM, New Delhi, India
e-mail: bhuvantnj7@gmail.com

B. Bhushan
Department of Computer Science and Engineering, Sharda University, Greater Noida, India
e-mail: bharat_bhushan1989@yahoo.com

J. Tavares et al. (eds.), *Emerging Technologies in Data Mining and Information Security*, Lecture Notes in Networks and Systems 164,
https://doi.org/10.1007/978-981-15-9774-9_78

provided by the consumer or even controls for their infrastructure, they turn toward the security being provided. The Internet is a group of many networks that can be easily accessed by any user around the globe in one way or the other. Thus, people and companies may reach a point on the Web without even caring for national or international boundaries, day or night, having said that with the ease and undemanding access to information comes the possibility that the precious information can be lost, stolen, changed, or misused [3, 4]. Many attacks may be launched at different layers. Active attacks are defined when the attacker stops the connection and then changes the information, whereas passive attack is defined as the one where the attacker intercepts the information with the sole purpose of reading and examining the information but not changing it [5]. If the information is stored electronically and is accessible on computers having networks, it is more dangerous to be stolen rather than the one written on paper and put in a cabinet of files [6]. Technical and security companies are depended too much on data protection standards internationally mainly ITIL, British Standard, and other international standards which are themselves. Test of these organizations for a longer period and their answers have been taken with an economic victory. Some of the most prominent attacks are attacks by dropping attacks, attacks by changing of message fields, attacks using impersonation, wormhole attacks, and attacks using fabrication [7, 8].

Therefore, it was mandatory to discover a method that addresses the flaws and loopholes and hence decrement the cost of the same. Data violation results in legal action and fines which are way too huge and may also destroy an organization's reputation. The main aim of protecting data from different threats is much more necessary today than it has been or ever will be [9].

In this paper, the main emphasis is on the data and information security. The remainder of the paper is organized as follows. Section 2 presents the background of security. Section 3 shows the different types of security models and gives a brief overview of threats and their countermeasures. Section 4 discusses the different risks involved, threats to security, and the importance of security, Sect. 5 discusses various challenges of security, and Sect. 6 concludes the paper by giving a summary.

2 Background

The importance of security is to protect the hardware from the threats that are present outside. Information security began just after the first mainframes were made [10]. It was first developed and used during Second World War where coincidentally the first modern computer was also created. Face recognition, different keys, and symbols were made for assigned personnel who were controlling the locations where the confidential information was being kept [11]. On the contrary, the security in the early days was basic and is made of only a simple document scheme where there were no jobs at that time because the main threat that time was physical theft of apparatus and other products. During this time of wars and all, there was no Internet or anything else to be worried about so security was solely focused on measures that were physically based which used to prevent access to people who did not have enough knowledge of the same topic and also on how to deal with the computer.

In the 1960s, the corporates began to be very protective about their computers, and also in the 1960s there was no network to panic about so security was basically concentrated on and around physical measures and also preventing access to folks who know how to work on a computer. Just to do this, passwords and multiple layers of security were added. During this time, the (ARPA) started inspecting the practical network communication which was made just to support the military's need to share information with those who need it [12]. In the next decade or so, ARPANET grew a lot in stature but it had a lot of loopholes which could be misused a lot. In around 1973, a scientist advised that there were many problems with ARPANET in terms of its security. Single user did not have enough controls to secure data against unauthorized users. There were no safety measures for the dial-up link. User identification and access were not existent. Phone numbers were distributed and were put up on the restrooms and different phone booths which made it easy for the hackers to use ARPANET. At that time all the focus was on the researching team who developed a system security called Multiplexed Information and Computing Service (MULTICS). Before 1970, after the structuring of the MULTICS project, many key figures had also developed a new operating system called UNIX. UNIX system did not plan passwords and security with MULTICS has had [13].

From 1960 to 1980, the world advanced to such an extent that everything became more connected, and also the viruses were advanced and the security systems could not keep up with the different innovative hacking techniques. During this time, the major point came when the network started to expand in many colleges, military, and govt. departments and also the ARPANET network was called the Internet and available to the public as the World Wide Web [14].

3 Security Model

3.1 Information Security Model

An information security model is the term used to explain the methods and different technologies made to preserve data. Its main focus is on collecting, storing, creating which are discussed below. It shows the relationship between the different processes as shown in Fig. 1 [15].

3.1.1 The Collection Process

This activity includes the different processes to collect and examine the different type of information that connects to the goals and finishes with the report. It helps in finding out data which are related to the structure of the corporation management,

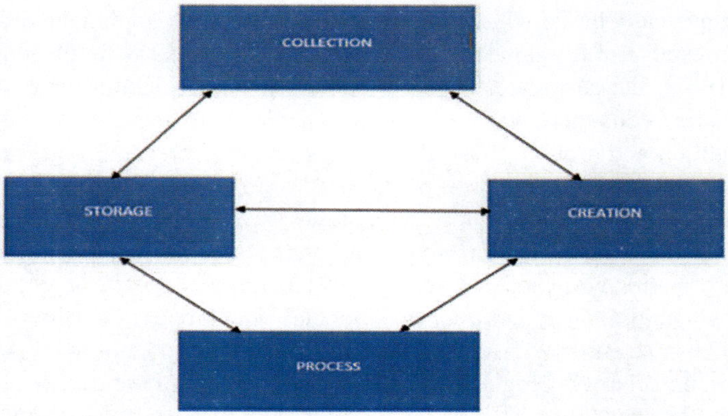

Fig. 1 Information security model

also understands the different threats, vulnerabilities, outcomes, and these things can only be done through building a synergy and a teamwork.

3.1.2 The Creation Process

This type of process includes different types of subprocess that generate a security scheme and rules that to be implemented for security management which ends with a policy statement. In general, the goals are specified in levels agreements and these agreements are seen as plans for different units.

3.1.3 Process

This activity has three subprocesses implementation: the process to put an agreed agreement in action; evaluation: this process is defined as the process to see and control the result of action; maintenance: to reconstruct the plan of operation.

3.2 Data Security

The data security model is depicted in Fig. 2 and various terms are defined below [16].

The key features of this type of security are: confidentiality, availability, and integrity also called the CIA triad. It shows the companies the way on how to keep data which is sensitive safe from the outside world and not let the data break out of the company. Threats and countermeasures of CIA triad are explained in Table 1.

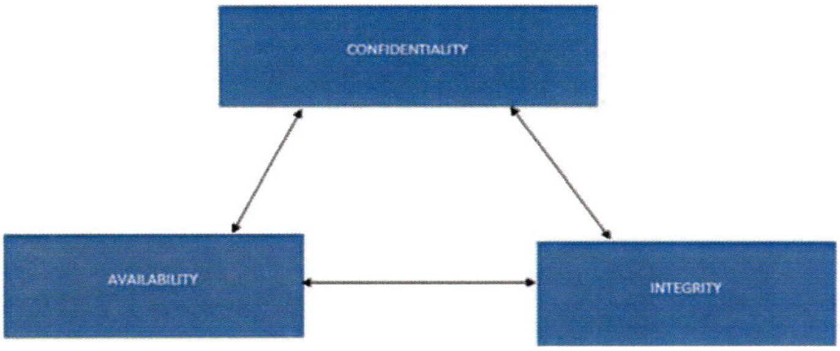

Fig. 2 Data security model

Table 1 Threats and countermeasures of CIA

	Threat sources	Countermeasures
Confidentiality	• Keeping an eye fixed on the network • Shoulder aquatics	• Encrypting data, the moment they are stored • By using the process of network padding • Using strict access control
Availability	• Device failure • DoS attacks • Non-human events	• Having proper backups to overcome system failure • IDS to follow the network traffic
Integrity	• Viruses • Logic bombs • Backdoors	• Hashing • Access control

3.2.1 Confidentiality

It is the property where the information is not available or open to different individuals or processes. It ensures that facts, figures, or particulars will solely be accessed by permitted members of the organization. The main way to avoid the confidentiality problem is by using encryption techniques to protect the data in such a way that even if the attacker gets access to the data, the attacker will not be able to decode it. Encryption standards are the Advanced Encryption Standard (AES) and Data Encryption Standard (DES). One more way to conserve the data is by using a virtual personal network tunnel.

3.2.2 Availability

It makes certain that the information or statistics available is accessible to please the wants of the organization. To make sure of the availability of data, the network management should maintain proper hardware and continuous upgrades and have a proper plan. Many attacks such as denial of service (DoS) or distributed denial of service (DDoS) can hinder a network and the resources may get unavailable as the network will be exhausted. This attack may prove costly to the corporate as many corporates are dependent on the network for business tools. Therefore, proper policies should be made to prevent these types of attacks.

3.2.3 Integrity

It guarantees that the information is in such an arrangement that it is true, correct, and also in its original purpose. The person who receives such information must only have the data the sender wanted him to have. The data can only be edited by the person who has the access by authorized people. Integrity is applied using different mechanisms such as data encryption and the other one being hashing. One of the most important things in integrity is that the loss of data might even occur due to non-human events such as server crash or another one being the electromagnetic pulse so it is advised to have a backup and enough systems to ensure integrity.

3.2.4 Non-Repudiation

In terms of information security, non-repudiation applies to having a formal contract, verbal agreement, or data transfer. The main objective of this method is to make sure that every person in a corporate bounded to a contract, or the parties that are required for communication or transfer of particular data will not be able to deny the authenticity of the signatures on their contracts or to prove whether they were the creator of the message or data required for transferring.

3.2.5 Authenctication

In security, authentication is defined as the process of confirming the identity of a folk or a particular device. One of the most basic instances is when a user enters a username and password on a site. While the above combination is a very ordinary way to check for authentication but many other types of authentication do exist such as passcode for unlocking a phone or a password to login to the computer.

4 Risk, Threats, and Importance of Security

This section discusses the various risks and threats that exist around security particularly those stored in systems and also talks about how important security is for us.

4.1 Risk Management

Management of risk is that the procedure of recognizing all the categories of threats and vulnerabilities affecting all the assets that were being employed by a company in achieving the goals set up by several organizations and conjointly to search out the countermeasures. Risk examination and investigation ways have their restrictions as a result of once incident happens, their extraordinariness and character bring about to unpredictable threats. There is loads of chance that something poor can happen that may cause loads of damage to a product which may cause a huge loss for the organization. A threat is some things that have the utmost potential to cause damage to be it imitation or act of nature. A risk analysis is taken out by a gaggle of employees who have data associated with business and different security areas [17].

In larger terms, the method of risk management process consists of firstly, identifying completely different assets and finding their values that embody everybody like folks, buildings, etc. Secondly, to conduct a threat judgment which incorporates war acts, accidents, etc. The only way to stop such things from happening is only by implementing strong policies and put them in proper use. Next is to assess different kinds of policies and standards that ease the danger management method. The last method is by recognizing different controls and then choosing and execute applicable controls. To search out a pleasant response associate, we want to contemplate the price of production and conjointly the worth of the plus [18].

Ex: Security for Libraries
As of today, libraries are only dependent and overly reliant on computers. Many libraries irrespective of the sizes have altogether dropped the use of cards for the high work saver machine called computers. Previously also some libraries used to have computers but were only accessible through the workstations but things are different now as most of the libraries now are accessible through the Web. The online searching of books and other databases that are created to store things like how many books are on the shelf, how many students have been registered in the database, how many books have been issued, who all has not returned the books to the library, etc. Such things should be only restricted to the teachers of the library and not for the end user.

An effective security in libraries should have: Teachers assigned for security tasks. On point training for all those involved in the security procedures and issues. Some policies should be made to make clear on how to deal with the security thefts. Security plans should be made based on the library is set up and proper measures to protect data integrity. Access levels should be made depending upon the faculty being in charge. These points are for any type of library that exists around the globe be it public library where anyone and everyone can sit and read or academic libraries which are situated in schools and colleges for students or corporate libraries for people working in different companies to complete a pending work [19].

4.2 Security Threats

New fears regarding security are appearing with every passing day from malware methods that can be installed on someone's computer, laptop, mobile phone, tabs, etc., without him or her even realizing it. It is one of those aims that trick the other worker to give his or her sensitive information to different types of viruses and also into some kind of theft attempts. Many-a-times the information which is leaked through the different kinds of threat assaults in the corporates are very tough to handle only because the protection methods that were installed in the respective machine are not correct or just not deemed good enough. The tough thing about managing the data properly is that the worker or the employee is not deemed qualified enough to manage information or data security. Occasionally the corporates do not hire workers or employees based on the qualification they have because the companies that hire workers for this particular job does not match with the skill set the interviewer to possess about the topic. So, it makes tougher for the workers to keep the companies' data safe and hence it makes attacks easy and steal the info if the workers do not have the right skill to conserve the sensitive information from getting leaked. The most difficult thing arises when in a corporate there is a miscommunication on vital topics which can also lead to cyberattacks. When the lack of understanding increases in terms of how to keep the information safe is when the hacker's attacks and information are at the biggest risk of being stolen [20].

4.3 Importance of Security

Information and data security are the practice of securing information while still giving access to all those who need it. Three ethics set a good foundation for a solid security system. In today's world where the Internet is reachable to every part of the globe, the protection of information is very mandatory and can be compared to how one protects his or her property. The main objective of providing security is to save the information from getting destroyed or unauthorized access. The reason we want our data to be safe and secure is that we do not want our valuables to be in somebody's else hands that we do not even know or trust. There are a lot of things we do not want outsiders to see be it our medical or financial records or the world seeing our emails or social media messages, our Internet passwords, our cards number, and even our banking information.

The security measures are very key in the place we live as it secures the information, allows the corporate's function, and also allows the shield operation of the implemented software on the companies' system. The main importance of security is: Having the assurance of his or her company's security, the need for workers with high skill set, allowance of funds for security keeping in mind the budget of the corporate without compromising the security of the firm, and disturbance in day-to-day life [21].

5 Challenges of Security

There could exist a variety of challenges within the endlessly dynamic surroundings that makes it onerous to utterly conserve our resources [22]. These embody the combining of organization and personal life, not the consistent implementation of policies, fragile links within the supply chain, possibility of insider threat, etc. [23].

5.1 *Combination of Organization and Private Life*

Net services while not having to pay further have created employees take its edges by victimization it for their use instead of doing it for the organization. For example, the employees could use the e-mail provided by the organization for a few personal or personal communication and lots of these employees should even be victimized by the mobile phones assigned by the organization that they use for their stuff. Several employees do not even have a desktop at the house and should use the portable computer issued by the organization for victimization system software like paying tax and on the opposite hand some employees bring their portable computer to figure which implies that the employees used the business assets that were being protected for his or her functions.

5.2 *Not Consistent Implementation of Policies*

Several organizations have not enforced their policies antecedently or have not done it in an exceedingly consistent means that conjointly depends on the position of the employee. The origin of such issues is once a security purpose tries to require down the felon, several corporations have underestimated the importance of implementing such policies.

5.3 *Fragile Link Within the Offer Chain*

Each trade or a corporation round the globe contains a service or a product that has its offer chain which can be a matter of concern on how the information is protected. In an exceeding production cycle of a product, the fragile data, either directly associated with the merchandise, the method and ways used or the organization themselves are spilled with completely different parties that conjointly include the vendor that at once takes the data out of the organization's hands and is subject to security attacks or cyberattacks.

5.4 Possibility of Insider Threat

The malicious inside leaks have clean intentions at the back of them, and are rarer, however, unintended threats to information might be an everyday occurrence, mainly because of the increasing variety of man or woman identities, or personalized devices being introduced to the workplace. These gadgets, which make use of Internet of things (IoT), are linked with unclear safety protocols in place. This then leads to records being either captured on such gadgets or leaked out to different events that will not be a part of the company.

6 Conclusion and Future Work

As everyone knows that today everyone is very much dependent on Internet, information, and even its security. Security as a term has become a very key figure in today's world. At this point, when the world is advancing at its peak several organizations have implemented and brought out a lot of new theories and software on how to protect data from different types of threats but not all are easy to implement. One of the major requirements an organization makes these software needs to be aware of some facts is that it should be easy to manage, also not tough to implement, and its cost should also be less. Therefore, the summary of my paper can be done in two major steps: Firstly, collecting the required data through analyzing the already existing system's situation to find out three things (the structure of management, the processing scenario in the organization, and the threats vulnerabilities and probabilities of occurrences). Secondly, creating the security plan which includes (policy statement the determination of responsibility, IT statement required, and the operational level agreement which considered the security policy).

References

1. B. Hauer, Data and Information leakage prevention within the scope of in- formation security. IEEE Access **3**, 2554–2565 (2015). https://doi.org/10.1109/access.2015.2506185
2. N. Smith, A history of first world war technology in 11 objects. Eng. Technol. **13**(11), 34–37 (2018). https://doi.org/10.1049/et.2018.1102
3. A. Arora, A. Kaur, B. Bhushan, H. Saini, Security concerns and future trends of internet of things, in *2019 2nd International Conference on Intelligent Computing, Instrumentation and Control Technologies (ICICICT)*. https://doi.org/10.1109/icicict46008.2019.8993222
4. F.A. El-Moussa, N. Linge, M. Hope, Active router approach to defeating denial-of-service attacks in networks. IET Commun. **1**(1),55 (2007). https://doi.org/10.1049/iet-com:20050441
5. D. Bradbury, Fighting ID theft. Eng. Technol. **10**(12), 60–63 (2015). https://doi.org/10.1049/et.2016.1207
6. S.R. Chhetri, S. Faezi, M.A.A. Faruque, Information leakage-aware computer-aided cyber-physical manufacturing. IEEE Trans. Inf. Forensics Secur. **13**(9), 2333–2344 (2018). https://doi.org/10.1109/tifs.2018.2818659

7. S.L.L. Correa, M.P. Mexas, Implementation of the service center accord-ing to best practices recommended by ITIL (the case of HU-UFJF). IEEE Latin Am. Trans. **16**(6), 1809–1816 (2018). https://doi.org/10.1109/tla.2018.8444403
8. S. Ji, T. Chen, S. Zhong, Wormhole attack detection algorithms in wireless network coding systems. IEEE Trans. Mob. Comput. **14**(3), 660–674 (2015). https://doi.org/10.1109/tmc.2014.2324572
9. I. Natgunanathan, A. Mehmood, Y. Xiang, G. Beliakov, J. Year-wood, Protection of privacy in biometric data. IEEE Access **4**, 880–892 (2016). https://doi.org/10.1109/access.2016.2535120
10. T.P. Layton, *Information Security*. ISO/IEC17799 ISBN: 9780849370878 (2006)
11. M.A. Abbas, H. Song, J.-P. Hong, Secure wireless communications in broadcast channels with confidential messages. IEEE Access **7**, 170525–170533 (2019). https://doi.org/10.1109/access.2019.2955603
12. L. Lopriore, Password capabilities revisited. Comput. J. **58**(4), 782–791 (2013). https://doi.org/10.1093/comjnl/bxt131
13. A. John, *"Spyware and Adware" the University of Calgary* (Springer Science + Business Media, USA, 2011). ISBN: 978-0-387-77740-5
14. B. Bhushan, G. Sahoo, Recent advances in attacks, technical challenges, vulnerabilities and their countermeasures in wireless sensor networks. Wirel. Pers. Commun. **98**(2), 2037–2077 (2017). https://doi.org/10.1007/s11277-017-4962-0
15. M.P.D. Silva, R.M. De Barros, Maturity model of information security for software developers. IEEE Latin Am. Trans. **15**(10), 1994–1999 (2017).https://doi.org/10.1109/tla.2017.8071246
16. G. Kumar, R. Saha, M.K. Rai, T.-H. Kim, Multi-dimensional security provision for secure communication in vehicular ad hoc networks using hierarchical structure and end-to-end authentication. IEEE Access **6**, 46558–46567 (2018). https://doi.org/10.1109/access.2018.2866759
17. P. Sinha, A.K. Rai, B. Bhushan, Information security threats and attacks with conceivable counteraction, in *2019 2nd International Conference on Intelligent Computing, Instrumentation and Control Technologies (ICICICT)*, 2019. https://doi.org/10.1109/icicict46008.2019.8993384
18. T. Varshney, N. Sharma, I. Kaushik, B. Bhushan, Architectural model of security threats and their countermeasures in IoT, in *2019 International Conference on Computing, Communication, and Intelligent Systems (ICCCIS)*, 2019. https://doi.org/10.1109/icccis48478.2019.8974544
19. M. Evans, Y. He, C. Luo, I. Yevseyeva, H. Janicke, E. Zamani, L.A. Maglaras, Real-time information security incident management: a case study using the IS-CHEC technique. IEEE Access **7**, 142147–142175 (2019). https://doi.org/10.1109/access.2019.2944615
20. H. Asgari, S. Haines, O. Rysavy, Identification of threats and security risk assessments for recursive internet architecture. IEEE Syst. J. **12**(3), 2437–2448 (2018). https://doi.org/10.1109/jsyst.2017.2765178
21. J. Knight, The importance of security cases: proof is good, but not enough. IEEE Secur. Priv. **13**(4), 73–75 (2015). https://doi.org/10.1109/msp.2015.68
22. X. Jing, Z. Yan, W. Pedrycz, Security data collection and da-ta analytics in the internet: a survey. IEEE Commun. Surv. Tutor. **21**(1), 586–618 (2019). https://doi.org/10.1109/comst.2018.2863942
23. A. Gauci, S. Michelin, M. Salles, Addressing the challenge of cyber security maintenance through patch management. CIRED Open Access Proc. J. **2017**(1), 2599–2601 (2017). https://doi.org/10.1049/oap-cired.2017.0252

Design and Implementation of Cryptographic Technique Employing Intricate Table and Rubik Rotation Algorithm

Rajdeep Chowdhury, Soma Sharma, Md Adil Alam, and Niharika

Abstract Malicious intrusions, executing hackers, and snowballing upshot of artificial intelligence have made network security as sturdy as aged boots. The indispensability of formulating cryptographic techniques would dent the safekeeping matters quite briskly. The paper demonstrates a cryptographic archetypal in aid of Rubik's cube scrambling and fragmentary ciphering the text, with the assistance of intricate table prior to scrambling, for concealing the treasured message, amended by means of brisk tackling with safekeeping matters and to have a bang at impending encounters. Elucidating explicitly, employment of a Rubik's cube of order 3×3 and an intricate table for ciphering the text, dependent on the proposed algorithm through scrambling procedure has been adhered. Escorted by the usage of keys through the ciphering process for endowing with innumerable stratum of safekeeping would inspire apiece end user to preserve their facts in innocuous repository. With enactment and implementation of the proposed cryptographic technique, dark Web could be sheltered with numerous strata of safekeeping, which in go would furnish the pro-stratum safekeeping to apiece individual and consequently benefiting the society in entirety. The fineness of the formulated ploy is split in the paper amid diverse segments. The afresh cryptographic modus operandi scheming is conversant in the attempt and credible in its incorporation.

R. Chowdhury (✉)
Academician and Author, Chinsurah, India
e-mail: dujon18@yahoo.co.in

S. Sharma · M. A. Alam · Niharika
JIS College of Engineering, Kalyani, India
e-mail: soma.sv99@gmail.com

M. A. Alam
e-mail: alam.adil55@gmail.com

Niharika
e-mail: niharikasingh12128@gmail.com

J. Tavares et al. (eds.), *Emerging Technologies in Data Mining and Information Security*,
Lecture Notes in Networks and Systems 164,
https://doi.org/10.1007/978-981-15-9774-9_79

Keywords Cryptography · Intricate table · Rubik rotation algorithm · Rubik · Scrambling · Crossover · Mutation · Embedding

1 Introduction

Contemporary setting of the Internet world is immensely engaged with the usage of deep Web as well as dark Web. Usual user remarkably employs surface Web and occasionally deep Web for individual concern, though globally unearthing digitization in apiece nook and cranny, yet contingency of deprivation has exaggerated [1].

To have a snap at these contingencies of data outbreak or manipulation by the unwanted individual, there is a necessity of safekeeping by means of cryptographic technique which would effortlessly confront all the challenges and preserve the data warehouse in innocuous custody [2–5]. Cryptography is a modus operandi of shielding information through employment of codes so that only those for whom the information is envisioned could read and process the same [2–10].

The cryptographic algorithm employed in the paper in aid with Rubik's cube scrambling has intellectualized the novelty of text ciphering and ensure uninterrupted traversing to the receiver's end. In the devised paper, a notion of ciphering the text through scrambling of Rubik's cube in a desired numeral of moves has been proposed [7, 11, 12].

Users have to enter the message desired to be encrypted and then further procedures would occur accordingly. In the primary intuition, plaintext entered by the user is converted into intricate text and then the same intricate text is ensued for embedding. Embedding is a method of placing the intricate text in the Rubik's cube according to the numbers scribbled on the subcubes, commencing from 1st position to 54th position. Post embedding, scrambling commences and the same takes place in desired numeral of moves, preferred by the user to obtain the ciphertext.

Long way off, the receiver would decipher the encrypted message through the inverse function of the keys. During ciphering procedure, keys are arbitrarily engendered as per the number of moves selected and hence these keys make the algorithm sturdy enough to endure the arduousness of the cryptographic algorithm.

2 Proposed Work

Flowchart of the comprehensive work entitling the procedure has been depicted underneath for effortlessness understanding (Fig. 1).

Devised intricate table to obtain the modified ASCII value for usage in the complete proposed work and implementation procedure have been portrayed in Table 1.

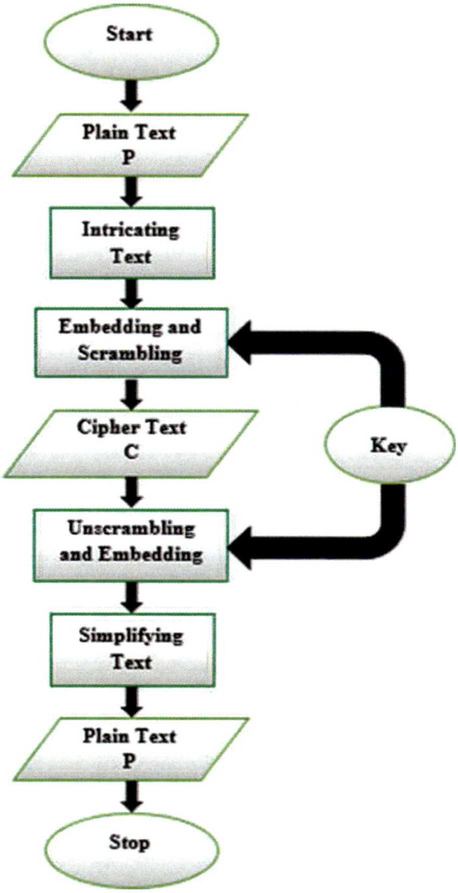

Fig. 1 Flowchart of the proposed work

2.1 Formulated Algorithm

Formulated algorithm could be effortlessly understood with the four expounded methodologies mentioned beneath, namely;

2.1.1. Intricating plaintext
2.1.2. Embedding plaintext
2.1.3 Ciphering modus operandi
2.1.4 Deciphering modus operandi

2.1.1 Intricating Plaintext

Plaintext engaged from the user is initially combined with the proposed algorithm.

Table 1 Intricate table for modified ASCII

Plain text	ASCII value	Position (4, 3, 2, 1) modified ASCII value				Plain text	ASCII value	Position (4, 3, 2, 1) modified ASCII value			
		4	3	2	1			4	3	2	1
(SP)	32	40	36	34	33	P	80	88	84	82	81
!	33	41	37	35	32	Q	81	89	85	83	80
“	34	42	38	32	35	R	82	90	86	80	83
#	35	43	39	33	34	S	83	91	87	81	82
$	36	44	32	38	37	T	84	92	80	86	85
%	37	45	33	39	36	U	85	93	81	87	84
&	38	46	34	36	39	V	86	94	82	84	87
‘	39	47	35	37	38	W	87	95	83	85	86
(	40	32	44	42	41	X	88	80	92	90	89
)	41	33	45	43	40	Y	89	81	93	91	88
*	42	34	46	40	43	Z	90	82	94	88	91
+	43	35	47	41	42	[	91	83	95	89	90
,	44	36	40	46	45	\	92	84	88	94	93
-	45	37	41	47	44	]	93	85	89	95	92
.	46	38	42	44	47	^	94	86	90	92	95
/	47	39	43	45	46	_	95	87	91	93	94
0	48	56	52	50	49	.	96	104	100	98	97
1	49	57	53	51	48	A	97	105	101	99	96
2	50	58	54	48	51	B	98	106	102	96	99
3	51	59	55	49	50	C	99	107	103	97	98
4	52	60	48	54	53	d	100	108	96	102	101
5	53	61	49	55	52	e	101	109	97	103	100
6	54	62	50	52	55	F	102	110	98	100	103
7	55	63	51	53	54	g	103	111	99	101	102
8	56	48	60	58	57	h	104	96	108	106	105
9	57	49	61	59	56	I	105	97	109	107	104
:	58	50	62	56	59	J	106	98	110	104	107
;	59	51	63	57	58	k	107	99	111	105	106
<	60	52	56	62	61	L	108	100	104	110	109
=	61	53	57	63	60	m	109	101	105	111	108
>	62	54	58	60	63	n	110	102	106	108	111
?	63	55	59	61	62	o	111	103	107	109	110
@	64	72	68	66	65	p	112	120	116	114	113
A	65	73	69	67	64	q	113	121	117	115	112

(continued)

Table 1 (continued)

Plain text	ASCII value	Position (4, 3, 2, 1) modified ASCII value				Plain text	ASCII value	Position (4, 3, 2, 1) modified ASCII value			
		4	3	2	1			4	3	2	1
B	66	74	70	64	67	r	114	122	118	112	115
C	67	75	71	65	66	s	115	123	119	113	114
D	68	76	64	70	69	t	116	124	112	118	117
E	69	77	65	71	68	u	117	125	113	119	116
F	70	78	66	68	71	v	118	126	114	116	119
G	71	79	67	69	70	w	119	128	115	117	118
H	72	64	76	74	73	x	120	112	124	122	121
I	73	65	77	75	72	y	121	113	125	123	120
J	74	66	78	72	75	z	122	114	126	120	123
K	75	67	79	73	74	{	123	115	128	121	122
L	76	68	72	78	75	|	124	116	120	126	125
M	77	69	73	79	76	}	125	117	121	128	124
N	78	70	74	76	79	~	126	118	122	124	128
O	79	71	75	77	78						

Step 1 Enter the plaintext.
Step 2 Retrieve its ASCII value as per the text input.
Step 3 Convert its ASCII value into binary and do crossover according to the proposed position.
Step 4 After obtaining the binary value post crossover operation, its ASCII value is calculated and the modified ASCII value is considered for ensuing procedure.
Step 5 Modified ASCII value endows with the intricate text for embedding.

2.1.2 Embedding Plaintext

Post obtaining the intricate text, the texts are cited to the subcubes of the Rubik's cube numbered from 01 to 54, right from 1st to 54th position.

Rubik's cube engaged here for embedding and scrambling purpose is of order 3 × 3. Mathematically, Rubik's cube is a permutation group, having six identical faces and apiece face having its own color. Apiece color is recurrent precisely nine times. Hence, the cube has an ordered list which has 54 elements.

Note: Once all the 54 subcubes are being embedded with primary 54 characters, then additional characters from plaintext would enter for embedding in the second round of scrambling in the equivalent Rubik's cube, with identical moves as employed during commencement (Fig. 2).

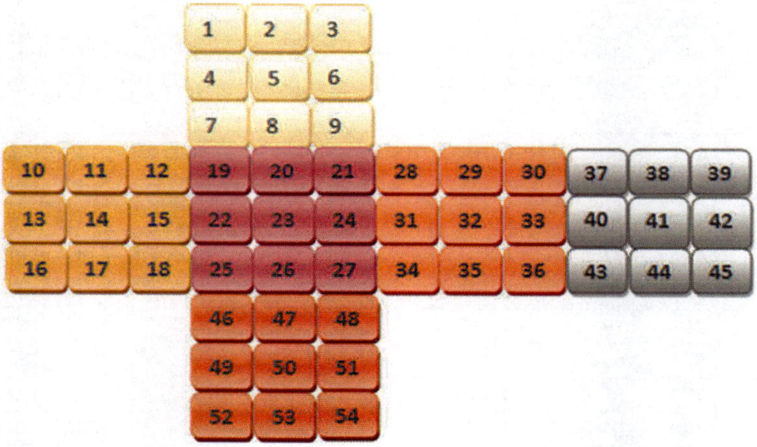

Fig. 2 Numbering of subcube faces in Rubik's cube

2.1.3 Ciphering Modus Operandi

Scrambling the faces according to the proposed algorithm would originate once embedding is accomplished. Prior to scrambling, a number of moves have to be entered, which would be utilized as key and which are randomly engendered. Randomly engendered keys would scramble the faces in a sequential order of moves and furnish with a sturdier ciphertext. Stratum of steps followed to obtain the ciphertext has boosted the safekeeping issues and furnished all sectors of the society an innocuous mode of transferring the data or employing the same.

2.1.4 Deciphering Modus Operandi

Deciphering entails inverse function of keys to unscramble the faces and backtracking all the methodologies employed during ciphering and embedding procedure to retrieve the plaintext.

2.2 Implementation

Implementation scheme could be simply perceived under the embellishment with an illustration of the four methodologies discussed in the proposed work.

2.2.1 Intricating Plaintext

Modified ASCII value = ASCII value after crossover
Crossover position = Sequential altering the position from 4th to 1st bit, that is 2^3rd to 2^0th position in the plaintext positioned in Rubik's subcubes, right from 1 to 54.

Table 1 could be demonstrated by means of the crafted illustration beneath:

Step 1 Let the character to be intricate be **'s'** from the plaintext.
Step 2 Corresponding ASCII value is 115.
Step 3 Binary of 115 is 01110011.

- *Crossover procedure*
- Position of plaintext **'s'** in the Rubik's cube is 3.
- In the laid position, crossover entails to be performed at 2nd bit as per the proposed 4th to 1st bit position laid in the sequence.
- Binary after crossover at 2nd position is 01110001.

Step 4 ASCII value is 113.
Step 5 As per the proposed modified ASCII value, intricate text for embedding would be **'q' (ASCII Value = 113)** (Table 2).

In alike means, all the text would be converted and the resultant of plaintext **'Vision@20'** would be '***^mqhgjB38.***'

Note: Post crossover, obtaining modified ASCII value of any character as 127 would lead to addition of +1 to that and ASCII value of 128 would be considered for further process and reverse of obtaining the equivalent character during deciphering would follow back-tracking procedure with position number.

Table 2 Plaintext to intricate text

Plaintext	ASCII value	Crossover	Modified ASCII value	Intricate text
V	86	4	94	^
i	105	3	109	m
s	115	2	113	q
i	105	1	104	h
o	111	4	103	g
n	110	3	106	j
@	64	2	66	B
2	50	1	51	3
0	48	4	56	8

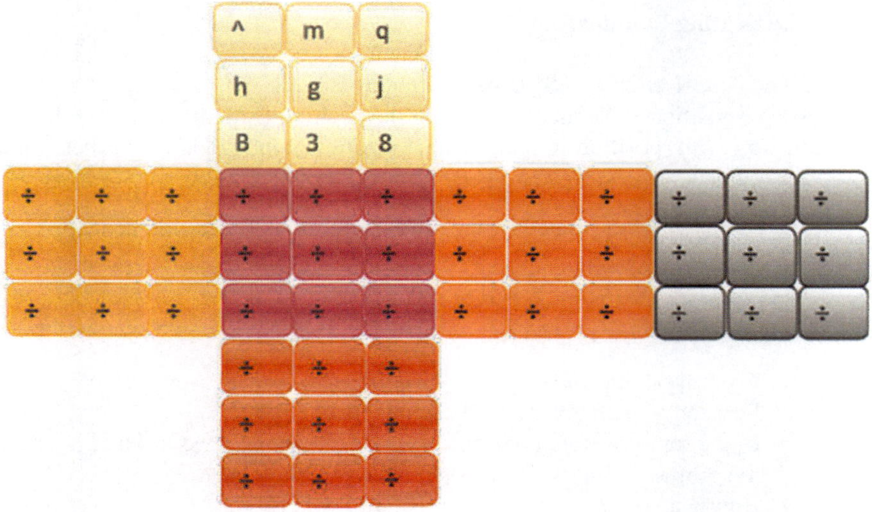

Fig. 3 Embedding process of intricate text

2.2.2 Embedding Plaintext

After obtaining the intricate text, post following the algorithm, modified text would be cited to the faces of subcubes chronologically, right from 1st to 54th position in the Rubik's cube for scrambling.

Here, the text to be cited on the faces of the subcubes are '***^mqhgjB38.***'

Figure 3 exhibits the embedding process as per the user input.

Note 1: Post embedding, if any block of Rubik's cube faces could not be embedded due to lack of plaintext, special character of ASCII value 246 would be cited in the blank spaces. (ASCII value of 246 is $\div$).

Note 2: If there are more characters than 54, then second round of embedding with equivalent numeral of moves would be performed, as entered during the commencement with identical cube of order 3×3.

2.2.3 Ciphering Modus Operandi

Ciphering procedure of the plaintext could be effortlessly understood in aid with the flow diagram portrayed in Fig. 4. Plaintext **'Vision@20'** has been employed to obtain the ciphertext. Ciphertext could be unalike as per the numeral of moves employed as keys, as the moves comprise the algorithm for scrambling and is randomly engendered, when the user enters the number of moves.

Exemplification of ciphering the plaintext **'Vision@20'** has been essentially performed and could be depicted through the screenshot in Fig. 5

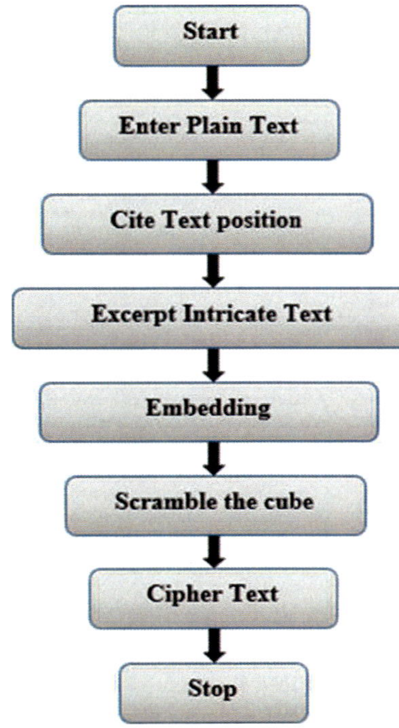

Fig. 4 Ciphering modus operandi portrayal

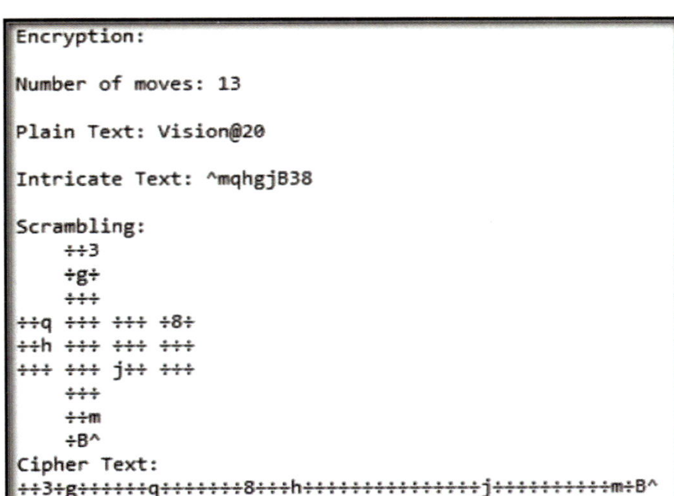

Fig. 5 Encryption process

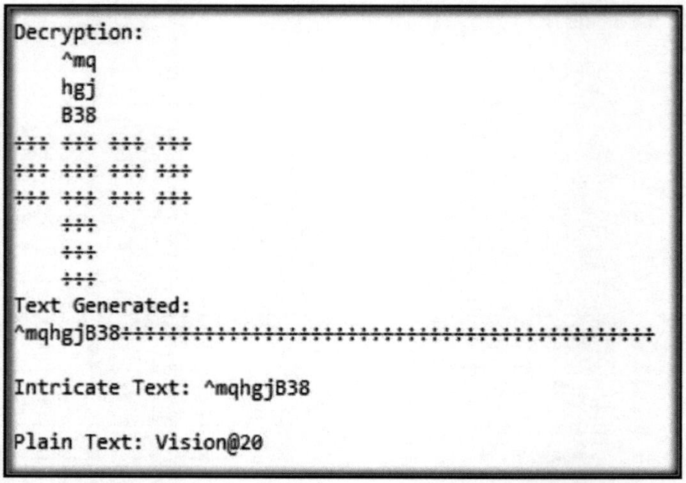

Fig. 6 Decryption process

2.2.4 Deciphering Modus Operandi

Unscrambling the faces employing inverse function of the keys and back-tracking the remaining process would effortlessly decipher the text to retrieve the plaintext.

Exemplification of deciphering the ciphertext has been essentially performed and could be depicted through the screenshot in Fig. 6.

3 Conclusion

The proposed cryptographic technique ensures proficient employment in diversified fields and is perceptively designated to dent the effects of uncertainty and ensure the user to preserve their data in safe hands. Employment of intricate table and scrambling of Rubik's cube for concealing the valuable text has made data warehouse impregnable and has furnished the cryptographic world a novel fangled modus operandi in ciphering and deciphering. Brisk accomplishment of acquiring keys while commuting with the proposed algorithm has proliferated the stratum of safekeeping, which in turn bring forth needle-point safekeeping stratum to apiece individual, as well as the society in entirety. Augmenting the safekeeping stratum and implementation of this impassable algorithm would enthrall the employment of the proposed cryptographic technique.

References

1. A. Kahate, Cryptography and network security, 4th edn. (Tata McGraw–Hill Publishing Company Limited, New Delhi, 2008). ISBN (10)–0-07-064823-9, ISBN (13)–978-007-06-4823-4
2. R. Chowdhury, O. Roy, S. Datta, S. Dasgupta, *Virtual Data Warehouse Model Employing Crypto–Math Modus Operandi and Intelligent Sensor Algorithm for Cosseted Transference and Output Augmentation. Knowledge Computing and Its Applications* (Springer, Singapore, 2018). ISBN (O)–978-981-10-6680-1, ISBN (P)–978-981-10-6679-5, pp. 111–129
3. R. Chowdhury, J. Khan, S. Sengupta, S. Dassgupta, S. Datta, M. De, Dynamic referencing data warehouse employing proposed time integrated seed based element engendered cryptographic algorithm, in *Proceedings of ACM Symposium on Women in Research 2016 (ACM–WIR 2016), ACM–International Conference Proceedings Series (ICPS)*, ACM's Women in Computing, Sri Aurobindo Institute of Technology, Indore, Paper ID–WIR–48, Published and Archived in ACM Digital Library. ISBN-978-1-4503-4278-0, pp. 142–149 (2016)
4. R. Chowdhury, S. Datta, S. Dasgupta, M. De, Implementation of central dogma based cryptographic algorithm in data warehouse for performance enhancement. Int. J. Adv. Comput. Sci. Appl. **6**(11), 29–34 (2015). ISSN (O)–2156 5570, ISSN (P)–2158 107X
5. R. Chowdhury, K. Dey, S. Datta, S. Shaw, Design and implementation of proposed drawer model based data warehouse architecture incorporating DNA translation cryptographic algorithm for security enhancement, in *Proceedings of International Conference on Contemporary Computing and Informatics*, IC3I 2014, Mysore, Published and Archived in IEEE Digital Xplore, ISBN–978-1-4799-6629-5, pp. 55–60 (2014)
6. R. Chowdhury, P. Chatterjee, P. Mitra, O. Roy, Design and implementation of security mechanism for data warehouse performance enhancement using two tier user authentication techniques. Int. J. Innov. Res. Sci. Eng. Technol. **3**(6), 165–172 (2014). ISSN (O)–2319 8753, ISSN (P)–2347 6710
7. R. Chowdhury, R. Bose, N. Sengupta, M. De, Logarithmic formula generated seed based cryptographic technique using proposed alphanumeric number system and Rubik rotation algorithm, in *Proceedings of 2012 International Conference on Communications, Devices and Intelligent Systems*, CODIS,Kolkata, Published and Archived in IEEE Digital Xplore. ISBN–978-1-4673-4700-6, pp. 564–567 (2012)
8. R. Chowdhury, A. Saha, A. Dutta, Logarithmic function based cryptosystem [LFC]. Int. J. Comput. Inf. Syst. **2**(4), 70–76 (2011). ISSN–2229 5208
9. R. Chowdhury, A. Saha, P. Biswas, A. Dutta, Matrix and mutation based cryptosystem [MMC]. Int. J. Comput. Sci. Netw. Secur. **11**(3), 7–14 (2011). ISSN–1738 7906
10. R. Chowdhury, S. Ghosh, Study of cryptology based on proposed concept of cyclic cryptography using cyclograph. Res. J. Eng. Technol. **2**(1), 17–20. ISSN (O)–2321 581X, ISSN (P)–0976 2973
11. E.R. Korf, Finding optimal solutions to Rubik's cube using pattern databases, in *Proceedings of American Association for Artificial Intelligence (AAAI–97)*, pp. 700–705 (1997)
12. J. Ewing, C. Kosniowski, *Puzzle it Out: Cubes, Groups and Puzzles* (Cambridge University Press, United Kingdom, 1982). ISBN (10)–0-52-128924-6, ISBN (13)–978-052-12-8924-5, p. 4

Impact of Routing Protocols and Artificial Intelligence Techniques on Ad hoc Networks in Critical Scenario

Meena Rao

Abstract Ad hoc networks play a crucial role in setting up of critical infrastructure in disaster or emergency situation. In today's world, where dealing with COVID 19 pandemic has become need of the hour, availability of critical data and transmitting information between various nodes can be lifesaving. The need for setting up an ad hoc network in a remote area or a hotspot point would help in sending and receiving information with least human intervention. To have a proper ad hoc network setup that relays critical data or information, it is necessary that the system provides good quality of service (QoS) parameters such as least error, good packet delivery ratio, and minimum delay. QoS parameters depend on routing protocols which transmit information between various nodes. Ad hoc networks mainly employ reactive routing protocols that provide routes between nodes as and when required. Applying artificial intelligence techniques with various routing protocols enhances monitoring and recognition of devices leading to better QoS parameters. This paper focusses on the study and application of various routing protocols in an ad hoc network setup in a critical area. Also, the advantage of the artificial intelligence technique on routing protocols is discussed.

Keywords Adhoc networks · Quality of service · Artificial intelligence

1 Introduction

Ad hoc networks are systems which can be set up anywhere, anytime, and with least infrastructure backup [1]. These networks are generally wireless and subsequently have nodes which can be temporary and can operate without much technological setup [2, 3]. The nodes or units are grouped to form a network which is dynamic.

M. Rao (✉)
Department of Electronics and Communication Engineering, Maharaja Surajmal Institute of Technology, C-4 Janakpuri, New Delhi 110058, India
e-mail: meenarao@msit.in

J. Tavares et al. (eds.), *Emerging Technologies in Data Mining and Information Security*, Lecture Notes in Networks and Systems 164,
https://doi.org/10.1007/978-981-15-9774-9_80

The nodes are dynamic in nature and also function as routers to decide how to receive and send packets. Setting up of the wireless ad hoc networks does not take much time. Moreover, the cost of setting up the system is generally quite reasonable. Further, the system does not have any requirement of any pre-existing infrastructure. All these factors have made the ad hoc networks suitable for being setup in disaster areas, emergency situations, and military operations. An emerging aspect where setting up of ad hoc networks for information transfer and data gathering can be looked into is in areas dealing with COVID 19. The transfer of information between various nodes or users, gathering of information from a particular node, and relaying it to central unit as well as sending the data from a centralized unit to a particular node is always very critical in emergency situation or relief operations.

The importance of having a robust ad hoc network is all the more important in areas that are affected by COVID 19 or for that matter any area facing a critical or an emergency situation. In hotspots or other areas, where there is a high requirement of information transfer as well as information gathering, setting up of wireless ad hoc networks can be thought of. Setting up of an efficient wireless ad hoc network in such an area would lead to less requirement of human intervention. Moreover, the network can be set up quite easily even in remote areas. Also, since the system does not involve any setup cost, it can be discontinued after its use over a certain period of time.

The setting up of an efficient wireless ad hoc network is useful only if it satisfies certain quality of service (QoS) parameters. If the system is being set up to gather and broadcast critical information, then reliability and efficiency of the system becomes very crucial. Hence, quality of service requirements play a very important role in achieving a high performance setup.

QoS parameters are directly dependent on the routing protocols that are employed by the system. In this paper, various ad hoc routing protocols, their performance, and effect or benefit of employing artificial intelligence with routing protocols will be discussed.

In Sect. 2, routing protocols are discussed, and Sect 3 discusses artificial intelligence techniques with routing protocols. Finally paper is concluded in Sect. 4.

2 Routing Protocols

Ad hoc networks communicate information packets between various nodes using various routing protocols. Routing protocols find out the routes between various nodes. The routes for information transfer can be directly between sender and receiver. Also, the routes can be between intermediate nodes as and when necessary. Most commonly used routing protocols that have been suggested in literature can be categorized as proactive and reactive routing protocols.

2.1 *Proactive or Table-Driven Routing Protocols*

Proactive routing protocols or table-driven routing protocols find and maintain links or routes with all the participant nodes. Maintaining route information to all the nodes is expensive and consumes a lot of bandwidth. Hence, proactive routing protocols are not found to be very suitable for ad hoc networks. Also, the routing information is maintained in the form of table which requires a lot of overheads. Moreover, there may be some nodes which are mobile and due to their changing location it is not very feasible to maintain a routing table. Since the routes are dynamically changing, not updating the routing tables may also cause loss of packets. This is the reason proactive routing protocols are not very popular in ad hoc network setups. When setting up ad hoc networks in critical areas wherever bandwidth and other resources are quite limited, employing proactive routing protocols is further ruled out.

2.2 *Reactive or On-Demand Routing Protocols*

Reactive routing protocols are suited to an on-demand system like ad hoc routing protocols. Here, the routes are found as and when necessary [4]. Since the route information is not maintained in the form of any tables, the overhead is reduced to a considerable extent. The technique is suitable for systems like ad hoc networks where bandwidth, infrastructure, etc., are limited. There are various ad hoc routing protocols that have been proposed in literature till now. Prominent among them are discussed below.

2.2.1 Dynamic Source Routing Protocol

The earliest reactive routing protocol is dynamic source routing (DSR). Here, the sender determines the complete route from source to destination. The packet also includes the route to the destination. Here, the intermediate nodes only forward the packets based on the predetermined route. Intermediate nodes do not make any routing decisions.

The advantage that DSR protocol offers is that a route is dynamically discovered by the source or host node when some information is to be sent. This arrangement is quite suitable for a bandwidth and resource constrained setup as routes are set up only when some information packets are to be transferred. Moreover, any dynamic change in the node positions can also be easily catered to as routes are discovered dynamically and on requirement basis.

Even though there are many advantages as compared to traditional routing protocols, still some problems persist. DSR protocol does not come up with a proper backup routing which should work in case of main route failure. This problem becomes all the more serious when the system has been setup in an emergency situation and transferring data is critical. The ad hoc networks if setup in a COVID 19 hotspot should have proper routing protocols to support any route failure condition.

2.2.2 Ad Hoc On-Demand Distance Vector Routing Protocol

Ad hoc on-demand distance vector (AODV) routing protocol also discovers routes as and when necessary. In AODV protocol also, a routing table is maintained and whenever any data packet has to be transmitted it is first looked up in the routing table. When the routes are not available in the routing table, route request (RREQ) packets are broadcasted by the source node. If routes are not available in the routing table, then a new route is searched for using route broadcasts. AODV also resulted in lot of routing overhead. Moreover, AODV protocol did not provide any solution to the problem of route failure. Hence, in any critical solution, route failure in AODV protocol would result in loss of packets. If applicable to any emergency area like a COVID 19 hotspot any loss of data or more delay in information transfer is not desirable.

2.2.3 Ad Hoc On-Demand Distance Vector Backup Routing (AODVBR) Protocol

AODVBR protocol works mostly on similar principles of AODV protocol. However, AODVBR protocol creates a mesh-like structure to provide an alternate route. During the route reply phase, the neighboring nodes hear the route reply packets (RREP). In this way, alternate path information is available. However, AODVBR protocol does not suggest a solution when both the primary and secondary route failures occur.

2.2.4 Ad Hoc On-Demand Distance Vector nth Routing Protocol (AODV nthBR) Protocol

AODV nthBR protocol proposes a solution of solving the problem of packet loss due to route failure by providing multiple backup routes [5]. Here, the nodes that are part of routing are selected based on their energy efficiency. On the failure of any one node, an alternate energy efficient node is selected. This technique helps in establishing an ad hoc network with least possibility of packet drop and better performance parameters compared to the above-mentioned reactive routing protocols.

AODV nthBR protocol provides a stable system as failure of the main as well as the next routes do not lead to route failure. Here when the primary route fails, the next nearest energy efficient node is selected for transferring data. In case the next node also fails, another node is searched based on the criteria of energy efficiency. Since route failure chances are less, QoS parameters automatically improve. This kind of a system is very relevant in a critical setup. In a critical setup, the nodes can

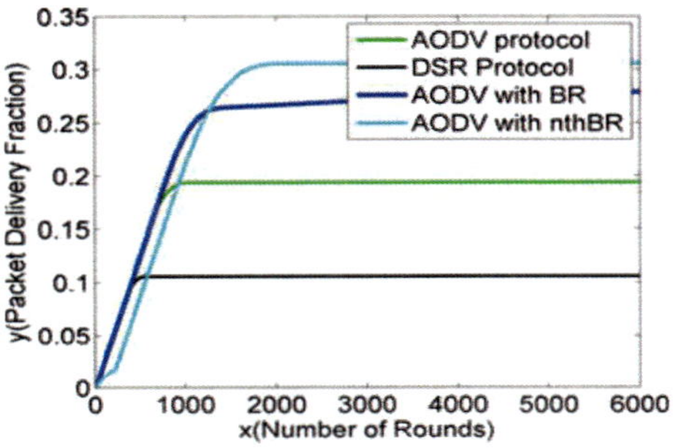

Fig. 1 Comparison of packet delivery fraction [6]

be patients or people who are being monitored and whose data has to be transferred to a base station which can be a hospital. So, in emergency situation, transferring this data on time with low packet drop rate and good QoS parameters is very important.

2.3 Comparison of Results Based on Quality of Service Parameters

A comparison of the results obtained after simulation of AODV, DSR, AODV, and AODV nthBR protocol is given below [6]. The results are simulated using MATLAB considering 60 nodes in the network and nodes moving at a speed of 20 m/s.

2.3.1 Packet Delivery Fraction

In Fig. 1, the packet delivery fraction is found to be maximum when AODV nthBR protocol is used for routing. This is because the packet loss is minimum as multiple backup routes are available on failure of the selected route. Also, the energy-efficient nodes are selected resulting in faster delivery of information and low loss. In critical or emergency situations where collecting and receiving information is very important, AODV nthBR protocol can be implemented.

2.3.2 End-to-End Delay

From Fig. 2, it can be seen that end-to-end delay is least when AODV nthBR protocol is used followed by AODV protocol, since emergency or critical areas demand least delay for information transfer. So, if an ad hoc network is set up in an area where information from COVID 19 patients has to be collected or maybe important information has to be sent from a centralized location to certain nodes in a COVID 19 hotspot area, then AODV nthBR protocol can be used as a routing protocol in the adhoc network.

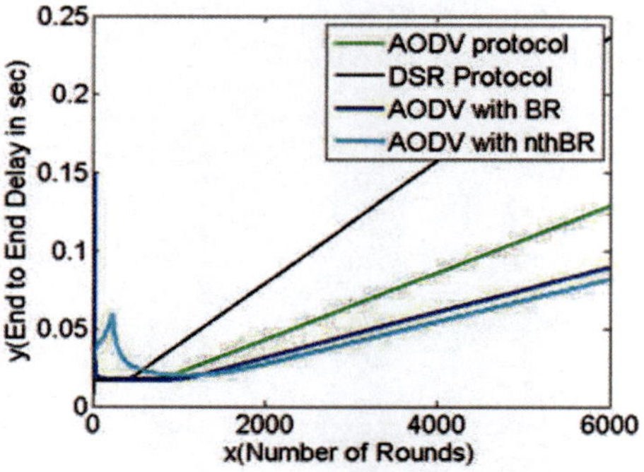

Fig. 2 End-to-end delay [6]

3 Artificial Intelligence Techniques with Routing Protocol

In this section, various artificial intelligence techniques that have been successfully implemented with various routing protocols will be discussed [7, 8]. Barbancho et al. have suggested that using artificial intelligence techniques result in better monitoring of nodes and observation of the nodes from the data.

For applying artificial intelligence techniques, directed diffusion and energy-aware routing protocols have been chosen by them. In directed diffusion technique, any information that is required is broadcasted to all the nodes from the base station. All the nodes that receive the request prepare a gradient toward the particular node from where it has received that request. This process is repeated until a gradient is set up from the source node to the broadcasting node. So, basically directed diffusion technique continues doing its task until the targeted node is reached and the targeted node is directly able to communicate with the base node. This approach works very well when one-to-one communication is required between a particular targeted node and centralized base station. If applied to an ad hoc network in a COVID 19 cluster, it can be used to gather data directly from the targeted node or person and information can be sent to the centralized base station. Also, any urgent or necessary information can be sent to the targeted node directly as well.

Another approach to apply artificial intelligence technique is using energy-aware routing. Here, many optimal routes are selected for information transfer. The selection of routes is based on the energy consumption of all the nodes. For information procurement and data transfer, different paths can be used at different instances. This way energy consumption of all the nodes is reduced. Depletion of energy is less at every node using this technique. So, there is a higher chance of sustainability of ad hoc network. Higher sustainability is a desirable factor especially in relief operations and monitoring of data in emergency or COVID 19 hotspot areas where resources are limited. Also, if these locations are at a remote area it would not be possible to provide infrastructure support at a very short notice. The artificial intelligence technique can be applied to AODV nthBR protocol as in this routing protocol nodes are selected based on their energy efficiency.

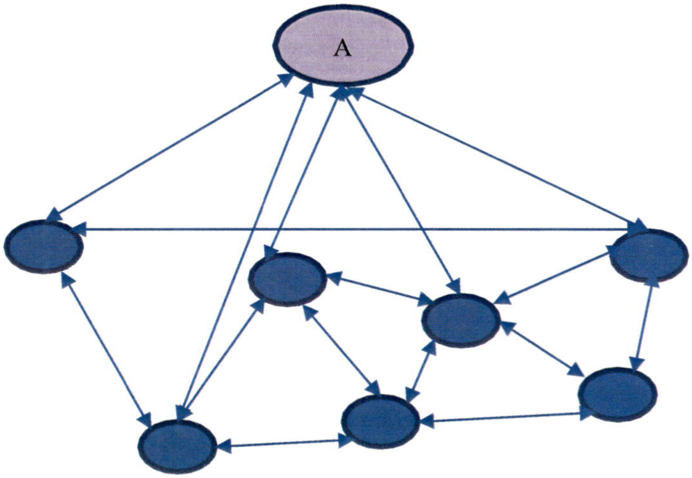

Fig. 3 Ad hoc network setup

Further, Singh et al. have worked on energy conservation and selection of cluster head [9]. They have proposed network partitioning using manageable clusters. This technique can be used especially in areas where a number of users or nodes are more but bandwidth is limited. Here, a self-organizing map (SOM) is chosen for preparing clusters. Managing clusters is easy and it also reduces the data traffic. Data values are transferred from the node to the base station or central node with the help of intermediate nodes. Hence, it saves bandwidth to a considerable extent. A possible sustainable model for an ad hoc network using artificial intelligence technique in routing protocols is shown in Fig. 3 Here, node A is the central node or base station which can be used to send information or collect important data from a COVID 19 hotspot area from various nodes or users. It can also be used to collect data from an area which is under critical or an emergency situation. All other nodes represent the users from where data is to be collected or any information is to be sent. The arrows represent the routing protocol on which artificial intelligence technique has been applied. Energy efficiency techniques can be particularly used for better QoS parameters.

4 Conclusion

In this paper, the need of having a robust routing protocol in critical setups that employ ad hoc network has been discussed. The existing routing protocols and its employability in a critical setup such as a COVID 19 area have been explored. Further, extension or applicability of artificial techniques on the routing protocols has been discussed and its suitability to ad hoc networks in critical setups has been deliberated. In future research, artificial intelligence techniques can be applied to routing protocols like AODV nthBR which choose the node for routing based on the energy efficiency criteria.

References

1. M. Rao, N. Singh, Quality of service enhancement in MANETs with an efficient routing protocol, in *Proceedings of the IEEE International Advance Computing Conference*, pp. 381–384(2014)
2. S. Ghalib, A. Kasem, A. Ali, Analytical study of wireless ad-hoc networks: types, characteristics, differences, applications, protocols, in *Proceedings International Conference on Futuristic Trends in Networks and Computing Technologies*, pp. 22–40 (2020)
3. Q.M. Yas, M. Khalaf, A trusted MANET routing algorithm based on fuzzy logic, in *Proceedings International Conference on Applied Computing to Support Industry: Innovation and Technology*, pp. 185–200 (2020)
4. A. Sharma, A. Singh, G. Lakshay: Comparative analysis of different routing protocols in vehicular ad-hoc network, in *Proceedings Amity International Conference on Artificial Intelligence*, pp. 404–410 (2019)
5. M. Rao, N. Singh, An improved routing protocol (AODVnthBR) for efficient routing in MANETs, in *Proceedings International Conference on Advanced Computing, Networking and Informatics*,pp.215–223 (2014)
6. M. Rao, N. Singh, Performance evaluation of AODV nthBR routing protocol under varying node density and node mobility for MANETs. Indian J. Sci. Technol. **8**(17), 1–9 (2015)
7. J. Barbancho, C. Leon C, F.J. Molina, A. Barbancho, Using artificial intelligence in routing schemes for wirelessnetworks.J.Comput.Commun.**30**(14–15), 45–49,2802–2811(2007)
8. Z.-J. Lee, S.-T. Cho, C.-Y. Lee, B.-Y. Peng,AODVwithintelligentpriorityflowschemeformulti-hop adhoc networks. Vietnam J. Comput. Sci. **3**, 259–264(2016)
9. M. Mittal, A. Singh, K. Kumar, Performance improvement in AODV routing protocol with artificial intelligence, in*Proceedings IEEE International Conference on Computing for Sustainable Development*, pp. 45–49(2016)

Algorithm for Inference of Gene Regulatory Networks Using Partial Least Squares Regression and Mutual Information

Nimrita Koul and Sunilkumar S. Manvi

Abstract Inference of gene regulatory networks refers to identification and localization of functional interactions among genes by application of computational analysis on genomic data. The gene expression data obtained from cDNA microarrays has been used often for such inference. This is an important problem as gene regulatory networks can help in diagnosis and prediction of certain diseases; it is a challenging problem because the data is very high dimensional, noisy, contains very few samples, and time series data is not easily available. The algorithms for inference of gene regulatory networks use the concepts from the fields of information theory, Boolean networks, regression, Bayesian networks, ordinary differential equations, etc. In this paper, we have proposed a method for inference of gene regulatory networks from cancer gene expression data using partial least squares regression to identify important genes, and thereafter, the mutual information is applied to identify the groups of interdependent genes. The results show that the proposed method can identify the biologically relevant networks.

Keywords Gene expression data · Gene regulatory networks · Mutual information · Partial least squares regression · Cancer classification

1 Introduction

Cells of organisms contain genetic material in the form of DNA and RNA structures in form of genes. A cell used proteins as structural material. A human cell contains around 24 thousand genes. Activity of certain genes is associated with incidence of some diseases like cancer. The genes don't act just in isolation, and they also

N. Koul (✉) · S. S. Manvi
REVA University, Bangalore, India
e-mail: emailnk1@gmail.com

S. S. Manvi
e-mail: ssmanvi@reva.edu.in

J. Tavares et al. (eds.), *Emerging Technologies in Data Mining and Information Security*, Lecture Notes in Networks and Systems 164,
https://doi.org/10.1007/978-981-15-9774-9_81

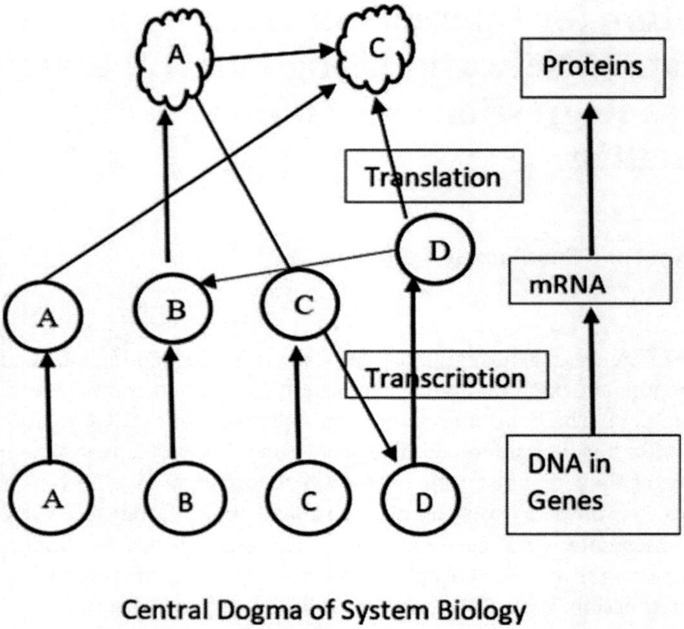

Fig. 1 Central dogma of systems biology

inter-regulate. The central dogma of system biology (Fig. 1) states that from DNA, the process of transcription produces mRNA which is converted into proteins by the process of translation. The proteins which are generated by activation or deactivation of one gene, called as source gene, can in turn lead to activation or deactivation of another gene, called target gene. This network of source gene, target gene and intermediate genes may lead to activation or deactivation of a gene which is an oncogene, i.e., is responsible for causing cancer. This network of mutual influence among genes is known as gene regulatory network. Certain proteins known as transcription factors bind to the sites on target genes and thereby activate or deactivate their expression. Thus, we see that there is a structured information flow within the cell from source gene to target gene through transcription factors. It is because of this, the research studies are dedicated to correctly inferring the gene regulatory networks. A gene regulatory network (GRN) shows the interactions and synergistic relations among groups of genes 2. Often these relations are directed from source to target via intermediate genes. Using the concepts of graph theory, these relations can be modeled in form of graphs where nodes represent genes and the edges represent their relationships. The edges can be labeled to show the strength of influence exerted by one gene on another. Their interactions lead to activation or deactivation of target genes. Reconstruction and analysis of GRNs can provide early information about a disease before the symptoms have manifested externally. They also help in identification of new targets for drugs to break the regulatory networks and cure the disease. With availability of large volumes of genomic data like the microarray gene expression data, next-generation sequencing data, the computational researchers have been able to devise many methods for computational inference of gene regulatory networks. Prominent approaches for inference of gene regulatory networks include information theoretic methods like mutual information, Boolean networks methods, Bayesian and dynamic Bayesian network methods, partial differential equation-based methods.

Background

High-throughput technologies such as cDNA microarrays and RNA-seq provide the expression values of entire genomes for an organism simultaneously under various experimental setups. Computational approaches can analyze this data and capture there relations with some assumptions. These methods can predict GRNs which need to be validated with clinical networks or with true underlying network called as Ground Truth or Gold Standard. Only few existing datasets have a corresponding ground truth. While inferring a GRN, we use gene expression values which are the concentrations of mRNA within the cell at the moment of measurement. Nodes of the graph represent gene expression values, edges represent regulatory effect exerted by one expression value on next in the space of gene expression values or concentration of mRNAs. Thus, rate of production of proteins from mRNA is regulated by value of node at source of incoming arrow. Rate of transcription is under the control of incoming edges. In computational approaches to construction of GRNs from gene expression, major simplification is applied and only the mRNA concentrations are used as inputs with the assumption that these concentrations of mRNA are representative of activity levels of the genes at source of an edge. Computational representation of a GRN is a graph. A graph is an ordered pair $G = (V, E)$ where V is the set of nodes and E is the set of edges. In the context of GRN inference problem, the set of nodes represents set of gene expression values of the genes in gene expression dataset, i.e., $V = x1, x2, \dots xn$, the set of edges, E, is the set of weights representing confidence in the relation between the genes at two ends of an edge. Often the graphs are represented as adjacency matrix. For a graph $G = (V, E)$, the adjacency matrix M is a $N * N$ matrix where N is number of nodes, V is set of all gene expressions in the gene expression dataset. Value in each cell in an adjacency matrix, '*aij*', is weight of the edge from node '*i*' to node '*j*'. Degree of a node is the number of other nodes that it is connected to. It is measured as in-degree and out-degree in case of directed graphs. Degree distribution measures the probability distribution of these degrees in a network and gives the probability of a node having certain number of links, e.g., $P(5)$ is the fraction of number of nodes in the network with exactly five links. Using degree distribution, we can classify networks, and the graph can be characterized as a feature vector. In scale-free networks, the degree distribution decays according to power law and the network has a small number of highly connected nodes called as hubs (Fig. 2).

In this paper, we have contributed as follows—1. We have proposed a new approach for inference of GRNs 2. We have implemented the approach using R and Python on leukemia gene expression dataset 3. We have provided comparative analysis with standard methods for GRN inference ARACNE and B3CNET. Rest

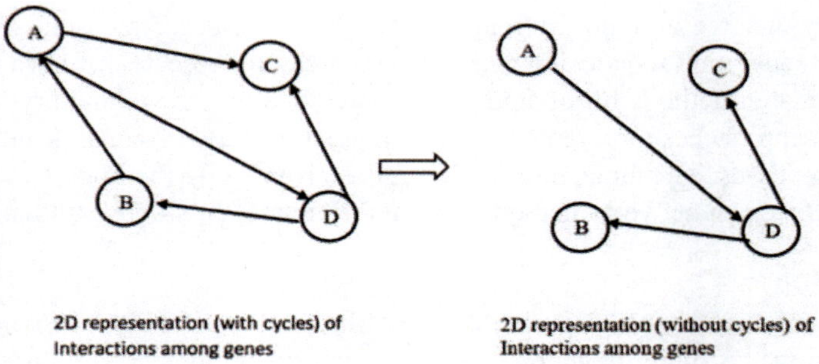

Fig. 2. 2D representation of GRNS

of the paper is organized as follows—Section 2 covers a brief survey of methods for GRN inference in literature, Sect. 2.1 presents the proposed approach, Sect. 3 presents the results and discussion on them, Section 4 concludes the paper.

2 Literature Survey

Algorithms for GRN inference are broadly classified as those based on supervised learning, unsupervised learning, and semi-supervised learning. GRN inference methods mostly used unsupervised learning. We can infer a directed or undirected graph based on the type of learning algorithm used. A directed graph, with an edge directed from node 'A' to node 'B', not only give the nodes involved in an interaction but also determines whether it is node 'A' that controls expression on node 'B' or vice versa. Authors in [1] were the first to discover the process of protein synthesis from DNA in cell. In [2], authors have discussed Bessel functions to capture the inter-regulatory relationships among data items. Authors in [3] have provided a detailed survey of three categories of methods used for inference of gene regulatory networks. In [4], authors have compared the performance of various methods used for computational gene regulatory network inference. Authors in [5] have developed a method known as ARACNE which is based on conditional mutual information. This method is one of the benchmark methods for inference of GRNs. They have used a method for removing cycles or transitive relations in graphs. Authors in [6] have used backward elimination for accurate inference of gene regulatory networks. In [7], authors have proposed an algorithm for reconstruction of the causal relations between the genes using information theoretic methods. Authors in [8] have presented a detailed analysis of a method known as C3NET for inference of gene regulatory networks. In [9], authors have used probabilistic graph method for inference of gene regulatory networks. Authors in [10] have applied Bayesian networks to infer gene regulatory relations among genes.

2.1 Proposed Approach

In this section, we present the proposed approach for inference of gene regulatory networks from microarray gene expression data. The problem of inference of GRN can be divided into three steps—finding set of differentially expressed genes, finding incoming edges to each differentially expressed gene, and adding these edges to an adjacency matrix. The algorithm we propose for inference of GRNs is given here -

Algorithm: GRN Inference.
Input: Microarray Gene Expression Data Set in.
Output: Graph in form of Adjacency matrix out.
Initialization:

1. Compute the set of deferentially expressed genes D using PLS
2. while not all elements in DEGs have been considered do
3. for the next gene 'g' in DEGs do
4. Compute mutual information with other genes in D, consider gene with highest MI as incoming edges to 'g'
5. the gene selected first, 'g' for analysis is the target gene, rest are the transcription factors
6. Make all incoming genes to 'g' as first column of adjacency matrix
7. end for
8. visualize the adjacency matrix
9. end while
10. return the graph

The algorithm was implemented in *R* using the functions from Bioconductor package on a Windows 10 machine.

3 Results and Discussion

In this section, we present the performance metrics of the proposed method in terms of area under ROC curve and area under precision recall curve as compared with two other methods—ARACNE and B3CNET on the Leukemia dataset presented originally in [11]. In Table 1, we have presented a comparison of area under ROC curve obtained while obtaining true positive rate of correctly identified edges by three

Table 1 Area under ROC curve comparison of proposed method with ARACNE and B3CNET for different number of DEGs Identified

No. of DEGs	Proposed method	ARACNE	B3CNET
10	0.66	0.68	0.73
20	0.64	0.60	0.67
30	0.57	0.56	0.63
40	0.52	0.45	0.54

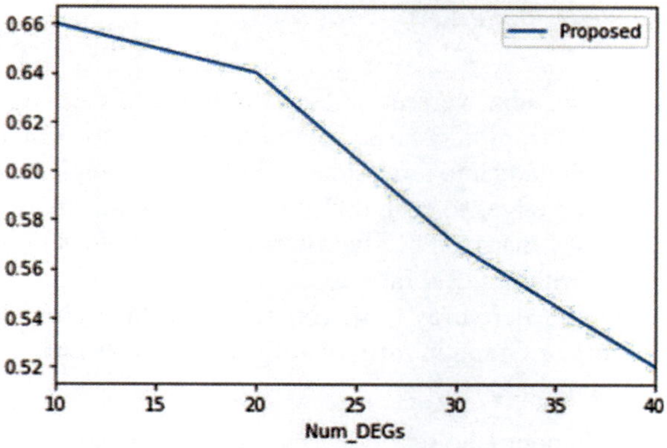

Fig. 3 AUROC proposed algorithm

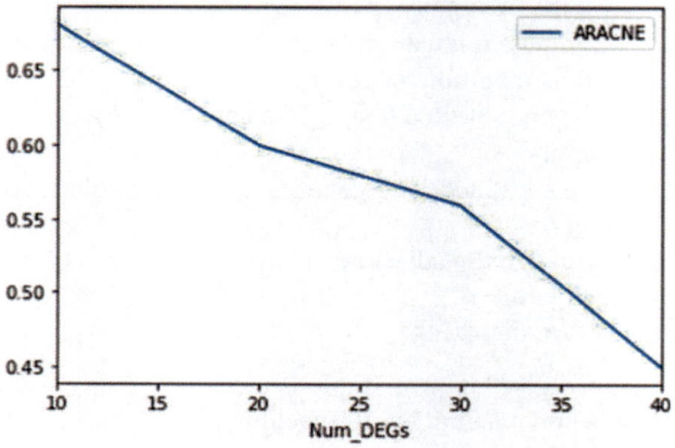

Fig. 4 AUROC ARACNE

methods including proposed method, ARACNE and B3CNET. Table 1 shows that the proposed method works at par with the standard methods which were originally designed for single cell data. Figures 3, 4 and 5 have diagrammatically presented a comparison of area under ROC curve in the proposed method and two existing methods on the leukemia dataset [11].

4 Conclusion

GRN inference problem is an important research field in systems biology, and many methods are being proposed in computational research for the same. In this paper, we proposed a new approach for inference of GRNS from microarray gene expression data. We tested our algorithm on the Leukemia microarray dataset and found that the networks we obtain are relevant with the ground truth as in gene ontology databases. As a future work, the algorithm can be tested on more number of datasets and on

Fig. 5 AUROC B3CNET

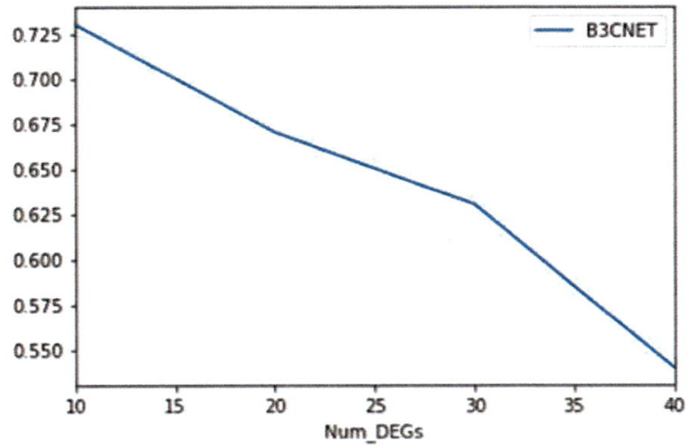

synthetic data to establish its robustness. Further study using gene ontology can be undertaken to explore more relevant results from the observed results.

Acknowledgements Authors would like to thank the Department of Science and Technology (DST), Government of India, for financial grant to carry out this research work under the scheme DST-ICPS 2019.

References

1. F. Crick, Central dogma of molecular biology. Nature **227**(5258), 561563 (1970)
2. G. Eason, B. Noble, I.N. Sneddon, On certain integrals of Lipschitz-Hankel type involving products of Bessel functions. Phil. Trans. Roy. Soc. London **A247**, 529–551 (1955)
3. S.R. Maetschke, P.B. Madhamshettiwar, M.J. Davis, M.A. Ragan, Supervised, semi-supervised and unsupervised inference of gene regulatory networks. Brief. Bioinform. **15**(2), 195211 (2014)
4. D. Marbach, R.J. Prill, T. Schaffter, C. Mattiussi, D. Floreano, G. Stolovitzky, Revealing strengths and weaknesses of methods for gene network inference. Proc Natl Acad Sci **107**(14), 628691 (2010)
5. A.A. Margolin, I. Nemenman, K. Basso, C. Wiggins, G. Stolovitzky, R.D. Favera, A. Califano, ARACNE: an algorithm for the reconstruction of gene regulatory networks in a mammalian cellular context. BMC Bioinform. **7** (2006)
6. P.E. Meyer, D. Marbach, S. Roy, M. Kellis, Information-theoretic inference of gene networks using backward elimination. BIOCOMP, Int. Conf. Bioinform. Comput. Biol, 700–705 (2010)
7. G. Altay, F. Emmert-Streib, Inferring the conservative causal core of gene regulatory networks. BMC Syst. Biol. **4**(1), 132–140 (2010)
8. G. Altay, F. Emmert-Streib, Structural influence of gene networks on their inference: analysis of C3NET. Biology Direct **6**(1), 31 (2011)
9. N. Friedman, Inferring cellular networks using probabilistic graphical models. Science **303**(5659), 799805 (2004)
10. N. Friedman, M. Linial, I. Nachman, D. Peer, Using Bayesian networks to analyze expression data. J. Comput. Biol. **7**(3), 601620 (2000)
11. Golub et al., Molecular classification of cancer: class discovery and class prediction by gene expression monitoring. Science **286**, 531–537 (1999)

Analyzing Automatic Object Classification to Query-Based Image Search Engine

Mansi Mahendru and Sanjay Kumar Dubey

Abstract As human can easily differentiate between similar objects of different texture and color, making our machine equally intelligent as human is a difficult task. A machine in today's time is doing excellent progress with textual data in terms of accuracy by applying several algorithms. The problem of getting good accuracy from image dataset is still not resolved with traditional technologies. This paper gives a brief survey for finding best algorithm suitable for developing query-based image search engine (QBIC) by using deep learning algorithm for object identification within complex images. QBIC system helps to find similar images by indexing and comparing the user query with images present in the database and retrieving the most important images on the top. The survey covers all the current research done related to similarity search and object classification.

Keywords QBIC · Object identification · Deep learning · Similarity search · Query

1 Introduction

An image speaks more than a text. Today everyone is having high-quality smartphones that are very much capable of taking high qualities photos and videos. Sometimes we are not interested in full image but are attracted toward some objects present in the image. We regularly do some kind of textual search on World Wide Web (WWW) but searching our area of interest by developing image similarity search engine is very interesting research area in today's time. Describing anything on

M. Mahendru (✉) · S. K. Dubey
Department of Computer Science and Engineering, Amity University Uttar Pradesh, Sec-125, Noida, U.P., India
e-mail: mansi1997mahendru@gmail.com

S. K. Dubey
e-mail: skdubey1@amity.edu

J. Tavares et al. (eds.), *Emerging Technologies in Data Mining and Information Security*, Lecture Notes in Networks and Systems 164,
https://doi.org/10.1007/978-981-15-9774-9_82

Google using textual search and by using keywords without any visual description is a real pain. So to deal with this problem, there is a necessity of developing the query by image content (QBIC) system. QBIC [1] consists of fetching back the most visually similar images for a given query by searching from database of images. If we have any image that consists of many kinds of objects, so here QBIC comes into existence as it is the strategy of automatically indexing the images by the visible features like shape, texture and color. In this paper, we are covering all the research done related to query-based search engine by using different methods and the results obtained by the author from their research. The first step of our research is to find which algorithm is best to make machine understandable in differentiating various objects. To build the first block, we will use a very interesting application of computer vision, i.e., object detection. There are many deep learning algorithms since 1960 when concept of pattern recognition comes into existence [2]. To extract similar images from web, we combine object detection algorithm [2] with visual search engine. The second block is to use the result obtained in the form of textual classification of objects into retrieval of stystically similar images. The main aim of this research paper is to examine the combination of object recognition with query-based image search engine to overcome from the issues with the image data in efficient way in less time.

2 Research Methodology

Under this section, the research framework is presented which shows key concepts in the terms of research questions and based on that what will be the direction of our analysis and how we will build on these choices at later stages.

2.1 Universal Research Questions Related to Combination of Query-Based Image Search Engine with Object Identification

Two general questions are framed based on the literature review performed.

RQ1. Which deep learning algorithm is best suited for extracting features from complex images?

RQ2. How the extracted features help the user to extract most similar images from the web browser?

Following is the process flow of the query-based search engine (Fig. 1).

A user will upload an image from or via mobile app, and then, QBIC system starts extracting features from the query image and applying the similarity function to compare the query features with the features of the images already indexed. From there, the most similar images are extracted from the web and retrieved to the user.

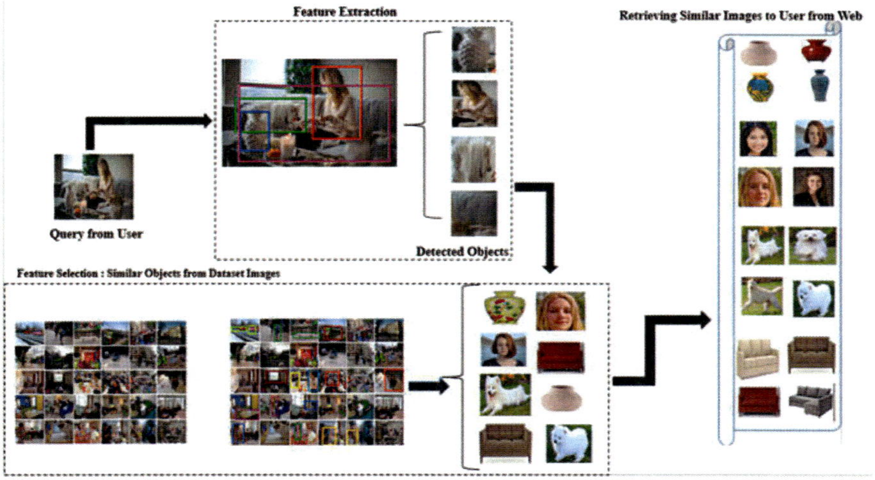

Fig. 1 Flow diagram of QBIC system

3 Related Work

Now days, working on image data is very interesting area of research. There has been a remarkable research on image data over the last years. This section includes the research done related to image retrieval system and what are the challenges faced by the researcher during their work. We have divided our literature survey into two parts. Under the first part, we have covered the work done related to the first building block, i.e., object detection over the last years. In the second part, we have covered all the research done on image retrieval system in the last years.

3.1 Classification of Objects

See Table 1.

3.2 Extraction of Similar Images

See Table 2.

Table 1 Analysis of literature review for classification of objects

Objective	Method or technique used	Dataset used	Analysis	References
Advancement in deep learning technology for 3D object detection using CNN and RNN	CNN and RNN	RGB-D object dataset	Object detection using both CNN and RNN gives more accurate and faster result than two-layer CNN	[3]
Applying CNN cascade for face detection	CNN cascade	Face detection dataset and benchmark dataset (FDDB)	The use of CNN in face detection achieves 14 FPS on CPU and 100 FPS on GPU for VGA images	[4]
Using CNN algorithm for identifying invariant image representation to compact the space acquired by the image	CNN algorithm	Benchmark image datasets	The use of CNN reduces the size of image and achieves the result in less than 100–500 region proposals per image	[5]
Applying RCNN for small object detection	Augment RCNN algorithm	Benchmark dataset	The use of RCNN improves the mAP by 29.8% than normal RCNN algorithm	[6]
Using fast RCNN for object detection	Fast RCNN on two networks VGG-net and SPP-net	PASCAL VOC 2012 dataset	Fast RCNN trains VGG-net 9× faster, test 213× faster, SPP-net 3× faster and tests 10× faster than RCNN algorithm	[7]
Applying faster RCNN algorithm for object detection	Faster RCNN algorithm	PASCAL VOC 2007 and 2012 dataset	Faster RCNN for VGG-net achieves 73.2% mAP on PASCAL VOC 2007 and 70.4% mAP on PASCAL VOC 2012 dataset	[8]

(continued)

Table 1 (continued)

Faster object detection using refining fast RCNN algorithm	Refining fast RCNN algorithm	Road view image dataset	Using refining fast RCNN improves the performance by merging both fast RCNN and faster RCNN into one block	[9]
Applying you only look once algorithm for real-time object detection	YOLO algorithm	PASCAL VOC 2007 and VOC 2012 dataset	Using YOLO and fast YOLO algorithm takes full images instead of region proposals and works more accurately than other deep learning algorithms	[10]
YOLO 9000 algorithm for object identification	YOLO 9000 algorithm	Image net classification and COCO dataset	YOLO 9000 is much stronger in detection more than 9000 classes for different objects	[11]
Developing YOLO-LITE for real-time object detection on portable devices	YOLO-LITE algorithm	PASCAL VOC and COCO dataset	YOLO–LITE achieves 33.81 and 12.26% mAP	[12]
YOLO_V3, modified YOLO algorithm for object detection	YOLO_V3 algorithm	COCO dataset	YOLO_V3 is slighter bigger but is 3 times faster than SSD model and achieves 28.2 mAP	[13]
Mini YOLO_V3 algorithm for object detection in embedded applications	Mini YOLO_V3 algorithm	MS COCO benchmark dataset	Size of mini YOLO_V3 is 23% of YOLO_V3, requires ½ detection time and achieves 50 of 52.1 mAP at 67 FPS	[14]
Applying FCNN for detecting remote sensing device using MIF-CNN	Multi-scale image block-level FCNN algorithm	BEIJING CAPITAL INTERNATIONAL and HONKONG INTERNATIONAL AIRPORT dataset	MIF-CNN can extract image of any size, and it improves the computing efficiency as compared to other algorithms	[15]

(continued)

Table 1 (continued)

Object detection algorithm for autonomous driving using Gaussian YOLO_V3	Gaussian YOLO_V3 algorithm	KITTI and BDD dataset	Gaussian YOLO_V3 improves the mAP by 3.09 and 3.5 as compared to original YOLO_V3	[16]

4 Results and Analysis

This section presents the solution to the above-mentioned research questions based on the analysis of previous work done in this area.

4.1 Deep Learning Algorithm Used for Extracting Objects from Complex Images.

According to the literature survey, the solution of this question is presented in the form of graph as shown in Fig. 2. The graph shows the mean average precision (mAP) of every deep learning algorithm that is used by the researchers in their work. Based on the below graph, it is recorded that YOLO family is performing better than RCNN family and YOLO_V3 is the latest version of YOLO algorithm which is more better, faster and accurate in detecting small objects which is the limitation of all previous deep learning algorithms. From the graph, YOLO_V2 has the highest mAP but it makes prediction only at the last layer due to which its mAP is more but accuracy is less than YOLO_V3.

4.2 Performance of Different Algorithm in Image-Based Search Engine

Figure 3 shows the performance of various algorithms used by the researchers in their past studies in building search engines. The best algorithm obtained from the above analysis is implemented in future to improve the visual search.

5 Conclusion and Future Scope

Firstly, this paper tries to find the pitfalls in various deep learning algorithms for object classification by analyzing the work done by many researchers and then analyzing the

Table 2 Analysis of literature review for extraction of similar images

Objective	Method or technique used	Dataset used	Analysis	References
Developing content-based image retrieval system using color and texture	Wavelet-based color histogram retrieval technique (WBCHIR)	WANG image database	Applying WBCHIR technique reduces the computational steps and only takes 5–6 min as whole indexing time	[17]
Developing content-based image retrieval system using KNN algorithm based on color, shape and texture	KNN algorithm	Image database consisting of 500 images	Applying KNN algorithm on CBIR extract features using color correlogram, color moment and color hsv by achieving precision of 0.86667 and recall 0.37143	[18]
Content-based image retrieval system for scaling invariant features	SIFT algorithm and K means algorithm	Caltech 101 and Li database	The use of two techniques makes the system fast and accurate and is able to manage large database effectively	[19]
New content-based image retrieval system to improve the image retrieval precision	Swarm optimization algorithm	Corel dataset	New CBIR system uses two features DCD and wavelet and curvelet and achieves 76.50% precision	[20]
Visual search at Pinterest	CNN algorithm	Images from Pinterest women's fashion category	They analyze that visual search can improve user engagement on web	[21]
Content-based image retrieval system using deep learning algorithm	CNN algorithm	Shopping shoes dataset and kitchen appliances dataset	Applying CNN on shoe dataset achieves 98.78% accuracy for 10th prediction and 98.44% accuracy for top 3 predictions	[22]

(continued)

Table 2 (continued)

Multimodal search engine for interior design	YOLO 9000 algorithm	IKEA dataset	Applying YOLO 9000 algorithm improves visual search component by 200%	[23]
Developing image-based search engine using memetic algorithm	Memetic algorithm	Corel dataset	Applying MA algorithm for retrieving similar images achieves 0.882 5 precision and 0.7002% recall	[24]
Image search engine using deep learning process	Deep belief neural network	Image database	Applying DBN achieves 98.6 and 96% accuracy	[25]

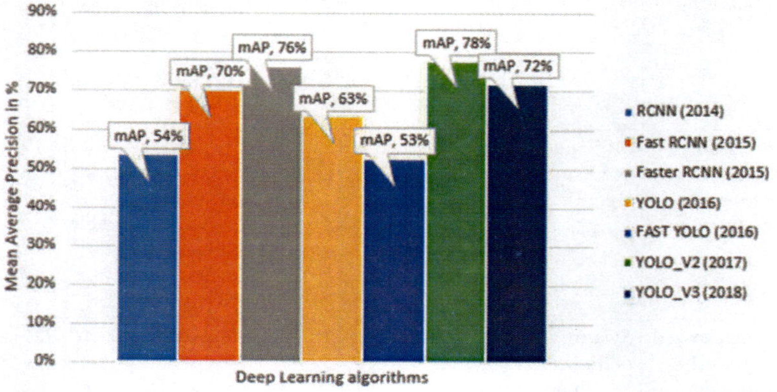

Fig. 2 Performance measurement of image-based search engines

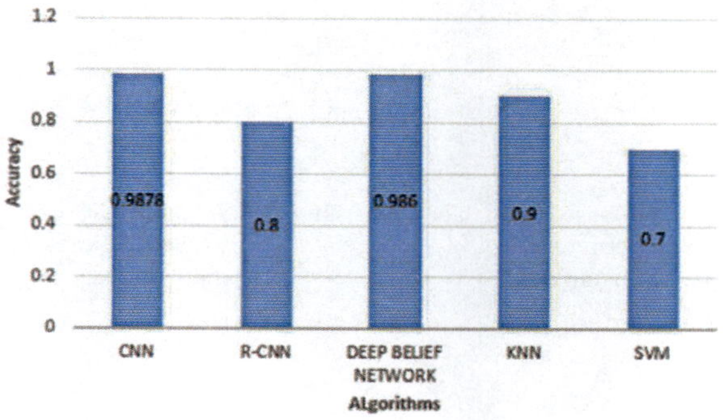

Fig. 3 Performance of various deep learning algorithm

various techniques used to build the visual search engine. By doing analysis on visual search engine, it is recorded that there are many challenges with image data that can only be solved by efficient deep learning algorithm. Paper presents the combination of object recognition and image-based search engine to explore complex problems that are not solved earlier. For this reason, analytical survey is done and questions are framed in this regard. The result of the research questions shows that effective deep learning algorithm is to be chosen in order to solve the problems with image data. Future work will include the improvement of accuracy in various deep learning algorithm and further to work on real images.

References

1. K. Ahmad, M. Sahu, M. Shrivastava, M.A. Rizvi, V. Jain, An efficient image retrieval tool: query based image management system**12**(1), 103–111 (2020)
2. C.K. Maharaja, C.K. Agrasen, S. Arora, The comparison between various object detection algorithms (2018)
3. R. Socher, B. Huval, B. Bath, C.D. Manning, A.Y. Ng, Convolutional-recursive deep learning for 3d object classification. Adv. Neural Inf. Process. Syst. 656–664 (2012)
4. H. Li, Z. Lin, X. Shen, J. Brandt, G. Hua, A convolutional neural network cascade for face detection, in *Proceedings of the IEEE Conference on Computer Vision and Pattern Recognition*, pp. 5325–5334 (2015)
5. K. Reddy Mopuri, R. Venkatesh Babu, Object level deep feature pooling for compact image representation, in *Proceedings of the IEEE Conference on Computer Vision and Pattern Recognition Workshops*, pp. 62–70 (2015)
6. C. Chen, M.Y. Liu, O. Tuzel, J. Xiao, R-CNN for small object detection, in *Asian Conference on Computer Vision* (Springer, Cham, 2016), pp. 214–230
7. R. Girshick, Fast R-CNN, in *Proceedings of the IEEE International Conference on Computer Vision*, pp. 1440–1448 (2015)
8. S. Ren, K. He, R. Girshick, J. Sun, Faster R-CNN: towards real-time object detection with region proposal networks. Adv. Neural Inf. Process. Syst. 91–99 (2015)
9. M.C. Roh, J.Y. Lee, Refining faster-RCNN for accurate object detection, in *2017 Fifteenth IAPR International Conference on Machine Vision Applications (MVA)* (IEEE, May 2017), pp. 514–517
10. J. Redmon, S. Divvala, R. Girshick, A. Farhadi, You only look once: unified, real-time object detection, in *Proceedings of the IEEE Conference on Computer Vision and Pattern Recognition*, pp. 779–788 (2016)
11. J. Redmon, A. Farhadi, YOLO9000: better, faster, stronger, in *Proceedings of the IEEE Conference on Computer Vision and Pattern Recognition*, pp. 7263–7271 (2017)
12. R. Huang, J. Pedoeem, C. Chen, YOLO-LITE: a real-time object detection algorithm optimized for non-GPU computers, in *2018 IEEE International Conference on Big Data (Big Data)* (IEEE, Dec 2018), pp. 2503–2510
13. J. Redmon, A. Farhadi, Yolov3: an incremental improvement. arXiv:1804.02767 (2018)
14. Q.C. Mao, H.M. Sun, Y.B. Liu, R.S. Jia, Mini-YOLOv3: real-time object detector for embedded applications. IEEE Access **7**, 133529–133538 (2019)
15. W. Zhao, W. Ma, L. Jiao, P. Chen, S. Yang, B. Hou, Multi-scale image block-level F-CNN for remote sensing images object detection. IEEE Access **7**, 43607–43621 (2019)
16. J. Choi, D. Chun, H. Kim, H.J. Lee, Gaussianyolov3: an accurate and fast object detector using localization uncertainty for autonomous driving, in *Proceedings of the IEEE International Conference on Computer Vision*, pp. 502–511 (2019)

17. M. Singha, K. Hemachandran, Content based image retrieval using color and texture. Signal Image Process **3**(1), 39 (2012)
18. P.A. Deole, R. Longadge, Content based image retrieval using color feature extraction with KNN classification. IJCSMC **3**(5), 1274–1280 (2014)
19. G.A. Montazer, D. Giveki, Content based image retrieval system using clustered scale invariant feature transforms. Optik **126**(18), 1695–1699 (2015)
20. S. Fadaei, R. Amirfattahi, M.R. Ahmadzadeh, New content-based image retrieval system based on optimised integration of DCD, wavelet and curvelet features. IET Image Proc. **11**(2), 89–98 (2016)
21. Y. Jing, D. Liu, D. Kislyuk, A. Zhai, J. Xu, J. Donahue, S. Tavel, Visual search at pinterest, in *Proceedings of the 21th ACM SIGKDD International Conference on Knowledge Discovery and Data Mining*, pp. 1889–1898, Aug 2015
22. S. Jain, J. Dhar, Image based search engine using deep learning, in *2017 Tenth International Conference on Contemporary Computing (IC3)* (IEEE, Aug 2017), pp. 1–7
23. I. Tautkute, A. Możejko, W. Stokowiec, T. Trzciński, L. Brocki, K. Marasek, What looks good with my sofa: multimodal search engine for interior design, in *2017 Federated Conference on Computer Science and Information Systems (FedCSIS)* (IEEE, Sept 2017), pp. 1275–1282
24. M.K. Alsmadi,An efficient similarity measure for content based image retrieval using memetic algorithm: an efficient similarity measure for content based image retrieval using memetic algorithm. Egypt. J. Basic Appl. Sci. **4**(2), 112–122. https://doi.org/10.1016/j.ejbas.2017.02.004 (2017)
25. R.R. Saritha, V. Paul, P.G. Kumar, Content based image retrieval using deep learning process. Cluster Comput. **22**(2), 4187–4200 (2019)

An Analytical Approach to Investigation of Asthma Diseases Using Soft Computing Techniques

Gautam Singh Yadav and Sanjay Kumar Dubey

Abstract Fresh air is required for the long survival of organisms, but pollution in cities rapidly increases due to industrial waste, burning fossil fuels, and agricultural activities. Lungs are the important organ in the body which helps in breathing and asthma is a respiratory disease in the lungs, and air pollution is the main cause of Asthma disease. As air pollution increases chances of respiratory disease can also increase rapidly. It needs to control both pollution and disease. The functioning of an organ gets changes during abnormality. Abnormality in the body negatively affects and includes sudden deaths. Due to the similar structure of different abnormalities, easily detection is not possible. So, doctors recommend various tests to the patients, and it consumes both time and money. Artificial intelligence has different techniques that can help in prediction. This paper presents a detail study on asthma severity detection techniques that were proposed by researchers and discuss analysis for severity detection through symptoms, questionnaire, and through automation.

Keywords Disease · Diagnosis · Prediction · Analysis · Soft computing

1 Introduction

Medical analysis of asthma disease examines in two ways questionnaire and clinical data. The severity of asthma disease can be confirmed based on symptoms and pattern detection. Physical examination of patients in asthma can be determined based on qualitative symptoms but clinical data, the asthma severity can be calculated with the help of air intake, cough, wheeze, and working activities [1]. Detection of the pattern is time consuming due to the similar structure of different diseases. In organism, lung

G. S. Yadav (✉) · S. K. Dubey
Department of Computer Science and Engineering, ASET, Amity University, Noida, India
e-mail: er.gautamtech96@gmail.com

S. K. Dubey
e-mail: skdubey1@amity.edu

J. Tavares et al. (eds.), *Emerging Technologies in Data Mining and Information Security*, Lecture Notes in Networks and Systems 164,
https://doi.org/10.1007/978-981-15-9774-9_83

[2] is responsible for breathing and polluted air may increase the severity. Disease detection in an early stage is helpful for a cure. Asthma severity can be detected in newborn to old age people, and it depends on the breathing of Air quality present in the environment [3]. According to the various report of pollution [4] control and health organization, pollution in cities rapidly increases through crops burning and fossil fuels. Algorithms can train a model from the prior dataset and predict the outcome using certain rules [5]. Symptoms are the treasure for the curing of any disease. In this paper, respiratory dysfunction in asthma, patients cannot survive for long without proper treatment.

2 Review Methodology

Many researchers designed an expert system based on cough, tightness in chest, wheeze, Asthma effect in night or morning, cold days, after exercise or in rest, fever, perennial problem, cough in sputum, sputum color, swelling ankles, particulate matter in city, smoking activity, asthma duration, FEV, VC, PEF, history of asthma, immunoglobulin E, overall use of quick-relief inhaler, did asthma bother your sleep?, diabetes, allergy, and working limitation of asthma diagnosis. Sound travels in an acoustic waves form. Khan et al. [6] design a system for the detection of resonated wheeze wave in analog-to-digital form. For capturing wheezing sound, create acoustic sensor with the help of metal foil like a parallel plate capacitor, and the gap between plates works as a pressure-sensitive dielectric material. The sensor works as a low-pass filter, connected with smartphone, attached to the chest. FFT used for sound frequency. Rabe et al. [7] design a tool for checking the effect of dupilmab on lung function in a patient who suffered from severe asthma. With the parameter like pre- and post-BD FEV1, FVC, FEV1/FVC, dupilmab steroid-sparing depends on FEV1. Matsuda et al. [8] used first wheeze and childhood asthma by tracking the viruses and cytokine/chemokine to distinguish asthma. Yeganeh et al. [9] design system for the estimation of particulate matter measurement based on the satellite-based model populated environment using ANFIS, SVM and BPANN, ANFIS give more accuracy RF, SVM and BPANN. Thukral et al. [10] design fuzzy-based system for detection of optimum result in asthma disease and observed accuracy of fuzzy system increased with clinical data, due to accuracy fuzzy techniques spread rapidly in last decades. Tartarisco et al. [11] used ANN, SVM, Bayesian network, fuzzy logic, and allergic disease detection. High accuracy results by ANN with maximum symptoms.

3 Related Work

Q1. Is it safe for the patient to suffer from the respiratory disease? What can be done to combat respiratory disease?

Q2. Can Disease have any impact on countries' economies? How it can be overcome?

Asthma is a chronic respiratory disease and not an age-related disease; Early stage asthma can be controlled by some precaution but not end. Late or severe asthma is very dangerous to patients. It needs regular checkup and doges. Asthma, symptoms, classification, precaution, and investigation are discussed below.

3.1 Analyze Asthma Through Questionnaire.

I. Is breathing problem associated with fever, cough, and wheeze?
II. Do you feel irritation and tightness in chest?
III. Does the problem seasonal, episodic, or paraxial Intermittent?
IV. Does the problem occur at rest, after exercise, morning, or night?
V. Does cough produce any sputum, sputum color (except white)?
VI. Does the problem occur unendly?
VII. Do you have any bump of feet and ankles?

3.2 Analysis on Asthma

Chronic inflammatory disease of the airways, disease impact changes with overtime due to the blocking of airways, the functioning of the lung is affected, the improper supply of oxygen in the body may have side effects like irritation. In attacks respiratory rate, breath sound [12] wheeze increases rapidly and in severe attacks narrow airways [13] pass the insufficient amount of air to the bodies and causes severe damage. Asthma classification, bronchial asthma are early-onset [14], atopic, allergic and extrinsic, increased levels of immunoglobulin E in the serum and Late-onset asthma: non-atopic, idiosyncratic and intrinsic, normal IgE [15] and other bronchial asthma includes non-allergic asthma, A etiopathogenesis, bronchial hyper-responsibleness, Inflammatory reaction with bronchial [16] wall are two main factors of pathogenesis. Allergen precaution and clinical features protection from allergens [17] are also a cure. The disease can enter into the body through various routes and should avoid fungal spores, animal dander, pollen, dust, milk, yeast, wheat, and fish, and it needs to take regular antibodies and prescribe drugs. Asthma is classified based on clinical features: episodic [18] asthma occurs in interval periods or spontaneous episodes, and characters are wheezing, its attacks are mild and severe [19]. Severe acute asthma causes unproductive cough; physical signs are central cyanosis, tachycardia, sweat, pulsus, and paradoxus. Chronic asthma indicators are chest tightness, wheeze occurs at night and breathlessness occurs on exertion. Investigation of bronchial Asthma can be identified by pulmonary function tests [20], peak expiratory flow (PEF), force expiratory volume (FEV), and vital capacity (VC). These estimate

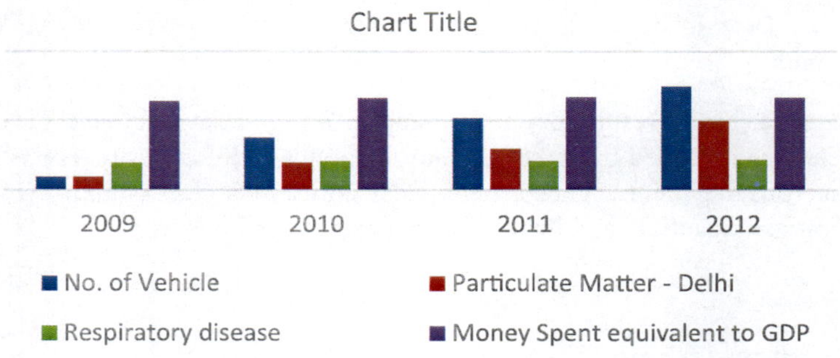

Fig. 1 Chart for the yearly increment of vehicle, PM, respiratory disease, and money spent

the degree of airflow obstruction, in adults, an increase in FEV of >12% or >200 mL from baseline 10–15 min after inhalation of 200–400 μg of salbutamol is considered a positive bronchodilator reversibility test. PEF changes daily in asthma and in the morning recorded lowest [21]. In histamine or methacholine provocation test, FEV falls more than 20%, mostly cough is the only symptom in asthma. Arterial blood gas during acute attack shows hypoxemia and hypocarbia. Controlled asthma has no limitation in exercise and activities, no exacerbation, symptoms occur not more than twice per week [22]. Asthma and vocal cord dysfunction can coexist, and misdiagnosis of VCD as asthma can be costly. India recorded a government debt equivalent to 69.62 percent of the country's gross domestic product in the 2019–2020 fiscal year [23], the lowest is 47.9 from 1980 to 2019. Asthma is dangerous at the age of 61 to 75 years of patients; these patients face different attacks related to the dysfunction of lungs. In 2016–2019 the number of patients reached from 22 to 53%, respectively [24] (Fig. 1).

Above graph shows yearly variation in external climate automatically and manually. Number of vehicles is directly responsible particulate matter or pollution change. Pollution increases the chances of respiratory disease. As the respiratory disease increases, the amount of money spent on cure also increases. According to the report of economic times, every year 1.2–1.4% of GDP money spent [25] on healthcare sector definitely affects the economy.

4 The Methodology Used for Asthma Analysis and Prediction

Many researchers and biologist used different techniques for asthma. Four major symptoms, cough, wheeze, shortness of breath, and tightness in chest, have been identified from different prior study. Wheeze sound was used by many researchers and biologist for the asthma diagnosis, so only major symptoms were selected for further research work. Identification of symptoms metrics for severity detection system (Table 1).

KNN is an instance-based learning technique which is used for classification of pattern recognition. In training through asthma dataset, it collects labeled instances. When a new query has to be classified, KNN algorithm selects *K* number of training instance near to query point. It follows Euclidian distance similarity to close points. If *K* points have majority of vote, it is assigned to the class of its nearest neighbor. If asthma symptom has similar behavior toward low, mid, high, it can be easily categorized with KNN [20]. Boosting method is one in which single "strong classifier" is generated by combining of different "weak classifiers." Performance of weak classifier is superior than random guessing but performs inadequately. AdaBoost is a classification technique which used labeled data for training of new classifier and determining the required weight for each when combined. Once training is complete of classifier, if performance is high then the assigned weight is also high. Linear combination of weak classifier provides final classification [20]. Troupe learning

Table 1 Asthma survey and classification

S. No.	Researcher	Symptom	Technique	Accuracy %
1	Prasadi et al. (2011)	Questionnaire, clinical data	C4.5, PSO, Bayesian network	84, 85, 82
2	Prasadi et al., (2011)	Questionnaire, clinical data	CSAMM, backpropagation	84, 83
3	Silveira et al. (2011)	Clinical data	Specialists, DT, NN, RF	91.2, 89.8, 89.5, 89.2
4	Silveira et al. (2011)	Clinical data	Gaussian SVM, linear SVM, NB, polynomial SVM	83, 76.2, 70.3, 58.4
5	Polat et al. [12]	Breath cycle	NN, GA	91, 93
6	Sihag et al. [4]	Particulate matter	SVM, RF, ANN	ANN
7	Sihag et al. [4]	Particulate matter	M5P, GP,	ANN
8	Sanak et al. (2015)	Anti-inflammatory drug (N-ERD)	ANN	92.90
9	Morten et al. (2016)	Wheeze	SVM	64.6, 83.5
10	Patel et al. (2013)	PEFR, FEV1, FVC, FEF	FL, ANN, ANFIS	78, 90, 94
11	Rocha et al. (2017)	Respiratory sound	Empirical rule-based SVM, ANN, KNN, GMM	ANN
12	Rajani et al. (2019)	BP, PA, age, BMI, smoke, diabetic, cholesterol	Fuzzy logic	99.3
13	Tartarisco et al. [11]	PROSPERO database (CRD42016038894)	ANN, SVM,	96, 80

provides solution based on decision tree. RF produces average of the individual trees in regression and in classification, class that is the mode of the classes output. Developer Breiman and Cutler used "bagging" and selection features for decreasing variance by averaging many noisy [11, 20], but approximately unbiased, models. Ideal candidate (tree) can acquire complex interaction structure and has low bias data. Variance of bagging is cut down in random selection by reducing correlation in tree. Clustering, detecting, outliers, and scaling of data can be easily done with RF. FDSC used pairwise dissimilarities for representation of objects symptoms and shows more complete representation than features. FDSC is pattern detection-based technique and easily catch the gap, dissimilarities in structures and images of asthma patients. It has also kernel-based approach which finds similarities. Method needs to obtain all pairwise distance as an $n * n$ dissimilarity matrix and Euclidean distance between i-th and jth object. Euclidian vector space must be equal dissimilarity space, and dimensional dissimilarity space represents training object used for training classifier. In dissimilarity space, test objects are represented as Euclidean distance to the object in the training set and applied to the trained classifier in this space [20]. Backpropagation NN has multiple hidden layers and has recursive property. Due to recursive, weight variation, and error tolerance nature, BPNN uses low memory. BPNN can be trained with minimum data. BPNN is used for the prediction of asthma disease. NN uses symptoms data that is allotted to neurons, and each neuron is connected to their respective weight. If the value of neuron passes to weight and its value is less than the activation function, then weight needs to be updated and again neuron transfers to weight; NN is a slow tool that takes more time in training to reach the error minimum. ANN detects pattern in dataset for the training of data. ANFIS is an ANN and fuzzy-based system. Fuzzy system can easily deal with group data and has learnability for nonlinear function approximation. ANFIS system is used for training the group numerical samples of complex function and used to classify pattern recognition and rule-based process control. ANFIS uses fuzzy data for training of neural networks. In asthma, symptoms have more than one value and system has ability to train itself with correct value with error tolerance. Particle swarm optimization is a population-based stochastic optimization technique which is similar to evolutionary calculation techniques like GA. PSO is initialized with random solution and searches for optimum solution by updating generation. Multilayer perceptron is a classical supervised, and each layer neuron is interconnected and received multiple inputs and connected with other weights. Multilayer perceptron network involves finding values, bias, and connection weights and reduces error function between desired and target result in the training dataset. Mean square error (MSE) is used as error function and has number of patterns, desired output, and target value. Network training is stopped when desired output and calculated value is equal and has minimum error tolerance value. Backpropagation network is used for adjustment of weights for minimization of error. Desire power in neural network is reduced for fixed size set of samples that has too large input. GA accuracy depends on feature selection [22] of factor. Every individual has solution in selection problem. Naturally if chromosome is selected its value is 1 else its value is 0, minimize erroring training data. Fitness function chosen for GA is ratio of number of input to total number of prognostic factors in training [23] in addition with MSE.

5 Issues in Asthma Diagnosis

i. Breathing rate in children and adults are different in asthma, and it needs to be observed carefully. Lack of standardization, result varied.
ii. Demographic data including weight, height, age [22], asthma severity may be different for the same group patient or the same severity for a different group.
iii. Asthma has several attacks like at night, early morning, and episodic. In asthma, two patients with the same symptoms not necessarily suffer the same attack.
iv. Asthma is an unended disease and can be controlled for a limited time using regular doses in the guidance of specialist but drugs overdose is not good.
v. A severity for different patients may vary due to symptoms (in mild asthma if the severity of symptoms increased then it can be moderate or severe).

6 Results and Discussion

Analysis of various algorithm and efficiency of model change with membership functions. CSAAM have 84.32, BPNN 82.21%, C4.5 83.83%, PSO 84.16% accuracy with considering above questionnaire. NN 89.5%, NB 70.3%, RF 89.2%, other ANFIS and BPANN results, accomplished that result varies as factors 0.9676 and 0.9536 for train and test phase. Categorizing of dimension (feature) from 46 to 18 provides similar performance in proposed, 95.54% of severity can be predicted by method. features, initial prognostic factors of used only 8% can achieve performance.

7 Conclusion

In this paper, we analyze severity of asthma and detection through major symptom using AI technique and their performance and also discuss our future working implementation model based on FIS and efficiency evaluation with NN and ANFIS. ANFIS results have high accuracy. Mean square error method is used for error minimization. ANFIS- and GA-based tools can easily detect asthma severity and cost efficient. Demand of automation rapidly increases in medical industry for cure of common disease and growth of the field. This research work done predicts severity to asthma, which can help in preventing the patients from unusual severe attacks and can help in reduction of test cost, time, and efficiency for the patients by predicting accuracy in severity. The research objectives are achieved with proper methodology.

References

1. B.D.C.N. Prasadl, P.E.S.N.K. Prasad, Y. Sagar, An approach to develop expert systems in medical diagnosis using machine learning algorithms (asthma) and a performance study. Int. J. soft Comput. (IJSC) **2**(1), 26–33 (2011)
2. P. Kozlik, J. Zuk, S. Bartyzel, J. Zarychta, K. Okon, L. Zareba, J.G. Bazan et al.,The relationship of airway structural changes to blood and bronchoalveolar lavage biomarkers, and lung function abnormalities in asthma. Clin. Experiment. Allergy (2020)
3. E. Chatzimichail, E. Paraskakis, A. Rigas, An evolutionary two-objective genetic algorithm for asthma prediction, in *2013 UKSim 15th International Conference on Computer Modelling and Simulation* (IEEE, 2013), pp. 90–94
4. P. Sihag, V. Kumar, F.R. Afghan, S.M. Pandhiani, A. Keshavarzi, Predictive modeling of PM 2.5 using soft computing techniques: case study—Faridabad, Haryana, India.Air Qual. Atmos. Health **12**, 1511–1520 (2019)
5. A. Sahoo, S. Samantaray, S. Bankuru, D.K. Ghose, Prediction of flood using adaptive neuro-fuzzy inference systems: a case study, in *Smart Intelligent Computing and Applications* (Springer, Singapore, 2020), pp. 733–739
6. Khan, S.M., Qaiser, N., Shaikh, S.F. and Hussain, M.M. Design Analysis and Human Tests of Foil-Based Wheezing Monitoring System for Asthma Detection. IEEE Transactions on Electron Devices, 67(vol no 1), pp.249–257 2019.
7. K.F. Rabe, P. Nair, J.F. Maspero, M. Castro, M.S. Rice, Y Deniz, P. Rowe, H.W. Staudinger, G. Pirozzi, B. Akinlade, N.M. Graham, The effect of dupilumab on lung function parameters in patients with oral corticosteroid-dependent severe asthma. Respirat. Med. X **2**, 100010 (2020)
8. S. Matsuda, M. Kato, T. Koike, Y. Kama, K. Suzuki, M. Enseki, H. Tabata, K. Hirai, Y. Yamada, H. Mochizuki, Differences in virus detection and cytokine profiles between first wheeze and childhood Asthma. Tokai J. Experiment. Clin. Med. **45**(1), 10–17 (2020)
9. B. Yeganeh, M.G. Hewson, S. Clifford, L.D. Knibbs, L.A. Morawska, Satellite-based model for estimating PM2. 5 concentration in a sparsely populated environment using soft computing techniques. Environ. Modell. Softw. 84–92 (2017)
10. S. Thukral, V. Rana, Versatility of fuzzy logic in chronic diseases: a review. Med. Hypotheses **122**, 150–156 (2019)
11. G. Tartarisco, A. Tonacci, P.L. Minciullo, L. Billeci, G. Pioggia, C. Incorvaia, S. Gangemi, The soft computing-based approach to investigate allergic diseases: a systematic review. Clin. Mole. Allergy **15**(1), 10 (2017)
12. I. Güler, H. Polat, U. Ergün, Combining neural network and genetic algorithm for prediction of lung sounds. J. Med. Syst. **29**(3), 217–231 (2005)
13. K. Porpodis, P. Katerina, T. Kontakiotis, I. Kioumis, D. Spyratos, D. Papakosta,Detection of asthma and rhinitis in furniture workers exposed to chemicals in the area of Thessaloniki Greece(2019)
14. B.M. Rocha, D. Filos, L. Mendes, I. Vogiatzis, E. Perantoni, E. Kaimakamis, P. Natsiavas et al.,A respiratory sound database for the development of automated classification, in *Precision Medicine Powered by pHealth and Connected Health* (Springer, Singapore, 2018), pp. 33–37
15. T.F. Carr, R. Altisheh, M. Zitt, Small airways disease and severe asthma. World Allergy Org. J. **10**(1), 20 (2017)
16. J. Finkelstein, I. Cheol Jeong, Machine learning approaches to personalize early prediction of asthma exacerbations. Ann. New York Acad. Sci. **1387**(1), 153 (2017)
17. L. Aliboni, F. Pennati, T.J. Royston, J.C. Woods, A. Aliverti, Simulation of bronchial airway acoustics in healthy and asthmatic subjects.PloS One **15**(2), e0228603 (2020)
18. A. Rodrigues, J.M. de Oliveira, K.C. Furlanetto, F.V.C. Machado, L.F. Belo, L.P. Schneider, Are the effects of high-intensity exercise training different in patients with COPD Vversus COPD+ asthma overlap? Lung **198**(1), 135–141 (2020)
19. X. Yang, Y. Guo, Data science for asthma study, in *Genomic Approach to Asthma* (Springer, Singapore, 2018), pp. 277–301

20. J.L. Amaral, A.J. Lopes, J. Veiga, A.C. Faria, P.L. Melo, High-accuracy detection of airway obstruction in asthma using machine learning algorithms and forced oscillation measurements. Comput. Methods Progr. Biomed. 113–125 (2017)
21. A. Silveira, C. Muñoz, L. Mendoza, Severe asthma exacerbations prediction using neural networks, in *International Conference on Engineering Applications of Neural Networks* (Springer, Cham, 2019), pp. 115–124
22. K.E. Tyrak, K. Pajdzik, E. Konduracka, A. Ćmiel, B. Jakieła, N. Celejewska-Wójcik, G. Trąd, A. Kot, A. Urbańska, E. Zabiegło, R. Kacorzyk, Artificial neural network identifies nonsteroidal anti-inflammatory drugs exacerbated respiratory disease (N-ERD) Cohort. Allergy (2020)
23. K.S. Burrowes, J. De Backer, R. Smallwood, P.J. Sterk, I. Gut, R. Wirix-Speetjens, S. Siddiqui et al.,Multi-scale computational models of the airways to unravel the pathophysiological mechanisms in asthma and chronic obstructive pulmonary disease (AirPROM).Interface Focus **2**, 20120057 (2013)
24. R.V. Veiga, H.J. Barbosa, H.S. Bernardino, J.M. Freitas, C.A. Feitosa, S.M. Matos, N.M. Alcântara-Neves, M.L. Barreto, Multiobjective grammar-based genetic programming applied to the study of asthma and allergy epidemiology. BMC Bioinform. **19**(1), 245 (2018)
25. https://tradingeconomics.com/india/government-debt-to-gdp. Last visited 22 Apr 2020

Cataract Eye Detection Using Machine Learning Models

Ankush Shetty, Kunal Bathija, and Priya R. L.

Abstract A cataract is clouding of the eye lens which results in decrease of vision. The existing systems are limited to use of small size image datasets resulting in lesser accuracy, and user-friendly application is not available. The proposed model is designed to use image classification models to differentiate a healthy eye and an eye with cataract. To classify types of images, it uses the VGG16 model. This model has 16 layers. The image is convoluted and pooled at each layer of the model. With the VGG16 model, the rural denizens will be able to determine whether a person is suffering from cataract or not, without consulting an ophthalmologist. The output is then classified accordingly. The proposed system provides a UI for the detection of cataract disease.

Keywords Cataract detection · Classification · VGG16 model for classification

1 Introduction

Cataract is a serious eye disease, which has affected over 65.2 million people worldwide. Cataract is produced due to clouding of the eye lens which results in blocking the light to go through the lens and project image on the retina. This results in partial transfer of the whole image to the brain which leads to blindness.

According to the National Blindness and Visual Impairment Survey India 2015–19 "The major reason for blindness is cataract for people of 50 years and above. Cataract is the root cause behind 66.2% blindness cases, 80.7% severe visual impairment cases

A. Shetty (✉) · K. Bathija · Priya R. L.
Department of Computer Engineering, Vivekanand Education Society's Institute of Technology, Hashu Advani Memorial Complex, RC Marg, Collectors Colony, Chembur, Mumbai, Maharashtra 400074, India
e-mail: 2017.ankush.shetty@ves.ac.in

K. Bathija
e-mail: 2017.kunal.bathija@ves.ac.in

Priya R. L.
e-mail: priya.rl@ves.ac.in

J. Tavares et al. (eds.), *Emerging Technologies in Data Mining and Information Security*, Lecture Notes in Networks and Systems 164,
https://doi.org/10.1007/978-981-15-9774-9_84

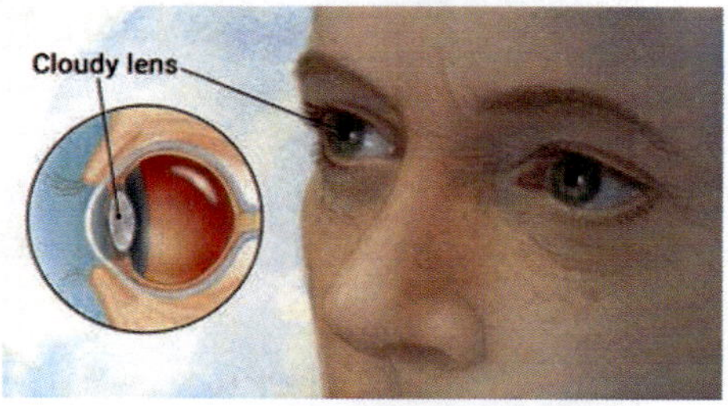

Fig. 1 Cataract visualization [2]

and 70.2% moderate visual impairment cases in the age group. India launched the National Programme for Control of Blindness in 1976 to reduce blindness frequency to 0.3% by 2020, but the estimated frequency of blindness still stands at 1.99%, 1.96% for severe visual impairment, 9.81% for moderate and moderate severe visual impairment at 11.77%, according to data released on October 2019." [1].

Figure 1 represents clouding of the lens which blocks the light to go through the lens resulting in partial or complete blindness depending upon the cataract formation on the lens.

The WHO released the World Vision Report in October 2019, which declared that eye care requires high cost, especially for the rural population, which gets majorly affected by visual impairment. This has resulted in expanding Universal Healthcare Coverage, including eye care services in it.

The experiment can make the future of classifying cataracts more easy, and ophthalmologists can perform appropriate operations to different categories of cataracts within a shorter time to cure patients suffering from cataracts. This project will help the ophthalmologists to identify the cataract diseases quickly and with more efficiency.

2 Literature Survey

The study of cataract has been going on for several years. The survey clearly indicates that the gist of the idea was that all the existing software uses limited dataset size which varies in a few hundreds. In most of those projects, they were concerned about detecting cataract using slit lamp images or fundus images. The graphical user interface is either missing or not user-friendly, i.e., a user-based cataract detection has not been made so far.

As discussed in the study [3, 4] used, only a few hundreds of images of fundus images require proper machinery as it is the back side of image thus having a drawback that a naive user cannot make use of it. The system is defined [5] using feature extraction of 5378 images classified by RNN algorithm. Few more studies [6, 7] classified the eye diseases into three types by using single perceptron training model. The system [3] obtained accuracy of 90.82%. This paper based on the data has produced the said accuracy, and the issue faced is limited data.

Table 1 Comparison between existing system

	Paper [8]	Paper [6, 7]	Paper [3]	Paper [5]	Paper [4]	Proposed system
Accuracy	96.1%	94.69%	90.82%	Only feature extraction	95.479%	Depends on data
Type of image used	Slit lamp	Slit lamp	Fundus image	Slit lamp	Fundus image	Regular eye images
Classification in	3 types	3 types	4 types	1 type	2 types	4 types
Algorithm used	SVM	Single perceptron training model	CNN	RNN, SVR	Decision tree, BPNN, SMO	CNN VGG16 model (16 layers)

In Table 1, we see that in most of the existing systems, datasets used as system input were slit lamp images or fundus images but not both. Accuracy of detecting cataract will increase one and only if both types of images are considered for the single eye. Various machine learning algorithms used in the previous models are SVM, CNN, RNN, SVR, BPNN, decision tree, etc.

Classification also varies among different systems as few of them have classified into two types or three types or others into four types.

3 Proposed System

The proposed system is used to detect the cataract disease using the machine learning models.

3.1 System Architecture

The detection model is depicted in Fig. 2 to define the flow of the proposed system. The first step to get access to the provided facilities is to complete the registration process as a patient. The registration process reads patient demographic information such as age, lifestyle, habits, smoking/alcohol, family history of cataract, had an eye injury or surgery, medications like steroids being used, and other diseases. Once the authentication of the user is successful, then the user will be transmitted to the home page which would have rudimentary information about various eye diseases and how to prevent them along with the symptoms of each eye disease.

Any novice user can also check whether he is suffering from cataract or not using the proposed system. All he/she has to do is to click an image of the eye and upload. Then using the VGG16 model, the image will be processed and detection of cataract

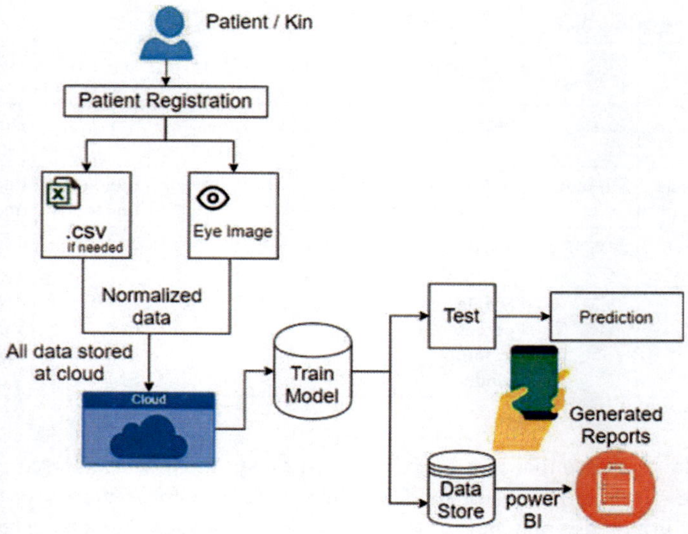

Fig. 2 System architecture of detection model

takes place. If the user is suffering from cataract, then he would be suggested to visit the nearest ophthalmologist along with the address of the clinic.

All the data along with the personal information and the data predicted from the eye image will be stored in the cloud database.

3.2 Methodology

The procedure of the proposed system used to detect cataract involves various subprocesses. Subprocesses involved in detecting cataract disease are depicted in modular diagram as shown in Fig. 3.

3.2.1 Load Image

The image on which users need to perform detection algorithms is first uploaded into the system. The image is taken with JPG extension and will be resized according to system requirement.

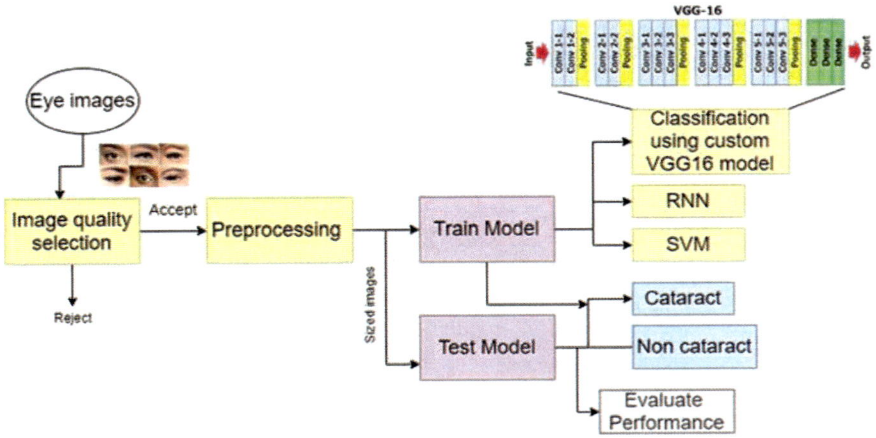

Fig. 3 Modular diagram

3.2.2 Preprocessing

Preprocessing is used to reduce noise [9] and improve desirable qualities in an image. The preprocessing techniques such as grayscaling are applied in the input image.

3.2.3 Gray Scaling

Before the image is fed into the model, it is necessary to initially change the color of the image for quick and efficient processing of the image. Hence, the grayscaling method is applied. The reason for using grayscaling is that it simplifies the algorithm and reduces the computational requirements because with color images it provides some extra information which would increase the time for training a model.

3.2.4 Segmentation

Finding a pupil from the image is quite difficult [10] so to make it easier OpenCV has introduced. The segmentation process is carried out using HoughCircles. HoughCircles are used to detect circles from the image. This function helps to find pupils from the image and then detect whether a person is suffering from cataract or not.

As represented in Fig. 4, it represents how the pupil can be detected from the image of an eye.

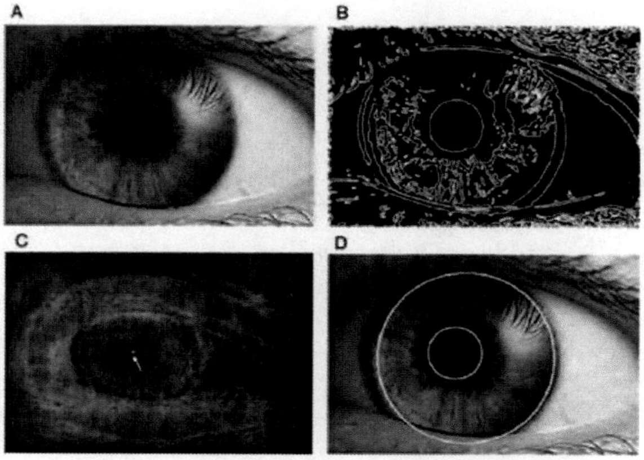

Fig. 4 HoughCircles algorithm [11]

3.2.5 Processed Image into the Trained Model

Once the model is trained properly using eye image datasets, the system obtains a preprocessed image which can be fed into the model for detecting whether a person is suffering from cataract or not.

3.2.6 Classification

After the detection of cataract disease, the proposed system helps to classify it to one of its types.

The types of cataract are:

- Nuclear sclerotic
- Cortical
- Posterior subcapsular.

The VGG16 model is used with 16 layers. It is considered to be one of the most effective vision model architectures till date and used in many real-time healthcare applications.

3.3 VGG-16 Model

VGG16 model is a convolutional neural network (CNN) architecture. VGG of VGG16 is an acronym for the Visual Geometry Group. The VGG16 model is used for image classification. The 16 in VGG16 comes from the fact that it has 16 layers that have weights. This model achieved 92.7% test accuracy in ImageNet, which consists of a dataset of over 14 million images which belongs to 1000 different classes. It supplants the AlexNet by replacing large kernel-sized filters with multiple 3 × 3 sized filters. The reason for using the VGG16 model is its simplicity to classify images which use only 3 × 3 convolutional layers which are placed on top of one another. A dense layer represents a matrix vector multiplication (assuming your batch size is 1).

Figure 5 shows various layers present in VGG16 model. This model consists of various convolutional layers, i.e., max pooling layers, activation layers, and fully connected layers. Figure 3 represents 13 convolutional layers, 5 pooling layers, and 3 dense layers which sum up to 21 layers but only 16 weight layers. The convolutional layer consists of 3 × 3 size whereas max pooling layer consists of 2 × 2 sized filters.

Figure 6 represents the architecture of the VGG16 model which displays the number of filters in each layer. Using Figs. 3 and 4, we find that there are 64 filter layers in Conv 1, 128 filter layers in Conv 2, 256 filter layers in Conv 3, and 512 layers in Conv 4 and Conv 5.

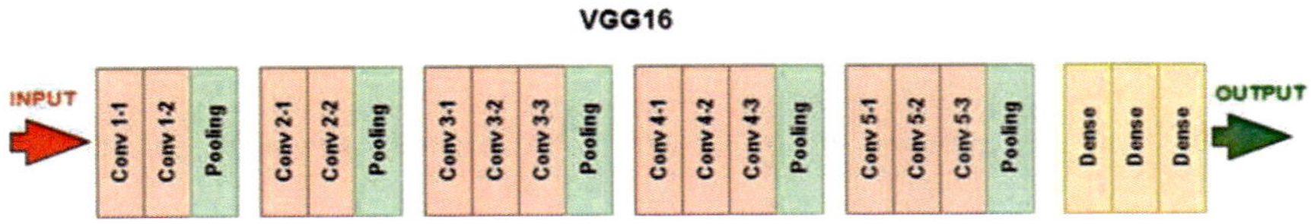

Fig. 5 VGG16 layer definition

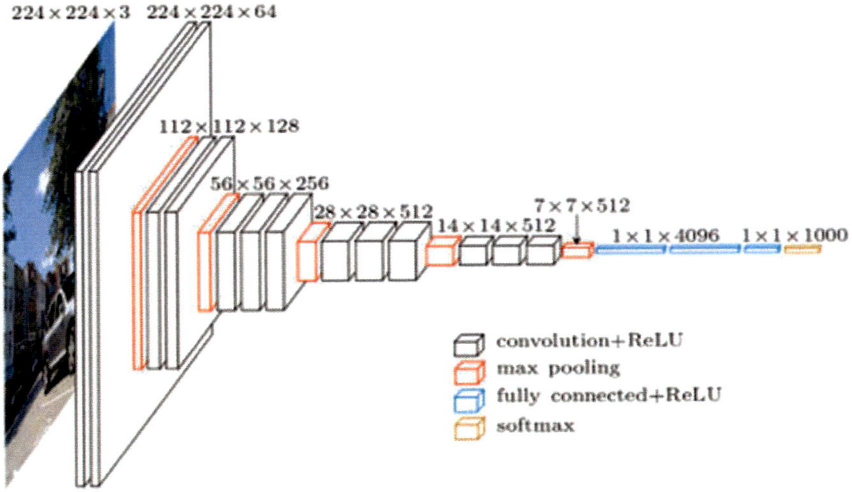

Fig. 6 VGG16 architecture [12]

4 Implementation

The proposed system is being carried out on Google Colab and it is using Python script to get all the cataract-related images from the Internet and it downloads those images. The VGG16 model is used for cataract detection using the downloaded images. As the images are of open type, the system uses all the images and thus resulting in loss again resulting in accuracy of 74.98%.

5 Result and Analysis

Initially we trained a model with 4–5 layers of pooling and convolution. The accuracy was close to 0.52 with loss at 0.735. We then trained the VGG model with weights pretrained and having 16 layers of pooling, convolution, and a dense layer at end. The accuracy came around 0.7 and loss in the order of e-4 on validation data. By varying epochs in between 4 and 30, we found the model maxed out accuracy at 7 epochs.

Figures 7 and 8 represent the model accuracy and model loss graphs, and the bifurcation of train and test is represented in the same figure.

Figure 9 represents the classification of eye images into cataract eye represented by [1.0] and regular eye represented by [0.1]. Figure 10 represents various layers of VGG16 model and gives brief description of those layer.

Fig. 7 Model accuracy

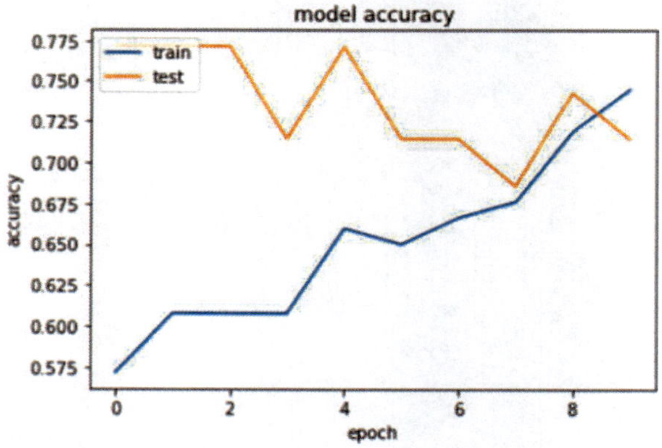

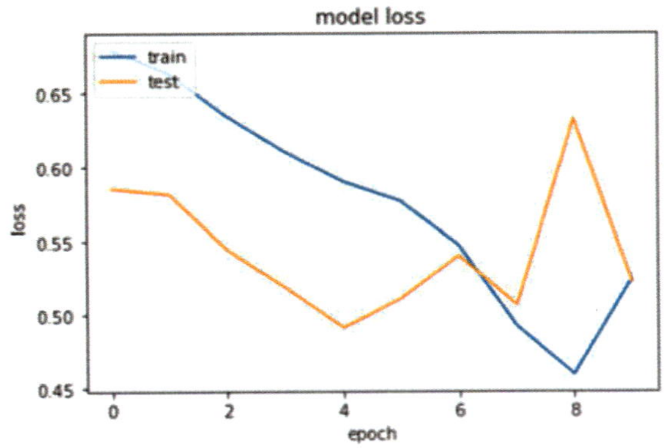

Fig. 8 Model loss

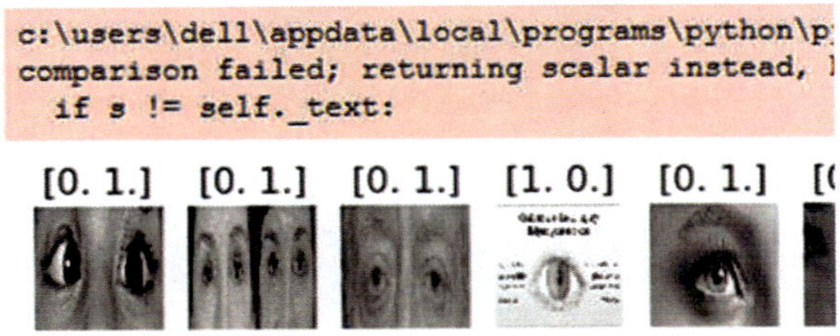

Fig. 9 Data being labeled by the model as cataract or non-cataract

6 Conclusion

A user interface for detecting cataract eye disease is thus developed and with sufficient dataset belonging to each stage of cataract, the proposed system will train the model not only to detect but also to identify the stage of cataract a person might be suffering.

Fig. 10 Weight parameters in each layer of model

Layer (type)	Output Shape	Param #
block1_conv1 (Conv2D)	(None, 224, 224, 64)	1792
block1_conv2 (Conv2D)	(None, 224, 224, 64)	36928
block1_pool (MaxPooling2D)	(None, 112, 112, 64)	0
block2_conv1 (Conv2D)	(None, 112, 112, 128)	73856
block2_conv2 (Conv2D)	(None, 112, 112, 128)	147584
block2_pool (MaxPooling2D)	(None, 56, 56, 128)	0
block3_conv1 (Conv2D)	(None, 56, 56, 256)	295168
block3_conv2 (Conv2D)	(None, 56, 56, 256)	590080
block3_conv3 (Conv2D)	(None, 56, 56, 256)	590080
block3_pool (MaxPooling2D)	(None, 28, 28, 256)	0
block4_conv1 (Conv2D)	(None, 28, 28, 512)	1180160
block4_conv2 (Conv2D)	(None, 28, 28, 512)	2359808
block4_conv3 (Conv2D)	(None, 28, 28, 512)	2359808
block4_pool (MaxPooling2D)	(None, 14, 14, 512)	0
block5_conv1 (Conv2D)	(None, 14, 14, 512)	2359808
block5_conv2 (Conv2D)	(None, 14, 14, 512)	2359808
block5_conv3 (Conv2D)	(None, 14, 14, 512)	2359808
block5_pool (MaxPooling2D)	(None, 7, 7, 512)	0
flatten (Flatten)	(None, 25088)	0
fc1 (Dense)	(None, 4096)	102764544
fc2 (Dense)	(None, 4096)	16781312
predictions (Dense)	(None, 1000)	4097000

Total params: 138,357,544
Trainable params: 138,357,544
Non-trainable params: 0

References

1. Dr. Rajendra Prasad Centre for Ophthalmic Sciences, AIIMS, New Delhi: National Blindness & Visual Impairment Survey India 2015–2019—A Summary Report
2. C.P. Niya, Automatic cataract detection and classification systems: a survey. TechS Vidya e-J. Res. **3** (2014–15), 28–3 (2015). ISSN 2322 - 0791
3. Y. Dong, Q. Zhang, Z. Qiao, J.-J.Yang, Classification of cataract fundus image based on deep learning, in 2017 IEEE International Conference on Imaging Systems and Techniques (IST) (2018)
4. N. Hnoohom, A. Jitpattanakul, Comparison of ensemble learning algorithms for cataract detection from fundus images, in 2017 International Computer Science and Engineering Conference (2018)
5. X. Gao, S. Lin, T.Y. Wong, Automatic feature learning to grade nuclear cataracts based on deep learning. IEEE Trans. Biomed. Eng. **62**(11) (2015)

6. R. Sigit, M.B. Satmoko, D.K. Basuki, Classification of cataract slit-lamp image based on machine learning, in 2018 International Seminar on Application for Technology of Information and Communication (iSemantic) (2018)
7. M.A.V. Patwari, M.D. Arif, M.N. Chowdhury, M. Nurul, A. Arefin, M.I. Ikhwanul, Detection, categorization, and assessment of eye cataracts using digital image processing, in The First International Conference on Interdisciplinary Research and Development (2011)
8. X. Qian, E. W. Patton, J. Swaney, Q. Xing, T.H. Zeng, Machine learning on cataracts classification using SqueezeNet, in IEEE 4th International Conference on Universal Village, UV2018, Session 2–4 (2018)
9. S. Loffe, Red eye detection with machine learning, in Proceedings 2003 International Conference on Image Processing: FUJIFILM Software, Califomia (2003)
10. S. Y. Gwon, C. W. Cho, H. C. Lee, W. O. Lee and K. R. Park, Robust eye and pupil detection method for gaze tracking. Int. J. Adv. Rob. Syst. (2013)
11. X. Wang, X. Du, The application of SVM in automatic Eye Position Algorithm, in 2009 2nd International Conference on ICT, vol. 1 (2009)
12. K. Simonyan, A. Zisserman + Visual Geometry Group, Very deep convolutional networks for large scale image recognition, in 2015 International Conference on Learning Representations, Department of Engineering Science, University of Oxford (2015)

Automated High Precision Path Annotations on Digital Map from Crowdsourced GPS Traces

Siddhartha Chatterjee, Saikat Samanta, and Ratna Mandal

Abstract Developing reliable and well-informed mobile phone-based transportation is followed by growing popularity and accessibility of digital maps, which has definitely simplified navigation. But that comes with severe drawbacks in the case of narrow unknown paths which have no place in these maps. The inevitability of poor path geometry precision and fine-grained scarcity knowledge is hampering a route guidance system. In this paper, we are offering an approach to construct high-precision maps from traces of crowd equipped with differential GPS receivers with updated existing path along with addition of new path segment scaled according to the exact width and shaped on electronic map. As the cost of GPS-enabled devices is decreasing rapidly, we assume that large amounts of GPS traces can be obtained with sufficient confidence to build up a new road section. Our approach is to make a server side application where the crowd traces will be uploaded through an user-friendly graphical interface and a fine-grained map standardized by human labeling will be generated. This paper describes an offline algorithm which generates an enhanced map by a certain time gap in a contiguous manner. Our new contribution is a spatial community detection of GPS coordinates from all uploaded traces subjected to a threshold value which is crucial to select as it varies with the width of path and volume of collected information from crowd.

S. Chatterjee (✉)
Department of Computer Science and Engineering, Gargi Memorial Institute of Technology, Kolkata, West Bengal 700144, India
e-mail: siddhartha.chatterjee31@gmail.com

S. Samanta
Department of Computer Science and Engineering, Bengal College of Engineering, Durgapur, West Bengal 713212, India
e-mail: s.samanta.wb@gmail.com

R. Mandal
Department of Computer Application, Institute of Engineering and Management, Kolkata, West Bengal 700091, India
e-mail: ratna.mandal.iem@gmail.com

J. Tavares et al. (eds.), *Emerging Technologies in Data Mining and Information Security*, Lecture Notes in Networks and Systems 164,
https://doi.org/10.1007/978-981-15-9774-9_85

Keywords GPS · Map matching · Annotation · Topology · Crowdsource

1 Introduction

Due to availability of digital maps Google [1], bing [2], OpenStreetMap, OpenHeatMap, etc., and several GPS-based applications like navigation, location tracking, route planning and recommendation, traffic analysis and forecasting, transport is now quite convenient throughout the globe. However, in many developing countries like India people still end up with several bitter travel experiences. Many a times people have a nightmarish experience when one hires a metered cab to reach to some unknown destination. In majority of the cases, one is left at the mercy of the driver who inevitably takes a much longer route, charging the passenger much more than the actual fare. Locating one's friend's house in a densely populated locality can also be a challenging task. The reason is although the digital maps of developing countries are not equipped with adequate information and are incomplete.

Maps in such developing areas hardly come up with details information of traffic direction, lanes details, permissible modes of transport, etc. Moreover, it is observed many of the path segments popularly traveled by pedestrians/two wheelers used to cut short the time and distance are found missing in either map view or satellite view. So these shortcuts are not taken into account by the subsequent algorithm suggestion of routes which are rarely followed in practice. As a consequence, these so-called shortest routes are not useable for navigation and tracking purpose as shown in Fig. 1.

Figure 1 illustrates an area in Kolkata, (a city in eastern India). The route from A to B is misguided but the reverse route from B to A is perfectly recommended. The reason behind this is that the route from B to A is labeled as one-way, whereas it is actually a two-way path.

Figures 2 and 3 show an area in a small town Durgapur in eastern India. Here, the shortcut pathways used by pedestrians and two-wheelers are partly visible in Google satellite view and are not available in map view. Thus, the actual path (red) followed is less than half of the recommended path (blue).

So, can these things be improved in practice? Is there any technological approach to do away with this issue, if we assume that people are able enough to afford at least a Java-based and GPS-enabled mobile phone at reasonable cost because the cost of these systems does decrease rapidly? Now who will feed this fine-grained information about existing and of entire territory? No doubt it will be the crowd; passing through the paths. Mao [3], Wu [4] have relied on crowdsourcing for definite confirmation about accuracy of map. So, the information will be collected from GPS traces of all travelers irrespective of his/her mode of transport. However, there is also a challenge with acquisition of information. Majority of common people are not so tech-savvy and also will not bother to waste any effort for acquisition purpose of information about his traversal. Hence, the process should incur zero cost as well as least effort. In addition, user-friendly graphical interface should be accessible for system interaction, so that general citizens feel free to participate and act as a

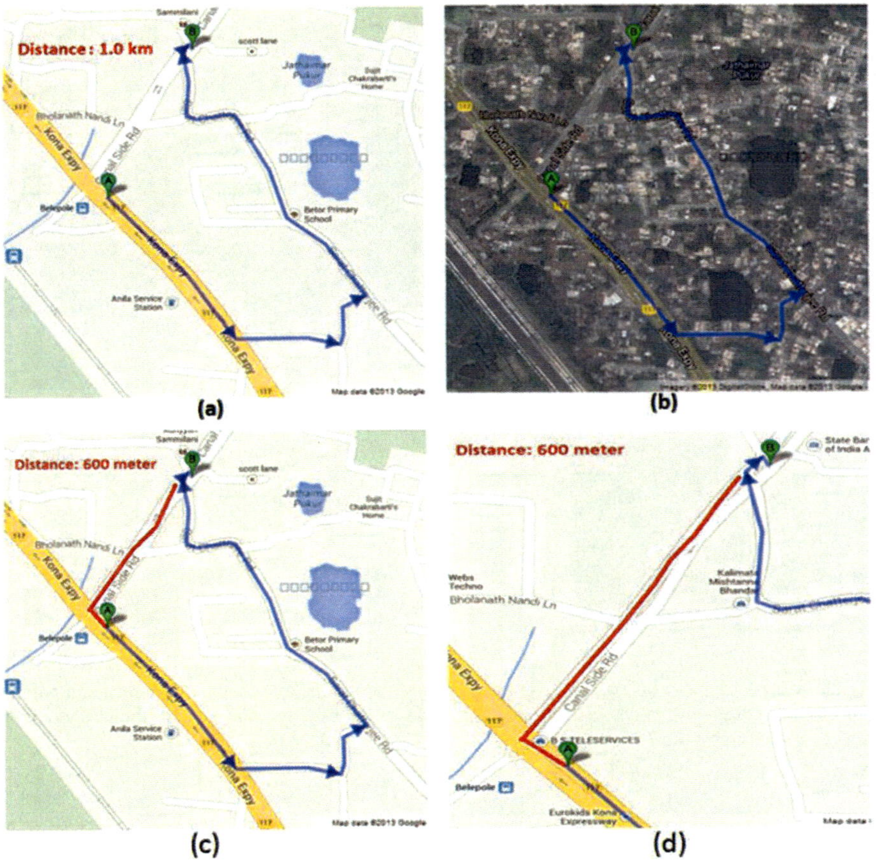

Fig. 1 **a** Wrongly recommended route. **b** Satellite view. **c** Short route which people will follow. **d** Zoomed view of alternate route

volunteer in the process. After the acquisition of information, those data are stored and kept in the system followed by proper authentication. For an individual, it may be a personalized system to upload the detail topology from the nearest landmark to his home in server for providing assistance to his friend on downloading the map from server. If he is even prior instructed the direction to home then also there are every chance to get confused among several lanes, by lanes or streets of a colony because no such information can guide him with respect to his current position. Things might be more complicated if he comes during any odd hour to have a complete path direction as nobody will be found on road at that time.

Therefore, how crowdsourced information can be authenticated, visualized, and be navigated?

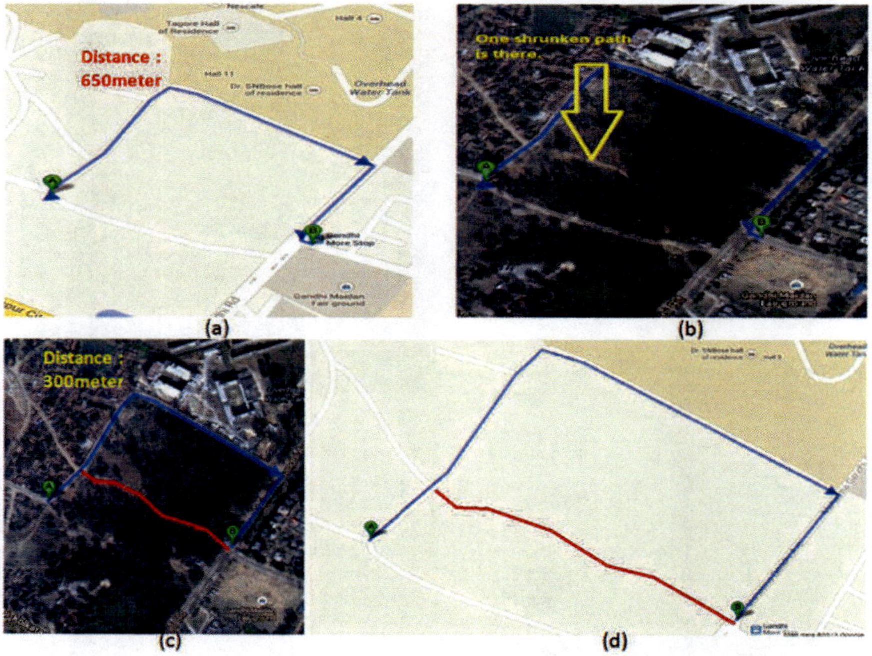

Fig. 2 **a** Invisible path on map. Recommended route is longer main road. **b** A shrunken path is visible in satellite view. **c** The path can be explored from crowdsource as shown in red. **d** Zoomed view of crowdsourced path

Generally, authentication process will be carried out by social domain but it is very much dependent on volume of data because crowdsourcing is a gradual process and will take some time to be well populated. So, initially due to the sparseness of data it cannot be guaranteed confidently and authentication may be some sort of a crucial issue.

Next comes the visualization task issue. Any standard digital map can be deployed here. Google Map is one of the most popular electronic maps. Our approach is to incorporate a layer on Google Map. There are several classical map-matching algorithms for basic purpose and these approaches are well described in section of literature study. But map-matching algorithms prove to be successful where a path exists on digital map. However, the problem becomes non-trivial where maps are incomplete. The paths either do not hold on map or not visible. Classical map-matching algorithm fails in such scenario where no base is available to match. Digital map provides the route in between two landmarks but visualization of route between any two intermediate points on a road segment extracting the actual route topology is non-trivial.

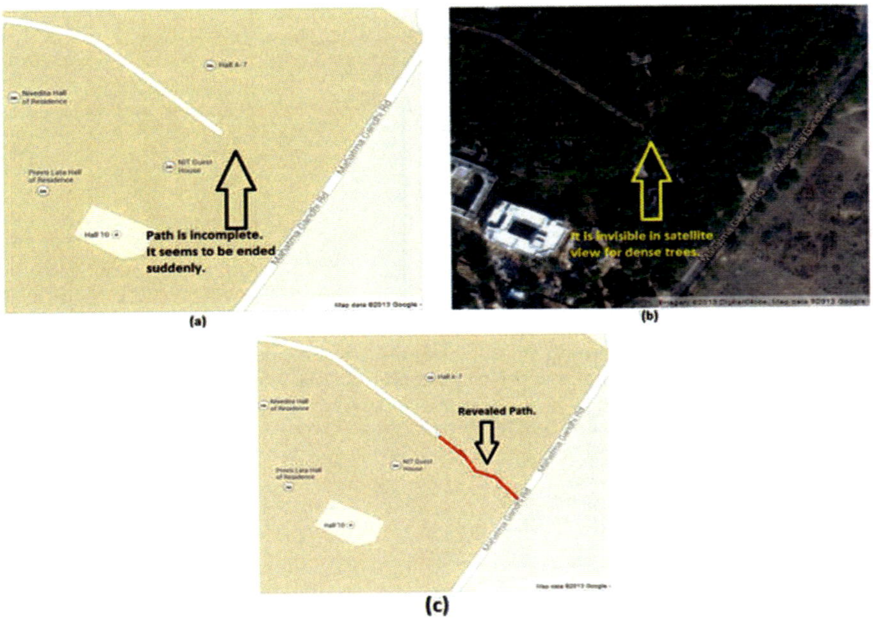

Fig. 3 **a** Invisible path segment on map with sudden end. **b** Invisible in satellite view for dense trees. **c** Actual path is drawn in red on map

This paper is primarily focused on map visualization through crowdsource which can be an added information and can be used for annotation and route recommendation.

The next focus is on navigation over this crowdsourced paths as well as on existing digital paths. How can one detect whether he/she is on right track or not? As new paths are built from multiple traces of GPS, who path to match when navigating the new route? One can think that, if there are multiple trails along the path then anyone can be matched as described in Fig. 4.

All trails from crowd should be taken into account to get the exact pattern rather than following a single one. If we think about navigation by using sensor reading without help of GPS then what modified information is required for these fine-grained map? Henceforth, only feeding the crowd information is not the end but the paths should have the directional/bearing information with respect to the North Pole for movement along with the magnitude value, i.e., distance. The paper remains organized according to different part discussed in the following. Section 2 summarizes the prior related work in detail. Section 3 provides the system architecture design. Section 4 closures the algorithm with description. Section 5 demonstrates the evaluation of our work with results and analysis. We roll up our paper and present the way ahead in Sect. 6.

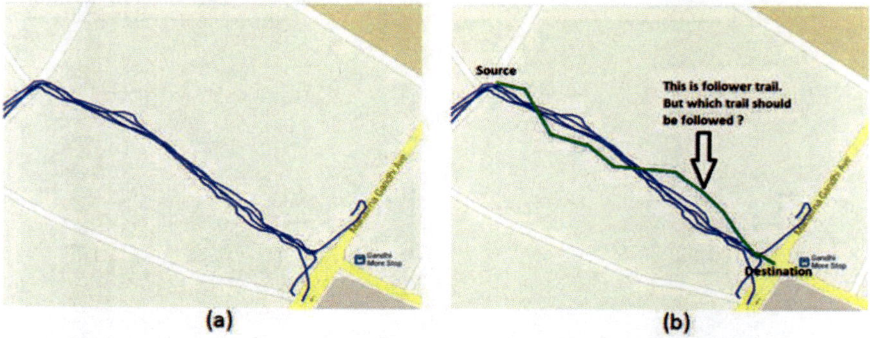

Fig. 4 **a** Multiple trails from crowd. **b** Which particular trail will be followed by new trail?

2 Related Literature

Inadequate digital maps are not necessary in automotive protection and convenience for any advanced applications. In 2002, Schroedl [5] provides a snapshot of a program that automatically produces digital road maps that are considerably more accurate and include lane layout details, including the number of lanes and their positions, as well as detailed intersection structures. Lane positions are determined by a spatial community detection algorithm to infer from scratch the maps connectivity structure, making the map slimming method entirely independent of the initial input map. Map-matching algorithm maps the GPS coordinates into a point of road segment. Online map-matching algorithm for GPS navigation means the matching is done based on currently received and past data. But the offline algorithm benefits from looking forward to the future stages [6]. From a literature study, it can be concluded that most of the real-time application prefers the online map matching supported by some sensors. But it produces inaccurate result due to the errors in sensor readings. Henceforth, offline approach is preferable where post-processing is possible because it gives less error than online one. The Marchal algorithm is a well-known offline algorithm which claims to be faster in computation rather than to be accurate in performance. Pereira [6] has modified the algorithm by setting a threshold value on the deviation between geographical point on map and its corresponding GPS point and checking its sustainability for subsequent matches at initial matching stage. Lou [7] has developed a map-matching system, which enables the GPS interface at long intervals to save resources. Biagioni [8] have categorized the map-inference algorithms into three classes and implemented those as open sources for a qualitative, quantitative, and comparative evaluation. The categorization is based on algorithmic foundations: (a) k-means algorithm, (b) trace merging, and (c) kernel density estimation (KDE). Griffin [9] has talked about standardizing a route by an offline map-matching algorithm. The algorithm decreases the number of points to be processed while preserving the accuracy of the routes returned by eliminating sensitive waypoints that result in incorrect matching of the map. Mao [3] has created an interactive map-matching machine truth-finder for exploring ground truth data with reduced human interaction costs.

Wu [4] introduces an approach of using statistical analysis system integrated with popular geospatial tools, statistical analysis framework for mapping GPS data on roadways and reducing a significant amount of GPS points to heat maps showing the frequency with which road segments are crossed.

From this literature study, we can conclude that, most of the authors have suggested an offline approach because an online algorithm is one that can process the input in a serial way, i.e., in order to feed the input to the algorithm, without making the entire input available from the beginning. By comparison, the entire problem data is given to an offline algorithm from the start and can solve the problem at hand. This prior work gives a clear view that a lot work is done to match a trail with available maps and some has targeted to do it with crowdsourcing. But almost all of them have raised the problem of all kinds transport convenience like navigation, traffic analysis, etc., on less-annotated and less-informative map.

3 System Overview

Motivated by the previously mentioned problem design, we are trying to formulate an interactive framework. The systems architecture is represented in Fig. 5.

Crowd—Pedestrians and vehicles equipped with a GPS-enabled PDA device.

Information acquisition—Collection of GPS traces from crowd and store in server.

Recommended path computation—Exploration of path from crowd recommendation.

Map generation—Fine-grained map generation with detail topological view.

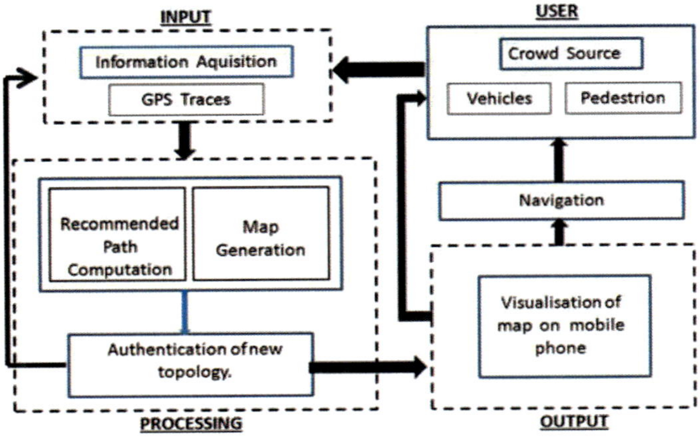

Fig. 5 System architecture

Authentication of new topology—The new path obtained from crowd needs to be authenticated by its social domain, i.e., the users. This part is important for the enhancement of the path and later processing.

Visualization of map on mobile phone—Now after the authentication the path will be displayed on the map scaled according to the existing path of map.

Navigation—Now we have a path on the map which is well authenticated and ready to be used for navigation. To build a navigation system, the crowdsourced data should be more informative because we intend to allow navigation with a low-end device at low energy and running cost. This navigation system will be made available to the user.

4 Algorithm

The objective is to approximately determine the width of the geographical path using numerous trails obtained from crowdsourcing. By crowdsourcing it implies collection of GPS trails obtained from large group of people or community using various modes of collection like by walking, riding a motorcycle, etc.

Speed is different for various modes or sources. So, for the sake of simplicity here we are considering only the trails gathered by walking. Numerous edges are formed with the coordinates lying in the proximity or threshold of each coordinate leading to formation of myriads of connected graph. These set of connected graphs are made closed using convex hull if the number of trails obtained is less than the optimum threshold, assumed to be 6. On the contrary when the threshold is more than or equal to optimum threshold then Louvain communities [10] are formed. Here, Louvain community detection is used as heuristic method based on modularity. In terms of computation time, it is shown to outperform all other established group detection methods. It breaks down the network into subunits or groups, collection of highly interconnected nodes. The consistency of the partitions that arise from these approaches is also calculated by the partitions so-called modularity. All the pairs of adjacent nodes form input to the Louvain algorithm which forms numerous communities based on the modularity. This community detection algorithm has not been extensively used in extracting topographical information from GPS coordinates as per our knowledge. Once the Louvain communities are formed, convex hull is created for every community. Convex hull is employed with an objective to close the connected graph formed in the process.

Algorithm 1: Generate Path Width
Input: *Crowd Sourced GPS Trails for a particular Geographical Path, Number of GPS Trails.*
Output: *Graph G (V, E , Set of Convex Hulls, Louvain communities.*
1. Store GPS co-ordinates for all the GPS Trails in a *Coordinate_List* array and Number of GPS Trails in a variable *Counter_trail*.
2. Assign *temp* variable with that value; which is set to detect a outlier co-ordinate.
3. Initialize *Threshold* value as the difference between *temp* and *Counter_trail* and greater than 2.
for *each ith co-ordinate in Coordinate_List* **do**
    for *each jth co-ordinate in Coordinate_List* **do**
        if *$i^{th} \neq j^{th}$ element* **then**
            Compute Distance between i^{th} and j^{th} element.
            if *Distance ≤ Threshold* **then**
                1. Include j^{th} coordinate in the ith Group.
                2. Add edge between j^{th} co-ordinate and the ith coordinate and create sub-graph of G.
            end
        end
    end
    if *Threshold ≤ Optimum_Threshold* **then**
        1: Reset *CrowdSource* variable.
        /*CrowdSource indicates the magnitude of GPS Trails*/
        2: *Polygon* = CreateConvexHull (i^{th} Group).
        /*CreateConvexHull generates a Convex Hull for each ith Group*/
        3: Display the *Polygon*.
    end
    else
        1. Set *CrowdSource* variable.
        2. create supergraph G combinning all subgraphs.
    end
end
if *CrowdSource is set* **then**
    Call Createcommunity(G) **/*Createcommunity applys Louvian community detection algorithm to community G communities*/**
end

Algorithm 2: Create_community
Input: *All Adjacent GPS Co-ordinates for each Co-ordinate i.e, GPSPoint_Graph.*
Output: *List of Pair of Co-ordinates i.e, Louvian Group, Louvian communities, Convex Hull for each community*
1. Initialize k as 0
2. List of Pair of Adjacent GPS Co-ordinates are stored in Louvian Group

for *each ith element in GPSPoint_Graph* **do**
    for *each jth element in GPSPoint_Graph* **do**
        1: append i^{th} and j^{th} element in pair to LouvianGroup[k]
        2: increment k by 1
    end
end

3. Louvaincommunities <- Call Form Louvaincommunity (LouvianGroup)
/*FormLouvaincommunity accepts all the pairs of Adjacent coordinate to form communities*/
4. Convex Hull is created for all the Louvian communities

for *each community in Louvaincommunities* **do**
    1: Polygon<-Call CreateConvexHull (community)
        /*CreateConvexHull generates a Convex Hull for each Louvian community*/
    2: Display the Polygon created in the above step
end

4.1 Description of Generate_Path_Width

All the GPS trails obtained are initially transformed into GPS coordinates and stored in a linear list. This linear list is to be used for creating graph having more than one component. Creation of graph is dependent on the threshold value, which refers to the limit used to determine whether there will be an edge between two GPS coordinates. It is defined as the maximum value between difference of maximum distance and number of GPS trails, and minimum threshold, assumed to be 2. Here, maximum threshold, for the convenience sake, is assumed to be 11 m which refers to the maximum width a geographical path can have. Now, for every coordinate, a connected graph is formed with the other coordinates, if those coordinates lie within the threshold value corresponding to that coordinate. This will eventually lead to formation of numerous overlapping connected graph as well as components. After formation of every connected graph, the threshold value is again checked and if it is less than optimum threshold then convex hull will be created for every connected graph. Conversely, if the threshold value is more than or equal to the optimum threshold then Louvain communities will be formed for every pair of coordinates incident on the same edge. This will be followed by creation of convex hull for each of the communities formed.

4.2 Description of Create_Community

All the groups or graph components are converted into edge representation, i.e., pair of coordinates incident on a same edge are formed. These pairs of coordinates are used as input to create set of communities using Louvain community formation algorithm. Convex hull is then created for each communities followed by displaying of the entire convex hull formed out of communities.

5 Result and Analysis

We present the experimental result in this section. We have two methodologies including two datasets. We collected our first dataset of road network around the campus and another one is outside the campus. Figure 6(1) shows three crowd-sourced trails on a particular path. In Fig. 6(2), threshold is 3 m and we observe that, a low threshold results a broken path, i.e., part of the path is significant to its topography or width is considered as an outlier.

Hence, the threshold must be increased. 6(3) show the same three crowdsourced trails of the particular path with clustering. Since important points in the graph are left out as outliers, community detection would also result in a heavily broken path. It may give some shape but these shapes would be separated by quite a distance and

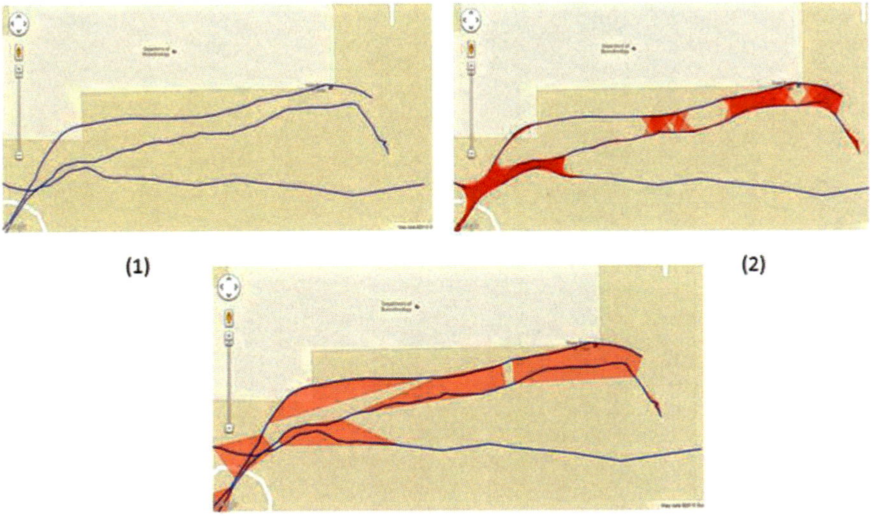

(1) (2) (3)

Fig. 6 Number of trails = 3 for threshold value 3

hence there would be ambiguity. Thus, we need to increase threshold and allow the path to grow to its proper shape.

In Fig. 7(1), there are three crowdsourced trails of a particular path. In Fig. 7(2), there are same three crowdsourced trails of the particular path. Here, threshold is set to 8. As compared to Fig. 6, we have noticed a significant improvement in the path since we have allowed the path to expand by increasing the threshold. From this figure, we have applied the convex hull algorithm to each vertex and its induced subgraph resulting in number of convex hulls. These convex hulls are nothing but the actual width of the path. Since we have only three trails, we can afford to create this many hulls. The complexity is low. In 7(3), the trail in the center represents the same three crowdsourced trails of the particular path. We can see that according to the three trails the lowermost one looks like an outlier or another path and it should be discarded. However, community detection would create a problem as even one point on this outlier path if considered would result it being a part of community formation. We create a convex hull on these communities and it may give us wrong results. The computation complexity is low but results are inaccurate.

Figure 8(1) shows eight crowdsourced trails on a particular path. Figure 8(2) of same eight crowdsourced trails on a particular path. We have applied the convex hull algorithm (without community detection) with a threshold of 3 m. We can see that as compared to the previous set where we had lesser number of trails, the width has increased significantly. We can see that the lowermost trail which is not a part of the shaded region is an outlier or may become part of the path more trails are added. We have formed a number of convex hulls (one for each vertex and its induced subgraph).

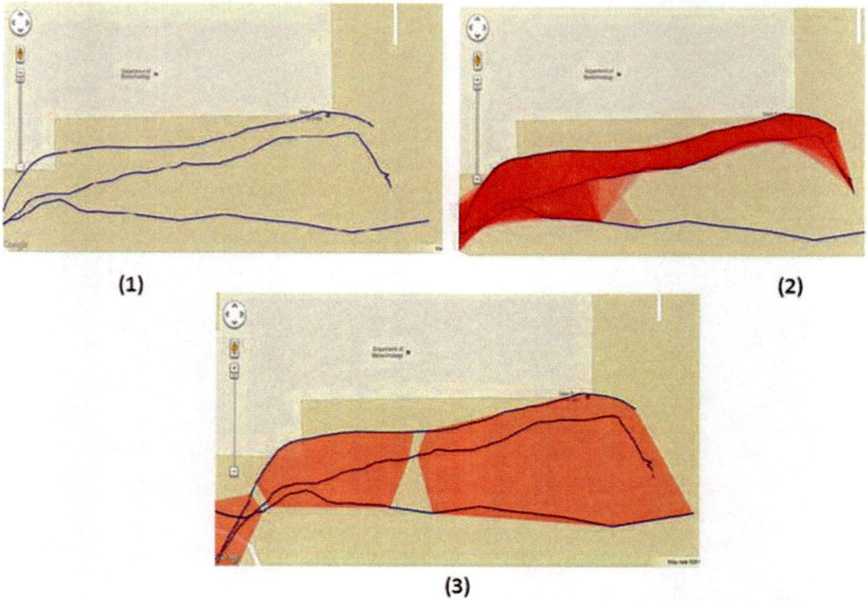

Fig. 7 Number of trails = 3 for threshold value 8

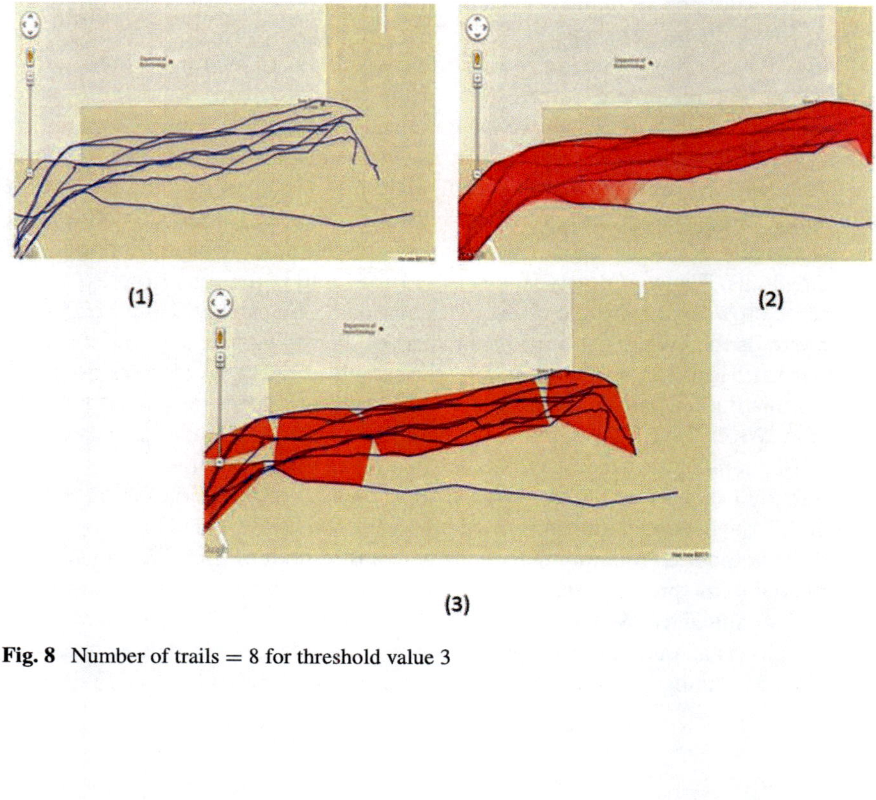

Fig. 8 Number of trails = 8 for threshold value 3

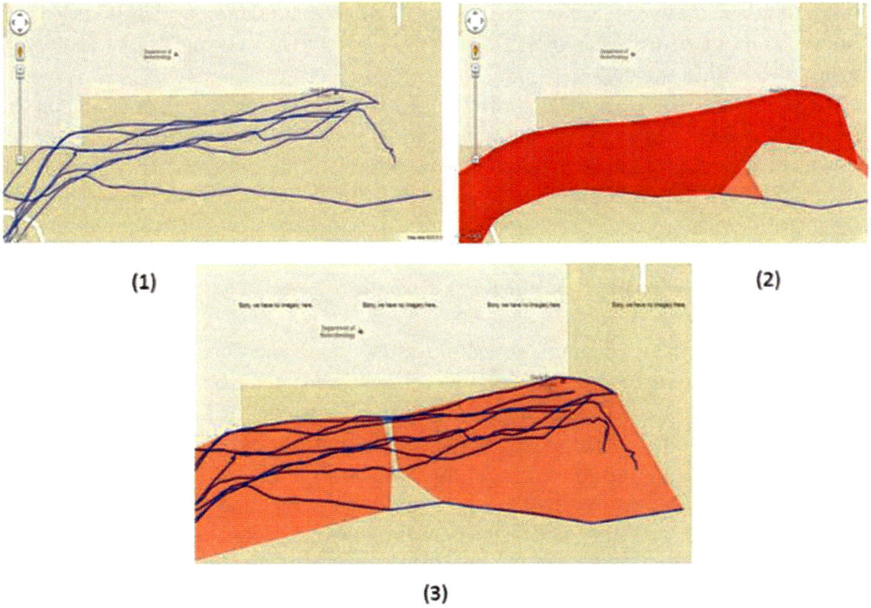

Fig. 9 Number of trails = 8 for Threshold value 8

Hence, computation complexity is high. We have taken a small threshold as the path is already heavily crowdsourced and its tendency to grow would reduce. 8(3) shows the same path with communities formed on the graph. We can see that community detection causes nodes to accumulate and hence we only require a few (in this case 5) convex hulls. Hence complexity reduces.

9(1) shows the trace of eight crowdsourced trails of a given path. 9(2) shows the trace of the same eight crowdsourced trails of the same path. We have used a threshold of 8, i.e., we have allowed path to expand heavily. Since we already have a path with a large width, giving freedom to expand with such a high threshold results in large number of outliers and an unusual increase in the width of the path. The convex hull created considers point's way outside even though the path has attained a large size. This results in a very wide path giving erroneous results. Hence threshold should be reduced. In 9(3), the third trail in the bottom is the same eight crowdsourced trail of the same path. There we have allowed a graph to be created based on threshold and applied community detection. Later we form a convex hull on those communities. It has a low computational cost. In this case, since we have such a large threshold, there are outliers in graph and community detection yields wrong results. Hence, we get wider path than expected. Hence, we must reduce the threshold so as to restrict the path from further growth.

The threshold can be considered as the limiting factor that controls the tendency of a path to grow. If we have a low threshold, we allow the path to expand by connecting to distant coordinates and including it as a part of the path. If the threshold is raised, then we start to restrict the path from further growth by taking only the nearby coordinates within the path.

The threshold would vary depending on the number of trails uploaded, i.e., on the strength of our dataset. A larger dataset means it is well crowdsourced while a small dataset means it is less crowdsourced. If we have a large number of trails or large dataset, then we reduce the threshold so as to restrict the path from further growth. In the result, we apply a clustering technique to reduce the computation overhead as we have a large dataset. When we have a small dataset or less number of trails, then we increase the threshold and allow the path to grow as clustering with small dataset may create wrong clusters.

Hence, we need an optimum threshold after which we consider a dataset well crowdsourced. This optimum threshold was observed to be 6, i.e., we considered that a large dataset would constitute of more than six trails. Trails less than six were found to be fewer crowdsourced and constitute a small dataset. Threshold is initially set to detect a GPS outlier. In our case, this was observed to be 10 m, i.e., GPS trail can give a maximum error of 10 m as compared to the original path. As the trace of that particular path gets enriched with more trails, we start decreasing the threshold. We have reduced the threshold by 1 for every GPS trail we upload except for first trail. When we upload the first trail then the threshold becomes the initial threshold. Initial threshold = 10 m. Threshold = 8 m. Number of trails uploaded = 3. Threshold = 3 m. Number of trails uploaded = 8.

The above continues until a point at which we assign a constant value of threshold no matter how many trails we get, i.e., we consider that path would grow very slowly at constant rate. This does not cause abnormalities since we already have such a large number of trails and shape would have become more or less constant. Hence, even if we allow a small constant threshold, the path would not change drastically. However, keeping such a small threshold would ensure that any change in the path in future would allow the change to take place (maybe slowly). According to our experiments, this constant threshold is attained when a minimum of nine trails are uploaded and the threshold attains a value of 2 m. We have taken another example and here we have four trails and Threshold = 6.

We can see that we get a perfect topology in case of Fig. 10(2). However, Fig. 10(3) has some outliers which have been included (in the initial part of the path) as we mentioned there would be since there are communities being formed for such a small dataset (less number of trails). Hence Fig. 10(2) is perfect for above example.

6 Conclusion and Future Work

In this paper, we seek to bring forward an interactive framework for uploading the crowdsource information on server for processing collected raw GPS trajectories. Generally, human asks other people who have traveled the place before for guidance to roam around any unknown place. Our system will implement the same cooperative

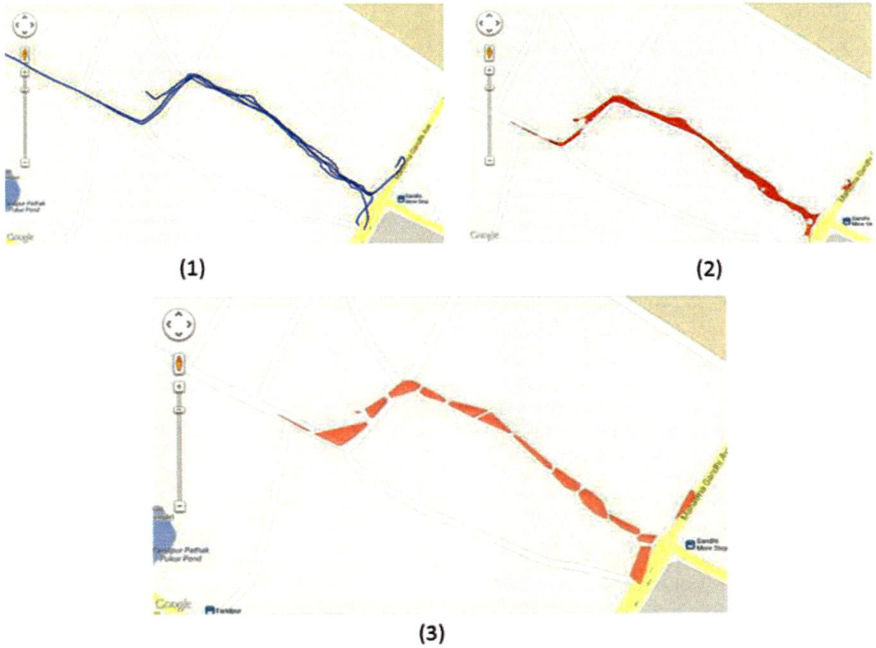

Fig. 10 Second dataset: number of trails = 4 for threshold value 6

human nature for creating a new trail. All GPS traces of travelers through a non-existing path on map will be collected through a cell-phone interface. We have figured out exact underlying pattern of path considering all footprints of all travelers along the path. Henceforth, the GPS coordinates are grouped; dense over a section of path. We have imposed a threshold value for this community formation. These subsequent communities will expose the continuous segments of path and thus the exact path pattern will be revealed. This newly exposed path will get enriched with more crowdsourcing gradually and this will be a continuous process running in server. The result shows that our algorithm nicely explores the unknown paths on map.

Crowdsourcing implies all modes of transports but in this paper we have taken only walking mode data into consideration. But, in future, our system will be able to process information from all modes of transport estimating different speed constraints.

To perform the task of navigation over this new map, it needs to be more informative and if we imagine of navigation only with compass sensor then how precise will be that information? Here, our target is to run the system at low-end devices with low execution cost. Hence, we may assume that only an Internet connection and compass sensor will be available as supportive. If we can store the topology with the directional value for a particular path segment, then it will be an easy task to walk along with compass only. We can add this informative layer on generated map. This segmentation approach will appear layerwise and user will get the more detail view as he will concentrate to the particular section.

References

1. http://www.bing.com/maps/
2. http://www.openstreetmap.org/
3. H. Mao, W. Luo, H. Tan, N. Xiao, *Exploration of Ground Truth from Raw GPS Data* (UrbComp12, Beijing, China, August 12, 2012). analytics.ncsu.edu/sesug/2012/RI-14.pdf
4. V.D. Blondel, J. Guillaume, R. Lambiotte, E. Lefebvre, Fast unfolding of communities in large etworks. J. Stat. Mech. Theor. Exp. **2008**(10) (1 Oct 2008)
5. S. Schroedl, K. Wagstaff, S. Rogers, P. Langley, C. Wilson, Mining GPS traces for map refinement. Int. J. Data Min. Knowl. Discovery **9**(1), 57–87 (2004)
6. F.C. Pereira, H. Costa, N.M. Pereira, An offline map-matching algorithm for incomplete map databases. Eur. Transp. Res. Rev. 107–124 (2009). (Springer)
7. Y. Lou, C. Zhang, Y. Zheng, map-matching for low-sampling-rate GPS trajectories, in ACM GIS '09 (Seattle, WA, USA, 4–6 Nov 2009), pp. 352–361
8. J. Biagioni, J. Eriksson, Inferring road maps from global positioning system traces. Transp. Res. Rec. J. Transp. Res. Board 61–71 (2012)
9. T. Griffn, Y. Huang, S. Seals, Routing-based map matching for extracting routes from GPS trajectories, in *Proceedings of the 2nd International Conference on Computing for Geospatial Research & Applications* (USA, 2011)
10. https://maps.google.co.in/maps/

Amplifier to Strengthen the Microvolt Signal for Biomedical Sensors

R. Chaitanya and R. Puviarasi

Abstract Aim of the project is to design an amplifier that is capable of taking a weak, low voltage input signal and strengthening it to provide a stable and amplified signal for better use. The project aims to function with biomedical sensor implants that are used to record and report on certain parameters of the human body. Considering a heart patient that requires a pacemaker to live, the amplifier would be capable of improving upon the quality of input to the pacemaker and thus help it provide a better grade of service for the patient. The design would thus have to be fabricated as miniature as possible to provide for smooth installation into the patient's current system and also need to provide a quality and stable output.

Keywords System · Device · Amplifier · CMRR · Gain · Noise · Sensor · Biomedical · Signal · Instrumentation amplifier · Impedance · Schematics

1 Introduction

Technology has grown immensely in every way of the world. The impact it has on the world over the past few decades has been exponential, and the requirement for additional growth is demanding. In today's society, there has been a great change in bio-electronics where sensors can be placed directly over the human body to perceive data through. There are varieties of sensors that are used for a different range of purposes, and the implementation of such technology is bound to grow as time moves forward.

R. Chaitanya · R. Puviarasi
Saveetha School of Engineering, Chennai, India
e-mail: chaitanyaravi739@gmail.com

R. Puviarasi (✉)
Saveetha Institute of Medical and Technical Sciences, Kuthambakkam, India
e-mail: puviarasi88@gmail.com

J. Tavares et al. (eds.), *Emerging Technologies in Data Mining and Information Security*, Lecture Notes in Networks and Systems 164,
https://doi.org/10.1007/978-981-15-9774-9_86

Many of the sensors being implemented are able to pin point their objective measurement values to certify at least 70% accuracy. However, in more critical cases, the measurement would need to be more accurate, and to do so, the implanted sensor would need to be able to take readings at the microvolt level. Consider an implanted sensor to check the sugar levels of patients affected by diabetes. The hypothetical sensor would be made to function through the mouth and be used to check the sugar levels through the saliva of the patient under observation periodically. It is also to be designed such that it would provide a detailed analysis and report on the patient's condition with respect to the time of the test. In such a scenario, the corresponding result will not be affected by any microvolt readings or fluctuations. However, if you were to consider other implants, such as neural implants, it would be highly affected by the voltage fluctuations or through the electromagnetic radiation that surrounds the patient. The implementation of such an amplifier to most devices would provide a better and higher quality output for the patient's condition. It would be used to improve upon the input fed into most sensors and reduce the error in the readings.

The main aim behind this project is to design a class A microvolt amplifier that is able to strengthen the weak input signal provided to it and provide a reasonable and stable output with minimal error and noise from the surrounds. The design is required to also be as small as possible without affecting its responsiveness to the biomedical sensor. It would be extremely useful to strength any weak signals or distorted signal and act as a filter to the background radiation that could potentially affect the results of most sensors. The amplifier could also be implemented into scanners to provide a higher resolution scan of the subject.

2 Literature Survey

In [1], R. F. Coughlin and F. F. Driscol discuss the characteristics of an operational amplifier and the working of it. The article provides details as to how the op-amp functions and the circuit configurations and classifications. The op-amp is a circuit very commonly used for VLSI design and can be used to design a variety of sequential and combinational circuits. Through the use of op-amps, the circuit parameters could be easily controlled. And due to its near infinite gain, in the case of CMOS op-amps, it provides us with insight as to how the low voltage amplifier could be designed.

In [2], Yu-Ming Hsiao, Miin-Shyue Shiau, Kuen-HanLi, Jing-Jhong Hou and Heng-Shou Hsu design a bio-amplifier using a cascaded network of amplifiers that consisted of multiple stages through the network and were able to amplify the provided input once through each stage and also provide a feedback loop from the output of each stage back into the first amplifier. In doing so, they were able to use a network of phase shifters and voltage controlled oscillators to stabilize the resultant output while also providing high CMRR.

In [3], Rakesh K. Lal, Maryam Shojaei-Baghini, Dinesh K. Sharma used network of n stage CMOS amplifiers that were able to take in a low voltage input and provide an amplified and stable signal for an output through repetitive amplification. The use of the CMOS amplifier is proved to be the best decision due to its low output impedance and high input impedance. The stages of the proposed network were also made such that it would be able to provide a constant and stable output without requiring any feedback. CMOS amplifier is as such used because of its high gain properties and is widely used in many applications.

In [4], Siegfried Selberherr, Member, Rui Martins and Francisco A. Vaz were able to fabricate a CMOS amplifier that was capable of providing very accurate readings for biomedical instrumentation. It proved to be very ideal for use in sensors that depended on accuracy.

In [5], A. P. Brokaw and M. P. Timko had theorized a new design for a monolithic instrumentation amplifier that was able to provide a maximum of 70% gain in its output without feedback.

In [6], Cameron Charles and Reid R. Harrison had proposed the way to design a low-noise low-power CMOS amplifier for neural recording applications. The proposed amplifier was able to work well and was used to detect any neural transmission through the use of a designed head set that was made through a network of capacitors with a low charge. Since the capacitor could respond well to the microvolt reading, it was capable of being used with an amplifier to take accurate readings.

In [7], Pado Augusto, Carlos A. dos Reis Fiho and Dal Fabbro design a CMOS instrumentation amplifier that is capable of performing microvolt amplification while providing a very high CMRR. This is done through the use of a resistor and transistor network that is used to boost the input impedance of the CMOS amplifier.

In [8], Khalil Najafi and Pedram Mohseni designed a fully integrated neural recording amplifier with DC input stabilization. The system uses headset designs that is seen in [6] and is capable of reading the low voltage of the neural transmissions of the brain for various purposes and record the commands through the different parts of the subject's mind.

In [9], K. D. Wise discusses the various types of bio-instrumentation amplifier designs for amplifying bio-signals fabricated using CMOS technologies. The bio-amplifiers play an important role in biomedical implant devices like hearing aids, pacemaker, etc. Those amplifiers are surveyed based on different techniques used for circuit design and fabrication methodologies.

In [10], bio-signals such as EEG, ECG, pulse rate, heartbeat rate are weak signal which have to be amplified for measurement and testing purposes in biomedical field. The process of amplification is achieved by using bio-amplifiers such as operational amplifiers as instrumentation amplifiers.

3 Methodology

In order to acquire highly efficient systems on chip (SoC) using CMOS, innovation is important to decrease the delay and increase the robustness, power utilization and surface area of the integrated circuits. This is a test investigational analysis of a class AB CMOS amplifier which is designed using various design layouts (serpentine, interdigitated and concentric). These techniques are accessed in function of delay area, product potency and amplifier characteristics such as common mode rejection ratio, offset, slew rate, electrical gain and power supply rejection ratio. In view of this test execution after effects of the class AB CMOS amplifier, the technique of using serpentine layout has proven to diminish the surface area roughly to 64% and decline in the consumption of power roughly to 39% with respect to the traditional technique. The serpentine layout technique could be utilized to enhance the electrical performance of their CMOS amplifiers in the design of SoC. The circuit schematics of a CMOS amplifier are shown in Fig. 1.

An instrumentation amplifier allows an engineer to adjust the gain of an amplifier circuit without having to change more than one resistor value. Compare this to the differential amplifier, which we covered previously, which requires the adjustment of multiple resistor values. The upper-left op-amp's negative feedback is causing the voltage to be equal to V1 at point 1 (top of Rgain). Similarly, in point 2 (bottom of Rgain), the voltage is kept to a value equal to V2. This sets up a decrease in voltage through Rgain equivalent to the difference in voltage between V1 and V2. This voltage drop induces a current through Rgain, and as the two input op-amp feedback loops draw no current, the same amount of current through Rgain has to flow through the two "R" resistors below and above it. Although this appears like a complex way of building a differential amplifier, it has clear advantages of providing extremely high input impedances on the V1 and V2 inputs (since they connect directly into the non-inverting inputs of their respective op-amps) and adjustable gain that can be controlled well by using a single resistor.

Figure 2 shows an example of an instrumentation amplifier.

The design of the system is to be implemented using a cascaded network of CMOS amplifiers that operate as instrumentation amplifiers as shown in [9]. The reason for doing so is to provide the best and most stable output signal from the amplifier. The design would also be capable of reading a very low voltage input signal and further amplifying it to be read by biomedical sensors such as pacemakers and insulin trackers. However, the drawbacks of the design are it could be made susceptible to any background EM radiation, and thus, it requires an additional filter design to operate smoothly. Based on the nature of the input signal, a low power,

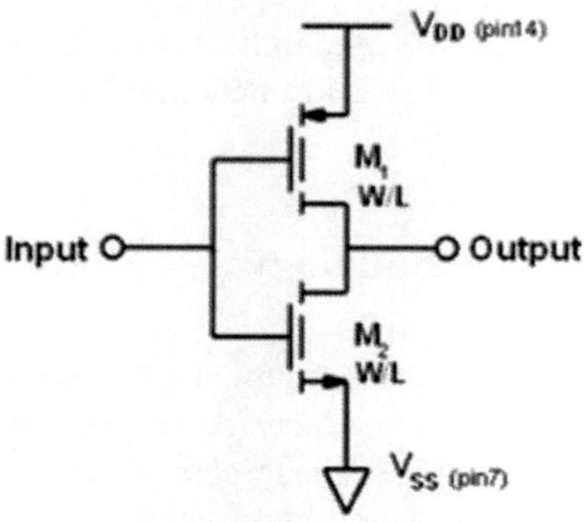

Fig. 1 Circuit schematic of a CMOS amplifier [3]

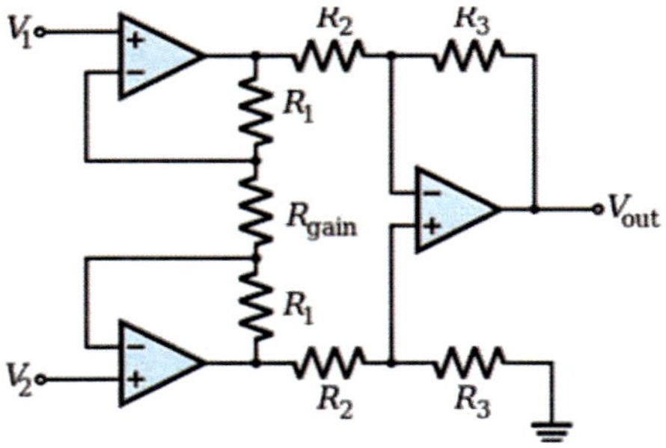

Fig. 2 Circuit schematic of an instrumentation amplifier

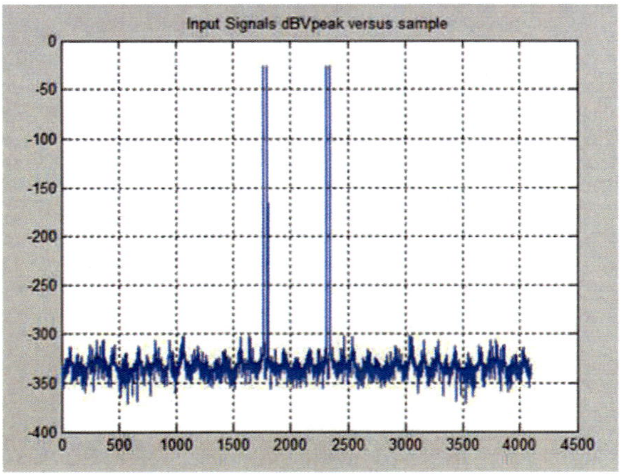

Fig. 3 Resultant filtered output waveform showing the maximum gain

low pass filter would seem ideal for it to function properly and smoothly. The input impedance and output impedance of the design could also be freely altered in order to optimize the circuit for other appliances as well. Thus, the circuit is to be analysed and implemented using the MATLAB software to project the returned gain of the output signal from the original weak input signal (Fig. 3).

4 Result

The prescribed circuit is thus simulated using the MATLAB software to show its significant gain through the different stages of the cascaded CMOS amplifier design as shown. The circuit is designed such that it is capable of establishing a feedback loop to the originating input waveform from each individual stage of the cascaded CMOS amplifier network as shown in Fig. 4.

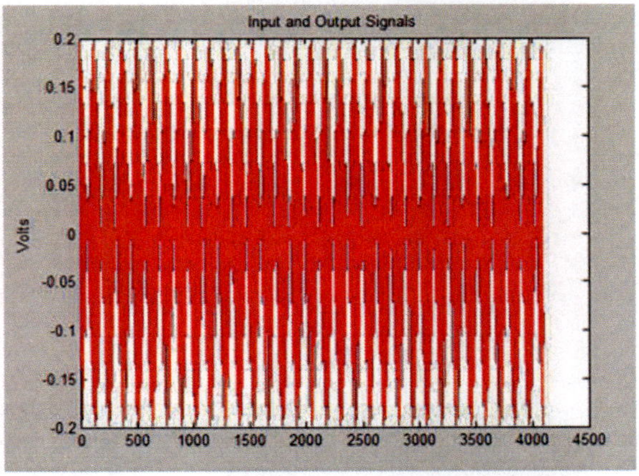

Fig. 4 Resultant input and output waveforms

When the signal is passed through the cascaded network of amplifiers, a portion of it is successfully amplified and thus produced at the output. The signal is however very susceptible to noise and shown below.

5 Discussion

As discussed through the literature survey, there were several different techniques that were used to design a stable and functioning bio-amplifier to be used in the medical field to amplify the resultant input signals of the biomedical sensors and implants. The increase in the input impedance of the proposed design seemed to be the most efficient way to improve upon the CMRR ratio of the amplifier, and additionally, it would provide for a stronger output signal. Doing so would thus improve the quality of the input signal to the sensors and would thus improve upon the resolution of their corresponding readings. The sensor was also able to function as optimum level through the implementation of a cascaded network of instrumentation amplifiers that were fed a feedback from the output signal. The signal produce was thus able to show a very high gain and CMRR ratio and less noise and interference (Fig. 5).

6 Conclusion

The circuit simulation of the bio-amplifier was able to take in a very low weak input signal and modulate and amplify it to acceptable voltage levels. The cascaded design of a network of instrumentation amplifiers with a feedback network to the main input pins was successful as when compared with the other prescribed methods. The design also offered very high CMRR. This is done also through the use of a

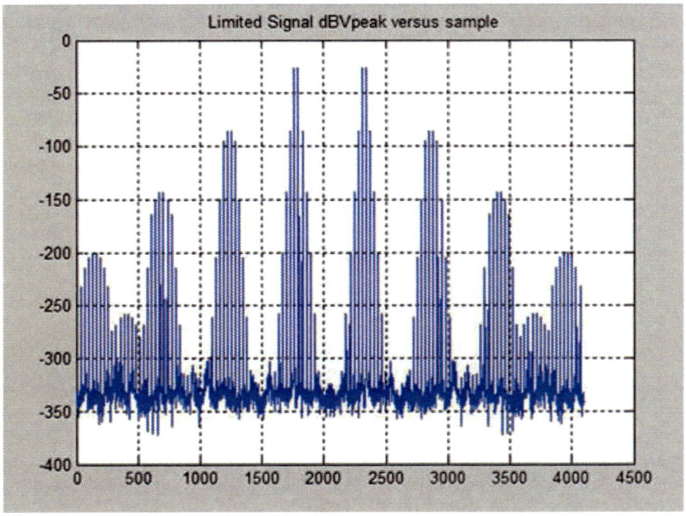

Fig. 5 Resultant unfiltered output waveform

resistor and transistor network to boost the input impedance of the CMOS amplifier to provide appropriate gain for the amplifier. The functioning also would have a very high signal-to-noise ratio (SNR). Thus, this is fit to function as a bio-signal amplifier that can provide a good quality and stable output for the biomedical sensors.

References

1. R.F. Coughlin, F.F. Driscol, *Operational amplifiers and Linear Integrated Circuits*, 2nd ed (1982)
2. Y.-M. Hsiao, M.-S. Shiau, K.-H. Li, J.-J. Hou, H.-S. Hsu, H.-C. Wuand, D.-G. Liu, *Design a Bio Amplifier with High CMRR*, vol. 2013 (Hindawi Publishing Corporation VLSI Design, 2013), Article ID 210265, 5p. (Department of Electronic Engineering, Feng Chia University, Taichung 40724, Taiwan)
3. R.K. Lal, M. Shojaei-Baghini, D.K. Sharma, An ultra low-power CMOS instrumentation amplifier for biomedical applications, in IEEE International Workshop on Biomedical Circuits & Systems. ISSN-0-7803-8665-5 (2004)
4. R. Martins, S. Selberherr, F.A. Vaz, A CMOS IC for portable EEG acquisition systems. IEEE Trans. Instrum. Measurement **47**(5), 1191–1195 (1998)
5. A.P. Brokaw, M. P. Timko An improved monolithic instrumentation amplifier. IEEE J. Solid-State Circ. **SC-10**, 417–423 (1975)
6. R.R. Harrison, C. Charles, A low-power low-noise CMOS amplifier for neural recording applications. IEEE J. Solid-State Circ. **38**(6) (2003)
7. P. Augusto, D. Fabbro, C.A. dos Reis Fiho, An integrated CMOS instrumentation amplifier with improved CMRR, in *Proceeding of the IS* Symposium on Integrated Circuits and System Design (SBCCI'OZ)* (2002)
8. P. Mohseni, K. Najafi, A fully integrated neural recording amplifier with DC input stabilization. IEEE Trans. Biomed. Eng. **51**(5) (2004)

9. K.D. Wise, A multi-channel microprobe for bio potential recording, Ph.D. dissertation, Stanford University, Stanford, CA, 1969
10. P.M. Van Peteghem, I. Verbauwhede, W.M.C. Sansen, Micro-power high-performance SC building block for integrated low-level signal processing. IEEE J. Solid-State Circ. **SC-20**, 837–844 (1985)
11. R. Pallas-Areny, J.G. Webster, Common mode rejection ratio for cascoded differential amplifier stages. IEEE Trans. Instrum. Measur. **40**(4), 677–681 (1991)
12. M. Shojaei-Baghini, R.K. Lal, D.K. Sharma, A Low-power and Compact Analog CMOS Processing Chip for Portable ECG Recorders (IEEE, 2005). ISSN-0-78 03-9162-4
13. J.-Y. Zhang, L. Wang, B. Li, Design of low-offset low-power CMOS amplifier for biosensor application. J. Biomed. Sci. Eng. 538–542 (2009)
14. T.H. Teo, G.K. Lim, D.S. David, K.H. Tan, P.K. Gopalakrishnan, R. Singh, Ultra low-power sensor node for wireless health monitoring system, IEEE International Symposium on Circuits and Systems (2007), pp 2363–2366
15. Y. Tseng, Y. Ho, S. Kao, C. Su, Low power front-end bio potential amplifier for bio signal recording. IEEE Trans. Biomed. Circ. Syst **6**(5) (2012)
16. M.J. Burke, D.T. Gleeson, A micropower dry-electrode ECG preamplifier. IEEE Trans. Biomed. Eng. **47**(2), 155–162 (2000)
17. C.-J. Yen, W.Y. Chung, M.C. Chi, Micro-power low-offset instrumentation amplifier IC design for biomedical system applications. IEEE Trans. Circ. Syst. I, Reg. Papers **51**(4), 691–699 (2004)
18. A. Goel, G. Singh, A novel low noise high gain CMOS instrumentation amplifier for biomedical applications. Int. J. Electr. Comput. Eng. (IJECE) **3**(4), 516–523 (2013). ISSN 2088-8708
19. M. Goswami, S. Khan, DC suppressed high gain active CMOS instrumentation amplifier for biomedical application, in Proceedings of IEEE International Symposium on Circuits and Systems (ISCAS'04). vol. 4 (2004), pp. 9–12
20. V. Ivanov, J. Zhou, I.M. Filanovsky, A 100-dB CMRR CMOS operational amplifier with single-supply capability. IEEE Trans. Circ. Syst. II, **54**(5), 397–401 (2007)

Automatic Street Light Controller Using LDR and Solar

C. H. Jyothi Priya and R. Puviarasi

Abstract Today we are planning a significant endeavor in which we can control the road lights with change of force of sunlight; i.e., as the force of daylight diminishes, power of light increments. For identifying the force, we utilize light ward resistor (LDR). This circuit is tried in different conditions, and it gives a flat out outcome. By and large road lights stay ON in morning time likewise because of some manual activity, which causes loss of vitality. For light activity, light ward resistor sensor is utilized in day and evening times. On the off chance that any individual or vehicle please the street, at that point the framework consequently sense this condition and work the light in higher force mode, and if no individual or vehicle out and about the will naturally goes diminish mode. This circuit is exceptionally valuable for sparing force and vitality via programmed control.

Keywords LDR · Solar cell · Light · Power supply

1 Introduction

In this paper, we ought to incline toward three modes, which is vitality age mode, vitality protection mode, and vitality application mode. In vitality age mode, the vitality is created with the assistance of sunlight-based board, in vitality preservation mode versatile applications and strategies are utilized with inserted framework. We have to spare some vitality since we rely upon for the most part on vitality assets like coal gaseous petrol cannot be supplanted. So we are utilizing sunlight-based panels for producing vitality. Rationing power is significant, superfluous occasions

C. H. Jyothi Priya
Saveetha School of Engineering, Chennai, India
e-mail: challajyothi665@gmail.com

R. Puviarasi (✉)
Saveetha Institute of Medical and Technical Sciences, Kuthambakkam, India
e-mail: puviarasi88@gmail.com

J. Tavares et al. (eds.), *Emerging Technologies in Data Mining and Information Security*, Lecture Notes in Networks and Systems 164,
https://doi.org/10.1007/978-981-15-9774-9_87

it ought to be turned off. Road light is of the significant force expending factor. The majority of the occasions road lights are ON significantly after the dawn in that time a great deal of vitality is getting waste. We are maintaining a strategic distance from this issue by having a programmed framework which turns the road lights ON and OFF. Every controller has a LDR which is utilized to recognize the encompassing light. In the event that the encompassing light is underneath the particular worth the lights are turned ON. Programmed road light controller is basic and influential idea, which utilizes transistor as a switch. By utilizing this framework, manual works are expelled totally. It naturally turns on the light when the daylight goes underneath the unmistakable area. It consequently goes OFF when the daylight comes.

There are more systems to control the road lights naturally and programmed road light controller utilizing LDR. In this paper, the analyst structuring a significant endeavor where we can control the road lights with change of force of daylight. For identifying the power, we utilize light ward resistor (LDR). For controlling, the force of light triac controlled circuit is utilized. In this structure, there is the essential of getting synchronized terminating beats for the door of the thermistor. Stage edge control conspire is utilized for get a sine wave by incorporated a stockpile voltage. The sign at yield terminal is synchronized with then heartbeat and is deferred from the inventory zero intersection signal. Triac instrumentation is utilized to give a voltage as a kind of perspective to a terminating circuit. This reference voltage is absolutely reliant on LDR obstruction whose worth change with the force of light. Automatic road light controller framework utilizing remote sensor systems. In this paper, the road light post frames the WSN hub. To recognize traffic out and about it utilizes Arduino Uno board and different sensors. To screen the status of the road lights raspberry pi3 PC is utilized as a web server. In view of the sensor signal, the lights are turned on during evenings when the traffic is available site page is created by utilizing HMTL and PHP. A savvy computerized road light controller and energetic outflow control system. This paper gives a keen and insightful system for sparing of non-sustainable wellsprings of vitality. What is more, the significant angle is that it manages the progression of traffic and the contamination levels at different places by checking the air quality controls and programmed road light control with various light forces. This paper microcontroller is utilized for controlling light by utilizing this framework manual activities are absolutely removed. If any vehicle is shown up, it naturally ON the light by utilizing sensor called infrared which detects the light. It naturally turns off when the vehicle is died. This paper controlling of power is likewise conceivable. The vitality utilization is additionally reduced. Automatic road light control by identifying vehicle movement in this paper photograph diodes and IR sensor are utilized to detect vehicle developments. Control rationales are actualized to control the lights dependent on vehicles and pedestrian and programmed road light control utilizing RTC and LDR. The light is regulated in this paper according to the timings customised in the system and the projects can be naturally changed by using continuous clock (RTC) and LDR to discern the light and it turns the light ON when there is no light according to the environmental changes.

2 Methodology

Following are the segments used to comprise this paper:

1. Solar board: It is a gadget which is utilized to change over the light vitality into electrical vitality. The get together of photovoltaic sun-oriented cells is having a pressed substance of photovoltaic module. It comprises of the photovoltaic exhibition of photovoltaic framework that produces and supplies sun-oriented power in certain applications.
2. Capacitor: It is utilized to decrease the clamor; furthermore, it is utilized in the filtration procedure.
3. Gem oscillator: It is an electronic oscillator which uses the mechanical resonation of a vibrating valuable stone of piezoelectric materials to make an electrical sign with a precise repeat. This repeat is generally used to screen time. To give a steady clock sign to computerized coordinated circuits, what's more, to balance out frequencies for radio transmitters and beneficiaries. In this paper, 12 MHz of gem oscillator is utilized.
4. Battery: It is a device provided to provide electrical gadgets such as spotlights and mobile phones containing at least one electrochemical cell with external associations. The electrical force is supplied, the cathode is positive, and the anode is in the negative terminal.
5. Voltage controller: It is intended to keep up a steady voltage level automatically. It might be a straightforward feed-forward plan. It might incorporate negative criticism control circles. It might utilize electronic segments. It might be utilized to manage at least one AC or DC voltage put together up with respect to the structure.
6. Infrared sensor: It is used to identify certain characteristics of its surroundings by perceiving infrared radiation. It is also prepared for assessing the glow sensors which are fit for evaluating the glow being released by a thing and perceiving development.
7. Resistor: It is a confined two-terminal electrical fragment that executes electrical resistance as a circuit segment. In most of the electronic circuits, resistors are used to diminish current stream, change signal levels, biased the dynamic components, to isolate voltages, and end transmission lines.
8. RELAY: It gives the selection between the controller and the gadget on the ground that has be probably aware gadgets may chip away at AC just as on DC at they get signals from microcontroller which deals with DC; thus, we require a hand-off to conquer any hindrance. The hand-off is incredibly helpful when you have to control a lot of flow or voltage with the little electrical sign.
9. LDR: ALDR is a photoconductor. It is one sort of resistor whose obstruction differs relying upon the measure of light falling on its surface. At the point when the light falls on the resistor, at that point the obstruction changes. These resistors are frequently utilized in numerous circuits where it is required to detect the nearness of light. These resistors have an assortment of capacities and opposition.

For example, when the LDR is in haziness, at that point it tends to be utilized to turn ON a light or to kill a light when it is in the light.

3 Working Principle

In this item, we utilized 89S5 microcontroller. The voltage of the microcontroller is 5 V DC. In that result, changing over the battery voltage into 5 V with the assistance of 5 voltage controller circuit, utilizing 7805 controller to manage the voltage into 5 volt DC. For filtration, one capacitor is ground from the controller. Here capacitor diminishes the commotion. The controller yield is associated with the pin 40 of the microcontroller legitimately. One of the gems is associated with the pin 18 and 19 of the microcontroller to give a swaying signal, 12 MHz precious stone is utilized for this reason. For diminishing the commotion, two capacitors are grounded from the crystal; in this paper, we are utilized (Fig. 1).

At the point when the LDR is in dim, then base voltage is null and the gatherer turns out to be increasingly positive. The controller senses this difference in voltage and switch on the yield drove which is associated with the port 0. Through the resistor in arrangement, the yield of the drove is associated with the port 0 by utilizing this rationale on the off chance that any vehicle crosses the principal infrared sensor, at that point all the LEDs are ON with high force and on the off chance that the traffic is proceed; at that point, the LED is on with the profoundly expanded power then previously. Also, in the event that any vehicle is absent out and about, at that point the LEDs are gleam with low power. In this process, we use 2 circuits of infrared sensor. This procedure should be possible by the light sensor (LDR). It is a light needy resistor. On the off chance that LDR is in dim; at that point, it offers a high obstruction and on the off chance that the light falls on the LDR; at that point, LDR offers a low opposition. In this paper, we are utilized the LDR with n-p-n transistor

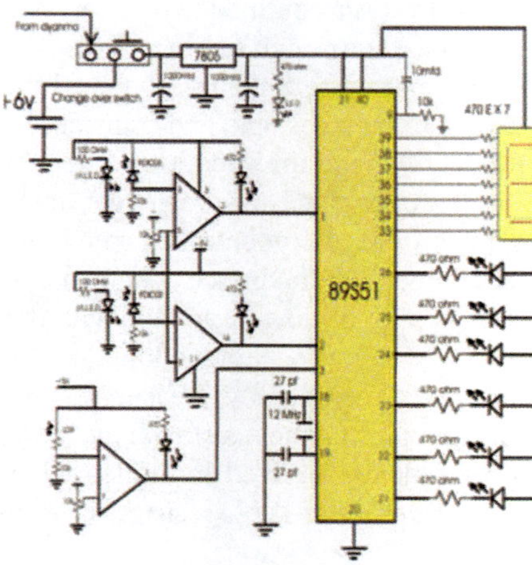

Fig. 1 Circuit diagram

circuit. The n-p-n transistor of the producer is associated with the ground, and the gatherer is associated with the pin 3 of the controller.

At the time, there is no base voltage when the LDR is in dim, and the turns out to be increasingly positive. The controller senses this difference in voltage and switch on the yield drove which is associated with the zero port. Through the resistor in arrangement, the yield of the drove is associated with the port 0 by utilizing this rationale on the off chance that any vehicle crosses the principal infrared sensor; at that point, all the LEDs are ON with high force and on the off chance that the traffic is proceed; at that point, the LED is on with the profoundly expanded power then previously. Also, in the event that any vehicle is absent out, and about, at that point the LEDs are gleam with low power. For these procedures, we utilize two infrared sensor circuits.

We utilize one photodiode circuit, when light falls on the photograph sensor the opposition of that sensor turns out to be low, and thus, negative voltage is applied to the controller. At the point when a vehicle crosses, the photodiode it obstruction turns out to be high, and thus, signal is change on the pin 2 of the controller as the controller sense this difference in signal on pin then every one of the lights is ON.

4 Result Analysis

The above graph tells that the outcomes for programmed road light framework by utilizing sun-oriented vitality. Figure 2 demonstrates that the daytime power is perfect (0). Then Fig. 3 demonstrates that the evening time with no vehicle, and it shows that the force is changes; for example, it improves when contrasted with daytime. In Fig. 4, when the street is loaded with traffic, the power of light is extremely high (100%).

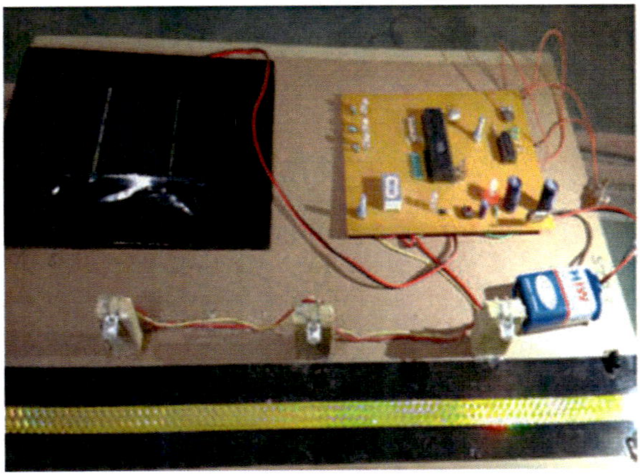

Fig. 2 Daytime-intensity of light is zero percent

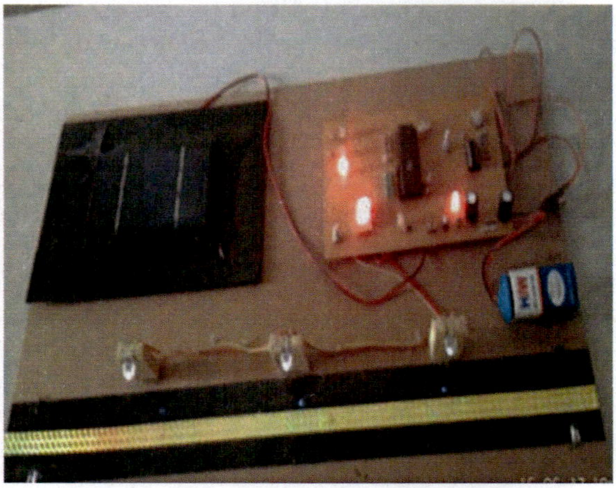

Fig. 3 Night-when no vehicles on the road intensity is very less

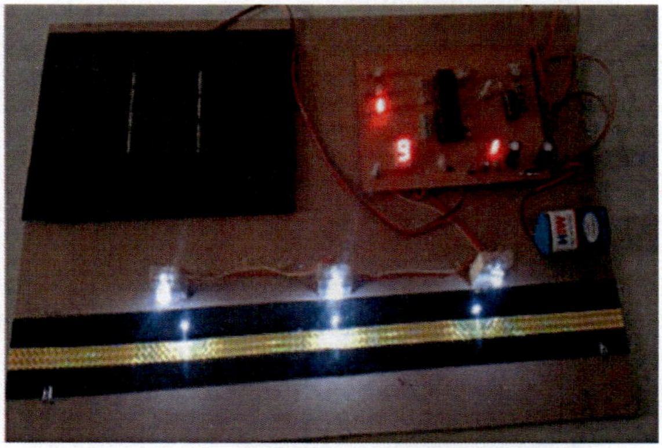

Fig. 4 Night time with vehicles on road (high light intensity). When full traffic in night (very high light intensity)

5 Conclusion

Programmed road light control framework dependent on sun-powered vitality is introduced in this paper. The exploratory framework shows that nearby planetary group is better answer for the lightning arrangement of the streets. Progressively number of activities are tried with this sustainable power source; this framework can likewise conquer the need of road lightning framework and interconnected and work with the electric influence grid. It is additionally helpful for sparing vitality and influence and furthermore sets aside cash for the operation in the night.

References

1. R.K. Kodali, S. Yerroju, Energy efficient smart street light, in IEEE conference 2017 (Department of Electronics and Communication Engineering, National Institute of Technology, Warangal, 2017)
2. O. Singh, *Solar LED Street Light System with Automatic Scheme* (Electrical Engineering Department, Greater Buddha University, Tushar Singh Sisodia Electrical Department, Greater Noida, 2017)
3. B. Roy, J. Datta, A smart street light intensity optimizer, in IEEE Conference/2017 (Department of Applied Electronics and Instrumentation Engineering, Guru Nanak Institute of Technology, Kolkata, W.B., India)
4. P.V. Parkash, D. Rajendra, Internet of Things based intelligence street lighting system for smart city. Int. J. Innov. Res. Sci. Eng. Technol. **5**(5) (2016)
5. D. Sunehra, Automatic street light control system using wireless sensor networks, in IEEE-2017 (SMIEEE, JNTU College of Engineering, Department of Electronics and Communication Engineering, Telangana, India, 2017)
6. K. Nirosha et al., Automatic street lights on/off application using IoT. Int. J. Mech. Eng. Technol. (IJMET) **8**(8), 38–47 (2017). (Article ID: IJMET_08_08_005)
7. F. Ramadhani, K.A. Bakar, M.-M. Gary Shafer, Optimisation of standalone street light system with consideration of lighting control. IEEE (2013)
8. N.S. Viswanath, S. PushpaBai, P. Rajesh, N.B. Mallikarjuna, IoT based green house monitoring system. Int. J. Electron. Commun. Eng. Technol. (IJECET) **6**(6), 45–47 (2015)
10. C. Kavitha, R. Gorrepotu, N. Swaroop, Advanced domestic alarms with IOT. Int. J. Electron. Commun. Eng. Technol. **7**(5), 77–85 (2016)
11. K. Prashanthet, Automated street lighting system using IoT. Int. J. Adv. Res. Ideas Innov. Technol. 1970–1973 (2018)
12. S.K. Cho, V. Dhingra, Street lighting control based on lon works power line communication, in IEEE International Symposium on Power Line Communications and Its Applications (Jeju City, 2–4 April 2008), pp. 396–398
13. J. Arthi, W. Lydiapreethi, B. Gunasundai, IOT based smart led street lighting system. IJRTI **2**(4). ISSN: 2456-3315
14. S. Bhosale, K. Gaware, P. Phalke, D. Wadekar, P. Ahire, IOT based dynamic control of street lights for smart city. Int. Res. J. Eng. Technol. (IRJET) **4** (2017)
15. S. Mishra, S. Gupta, S. Singh, T. Tiwari, A. Mohan, Arduino based led street light auto intensity control system. Int. J. Adv. Res. Eng. **3**(4) (2016)
16. S. Arkade, A. Mohite, R. Sonawane, V. Patil, IoT based street lights for smart city. Int. J. Res. Appl. Sci. Eng. Technol. (IJRASET) **4**((XII) (2016)
17. B. Abinaya, S. Guru Priya, *IOT Based Smart and Adaptive Lighting in Street Lighting* (Department of Information Technology Sri Sairam Engineering College), 2nd International Conference on Computing and Communications Technologies (ICCCT) (2017)
18. F. Leccese, M. Cagnetti, D. Trinca, A smart city application: a fully controlled street lighting Isle based on Raspberry-Pi Card, a ZigBee sensor network and WiMAX. Sensors **14**, 24408–24424 (2014). https://doi.org/10.3390/s141224408
19. A.M. Sindhu, J. George, S. Roy, J. Chandra, Smart streetlight using IR sensors. IOSR J. Mob. Comput. Appl. (IOSR-JMCA) **3**(2). e-ISSN: 2394-0050, P-ISSN: 2394-0042
20. P.V. Parkash, D. Rajendra, Internet of Things based intelligent street lighting system for smart city. Int. J. Innov. Res. Sci. Eng. Technol. **5**(5) (2016)
21. R. Salvi, S. Margaj, K. Matel, B. Aher, Smart street light using Arduino uno microcontroller. Int. J. Innov. Res. Comput. Commun. Eng. **5**(3) (2017)

EEG Processing for Computer-Aided Seizure Detection Using Artificial Neural Networks

P. Sai Ankith and M. K. Mariam Bee

Abstract Electroencephalography (EEG) is a critical gadget for making inquiries concerning the human cerebrum job, which obtains the tumultuous nature and also moreover quantum dives and additionally non-fixed practices of subconscious flags generally made use of for the recognition of epilepsy and seizure. The trouble is to build as well as make up signal handling calculations that leave these unpretentious subtleties and in addition utilize it for analysis, following and also moreover treatment of individuals with epilepsy. The suggested strategy had accomplished amazing effectiveness with the typical exactness of 99.4%, the regular selectivity of 91.1%, the typical level of affectability of 92.1%, the normal particularity of 99.5%, and also furthermore the traditional disclosure expense of 95.8%.

Keywords Electroencephalography (EEG) · Epilepsy · Seizure · Artificial neural networks (ANN)

1 Introduction

Trick exam of epilepsy bases on the experiences from a viewer. Commonly, by charming appraisal of EEG recording, epilepsy can be found. Also, with a couple check out, there still exist different difficulties in the intro of epilepsy. Psyche—COMPUTER structure UI is made use of to select the EEG signals. The fundamental part of this user interface is steel terminals, which are used for assessing the electrical renovation of the mind from the head surface area. Generally, evaluations are made by a cathode organizing strategy called the comprehensive 10–20 framework [1] EEG signals have actually unfathomably lessened plentifulness (in μv levels) worths. To

P. S. Ankith (✉) · M. K. Mariam Bee
Department of Electronics and Communication Engineering, Saveetha School of Engineering (SIMATS), Kuthambakkam, India
e-mail: ankithpulavarthi123@gmail.com

M. K. Mariam Bee
e-mail: mariamresearch09@gmail.com

J. Tavares et al. (eds.), *Emerging Technologies in Data Mining and Information Security*, Lecture Notes in Networks and Systems 164,
https://doi.org/10.1007/978-981-15-9774-9_88

have the decision to unravel these signs, they should certainly be reinforced by EEG tools. This raised EEG signal is used as a paper in continuing on kind. The recorded EEG signals take after the multichannel seismograph report. Analysis of EEG stories is done by celebrities that attended to significant master at the present time.

The five basic wave types in EEG signals are: alpha (8–13 Hz), beta (13–30 Hz), delta (0–4 Hz), gamma (30–100 Hz), also as theta (4–7 Hz) [2, 3]. EEG signals are not uncertain and also their ampleness, phase, and frequencies alter on and on. Along these lines, these signs are astonishingly hard to decode. EEG details is clearly checked by professionals [3].

Electroencephalogram (EEG) as a depictive indicator including nuances of the electric exercises made by the cerebral cortex afferent neuron has actually been just one of the most used indications to deductively diagram brain attempts. Furthermore, the intro of epileptic structure launches in the EEG is a significant sector in the searching for of epilepsy [4]. Throughout the past different decades, an affirmation of approaches has been suggested for the automated affirmation of epileptic seizures. These estimations and all points considered included time-space, ordinariness area name, time-rehash area comparably as nonlinear methods [5] as the most long-lasting sign handling method, the Welch horrible evaluation [6], and also the stage slant record [7] had achieved wonderful outcomes. Previous appraisals suggested that the EEG signals were nonlinear and non-stationary optional strategies, so the link estimate [8], Lyapunov kind [9], also as spoil [10] had actually been brought into the utilitarian analysis of epileptic seizure recognizing confirmation. What is more, the wavelet change had in like means been made use of considering the way that the multiscale assessment might bring an additional out of the case new viewpoint to the EEG assessment [11, 12]. The square delineation of line of work is showed up in Number 1.

2 Materials and Methods

2.1 Materials

Currently, the University of Bern11 open accessibility of the EEG dataset is used for the set order of EEG signals (22). Now, central EEG signals as well as 175 non-focal EEG signals are made use of Number 1 Image of main as well as non-central EEG signals.

2.2 Methods

This evaluation recommends an added approach in the choice of epilepsy that is not exactly comparable to the techniques presented in the past tests. The suggested lotion method offered accuracy outcomes that climb above to anything any type of kind of shown in the discovering Transforming complying with is a short framework of our proposed technique. It is possible to hold this view with the Fourier adjustment methodology, which is among the most crucial likewise as liberal components of indication dealing with (Figs. 1 and 2).

On the off opportunity that we require to acquire the advantages of this change with a made sure respected classifier, we have to present the online and authentic evaluations of the unforeseeable number to the information autonomously. When twisted numbers are cherished, 2-D information will certainly be shown up to the

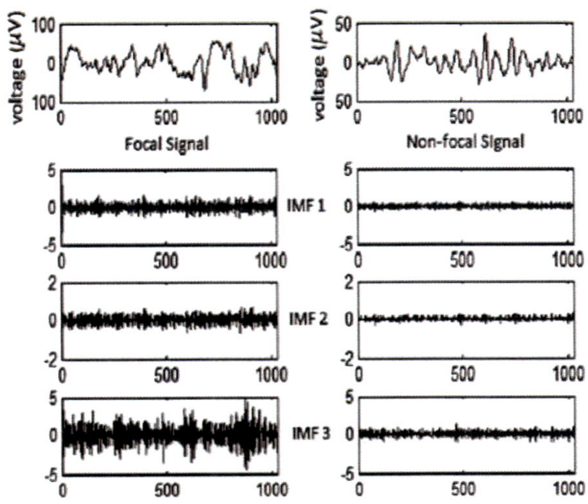

Fig. 1 Illustration of EEG signals

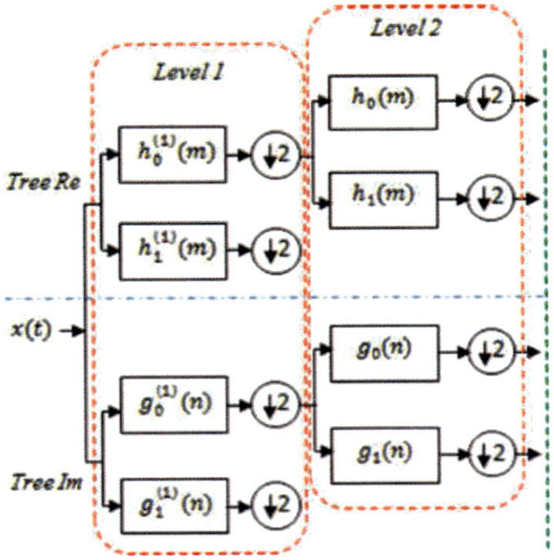

Fig. 2 DTCWT with filters

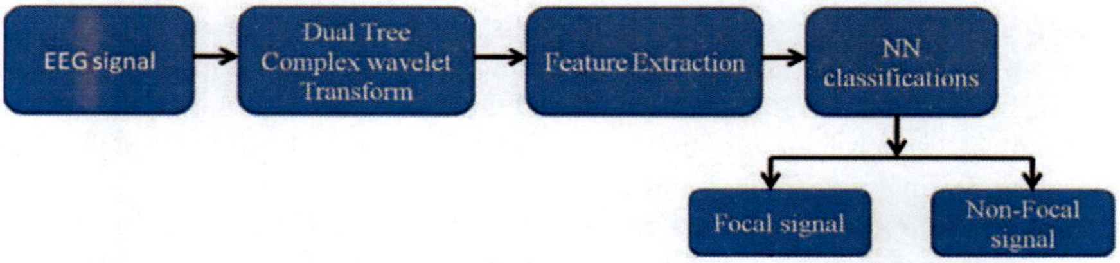

Fig. 3 Block diagram of the work

framework as 1-D details considering that confounded numbers have both certifiable in a similar way as unique parts.

This is appeared in substantially even more details in sticking to territories that deal with the upsides of these figuring's while exposing complex-respected evaluations. Considering this info, a framework that collaborates with caught numbers was concentrated so worrying managing the EEG appreciates in high precision rates. Doubletree workplace wavelet adjustment (DTCWT) has actually been used in numerous degrees for entailing the removal as well as quantifiable highlights are received from the obtained confounding popular area vector. A relative number of highlights for each and also every arrangement of EEG details is accustomed as a fragment with the ANN. The limit behind utilizing the DTCWT-ANN go across type model is understood dealing with the issue with an estimate that limits entirely in the questionable area. In this way, it is possible to see the impacts of the complex-viewed procedures similarly as splitting up them and the real all over related to strategies. Number 2 is the doubletree complex distinct WT with channels for the authentic area as well as narrative little the DTCWT coefficients.

Figure 3 reveals the advised epilepsy medical problem location making use of a fifty percent kind representation strategy. The changed recognizable proof of epilepsy medical troubles is recommended right now a hybrid course of action calculation. The suggested technique integrates gas mileage setup, including evacuation, as well as request stages. The EEG signals are shabby making use of double tree complicated wavelet change (DTCWT) as well as later on includes are separated from these coefficients. These features are after that arranged to utilize an ANN representation method to sort out the EEG signals right into either focal or non-focal EEG signals [13].

3 Feature Extraction

3.1 Maximum and Minimum Value

The design biggest equally as test least additionally called the greatest understanding as well as littlest observing is the evaluations of the very best equally as least parts of an instance. The base, just as the suitable worth, is the first and last demand insights (frequently suggested $X(1)$ as well as $X(n)$ separately, as an example, dimension of n).

3.2 Mean

The true mean indicates the mean or common that is used to obtain the main fondness of the info concerned. It is directed by including all the data centers around an individuals and after that separating the essentially by the amount of core passions. The occurring number is recommended as the criterion or the mean

$$\text{Mean } (\mu) = \frac{1}{N} \sum_{(n=1)}^{N} (X(n))$$

where xn is the examination EEG indicator and also I is the example at different unfathomable interims in the EEG signal. The absolute array of models in the assessment EEG signal is portrayed as N.

3.3 Median

The mean naturally with an unmistakable arbitrary variable trusts whether the array of terms in the dissemination is even or unusual. If that the amount of terms is unusual, after that the normal is the estimate of the term in the facility. This is the worth with completion objective that the assortment of terms having values above or equal to it amounts the selection of terms having worth's not specifically or comparable to it. On the off opportunity that the assortment of terms is likewise, after that the middle is the regular of the two terms between, with completion objective that the quantity of terms having worth's greater than or comparable to it balances as the variety of terms having worth's not specifically or comparable to it.

$$\text{Mediam} = ((N+1)/2)^1.$$

$$= \left(\left(\frac{N}{2}\right)^{\text{th}} \text{value} + \left(\frac{N}{2}+1\right)^{\text{th}} \text{value}\right) \Big/ 2$$

The regular appropriation with a constant arbitrary variable is the worth m to such an extent that the likelihood most likely to the very least 1/2 (half) that a haphazardly picked element on the element will favorably be not exactly or equal to m, equally as the possibility remains in any type of occasion 1/2 that a discretionarily selected factor on the aspect will certainly be a lot more significant than or equivalent to *m*.

3.4 Standard Deviation

The standard deviation of an abstract variable, organized masses, informational index, or chance movement is the square facility of its difference. This acceptance of a standard deviation is routinely called the "common botch" of the action or "typical mix-up of the suggest" when suggesting a mean (22)

$$\sigma = \sqrt{\frac{\sum (x - \mu)^2}{N}}$$

The SD of the EEG examination signal is defined by the going along with problem:

4 ANN Classifier

It is orchestrated into a winding NN and also manages sending back increasing NNs. The all-encompassing NN has no remarks for establishing up the occasions of EEG indications and it cannot be utilized right now for the representation of EEG signals for epilepsy discovery by non-partisanship of the non-attendance of responses when the mistake price is horrendous. This control of this winding NN is bewildered by feed in advance back propagation NNs, which have investigation furthermore as the made result is an exam of the info when there is a horrible (Fig. 4).

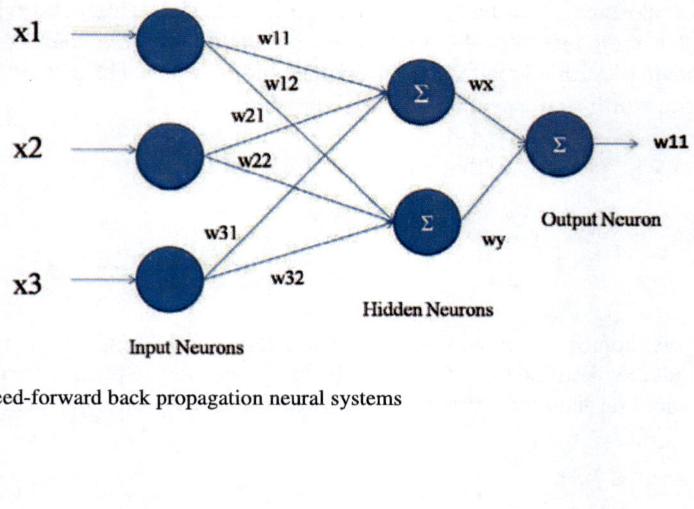

Fig. 4 Feed-forward back propagation neural systems

Starting now and also into the not so remote, comparative cut-off factors are expelled from the test EEG signal, in addition, these buildings are confined by the self-governing NN need mode close with qualified trademark strategies. The favorable worth addresses that the examination EEG signal is non-focal and the troublesome worth tends to that the examination EEG signal is main (22).

5 Results and Discussion

Regarding feasible is used for hallmark extraction and when it is sorted out with linear discriminant assessment classifier, provided the criteria like performance index, precision, level of sensitivity, uniqueness, high-quality worth as well as dead time, the typical end results are appeared in Table 3. The mathematical solutions for the efficiency index (PI), level of sensitivity, individuality as well as precision are given as sticks to

$$\text{PI} = \left(\frac{\text{PC} - \text{MC} - \text{FA}}{\text{PC}}\right) \times 100$$

where perfect classification = COMPUTER, missed classification = MC, and also the false alarm = FA. The level of sensitivity, uniqueness, and accuracy actions are revealed mathematically by the complying with

$$\text{Sensitivity} = \frac{\text{PC}}{\text{PC} + \text{FA}} \times 100$$

$$\text{Specificity} = \frac{\text{PC}}{\text{PC} + \text{MC}} \times 100$$

$$\text{Accuracy} = \frac{\text{Sensitivity} + \text{Specificity}}{2} \times 100$$

The quality value Q_V is mathematically explained as follows

$$Q_v = \frac{C}{(R_{\text{fa}} + 0.2) * \left(T_{\text{dly}} * P_{\text{dct}} + 6 * P_{\text{msd}}\right)}$$

where C = scaling regular,

R_{fa} = number of phony alarm system obtained per collection,

T_{dly} = the regular deferment of the beginning tactical plan like a blaze

P_{dct} = the price (%) of perfect tactical plan as well as

P_{msd} = the rate (%) of perfect risk degree missed

The suggested crossbreed characterization technique for the class of focal signals in a similar way as non-focal signals achieved 98% of the degree of affectability, 99% of unequivocality, and also moreover 99.3% of accuracy, as showed up in Table 1. The discussion success will certainly stabilize loved one with the amount of examination

Table 1 Performance assessment of the proposed blend course of action procedure (for testing 350 EEG signals)

Parameters	Experimental results (%)
Sensitivity	98
Specificity	99
Accuracy	99.3

EEG signals. In this way, a total of 100 EEG signals (50 focal likewise as 50 non-focal signals) is analyzed by the recommended crossbreed procedure at today time.

6 Conclusions

The mechanized place and also power evaluation of epilepsy health issues are recommended now a mix order equation. The advised epilepsy discovery framework discovers the 50 main EEG flags as main EEG flags out of 50 main EEG signals. The suggested epilepsy acknowledgment framework determines the 48 non-focal EEG flags as non-focal EEG flags out of 50 non-focal EEG signals. The suggested mix category approach for the classification of main signals just as non-focal signals accomplished 98% of the level of affectability, 99% of particularity, and also 99.3% of precision.

References

1. M. Peker, B. Sen, D. Delen, A novel method for automated diagnosis of epilepsy using complex-valued classifiers. IEEE J. Biomed. Health Inf. **20**(1) (2016)
2. B. Sen, M. Peker, F.V. Celebi, A. Cavusoglu, A relative report on arrangement of rest organize dependent on EEG signals utilizing highlight choice and grouping calculations. J. Prescription. Syst. **38**(18) (2014)
3. B. Senand, M. Peker, Novel methodologies for mechanized epileptic finding utilizing FCBF feature choice and arrangement calculations. Turk. J. Electr. Eng. Comput. Sci. **21**, 2092–2109 (2013)
4. N. Sadati, H.R. Mohseni, A. Maghsoudi, Epileptic seizure detection using neural fuzzy networks, in 2006 IEEE International Conference on Fuzzy Systems Sheraton Vancouver Wall Center Hotel (Vancouver, BC, Canada, 16–21 July 2006)
5. U.R. Acharya et al., Computerized EEG examination of epilepsy: a survey. Knowl. Based Syst. **45**, 147–165 (2013)
6. K. Polat, S. Güneş, Grouping of epileptiform EEG utilizing a half and half framework dependent on choice tree classifier and quick Fourier change. Appl. Math. Comput. **187**(2), 1017–1026 (2007)
7. P. Rana et al., Seizure discovery utilizing the stage incline record and multichannel ECoG. IEEE Trans. Biomed. Eng. **59**(4), 1125–1134 (2012)
8. Y.C. Lai et al., Connection measurement and autocorrelation changes in epileptic seizure elements. Phys. Fire up. E Stat. Nonlin. Delicate Matter Phys. **65**(3), 031921 (2002)
9. N. Guler et al., Intermittent neural systems utilizing Lyapunov types for EEG signals arrangement. Expert Syst. Appl. **29**(3), 506–514 (2005)

10. Y. Melody, P. Liò, another methodology for epileptic seizure recognition: test entropy based component extraction and extraordinary learning machine. J. Biomed. Sci. Eng. **3**(06), 556–567 (2010)
11. Y.U. Khan, J. Gotman, Wavelet based programmed seizure recognition in intracerebral electroencephalogram. Clin. Neurophysiol. **114**(5), 898–908 (2003)
12. T. Kalayci, O. Ozdamar, Wavelet preprocessing for mechanized neural system location of EEG spikes. IEEE Eng. Medications. Biol. Mag. **14**(2), 160–166 (1995)
13. M.K. Mariam Bee, K. Vidhya, A mechanized system for the characterization of central and nonfocal EEG signals utilizing a half breed order approach. Int. J. Imaging Syst. Technol. 1–7 (2019). https://doi.org/10.1002/ima.22360. (Wiley Periodicals, Inc.)
14. V. Dinkelacker, V.- H. Nguyen-Michel, L. Thivard, V. Navarro, C. Adam, O. Pallanca, I. Arnulf, I feel my arm shaking: incomplete cataplexy mixed up for drug-safe central epilepsy. Sleep Med. (2017)
15. L.D. Iasemidis, Epileptic seizure forecast and control. IEEE Trans. Biomed. Eng. **50**(5), 549–558 (2003)
16. M. Ceylan, R. Ceylan, Y. Özbay, S. Kara, Utilization of complex discrete wavelet change in characterization of doppler signals using complex-esteemed counterfeit neural system. Artif. Intell. Medications. **44**(1), 65–76 (2008)
17. M.K. Mariam Bee, K. Vidhya, An automated methodology for the classification of focal and nonfocal EEG signals using a hybrid classification approach. Int J Imaging Syst. Technol. 1–7. https://doi.org/10.1002/ima.22360
18. D. Buck, G.A. Bread Cook, A. Jacoby, D.F. Smith, D.W. Chadwick, Patients' experiences of injury because of epilepsy. Epilepsia **38**(4), 439–444 (1997)
19. N. Aydin, F. Marvasti, H. S. Markus, Embolic doppler ultrasound signal discovery utilizing discrete wavelet change. IEEE Trans. Inf. Technol. Biomed. **8**(2), 182–190 (2004)
20. G. Strang, T.Q. Nguyen, *Wavelets and Filter Banks* (Wellesley-Cambridge, Wellesley, MA, USA, 1998)
21. J.M. Lina, Picture preparing with complex Daubechies wavelets. J. Math. Imag. Vis. **7**, 211–223 (1997)
22. I.W. Selesnick, R.G. Baraniuk, N. Kingsbury, The double tree complex wavelet change. IEEE Signal Process. Mag. **22**(6), 123–151 (2005)

An Automated Method for Epileptic Seizure Detection Using Support Vector Machine

P. Sai Ankith and M. K. Mariam Bee

Abstract It is very essential for the precise mechanized discovery of epileptic seizures in enduring electroencephalogram (EEG) chronicles. This work gives a novel procedure to programmed epilepsy seizure revelation dependent on EEG assessment that controls the hidden non-straight nature of EEG information. The exhibition of the subsequent framework is checked with open datasets, recently refined to kill antiquities or outside interruptions, anyway in like manner with select datasets recorded under reasonable and non-perfect working conditions. Utilizing open datasets takes into account examination capacities while the selective one uncovers the proficiency of the framework under reasonable conditions of clamor, antiques, just as signs of different amplitudes. The electroencephalogram (EEG) is a symbolic sign which contains bits of information about the mind. The waveform can contain supportive data with respect to its state. Eventually, the human watcher cannot straight screen this unnoticeable information. What is more, considering that history signals are incredibly passionate, the signs and symptoms may show up aimlessly while scaling. Thus, the EEG signal parameters, expelled and assessed utilizing PC systems, are particularly significant in end. The inspiration driving this action is to arrange EEG

P. Sai Ankith (✉) · M. K. Mariam Bee
Department of Electronics and Communication Engineering, Saveetha School of Engineering (SIMATS), Chennai, Tamil Nadu, India
e-mail: ankithpulavarthi123@gmail.com

M. K. Mariam Bee
e-mail: mariamresearch09@gmail.com

J. Tavares et al. (eds.), *Emerging Technologies in Data Mining and Information Security*, Lecture Notes in Networks and Systems 164,
https://doi.org/10.1007/978-981-15-9774-9_89

information from customary and besides epileptic subjects. Hence, a support vector device (SVM) to perceive regular similarly as epileptic EEG signals is proposed.

Keywords EEG · Focal · Non-focal · SVM · Wavelet transform · Classifier

1 Introduction

EPILEPTIC takeovers are results of temporary furthermore amazing electricity agitating impacts of that cerebrum. About one in every 100 people in their lives will encounter a seizure every once in a while [1] But the location of an epileptic seizure is not evident, and its approach is not fully observed at this point in addition. Electroencephalogram (EEG) as an indicator of the subtleties of electrical exercises performed by the cortex afferent neuron was one of the most commonly used signs of clinical obstruction. Also, the divulgence of epileptic typical EEG discharges is an integral portion of the clinical end of epilepsy. The distinctive evidence of epilepsy that consolidates EEG visual testing represents the spikes similarly as seizures are typically dull, particularly with respect to large annals. In addition, bio-signals are unimaginably unique, so banter is conceivable on a comparable record; hence, the EEG signal judgments drawn out similarly as analyzed using PCs are truly important in diagnostics.

The early strategies of modified EEG getting ready relied upon Fourier change. This strategy relies upon earlier monitoring that the EEG broaden has some trademark waveforms that drop in a general sense inside four ordinariness gatherings. Such techniques have checked to be valuable for various EEG representations, yet fast Fourier change (FFT) fights with a giant level of affectability disturbances. Parametric frameworks for impact run estimation, for example, autoregressive (AR), decreases the nebulous vision episode issues what is more gives much better regularity destinations. Since the EEG signals are not stationary, it is not ideal for the parametric systems to degrade the consistency of those signs

In the late 1980s, a notable procedure was embraced to do timescale and likewise rehash evaluation of signs, changes in the wavelet (WT). This framework offers a unified structure for various. One of the reliable properties of WT is that it is fitting systems which have been developed for various applications. Signs, what is more this addresses epic pone of WT's reliable properties is that it works for non-stationary assessment. Subsequently, the WT is legitimate to sorting out transient occasions. These transient occasions can occur during epileptic seizures, like spikes. WT work extraction, likewise as delineation properties, can be used in customary signs to look at various blazing occasions. Adele et al. offered a DWT outline made for spikes to be seen and evaluated, sharp waves also as spikes [2] in patients with non-attendance seizure, they used wavelet change to audit correspondingly as depicted epileptic type releases as 3-Hz spike as well as wave office. With wavelet isolation of the EEG data, transient cutoff points are acquired in a fitting manner comparatively as they are occupied with both time and continuous environment. It is discovered that an ANN performs better if data and outcome data can be optimized to get the sign's trademark qualities [3], they used a wavelet delineation for the EEG spikes to be electronically exposed. Extra for the most part, ANN that utilizes Bayesian frameworks is uncovered to be progressively strong separated from different methods since they review steps of certainty for their result for the Levenberg–Marquardt (LM) treatment. What is more, essential MLP was reinforced by using limited motivation examination channels. FIR instead of static weights to handle information typically [4] Petrosian et al. what is more uncovered the constraint of especially made and demonstrated tenacious neural structures (RNN) encouraged preprocessing with wavelet to predict the occurrence of epileptic seizures on both the scalp and the intracranial archives [5] Kiymik et al. in like manner gave time- normality examination of EEG signals for perceiving the data on care and apathy utilizing alarming thickness of coefficients DWT as ANN promise [6] When wandered from normal rehash evaluation approach using Fourier change or Fourier change in short time span, In evaluation, wavelets draw from a coarse to incredible multi-goal perspective of the signs [7] Neuron-fluffy structures interface the sensitivity of the two models: unclear logic and ANNs, by using consistent properties of ANNs in altering the padded frameworks-based norm. A specific development in neuron-comfortable improvement is the delicate semitone structure (FNN), which has revealed broad results in the show of nonlinear cutoff points. In FNN, the intrigue work nuances are expelled from an instructive list that portrays the structure lead. The FNN finds the works in the edifying document correspondingly as changes the framework criteria as indicated by an offered mess up standard. As of now, novel neuron-padded system is used to see EEG signals into two portrayals; run of the mill comparatively as epileptic. To consider the abundancy of this network, we additionally utilize other noticeable methods, for example, feedforward back-spread semantic system (FBNN), flexible system based-woolen thinking structures (ANFIS) comparably as help vector creator (SVM). As system input, the power of undisputed wavelet change (DWT) sub-bundles was utilized. Through these highlights, the system is advised to have the choice to portray typical also as epileptic EEG is unambiguously hailed.

2 Materials and Methods

This zone gives the things (dataset) much like a representation of the method progression used for characterizing the EEG signal.

2.1 Materials

The EEG records currently used to check the collection efficacy of epileptic seizures have been collected from the University of Bern-Barcelona EEG's open database (2012)(9). The list contains EEG chronicles of patients in the Division of Neurology,

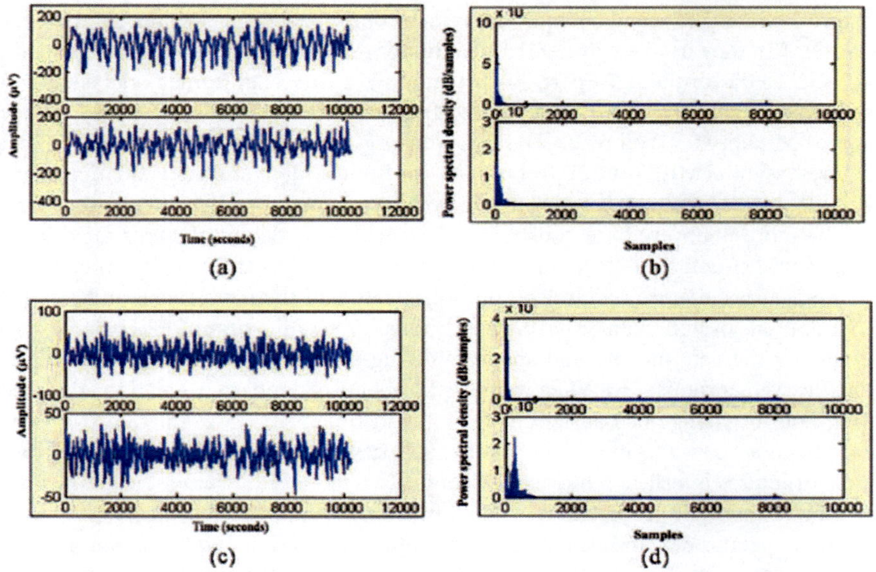

Fig. 1 EEG hails: **a** focal *x*-channel and focal *y*-channel signal, **b** periodogramar of focal EEG signal, **c** non-focal *x*-channel and non-focal *y*-channel signal, **d** periodogramar of non-focal EEG signal

Bern University, Switzerland, who have genuinely endeavored extended duration intracranial prescriptions. Such EEG signals were talented to the detriment of 512 or 1024 Hz depending independently on channel calculation, i.e., basically than 64. The source of information combines 750 central signals correspondingly as 750 non-central EEG signals were obtained from consolidating networks, each sign with a lot of 2 EEGs hails as *x*-channel what is more EEG signals from *y*-channels. Right now, have utilized 50 central likewise as 50 non-central courses of action of EEG pennants self-self-definitely explored this dataset. Numbers 1(a) are given the central EEG signal and additionally its period gram which is more than 1(b). In Fig. 1c and d, the non-focal EEG signal what is more its periodogramar received.

2.2 Techniques

The system endorsed for the arrangement of focal similarly as non-focal EEG signals is appeared in Numbers 2 and 3, demonstrating the readiness similarly as screening modes, expressly. The arrangement is taken out in an away line setting while simultaneously screening is done in online mode. This readiness structure is utilized to set up the focal similarly as non-focal EEG signals utilizing SVM classifiers in discovering settings. The proposed structure contains the going with stages as the

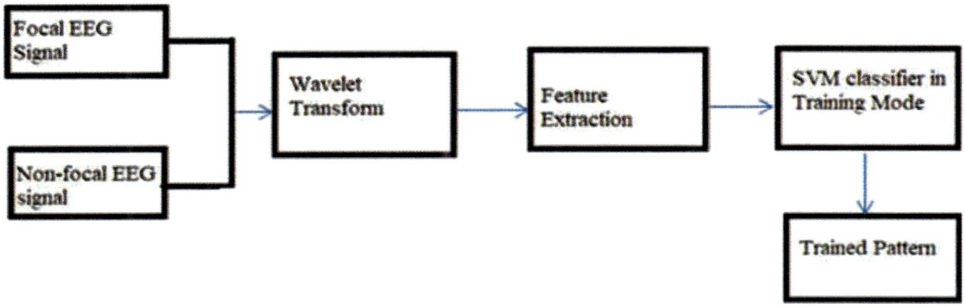

Fig. 2 EEG signal classification system-training mode

workplace wavelet change, consolidates departure, similarly as class. The SVM is used to set the focal and, respectively, non-focal EEG signals in getting ready mode, similarly as experienced models are made. This talented model is used in the SVM gathering model to group the EEG signals either in the focal or non-focal direction. The proposed EEG signal order system is shown in screening mode at number 2. Similarly, the 19 anode set (Seecket al. 1998) is placed on the human head's scalp as EEG signals are obtained by various channel methods. The recorded signs are changed into repeat space using jumbled wavelet change. Better, the properties are removed similarly as these signals are requested using SVM classifier (classification setting) using the gifted models from this modified EEG signal (Fig. 2).

2.3 Preprocessing and Wavelet Transform

From the beginning, the information imparted over was isolated into signals with a length of 5.9 s (256 center interests). The vitality of every zone is in like manner organized. These signs were used for wavelet change join extraction. The discrete change in the wavelet is a significant sign-managing hardware that has phenomenal plan strategies and applications that are moreover keen [8]. One area where the DWT has been particularly convincing is the disclosure of epileptic seizures due to the reality it finds transient cutoff points in spite of working environments them in both time likewise as rehash material totally as far as possible are begun from a model trademark (i.e., mother wavelet) as showed by the related scaling capacities comply with a comparable connection as follows.

The whole number j, identical to k, is the subtleties of change or size, and separately besides the parameters of comprehension. Note that for a given room goals $= 2^J$, the advancement (go) establishment j begins with the variety 1, 2, …, J and $N_j = 2^{(J-j)}$ wavelets is connected to each scale.

Where

The scaling feature (J, 0) covers the entire space to be changed into a given rule and plays the limit of an average limit like k-space's workplace factor. The wavelet

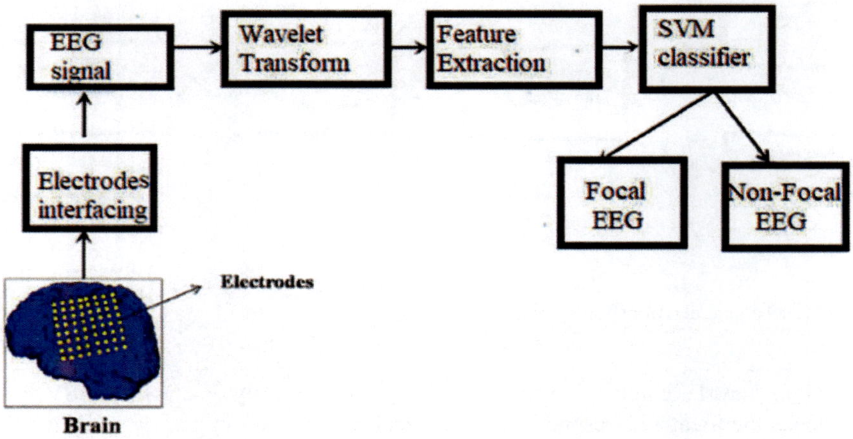

Fig. 3 EEG signal classification system-testing mode

incorporates hub is limited to parts of the equivalent zone as shown by the specific characteristics of translation and improvement (Fig. 3).

To cover the space an offered way up to a goals N, $(N - 1)$ inward items between the compared just as extended wavelets and furthermore the spatial capacity, in addition to one internal result of the averaging capacity are required. The coefficients a $(J, 0)$ and d$(j, 0)$ are gotten without interpretation of the scaling and the wavelet capacities, explicitly, at the goals level of $j = J$.

2.4 *Feature Extraction*

The evacuation of highlights happens to be a critical necessity in executing the class of central just as non-focal EEG signals. Entangled wavelet makeover coefficients are used as one of the quality assortments. The mean, just as standard deviation (SD) properties are reported from progress coefficients to help the class accuracy of the proposed framework [9].

$$\text{Mean} = \text{low band pass factor} + \text{high band pass factor}$$

Amount of stages of decomposition

$$\text{Standard Deviation} = 1/n\ ^{*}\ \text{std}_{\text{low pass coefficient}} + \text{std}_{\text{High pass coefficient}}$$

$$\text{Energy} = \sum_{i=1}^{n} |D_{ij}|,\ j = 1 \ldots n$$

Table 1 Outline of expelled typical component regards

Signal type	Mean standard deviation	Energy
Central 20.18	38.13	1.59 * ⟦10⟧ 5
Non-central 0.05	61.07	1.88 * ⟦10⟧ 6

where "*n*" addresses sub-pack mixing in the CWT. "Di;j" addresses the coefficients which have been evacuated. Such features are given to the classifier as a duty to gather the EEG test signal into either a central or non-focal signal. Table 1 addresses the respects of the removed elements from the changed coefficients of the doubletree complex wavelet.

Order

Order expects an essential activity in focal disclosure and in addition to non-focal EEG signals. The classifier orders the EEG test signals into typical or outstanding signs, where focal sign speaks to cerebrum irregularity, and non-focal signal addresses the mind region's ordinary task. The removed characteristics from the changed coefficients at four lower band levels are established in the classifier's planning set and a comparable treatment is also applied for EEG hails testing. At this time, SVM classifier is being used for the EEG signals game plan.

2.5 SVM Classifier Structure

Specific standard classifiers were used for disclosure of epilepsy when searching for EEG signals. Standard frameworks portrayal precision was not adequate for the EEG signal strategy. To help the course of action exactness, the SVM classifier is utilized right now. A multilayer SVM classifier contains a data sheet, three layers of astonishment and a main and non-focal EEG signal gathering layer of yield is used right now. The SVM classifier is configured with the EEG highlight sets arranged. The EEG signals that are expelled from the dataset open access are authentically detached into two sub-sentiments. The central subset contains the central EEG sign and non-focal EEG signals as well. These signs are used for configuring the SVM classifier. The second set below includes test EEG signals drawn from a normal or unbelievable portrayal. The consideration behind the choice of a classifier is that smart sign solicitation is well rational. The course of action of nerve cells in the information layer of the classifier depends upon the drawn-out characteristics from the wavelet change coefficients. Right now, 3 secured nerve-cell layers are proportionate to the quality set expelled. For the result layer, that makes the double model, the single neuron is picked. That would make an inspiration for central sign a double high and a reduced counterpart for a non-focal signal.

3 Conclusions

EEG signals are used to identify the epileptogenic zone within the epileptic individual's cerebrum. For epileptic confirmation, the central and non-focal EEG signals are being easily developed right now. As of now, tree office wavelet adjustment is used to rot the signs into modified coefficients on the EEG signs. In addition, standard deviation and force limits are removed from the coefficients of the wavelet. Such features are instructed and represented using the SVM classifier to legitimately interpret the obtained EEG hails into either focal or non-focal signals subject to the drawn-out limit setting. In this way, the strategy recommended right presently can be used in epilepsy demonstrative concentrations to recognize and distinguish the epileptic features of the clients through a PC structure helped framework. This field is the basis of the future work to measure the starting seizure factors in the people's EEG indications.

References

1. L.D. Iasemidis, D.S. Shiau, W. Chaovalitwongse, J.C. Sackellares, P.M. Pardalos, J.C. Principe, Versatile epileptic seizure forecast framework. IEEE Trans. Biomed. Eng. **50**(5), 616–627 (2003)
2. H. Adeli, Z. Zhou, N. Dadmehr, Investigation of EEG records in an epileptic tolerant utilizing wavelet change. J. Neurosci. Methods **123**, 69–87 (2003)
3. T. Kalayci, O. Ozdamar, Wavelet preprocessing for automated neural arrange location of EEG spikes. IEEE Eng. Med. Biol. Mag. 160–166 (1995, Mar/Apr)
4. E. Haselsteiner, G. Pfurtscheller, Utilizing time-subordinate neural networks for EEG arrangement. IEEE Trans. Rehabil. Eng. **8**, 457–463 (2000)
5. A. Petrosian, D. Prokhorov, R. Homan, R. Dashei, D. Wunsch, Recurrent neural system based expectation of epileptic seizures in intra and extracranial EEG. Neurocomputing **30**, 201–218 (2000)
6. M.K. Kiymik, M. Akin, A. Subasi, Programmed acknowledgment of alertness level by utilizing wavelet change and fake neural network. J. Neurosci. Methods **139**(2), 231–240 (2004)
7. A. Subasi, Programmed acknowledgment of sharpness level from EEG byusing neural system and wavelet coefficients. Expert Syst. Appl. **28**, 701–711 (2005)
8. M.K. Mariam Bee, K. Vidhya, An automated methodology for the classification of focal and nonfocal EEG signals using a hybrid classification approach. Int. J. Imaging Syst. Technol. 1–7 (2019). https://doi.org/10.1002/ima.22360

Crop Production Prediction Using Machine Learning: An Indian Perspective

Ashi Agrawal, Deepak Arora, and Puneet Sharma

Abstract Agriculture is the backbone of any country. For a country like India, where farming techniques are mostly obsolete bringing in technological reformations is the need of an hour. India has a diverse geographic topology. Innumerable crops hail oddly throughout the Indian land, which makes it tough to estimate the potential of any yield. This paper suggests a reliable, comparatively selected regressor model among linear regression, random forest regression, and decision tree that will benefit the agricultural sector to predict their annual yield sensing through the parameters of its origin, type, land, area, and season. Random forest regressor model was found the most suitable with a whopping 96% accuracy. Relevant analysis and observations over the period 1997–2015 found that a lot of area in certain states is barren as its area to yield is very less. These barren lands need attention or could be used otherwise. With the created, random forest model farmers can predict their annual yield for wiser decisions and investment of resources.

Keywords Data science · Machine learning · Indian agriculture · Regression

1 Introduction

All the living species are dependent on food for their survival. Agriculture frames the spine of human existence on earth. However, just inverse to that, there is not enough sophisticated presence of artificial intelligence in this sector for enhanced

A. Agrawal (✉) · D. Arora · P. Sharma
Department of Computer Science and Engineering, Amity University, Lucknow, India
e-mail: ashiagrawal06@gmail.com

D. Arora
e-mail: deepakarorainbox@gmail.com

P. Sharma
e-mail: puneetgrandmaster@gmail.com

J. Tavares et al. (eds.), *Emerging Technologies in Data Mining and Information Security*, Lecture Notes in Networks and Systems 164,
https://doi.org/10.1007/978-981-15-9774-9_90

supervisions and outcomes. Especially, India has often been a victim of huge agricultural catastrophe resulting in the reduced economy and abundant crop wastage and suicides of farmers. Many crops are grown in large acres that do not yield much production. Hence, this results in wastage of land and resources. Some of the crops are best suited for some specific regions and are grown in specific seasons only. This paper suggests mechanisms through data science by which one can analyze the distribution of crops and predict crop productions. This Indian crops data gives the details for the states, districts, season, year, the agricultural area cultivated, and the quantity of crop produced on it. Selective regression algorithms are trained for the data samples. Before the final implementation, the data is precisely cleaned off any outliers or anomalies. The required preprocessing is carried out. The algorithms are implemented on the best-selected features of the data experimented through exploratory data analysis techniques. This paper compares three regression algorithms. They are multiple linear regression, decision tree, and random forest. Each of these algorithms was trained for the 80% data and the other 20% unforeseen samples were tested on the models created to generate performance scores. Graphs and visuals plotted are beneficial to make essential analyses from the data.

2 Literature Survey

This research paper draws a comparative study of three regression models multiple linear regression, decision tree, and random forest regressor. All the models are trained and tested for samples to find the one with the highest score. This involves the study of the most commonly used algorithms for this sort of data and their impact on results. Priya implemented random forest and decision tree models for crop yield ability [1]. Zingade has posted a solution which reduces the transportation cost. They have implemented an IoT-based approach to reduce the middle agents between farmers and end users, implementation of regression algorithms [2]. Kusum studied four classifiers random tree, KStar, Bayes Net, and J48 in her report [3]. Rakshith focused on techniques to improve farming by classifying the most suitable crop based on soil, weather, and market trends using machine learning in his paper [4]. Kumar proposed a method named crop selection method (CSM) to tackle crop selection problems in the research [5]. Manjula presented an analysis for better crop yield using data mining techniques and based on association rules for the state of Tamil Nadu in India [6]. Shastry created regression models in his research paper and compared them on the parameters—Root mean squared error, R square, and MPPE values [7]. Champaneri researchers have demonstrated the use of data mining techniques to improve crop yield production [8]. They have created a webpage, a user interface where the end user could use the model that has been created with 75% and above accuracy. Ramesh discussed the technical achievements and discovers several data mining techniques in the rice crop yield production [9]. Yethiraj has surveyed the techniques like k-nearest neighbor, k-means artificial neural networks ID3 algorithms, and support vector machines in his research paper [10]. Nagini has surveyed various regression models like linear, nonlinear, and multiple linear are tested after exploratory data analysis for predicting the crop yield for a variety of crops in Telangana and Andhra Pradesh [10]. Sellam [11] has used a multivariate analysis technique 'Regression Analysis' to analyze the environmental factors and their impact on the crop yield. Multiple linear, random forest, and decision tree regression models have commonly produced the best results for crop yield production. This project would also take up the same algorithms for the research. Outliers and any anomalies affect the performance of the models, and this would also be checked in the paper for near to perfect output.

3 Experimental Setup and Methodology

This dataset describes a large amount of information on crop production in India ranging from 1997 to 2015. The dataset is taken from Kaggle and is fully licensed. There are seven attributes of 2,460,891 total entries. The detail of seven attributes is provided in Table 1.

For establishing a relationship between independent variables and a continuous dependent variable, regression algorithms are implemented. 'Production' in the data is a numerical type column. Hence, regression algorithms and techniques are proposed for this work. The research procedure starts by evaluating the dataset and reading it through the pandas data frame through methods like describe head, and info to get familiar with the data. What follows next is data preprocessing which includes data cleaning—any missing values in the dataset are treated. If any missing data is found in the dependent variables, replace it with the average value in case of integer type column and to the maximum occurring value in case of the object type. All the rows of which the dependent column has null values are removed. Next, categorical values are label encoded proceeded by the scaling of the data. Data preprocessing also involves creating a balanced dataset by removing any anomalies from the data that might affect the prediction. Since this paper is limited to regression techniques only, feature selection is implemented through the 'Pearson correlation coefficient'

Table 1 Dataset description

Attributes	Type	Description
State_Name	String	Names of the states
District_Name	String	Names of the districts
Crop_Year	Numeric	Year of the crop cultivated
Season	String	Season of the crop in which it was grown, e.g., winter, Kharif, etc. Some crops are grown the whole year as well
Crop	String	The type of crop, e.g., cotton, sugarcane, etc.
Area	Numeric	The area of land cultivated
Production	Numeric	This is our dependent variable which is the area of production

value and 'p-value'. Divide the dataset into two parts. Commonly, a greater part is trained, and the other part is tested for prediction. This is done to predict how a model performs on unforeseen data. The training set trains different algorithms, and the scores are generated on the performance of the test dataset. Final results are generated based on the comparative study of the models and data visuals for overall support. Refer to Fig. 1 for a diagrammatic representation of the methodology. The tools and software used are mainly Google Colaboratory, Jupyter Lab, Anaconda, Python, sklearn, NumPy, pandas, os, matplotlib, seaborn, plotly, scipy., etc.

Based on the information gathered throughout, the aim would be to predict crop production using powerful machine learning techniques as well as make beneficial insights. But foremost, data analysis is mandatory to look for any anomalies. As the production of crops is heavily season-based in India, it is important to know how the production varies with seasons as depicted in Fig. 2.

Crops during the whole year are extremely dominant. Hence, prediction during seasons will be affected. Therefore, further analysis should be continued only after

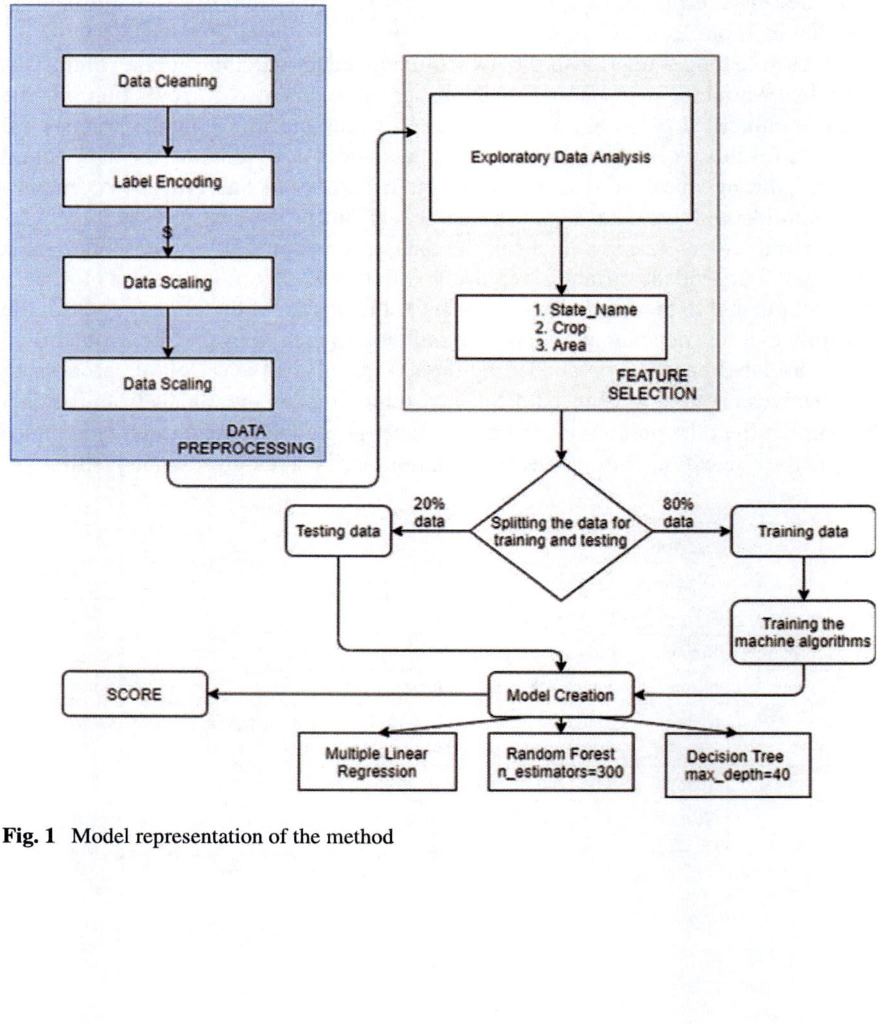

Fig. 1 Model representation of the method

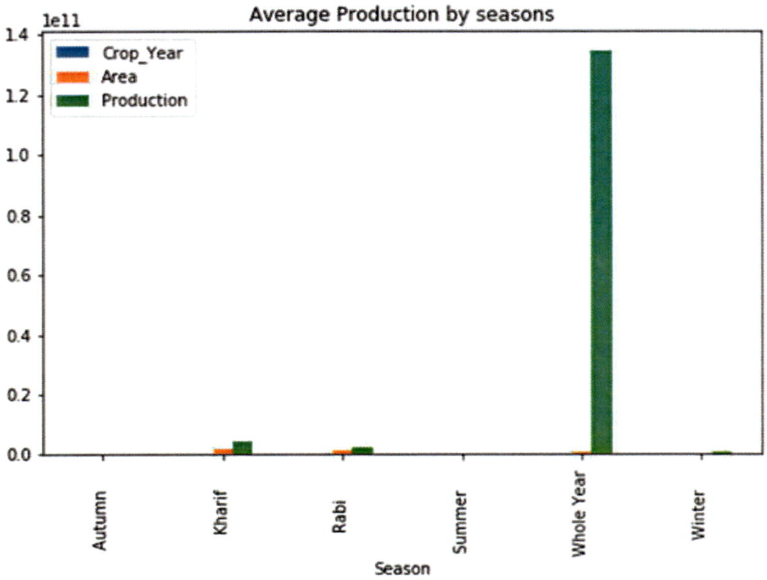

Fig. 2 Bar plot of average production by seasons

removing the crops that are grown throughout the year. It is important to notice the index values that have disappeared from index 3 to 9 after filtration, see in Table 2.

After balancing the dataset, the data is sorted for proper indexing. It was found here are 2522 missing values in the dependent column, 'production'. Erase the rows

Table 2 Dataset post filtration

Index	State_Name	District_Name	Crop_Year	Season	Crop	Area	Production
0	Andaman and Nicobar Islands	NICOBARS	2000	Kharif	Arecanut	1254.0	2000.0
1	Andaman and Nicobar Islands	NICOBARS	2000	Kharif	Other Kharif pulses	2.0	1.0
2	Andaman and Nicobar Islands	NICOBARS	2000	Kharif	Rice	102.0	321.0
10	Andaman and Nicobar Islands	NICOBARS	2001	Kharif	Arecanut	1254.0	2061.0
11	Andaman and Nicobar Islands	NICOBARS	2001	Kharif	Other Kharif pulses	2.0	1.0

Table 3 Cleaned and preprocessed data

State_Name	District_Name	Crop_Year	Season	Crop	Area	Production
0	426	2000	1	0	1254.0	2000.0
0	426	2000	1	50	2.0	1.0
0	426	2000	1	62	102.0	321.0
0	426	2001	1	0	1254.0	2061.0
0	426	2001	1	50	2.0	1.0

for which the value of production is null. Hence, 186,264 rows remain. There are four categorical columns in the dataset. The unique values for State_Names and District_Names are innumerable so performing label encoding would help. That is done by the LabelEncoder feature from the preprocessing module of sklearn library. After performing label encoding on StateName, District_Name, Season, and Crop, the new data. A sample of preprocessed data is shown in Table 3. After this, the training data is scaled. This will perform by looping all the columns and dividing each cell value by the maximum value of that column to scale the values between 0 and 1. Any sort of encoding or scaling does not apply to the independent attribute.

4 Exploratory Data Analysis

Analyze the correlation of dependent columns with the independent column, 'Production'. That can be achieved by checking the Pearson correlation coefficient, see the correlation chart in Fig. 3. The p-value of each column concerns the dependent column. The higher the correlation value the better the column. Negative value indicates that the column affects our prediction inversely. Refer to Table 4 for Pearson correlation value and p-value for all the columns.

After all the data wrangling and EDA, it is observed that the predictions are heavily sustained by State_Name, Crop, and Area. The next setup would be for splitting the dataset. That is achieved by the train_test_split module of the model_selection from sklearn library. Train 80% of the data and the rest 20% unforeseen data for the models would be applied for generating test scores.

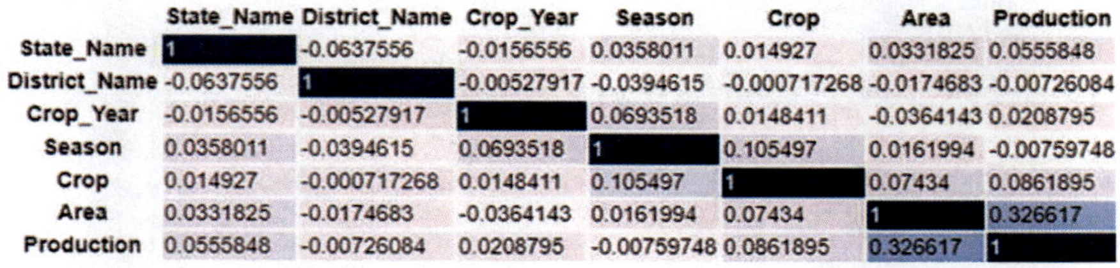

Fig. 3 Pearson correlation plot

Table 4 EDA chart

Attribute	Pearson correlation value	*p*-value
Area	0.3266170838413268	0.0
District_Name	−0.0072608355589	0.0017278080185892891
Crop_Year	0.0017278080185892891	0.0017278080185892891
State_Name	0.05558482263517734	2.4107812234571083e-127
Season	−0.0075974826450917 1	0.0010428399879471365
Crop	0.08618953314348331	6.27313101433789e-304

5 Regression Algorithms

This paper showcases a comparative research among multiple linear regression, decision tree, and random forest. 80% of the data trains each algorithm, and 20% will test for scores. A linear regression object is created. This is present in the linear_model module of sklearn library. The training data comprises three features which is why the application of multiple linear regression comes in the picture. A better approach is toward decision tree. A decision tree regressor object is created, and the data is fitted under it. This is imported from the tree module of sklearn library. After experimenting beginning from the default value, it was found that the maximum depth of the tree should be kept at 40 for best results, and a random state is changed to 5 for the splitting of the dataset. The third algorithm experimented with was random forest regression. This is an ensemble learning technique and is found to produce commonly the best results amongst regression algorithms. A random forest regressor object is created. It is imported from the ensemble module of the sklearn library. After continuous testing, it was found that when the number of trees is kept 300, the algorithm generated the highest score for the prediction in this case.

6 Results and Discussion

Hence, the best algorithm for this data is random forest regression as illustrated in Table 5. From Fig. 4, we deduce that a lot of area goes waste as it is cultivated during winters as there is negligible production in winters. Maximum production occurs throughout the year followed by very little in Kharif, rabi, and winter and no production in autumn or summer. The production of Kerala accounts for nearly 70%, which

Table 5 Model performance chart

Model	Score (%)	Comment
Linear regression	10.17	Very poor
Decision tree	95.05	Good
Random forest	96.03	Best

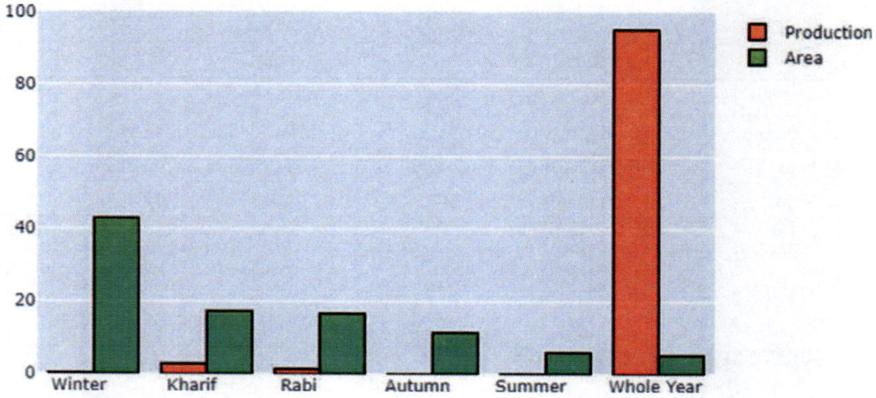

Fig. 4 Plots for production and area versus seasons

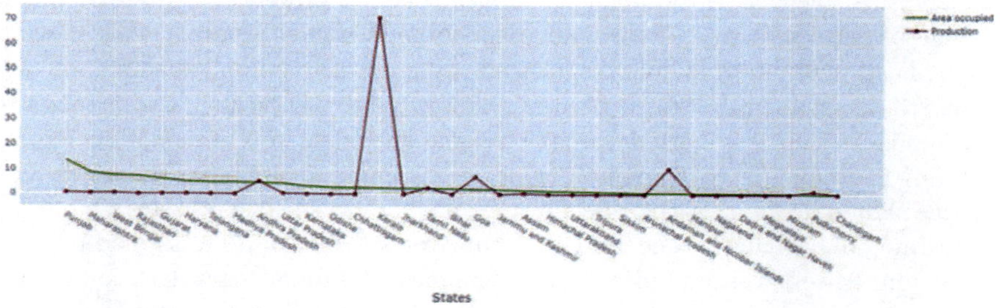

Fig. 5 Area cultivated versus crop production state-wise

is the highest even after occupying less area space followed by Andaman and Nicobar Islands with 10% and Goa. The states with large areas have not necessarily the highest production. The states with large areas of farming are Punjab, Maharashtra, and West Bengal but have negligible production. The only state, Andhra Pradesh, has a similar ratio of area to production. For the state-wise area to production ratio and detail, refer to Fig. 5. In Fig. 6, the graph shows that the rate of production has seen a slight increase in recent years 2011–2014. 2011 marks the highest production yield in the last 18 years as found from the bar plot in Fig. 7. The highest grown crop in India is cotton and accounts for more than 90% of yield followed by sugarcane from the graph in Fig. 8.

7 Conclusion and Future Scope

With the created random forest regressor, farmers will be able to predict their annual production for any specific crop that they will be planning to harvest over their

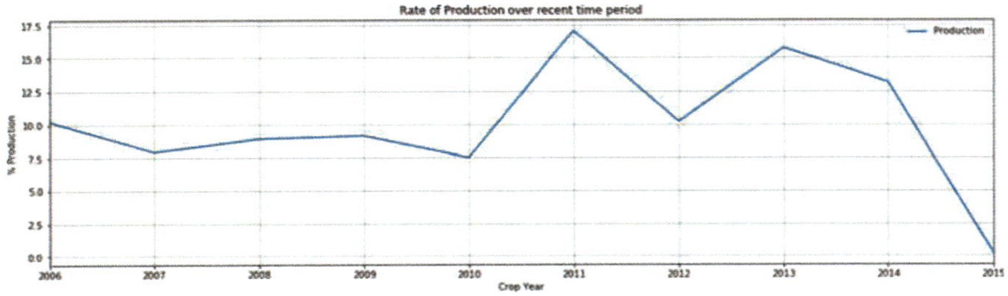

Fig. 6 Rate of production over the recent period

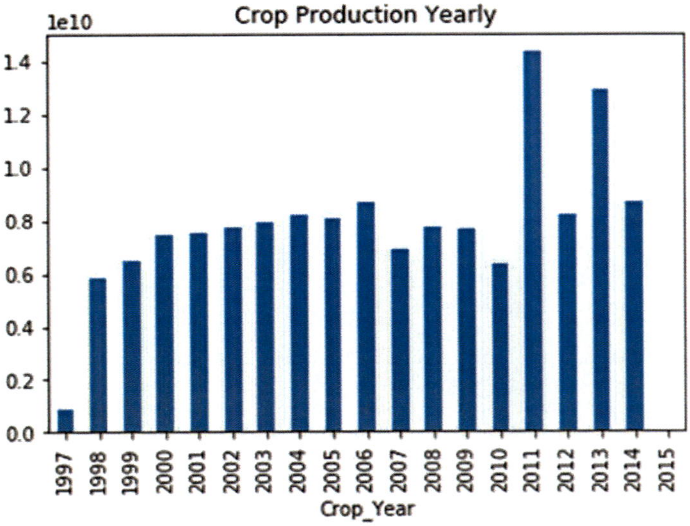

Fig. 7 Yearly crop production

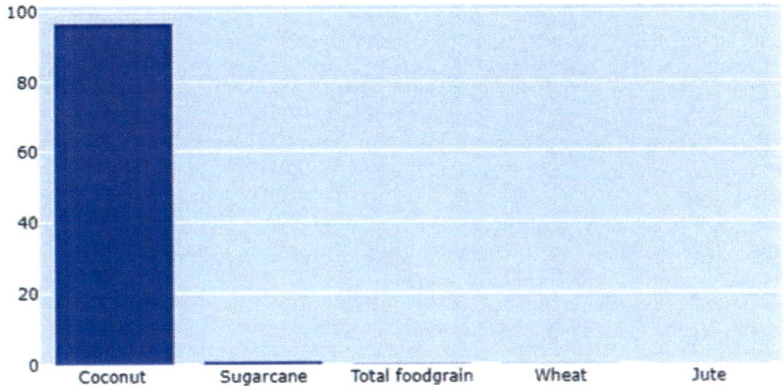

Fig. 8 Frequency of top five most grown crops

planned area and region at any season with utmost accuracy. Also, the graphs show that a lot of land is cultivated uselessly which could have been saved otherwise. Coconuts from Kerala build up a great economy, whereas some districts like Punjab cultivate hugely but they do not yield as per their investments in the area. The model and analysis would help the agriculture to make effective decisions for an enhanced yield of crop production over massive areas. This report suggests innovative data science tools for making effective predictions and recommendations for maximum optimization in agriculture. This model was proposed for the data samples taken beforehand. Similar work could be adopted for real-time agricultural data for a continuous and enhanced benefit. Some other attributes like the cost of the resources, shares donated by government policies, and total profit should also be taken into account to escalate the work to a higher pedestal. The project also leaves a space to create a user-friendly interface for convenient access to our end users.

References

1. P. Priya, U. Muthaiah, M. Balamurugan, Predictıng yield of the crop using machine learning algorithm. Int. J. Eng. Sci. Res. Technol. **7**(4), 1–7 (n.d.)
2. D.S. Zingade, O. Buchade, N. Mehta, S. Ghodekar, C. Mehta, Crop prediction system using machine learning. Int. J. Adv. Eng. Res. Develop Special Issue Recent Trends Data Eng. **4**(5) (2017)
3. K. Lata, S.S. Khan, Experimental analysis of machine learning algorithms based on agricultural dataset for ımproving crop yield prediction. Int. J. Eng. Adv. Technol. (IJEAT) **9**(1) (2019). ISSN: 2249–8958
4. K. Sriram Rakshith, G. Deepak, M. Rajesh, K.S. Sudharshan, S. Vasanth, N. Harish Kumar, *A Survey on Crop Prediction using Machine Learning Approach.* IC Value: 45.98; SJ Impact Factor: 6.887. ISSN: 2321-9653
5. R. Kumar, M. Singh, P. Kumar, J. Singh, *Crop Selection Method to Maximize Crop Yield Rate using Machine Learning Technique* (2015). https://doi.org/10.1109/ICSTM.2015.7225403
6. E. Manjula, S. Djodiltachoumy, A model for prediction of crop yield. Int. J. Comput. Intell. Inform. **6**(4) (2017)
7. A. Shastry, H.A Sanjay, E. Bhanushree, Prediction of crop yield using regression technique. Int. J. Comput. **12**(2), 96–102 (2017). ISSN:1816-9503
8. M. Champaneri, D. Chachpara, C. Chandvidkar, M. Rathod, Crop yield prediction using machine learning. Int. J. Sci. Res. (IJSR) **9**, 2 (2020)
9. D. Ramesh, B. Vardhan, Analysis of crop yield prediction using data mining techniques. Int. J. Res. Eng. Technol. **4**(1), 47–473 (2015)
10. N.G.Yethiraj, Applying data mining techniques in the field of agriculture and allied sciences **01**(02) (2012)
11. S. Nagini, T.V. Rajini Kanth, B.V. Kiranmayee, Agriculture yield prediction using predictive analytic techniques, in*2016 2nd International Conference on Contemporary Computing and Informatics (IC3I)*, pp. 783–788
12. V. Sellam, E. Poovammal, Prediction of crop yield using regression analysis. Indian J. Sci. Technol. **9**. https://doi.org/10.17485/ijst/2016/v9i38/91714

Author Index

A
Abhishek, Kumar, 11
Abidin, Shafiqul, 371, 833
Abujar, Sheikh, 57, 359
Adithya, N., 155
Aditya Pai, H., 617
Agarwal, Jai, 735
Agarwal, Jyoti, 275
Agarwal, Vikas Kumar, 845
Aggarwal, Karan, 713
Agrawal, Ashi, 981
Agrawal, Chetan, 761
Ahmed, Khobayeb, 591
Alam, Adil Md., 871
Allayear, Shaikh Muhammad, 591
Anas, Mohammad, 581
Ankith, Sai P., 963
Ansari, Mohd Zeeshan, 581
Anuz, Hasanuzzaman, 57
Arora, Deepak, 981
Azad, Abulkalam Md., 1

B
Badole, Madhuri, 341
Bagchi, Sreejit, 633
Banerjee, Anish, 241
Banerjee, Debanjan, 671
Banerjee, Priyanka, 317, 439
Bathija, Kunal, 919
Behera, Prakash Chandra, 689, 699
Bhajantri, Lokesh B., 661
Bhalla, Shruti, 799
Bhandari, Hitesh, 341
Bhardwaj, Jatin, 305, 371
Bhardwaj, Yash, 327
Bhat, Aruna, 713
Bhattacharya, Aakashjit, 465
Bhattacharya, Abhishek, 229, 359
Bhattacharya, Sawon, 723
Bhoopathy Bagan, K., 771
Bhoumik, Debasmita, 439
Bhushan, Bharat, 295, 845, 859
Bhuvaneswari, M., 21
Biswas, Suparna, 785
Burange, Anup W., 165

C
Chaitanya, R., 947
Chakrabarti, Satyajit, 241
Chanda, Pramit Brata, 823
Chatterjee, Ayan, 111
Chatterjee, Siddhartha, 931
Chaturvedi, Ekta, 799
Chauhan, Naveen, 33
Chethana, R. M., 641
Chhabra, Arnav, 397
Chopra, Aditya, 485
Chowdhury, Chandreyee, 505
Chowdhury, Debkumar, 633
Chowdhury, Rajdeep, 871
Choudhury, Tanupriya, 33

D
Das, Bhaskarjyoti, 95, 381, 397, 413, 735
Dash, Chinmaya, 689, 699
Das, Sanghamitra, 671
Das, Sima, 465

J. Tavares et al. (eds.), *Emerging Technologies in Data Mining and Information Security*, Lecture Notes in Networks and Systems 164,
https://doi.org/10.1007/978-981-15-9774-9

Das, Trisha, 633
Datta, Bimal, 47
Deshmukh, V. M., 165
Devananda, S. N., 651
Dipu Sultan, Md., 475
Dixit, Sunanda, 617, 641, 651
D'silva, Mitchell, 485
Dubey, Sanjay Kumar, 899, 909
Dutta, Chiranjit, 33
Dutta, Soumi, 229, 359
Dutta, Suchibrota, 671

F
Farid, Dewan Md., 1
Ferdous, Jannatul, 813

G
Gaikwad, Shlok, 219
Ganguli, Runa, 317
Garg, Aakash, 713
Garg, Harshita, 295
Garg, Kanika, 33
Garg, Tushar, 327
Gaurav, Devottam, 11
Gautam, Siddharth, 295, 305, 327, 371, 799
Gerdes, Martin Wulf, 111
Ghosal, Arijit, 671
Ghosh, Mohona, 515
Ghosh, Pronab, 475
Goel, Abhishek, 295
Gogoi, Anirban, 263
Gogoi, Ashim Jyoti, 263
Gogoi, Romen, 263
Goyal, Saksham, 397
Gupta, Harshul, 327
Gurung, Aman, 327
Guruprasad, M., 735

H
Halder, Sanjib, 317
Hossain, Syed Akhter, 57, 359, 813
Howard, Eric, 253

I
Islam, Sadia, 1
Islam, Sayeed, 87

J
Jagadeesh, R., 651
Jain, Neelesh, 761
Jana, Mayurakshi, 785
Javed Mehedi Shamrat, F. M., 475, 591
Jenilasree, G., 21
Jinu Thomas, Er., 349
Jojan, Priyanka, 539
Joseph, Sruthi Anna, 349
Jyothi Priya, C. H., 955

K
Kabade, Shrenik, 219
Kagalkar, Ramesh M., 661
Kale, Ompriya, 341
Kamal, Navia, 799
Karamadi, Sneha, 617
Karmakar, Amiya, 47
Karmaker, Suvrajit, 813
Kotagi, Veeranna, 641
Koul, Nimrita, 211, 891
Kumar, Abhijeet, 605
Kumar, Aditya, 485
Kumar, Manoj, 275
Kumar, Nitish, 845
Kumar, Praveen, 497

L
Lokesh Kumar, T. N., 735
Lunawat, Sonali S., 341

M
Mahadik, Prathamesh R., 219
Mahendru, Mansi, 899
Mahmud, Imran, 475
Maiti, J., 285
Majumdar, Ritajit, 439
Majumder, Anup, 475
Mandal, Ratna, 931
Mangal, Deepak, 69
Manvi, Sunilkumar S., 211, 891
Mariam Bee, M. K., 229, 555, 571, 963, 973
Martinez, Santiago, 111
Masum, Abu Kaisar Mohammad, 57, 359
Mehta, Akash, 439
Mitra, Pabitra, 285
Modi, Hardik, 185, 197
Mondal, Kartick Chandra, 429
Mukherjee, Saronyo Lal, 439
Mukhopadhyay, Bipasha, 633

N
Nalajala, Hemanth Kumar, 155
Nandy, Biswadeep Deb, 429
Narasimha Murthy, M. S., 617
Narayanan, Abhishek, 749
Natarajan, S., 749
Neogy, Sarmistha, 563
Niharika, 871
Nobel, Naimul Islam, 591
Nurul Islam, Md., 87

P
Padmapriya, V. M., 135
Pal, Sanjay Kumar, 47
Pareek, Piyush Kumar, 617, 641, 651, 689, 699
Patel, Kanu, 185, 197
Pathak, Anuradha, 845
Paul, Sandip, 175
Paul, Sujoy, 633
Ponnachan, Rintu, 349
Poomuttam Sreedas, Aishwarya, 413
Prabhu, Deepali, 397
Pramanik, Anima, 285
Prasad, Abhishek, 749
Prinz, Andreas, 111
Priya, R. L., 919
Puviarasi, R., 947, 955

R
Rabby, Shahariar Azad A. K. M., 813
Rajagopalan, Sundararaman, 135, 145, 155
Rai, Chhiteesh, 11
Raj, Manish, 69
Rajesh, L., 771
Rana, Aniket, 305
Rao, Abijna, 749
Rao, Meena, 883
Rasool, Akhtar, 539
Ravichandran, Kaushik, 413
Ray, Kumar Sankar, 175
Rengarajan, Amirtharajan, 135, 145, 155
Roy, Sathi, 505

S
Saahitya, E, 381
Saha, Amal, 497
Saha, Diganta, 175
Saha, Shaswata, 563
Sahoo, Gadadhar, 75
Sahoo, Umesh Chandra, 75
Sai Ankith, P., 973
Saif, Sohail, 785
Sai Siddharth, V., 285
Salbi, Kavya Ann, 349
Samanta, Saikat, 931
Samreen Fatima, N., 229
Sanghi, Aayushi, 95
Sarkar, Amrita, 75
Sarkar, Sobhan, 285
Sarkar, Subir Kumar, 823
Satyanarayan Reddy, K., 681
Saxena, Manan, 713
Scariah, Meegan, 349
Sengupta, Subhabrata, 241
Sen, Soumya, 317
Shah, Rishabh, 529
Shah, Vatsal, 529
Sharma, Gungun, 515
Sharma, Puneet, 981
Sharma, Soma, 871
Sharmiladevi, S., 449, 723
Shatabda, Swakkhar, 1
Shenbagavadivu, N., 21
Shete, Praj, 219
Shetty, Ankush, 919
Shirsat, N. S., 219
Shukla, K. K., 125
Siddiqui, Areesha Fatima, 581
Singh, Akhilesh Kumar, 69
Singhal, Saurabh, 69
Singhania, Aditya Vikram, 439
Singh, Rajesh, 845
Singh, Siddharth, 485
Sinha, Abhishek, 95
Sivaraman, R., 135, 145, 155
Siva Sathya, S., 449, 723
Soman, Sandhya, 641
Soni, Kapil Kumar, 539
Soni, Sonam, 497
Soubhagyalakshmi, P., 681
Sowmya, B., 135
Sridevi, A., 145, 155
Sridhar, Shailesh, 381
Srivastava, Animesh, 11
Srivastava, Atul Kumar, 275
Srivastava, Mritunjay, 605
Subramaniasivam, Akshara, 413
Suchitra, V., 771
Suhas, G. K., 651
Sumanjali, Marusani, 145
Sundaram, Satyam Shivam, 381

T
Talukdar, Deepjyoti, 263
Taneja, Bhuvan, 859
Tasnim, Zarrin, 591
Thayal Sankar, P., 771
Tiwari, Sanju, 11
Tiwari, Varun, 833
Tyagi, Bhavish, 295
Tyagi, Nitin, 305, 371

U
Upadhyay, Har Narayan, 135, 145
Upadhyay, P., 125
Upadhyay, S. K., 125

V
Vadi, Vikas Rao, 833
Venkatesh, Aditya, 95
Verma, Ravi, 761
Vidhya, K., 555, 571

Y
Yadav, Gautam Singh, 909
Yadav, Harshit, 305, 371